Brazilian Sand Flies

Elizabeth F. Rangel • Jeffrey J. Shaw
Editors

Brazilian Sand Flies

Biology, Taxonomy, Medical Importance and Control

Springer

Brazilian Ministry of Health
FIOCRUZ
Oswaldo Cruz Foundation

Editors
Elizabeth F. Rangel
Instituto Oswaldo Cruz
Fundação Oswaldo Cruz
Rio de Janeiro, Brazil

Jeffrey J. Shaw
Instituto de Ciências Biomédicas
Universidade de São Paulo
São Paulo, Brazil

ISBN 978-3-030-09260-3 ISBN 978-3-319-75544-1 (eBook)
https://doi.org/10.1007/978-3-319-75544-1

Printed on acid-free paper

This Springer imprint is published by the registered company Springer International Publishing AG part of Springer Nature
The registered company address is: Gewerbestrasse 11, 6330 Cham, Switzerland

There are people who pass and leave their mark. However, there are those who go far beyond leaving a simple milestone—whose remarkable genius gives us a legacy upon which the future is built.

It is in this vein that we express our deep respect, gratitude and admiration to all those dedicated pioneers who have studied phlebotomines. This book is founded on their publications, especially those of Brazil's entomological icons Leônidas de Mello Deane, Oswaldo Paulo Forattini, Amilcar Viana-Martins, Alda Lima Falcão, Otávio Mangabeira Filho, Reinaldo Maurício Golbert Damasceno and Mauro Pereira Barretto.

Foreword

Entomology, the branch of zoology concerned with the study of insects, was not considered a science until the nineteenth century, when the first entomological encyclopedia, *Introduction to Entomology* (1856), was published by the British entomologists William Kirby, widely considered the founder of entomology, and his colleague William Spence. Arthropods are the cornerstone in the transmission of a significant number of diseases of concern in public health and veterinary medicine, among them leishmaniasis, and its sand fly vectors are of outstanding importance.

The earliest fossil specimens of sand flies are found in amber from the Lower Cretaceous (135–120 million years). They closely resemble some current species belonging to the Hertigiini tribe, which contains some of the most primitive sand fly species. This means that sand flies most probably existed at the end of the Jurassic period 140–135 million years ago, and that the subfamily Phlebotominae possibly appeared before the formation of the Atlantic Ocean 190 million years ago. This is particularly important when considering their evolution and the parasites they transmit.

When in 1691 the Italian naturalist Philippo Bonanni published the first known description of a sand fly in his *Micrographia Curiosa*, which he called "*Culex minimus*", he could never have imagined the extraordinary importance that these small Diptera would have a couple of centuries later. Sand flies are Nematocera dipterans of great medical and veterinary importance because of the hematophagic habits of the females that are the invertebrate hosts—vectors—of several pathogens (bacteria, parasites and viruses) that infect humans and animals. Among these pathogens are *Leishmania* species that have caused visceral, cutaneous and mucocutaneous leishmaniasis in humans in 98 countries around the world.

The oldest description of cutaneous leishmaniasis is thought to be that which has been translated from a clay tablet in the library of the Assyrian king Ashurbanipal in Nineveh (today Mosul, Iraq) written two or three millenniums BC in Acadian language. The natural history of cutaneous leishmaniasis was described by Avicenna (980–1037) and Gorgani (1042–1136; Jorjani, Iran), who named it "*pashé gazidegui*" or "bitten by a mosquito", anticipating by centuries the role that insects play in transmitting diseases as shown by both Patrick Manson and Carlos Finlay at the end of the 1800s. We found a very intriguing painting by the Spanish bishop Martinez de Compañón from the1800s showing a Peruvian *criollo* with skin ulcers on the leg and left cheek and a destroyed nose, a typical sign of mucocutaneous leishmaniasis. Besides its graphic value, it is fascinating that there is an association between the disease and being bitten, represented by a small bird and a stick-insect on the ulcers.

Leishman, Donovan, Wright, Lindenberg and Carini independently identified the parasite causing leishmaniasis, which in 1903 was given the generic name *Leishmania* by Ronald Ross. Paul Ehrlich was the first person to use heavy metals, such as sodium antimony tartrate, to treat human African trypanosomiasis and syphilis; it was first used for cutaneous and mucocutaneous leishmaniasis in Brazil by Vianna and was later used in Italy for visceral leishmaniasis. In the 1920s, Marzinovsky and Schurenkowa infected volunteers and proved the pathogenicity of *Leishmania*. Finally, in 1928 Major H.E. Shortt infected sand flies by feeding them on a case of post-kala-azar dermal leishmaniasis from Assam, India.

The oldest records of cutaneous and mucocutaneous leishmaniasis in America are numerous pre-Inca (around 1400s) pottery representations in Peru and Ecuador, named "huacos". The accuracy of these representations is simply shocking. Conclusive proof has been found in macrophages parasitized with *Leishmania* after immunohistological examinations of the mummy of a 6-year-old girl dated 800 BC, proving that leishmaniasis existed before the arrival of Europeans in America. In fact, all of the parasites that cause cutaneous and mucocutaneous leishmaniasis existed before man's presence in the Americas; they are therefore zoonotic diseases that depend on contact with environments in which there are enzootic cycles, and thus infected sand flies. However, *Leishmania infantum*, which causes visceral leishmaniasis, was imported with infected dogs during the Spanish and Portuguese colonization of America in the 1700s. Its primary vector is now the permissive vector (transmitting different *Leishmania*) *Lutzomyia longipalpis*, described by Lutz and Neiva in 1912. It is certainly the best studied sand fly species in the Neotropical region because, among other things, its pervasive nature means this potentially fatal disease has become endemic throughout South and Central America in a very short period of time. The first autochthonous case of visceral leishmaniasis in South America was reported in 1913 by Migone in Paraguay, but it probably became endemic long before this.

To date, around 1000 sand fly species have been identified throughout the world and more than 500 of these are found in the Neotropical region, approximately 20 species of which are considered proven vectors of leishmaniasis in the Americas. The abundance of sand fly species in the Americas still represent a fascinating challenge for entomologists today.

Great advances have been made in our comprehension of arthropod-borne diseases since late1900s. An accurate understanding of the insects involved in the transmission of these diseases is an essential key to designing and implementing control programs. However, in the case of the leishmaniases there are still fundamental epidemiological knowledge gaps related to transmission, which impedes the design of adequate control programs or may lead to their failure. For instance, in the 1970s it was wrongly assumed that the malaria control campaign in India would be sufficient to eliminate leishmaniasis as a collateral benefit. Its failure alone emphasizes that there is no doubt that the study of sand flies is a discipline in its own right. The accurate identification of sand fly species and complexes of species, exhaustive studies of their biology associated with changes in the environment in which they occur, precise knowledge of the mechanisms regulating the intra-vectorial development of *Leishmania*, as well as transmission, and the eco-epidemiological factors that influence the transmission are key factors to defining control programs. The dynamics of sand fly biology are also intimately linked to research on domestic and wild reservoirs that assess synanthropy between sylvatic and domestic cycles.

The book that, dear reader, is in your hands today is a precious jewel of knowledge, reflecting decades of passion for fieldwork and meticulous effort in the laboratory by the authors. Eleven chapters navigate the importance of sand flies in the New World, with emphasis on Brazil, from biology to control. This is a book of paramount importance in that it takes a modern approach to the control of leishmaniasis—a public health problem affecting the poorest among the poor in our societies—and is a solid piece of knowledge allowing Latin American scientists to accentuate their knowledge of the fascinating natural history of a unique disease by understanding the biology of its vector better.

Jorge Alvar
Head of the Leishmaniasis Program
Drugs for Negleted Diseases Initiative
Geneva, Switzerland

Preface

Since the publication of *Flebotomíneos do Brasil* in 2003, we have unfortunately lost some eminent scientists who contributed enormously to our present understanding of this fascinating and diverse group of insects in Brazil. We have included references to their obituaries at the end of this preface. We had the good fortune to know all of them personally and over the years each one of them became a cherished friend. Because of this it is emotionally difficult to write about them as memories flood in, bringing them back to life.

Our first loss was Adelson de Souza (Adelson Alcimar Almeida de Souza, 1942–2010). He took over Wellcome Parasitology Unit's sand fly laboratory in the Instituto Evandro Chagas's from Lee Ryan, who was the last permanent expatriate entomologist funded by the Wellcome Trust. Amongst Adelson's specialties were football and field work, both of which he took very seriously. With his field team he undertook extensive work throughout Amazonia and isolated many trypanosomatids. These cultures are a treasure to us all and studying them has opened many new avenues showing how important sand flies are as vectors of trypanosomatids of different vertebrate phyla. Seven years after his death his field collections have led to publications thanks to the efforts of Thiago Vasconcelos dos Santos. He was always good humored, which made field trips with him so enjoyable.

Italo Sherlock (Italo Rodrigues de Araujo Sherlock, 1936–2009) began his scientific career at the age of 14 with Maria and Leonidas Deane when they were working on visceral leishmaniasis in Sobral, Ceará State. They immediately recognized his intelligence and aptitude for entomology. He went with them when they returned to São Paulo, and Leonidas managed to get him enrolled in an entomology course even though he was underage for it. He moved to Rio de Janeiro and while visiting the Instituto Oswaldo Cruz he met Otávio Mangabeira Filho who invited him to go with him to Bahia, which became his home for the rest of his life. Even though he qualified as an ear nose and throat specialist he never abandoned entomology. He described many sand fly species and his work on the reservoirs and the vector of visceral leishmaniasis in Bahia remains a reference for us all. Italo's generous nature was much appreciated by those who knew him.

Bob Killick-Kendrick (Robert René Killick-Kendrick, 1929–2011) was known worldwide for his expertise on sand fly biology. He studied them in many parts of the world, especially in the Mediterranean region but also in Brazil. He was very influential in promoting the use of insecticide-impregnated collars for the control of canine visceral leishmaniasis, which is one of the most effective control measures. Bob first visited Brazil in 1976 and accompanied us on a field trip to collect infected sand flies in the Serra das Carajás, notorious as being highly endemic for *L.* (*Viannia*) *braziliensis*. His objective was to take back some infected flies, but besides this he took back his own strain as he became infected during the field work. He realized the potential of *Lutzomyia longipalpis* as a laboratory model and established the first colony of this species outside Brazil from eggs of Lapinha Cave flies. Bob had a warm and generous personality and always emphasized the crucial importance of well-planned field work, which he himself followed throughout his life.

Almério Gomes (Almério de Castro Gomes, 1942–2012) began his research career studying sand flies as a student under the guidance of Osvaldo Paulo Forattini. Almério's enthusiasm during his lectures was infectious, as was his passion for perfection. His good humor was always appreciated when discussions at meetings became heated. He produced many papers on the ecology of São Paulo State's sand fly fauna that evolved into the creation of technical guidelines for the control of cutaneous and visceral leishmaniasis. He was dedicated to promoting vector control and moved from studying sand flies to *Aedes aegypti* as Brazil's dengue epidemic expanded. Almério became a respected expert and consultant on the control of dengue fever, travelling extensively as an advisor throughout Latin America.

Alexandre Peixoto (Alexandre Afranio Peixoto, 1963–2013) was a brilliant young evolutionary geneticist who died tragically in a rafting accident. He was renowned for his easy-going nature, his generosity and was admired by his students. His application of the genetic analytical methods developed for *Drosophila* revolutionized our understanding of the speciation within the *Lutzomyia longipalpis* complex. He and his group, based in the Instituto Oswaldo Cruz, found that the male's courtship song showed the greatest variation between different geographical populations, suggesting that it potentially identified incipient species. Subsequent analysis was extended to 21 nuclear genes. Over a period of 6 years, his group and that of Richard Ward produced eight articles on the genetics of the *Lu. longipalpis* complex.

Richard Ward (Richard Douglas Ward, 1944–2015) began his entomological career working as a volunteer on the control of Culicoides in the Caribbean Cayman Islands, where he met his wife Mary. He returned briefly to England and came to Brazil in 1970, working at the Instituto Evandro Chagas's Wellcome Parasitology Unit. He left 7 years later to take up a post as reader at Liverpool University but returned to Brazil many times during collaborative research projects on *Lutzomyia longipalpis*. His hallmark was finding practical solutions. He successfully colonized *Lu. longipalpis* in the early 1970s using females captured in Ceará and developed a simple artificial feeding method that led to the first successful transmission of visceral leishmaniasis by this sand fly. Richard's fascination by the one and two spots of the male *Lu. longipalpis* opened the door to its genetic complexity. Later he championed the use of pheromones in control strategies. Richard was very generous, good-humored and well-known for caring about and helping his students.

María Cristina Ferro de Carrasquilla (1947–2015) was born in Ipiales, Colombia and went to school in Madrid, Spain. She graduated in Microbiology from the Los Andes University in Colombia in 1969. As a child she was passionately interested in insects, a passion that continued throughout her life. After gaining her MSc at the London School of Hygiene and Tropical Medicine she returned in 1976 to join the entomology laboratory of the Colombia's National Health Institute in Bogota. This is when she began her lifelong study of sand flies, and she became the laboratory's leader in 1994. Her research on sand flies spans from discovering new species and evaluating vectorial capacity to studying the environmental and ecological variations affecting their biology, all of which are crucial in understanding the epidemiology of leishmaniasis in Colombia and its neighbors. The scope of her work extended to other insect groups of medical importance, especially mosquitoes involved in the transmission of Venezuelan Equine Encephalitis and yellow fever. Maria Cristina loved her work and was loved and respected by those who worked with her.

Bruce Alexander (John Bruce Alexander Flynn, 1956–2016) was born and died in Edinburgh, Scotland. He gained his MSc under Bob Killick-Kendrick and his PhD studying sand flies in Colombia at Florida University with David Young. A full-blown sand fly pedigree. Between 1986 and 2000 he worked in Minas Gerais, initially at the Federal University and latterly in FIOCRUZ's René Rachou Research Institute in Belo Horizonte. He returned to Liverpool's School of Tropical Medicine in 2005 where he became involved in relating sand fly distribution to plants and chickens and their resistance to insecticides. In Australia he was involved in implicating midges as vectors of the kangaroo leishmania. With his wife Cristina, Bruce established Xeroshield, which aimed at finding practical solutions for insect control. Bruce had a dry sense of humor and was renowned for his passionate and argumentative nature in discussions. He was also an enthusiastic amateur ornithologist and a gifted artist. His exceptional enthusiasm is reflected by the fact that, even though he was seriously ill, he left Edinburgh to give a talk in Liverpool days before his death.

Dora Feliciangeli (Maria Dora Feliciangeli de Piñero, 1940–2017) was born in Borgomanero in Novara, Northern Italy and was one of Venezuela's most notable medical entomologists who carried the flag of medical entomology wherever it was needed, which thankfully included leishmaniasis. She was an enthusiastic teacher and tireless researcher who coordinated many programs of great public health importance. Her participation in the study of canine visceral leishmanias on the island of Margarita resulted in reducing the disease's incidence. Her extensive studies on sand fly biology helped to understand the complexity and many epidemiological facets of leishmaniasis in Venezuela. Dora's attention to detail and scientific courage are exemplified by her recognition of a new species of the *longipalpis* complex, which she described and named *Lutzomyia pseudolongipalpis*, with her student and long-term collaborator Jazzmin Arrivillaga.

Ralph Lainson (1927–2015) the editor of this book's 2003 edition along with Elizabeth Rangel. After a short illness he passed away in Belém, Pará State at the age 88, but his scientific legacy lives on as one of world's outstanding parasitologists. His enthusiasm continued right up until his death—a hallmark of this remarkable scientist whose academic interests spanned many groups of parasitic protozoa and their vectors. He impressed people by his simplicity, good humor and generous nature, but this masked a shrewd, focused intelligence that was the secret of his success. Entomology was the special subject of his bachelor's degree, which he obtained from Brighton Technical College in 1951 (now Sussex University). Although the major passion of his professional life was parasitic protozoa he was an enthusiastic amateur lepidopterist. However, the vectors of many parasitic protozoa are insects, so they became as important to him as the parasite For a very simple reason: he wanted to untangle the parasite's life cycle, which is impossible without studying its vector(s). Sand flies were no exception. He was never the lead investigator of projects involving sand fly biology, but he gave them his full support, which is reflected by the four entomologists who worked under his leadership in the Wellcome Parasitology Unit. This, however, was not the case when the parasite was involved. He spearheaded the experiments that led to the first experimental transmission of *Leishmania infantum* by *Lu. longipalpis*. In addition, he used sand flies to compare the development of different *Leishmania* species, which led to taxonomic revisions of the genus *Leishmania* that culminated in the creation of the genus *L. (Viannia)*. His enthusiasm led to discovery of the epidemiological importance of many sand fly species in Amazonia.

For reasons beyond our control there has previously never been an English version of *Flebotomíneos do Brasil*. At the time of its publication Ralph was disappointed by this, so we are very happy that thanks to the interest of our publisher, Springer, a completely updated English version is now available.

São Paulo, Brazil — Jeffrey J. Shaw
Rio de Janeiro, Brazil — Elizabeth F. Rangel

References

Anon (2012) Obituário: Almério de Castro Gomes (*1942 †2012). Braz Soc Trop Med News Lett. http://www.sbmt.org.br/portal/noticias-178/

Genta FA, Diaz-Albiter HM, Salgueiro P, Gomes B (2016) Control of vector-borne human parasitic diseases. Biomed Res Int 2016:1014805. (Tribute to Dr. Bruce Alexander).

Klaczko LB, Bruno RV (2013) Alexandre Afranio Peixoto (1963–2013). Gen Mol Biol 36(2):i.

Lainson R, Baker JR (2011) Robert René Killick-Kendrick MPhil, PhD, DSc, FSB 20 June, 1929–22 October, 2011. Rev Pan-Amaz Saude 2:21–55.

Leite G (2009) Italo Rodrigues de Araujo Sherlock (1936–2009). Rev Soc Bras Med Trop 42:360.

Olano VA (2015) María Cristina Ferro de Carrasquilla. Biomédica 35:444–445.

Reisz M (2015) Obituary: Richard Ward, 1944–2015. https://www.timeshighereducation.com/people/richard-ward-1944-201

Sanchez MA (2017) Obituary (Dora Feliciangeli). Leish-L. http://lineu.icb.usp.br/pipermail/leish-l/2017-November/002817.html

Shaw J (2015) In memory of Ralph Lainson - a parasitologist of amazing abilities. Acta Trop 150:224–226.

Watts G (2015) Ralph Lainson. Lancet 386(9988):24.

Contents

Contributors

Rafaella Albuquerque Communicable Diseases General Office, Ministry of Health, Brasília, Brazil

Margarete M. S. Afonso Instituto Oswaldo Cruz, Fundação Oswaldo Cruz, Rio de Janeiro, Brazil

Beatriz G. Brazil Instituto Oswaldo Cruz, Fundação Oswaldo Cruz, Rio de Janeiro, Brazil

Reginaldo P. Brazil Instituto Oswaldo Cruz, Fundação Oswaldo Cruz, Rio de Janeiro, Brazil

Bruno M. Carvalho Instituto Oswaldo Cruz, Fundação Oswaldo Cruz, Rio de Janeiro, Brazil

Wagner Alexandre Costa Instituto Oswaldo Cruz, Fundação Oswaldo Cruz, Rio de Janeiro, Brazil

Simone M. Costa Instituto Oswaldo Cruz, Fundação Oswaldo Cruz, Rio de Janeiro, Brazil

Ana Cecília Cruz Instituto Evandro Chagas, Secretaria de Vigilância em Saúde, Ananindeua, Pará, Brazil

Pedro Fernando da C. Vasconcelos Instituto Evandro Chagas, Secretaria de Vigilância em Saúde, Ananindeua, Pará, Brazil

Gustavo M. de Aguiar Instituto Oswaldo Cruz, Fundação Oswaldo Cruz, Rio de Janeiro, Brazil

Amélia T. de Rosa University of Texas Medical Branch, Galveston, TX, USA

Instituto Evandro Chagas, Secretaria de Vigilância em Saúde, Ananindeua, Pará, Brazil

Vanessa C. de Freitas Instituto de Pesquisas René Rachou, Fundação Oswaldo Cruz, Belo Horizonte, Minas Gerais, Brazil

Júlia dos S. Silva Instituto Oswaldo Cruz, Fundação Oswaldo Cruz, Rio de Janeiro, Brazil

Eunice A. B. Galati Faculdade de Saúde Pública, Universidade de São Paulo, São Paulo, Brazil

Ralph Lainson (Deceased) Instituto Evandro Chagas, Belém, Brazil

Ana Nilce S. Maia-Elkhoury Communicable Diseases and Environmental Determinants of Health (CDE), VT, Pan American Health Organization (PAHO), Duque de Caxias, Rio de Janeiro, Brazil

Carolina C. Monteiro Instituto de Pesquisas René Rachou, Fundação Oswaldo Cruz, Belo Horizonte, Minas Gerais, Brazil

Paulo F. P. Pimenta Instituto de Pesquisas René Rachou, Fundação Oswaldo Cruz, Belo Horizonte, Minas Gerais, Brazil

Ana Clara M. A. Pires Instituto de Pesquisas René Rachou, Fundação Oswaldo Cruz, Belo Horizonte, Minas Gerais, Brazil

Elizabeth F. Rangel Instituto Oswaldo Cruz, Fundação Oswaldo Cruz, Rio de Janeiro, Brazil

Paul D. Ready London School of Hygiene and Tropical Medicine, and Natural History Museum, London, UK

Oscar D. Salomón National Institute of Tropical Medicine, Ministry of Health, Puerto Iguazú, Argentina

Nágila Francinete C. Secundino Instituto de Pesquisas René Rachou, Fundação Oswaldo Cruz, Belo Horizonte, Minas Gerais, Brazil

Jeffrey J. Shaw Instituto de Ciências Biomédicas, Universidade de São Paulo, São Paulo, Brazil

Nataly A. Souza Instituto Oswaldo Cruz, Fundação Oswaldo Cruz, Rio de Janeiro, Brazil

Vanessa R. Vieira Instituto Oswaldo Cruz, Fundação Oswaldo Cruz, Rio de Janeiro, Brazil

Felipe M. Vigoder Instituto de Biologia, Universidade Federal do Rio de Janeiro, Rio de Janeiro, Brazil

Maurício Luiz Vilela Instituto Oswaldo Cruz, Fundação Oswaldo Cruz, Rio de Janeiro, Brazil

Adriana Zwetsch Instituto Oswaldo Cruz, Fundação Oswaldo Cruz, Rio de Janeiro, Brazil

Sand Flies: Medical Importance

Wagner Alexandre Costa and Nataly A. Souza

Phlebotomine sand flies (Diptera, Psychodidae, Phlebotominae) are small Diptera that appeared in the Cretaceous, approximately 120–135 million years ago (Hennig 1972; Azar et al. 1999); they represent a group of approximately 1000 known species around the world, of which 530 were found in the Americas (Shimabukuro et al. 2017). They carry trypanosomatid protozoans, including species from the *Leishmania* genus. These can affect human health, contaminate domestic animals, and make wild mammals into reservoirs for infectious diseases (Shaw et al. 2003; Brazil and Brazil 2014). It is estimated that 98 species of phlebotomine sand flies are possible natural vectors of *Leishmania* spp. (Killick-Kendrick 1999; Galati 2003; WHO 2010a). This is due to specific characteristics of their biology such as anthropophilia; being naturally infected by the same *Leishmania* that circulates among humans; and their spatial distribution, which coincides with likely places of infection.

These Diptera are holometabolous insects. The adult forms are considered to be cryptozoans because of their thin integument (exoskeleton) and their being extremely sensitive to variations in temperature and humidity. Natural shelters for adults include hollow spaces in tree trunks, fallen leaves, grottos or animal holes, and large treetops. In environments with human occupation or interference, artificial shelters include shaded and humid areas, other than domestic animals shelters such as hen houses, pigsties, and barns. The interior of houses can also be cited here (Aguiar et al. 1987; Killick-Kendrick 1999; Brazil and Brazil 2003). Immature phlebotomine forms develop in decomposing organic matter, particularly vegetable matter.

Both males and females feed on carbohydrates, especially aphids' nectar and secretions. They normally leave their natural shelters to feed at night and/or at dusk. Hematophagic activity, however, may take place throughout the night period, from

W. A. Costa (✉) · N. A. Souza
Instituto Oswaldo Cruz, Fundação Oswaldo Cruz, Rio de Janeiro, Brazil
e-mail: wagnerac@ioc.fiocruz.br

E. F. Rangel, J. J. Shaw (eds.), *Brazilian Sand Flies*,
https://doi.org/10.1007/978-3-319-75544-1_1

dusk to dawn. Only females are hematophagous, a required condition for their ovaries to mature. Because of their feeding habits, females are infected when they bite *Leishmania* spp. reservoirs and may later transmit infectious forms of the disease by biting humans and other mammals (Ministério da Saúde do Brasil 2017; Brazil and Brazil 2003).

In Brazil, as in the whole Every American, South, Central and North Continent, the species of phlebotomines that are responsible for spreading *Leishmania* spp. belong to the following genera: *Bichromomyia*, *Lutzomyia*, *Migonemyia*, *Nyssomyia*, *Pintomyia*, *Psychodopygus* and *Trichophoromyia* (Galati 2003; Marcondes 2007).

Current policies in Brazil establish that the control of leishmaniasis is the responsibility of the country's public health system (*Sistema Único de Saúde* [SUS]). According to the Health Surveillance Department's 'National Program of Leishmaniasis', Municipal Health Secretariats supported by State Health Secretariats are responsible for organizing the basic network of patient care and for implementing actions to combat the disease's vector (Ministério da Saúde do Brasil 2014, 2017).

The success of vector surveillance and control actions for leishmaniasis as planned by the Ministry of Health remains a great challenge in Brazil. These are illnesses with a complex eco-epidemiology, with different transmission cycles, unique links in their epidemiological chain and which can occur in various Brazilian biomes.

Considering this context and the diversity of *Leishmania* spp. vectors, entomologic surveillance will still need to collect more precise data on phlebotomine sand flies, including information for their correct identification and on their incidence in likely places of infection. Regarding incidence, both home and wild environments should be considered in order to obtain a full understanding of the local epidemiology. Another relevant factor to take into account is behavioral change in local vectors provoked by environmental changes. Important gaps in knowledge on phlebotomine vectors still exist, and they need to be addressed.

Leishmaniasis

Leishmaniasis is among the diseases with a higher impact on human health. Contributing to this scenario are environmental changes, which can be related to global climatic changes and impact on specific ecosystems. This context, however, becomes more complex as alterations in the environment can also be aggravated by the socioeconomic and living conditions of populations in risk areas for leishmaniasis. These conditions are important determinants in the process of disease transmission (Ministério da Saúde do Brasil 2017; Rangel et al. 2014).

The different forms of leishmaniasis are distributed worldwide, with an estimated 350 million people living in endemic regions. These illnesses are among the six most important infectious diseases in the world, which can be explained by their high detection coefficient, high number of related deaths and deformities, as well as their presence in 98 countries. Approximately 15 million people around the world are infected by leishmaniasis and endemic areas include the Americas, Africa, Asia and Southern Europe (Desjeux 2004; Malafaia 2009; WHO 2010a; Alvar et al. 2012; Ministério da Saúde do Brasil 2014, 2017).

Approximately 1.6 million new cases of these diseases are estimated to emerge each year, of which 500,000 are visceral leishmaniosis and 1.1 million are cutaneous or mucocutaneous leishmaniasis. Ninety percent of the cases of visceral leishmaniosis are concentrated in Bangladesh, Brazil, Ethiopia, India, Nepal and Sudan, 90% of the cases of cutaneous leishmaniosis can be found in Afghanistan, Saudi Arabia, Algeria, Brazil, Iran, Peru, Syria and Sudan, and 90% of the mucocutaneous cases are from Bolivia, Brazil and Peru. Only 600,000 of this total estimated number of 1.6 million cases are reported (Alvar et al. 2012; WHO 2010a). Every year, 500,000 new cases of leishmaniasis emerge, 60,000 related deaths occur and 200 million individuals are found to be at risk of developing the disease. Thus, the actual global leishmaniasis-related mortality is likely to be even higher than this estimate (de Oliveira et al. 2012).

Leishmaniasis presents a range of clinical manifestations, which are associated with infection from different types of *Leishmania*. In the Americas, they are distributed from the southern United States to the north of Argentina (Ministério da Saúde do Brasil 2014, 2017; WHO 2010a).

In Latin America, the disease has been described in at least 12 countries, although 90% of the cases are concentrated in Brazil. These cases generally occur in rural and peri-urban areas, commonly affected by elements of vulnerability, such as lack of basic sanitation services and garbage collection. These factors feed upon themselves, and generate a poverty–disease cycle (Alvar et al. 2012; WHO 2010a).

In Brazil, the leishmaniases are considered emerging endemic diseases in clear territorial expansion. They are included in the Ministry of Health's "System of Diseases with Compulsory Notification" and cases have been reported in all states of the country. Brazil is considered to have the biggest prevalence of leishmaniasis in the Americas, both for visceral and cutaneous forms (Ministério da Saúde do Brasil 2014, 2017; WHO 2010a). The importance of the national public health context for the spread of these diseases has increased significantly over recent years. This is mainly due to urbanization processes and to changes in the natural habitats of the species involved in the transmission cycle of leishmaniasis. The scenario for these illnesses in Brazil is therefore complex, and difficulties related to their control and the elimination of risk factors should always be taken into account (Toledo et al. 2017).

Migration flows should also be considered as an important social component influencing on the epidemiology of leishmaniasis. These phenomena have been observed in the north-east, center-east, south-east and south regions of the country, including cases of in-house transmission and cases in urban environments (Rangel 1995; Rangel and Lainson 2009; WHO 2010a).

American cutaneous leishmaniasis (ACL) is associated with a range of dermotropic leishmania (of the sub-genera *Leishmania* and *Viannia*) and manifests itself clinically through localized skin lesions, multiple lesions, cutaneous–mucosal and cutaneous–diffuse lesions. The latter tends to be very aggressive and it mainly develops in immunosuppressed patients (Ministério da Saúde do Brasil 2017; WHO 2010a).

Combined infection by *Leishmania* and HIV is an emerging problem that requires urgent attention (Lainson and Shaw 1972; Silveira et al. 1987; Lainson and Shaw 1989; Grimaldi et al. 1989; Lainson et al. 1994; WHO 2010a).

Throughout several regions in Brazil, the transmission of ACL occurs in deforested areas (Ministério da Saúde do Brasil 2017). This has mainly been concentrated in sites where houses are close to secondary forests (Lima et al. 1988; Sabroza 1981).

In the case of American visceral leishmaniasis (AVL), the parasite mainly develops in cells of the mononuclear phagocytic systems of the spleen, liver, bone marrow and lymphoid tissues. If the disease is not treated, it may lead to death (WHO 2010a), with cases concentrated in Venezuela, Bolivia and Brazil (Ministério da Saúde do Brasil 2014).

This disease has moved from being endemic in exclusively rural areas to, in recent decades, important numbers of AVL being recorded in urban sites, including state capitals, and both large- and medium-sized cities (Ministério da Saúde do Brasil 2014; Vilela et al. 2014; WHO 2010a). Today, the main challenge to fighting AVL lies precisely in the urbanization of its main vector: *Lutzomyia longipalpis*.

Leishmaniases as Neglected Diseases

Neglected diseases affect the lives of 1 billion people around the world and threaten the health of millions of others. They generate a substantial burden in terms of health, economic and social conditions. Neglected diseases represent a group of old illnesses that have afflicted humanity for centuries which proliferate in impoverished environments within tropical areas of the world. In many cases, these diseases remain hidden, concentrated in remote rural environments, but also in urban communities (Araújo-Jorge et al. 2014).

As a part of the group of neglected diseases, not only do leishmaniases prevail in contexts affected by poverty, but they also contribute to maintaining inequalities. A notable geographic spread of these diseases can be observed, which is related to numerous environmental, social, political and historic conditions. With these changes, the relationship between the components of the diseases' transmission chain in urban environments has become much more complex (WHO 2010b; Ministério da Saúde do Brasil 2014, 2017; Rangel et al. 2014).

Due to global climatic and environmental changes, uncertainties regarding nature and its impact on the scale of local ecosystems add to the complexities of the new reality of urban Brazil. In that sense, new barriers to fighting leishmaniasis appear (Costa 2016).

Considering the epidemiologic scenario for these diseases, vector control through an integrated program—one that contemplates the interface between health and the environment—should be a priority measure. A program with this approach would focus on local communities' quality of life, aiming to mitigate social inequalities.

This approach to vector control—where epidemiologic surveillance can subsidize more targeted action—requires rational decision-making processes that optimize the use of available resources. These actions should be undertaken under the perspective of public health, using integrated interventions that are based on disease control tools, aiming to prevent mortality, reduce morbidity and interrupt the transmission cycle of the disease (WHO 2010b).

Bartonellosis

Bartonellosis, which has not yet been reported in Brazil, and is largely unknown by Brazilian health professionals, is also transmitted by phlebotomine sand flies. However, it has been observed near the Brazilian border and its impact might be aggravated by the development model adopted in the Amazon forest. The infection is caused by the bacteria *Bartonella bacilliformis*, the only known vectors of which are phlebotomines from the *Lutzomyia* genus (Alexander 1995). Townsend (1913, 1914) was the first to suggest that phlebotomines could act as vectors for *B. bacilliformis*, based on evidence derived from the presence of phlebotomine sand flies in areas where the disease occurred. Also known as "Carrion disease", Bartonellosis was originally observed in the inter-Andian valleys of Peru, Colombia and Ecuador. The following species of phlebotomines are suggested as potential vectors: *Lutzomyia (Helcocyrtomyia) peruensis*, *Pintomyia (Pifanomyia) columbiana* and *Pi. (Pif.) verrucarum* (Sherlock 1962; Vargas and Perez 1985; Gray et al. 1990; Galati 2003; Marcondes 2007).

Studies have presented evidence that *Pi. verrucarum* is the most likely vector of *B. bacilliformis* in the Rimac Valley in Peru, although there are other species of anthropophilic phlebotomine sand flies in the area, such as *Lu. noguchii* and *Lu. peruensis*, which could also be involved in the local epidemiology.

Pi. verrucarum does not seem to exist in Ecuador and the identification of a possible vector is still pending (Alexander et al. 1992; Young and Duncan 1994; Maroli et al. 2013).

The most likely vector in Colombia is *Pi. columbiana*, which is intimately related to *Pi. verrucarum*. This species is highly anthropophilic and can be found in all areas of Colombia in which Bartonellosis outbreaks have occurred (Gamarra 1964). Outbreaks continue to be notified in areas in which *B. bacilliformis*, and not *Lu. Verrucarum*, is endemic. This leads to the conclusion that other phlebotomines, or even other arthropods, might be acting as vectors in this coastal region (Alexander 1995; Maroli et al. 2013).

The possibility of Bartonellosis entering Brazil through the Amazon forest raises many concerns. The spread of the disease could be accelerated by the process of increased anthropic action that is occurring in the region. The south-western portion of the Brazilian Amazon possesses favorable conditions for the emergence of Bartonellosis, with environmental changes and migration flows affecting the region (http://www.pragas.com.br/noticias/destaques/infeccao_bacteriana.php). Due to environmental impact, some species of phlebotomine sand flies from Brazil could become apt to transmission.

For a better understanding of the epidemiology of Bartonellosis, it is essential to improve surveillance of this disease in countries that share a border with Brazil. Because Bartonellosis is directly related to deforestation and migration, the surveillance tools used need to be the most precise available, such as modelling for complex systems in which phlebotomine sand flies and possible reservoirs are included, and the environment being seen as a socio-natural, dynamic and complex element. The results of these studies may point to possible phlebotomine vectors, as well as assisting in strategic planning for border surveillance.

References

Aguiar GM, Vilela ML, Lima RB (1987) Ecology of the sandflies of Itaguaí, an area of cutaneous leishmaniasis in the State of Rio de Janeiro. Food preferences (Diptera, Psychodidae, Phlebotominae). Mem Inst Oswaldo Cruz. 82(4):583–584

Alexander BA (1995) review of bartonellosis in Ecuador and Colombia. Am J Trop Med Hyg. 52:354–359

Alexander JB, Takaoka H, Eshita Y, Gomez EA, Hashiguchi Y (1992) New records of phlebotomine sandflies (Diptera: Psychodidae) from Ecuador. Mem Inst Oswaldo Cruz. 87:123–130

Alvar J, Vélez ID, Bern C, Herrero M, Desjeux P, Cano J, Jannin J, den Boer M (2012) Who Leishmaniasis Control Team. Leishmaniasis worldwild and global estimates of this incidence. Plos One. 7(5):e35671

Araújo-Jorge T, Matraca M, Neto AM, Trajano V, d'Andrea P, Fonseca A (2014) Doenças Negligenciadas, Erradicação da Pobreza e o Plano Brasil sem Miséria. In: O Brasil sem Miséria. Editora Ministério do Desenvolvimento Social, Brasília, pp 1–23

Azar D, Nel A, Solignac M, Paicheler J-C, Bouchet F (1999) New genera and species of psychodoid flies from the Lower Cretaceous amber of Lebanon. Palaeontology. 42:1101–1136

Brazil RP, Brazil BG (2003) Biologia de flebotomíneos neotropicais. In: Rangel EF, Lainson R (eds) Flebotomíneos do Brasil. Fiocruz, Rio de Janeiro, pp 257–274

Brazil RP, Brazil BG (2014) Vetores na Leishmaniose Tegumentar Americana. In: Leishmanioses do Continente Americano. Editora FIOCRUZ, Rio de Janeiro, pp 193–200

Costa WA (2016) Ações Integradas de Vigilância e Controle de Vetores de Agentes Etiológicos das Leishmanioses. Experiência com Comunidades de Área Endêmica de Leishmaniose Tegumentar Americana, Campus FIOCRUZ Mata Atlântica, Jacarepaguá, Rio de Janeiro, RJ, Brasil. Ministério da Saúde - Fundação Oswaldo Cruz – Instituto Oswaldo Cruz Doutorado em Programa de Pós-Graduação de Biodiversidade e Saúde, p 122

de Oliveira GC, Carneiro LM, Menoni SMF, Jurado SR (2012) Interrelação entre Leishmaniose Visceral e Saneamento Básico. CONVIBRA. 3–7. Disponível em. http://www.convibra.org/upload/paper/2013/72/2013_72_7664.pdf

Desjeux P (2004) Leishmaniasis: current situation and new perspectives. Comp Imm Microb Infec Dis. 27:305–318

Galati EAB (2003) Morfologia e taxonomia: classificação de Phlebotominae, pp 23–51. In: Rangel EF, Lainson R (eds) Flebotomíneos do Brasil. Editora Fiocruz, Rio de Janeiro, pp 368

Gamarra A (1964) Geografia medica de la verruga peruana. Verruga peruana en la zona fronteriza de Canchis-Chinchipe. Rev Sanid Polic 24:50–53

Gray GC, Johnson AA, Thornton SA, Smith WA, Knobloch J, Kelley PW, Escudero LO, Huayda MA, Wignall FS (1990) An epidemic of Oroya fever in the Peruvian Andes. Am J Trop Med Hyg 42(3):215–221

Grimaldi G, Tesh BR, Mc Mahon-Pratt D (1989) A review of the geographic distribution and epidemiology of leishmaniasis in the New World. Am J Trop Med Hyg. 41:687–725

Hennig W (1972) Psychodidae (Phlebotominae), mit einer kritischen Übersicht über das phylogenetische System der familie und die bisher beschriebenen Fossilien (Diptera). Stuttgarter Beiträge Naturk. 241:1–69

Killick-Kendrick R (1999) The biology and control of phlebotomine sand flies. Clin Dermatol. 17(3):279–289

Lainson R, Shaw JJ (1972) Leismaniasis in the New World: Taxonomic problems. Br Med Bull. 28:44–48

Lainson R, Shaw JJ (1989) *Leishmania* (*Viannia*) *naiffi* sp.n. a parasite of the armadillo, *Dasypus novemcinctus* (L.) in Amazon Brazil. Ann Parasitol Hum Comp. 64:3–9

Lainson R, Shaw JJ, Souza AAA, Silveira FT, Braga RR, Ishikawa EAY (1994) The dermal leishmaniasis of Brazil, with special reference to the eco-epidemiology of the disease in Amazonia. Mem Inst Oswaldo Cruz. 89:435–443

Lima LC, Marzochi MCA, Sabroza PC, Souza MA (1988) Observações sobre a leishmaniose tegumentar, cinco anos após profilaxia. Rev Saúde Públ S Paulo. 22(1):73–77

Malafaia G (2009) A importância do encorajamento de estudos sobre as doenças tropicais negligenciadas. Rev Soc Bras Med Trop. 42(5):609–610

Marcondes CB (2007) A proposal of generic and subgeneric abbreviations for Phlebotomine sand flies (Diptera: Psychodidae: Phlebotominae) of the world. Entomol News. 118:351–356

Maroli M, Feliciangeli MD, Bichaud L, Charrel RN, Gradoni L (2013) Phlebotomine sandflies and the spreading of leishmaniases and other diseases of public health concern. Med Vet Entomol. 27(2):123–147

Ministério da Saúde do Brasil (2014) Manual de Vigilância e Controle da Leishmaniose Visceral. Secretaria de Vigilância em Saúde (SVS). Editora MS, Brasília, p 120

Ministério da Saúde do Brasil (2017) Manual de Leishmaniose Tegumentar. Secretaria de Vigilância em Saúde (SVS). Editora MS, Brasília, p 177

Rangel EF (1995) Tropical diseases, society and the environment. SAREC Documentation/TDR, pp 103–110

Rangel EF, Lainson R (2009) Proven and putative vectors of American cutaneous leishmaniasis in Brazil: aspects of their biology and vectorial competence. Mem Inst Oswaldo Cruz. 104(7):937–954

Rangel EF, Costa SM, Carvalho BM (2014) Environmental changes and the geographic spreading of American cutaneous leishmaniasis in Brazil. In: Claborn D (ed) Leishmaniasis – trends in epidemiology, diagnosis and treatment. InTech, Rijeka, pp 1–25

Sabroza PC (1981) O domicílio como fator de risco na leishmaniose tegumentar americana: estudo epidemiológico em Jacarepaguá, município do Rio de Janeiro [Dissertação de Mestrado]. Escola Nacional de Saúde Pública Sérgio Arouca, Fundação Oswaldo Cruz, Rio de Janeiro, p 200

Shaw J, Rosa AT, Souza A, Cruz AC (2003) Os flebotomíneos brasileiros como hospedeiros e vetores de determinadas espécies. In: Rangel EF, Lainson R (eds) Flebotomíneos do Brasil. FIOCRUZ, Rio de Janeiro, pp 337–352

Sherlock IA (1962) Note on natural breeding places of Phlebotomus in Salvador, Bahia (Diptera, Psychodidae). Rev Bras Biol. 22:125–127

Shimabukuro PHF, de Andrade AJ, Galati EAB (2017) Checklist of American sand flies (Diptera, Psychodidae, Phlebotominae): genera, species, and their distribution. Zookeys. 8(660):67–106

Silveira FT, Shaw JJ, Braga RR, Ishikawa EEA (1987) Dermal leishmaniasis in the amazon region of Brazil: *Leishmani* (*V.*) *lainsone* sp.n., a new parasite from the state of Pará. Mem Inst Oswaldo Cruz. 82:289–292

Toledo CRS, de Almeida AS, Chaves SAM, Sabroza PC, Toledo LM, Caldas JP (2017) Vulnerabilidade à transmissão da leishmaniose visceral em área urbana brasileira. Rev S Publ. 51:49

Townsend CHT (1913) Preliminary characterization of the vector of verruga, *Phlebotomus verrucarum* sp. nov. Insecutor Inscitiae Menstruus. 1:107–109

Townsend CHT (1914) Resumen de las labores en le Peru sobre él *Phlebotomus verrucarum* y su agencia en la trasmision de lá verruga. An Zool Aplicada (Lima) 1:44–64

Vargas CM, Perez E (1985) La enfermidade de Carryon y leishmaniasis andina em la region de Conchucos, Distrito de Chavin, San Marcos e Huantar, Provincia de Huari. Departamento de Ancash. Diagnótico. 16:5–12

Vilela ML, Afonso MM, Costa SM, Costa WA, Rangel EF (2014) *Lutzomyia (Lutzomyia)* longipalpis: fatores associados ao processo de expansão e urbanização da leishmaniose visceral americana. In: Leishmanioses do Continente Americano. Editora FIOCRUZ, Rio de Janeiro, pp 183–192

WHO (2010a) Control of the leishmaniases: report of a meeting of the WHO Expert Committee on the Control of Leishmaniases, Geneva. WHO Technical Report Series, n° 949, pp 1–202

WHO (2010b) Trabalhando para superar o impacto global de doenças tropicais negligenciadas. Primeiro relatório da OMS sobre doenças tropicais Negligenciadas, Geneva, pp 1–188

Young DG, Duncan MA (1994) Guide to the identification and geographic distribution of Lutzomyia sand flies in Mexico, the West Indies, Central and South America (Diptera: Psychodidae). Memoirs of the American Entomological Institute no. 54. Associated Publishers, Gainesville, p 881

Phlebotominae (Diptera, Psychodidae): Classification, Morphology and Terminology of Adults and Identification of American Taxa

Eunice A. B. Galati

- SUMMARY
- **Classification of Phlebotominae**
- History
- Classification of the species; stages and sexes described and geographical distribution
- Siglas of the American countries and Brazilian states
- Figures 1-4. Cladograms of Phlebotominae
- Figure 1. Tribes and subtribes
- Figure 2. Sergentomyiina: genera, subgenera and series of species
- Figure 3 Lutzomyiina: genera, subgenera, groups of species and series of species
- Figure 4. Psychodopygina: genera, subgenera and groups of species
- **Morphology and Terminology and Identification of the American taxa**
- Morphology and Terminology of adults
- Identification of the American taxa
  - Psychodidae: characterization
  - Phlebotominae: characterization
- Identification keys:
- Subtribes
  - HERTIGIINA: genera
  - BRUMPTOMYIINA: genera
  - SERGENTOMYIINA: genera, subgenera and series of species
  - LUTZOMYIINA: genera, subgenera, groups and series of species
  - PSYCHODOPYGINA: genera, subgenera, groups and series of species

E. A. B. Galati (✉)
Faculdade de Saúde Pública, Universidade de São Paulo, São Paulo, Brazil
e-mail: egalati@usp.br

E. F. Rangel, J. J. Shaw (eds.), *Brazilian Sand Flies*,
https://doi.org/10.1007/978-3-319-75544-1_2

- Species
  - *Edentomyia piauiensis*
- HERTIGIINA
  - *Warileya*
  - *Hertigia*
- BRUMPTOMYIINA
  - *Brumptomyia*
  - *Oligodontomyia*
- SERGENTOMYIINA
  - *Deanemyia*
  - *Micropygomyia*
    - *(Silvamyia)*
    - *(Sauromyia)*
      - Atroclavata Series
      - Oswaldoi Series
    - (*Coquillettimyia*)
      - Chiapanensis Series
      - Vexator Series
    - (*Micropygomyia*)
      - Cayennensis Series
      - Pilosa Series
    - *Incertae Sedis*
      - *Micropygomyia xerophila*
- LUTZOMYIINA
  - *Sciopemyia*
  - *Lutzomyia*
    - (*Helcocyrtomyia*): series
      - Osornoi Series
      - Peruensis Series
      - Sanguinaria Series
    - (*Castromyia*)
    - (*Lutzomyia*)
    - (*Tricholateralis*)
    - *Incertae Sedis*

    - *Lutzomyia chotensis*
    - *Lutzomyia ignacioi*
    - *Lutzomyia manciola*
    - *Lutzomyia ponsi*
    - *Lutzomyia tanyopsis*

- *Migonemyia*
  - (*Migonemyia*)
  - (*Blancasmyia*)
- *Pintomyia*
  - Isolated Species: *Pintomyia* (*Pifanomyia*) *nuneztovari*
  - Isolated Species: *Pintomyia* (*Pifanomyia*) *diamantinensis*
  - (*Pintomyia*)
  - (*Pifanomyia*)
    - Evansi Series
    - Monticola Series
    - Pacae Series
    - Pia Series
    - Serrana Series
    - Townsendi Series
  - Verrucarum Series
  - *Incertae Sedis*
    - *Pintomyia diamantinensis*
    - *Pintomyia maracayensis*
    - *Pintomyia nuneztovari*
    - *Pintomyia rangeliana*
    - *Pintomyia* sp. de Anchicaya
    - *Pintomyia naiffi*
- *Dampfomyia*
  - (*Dampfomyia*)
  - Grupo Delpozoi
  - (*Coromyia*)
  - Incertae Sedis
    - *Dampfomyia caminoi*
- *Expapillata*
- *Pressatia*
- *Trichopygomyia*
- *Evandromyia*
  - (*Aldamyia*)
  - (Evandromyia)

- Species of the Americas
- Tribes
- Subtribes
- Genera
- Subgenera of the Americas
- Groups of species of the Americas
- Series of species of the Americas

- **Indices of Synonyms junior:**
  - Species of the Americas
  - TribeGenera of the Americas
  - Genera of Americas
  - Subgenera of the Americas

Classification of Phlebotominae

The development of our knowledge of the evolution of organisms has been highly dynamic and has led to alterations in the systematics and consequently in the taxonomy of living beings. These alterations certainly cause problems for beginners in the field or those who are dedicated to other branches of science and need this knowledge, especially when the modifications made take place at the genus level, because they imply changes in the nomenclature of the species (Lewis et al. 1977; Artemiev 1991). However, this is an inevitable aspect of science that serves as a stimulus in the search for a system that will truly reflect the affinities among the taxa and contribute to the understanding of the relationship between organisms, which is especially important in the case of vectors and parasites. Thus, we judged it opportune to present a historical sketch of the classification of the Phlebotominae in order to provide the basic elements for those who have to deal with changes of nomenclature.

Barretto (1961) and Duckhouse (1973), in their systematic studies of the American subfamilies and genera, presented revisions of Psychodidae. Lewis et al. (1977) also greatly developed this theme, including a bibliography of the 18th and 19th centuries, of difficult access in approaching the question of the priority of the name Psychodidae or Phlebotomidae raised by Abonnenc and Leger (1976a). They concluded that the former has priority, as it was used by Newman (1834) as Psychodites. The principal aspects of these revisions and other more recent ones that relate to Phlebotominae, mainly those of the Americas, are presented below.

Bibio papatasi Scopoli, 1786, was the first Psychodidae species based on a female from the Lombardy region of northern Italy. Scopoli maintained the second name of the species by which the insect was popularly known (Scopoli 1786; Grassi 1907).

Psychoda Latreille, 1796, was the first genus of Psychodidae described, without designation of the type species, though compared with *B. papatasi*.

Rondani (1840) proposed the genus *Flebotomus*, with the type species *F. papatasi* (Scopoli, 1786), and placed it in the tribe that he called Flebotomidae, in the Flebotominae family. Rondani (1843) included the family Flebotominae in the tribe Tipulidae and described *Hebotomus minutus* (probable typographical error for *Flebotomus*). Agassiz (1846) in the "*Nomenclatoris Zoologici Index Universalis*" amended *Flebotomus* to *Phlebotomus*.

Bigot (1854) first included *Phloebotomus* (sic), *Psychoda*, *Nemapalpus* Macquart, 1838, *Trichomyia* Haliday, 1839, and the fossil genus *Posthon* Loew, 1845, in Psychodidae.

Coquillett (1907) described the first American sandflies, *Flebotomus vexator* and *F. cruciatus*, and disagreed with the change of name to *Phlebotomus*.

Newstead (1911) proposed the first division of the *Phlebotomus*, using the setae that covered the abdomen as differential characteristics. In 1914, he suggested a division based on the characteristics of the male genitalia as well as the wing vein pattern, palpal and antennal formula.

França (1919) studied the morphology of the male genitalia in detail and suggested several ratios for phlebotomine identification; he proposed the first division of the European and African species of *Phlebotomus* into two subgenera, *Phlebotomus*, *s. str.* and *Newsteadia* (without type species designation), with the inclusion of the species *P. sergenti* Parrot, 1913; *P. ingrami* Newstead, 1914; *P. minutus*; *P. simillimus* Newstead, 1914; *P. perniciosus* Newstead, 1911; and *P. mascittii* Grassi, 1908.

França and Parrot (1920) maintained the above division. They replaced the name *Newsteadia* – already in use – by *Sergentomyia* (still without type species designation) and presented the diagnosis of both the subgenera on the basis of the male genitalia and commented on the existence of an intermediate group – which could include *P. malabaricus* Annandale, 1910 – between the two, forming a third subgenus, *Neophlebotomus*.

França (1920) reviewed the three Brazilian species described by Lutz and Neiva (1912) – *P. longipalpis*, *P. intermedius* and *P. squamiventris* – and described *P. migonei* França, 1920, from Paraguay. He commented that among known American species, the palpal formula with the fifth segment shorter than the third had not been observed among Old World species. Further, of those with the fifth palpal segment longer than the third, *P. longipalpis* was the most interesting species, because its palpal formula was similar to that of *P. papatasi*; however, its genitalia are completely different from those of the other species of the genus *Phlebotomus*, thus justifying its designation as the species-type of a new subgenus *Lutzia* França, 1920. Then beyond this subgenus, he adopted the two others he had previously proposed in 1919, *Phlebotomus s. str.* and *Newsteadia* França, 1919, as members of the genus *Phlebotomus*; however, as the name *Newsteadia* was already in use, he changed it to *Sergentomyia* França, 1920, and designated *P. minutus* Rondani, 1843, as its type species. The differentiation of the three subgenera was based on the male genitalia. He did not mention *Neophlebotomus*.

França and Parrot (1921), in a new classification based on the wing indices and male genitalia, accepted the three former subgenera and created another two,

Brumptomyia (without the designation of the type species), which included *P. brumpti* Larrousse, 1920, and *P. vexator*, New World species, and *Prophlebotomus* (without the designation of the type species) for *P. perturbans* Meijere, 1908, *P. minutus*, *P. antennatus* Newstead, 1920, and *P. fallax* Parrot, 1920. The American species described previously – with the exception of *P. cruciatus*, known only by the female – were included in *Sergentomyia*.

Alexander (1920) created the subfamily Bruchomyiinae for *Bruchomyia*, gen. n., with the type species *Bruchomyia argentina* Alexander, 1920. Edwards (1921) created the subfamily Nemopalpinae for the genus *Nemopalpus* Macquart, 1838, which had previously been considered among the Phlebotominae. Tonnoir (1922) included *Nemopalpus* and *Bruchomyia* in Phlebotominae; in this Tonnoir was followed by Alexander (1928, 1929). This latter author, in 1940, described the genus *Eutonnoiria* (type species: *Bruchomyia edwardsi* Tonnoir, 1839) for Africa.

França (1924) replaced the name *Lutzia*, already used for Culicidae, with *Lutzomyia*. Dyar and Nunez-Tovar (1926/1927) suggested replacing *Lutzia* with *Fransaia*, Cordero et al. (1928), with *Lutziomyia*; and Strand (1932), with *Lutziola*.

Sinton (1928) rejected França and Parrot's (1921) classification. He alleged that the basic character used, the wing index (*alpha/beta*), was so highly variable that closely related species could be classed as distinct subgenera. He separated the Old World species into three divisions based on Newstead's observations, on the setae that covered the abdominal tergites, and on those of Adler and Theodor (1926), on the spermathecae. The first division included the species with erect setae and ringed spermathecae; the second, those with recumbent setae and smooth spermathecae; and the third, the intermediate species (Costa Lima 1932, Theodor 1948).

Dyar (1929), updating knowledge of the American phlebotomines, made the following commentary on the subdivisions proposed by França and Parrot (1920, 1921), "While these subdivisions are not of much importance, it may be convenient to make use of them". He adopted *Brumptomyia* (designating *P. brumpti* as type species), *Lutzomyia, Neophlebotomus* (he accepted *P. malabaricus* Annandale, 1910, as type species, thus designated by Larrousse, 1921, in an unpublished thesis presented to the Paris Medical School) and proposed the creation of *Shannonomyia* (type species: *P. panamensis* Shannon, 1926). The American species considered by França and Parrot (1921) were included under *Neophlebotomus*. He designated *P. minutus* as the type species of *Prophlebotomus* França and Parrot, 1921, which led him necessarily to include it as synonymous with *Sergentomyia*.

Adler and Theodor (1929) proposed that the phlebotomines should constitute a family, as suggested by Walker (1851), because this was the only group whose females present jaws; however, Tonnoir (1933) disagreed with those authors when he described the genus *Horaiella* (type species: *H. prodiginosa* Tonnoir, 1933), whose females also have jaws as well as other characteristics that serve as links between the phlebotomines and other psychodides. He criticized the creation of monogeneric families and included the genus among the Trichomyiinae.

Nitzulescu (1931) considered that França and Parrot's 1921 classification did not obey the natural division of the genus *Phlebotomus*. He brought together just those internal morphological characteristics studied by Adler and Theodor (1926, 1929) and divided the genus into five subgenera: (1) *Phlebotomus*, *s. str.* and four new ones: (2) *Larroussius* (type species: *P. major* Annandale, 1910), (3) *Adlerius* (type species: *P. chinensis* Newstead, 1916), (4) *Sintonius* (type species: *P. hospittii* Sinton, 1924) and (5) *Brumptius* (type species: *P. minutus*) – the first three without cibarial armature and the last two with it, and the first four with ringed spermathecae and the fifth with smooth spermathecae.

Theodor (1932) subdivided the Phlebotominae into three tribes (without naming them), two for the Old World species and one for America. The tribes were subdivided into species groups.

Costa Lima (1932) was of the same opinion as Dyar (1929) as regards placing *Prophlebotomus* as synonymous with *Sergentomyia* and in the revalidation of *Neophlebotomus*; in the revision of the American species he created the subgenus *Pintomyia* (type species: *Phlebotomus fischeri* Pinto, 1926); the others were not included in any other subgenus of *Phlebotomus*.

Mangabeira (1941a, b, c) adopted the subgenera *Lutzomyia, Brumptomyia, Shannonomyia, Pintomyia* and described three new ones, namely, *Evandromyia* (type species: *Flebotomus infraspinosus* Mangabeira, 1941), *Psychodopygus* (type species: *F. unisetosus* Mangabeira, 1941) and *Viannamyia* (type species: *F. tuberculatus* Mangabeira, 1941). In 1942, he further described *Pressatia* (type species: *F. triacanthus* Mangabeira, 1942) and *Castromyia* (type species: *Phlebotomus castroi* Barretto & Coutinho, 1941).

Dampf (1944) accepted Dyar's (1929) opinion as regards *Neophlebotomus* and included *Prophlebotomus* and *Brumptius* as synonymous with *Sergentomyia*. Concerning the American species, "Dampf appears to write as if he accepted various previously created subgenera, among them being *Brumptomyia*, *Shannonomyia*, *Castromyia* and *Pintomyia*", and expressed the opinion that the phlebotomines should constitute a family distinct from the psychodides (Barretto 1955).

Addis (1945) created the American subgenus *Dampfomyia* (type species: *Phlebotomus anthophorus* Addis, 1945).

Theodor (1948) adopted the category of subfamily for the phlebotomines and, fundamentally using the cibarial armature, first proposed the division of the Phlebotominae into four genera, namely, *Phlebotomus* and *Sergentomyia* for the Old World species and *Brumptomyia* and *Lutzomyia* for those of the New World. He considered *Prophlebotomus* and *Neophlebotomus* to be synonymous with *Sergentomyia*.

Hertig (1948) described the genus *Warileya* (type species: *W. phlebotomanica* Hertig, 1948) and Fairchild (1949) *Hertigia* (type species: *H. hertigi*, Fairchild, 1949), both of the neotropical region.

Barretto (1950a), discussing the authorship of the genus *Phlebotomus*, attributed by the majority of authors to Rondani (1840), considered Rondani and Berté as the authors of this genus, because the authorship was clearly mentioned on page 12 of the original publication. He analyzed the polemical question regarding the change of the spelling of *Flebotomus* to *Phlebotomus* raised by Coquillett (1907), and concluded that a spelling error had occurred, such that *Phlebotomus* was the correct form.

The International Commission on Zoological Nomenclature (ICZN) (1950) agreed "(1) to amend to *Phlebotomus* the generic name originally published by Rondani, 1840 *Flebotomus* (Classe Insecta, Order Diptera)" and (2) "to place the generic name *Phlebotomus* Rondani, 1840 (type species: *Bibio papatasi* Scopoli, 1786, by monotypy) on the *Official List of Generic Names in Zoology* and the trivial name *papatasi* Scopoli, 1786 (as originally published in the binomial combination *Bibio papatasi*), on the *Official List of Specific Trivial Names in Zoology*". These decisions were published as Opinion 256 (ICZN, 1954).

Parrot (1951) did not accept the promotion of *Phlebotomus s. lat.* to the subfamily Phlebotominae, nor its division into genera. He argued that as the genus *Phlebotomus*, endowed with extreme plasticity, was still undergoing a process of evolution and also because phlebotomology was a young science, the adoption of a minimum of formal categories was recommendable. He disagreed with placing *Prophlebotomus* and *Neophlebotomus* as synonymous with *Sergentomyia*, justifying his position by affirming that the drawing of *Sergentomyia*'s genitalia presented by França and Parrot (1920) completed the definition of the subgenus and that this drawing unquestionably represented the genitalia of *P. perniciosus* Newstead, so this species would be the type "indicated" for the subgenus *Sergentomyia.* He concluded that (1) the designation of *P. minutus* as type species of *Sergentomyia* proposed by França 1920 was incorrect; (2) *Prophlebotomus*, with the type species *P. minutus*, was a valid subgenus, *Sergentomyia* (type species: *P. perniciosus*) and *Larroussius* (type species: *P. major*) being synonymous with it; (3) *Neophlebotomus* was considered a *nomen nudum*.

Barretto (1955) judged that despite the accumulated knowledge of the phlebotomines, it would at that time be difficult to propose a rational and satisfactory classification. On the other hand, the distribution in groups would be advantageous because of the great number of species (166 for the Old World and 199 for the Americas, at that point). However, he would prefer a classification with few categories. He was in accordance with Theodor (1948) as to the elevation of some subgenera to the genus level, considering its practical utility; however, he disagreed with the classificationTheodor proposed for the Western hemisphere phlebotomines. Barretto argued that the phlebotomines of this latter hemisphere included not only the genera *Brumptomyia* and *Lutzomyia*; there was "at least one highly homogenous group of species closely related to an exotic group of *Sergentomyia*". The greater part of the American species would be included in this latter subgenus. *Lutzomyia* would be restricted, at the most, to a reduced number of species; however, the general characteristics of these species were those of the genus *Sergentomyia*, though they differed slightly from those of its type species. Thus he considered the American phlebotomines as consisting of the genera *Brumptomyia*, *Warileya* and *Sergentomyia*, this last having *Lutzomyia* as its synonym. He made no reference to *Hertigia*. He disagreed with Parrot's (1951) designation of *P. perniciosus* as the type species of *Sergentomyia* and also considered *Neophlebotomus* a *nomen nudum*. He argued that, in accordance with the rules of the ICZN, the designation of *P. malabaricus* as type species proposed by Larrousse (1921) and accepted by Dyar (1929) was incorrect, as França and Parrot (1920), when proposing this subgenus, doubted the correctness of its inclusion. They agreed with Dyar (1929) and Theodor (1948) in considering *Prophlebotomus* as a synonym of *Sergentomyia*. Finally they considered *Brumptius* as synonymous with *Sergentomyia*.

Fairchild (1955) divided Psychodidae into the subfamilies Psychodinae, Trichomyiinae and Phlebotominae. This last was divided into two tribes, namely, Bruchomyiini, with the genera *Nemopalpus*, *Bruchomyia* and *Eutonnoiria*, and Phlebotomini with the genera *Phlebotomus*, *Warileya* and *Hertigia*. On the basis of male genitalia he divided *Phlebotomus* into five subgenera: (1) *Phlebotomus* and (2) *Sergentomyia* containing Old World species and (3) *Psychodopygus*, (4) *Viannamyia* and (5) *Brumptomyia* containing New World species. The American species with short palpi were included in *Viannamyia* and in *Psychodopygus* (= *Shannonomyina* Pratt, 1947); the latter was divided into two species groups, Intermedius and Panamensis. The other species were included in *Brumptomyia*, being divided into nine species groups, some of them in series of species: (1) the Brumpti group, with the series (1.1) Brumpti and (1.2) Vexator; (2) the Vespertilionis group; (3) the Triacanthus group (= *Pressatia*), with the series (3.1) Triacanthus and (3.2) Fischeri; (4) the Anthophorus group (= *Dampfomyia*); (5) the Cayennensis group; (6) the Shannoni group; (7) the Cruciatus group, with 11 series of species: (7.1) Cruciatus, (7.2) Verrucarum, (7.3) Migonei, (7.4) Walkeri, (7.5) Castroi (= *Castromyia*), (7.6) Atroclavatus, (7.7) Baiyti, (7.8) Longipalpis (= *Lutzomyia*), (7.9) Infraspinosus (= *Evandromyia*), (7.10) Castanheirai and (7.11) Servulolimai; (8) the Longispinus group and (9) the Alphabeticus group.

Barretto (1961) included in Psychodidae the subfamilies Bruchomyiinae and Phlebotominae, in addition to others.

Barretto (1962) accepted for the American phlebotomines the genera *Warileya, Brumptomyia* and *Lutzomyia*. This last was divided into 15 subgenera, seven of which had previously been proposed: (1) *Lutzomyia s. str.*, with the groups: (1.1) Longipalpis and (1.2) Cavernicola; (2) *Pintomyia*; (3) *Evandromyia*; (4) *Psychodopygus*, with seven groups: (4.1) Unisetosa, (4.2) Geniculata, (4.3) Bispinosa, (4.4) Panamensis, (4.5) Davisi, (4.6) Arthuri and (4.7) Matosi; (5) *Viannamyia*; (6) *Pressatia*; (7) *Dampfomyia* and eight new subgenera; (8) *Micropygomyia* (type species: *Phlebotomus cayennensis* Floch & Abonnenc, 1941); (9) *Sciopemyia* (type species: *P. nordestinus* Mangabeira, 1942); (10) *Helcocyrtomyia* (type species: *P. peruensis* Shannon, 1929), which was subdivided into the groups (10.1) Peruensis, (10.2) Oswaldoi, (10.3) Vexatrix and (10.4) Sanguinarius; (11) *Trichophoromyia* (type species: *P. ubiquitalis* Mangabeira, 1942), with three groups: (11.1) Ubiquitalis, (11.2) Brachypygus and (11.3) Ininii; (12) *Coromyia* (type species: *P. vespertilionis* Fairchild & Hertig, 1947), with five groups: (12.1) Vespertilionis, (12.2) Vesicifera, (12.3) Christophei, (12.4) Serrana and (12.5) Verrucarum; (13) *Trichopygomyia* (type species: *P. longispinus* Mangabeira, 1942); (14) *Nyssomyia* (type species: *P. intermedius* Lutz & Neiva, 1912); and (15) *Psathyromyia* (type species: *P. shannoni* Dyar, 1929), subdivided into three groups: (15.1) Shannoni, (15.2) Volcanensis and (15.3) Aragaoi.

Ortiz and Scorza (1963) created the subgenus *Pifanomyia* (type species: *Flebotomus serranus* Damasceno & Arouck, 1949) for an American group of species.

Rohdendorf (1964) raised the phlebotomines to family level, arguing that their females are hematophagous and their larvae live in an environment distinct from that of other psychodides.

Theodor (1965) clarified the principles that had guided his 1948 classification of the Old World species in order to justify his disagreements with Fairchild (1955) and proposed a new classification for the New World phlebotomines. He highlighted the importance of the use of the cibarium and male and female genitalia in distinguishing the groups and adopted the genera *Hertigia, Warileya, Brumptomyia* and *Lutzomyia*. This last he divided into the eight subgenera previously proposed by Barretto (1962) and 16 groups of species. The subgenera were (1) *Lutzomyia, s. str.*, subdivided into two series: (1.1) Longipalpis and (1.2) Renei; (2) *Psychodopygus,* with the series (2.1) Unisetosa, (2.2) Panamensis and (2.3) Arthuri; (3) *Evandromyia*; (4) *Viannamyia*; (5) *Castromyia*; (6) *Pressatia*; (7) *Pintomyia* and (8) *Dampfomyia*. The groups of species were (1) Cruciata; (2) Migonei, with two series: (2.1) Migonei and (2.2) Evandroi and six species of doubtful classification; (3) *Verrucarum*, with two series of species: (3.1) Verrucarum and (3.2) Serrana (= *Pifanomyia*); (4) *Vexatrix* (= *Helcocyrtomyia, partim*), with two series of species: (4.1) Vexatrix and (4.2) Peruensis; (5) Vespertilionis (= *Coromyia*), with two series of species: (5.1) Vespertilionis and (5.2) Deleoni; (6) Intermedia (= *Nyssomyia*); (7) Lanei; (8) Auraensis (= *Trichophoromyia, partim*); (9) Aragaoi (= *Psathyromyia, partim*), with two series of species: (9.1) Aragaoi and (9.2) Brasiliensis; (10) Cayennensis (= *Micropygomyia*), with two series of species: (10.1) Cayennensis and (10.2) Chiapanensis; (11) Oswaldoi (= *Helcocyrtomyia, partim*); (12) Shannoni (= *Psathyromyia, partim*); (13) Longispina (= *Trichopygomyia*); (14) Castanheirai (= *Trichophoromyia, partim*); (15) Pilosa; and (16) Baityi. In addition to these he considered 9 species as isolated and 11 as being of doubtful classification or inadequately described.

Martins and Silva (1965) proposed the subgenus *Barretomyia* (type species: *Phlebotomus tupynambai* Mangabeira, 1942), correcting its name to *Barrettomyia* in 1968.

Forattini (1971, 1973) accepted the subfamilies Phlebotominae and Bruchomyiinae. *Hertigia*, as Barretto (1961) had proposed, was included in Bruchomyiinae. Forattini considered the genera *Warileya* and *Brumptomyia* and that the species that were included in *Lutzomyia* by Theodor (1948, 1965) and Barretto (1962) were divided into five genera: (1) *Lutzomyia*, (2) *Psychodopygus*, (3) *Pintomyia*, (4) *Pressatia* and (5) *Viannamyia*. The genus *Lutzomyia* was subdivided into four subgenera: (1) *Lutzomyia, s. str.* (= *Evandromyia*; *Helcocyrtomyia, partim*; *Trichopygomyia, partim*; *Coromyia, partim* and *Pifanomyia)*; (2) *Coromyia*; (3) *Trichopygomyia* (= *Sciopemyia, partim*; *Helcocyrtomyia, partim* and *Psathyromyia, partim*); (4) *Barrettomyia* (= *Castromyia*). Several species were considered as *incertae sedis*. The genus *Psychodopygus* was subdivided into two subgenera: (1) *Psychodopygus* (= *Shannonomyina* Pratt) and (2) *Trichophoromyia* (= *Nyssomyia*, *Psathyromyia, partim*; *Sciopemyia, partim* and *Helcocyrtomyia, partim*). This classification was not approved because the majority of the species included were grouped on the basis of artificial characteristics (Young and Fairchild 1974; Lewis et al. 1977).

Hennig (1972), in a phylogenetic study of Psychodidae, added Phlebotominae and Bruchomyiinae to the "Unterfamiliengruppe" Phlebotomoinea. The genus *Phlebotomites* (type species: *Phlebotomites brevifilis* Hennig, 1972) was created to accommodate two fossil species of the Cretaceous period found in Lebanese amber. In his classification this genus is included in an intermediate position between *Warileya* and Phlebotominae, *s. str.* (*Sergentomyia*, *Phlebotomus*, *Lutzomyia* and *Brumptomyia*).

Duckhouse (1973) considered Phlebotominae and Bruchomyiinae as belonging to the Psychodidae.

Young and Fairchild (1974) proposed a classification very similar to that of Theodor (1965), although with some modifications in the genus *Lutzomyia*: (1) five groups of species were created: (1.1) Delpozoi, for two species considered isolated; (1.2) Gasparviannai, for species that belonged to the groups Cruciata, *partim* and Migonei, *partim*; (1.3) Rupicola, for *L. rupicola* Martins, Godoy & Silva, 1962 and *L. correalimai* Martins, Coutinho & Luz 1970, both considered of doubtful position; (1.4) Dreisbachi, for species that belonged to the groups Castanheirai, *partim* and Auraensis, *partim*; (1.5) Atroclavata, for *L. atroclavata* and *L. venezuelensis* (Floch & Abonnenc, 1948) (= *P. zuliaensis* Floch & Abonnenc, 1948) both considered isolated; (2) the Cruciata group was not considered and its species included in the series Cruciata of the subgenus *Lutzomyia*; (3) the Migonei group was amplified and the name of one of its series modified, being then composed as follows: (3.1) the Migonei series, (3.2) the Walkeri series (= the Evandroi series of Theodor) and (3.3) the Costalimai series (= *Castromyia, partim* of Theodor); (4) the subgenus *Evandromyia* was divided into two series: (4.1) Infraspinosa and (4.2) Monstruosa; (5) *Trichophoromyia* Barretto was adopted (species of the Castanheirai group, *partim* and of Theodor's Auraensis group were included); (6) *Nyssomyia* Barretto, 1962 (= Theodor's Intermedia group) was adopted; (7) *L. trinidadensis* (isolated species in Theodor's classification) was included in the Oswaldoi group and (8) *L. castroi* (type species of *Castromyia*) was considered as an isolated species.

Abonnenc and Leger (1976a) adopted the category of family for the phlebotomines. They divided Phlebotomidae into three subfamilies: Euphlebotominae, Neophlebotominae and Disphlebotominae; as these were in disagreement with the rules of the ICZN they later changed them to Phlebotominae, Lutzomyiinae and Hertigiinae, respectively (Abonnenc and Leger 1976b).

Lewis et al. (1977), for the American species, followed Young and Fairchild's (1974) system, with few modifications: *Hertigia* was considered as a subgenus of *Warileya*, and they created the group Saulensis, considered an isolated species in this latter classification.

Martins et al. (1978) proposed a classification similar to the two preceding ones. However, with the exception of *Castromyia* and *Sciopemyia,* these authors adopted the subgenera previously proposed. The species of the former were included in the Amarali group and those of the latter together with those of the Rupicola group were considered as isolated; the Vexator group was considered the Vexator series of the subgenus *Helcocyrtomyia*. These authors adopted the Cruciata group. The Migonei group became more restricted; the Walkeri series was included as a group of species; the taxa of the Aragaoi group of other classifications were included in the Brasiliensis group; the Baityi group was denominated the Gorbitzi group; the Pilosa group was denominated the Chassigneti group; the Saulensis group was included as the Saulensis series of the subgenus *Coromyia,* and *Hertigia* was considered a genus.

Vargas (1978) proposed the creation of two subgenera for the genus *Lutzomyia*, *Forattiniella* (type species: *Phlebotomus lutzianus* Costa Lima, 1932) and *Aguayoi* (type species: *Lutzomyia dispar* Martins & Silva 1963).

Ready et al. (1980), based on the characteristics of adults, eggs, larvae and ecological niche, considered *Psychodopygus, s. str.* a distinct genus, just as Forattini (1971, 1973) did. Ryan (1986) also adopted *Psychodopygus* as a genus in a review of phlebotomines of Pará state, Brazil.

Young and Arias (1984) created the Microps group in the genus *Lutzomyia.*

Leng (1987) described a new genus of Phlebotominae for Asia called *Chinius* (type species *C. junlianensis* Leng 1987) with primitive characteristics.

Artemiev (1991) divided the Phlebotominae into two tribes: Idiophlebotomini (type genus: *Idiophlebotomus* Quate & Fairchild, 1961) and Phlebotomini (type genus *Phlebotomus* Rondani & Berté, 1840). The former was subdivided into subtribes (1) Idiophlebotomina with five genera – (1.1) *Idiophlebotomus*; (1.2) *Spelaeophlebotomus* (type species: *P. gigas* Parrot & Schwetz, 1937); (1.3) *Phlebotomites* Hennig, 1972 type species: *Phlebotomites brevifilis* Hennig, 1972); (1.4) *Phlebotomiella* Meunier, 1905 (type species: *Phlebotomiella tipuliformis* (Meunier, 1905) and (1.5) *Chinius* Leng, 1987 – and (2) the Hertigiina for the Americas, with the genera *Hertigia* and *Warileya*. The tribe Phlebotomini was subdivided into five subtribes – three for the Old World: (1) Phlebotomina, (2) Spelaeomyiina (genus type: *Spelaeomyia* Theodor, 1948) and (3) Sergentomyiina (type genus: *Sergentomyia* França & Parrot, 1920); one for the Australian region: Australophlebotomina (type genus: *Australophlebotomus* Theodor, 1948) and one for the Americas: Brumptomyiina (type genus *Brumptomyia* França & Parrot, 1921). Phlebotomina was divided into two genera, one of them, *Phlebotomus,* into nine subgenera, Spelaeomyiina, with only one genus and Sergentomyiina with three genera, one of them, *Sergentomyia,* with two subgenera. Brumptomyiina was divided into 10 genera: (1) *Brumptomyia*; (2) *Isolutzomyia* (type species: *Lutzomyia cirrita* Young & Porter, 1974); (3) *Psychodopygus*, with three subgenera: (3.1) *Psychodopygus*, *s. str.*, (3.2), *Shannonomyia* Dyar, 1929 and (3.3) *Eupsychodopygus* (type species: *Flebotomus. arthuri* Fonseca, 1936); (4) *Nyssomyia*, with two subgenera: (4.1) *Nyssomyia, s. str.* and (4.2) *Bichromomyia* (type species: *Flebotomus flaviscutellatus* Mangabeira, 1942); (5) *Trichophoromyia*; (6) *Psathyromyia*, with three subgenera: (6.1) *Psathyromyia, s. str.* with three groups: (6.1.1) Shannoni, (6.1.2) Volcanensis and (6.1.3) Lanei; (6.2) *Oophoromyia*, subgen. n. (type species: *Phlebotomus aragaoi* Lima, 1932), with two groups: (6.2.1) Aragaoi and (6.2.2) Brasiliensis; (6.3) *Xiphomyia* (type species: *Phlebotomus aclydifera* Fairchild & Hertig, 1952); (7) *Viannamyia*; (8) *Pintomyia*; (9) *Lutzomyia* with eight subgenera: (9.1) *Lutzomyia, s. str.*, with three groups: (9.1.1) Longipalpis, (9.1.2) Cavernicola and (9.1.3) Lichyi; (9.2); *Helcocyrtomyia*, with three groups: (9.2.1) Peruensis, (9.2.2) Scorzai and (9.2.3) Cruciata; (9.3) *Barrettomyia*, with four groups: (9.3.1) Tupynambai, (9.3.2) Migonei, (9.3.3) Evandroi and (9.3.4) Baityi; (9.4) *Coromyia*, with six groups: (9.4.1) Vespertilionis, (9.4.2) Serrana, (9.4.3) Vesicifera, (9.4.4) Verrucarum, (9.4.5) Monticola and (9.4.6) Saulensis; (9.5) *Evandromyia*, with two groups: (9.5.1) Infraspinosa, stat. n. and (9.5.2) Monstruosa; (9.6) *Pressatia*; (9.7) *Trichopygomyia*; (9.8) *Dampfomyia*; and (10) *Micropygomyia*, with three subgenera: (10.1) *Micropygomyia*, *s. str.* with three groups: (10.1.1) Cayennensis, (10.1.2) Chiapanensis and (10.1.3) Atroclavata; (10.2) *Sciopemyia*, with four groups: (10.2.1) Microps, (10.2.2) Pilosa, (10.2.3) Gasparviannai and (10.2.4) Delpozoi; (10.3) *Sauromyia* (type species: *Flebotomus oswaldoi* Mangabeira, 1942), with three groups: (10.3.1) Oswaldoi, (10.3.2) Trinidadensis and (10.3.3) Vexator.

Williams (1993) proposed promoting the phlebotomines to the Phlebotomidae family, with two subfamilies: Phlebotominae, *s. str.* and Bruchomyiinae.

Young and Duncan (1994), in a revision of the genus *Lutzomyia* – excepting the species of Canada and United States – followed basically the classification described by Lewis et al. (1977). However, they adopted the subgenera *Coromyia*, *Psathyromyia* and *Sciopemyia.*

Galati (1995) proposed a classification for Phlebotominae, with emphasis on the American species, with a phylogenetic approach in which Phlebotominae was divided into the tribes Hertigiini and Phlebotomini. The former consisted of two subtribes, Hertigiina for the Americas and Idiophlebotomina, including species of the Old World and the Australian region. Phlebotomini was divided into six subtribes, namely, Phlebotomina, Australophlebotomina, Brumptomyiina, Sergentomyiina, Lutzomyiina and Psychodopygina, with the American species distributed among the last four.

Galati et al. (2003) described the monospecific genus *Edentomyia* Galati, Andrade Filho, Silva & Falcão, 2003, with the type species *E. piauiensis* Galati, Andrade-Filho, Silva & Falcão, 2003, found in caves in the Brazilian state of Piauí, with characteristics that permit its inclusion in Phlebotomini but not in any of its subtribes considered in Galati's classification (1995).

Based on two fossil specimens of Cretaceous Lebanese amber, Azar et al. (1999) described two genera: *Libanophlebotomus* and *Mesophlebotomites*; Poinar (2004) described the genus *Palaeomyia* from a fossil specimen found in Cretaceous Burmese (Myanmar) amber, and Solórzano and Kraemer (2009) described the genus *Phlebotoiella* from Vastan (Indian amber). However, on the basis of the characteristics described for the four genera, it was not possible to include them in either of the two tribes considered in Galati's classification. Malak et al. (2013) recently described three species from Cretaceous Burmese (Myanmar) amber and included them in the *Phlebotomites* genus, described by Hennig (1972) on the basis of species found in Lebanese amber.

The artificial taxa *Phlebotomus breviductus* Barretto & Coutinho, 1950, and *Phlebotomus oliverioi* Barretto & Coutinho, 1941, each described on the basis of two species, were proposed the exclusion from the provisions of the ICZN (Articles 17.2 and 23.8) (Andrade et al. 2013, 2014).

The cladograms of Galati's (1995) classification are partially reproduced here (Figs. 1, 2, 3, and 4), as well as a classification by sequencing, with the inclusion of the series of species of the genus *Psychodopygus* and those of the subgenus *Helcocyrtomyia* and the genus *Edentomyia*. In the classification of the 1,008 taxa of the group of species (species and subspecies) described [977current and 31 fossil

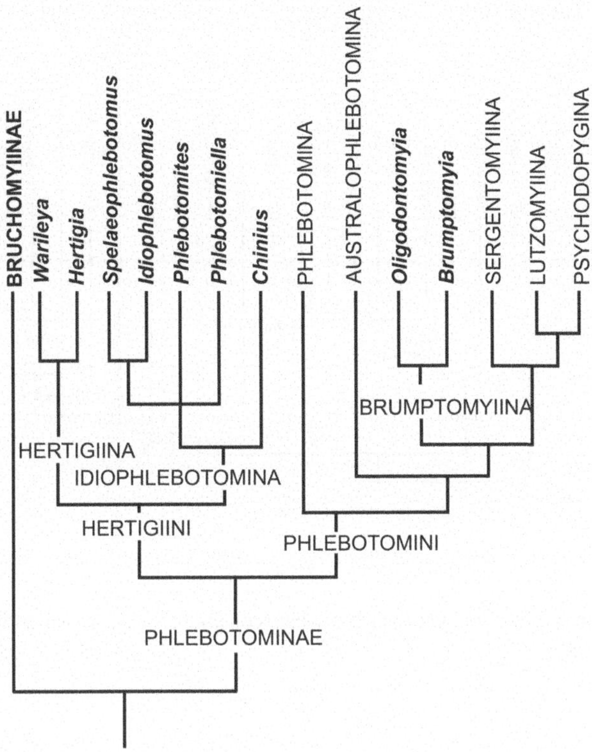

Fig. 1 Cladogram of Phlebotominae: tribes and subtribes, with the inclusion of genera in Hertigiini and Brumptomyiina

Fig. 2 Cladogram of Phlebotominae: genera, subgeneras and series of species of Sergentomyiina

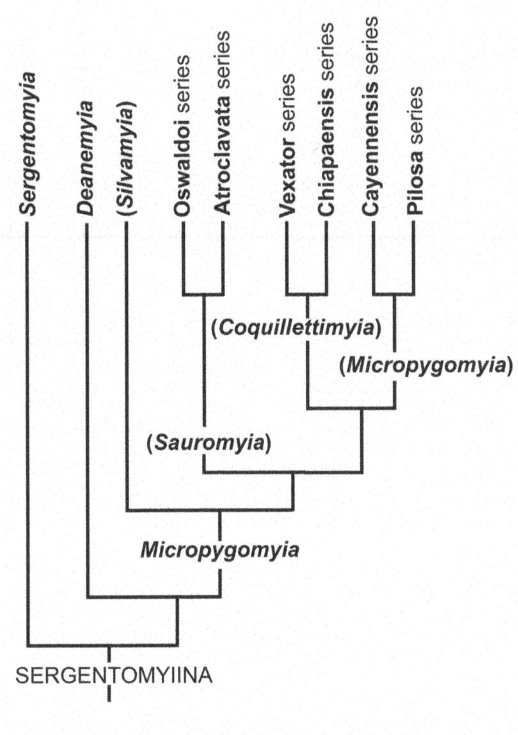

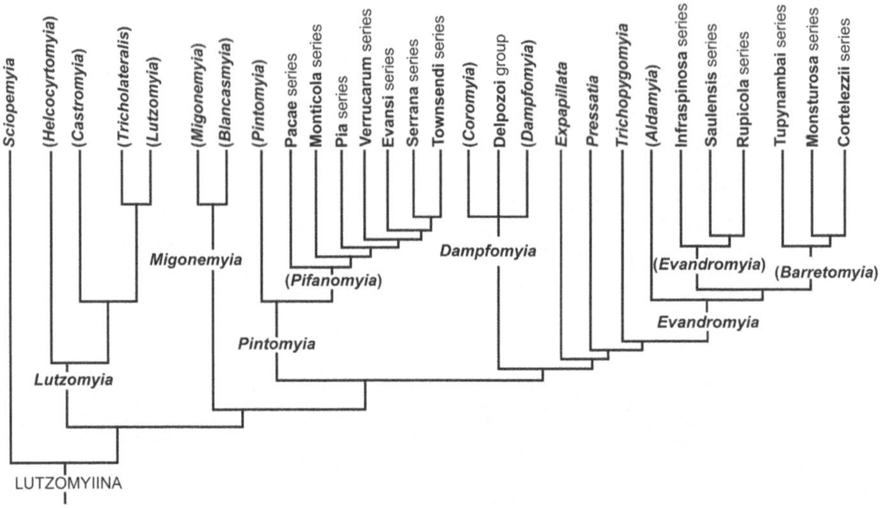

Fig. 3 Cladogram of Phlebotominae: genera, subgenera, group of species and series of species of Lutzomyiina

Fig. 4 Cladogram of Phlebotominae: genera, subgenera, groups and series of Psychodopygina. *Xiphomyia* (a pre-occupied name) has been replaced by *Xiphopsathyromyia* by Ibañes et al. (2015)

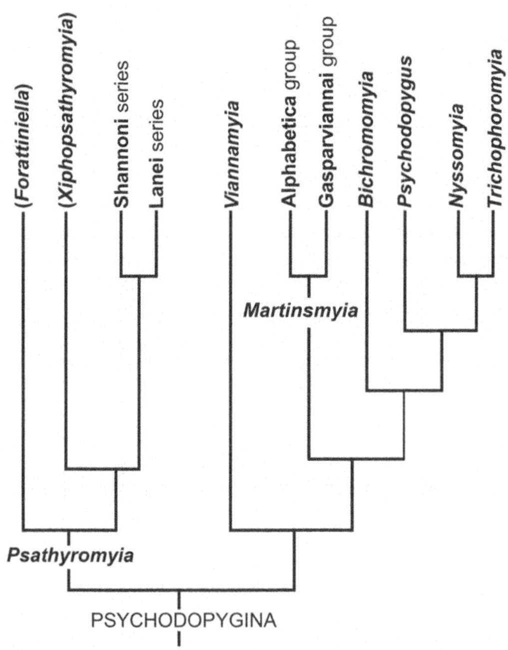

(+)], those of the Americas are emphasized, totaling 539 species/subspecies, 522 current and 17 fossil species, which are listed in alphabetical order in the respective supra-specific categories (genera, subgenera, groups and series of species) because no overall study has been undertaken of the phylogenetic relationship between them.

Artemiev (1991) and Galati (1995) proposed in their classifications several new taxa in the category of family and genus groups. However, Artemiev only listed the taxa without giving any diagnosis that differentiated them from the closest taxa, which conflicts with Article 13.1.1 of the Code of Zoological Nomenclature and cites no references from which the information might be recovered (Article 13.1.2). Galati (1995), although giving sets of characteristics that permit the differentiation of each taxon, presents the characteristics in the form of numbers (0,1, 2, 3, . . .) instead of describing them in words as recommended by Article 13.1.1. Thus both classifications are in disagreement with the rules of the Code of Zoological Nomenclature (ICZN 1999), making these names unavailable. However, the ICZN 2017; considers that such names may become available if a subsequent author publishes them again by fulfilling the availability criteria. But in this case, the authorship is no longer attributed to the author who originally proposed the name but rather to the author(s) who made the name available. For the American taxa the author and date became Galati (2003b) who provides this information in her identification keys.

Phylogenetic Classification of Phlebotominae (Diptera, Psychodidae)

The American groups are emphasized. Stages are presented as follows: E = egg, L = larva, P = pupa and sexes of the adults "m" (male) and "f" (female) described and the geographical distribution by country, and Brazilian states (in parentheses). The underlined country or Brazilian state represents that of the type locality. The full names of the countries and Brazilian states are shown in Table 1.

PHLEBOTOMINAE Rondani, 1840

HERTIGIINI Abonnenc & Leger, 1976

HERTIGIINA (Type genus: *Hertigia* Fairchild, 1949)

Warileya Hertig, 1948 (Type species: *Warileya phlebotomanica* Hertig, 1948)

euniceae Fernández, Carbajal, Astete & Wooster, 1998 (mf) - PE.
fourgassiensis Le Pont & Desjeux, 1984 (mf) - GF.
leponti Galati & Cáceres 1999 (m) - PE.
lumbrerasi Ogosuku, Perez, Davies & Villaseca, 1996 (mf) - PE.
nigrosaccula Fairchild & Hertig, 1951 (mf) - PA, CO.
phlebotomanica Hertig, 1948 (mf) - PE, EC.
rotundipennis Fairchild & Hertig, 1951 (mf) - CR, PA, CO, PE, BO.
yungasi Velasco & Trapido, 1974 (m) - BO.

Hertigia Fairchild, 1949 (Type species: *Hertigia hertigi* Fairchild, 1949)

hertigi Fairchild, 1949 (mf) - CR, PA.

BRUMPTOMYIINA Galati 2003 (Type genus: *Brumptomyia* França & Parrot, 1921)

Brumptomyia França & Parrot, 1921 [Type species: *Brumptomyia brumpti* (Larrousse, 1920)]

angelae Galati, Santos & Silva, 2007 (m) - BR (PR).
avellari (Costa Lima, 1932) (ELPmf) - PA, CO, VE, PE, BO, BR (RR, PA, RO, AC, TO, MA, PI, BA, MG, ES, RJ, SP, PR, GO, MT, MS,DF), AR, PY.
beaupertuyi (Ortiz, 1954) (mf) - CO, VE.
bragai Mangabeira & Sherlock, 1961 (m) - BR (BA, ES, SP).
brumpti (Larrousse, 1920) (Pmf) - BO, BR (AM, PA, MA, TO, MG, RJ, SP, PR, SC, MT, MS), AR.
cardosoi (Barretto & Coutinho, 1941) (mf?) - BR (MG, ES, RJ, SP, PR).
carvalheiroi (Shimabukuro, Marassá & Galati, 2007) (mf) - BR (SP).
cunhai (Mangabeira, 1942) (mf) - HN, BR (AP, PA, RO, BA, MG, ES, RJ, SP, PR, RS, MT, MS).
devenanzii (Ortiz & Scorza, 1963) (mf) - VE.
figueiredoi Mangabeira & Sherlock, 1961 (m) - BR (BA, ES).
galindoi (Fairchild & Hertig, 1947) - (mf) - NI, CR, PA, CO, EC, PE, BO.
guimaraesi (Coutinho & Barretto, 1941) (mf) - CO, BR (MG, ES, RJ, SP, PR), AR, PY.
hamata (Fairchild & Hertig, 1947) (mf) - MX, BZ, PA, CO, EC, PE.
leopoldoi (Rodriguez, 1953) (mf) - BZ, PA, CO, EC.
mangabeirai (Barretto & Coutinho, 1941) (m) - BR (PA, MG, SP, PR).
mesai Sherlock, 1962 (mf) - MX, BZ, HN, CO, BR (PR, MS, SP) (Revalidated: Ibáñez-Bernal 1999).
nitzulescui (Costa Lima, 1932) (mf) - BR (PE, MG, ES, RJ, SP, PR, SC, RS, MS).
orlandoi Fraiha, Shaw & Lainson, 1970 (m) - BR (MT).
ortizi Martins, Silva & Falcão, 1971 (mf?) - BR (MG, SP, PR).
pentacantha (Barretto, 1947) (mf) - CO, EC, PE, BO, BR (PA, AC, RO, MT).
pintoi (Costa Lima, 1932) (mf) - VE, SR, GF, BO, BR (RR, AM, PA, RO,MA, MG, RJ, SP, MT, MS, DF), AR.
quimperi Galati & Cáceres, 1999 (m) - PE.
spinosipes (Floch & Abonnenc, 1943) (f) - PA, GF, BR (RR, PA).
travassosi (Mangabeira, 1942) (ELPmf) - PA, SR, GF, BR (AP, RR, PA, RO, MA, MG).
troglodytes (Lutz, 1922) (mf) - PE?, BR (MG, ES, RJ, SP, SC).

virgensi Mangabeira & Sherlock, 1961 (m) - BR (BA).

Oligodontomyia Galati, 2003 [Type species: *Ol. oligodonta* (Young, Pérez & Romero, 1985)]

isopsi (Leger & Ferte, 1996) (mf) - CH.
oligodonta (Young, Pérez & Romero, 1985) (mf) - PE.
toroensis (Le Pont, Torrez-Espejo & Dujardin, 1997) (mf) - BO.

SERGENTOMYIINA Galati 2003 (Type genus: *Sergentomyia* França & Parrot, 1920)

Deanemyia Galati, 2003 [Type species: *Deanemyia samueli* (Deane 1955)]

appendiculata (Martins, Falcão & Silva, 1961) (m) - BR (MG).
derelicta (Freitas & Barrett, 1999) (mf) - BR (PA).
maruaga (Alves, Freitas & Barrett, 2008) (f) - BR (AM).
ramirezi (Martins, Falcão, Silva & Miranda-Filho, 1982) (mf) - BR (MG), BO.
samueli (Deane, 1955) (mf) - BR (PA, MA, PI, CE, RN).

Micropygomyia Barretto, 1962 [Type species: *Micropygomyia oswaldoi* (Mangabeira, 1942)]

(***Silvamyia***) Galati, 2003 [Type species: *Micropygomyia* (*Silvamyia*) *acanthopharynx* (Martins, Falcão & Silva, 1962)]

acanthopharynx (Martins, Falcão & Silva, 1962) (mf) - BR (RO, TO, MG, GO, MT, MS, DF).
echinatopharynx Andrade Filho, Galati, Andrade & Facão, 2004 (mf) - BR (TO).

(***Sauromyia***) Galati 2003 [Type species: *Micropygomyia* (*Sauromyia*) *oswaldoi* (Mangabeira, 1942)]

Oswaldoi series Barretto, 1962

capixaba (Dias, Falcão, Silva & Martins, 1987) (mf) - BR (RN, PE, BA, MG, ES, RJ).
dereuri (Le Pont, Matias, Martinez & Dujardin 2004 (mf) - BO.
+*dorafeliciangeli* Andrade Filho, Galati & Brazil, 2009 (m) - DO (Fossil, amber, Miocene).
ferreirana (Barretto, Martins & Pellegrino, 1956) (mf) - BR (MG, ES, RJ, SP, PR, MS, RS).
syn. *Lutzomyia borgmeieri* Martins, Falcão & Silva, 1972 (Galati et al. 2002).
huacalquensis (Le Pont, Matias, Martinez & Dujardin 2004 (mf) - BO.
longipennis (Barretto, 1946) (mf) - PE, BR (RR, AM, AP, PA, RO, TO, MA, AC, MG, RJ, SP, PR, GO, MT, MS, DF).

machupicchu (Martins, Llanos & Silva, 1975) (m) - PE.
oswaldoi (Mangabeira, 1942) (OLPmf) - BO, BR (AP, PA, RO, TO, MA, PI, CE, RN, PE, BA, MG, RJ, GO, MT, MS), AR.
+*paterna* (Quate, 1963) (m) - MX (Chiapas, Simojovel - Fossil: Oligocene/Miocene)
peresi (Mangabeira, 1942) (mf) - GF, BO, BR (RR, AM, PA, AC, RO, TO, MA, PI, CE, RN, PE, MG, RJ, GO, MT, MS), AR.
petari Galati, Marassá & Gonçalves-Andrade, 2003 (mf) - BR (SP).
pratti (Vargas & Diaz-Nájera, 1951) (m) - MX.
pusilla (Dias et al. 1986) (mf) - GF, BR (AP, RR, AM, PA, RO, MA, MT).
syn. *Phlebotomus* sp. de Saul Floch & Abonnenc, 1944 (Dias et al. 1986).
quechua (Martins, Llanos & Silva, 1975) (mf) - PE.
quinquefer (Dyar, 1929) (mf) - BO, BR (MA, CE, RN, PE, AL, BA, MG, ES, RJ, PR, GO, MT, MS), PY, AR.
syn. *Flebotomus rickardi* Costa Lima, 1936 (Fairchild & Hertig 1957).
rorotaensis (Floch & Abonnenc, 1944) (mf) - PA, CO, VE, SR, GF, PE, BR (AP, RR, AM, PA, RO, TO, MA, PI).
syn. *Phlebotomus* sp. de Rorota Floch & Abonnenc, 1941 (Floch & Abonnenc 1952)
saccai (Feliciangeli, Ramírez Pérez & Ramirez, 1989) (mf) - VE.
sp. 2 de Araracuara (Morales & Minter, 1981) (m) - CO.
trinidadensis (Newstead, 1922) (mf) - MX, BZ, GT, HN, NI, CR, PA, CO, VE, TT, SR, GF, EC, PE, BO, BR (RR, AP, AM, PA, AC, RO, TO, MA, CE, RN, BA, MG, MT).
syn. *Phlebotomus baduelensis* Floch & Abonnenc, 1942 (Fairchild & Hertig 1948).
syn. *Phlebotomus yucatanensis* Galliard, 1934 (Fairchild & Hertig 1948).
syn. *P. yucatanensis* var. *baduelensis* Floch & Abonnenc, 1941 (Fairchild & Hertig 1948).

villelai (Mangabeira, 1942) (mf) - BR (PA, AC, RO, TO, MA, PI, CE, RN, SE, BA, MG, GO, MT, MS) [Revalidated (Galati 2003)][1].

syn. *Lutzomyia goiana* Martins, Falcão & Silva, 1962 (Galati 2003).

vonatzingeni Galati, 2007 (mf) - BR (PA, TO, MS, MG).

zikani (Barretto, 1950) (mf) - BR (ES, PA).

Atroclavata Series Fairchild, 1955

atroclavata (Knab, 1913) (mf) - CR, PA, CO, VE, TT, VI, GP, MQ.

syn. *Phlebotomus guadeloupensis* Floch & Abonnenc 1945 (Fairchild & Hertig 1948).

syn. *Phlebotomus tejerae* Larrousse, 1921 (Dyar & Nuñez-Tovar 1926/1927).

venezuelensis (Floch & Abonnenc, 1948) (mf) - CO, VE.

syn. *Phlebotomus zuliaensis* Floch & Abonnenc, 1948 (Pifano et al. 1962).

(***Coquillettimyia***) Galati, 1995 [Type species: *Micropygomyia* (*Coquillettimyia*) *vexator* (Coquillett, 1907)]

Vexator Series Fairchild, 1955

apache (Young & Perkins, 1984) (mf) - US.

oppidana (Dampf, 1944) (mf) - CA, US, MX.

vexator (Coquillett, 1907) (mf) - CA, US, MX.

syn. *Phlebotomus vexator occidentis* Fairchild & Hertig, 1957 (Young & Perkins, 1984).

vindicator (Dampf, 1944) (mf) - MX.

Chiapanensis Series Theodor, 1965

californica (Fairchild & Hertig, 1957) (mf) - US.

chiapanensis (Dampf, 1947) (mfELP) - MX, SV, HN, NI, CR, PA.

stewarti (Mangabeira & Galindo, 1944) (mf) - US, MX.

(***Micropygomyia***) *s. str.* Barretto, 1962

[1] This was proposed on the basis of the comparison of the type male of *Mi. goiana* and three males from Piauí state identified by one of the authors of the species (Falcão A.L.) with seven specimens captured in VI.1940 by Mangabeira in the type locality of *Ph. villelai*. Three other males from Aurá, Belém, Pará, captured by Damasceno, VIII. 1940, and also identified as *Ph. villelai* by Mangabeira, 1942, belong to an undescribed species. All these specimens are deposited in the Museu de Zoologia of São Paulo University (Galati 2003a, b, p. 33).

Cayennensis Series Fairchild, 1955

absonodonta (Feliciangeli, 1995) (mf) - PE, VE.

ancashensis Galati & Cáceres, 2007 (m) - PE.

cayennensis cayennensis (Floch & Abonnenc, 1941) (mfLP) - MX, BZ, SV, HN, NI, CR, PA,CO, VE, TT, GF, EC, PE, BR (RR, AM, PA, RO).

cayennensis braci (Lewis, 1967) (mf) - KY.

cayennensis cruzi (Gonzales & Garcia, 1981) (mf) - CU.

cayennensis hispaniolae (Fairchild & Trapido, 1950) (mf) - DO, HT.

cayennensis jamaicensis (Fairchild & Trapido, 1950) (mf) - JM.

cayennensis maciasi (Fairchild & Hertig, 1948) (mf) - MX, GT, BZ.

cayennensis puertoricensis (Fairchild & Hertig, 1948) (mf) - PR.

cayennensis viequesensis (Fairchild & Hertig, 1948) (mf) - PR, VI.

ctenidophora (Fairchild & Hertig, 1948) (f) - MX.

cubensis (Fairchild & Trapido, 1950) (mf) - US, CU.

duppyorum (Fairchild & Trapido, 1950) (mf) - JM.

durani (Vargas & Diaz-Nájera, 1952) (mf) - MX, HN, SV.

farilli (Vargas & Diaz-Nájera, 1959) (f) - MX.

hardisoni (Vargas & Diaz-Nájera, 1952) (mf) - MX.

lewisi (Feliciangeli, Ordoñez & Férnandez, 1984) (mf) - VE.

micropyga (Mangabeira, 1942) (mf) - CR, PA, CO, VE, TT, GF, EC, PE, BO, BR (RR, AP, AM, PA, TO, MA, AC, RO, MG, GO, MT).

schreiberi (Martins, Falcão & Silva, 1975) (mf) - BR (CE, PB, PE, BA, MG, ES, RJ, SP).

wirthi (Vargas & Dias-Nájera, 1951) (f) - MX.

yencanensis (Ortiz, 1965) (mf) - CO, VE.

Pilosa Series Theodor, 1965

chassigneti (Floch & Abonnenc, 1944) (mf) - SR, CO, GF, BR (AP, AM).

mangabeirana (Martins, Falcão & Silva, 1963) (mf) - BR (RR).

pilosa (Damasceno & Causey, 1944) (mf) - CR, PA, CO, VE, TT, GR, BR (AM, PA, AC, RO, MA).

Incertae sedis

+*Micropygomyia brandaoi* Andrade Filho, Galati, Falcão & Brazil, 2008 (m) - DO. (Fossil, amber, Miocene).
Micropygomyia xerophila (Young, Brenner & Wargo, 1983), *comb. n.* (mf) - US.

LUTZOMYIINA Abonnenc & Leger, 1976 (Type genus: *Lutzomyia* França, 1924)

Sciopemyia Barretto, 1962 [Type species: *Sciopemyia sordellii* (Shannon & Del Ponte, 1927)]

fluviatilis (Floch & Abonnenc, 1944) (mf) - GF, BR (AP, PA, AC, RO, MA, PE).
microps (Mangabeira, 1942) (mf) - BR (PA, TO, MA, BA, MG, ES, RJ, SP, SC).
nematoducta (Young & Arias, 1984) (mf) - CO, BR (AM).
pennyi (Arias & Freitas, 1981) (m) - BR (AM).
preclara (Young & Arias, 1984) (mf) - CO, PE, BO, BR (AM, AC).
syn. *Lutzomyia* sp. near *L. microps* Young et al. 1985 (Young & Duncan 1994).
servulolimai (Damasceno & Causey, 1945) (mf) - PE, BO, BR (AM, PA, AC, RO, MA, CE, MT, DF).
sordellii (Shannon & Del Ponte,1927) (mf) - CR, PA, CO, VE, TT, GF, EC, PE, BO, BR (RR, AP, AM,PA, AC, RO, TO, MA, PI, CE, RN, PE, ES, RJ, SP, PR, MG, GO, MT, MS, DF), AR.
syn. *Phlebotomus longicornutus* Floch & Abonnenc, 1943 (Barretto 1946a).
syn. *Phlebotomus nordestinus* Mangabeira, 1942 (Young & Morales 1987).
vattierae (Le Pont & Desjeux, 1992) (mf) - PE, BO, CO, BR (AC).

Lutzomyia França, 1924 [Type species: *Lutzomyia longipalpis* (Lutz & Neiva, 1912)]

syn. *Françaia* Dyar & Nuñez-Tovar, 1926/1927 (new name for *Lutzia, non* Theobald, 1903).
syn. *Lutzia* França, 1920, *non* Theobald, 1903.
syn. *Lutziola* Strand, 1932 (new name for *Lutzia, non* Theobald, 1903).
syn. *Lutziomyia* Cordero, Vogelsang & Cossio, 1928 (new name for *Lutzia, non* Theobald, 1903).

(***Helcocyrtomyia***) Barretto, 1962 [Type species: Lutzomyia (*Helcocyrtomyia*) *peruensis* (Shannon, 1929)]

Sanguinaria Series Barretto, 1962

adamsi Fernandez, Galati, Carbajal, Wooster & Watts, 1998 (mf) - PE.
botella (Fairchild & Hertig, 1961) (f) - PA.
caceresi Le Pont, Matias, Martinez & Dujardin 2004 (m) - BO.
cirrita Young & Porter, 1974 (mf) - CO.
gonzaloi Ogusuku Canales & Pérez, 1997 (mf) - PE.
guderiani Torrez-Espejo, Cáceres & Le Pont, 1995 (mf) - PE, BO.
hartmanni (Fairchild & Hertig, 1957) (mf) - CR, PA, CO, EC, PE.
kirigetiensis Galati & Cáceres, 1992 (mf) - PE
monzonensis Ogusuku Canales & Pérez, 1997 (mf) - PE.
sanguinaria (Fairchild & Hertig, 1957) (mf) - HN, NI, CR, PA, CO.
scorzai (Ortiz, 1965) (mf) - VE, CO, PE.
sp. de Pichinde Young,1979 - (mf) - CO.
tolimensis Carrasquilla, Munstermann, Marín, Ocampo & Ferro, 2012 (mf) - CO.
tortura Young & Rogers, 1984 (mf) - CO, EC, BO.
velezi Bejarano, Vivero & Uribe, 2010 (m) - CO.

Osornoi Series Galati & Cáceres, 1994

caballeroi Blancas, Cáceres & Galati, 1989 (ELmf) - PE.
castanea Galati & Cáceres, 1994 (mf) – PE, EC.
ceferinoi (Ortiz & Alvarez, 1963) (mf?) - CO, VE. (f? see Galati & Cáceres 1994.
erwindonaldoi (Ortiz, 1978) (mf?) - CO, VE. (f? see, Galati & Cáceres 1994).
herreri Galati & Cáceres, 2003 (mf) - PE.
imperatrix (Alexander, 1944) (f) - PE.
larensis Arredondo, 1987 (mf) - VE.
munaypata Ogusuku, Chevarria, Porras & Pérez, 1999 (mf) - PE.
osornoi (Ristorcelli & Van Ty, 1941) (mf) - CO, EC, PE, BO.
syn. *Phlebotomus montoyai* Sherlock, 1962 (Young & Porter 1974).

quillabamba Ogusuku, Chevarria, Porras & Pérez, 1999 (mf) - PE.
rispaili Torrez-Espejo, Cáceres & Le Pont, 1995 (m) - BO, PE.
strictivilla Young, 1979 (mf) - CO, VE, EC.
wattsi Fernández, Carbajal, Astete & Wooster, 1998 (m) - PE.

Peruensis Series Barretto, 1962
ayacuchensis Cáceres & Galati, 1988 (mf) - EC, PE.
blancasi Galati & Cáceres, 1990 (mf) - PE.
chavinensis Pérez & Ogusuku, 1999 (mf) - PE.
galatiae Le Pont, Martinez, Torrez-Espejo & Dujardin, 1998 (mf) - BO.
noguchii (Shannon, 1929) (ELmf) - PE.
pallidithorax Galati & Cáceres, 1994 (mf) - PE.
peruensis (Shannon, 1929) (ELmf) - PE, BO.
pescei (Hertig, 1943) (mf) - PE.
tejadai Galati & Cáceres, 1990 (ELmf) - PE.

Incertae sedis
Lutzomyia vargasi (Fairchild & Hertig, 1961 (m) - MX.
Lutzomyia infusca Porter & Young, 1999 (mf) - GT.

(***Castromyia***) Mangabeira, 1942 [Type species: *Lutzomyia* (*Castromyia*) *castroi* (Barretto & Coutinho, 1941)]
amarali (Barretto & Coutinho, 1940) (mf) - BR (MG, ES, RJ, SP, PR).
syn. *Lutzomyia diacantha* Martins & Silva, 1965 (Martins et al. 1978).
caligata Martins, Falcão & Silva, 1965 (m) - BR (RO).
castroi (Barretto & Coutinho, 1941) (m) - BR (SP).

(***Tricholateralis***) Galati, 2003 [Type species: *Lutzomyia* (*Tricholateralis*) *cruciata* (Coquillett, 1907)]
araracuarensis Morales & Minter, 1981 (m) - CO, BR (AM).
carvalhoi (Damasceno, Causey & Arouck, 1945) (mf) - GF, BR (AP, PA, RO).
cruciata (Coquillett, 1907) (ELPmf) - US, MX, BZ, GT, HN, SV, NI, CR, PA, BR (MT?).

cultellata (Freitas & Albuquerque, 1996) (m) - PE, BR (AM). (n. comb.)[2].

diabolica (Hall, 1936) (mf) - US, MX.

evangelistai Martins & Fraiha, 1971 (mf) - CO, PE, BO, BR (AP, AM, PA, RO, AC, MT).

falcata Young, Morales & Ferro, 1994 (mf) - CO, EC, BR (AM, MT).

flabellata Martins & Silva, 1964 (m) - BO, BR (AM, AC, RO, MT).

gomezi (Nitzulescu, 1931) (ELPmf) - MX, HN, SV, NI, CR, PA, CO, VE, TT, GF, EC, PE, BO, BR (RR, AP, AM, PA, AC, RO, TO, MA, MT, GO, BA).

syn. *Flebotomus (Brumptomyia) suis* Rozeboom, 1940 (Barretto 1946b).

syn. *Phlebotomus japignyi* Floch & Abonnenc, 1944 (Fairchild & Hertig 1948).

legerae Le Pont, Gantier, Hue & Valle, 1995 (mf) - NI.

maesi Le Pont, Ibáñez–Bernal & Fuentes, 2011 (mf) - NI.

marinkellei Young, 1979 (mf) - CO, BR (AM, RO, AC, MT).

sherlocki Martins, Silva & Falcão 1971 (mf) - CO, EC, PE, BO, BR (RR, AM, PA, AC, RO, TO, MT).

spathotrichia Martins, Falcão & Silva, 1963 (mf) - EC, GF, BO, BR (RR, AP, AM, PA, RO, MA, MT).

syn. *Lutzomyia eliensis* Le Pont & Desjeux, 1983 (Lebbe et al. 1987).

(*Lutzomyia*) *s. str.*

syn. (*Aguayoi*) Vargas, 1978 (Espécie-tipo: *Lutzomyia dispar* Martins & Silva, 1963) (Galati 1995).

alencari Martins, Souza & Falcão, 1962 (mf) - BR (MG, ES, RJ).

almerioi Galati & Nunes, 1999 (mf) - BR (MS, SP).

---

[2]R. Fernandes (NAMRID) captured a specimem in Peru and had difficulties identifyng it because many of its characteristics were different from those of *Psathyromyia*, in which the species was included by its authors (*in* Barrett et al. 1996) and adopted by Galati (2003a, b). However, it was observed in the specimen captured by R. Fernandes the presence of setae on the abdominal pleurae, an autapomorphy of the subgenus *Tricholateralis*, beyond other typical characteristics of this group of insects.

battistinii (Hertig, 1943) (mf) - PE.
bicornuta (Blancas & Herrer, 1959/1960) (mf) - PE.
bifoliata Osorno-Mesa, Morales, Osorno & Hoyos, 1970 (mf) - CO.
cavernicola (Costa Lima, 1932) (mf) - BR (TO, MG, GO).
cruzi (Mangabeira, 1938) (mf) - BO, BR (GO, MT, MS).
dispar Martins & Silva, 1963 (mf) - BR (MA, PI, SP, GO, MT, MS).
elizabethrangelae Vilela, Azevedo & Godoy, 2015 (mf) – BR (TO).
falquetoi Pinto & Santos, 2007 (m) - BR (ES).
fonsecai (Costa Lima, 1932) (fm) - BO. (nov. comb; Galati et al. 2011).
forattinii Galati, Rego, Nunes & Teruya, 1985 (mf) - BO, BR (MS).
gaminarai (Cordero et al. 1928) (mf) - BR (PR, RS), PY, UY.
ischnacantha Martins, Souza & Falcão, 1962 (mf) - BR (MG, DF).
ischyracantha Martins, Falcão & Silva, 1962 (mf) - BR (MG, RJ).
lichyi (Floch & Abonnenc, 1950) (ELmf) - CR, PA, CO, VE, TT, GF, EC, PE, BR (RR, MT).
 syn. *Phlebotomus foliatus* Mirsa & Ortiz, 1952 (Fairchild & Hertig 1952).
 syn. *Phlebotomus vexillarius* Fairchild & Hertig, 1952 (Floch & Kramer 1965).
longipalpis (Lutz & Neiva, 1912) (ELPmf) - MX, GT, HN, SV, NI, CR, PA, CO, VE, BO, BR (RR, AP, PA, AC, RO, TO, MA, PI, CE, RN, PB, PE, AL, SE, BA, MG, ES, RJ, SP, GO, MT, MS, DF, PR, RS), AR, PY, UY.
 syn. *Phlebotomus almazani* Galliard, 1934 (Fairchild & Hertig 1958).
 syn. *Phlebotomus otamae* Nuñez-Tovar, 1924 (Dyar & Nuñez-Tovar 1926/1927).
matiasi Le Pont & Mollinedo, 2009 (m) - BO.
pseudolongipalpis Arrivillaga & Feliciangeli 2001 (L mf) - VE.
renei (Martins, Falcão & Silva, 1957) (ELPmf) - BR (MG, GO, MS).
souzalopesi Martins, Silva & Falcão, 1970 (mf) - BR (ES).

Incertae sedis

Lutzomyia chotensis Galati, Caceres & Zorrila, 2003 (m) - PE.
Lutzomyia ignacioi Young, 1972[3] (mf) - VE, CO.
Lutzomyia manciola Ibáñez-Bernal, 2001[4] (f) - BZ.
Lutzomyia ponsi Perruollo, 1984[5] (f) - VE.
Lutzomyia tanyopsis Young & Perkins, 1984[6] (f) - US.

Migonemyia Galati, 2003 [Type species *Migonemyia migonei* (França, 1920)]

(*Migonemyia*) *s. str.*

migonei (França, 1920) (ELPmf) - CO, VE, TT, PE, BO, BR (AP, AM, PA, AC, RO,TO, MA, CE, RN, PB, PE, AL, BA, MG, ES, RJ, SP, PR, SC, RS, MT, MS, GO), AR, PY.
syn. *Phlebotomus araozi* Paterson & Shannon, 1926 (Dyar, 1929).
syn. *Phlebotomus rangeli* Nuñez-Tovar, 1924 (Dyar & Nuñez-Tovar 1926/1927).
rabelloi (Galati & Gomes, 1992) (mf) - BR (SP).
vaniae Galati, Fonseca & Marassá, 2007 (mf) - BR (SP).

---

[3]In the classification of Galati (2003a, b) this species was inserted in *Psathryromyia* (*Incertae sedis*). However, two male and two female paratypes (deposited at the Entomological Collection - Smithsonian Institution/Walter Reeed Biosystematic Unit, Suitland, MD, USA) were examined by Sábio P.B. (Personal information) who observed in these specimens the presence of the ventro-cervical sensilla which excludes it from *Psathryromyia*. The presence of the papilla in fIII was also observed, therefore, its insertion in Sergentomyiina should be excluded. On the other hand, three characters observed: absence of setae on the anterior region of the katepisternum, ascoids with reduced posterior spur and ringed spermathecae are synapomorphies shared with some of the subgenera of *Lutzomyia, Castromyia*, *Tricholateralis* and *Lutzomyia*. However, it does not have characters which permits its insertion into any of these subgenera.

[4]The insertion in *Lutzomyia* is provisional. No characters of the female were described which make it possible to include it more precisely in a given genus, and the male is not known either. The inclusion of *Lu. manciola* in *Sciopemyia* was suggested by Ibáñez-Bernal (2001). However, this genus presents a short head and labrum-epifaringe, so that the sum of both is smaller than that of fI + fII, which are long; also, there is no information on the ventro-cervical sensillae, which are absent in *Sciopemyia* (Galati 2003, 36).

[5]The insufficient description of the characters of this species and the similarity of their epermathecae with those of *Lu. ignacioi* led us to include it together with this species. See the previous note.

[6]Although *Lu. tanyopsis* presents ascoids with posterior spur just as *Psathyromyia*, this characteristic represents a plesiomorphic state that alone does not allow its inclusion in this genus; moreover, its very long clypeus also distinguishes it from the species of this group. On the other hand, the species of Lutzomyiina and Sergentomyiina do not present ascoids with posterior spur and those of this latter subtribe present another palpal formula (Galati 2003a, b, p.36).

(***Blancasmyia***) Galati, 2003 [Type species: *Migonemyia (Blancasmyia) gorbitzi* (Blancas,1955)]

bursiformis (Floch & Abonnenc, 1944) (mf) - CO, VE, GF EC, BR (RR, AM, PA, MA, PI, SP, GO, MT, MS).

syn. *Flebotomus baityi* Damasceno, Causey & Arouck, 1945, (Galati 2003)[7].

cerqueirai (Causey & Damasceno, 1945) (mf) - CO, PE, BR (AM, PA, RO, BA).

gorbitzi (Blancas, 1959/1960) (mf) - CR, PA, CO, EC, PE.

syn. *Phlebotomus* sp. M, Hanson, 1961 (Martins et al. 1978).

syn. *Phlebotomus hansoni* Fairchild & Hertig, 1961 (Christensen & Rutledge 1973).

moucheti (Pajot & Le Pont, 1978) (mf) - PE, GF, BR (AM).

Pintomyia Costa Lima, 1932 [Type species: *Pintomyia fischeri* (Pinto 1926)]

(***Pintomyia***) ***s. str.***

bianchigalatiae (Andrade Filho, Aguiar, Dias & Falcão, 1999) (mf) - BR (ES, RJ, SP, MG, PR), AR.

christenseni (Young & Duncan, 1994) (mf) - PA, CO, VE, TT, BR (AP, RR, AM, PA, AC, RO, TO, MA, GO, MG, SP, PR, MT, MS, DF).

damascenoi (Mangabeira, 1941) (mf) - CO, SR, GF, BR (RR, AP, AM, PA, AC, RO, MA, BA, MG, GO, MT, MS).

syn. *Phlebotomus spinosus* Floch & Abonnenc, 1942 (Young & Duncan 1994).

fischeri (Pinto 1926) (ELPmf) - VE, PE, BO, BR (CE, PE, BA, ES, RJ, SP, PR, SC, RS, MG, GO, MT, MS, DF), AR, PY.

gibsoni (Pifano & Ortiz, 1972) (f) - VE.

kuscheli (Le Pont et al. 1998) (mf) - BO, BR (MS).

mamedei (Oliveira, Afonso, Dias & Brazil, 1994) (f) - BR (RJ, MG, ES, MS).

pessoai (Coutinho & Barretto, 1940) (ELPmf) - BR (BA, MG, ES, RJ, SP, PR, SC, RS, GO, MS), AR, PY.

---

[7] The proposal was based on the capture of two males of *Mg. baityi* (Damasceno, Causey & Arouck, 1945) along with 3 females identical to *Mg. bursiformis* (Floch & Abonnenc 1944) in experimental chicken in Brejão Farm, municipality of Guaira, State of São Paulo, IV.1976 and in the absence of other species that were not known by both sexes (Gomes et al. 1978) (Galati 2003a, b, pg. 37).

(***Pifanomyia***) Ortiz & Scorza, 1963 [Type species: *Pintomyia* (*Pifanomyia*) *serrana* (Damasceno and Arouck 1949)]

Pacae Series Galati, 2003

gruta (Ryan, 1986) (mf) - BR (PA, RO).

pacae (Floch & Abonnenc, 1943) (mf) - SR, GF, BR (RR, AP, AM, PA, RO, MA).

Monticola Series Galati, 2003

misionensis (Castro, 1959) (mf) - BR (TO, BA, ES, RJ, SP, PR, RS, MG, GO, MS), AR, PY.

syn. *Lutzomyia coelhoi* Coelho, Falcão & Falcão, 1967 (*nomen nudum*, Martins et al. 1978).

syn. *Phlebotomus* sp. Bejarano & Duret, 1950 (Castro 1959).

monticola (Costa Lima, 1932) (ELPmf) - PE, BR (PA, BA, MG, ES, RJ, SP, PR, SC, RS, MS), AR, PY.

syn. *Lutzomyia paulwilliamsi* Martins, Falcão & Silva, 1977 (Young & Duncan 1994).

Pia Series Galati, 2003

emberai (Bejarano, Duque & Vélez, 2004) (f) - CO.

limafalcaoae Wolff & Galati, 2002 (mf) - CO.

pia (Fairchild & Hertig, 1961) (mf) - CR, PA, CO, VE, PE, BO.

reclusa (Fernández & Rogers, 1991) (mf) - PE.

suapiensis (Le Pont, Torrez-Espejo & Dujardin, 1997 (f) - PE, BO.

tihuiliensis (Le Pont, Torrez-Espejo & Dujardin, 1997) (f) - PE, BO, CO.

tocaniensis (Le Pont, Torrez-Espejo & Dujardin, 1997) (f) - PE, BO.

torrealbai (Martins, Fernandez & Falcão, 1979) (m) - VE.

valderramai (Cazorla, 1988) (m) - VE.

Verrucarum Series Fairchild, 1955

andina (Osorno, Osorno-Mesa & Morales, 1972) (mf) - CO.

antioquiensis Wolff & Galati, 2002 (m) - CO.

aulari (Feliciangeli, Ordoñez & Manzanilla, 1984) (mf) - VE.

cajamarcensis (Galati , Cáceres & Le Pont, 1995) (mf) - PE.

columbiana (Ristorcelli & Van Ty, 1941) (mf) - CO.

syn. *Phlebotomus monticolus* var. *incarum* Ristorcelli & Van Ty, 1941 (Rozeboom 1947).

deorsa (Pérez, Ogusuku, Monje & Young, 1991) (m) - PE.
disiuncta (Morales, Osorno & Osorno-Mesa, 1974) (mf) - CO.
itza Ibáñez-Bernal, May-UC & Rebollar-Tellez, 2010 (m) - MX.
moralesi (Young, 1979) (mf) - CO.
verrucarum (Townsend, 1913) (ELmf) - PE.

Evansi Series Galati, 2003

evansi (Nuñez-Tovar, 1924) (mf) - MX, GT, HN, SV, NI, CR, CO, VE, PE.
maranonensis (Galati, Cáceres & Le Pont, 1995) (mf) - EC, PE.
nevesi (Damasceno & Arouck, 1956) (mf) - EC, CO, PE, BO, BR (PA, AC, RO, MT, MA).
ovallesi (Ortiz, 1952) (mfLP) - MX, BZ, GT, HN, NI, CR, PA, CO, VE, TT.

Serrana Series Barretto, 1962

boliviana (Velasco & Trapido, 1974) (mf) - BO.
christophei (Fairchild & Trapido, 1950) (mf) - DO, HT.
diazi (Gonzales & Garcia, 1981) (mf) - CU.
guilvardae (Le Pont et al. 1998) (m) - BO.
novoae (Gonzales & Garcia, 1981) (mf) - CU.
odax (Fairchild & Hertig, 1961) (mf) - GT, HN, NI, CR, PA, VE, GF, BR (RO, MA).

syn. *Lutzomyia dubia* Martins, Falcão & Silva, 1965 (Forattini 1973).

oresbia (Fairchild & Hertig, 1961) (mf) - CR, PA.
orestes (Fairchild & Trapido, 1950) (mf) - CU, KY, BR (MA?).
ottolinai (Ortiz & Scorza, 1963) (mf) - VE.
piedraferroi (León, 1971) (m) - GT.
robusta (Galati, Cáceres & Le Pont, 1995) (mf) - EC, PE.
salomoni Fuenzalida & Quintana, 2017 (mf) - AR.
serrana (Damasceno & Arouck, 1949) (mf) - MX, BZ, HN, NI, GT, CR, PA, CO, VE, GF, EC, PE, BO, BR (AP, RR, AM, PA, AC, RO, MT,MA, BA, ES, MG, RJ).

syn. *Phlebotomus guayasi* Rodriguez, 1956 (Fairchild & Hertig 1961a).

torresi (Le Pont & Desjeux, 1991) (mf) - BO, AR.

Townsendi Series Galati, 2003

amilcari (Arredondo, 1984) (mf) - VE.

longiflocosa (Osorno-Mesa, Morales, Osorno & Hoyos, 1970) (mf) - CO.

nadiae (Feliciangeli, Arredondo & Ward, 1992) (mf) - VE.

+*paleotownsendi* Andrade Filho, Falcão, Galati & Brazil, 2006 (m) - DO (Fossil, amber, Miocene).

+*paleotrichia* Andrade Filho, Brazil, Falcão & Galati, 2007 (m) -DO (Fossil, amber, Miocene).

quasitownsendi (Osorno, Osorno-Mesa & Morales, 1972) (mf) - CO.

sauroida (Osorno-Mesa, Morales & Osorno, 1972) (mf) - CO, VE.

spinicrassa (Morales, Osorno-Mesa, Osorno & Hoyos, 1969) (mf) - CO, VE.

torvida (Young , Morales & Ferro, 1994) (mf) - CO.

syn. *Lutzomyia* sp. de Reventones, Ferro & Morales, 1988 (Young & Duncan 1994).

townsendi (Ortiz, 1959) (mf) - CO, VE.

youngi (Feliciangeli & Murillo, 1985) *in* Murillo & Zeledón, 1985 (mf) - CR, CO, VE.

Incertae sedis

+*Pintomyia* (*Pifanomyia*) *adiketis* Poinar, 2008 (f)) - DO (Fossil, amber, Mioceno).

+*Pintomyia (Pifanomyia) bolontikui* Ibáñez-Bernal, Kraemer, Stebner & Wagner, 2013 (m) - MX (Fossil, amber Simojovel, Chiapas, Miocene).

+*Pintomyia* (*Pifanomyia*) *brazilorum* Andrade Filho, Galati & Falcão 2006 (m) - DO (Fossil, amber, Miocene).

Pintomyia (Pifanomyia) diamantinensis (Barata, Serra e Meira & Carvalho,2012) *comb. nov.* (mf) - BR (MG).

+*Pintomyia (Pifanomyia) dissimilis* Andrade Filho, Serra e Meira, Sanguinette & Brazil, 2009 (m) - DO (Fossil, amber, Miocene).

+*Pintomyia (Pifanomyia) dominicana* Andrade Filho, Galati & Brazil, 2009 (m) - DO (Fossil, amber, Miocene).
+*Pintomyia* (*Pifanomyia*) *falcaorum* Brazil & Andrade Filho, 2002 (m) -DO (Fossil, amber, Miocene).
+*Pintomyia* (*Pifanomyia*) *filipalpis* (Peñalver & Grimaldi, 2005) (m) -DO (Fossil, amber, Miocene).
+*Pintomyia (Pifanomyia) killickorum* Andrade Filho & Brazil 2004 (m) -DO (Fossil, amber, Miocene).
Pintomyia (*Pifanomyia*) *maracayensis* (Nuñez-Tovar, 1924)[8] (m) - VE.
+*Pintomyia (Pifanomyia) miocena* (Peñalver & Grimaldi, 2005) (m) - DO (Fossil, amber, Miocene).
Pintomyia (Pifanomyia) naiffi (Freitas & Oliveira, 2013), *n. comb.* (m) -BR (AC).
Pintomyia (Pifanomyia) nuneztovari (Ortiz, 1954) (mf) - GT, HN, PA, CO, VE, PE, BO.

syn. *Lutzomyia nuneztovari anglesi* Le Pont & Desjeux, 1984 (Young & Duncan 1994).

+*Pintomyia (Pifanomyia) paleopestis* (Peñalver & Grimaldi, 2005) (m) -DO (Fossil, amber, Miocene).
Pintomyia (Pifanomyia) rangeliana (Ortiz, 1953) (mf) - PA, CO, VE, TT.
Pintomyia sp. de Anchicaya (Young, 1979) (m) -CO, EC.
+*Pintomyia (Pifanomyia) succini* (Peñalver & Grimaldi, 2005) (m) -DO (Fossil, amber, Miocene).

Dampfomyia Addis, 1945 [Type species: *Dampfomyia anthophora* (Addis, 1945)]

syn. (*Anthophorus*) León, 1971 (the subgenus was characterized and the type species was not designated) (Galati 2003).

(***Coromyia***) Barretto, 1962 [Type species *D.* (*Coromyia*) *vespertilionis* (Fairchild & Hertig, 1947)]

[8] Male insufficiently described; however, the characteristicss of the genitalia designed suggest that the species belonged to the series Townsendi (see identification keys for series of species).

aquilonia (Fairchild & Harwood, 1961) (mf) - CA, US.
beltrani (Vargas & Díaz-Nájera, 1951) (mfELP) - MX, HN.
deleoni (Fairchild & Hertig, 1947) (mf) - MX, BZ, GT, HN, SV, CR.
disneyi (Williams, 1987) (mf) - BZ, GT.
syn. *Lutzomyia beltrani* "Belize Form" Williams, 1976 (Williams 1987).
isovespertilionis (Fairchild & Hertig, 1958) (mf) - CR, PA, CO.
steatopyga (Fairchild & Hertig, 1958) (mf) - MX.
vesicifera (Fairchild & Hertig, 1947) (mf) - NI, CR, PA.
vespertilionis (Fairchild & Hertig, 1947) (mf) - NI, CR, PA, CO, EC.
viriosa (Fairchild & Hertig, 1958) (mf) - CR, PA.
zeledoni (Young & Murillo, 1984) (mf) - HN, NI, CR.

Delpozoi Group Young & Fairchild, 1974
delpozoi (Vargas & Díaz-Nájera, 1953) (mf) - MX, BZ, GT.
inusitata (Fairchild & Hertig, 1961) (mf) - MX.
sp. de Suchitepequez (Young & Duncan, 1994) (mf) - GT.
syn. *Lutzomyia piedraferroi* Young & Duncan, 1994 *non* León, 1971 (Galati 2003).

(*Dampfomyia*) *s. str.*
anthophora (Addis, 1945) (ELPmf) - US, MX, NI.
atulapai (León, 1971) (mf) - MX, GT, SV.
dodgei (Vargas & Diaz-Nájera, 1953) (mf) - MX, SV. Revalidated by Young & Duncan (1994).
insolita (Fairchild & Hertig, 1956) (mf) - CR, PA.
syn. *Phlebotomus rubidulus* Fairchild & Hertig, 1956 (Christensen & Rutledge 1973).

leohidalgoi (Ibáñez-Bernal, Hernández-Xoliot & Mendoza, 2006) (mf) -MX.
permira (Fairchild & Hertig, 1956) (mf) - MX, BZ, GT.
syn. *Phlebotomus tikalaensis* León, 1971 (Young & Duncan 1994).
rosabali (Fairchild & Hertig, 1956) (mf) - CR, PA, CO.

Incertae sedis

Dampfomyia caminoi (Young & Duncan, 1994) (mf) - MX.

Expapillata Galati, 2003 [Type species: *Expapillata firmatoi* (Barretto, Martins & Pellegrino, 1956)]

cerradincola (Galati, Nunes, Oshiro & Dorval, 1995) (mf) -BO, BR (TO, MG, MT, MS).

firmatoi (Barretto, Martins & Pellegrino, 1956) (mf) - BR (MG, ES, RJ, SP, PR, SC, RS), AR.

Pressatia Mangabeira, 1942, [Type species: *Pressatia triacantha* (Mangabeira, 1942)]

calcarata (Martins & Silva, 1964) (mf) - VE, PE, BO, BR (AC, RO).

camposi (Rodríguez, 1950) (mf) - NI, CR, PA, CO, EC.

syn. *Phlebotomus acanthobasis* Fairchild & Hertig, 1952 (Fairchild & Hertig 1958).

syn. sp. O Floch & Abonnenc, 1953 (Martins et al. 1978).

choti (Floch & Abonnenc, 1941) (Pmf) - CO, EC, PE, BO, SR, GF, BR (AP, RR, AM, PA, AC, RO, TO, MA, PE, AL, SE, BA, MG, ES, SP, MT).

duncanae (Le Pont, Martinez, Torrez-Espejo & Durjardin, 1998) (mf) - BO, CO, PE.

syn. *Lutzomyia (Pressatia)* sp. 1 Young & Morales 1987 (Le Pont et al. 1998).

dysponeta (Fairchild & Hertig, 1952) (mf) - CR, PA, CO, VE, EC, BR (MT).

equatorialis (Mangabeira, 1942) (m) - GF, BR (PA, MG, ES).

triacantha (Mangabeira, 1942) (ELPmf) - CO, EC, VE, GF, PE, BR (RR, AM, PA, AC, RO, MA, MT).

trispinosa (Mangabeira, 1942) (m) - PE, GF, BR (RR, AP, AM, PA, RO, MA, SP?).

Trichopygomyia Barretto, 1962 [Type species: *Trichopygomyia longispina* (Mangabeira, 1942)]

conviti (Ramirez-Pérez, Martins & Ramirez, 1976) (mf) - CO, VE, BR (AM).

dasypodogeton (Castro, 1939) (mf) - BO, BR (RR, AM, PA, AC, RO, TO, MT).

depaquiti (Gantier, Gaborit & Rabarison, 2006) (m) -GF.

elegans (Martins, Llanos & Silva, 1976) (mf) -PE, BR (AC).

ferroae (Young & Morales, 1987) (mf) - CO.

gantieri (Le Pont & Desjeux, 1987) (mf) - BO.

longispina (Mangabeira, 1942) (ELmf) - CO, VE, GF, BR (RR, AM, PA, RO, PE, BA, MG, ES, RJ, MT).
martinezi (Young & Morales, 1987) (mf) - CO.
pinna Feliciangeli, Ramirez-Pérez & Ramirez, 1989) (mf) - VE, BR (RR, AM).
ratcliffei (Arias, Ready & Freitas, 1983) (m) - BR (AM)
rondoniensis (Martins, Falcão & Silva, 1965) (m) -BO, BR (AM, RO, MT).
trichopyga (Floch & Abonnenc, 1945) (mf) - SR, GF, BR (RR, AP, AM, PA, RO, BA).
triramula (Fairchild & Hertig, 1952) (LPmf) -MX, BZ, GT, CR, PA, CO, EC.
turelli (Fernández, Galati, Carbajal & Watts, 1998) (mf) - PE.
wagleyi (Causey & Damasceno, 1945) (mf) - CO, VE, BO, BR (AM).
witoto (Young & Morales, 1987) (m) - CO, EC.

Evandromyia Mangabeira, 1941 [Type species: *Evandromyia infraspinosa* (Mangabeira, 1941)]

(*Aldamyia*) Galati, 2003 [Type species: *Evandromyia* (*Aldamyia*) *walkeri* (Newstead, 1914)]

aldafalcaoae (Santos, Andrade Filho & Honer, 2001) (mf) - BR (MS), AR.
andersoni (Le Pont & Desjeux, 1988) (mf) - BO, BR (AC, MT).
apurinan Shimabukuro, Figueira & Silva, 2013 (mf) -BR (AM).
bacula (Martins, Falcão & Silva, 1965) (mf) BO, BR (AM, PA, AC, RO, TO, MA, MG, PR, GO, MT, DF).
carmelinoi (Ryan et al. 1986) (mf) - BR (RR, PA, TO, MA, PI, SP, PR, MG, GO, MS, MT).
syn. *Lutzomyia* sp. 222.12 Ryan, 1986 (Young & Duncan 1994).
dubitans (Sherlock, 1962) (mf) - CR, PA, CO, VE, TT, BR (RR, AM, MT).
syn. *Phlebotomus marajoensis* Fairchild & Hertig, 1961, *non* Damasceno & Causey, 1944 (Feliciangeli 1985).
evandroi (Costa Lima & Antunes, 1936) (mf) - BR (RR, AM, AP, PA, AC, RO, TO, MA, PI, CE, RN, PB, PE, AL, BA, ES, PR, MG, GO, MT, MS, DF), AR.
hashiguchii Leon, Teran, Neira & Le Pont, 2009 (m) - EC.

lenti (Mangabeira, 1938) (ELPmf) - SR, BO, BR (PA, TO, MA, PI, CE, RN, PB, PE, AL, BA, MG, ES, RJ, SP, PR, GO, MT, MS, DF).

syn. *Lutzomyia lentioides* Forattini, 1971; new name for *P. pinottii* Lucena, *non P. pinottii* Damasceno & Arouck (Martins et al. 1978).

syn. *Phlebotomus pinottii* Lucena, 1960, *non* Damasceno & Arouck, 1956 (Forattini 1973).

orcyi Oliveira, Sanguinette, Almeida & Andrade Filho, 2015 (mf) -BR (MS).

sericea (Floch & Abonnenc, 1944) (mf) - EC, CO, VE, SR, GF, BR (RR, AP, AM, PA, AC, RO, CE, BA, ES, MT).

syn. *Flebotomus deanei* Damasceno, Causey & Arouck, 1945 (Floch & Abonnenc 1952).

sp. de Baduel (Floch & Abonnenc, 1945) (fm) - GF, CO, SR, BR (AP, PA?, RO?, MA?).

termitophila (Martins, Falcão & Silva, 1964) (mf) -BO, BR (PA, AC, RO, TO, MA, PI, AL, BA, MG, ES, RJ, SP, GO, MS, MT), AR.

walkeri (Newstead, 1914) (mf) - PA, CO, VE, TT, GF, EC, PE, BO, BR (AP, RR, AM, PA, AC, RO, TO, MA, PI, CE, RN, PB, AL, PE, MG, RJ, GO, MT) PY.

syn. *Phlebotomus marajoensis* Damasceno & Causey, 1944 (Forattini 1973).

syn. *Phlebotomus gasti* Sherlock, 1962 (Young 1979).

williamsi (Damasceno, Causey & Arouck, 1945) (mf) - VE, PE, BR (RR, AP, AM, PA, AC, RO).

(*Evandromyia*) *s. str.*

Infraspinosa Series

begonae (Ortiz & Torrez, 1975) (mf) - CO, VE, BR (RR, AM, PA, RO, TO, MT).

bourrouli (Barretto & Coutinho, 1941) (mf) - BO, BR (AM,AP,AC,PA, RO, MA, SP, GO, MT, MS).

brachyphalla (Mangabeira, 1941) (mf) - GF, BR (AP, PA, MA, RO, TO).

georgii (Freitas & Barrett, 2002) (mf) - BR (RR, AM, PA, AC, RO).

infraspinosa (Mangabeira, 1941) (mf) - CO, VE, PE?, SR, GF, BO, BR (RR, AP, AM, PA, AC, RO, MA, MT).

inpai (Young & Arias, 1977) (mf) - VE, BR (RR, AP, AM, RO, PA).

ledezmaae Leon, Teran, Neira & Le Pont, 2009 (m) - EC.
pinottii (Damasceno & Arouck, 1956) (mf) - VE, GF, BR (PA, RO, TO, MA).
Lutzomyia arouckii Barretto, 1962; new name for *F. pinottii*
Damasceno & Arouck [(unnecessary new name (Young & Arias 1977)].
sipani (Fernández, Carbajal, Alexander & Need, 1994) (m) -CO, PE, BR (AM).
tarapacaensis (Le Pont, Torrez-Espejo & Galati, 1997) (m f) -BO, BR (PA, AC).

Saulensis Series Lewis et al. 1977
saulensis (Floch & Abonnenc, 1944) (mfL) - CR, PA, CO, VE, GF, EC, PE, BO, BR (AP, RR, AM, PA, AC, RO, TO, MA, PI, CE, GO, MT, MS, DF).
syn. *Phlebotomus pinealis* Floch & Abonnenc, 1944 (Fairchild & Hertig, 1958)
wilsoni (Damasceno & Causey, 1945) (mf) - BR (AM, AC, RO, MA, MT).

Rupicola Series Young & Fairchild, 1974
correalimai (Martins, Coutinho & Luz, 1970) (mf) - BR (SP, PR, RS).
gaucha Andrade Filho, Souza & Falcão, 2007 (f) - BR (RS).
grimaldii Andrade Filho, Pinto, Santos & Carvalho, 2009 (mf) - BR (ES).
rupicola (Martins, Godoy & Silva, 1962) (mf) - BR (RJ, SP, MG).
tylophala Andrade & Galati, 2012 (mf) -BR (MG).

(*Barrettomyia*) Martins & Silva, 1968 [Type species: *Evandromyia* (*Barrettomyia*) *tupynambai* (Mangabeira, 1942)]

Monstruosa Series Lewis et al. 1977
monstruosa (Floch & Abonnenc, 1944) (mf) - CO, VE, SR, GF, BR (RR, AP, AM, PA, AC, RO, MA, GO, MT).
syn. *Phlebotomus falciformis* Floch & Abonnenc, 1944 (Fraiha et al. 1970).
teratodes (Martins, Falcão & Silva, 1964) (mf) - BR (TO, MA, MG, GO, MT, MS, DF), PY.

Tupynambai Series

bahiensis (Mangabeira & Sherlock, 1961) (mfP) - BR (BA).
callipyga (Martins & Silva, 1965) (mf) - BR (MG, ES).
costalimai (Mangabeira, 1942) (mf) - BR (BA, MG, ES, RJ).
petropolitana (Martins & Silva, 1968) (mf) - BR (MG, ES, RJ, SP).
tupynambai (Mangabeira, 1942) (mfLP) - BR (PB, AL, PE, BA, ES, RJ, MG).

Cortelezzii Series Galati, 2003

cortelezzii (Brèthes, 1923) (mf) - PE, BO, BR (AC, TO, MA, CE, BA, MG, ES, RJ, SP, PR, RS, MT, MS,GO), AR, PY, UY.
corumbaensis (Galati, Nunes, Oshiro & Rego, 1989) (mf) -BO, BR (TO, MA, AL, MT, MG, GO, MS, DF), AR.
sallesi (Galvão & Coutinho, 1939) (mf) - EC, PE, BO, BR (AC, TO, MA, CE, RN, PE, AL, SE, BA, MG, ES, RJ, SP, PR, MT, MS, GO, DF), AR, PY.
spelunca Carvalho, Brazil, Sanguinette & Andrade Filho, 2011 (mf) - BR (MG).

Incertae sedis

Evandromyia (Barrettomyia) edwardsi (Mangabeira, 1941) (mf) - BR (MA, BA, MG, ES, RJ, SP, PR, SC).

PSYCHODOPYGINA Galati, 2003 (Type genus: *Psychodopygus* Mangabeira, 1941)

Psathyromyia Barretto, 1962 [Type species: *Psathyromyia shannoni* (Dyar,1929)]

(***Forattiniella***) Vargas, 1978 [Type species: *Psathyromyia* (*Forattiniella*) *lutziana* (Costa Lima, 1932)]

syn. (*Oophoromyia*) Artemiev, 1991 [Type-species: *Pa.* (*Oophoromyia*) *aragaoi* (Costa Lima, 1932)] (Galati 1995)

abunaensis (Martins, Falcão & Silva, 1965) (mf) - EC, CO, PE, BO, BR (AM, AC, RO).
antezanai (Le Pont, Dujardin, Mouchet & Desjeux, 1990) (mf) - BO.
aragaoi (Costa Lima, 1932) (mf) - CR, PA, CO, VE, TT, GF, EC, PE, BO, BR (RR, AP, AM, PA, AC, RO, TO, MA, PE, BA, MG, RJ, SP, PR, GO, MT, MS, DF), PY.

syn. *Phlebotomus heckenrothi* Floch & Abonnenc, 1942 (Forattini 1973).

barrettoi barrettoi (Mangabeira, 1942) (mf) - CO, TT, SR, GF, EC, PE, BO, BR (RR, AP, AM, PA, RO, PE, BA, MG, RJ, GO, MT).

syn. *Phlebotomus* sp. de Maripa Floch & Abonnenc, 1946 (Young 1979).

barrettoi majuscula (Young, 1979) (mf) - HN, SV, NI, CR, PA, CO, EC.

brasiliensis (Costa Lima, 1932) (LPmf) - GF, PE, BR (RR, AM, PA, AC, RO, TO, MA, CE, AL, PE, BA, MG, RJ, SP, GO, MT, DF).

campograndensis (Oliveira, Andrade Filho, Falcão & Brazil, 2001) (mf) - GF, BR (RR, AP, AM, MS), AR.

syn. *Phlebotomus* sp. n° 768 Floch & Chassignet, 1948 (Galati 2003).

carpenteri (Fairchild & Hertig, 1953) (mfL) - MX, BZ, CR, PA, CO.

castilloi Leon, Mollinedo & Le Pont, 2009 (m) - EC, BO, GF.

coutinhoi (Mangabeira, 1942) (m) - PE,BO, BR (AM, PA, RO, MT).

elizabethdorvalae Brilhante, Sábio & Galati, 2016 (mf) - BR (AC).

inflata (Floch & Abonnenc, 1944) (mf) - GF, BO, BR (AP, AM, PA, RO, MT).

lutziana (Costa Lima, 1932) (mf) - CO, VE, SR, GF, PE, BO, BR (RR, AP, AM, PA, AC, RO, TO, MA, BA, MG, ES, RJ, SP, GO, MT, MS, DF), PY.

syn. *Phlebotomus* sp. de Cayenne, Floch & Abonnenc, 1945 (Forattini 1973).

naftalekatzi (Falcão, Andrade Filho, Almeida & Brandão-Filho, 2000) (mf) - BR (PE).

pascalei (Coutinho & Barretto, 1940) (mfP) - BR (TO, BA, MG, ES, RJ, SP, PR, RS), AR.

pradobarrientosi (Le Pont, Matias, Martinez & Dujardin, 2004 (M) - BO, BR (AC, MS).

runoides (Fairchild & Hertig, 1953) (mfL) - CR, PA, CO, EC, PE, BR (RR, AM, AC, RO, TO, MG, MT).

+*schleei* (Peñalver & Grimaldi, 2005) (m) - DO (Fossil, amber - Miocene).

texana (Dampf, 1938) (mf) - US, MX.

(***Xiphopsathyromyia***) Ibánez-Bernal & Marina, 2015 [Type species: *Psathryromyia* (*Xiphopsathyromyia*) *aclydifera* (Fairchild & Hertig, 1952)]

new name for (*Xiphomyia*) Galati 2003 *non* Townsend, 1917.

aclydifera (Fairchild & Hertig, 1952) (LPmf) - MX, BZ, HN, GT, NI, CR, PA, CO, EC, BO.

dreisbachi (Causey & Damasceno, 1945) (mf) - CO, EC, VE, SR, GF, PE, BO, BR (AP, AM, RR, AM, PA, AC, RO, MT).

syn. *Phlebotomus* sp. de Crique Anguille Floch & Abonnenc, 1945 (Floch & Abonnenc 1952).

hermanlenti (Martins, Silva & Falcão, 1970) (mf) - BR (TO, MA, MG, GO, MT, MS, SP).

ruparupa (Martins, Llanos & Silva, 1976) (mf) - PE, BO.

(***Psathyromyia***) *s.str.*

Lanei Series Theodor,1965

digitata (Damasceno & Arouck, 1950) (m) - BR (BA).

lanei (Barretto & Coutinho, 1941) (ELPmf) - BR (BA, MG, ES, RJ, SP, PR, SC, RS, MS), AR, PY.

pelloni (Sherlock & Alencar, 1959) (mf) - BR (BA, MG, ES, RJ, SP, PR, SC).

Shannoni Series Fairchild, 1955

abonnenci (Floch & Chassignet, 1947) (mf) - PA, CO, VE, SR, GF, EC, PE, BO, BR (AP, RR, AM, PA, AC, RO, TO, MA, CE, GO, MT).

baratai Sábio, Andrade & Galati, 2015 (mf) - BR (SP, MG, PR), AR.

barretti Alves & Freitas, 2015 (mf) - BR (AM).

bigeniculata (Floch & Abonnenc, 1941) (mf)[9] - GF, BR (AP, AM, PA, RO?, AC,TO? MA?, PI?, CE?, RN?, PB? PE? BA, MG, ES, RJ?, SP, MT, MS, GO?, DF, PR?, SC?, RS), AR, PY? Revalitaded from the synonymy of *Pa. Shannoni* by Sábio et al. (2014).

syn. *Phlebotomus microcephalus* Barretto & Duret, 1953 (Sábio et al. 2016).

[9] Their occurrence in the Brazilian states is with question mark as well as Paraguay yet need confirmation, because the species were reported as *Pa. shannoni*.

campbelli (Damasceno, Causey & Arouck, 1945) (mf) - CO, VE, GF, PE, BO, BR (RR, AM, PA, AC,RO, TO, MT, MS, GO).

cratifer (Fairchild & Hertig, 1961) (mf) - MX, BZ, HN, CR, PA.

dasymera (Fairchild & Hertig, 1961) (Lmf) - MX, BZ, NI, CR, PA, CO, VE, EC, BR (AM, RO, TO, MT).

dendrophyla (Mangabeira, 1942) (mf) - CO, VE, SR, GF, EC, PE, BO, BR (RR, AP, AM, PA, AC, RO,TO, MA, MT, MG).

guatemalensis (Porter & Young, 1986) (mf) - GT.

lerayi (Le Pont, Martinez, Torrez-Espejo & Dujardin, 1998) (m) - BO, CO.

limai (Fonseca 1935) (ELPmf) - BR (MG, ES, SP). Revalidated by Sábio et al. (2014).

    syn. *Phlebotomus pestanai* (Barretto & Coutinho, 1941) (Sábio et al. 2014).

punctigeniculata (Floch & Abonnenc, 1944) (mf) - PA, CO, VE, SR, GF, EC, PE, BO, BR (RR, AM, PA, AC, RO, TO, MA, MT, MS GO, SP), AR.

    syn. *Phlebotomus christophersoni* Damasceno & Causey, 1944 (Fairchild & Hertig 1950).

ribeirensis Sábio, Andrade & Galati, 2014 (mf) - BR (SP).

scaffi (Damasceno & Arouck, 1956) (mf) - CO, SR, GF, PE, BO, BR (RR, AP, AM, PA, AC, RO, MA, MT).

shannoni (Dyar, 1929) (mfELP)[10] - US, MX, BZ, GT, HN, NI, CR, PA, CO, VE.

soccula (Fairchild & Hertig, 1961) (m) - CR, PA.

souzacastroi (Damasceno & Causey, 1944) (m) - BR (AM).

undulata (Fairchild & Hertig, 1950) (mf) - MX, GT, BZ, HN, SV, CR, PA, CO, EC, BO.

    syn. *Phlebotomus humboldti* Vargas & Díaz-Nájera, 1959 (Rosabal & Trejos 1964).

[10]There is a wide geographical distribution indicated for this species, possibly, its actual occurrence is limited to the trans-Andean and Andean regions. In countries with trans-Andean, Andean and cis-Andean areas, studies are needed to identify the limit of its distribution. In cis-Andean areas where *Pa. shannoni* has been identified, it is possible that most of the specimens belong to *Pa. bigeniculata.*

volcanensis (Fairchild & Hertig, 1950) (mf) - CR, PA, BO.

Incertae sedis

Psathyromyia maya Ibáñez-Bernal, May-UC & Rebollar-Tellez, 2010 (f) - MX.
Psathyromyia pifanoi Ortiz, 1972 (Martins et al. 1978) (mf) - CO, VE, PE, BR (AM, PA, AC).
syn. *Lutzomyia cuzquena* (Martins, Llanos & Silva, 1975) (Sábio et al. 2016).

Viannamyia Mangabeira, 1941 [Type species: *Viannamyia tuberculata* (Mangabeira, 1941)]

caprina (Osorno-Mesa, Morales & Osorno, 1972) (mf) - HN, NI, CR, PA, CO, PE.
fariasi (Damasceno, Causey & Arouck, 1945) (m) - GF, BR (AM, PA).
furcata (Mangabeira, 1941) (L? P? mf) - CR, CO, VE, GF, EC, PE, BO, BR (RR, AP, AM, PA, AC,RO, TO, MA, PE, BA, MT, MS).
syn. *Phlebotomus arborealis* Floch & Abonnenc, 1944 (Barretto 1962).
tuberculata (Mangabeira, 1941) (mf) - PA, CO, VE, SR, GF, PE, BO, BR (RR, AP, AM, PA, AC, RO, TO, MA, PE, BA, MT).
syn. *Phlebotomus* sp. X, Floch & Abonnenc, 1944 (Barretto 1946a).
syn. *Lutzomyia munangai* Wijers & Huisenga, 1967 (Lewis 1975).

Martinsmyia Galati, 2003 [Type species: *Martinsmyia alphabetica* (Fonseca, 1936)]

Alphabetica Group Fairchild, 1955

alphabetica (Fonseca, 1936) (ELPmf) - BR (SP, PR, SC, RS), AR, PY.
brisolai (Le Pont & Desjeux, 1987) (mf) - BO, BR (MT).
minasensis (Mangabeira, 1942) (mf) - BR (TO, MG, RJ).
mollinedoi (Le Pont & Desjeux, 1991) (mf) - BO.
oliveirai (Martins, Silva & Falcão, 1970) (mf) - BR (PA, TO, MA, PI, MG, GO, MT, MS).
pisuquia (Ogusuku, Guevara, Revilla, Inga & Pérez, 2001) (mf) - PE.
quadrispinosa (Floch & Chassignet, 1947) (m) - GF.

syn. *Phlebotomus* sp. 1 de Baduel Floch & Abonnenc, 1947 (Floch & Abonnenc 1952).

reginae Carvalho, Brazil, Sanguinette & Andrade Filho, 2010 (MF) - BR (TO).

waltoni (Arias, Freitas & Barrett, 1984) (mf) - BR (RO).

Gasparviannai Group Young & Fairchild, 1974

cipoensis (Martins, Falcão & Silva, 1964) (mf) - BR (MG).

gasparviannai (Martins, Godoy & Silva, 1962) (mf) - BR (BA, MG, ES, RJ).

Bichromomyia Galati, 2003 [Type species: *Bichromomyia flaviscutellata* (Mangabeira, 1942)]

flaviscutellata (Mangabeira, 1942) (E L Pmf) - CO, VE, TT, SR, GF, EC, PE, BO, BR (RR, AP, AM, PA, AC, RO, TO, MA, CE, BA, ES, RJ, SP, MG, MS, MT, GO, DF).

syn. *Phlebotomus apicalis* Floch & Abonnenc, 1943 (Barretto 1946a).

inornata (Martins, Falcão & Silva, 1965) (m) - BO, BR (AM, RO, MA).

olmeca olmeca (Vargas & Díaz-Nájera, 1959) (mf) - MX, BZ, GT, HN, NI, CR.

olmeca bicolor (Fairchild & Theodor, 1971) (mfELP) - CR, PA, CO, VE, EC, PE, BR (RR, AM, PA, AC, MT).

olmeca nociva (Young & Arias, 1982) (mf) - PE, BR (AM, PA, MA, MT).

reducta (Feliciangeli, Ramirez Pérez & Ramirez, 1988) (mf) - CO, VE, PE, BR (AM, AC, RO).

Psychodopygus Mangabeira, 1941 [Type species: *Psychodopygus chagasi* (Costa Lima, 1941)]

Arthuri Series Barretto, 1962

syn. (*Eupsychodopygus*) Artemiev, 1991 [Espécie-tipo: *Psychodopygus* (*Eupsychodopygus*) *arthuri* (Fonseca, 1936)].

arthuri (Fonseca, 1936) (LPmf) - BR (MA, RJ, SP).

lloydi (Antunes, 1937) (mf) - BR (MA, MG, RJ, SP, PR).

syn. *Flebotomus rachoui* Damasceno & Arouck, 1956 (Galati 1981)

matosi (Barretto & Zago, 1956) (mf) - BR (BA, MG, ES, RJ).

Davisi Series Barretto, 1962

amazonensis (Root, 1934) (mf) - CO, VE, SR, GF, TT, EC, PE, BO, BR (RR, AP, AM, PA, AC, RO, MA, MT).

syn. *Lutzomyia robini* Abonnenc, Arias, Léger & Young, 1980) (Lebbe et al. 1987).

claustrei (Abonnenc, Léger & Fauran, 1979) (mf) - CO, VE, SR, GF, PE, BO, BR (RR, AP, AM, PA, RO, TO, MA, PE, MT, MS).

davisi (Root 1934) (Lmf) - CO, VE, SR, GF, EC, PE, BO, BR (RR, AP, AM, PA, AC, RO, MA, BA, MG, ES, RJ, GO, MS, MT).

syn. *Phlebotomus rooti* Mangabeira, 1942 (Martins et al. 1978).

parimaensis (Ortiz & Álvarez, 1972) (f) - VE.

Panamensis Series

syn. (*Shannonomyia*) Dyar, 1929 (Type species: *Phebotomus* (*Shannonomyia*) *panamensis* Shannon, 1926, *non* Alexander, 1920).

syn. (*Shannonomyiina*) Pratt, 1947, new name for *Shannonomyia*, *non* Alexander, 1920 (Pratt 1947).

ayrozai (Barretto & Coutinho, 1940) (mf) - PA, CO, VE, TT, GF, EC, PE, BO, BR (RR, AP, AM, PA, AC, RO, MA, BA, MG, ES, RJ, SP, PR,SC, MT).

syn. *Lutzomyia tintinabula* Christensen & Fairchild, 1971 (Young 1979).

syn. *Lutzomyia* sp. de Turure Lewis, 1975 (Lebbe et al. 1987).

carrerai carrerai (Barretto, 1946) (mf) - CO, VE, EC, PE, BO, BR (RR, AM, PA, AC, RO, MA, MT, BA, MG, RJ).

syn. *Sergentomyia pessoana* Barretto, 1955 (Martins et al. 1978).

carrerai thula (Young, 1979) (mf) - HN, CR, PA, CO, EC.

fairchildi (Barretto, 1966) (mf) - BR (MG, ES, RJ).

hirsutus hirsutus (Mangabeira, 1942) (mf) - CO, SR, GF, EC, PE, BO, BR (RR, AP, AM, PA, AC, RO, MA, BA, MG, ES, RJ, SP, MT).

syn. *Phlebotomus colasbelcouri* Floch & Chassignet, 1947 (Barretto 1953).

syn. *Phlebotomus* sp. II de Baduel, Floch, 1947 (Martins et al. 1978).
syn. *Phlebotomus* sp. C. Velasco, 1973 (Young 1979).
hirsutus nicaraguensis (Fairchild & Hertig, 1961) (mf) - NI, PA, BR (MT?).
joliveti Le Pont, Leon, Galati & Dujardin, 2009 (m) - GF.
llanosmartinsi Fraiha & Ward, 1980 (mf) - PE, BO, BR (AM, AC, RO, MT).
nocticolus (Young, 1973) (mf) - MX, PA, CO, GF, EC, PE, BO.
panamensis (Shannon, 1926) (ELPmf) - MX, BZ, GT, HN, NI, CR, PA, CO, VE, SR?, GF, EC, PE, BR (RR).
paraensis (Costa Lima, 1941) (ELmf) - CO, VE, SR, GF, EC, PE, BO, BR (RR, AP, AM, PA, AC RO, MA, RJ?, MT).
recurvus (Young, 1973) (mf) - PA, CO.
yasuniensis Leon, Neira & Le Pont, 2009 (m) - EC.
yucumensis (Le Pont, Caillard, Tibayrenc & Desjeux, 1986) (mf) - PE, BO, BR (AC, RO).

Guyanensis Series Barretto, 1962
corossoniensis (Le Pont & Pajot, 1978) (mf) - CR, PA, SR, GF, BR (AM, PA, AC, RO, MA).
dorlinsis (Le Pont & Desjeux, 1982) (m) - GF.
francoisleponti Zapata, Depaquit & León 2012 (mf) - EC. BR (AC).
geniculatus (Mangabeira,1941) (mfLP) - BZ, GT, NI, CR, PA, CO, VE, GF, EC, PE, BO, BR (AP, AM, PA, AC, RO, MA, MT, ES, RJ, SP, PR).
guyanensis (Floch & Abonnenc, 1941) (f) - SR?, GF, BR (MT).
lainsoni Fraiha & Ward, 1974 (mf) - PE, BO, BR (PA, AC, RO, MT, GO).
luisleoni Leon, Mollinedo & Le Pont, 2009 (m) - EC.
sp. de Trés Esquinas (Young, 1979) (f) - CO.

Chagasi Series Barretto, 1962
bernalei (Osorno-Mesa, Morales & Osorno, 1967) (mf) - CO, VE, BO, BR (AM).
chagasi (Costa Lima, 1941) (mf) - CO, VE, PE, BR (AP, RR, AM, PA, AC, RO, MT).

syn. *Flebotomus* (*Psychodopygus*) *unisetosus* Mangabeira, 1941 (Martins et al. 1968).

complexus (Mangabeira, 1941) (Emf) - BO, BR (AM, PA, RO, MA, MT).

douradoi (Fé, Freitas & Barrett, 1998) (mf) - BR (AM).

fairtigi (Martins, 1970) (mf) - CO.

syn. *Phlebotomus* (*Shannonomyina*) *squamiventris* Fairchild & Hertig, 1951 *non* Lutz & Neiva, 1912 (Martins 1970).

killicki (Feliciangeli, Ramirez-Pérez & Ramirez, 1988) (mf) - VE.

leonidasdeanei Fraiha, Ryan, Ward, Lainson & Shaw, 1986 (mf) - BR (PA).

syn. *Psychodopygus* sp. no. 401.63, Fraiha et al., 1978 (Fraiha et al. 1986).

squamiventris squamiventris (Lutz & Neiva, 1912) (mf) - VE, GY, PE, BR (RR, AM, PA, MA, MT).

squamiventris maripaensis (Floch & Abonnenc, 1946) (mf) - SR, GF, BR (AP, RR, AM, PA).

wellcomei Fraiha, Shaw & Lainson, 1971 (mfELP) - VE, BR (AM, PA, MA, RN, CE, PE).

Bispinosus Series Barretto, 1962

bispinosus (Fairchild & Hertig, 1951) (mf) BZ, HN, GT, NI, CR, PA, SR, CO, GF, EC, BR (AP, AM, PA, AC, RO, BA, MT).

Nyssomyia Barretto, 1962 [Type species: *Nyssomyia intermedia* (Lutz & Neiva, 1912)

anduzei (Rozeboom, 1942) (mfL) - CR, PA, VE, GF, PE, BR (RR, AM, AP, PA, AC, RO, MA, MT, BA).

antunesi (Coutinho, 1939) (Emf) - CO, VE, TT, SR, GF, PE, BO, BR (RR, AP, AM, PA, AC, RO, TO,MA, MT, MS).

syn. *Phlebotomus balouroensis* Floch & Abonnenc, 1944 (Barretto 1946a).

syn. *Phlebotomus intermedius* var. *acutus* Floch & Abonnenc, 1942, *partim* (Barretto 1946a).

syn. *Phlebotomus machicouensis* Floch & Abonnenc, 1942 (Theodor 1965).

bibinae (Léger & Abonnenc, 1988) (f) - GF.

delsionatali Galati & Galvis, 2012 (mf) - BR (MT).

edentula (León, 1971) (mf) - GT, HN, CR, PA.

elongata (Floch & Abonnenc, 1945) (m) - GF.

syn. *Phlebotomus intermedius longiductus* Floch & Abonnenc, 1941 (Floch & Abonnenc 1952).

syn. *Phlebotomus longiductus* Floch & Abonnenc, 1944 (Barretto 1950b).

fraihai (Martins, Falcão & Silva, 1979) (mf) - PE, BO, BR (RR, AM, PA, AC, RO, MT, BA, ES). Revalidated by Godoy & Galati (2016).

hernandezi (Ortiz, 1965) (mf) - CO, VE.

intermedia (Lutz & Neiva, 1912) (ELPmf) - BR (TO, MA, PI, RN, PB, PE, AL, SE, BA, MG, ES, RJ, SP, PR, GO, MS, MT, DF).

syn. *Phlebotomus lutzi* Manson-Bahr, 1925 (Barretto & Pessôa 1946).

syn. *Phlebotomus mazzai* Paterson, 1926 (Dyar 1929).

neivai (Pinto, 1926) (mf) - BO, BR (PA, SP, PR, SC, RS, MG, MS, GO, DF), AR, PY. Revalidated by Marcondes (1996).

richardwardi (Ready & Fraiha, 1981) (Emf) - CO, EC, PE, BO, BR (RR, AP, AM, PA, AC, RO, TO, MA, MT).

syn. *Lutzomyia* sp. 260.44, Ward & Ready 1975; Lainson et al. 1976 (Ready & Fraiha 1981).

shawi (Fraiha, Ward & Ready, 1981) (mf) - CO, PE, BO, BR (PA, AC, RO, MT).

syn. *Lutzomyia* sp. 260.43 Ward et al. 1973, Ward & Ready 1975, Fraiha et al. 1978 (Fraiha et al. 1981).

singularis (Costa Lima, 1932) (f) - BR (SP).

sylvicola (Floch & Abonnenc, 1945) (m) - GF, BR (PA, AC).

syn. *Phlebotomus sylvestris* Floch & Abonnenc, 1944 *non* Sinton, 1924 (Floch & Abonnenc 1945).

trapidoi (Fairchild & Hertig, 1952) (Lmf) - GT, HN, NI, CR, PA, CO, EC.

umbratilis (Ward & Fraiha, 1977) (Lmf) - CO, VE, SR, GF, PE, BO, BR (RR, AP, AM, PA, AC, RO, MA, CE, PE, MT).

syn. *Lutzomyia* sp. 260.31 Ward, 1973 (Young 1979).

urbinattii Galati & Galvis, 2012 (mf) - BR (PA, MT).

whitmani (Antunes & Coutinho, 1939) (ELPmf) - SR, GF, PE, BO, BR (AP, AP, AM, PA, AC, PA, AC,RO, TO, MA, PI, CE, RN, PB, PE, AL, SE, BA, MG, ES, RJ, SP, PR, RS, GO, DF, MT, MS), AR, PY.

syn. *Phlebotomus acutus* Floch & Abonnenc, 1942 (Floch & Abonnenc 1952).

syn. *Phlebotomus intermedius acutus* Floch & Abonnenc, 1941 *partim* (Floch & Abonnenc 1952).

ylephiletor (Fairchild & Hertig, 1952) (LPmf) - MX, BZ, GT, HN, NI, CR, PA, CO, EC.

yuilli yuilli (Young & Porter, 1972) (Emf) - CO, VE, EC, PE.
yuilli pajoti (Abonnenc, Léger& Fauran 1979) (mf) - CO, SR, GF, PE, BR (AP, PA, RO).
syn. *Phlebotomus* sp. de Souvenir Floch & Abonnenc, 1944 (Abonnenc et al. 1979).

Trichophoromyia Barretto, 1962 [Type species: *Trichophoromyia ubiquitalis* (Mangabeira, 1942)]
acostai (Llanos, 1966) (m) - PE.
syn. *Phlebotomus townsendi* Llanos, 1964 *non* Ortiz, 1960 (Llanos 1966).
adelsonsouzai Santos, Silva, Barata, Andrade & Galati, 2013 (mf) - BR (PA).
arevaloi Galati & Cáceres, 1999 (m) - PE.
auraensis (Mangabeira, 1942) (mf) - CO, VE, SR, PE, BO, BR (AM, PA, AC, RO, MT).
beniensis (Le Pont & Desjeux, 1987) (mf) - BO.
bettinii (Feliciangeli, Ramirez Pérez & Ramirez, 1988) (mf) - CO, VE.
brachipyga (Mangabeira, 1942) (mf) - GF, BR (AP, RR, AM, PA, AC, RO).
castanheirai (Damasceno, Causey & Arouck, 1945) (mf) - BR (AM, PA, RO).
cellulana (Young, 1979) (mf?) - CO, EC.
clitella (Young & Pérez, 1994) (m) - PE, BR (MT).
syn. *Lutzomyia* (*Trichophoromyia*) sp. 2, Young Pérez & Romero, 1985 (Young & Pérez 1994 *in* Young & Duncan 1994).
dunhami (Causey & Damasceno, 1945) (m) - BR (AM).
eurypyga (Martins, Falcão & Silva, 1963) (mf) - VE, BR (RR, AM, PA, RO).
flochi (Abonnenc & Chassignet, 1948) (m) - GF, BR (AM, AC, RO).
gibba (Young & Arias, 1994) (m) - BR (AM).
howardi (Young, 1979) (mf) - CO, PE?, BR (MT).
incasica (Llanos, 1966) (m) - PE.
new name for *Phlebotomus adleri* Llanos, 1964 *non* Theodor, 1963 (Llanos 1966).
ininii (Floch & Abonnenc, 1943) (mf) - SR, GF. BR (AP, AC).
syn. *Phlebotomus* sp. Floch & Abonnenc, 1942 (Martins et al. 1978).
syn. *Phlebotomus* sp. B du Gallion, Floch, 1943 (Barretto 1947).
lopesi (Damasceno, Causey & Arouck, 1945) (m) - BR (AM).
loretonensis (Llanos, 1964) (m) - PE, BR (RO).
meirai (Causey & Damasceno, 1945) (m) - BR (AM).

melloi (Causey & Damasceno, 1945 (m) - SR, BR (AM, PA, AC) (Revalidated by Young &Duncan (1994).
napoensis (Young & Rogers, 1984) (mf) - EC.
nautaensis (Fernandez, Lopez, Cardenas & Requena, 2015) (m) - PE
nemorosa (Young & Pérez, 1994) (m) - PE.
syn. *Lutzomyia (Trichophoromyia)* sp. 1 Young, Pérez & Romero, 1985 (Young & Perez 1994 *in* Young & Duncan 1994).
octavioi (Vargas, 1949) (m) - PE, BO, BR (RR, AM, PA, RO, MT).
new name for *Phlebotomus affinis* Mangabeira, 1942 non Theodor, 1933 (Vargas 1949)
omagua (Martins, Llanos & Silva, 1976) (mf) - PE.
pabloi (Barreto, Burbano & Young, 2002) (mf) - CO. EC.
pastazaensis (Fernandez, Carbajal, Alexander & Need, 1993 (mf) - PE.
readyi (Ryan, 1986) (mf) - BR (AM, PA, RO).
reburra (Fairchild & Hertig, 1961a, b) (mf) - CR, PA, CO, EC.
reinerti (Young & Duncan, 1994) (m) - BR (PA).
rostrans (Summers, 1912) (mf) - BR (AM).
ruifreitasi Oliveira, Teles, Medeiros, Camargo & Pessoa, 2015 (m) - BR (AC).
ruii (Arias & Young, 1982) (mf) - CO, BR (RR, AM, PA, RO, MT).
saltuosa (Young, 1979) (m) - CO.
sinuosa (Young & Duncan, 1994) (m) - PE.
sp. 1. de Araracuara (Morales & Minter, 1981) (mf) - CO.
ubiquitalis (Mangabeira, 1942) (mf) - CO, VE, SR, GF, EC, PE, BO, BR (RR, AP, AM, PA, AC, RO, MA, MT)
syn. *Phlebotomus basispinosus* Barretto & Coutinho, 1943 (Theodor 1965).
syn. *Phlebotomus cauchensis* Floch & Abonnenc, 1943 (Barretto 1950a, b).
velascoi (Le Pont & Desjeux, 1992) (mf) - BO.
viannamartinsi (Sherlock & Guitton, 1970) (mf) - BR (MA, BA).
wilkersoni (Young & Rogers, 1984) (mf) - EC.

SUBTRIBE UNDETERMINED

Edentomyia Galati, Andrade Filho, Silva & Falcão, 2003 (Type species: *Edentomyia piauiensis* Galati, Andrade Filho, Silva & Falcão, 2003)
piauiensis Galati, Andrade Filho, Silva & Falcão, 2003 (mf) - BR (PI, PA).

Table 1 Abbreviations of American countries and Brazilian states

American Countries		Brazilian States	
AR	Argentina	(AC)	Acre
BO	Bolivia	(AL)	Alagoas
BR	Brazil	(AM)	Amazonas
BZ	Belize	(AP)	Amapá
CA	Canada	(BA)	Bahia
CH	Chile	(CE)	Ceará
CO	Colombia	(DF)	Distrito Federal
CR	Costa Rica	(ES)	Espírito Santo
CU	Cuba	(MA)	Maranhão
DO	Dominican Republic	(MG)	Minas Gerais
EC	Ecuador	(MS)	Mato Grosso do Sul
GF	French Guyana	(MT)	Mato Grosso
GP	Guadalupe	(PA)	Pará
GT	Guatemala	(PB)	Paraíba
GY	Guyana	(PE)	Pernambuco
HT	Haiti	(PI)	Piauí
HN	Honduras	(PR)	Paraná
JM	Jamaica	(RJ)	Rio de Janeiro
KY	Cayman Islands	(RN)	Rio Grande do Norte
MQ	Martinique	(RO)	Rondônia
MX	Mexico	(RR)	Roraima
NI	Nicaragua	(RS)	Rio Grande do Sul
PA	Panama	(SC)	Santa Catarina
PE	Peru	(SE)	Sergipe
PR	Puerto Rico	(SP)	São Paulo
PY	Paraguay	(TO)	Tocantins
SR	Suriname		
SV	El Salvador		
TT	Trinidad and Tobago		
US	United States of America		
UY	Uruguay		
VE	Venezuela		
VI	Virgin Islands		

Morphology and Terminology of Adults and Identification of the American Taxa

Morphology and Terminology of Adults

External and internal structures which resist the process of clarification have been described for male and female phlebotomines. It is suggested that the reader consult Jobbling (1987) who produced an excellent study of male and female *Phlebotomus papatasi* for other characteristics of their internal anatomy.

Beyond the characteristics traditionally used for descriptions as well as in studies of the revision of these Diptera and in the proposal for standardized descriptions made by the CIPA group (1991), others, resulting from intense morphological study with an evolutionary focus, have been introduced to develop a phylogenetic approach to the group.

The insects were mounted in lateral and/or dorsal-lateral positions for the observation of the structures. Emphasis is given, as regards the head, to the characteristics of the appendages. The characteristics of the cervix are described here for the first time and this allows subdivision of the species into groups. Concerning the thorax, beyond the characteristics traditionally used regarding wings, legs and some seta tufts, the importance of other elements of chetotaxy, furca and sutures are brought out. As for the abdomen, information regarding seta covering and the presence of papillae on the tergites of the males has been expanded. As regards some of the common characteristics presented in descriptions of females, but also mainly of male genitalia, a new terminology has been adopted for them, approximating more closely to that used for Diptera in general.

As for the female genitalia, some new structures on the VIII and IX tergites, useful in distinguishing species and groups, will be presented here.

The points of reference for orientation as to what is basal or distal (apical) vary. Thus as regards the head, the characteristics to be found on the dorsal side take the frontoclypeal suture as their basal point of reference; in the ventral view, the basal point is the occipital foramen, and for the internal structures, it is the limit with the bucal appendages. As for the thorax, the basal and ventral parts relate to the sternum, and as regards the abdomen, the basal parts relate to the thorax. For the female, the spermathecal ducts, spermathecae and terminal knob take the genital chamber as the basal point of reference. As for what is internal and external, for the head, the internal is that which approximates to the bucal appendages and for the genitalia, to the parameral sheath.

Head

Externally, the cephalic capsule (Figs. 5, 6 and 7) is formed by the **compound eyes** which may occupy the larger part of the head, with the gena situated laterally and before the eyes. The **front**, composed of the part which extends from the vertex to

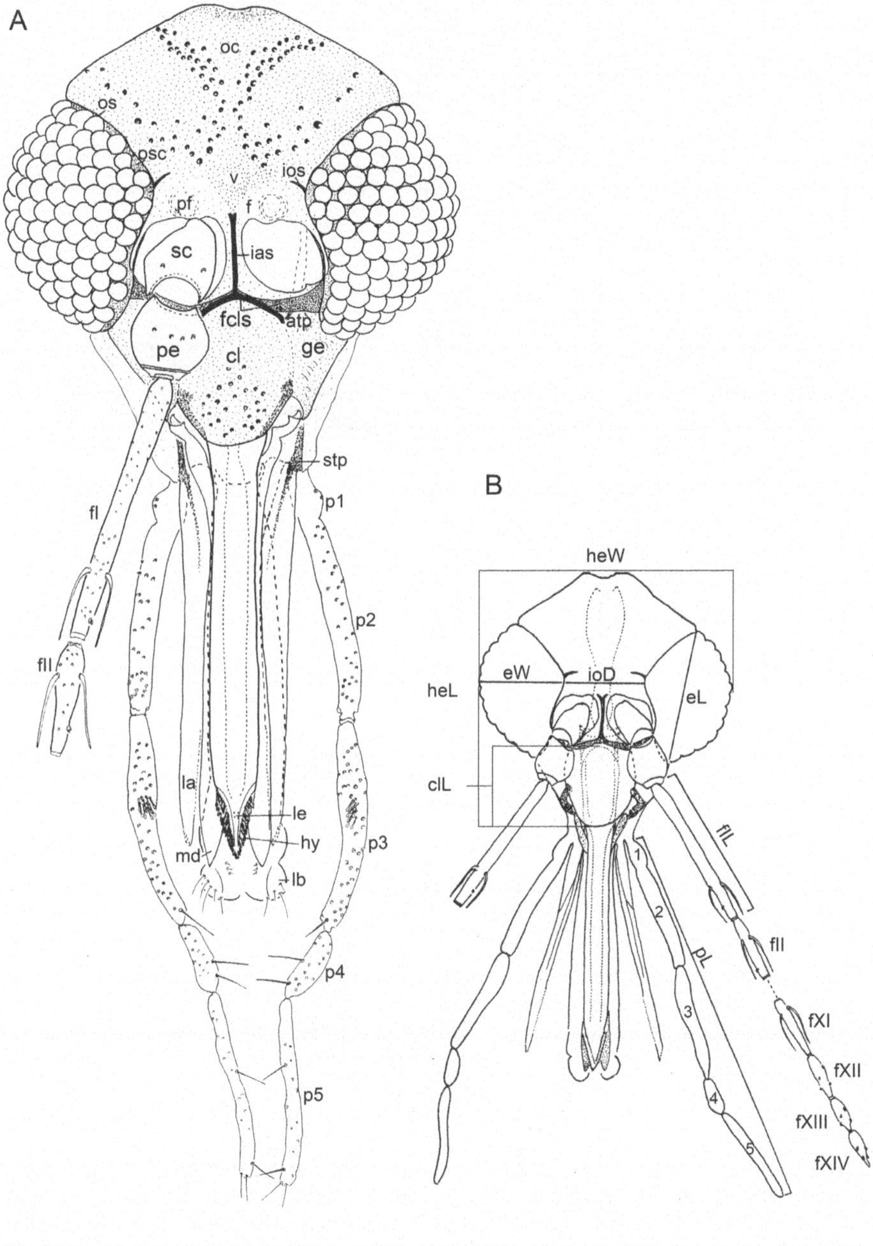

Fig. 5 Dorsal view of the head and its appendices of a Phlebotomine female. **A** – characteristics: **atp** – anterior tentorial pit; **cl** – clypeus; **f** – front; **fcls** – frontoclypeal suture; **fI** – 1st flagellomere; **fII** – 2nd flagellomere; **ge** – gena; **hy** – hypopharynx; **ias** – interantennal suture; **ios** – interocular suture; **la** – lacinia of maxilla; **lb** – labium; **le** – labrum-epipharynx; **md** – mandible; **oc** – occiput; **os** – ocular suture; **osc** – ocular sclerite; **pe** – pedicel; **pf** – protuberance of the front; **p1** – 1st palpal segment; **p2** – 2nd palpal segment; **p3** – 3rd palpal segment; **p4** – 4th palpal segment; **p5** – 5th palpal segment; **sc** – scape; **stp** – stipe of maxilla; **v** – vertex; **B** – measurements most frequently used: **eL** – eye length; **eW** – eye width; **clL** – clypeus length; **fIL** – 1st flagellomere length; **fXI** – 11th flagellomere; **fXII** – 12th flagellomere; **fXIII** – 13th flagellomere; **fXIV** – 14th flagellomere; **heL** – head length; **heW** – head width; **ioD** – interocular distance; **pL** – palpus length. *Nyssomyia intermedia*

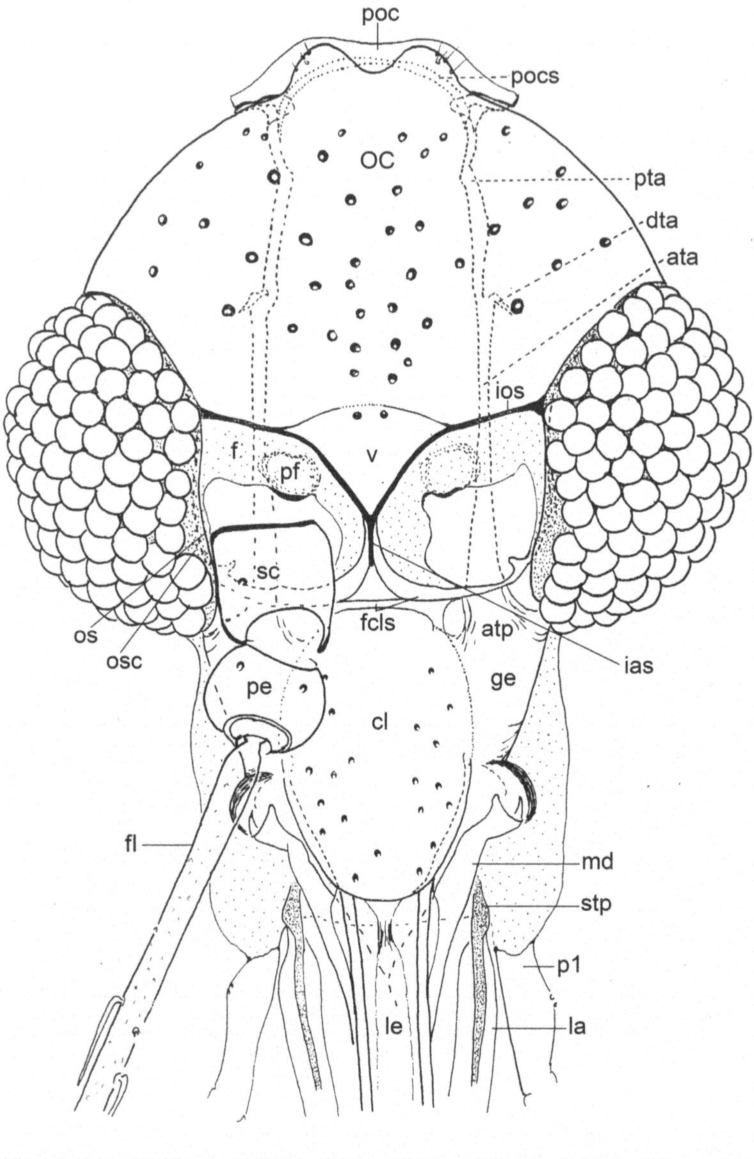

Fig. 6 Dorsal view of the head of a Phlebotomine female: **ata** – anterior tentorial arm; **atp** – anterior tentorial pit; **cl** – clypeus; **dta** – dorsal tentorial arm; **fl** – 1st flagellomere; **f** – front; **fcls** – frontoclypeal suture; **ge** – gena; **ias** – interantennal suture; **ios** – interocular suture; **la** – lacinia of the maxilla; **le** – labrum-epipharynx; **md** – mandible; **oc** – occiput; **os** – ocular suture; **osc** – ocular sclerite; **pe** – pedicel; **pf** – protuberance of the front; **poc** – postocciput; **pocs** – postoccipital suture; **pta** – posterior tentorial arm; **p1** – 1st palpal segment; **sc** – scape; **stp** – stipe; **v** – vertex. *Warileya lumbrerasi*

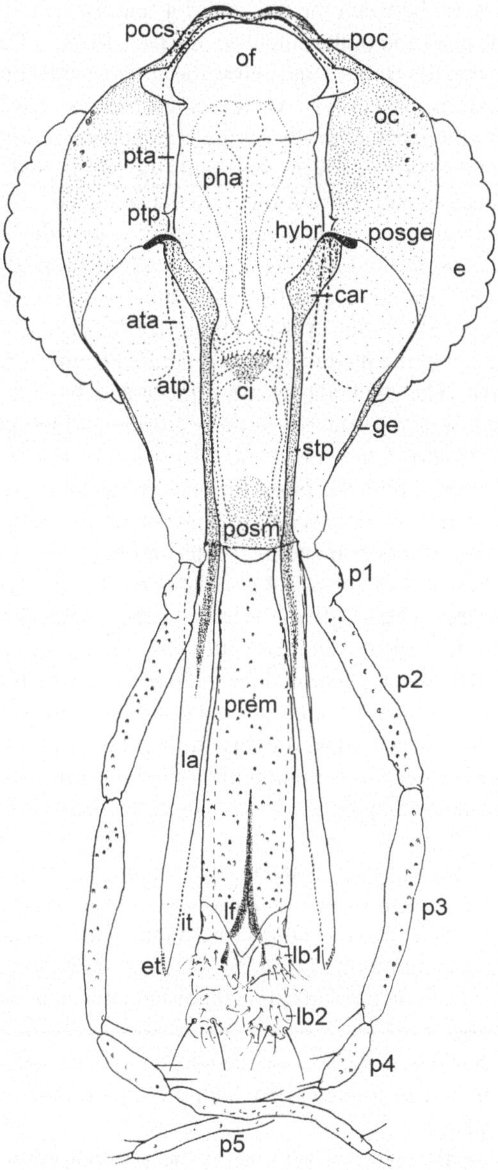

Fig. 7 Ventral view of the head of a Phlebotomine female: **atp** – anterior tentorial pit; **ata** – anterior tentorial arm; **car** – cardo; **ci** – cibarium; **e** – eye; **et** – external teeth of lacinia; **ge** – gena; **lf** – labial furca; **hybr** – hypostomal bridge; **it** – internal teeth of lacinia; **la** – maxillary lacinia; **lb1** – labellum 1; **lb2** – labellum 2; **oc** – occiput; **of** – occipital foramen; **pha**- pharynx; **poc** – postocciput; **pocs** – postoccipital suture; **posge** – postgena; **posm** – postmentum; **prem** – prementum; **pta** – posterior tentorial arm; **ptp** – posterior tentorial pit; **p1** – 1st palpal segment; **p2** – 2nd palpal segment; **p3** – 3rd palpal segment; **p4** – 4th palpal segment; **p5** – 5th palpal segment; **stp** – stipe. *Nyssomyia intermedia*

the frontoclypeal suture, between the two anterior tentorial pits – morphologically, the area dorsal to the insertion of the antennae, the post-front, is frequently called the front and the area above the clypeus and below the antennae, the pre-front, is referred to as the **face** (McAlpine 1981). The **vertex** is not always clearly defined, as it is limited anteriorly by the front, laterally by the eyes and posteriorly by the **occiput**, and the upper half of the **post-cranium** represents the whole posterior surface of the head capsule. The occiput merges ventrally into the postgenae. This latter sclerite is the part of the postcranium situated laterally and below the occipital foramen. Ventrally the head capsule is closed by a sclerite (**hypostomal bridge**) resulting from the median approximation and fusion of the postgenae, in agreement with Crampton (1942), *apud* McAlpine (1981). The narrow sclerite is situated dorsally to the occipital foramen, and the **postocciput** is separated from the postcranium by the **postoccipital suture.** The appendages are represented by the antennae and the mouthparts. The endoskeleton consists of the tentorium, which presents three pairs of arms, namely, the anterior, the dorsal and the posterior. The anterior arms spring from the anterior tentorial pits, the posterior ones from the posterior tentorial pits, and the dorsal ones, which are believed to be a prolongation of the anterior arms, are situated at the junction of this arm and the posterior one.

The **eyes** (Figs. 5, 6 and 7) consist of a variable number of eye facets which are also of variable diameter. The length of the eye may be greater than, equivalent to or less than half that of the capsule, and the width also varies, thus affecting the internal ocular distance. The eye is separated from the sclerite which is adjacent to it by the ocular suture; the narrow area situated between this suture and the ommatidia is the ocular sclerite. In a dorsal view, roughly in the middle of its ocular suture, the **post-frontal suture** begins and is commonly called the **interocular suture** in the phlebotomines. This latter may be longer or shorter and be united or not to the interantennal suture.

The **vertex** can be seen in Figs. 5 and 6. In the majority of phlebotomine species the limits of this sclerite are not evident. However, in some species of *Warileya* it is clearly outlined by the post-frontal sutures (interocular suture) and the vertex occipital suture, and presents some hairs (Fig. 6). In some species of other groups it is possible to see it by virtue of a lack of pigmentation in the area corresponding to the vertex occipital suture.

The **front** (Figs. 5 and 6) is the area situated dorsal to the insertion of the antennae, delimited by the interocular sutures and which presents protuberances that may be more or less developed.

The **face** is a greatly reduced sclerite in the phlebotomines. It is delimited dorsally by the area of insertion of the antennae, ventrally by the frontoclypeal suture and laterally by the eyes.

The **clypeus** (Figs. 5 and 6) is situated between the face and the labrum. Dorsally, it is limited by the frontoclypeal suture and anteriorly by the clypeolabral articulation. Its length is variable as regards the head, and hairs may cover its surface totally or partially and it is rarely bare.

The regions lateral to the clypeus and between the inferior edge of the eye and the subcranial cavity are the **gena** (Figs. 5, 6 and 7) that are united to the **postgena**

(Fig. 7) situated laterally in the ventral region of the head. The clypeogenal suture separates the gena from the clypeus. This suture in some groups of phlebotomines is well marked all along the lateral edge of the clypeus and in others only on the anterolateral edge of the clypeus. The mandibles are inserted in the anterior region of the gena.

The **occiput** (Figs. 5, 6 and 7), in its dorsal area, presents a hairy covering that may occupy its whole surface (Fig. 6) or be disposed as an *x* or in arrow shape (Fig. 5).

The mouth parts form a tube, the **proboscis**, with three unpaired and two paired pieces. The former are the **labrum-epipharynx**, in a dorsal position, the **labium** in a ventral position and the **hypopharynx** between them. The paired pieces, also situated between the labro-epipharynx and the labium, in an anterolateral position, are the **mandibles** and posterolaterally, the **lacinia of the maxillae**.

The **labrum** arises from a single preoral lobe (Snodgrass 1944, *apud* McAlpine 1981). Its ventral surface is denominated the **epipharynx**, despite being an integral part of the labrum. A small sclerite, called the **torma**, occurs laterally at the base of the epipharynx and is connected with the clypeus at the lateral terminations of the labroclipeal suture. Externally, the labrum is joined to the clypeus by the clypeo-labral suture and internally, the epipharynx forms the top of the food canal (McAlpine 1981). The **labrum-epipharynx**, blade-shaped, has two or more pairs of spiniform sensilla (apical sensilla) at its apex, spiniform teeth implanted apicolaterally in the dorsal surface and teeth with a slightly more laminated aspect in the ventral one (Fig. 8). Four spiniform sensilla are present in the preapical region of the food canal and others, also spiniform, in variable number may occur throughout the food canal.

The **hypopharynx**, also blade-shaped, contains the salivary canal. It presents many teeth, like a slender and deep fringe in males on the apicolateral edge, but in

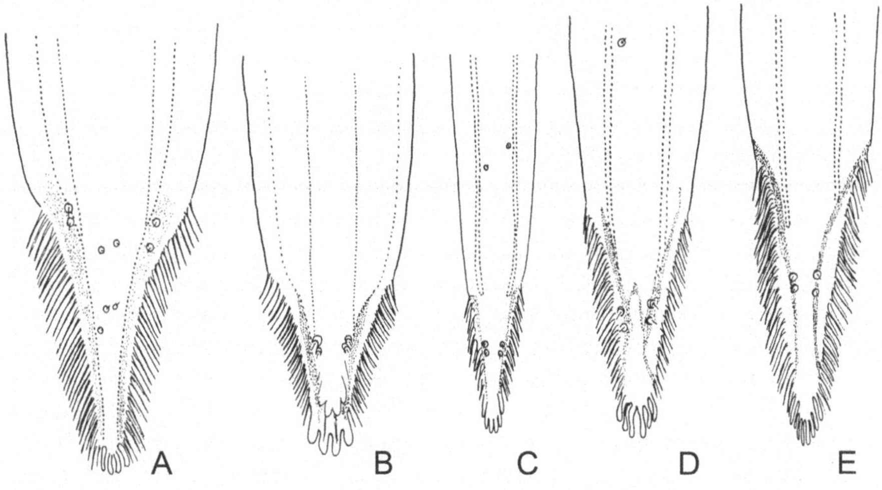

Fig. 8 Apical region of the labrum-epipharynx of phlebotomine females. **A** – *Warileya phlebotomanica*; **B** – *Micropygomyia vexator*; **C** – *Sciopemyia microps*; **D** – *Lutzomyia longipalpis*; **E** – *Lutzomyia peruensis*

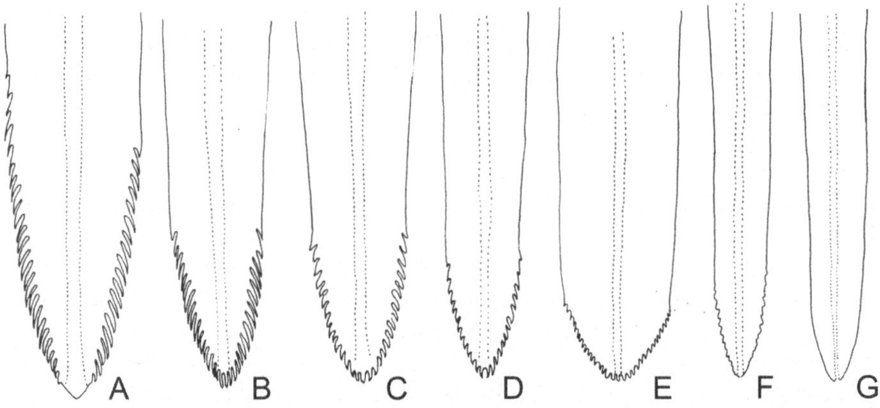

Fig. 9 Apical region of the hypopharynx of phlebotomine females. **A** – *Warileya phlebotomanica;* **B** – *Nyssomyia intermedia;* **C** – *Lutzomyia longipalpis;* **D** – *Trichopygomyia longispina;* **E** – *Micropygomyia vexator;* **F** – *Sciopemyia sordellii;* **G** – *Micropygomyia cayennensis*

the females the teeth vary in size, from slender and deep to wide and flattened, like those of a saw (Fig. 9-F, G), though sometimes they are altogether absent. Species of Sergentomyiina, with their atrophied hypopharynx, whether with flattened or no teeth, have been associated with feeding on cold-blooded animals (Hennig 1972; Lewis 1975).

The **labium** (Figs. 7 and 10) arises from the second pair of the united maxillae and consists of the proximal **post-mentum** and the distal **pre-mentum** (teca). The post-mentum is reduced and the main sclerite is the pre-mentum. At its distal extremity this latter bears the first and second labella derived from a pair of bi-segmented labial palpi (McAlpine 1981). On the median longitudinal region of the pre-mentum occur sutures that may be united close to the labella, forming the **labial fork** (Figs. 7 and 10-A, C), though in some species (*Lutzomyia dispar* and *Lutzomyia fonsecai*) or species of *Deanemyia* and practically all species of *Sergentomyia,* this does not happen (Fig. 10-B).

The **mandibles** are functional only in the females. They are absent in the majority of the males, but at least in one species of *Warileya*, though well less developed than in the females, they are present (Galati and Cáceres 1999). Their articulations occur at the mandibular joint situated in the antero-lateral part of the clypeus/gena (Figs. 5 and 6). They are flattened blade-like pieces and their apex may be more or less pointed (Fig. 11). Highly saw-like teeth are present on their internal apical edge.

The **maxillae** are shown in Figs. 7, 12 and 13. Each maxilla is formed of a basal region, the **cardo** that is united to the stem-like **stipes**. This supports the **lacinia** and the **maxillary palpus**, formed of five segments (Cumming and Wood 2009). The **lacinia** may present a set of **teeth** on the apical external edge (**external teeth**), disposed longitudinally (Figs. 12-B, D) or transversally (Fig. 12-E) to the edge; they may also be in a single row (Fig. 12-B) or in two longitudinal rows (Figs. 12-C, D). Teeth are also present on the internal edge (**internal teeth**), more or less close to the

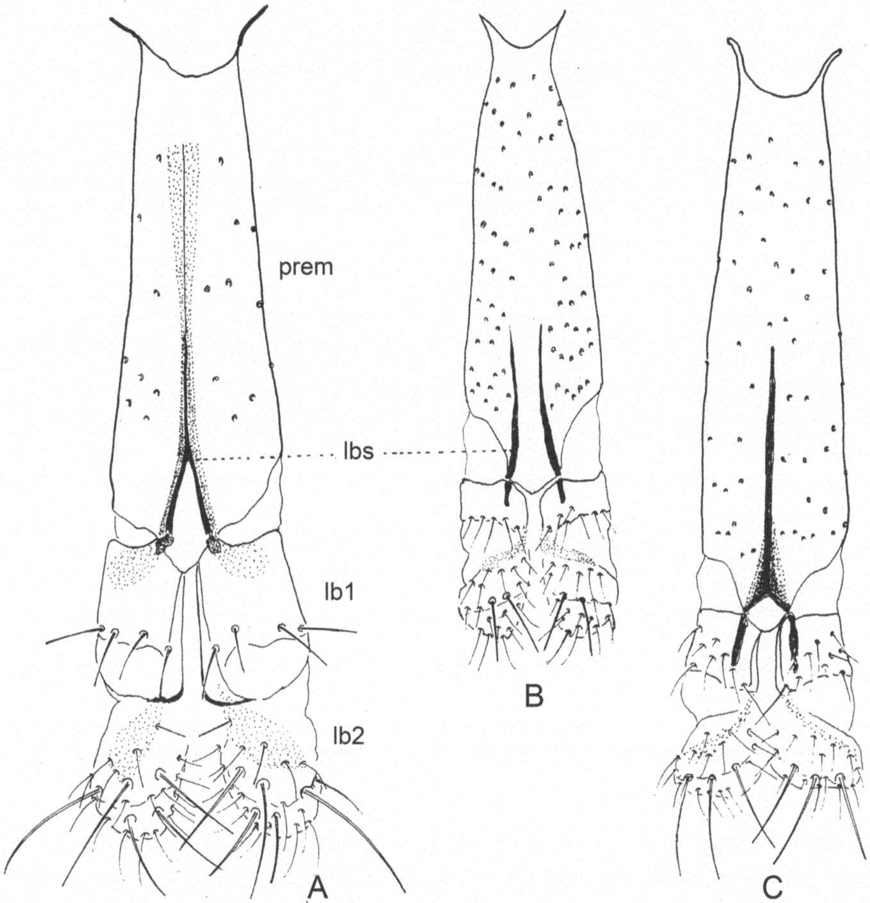

Fig. 10 Labium of phlebotomine females in ventral view: **lb1** – labellum I; **lb2** – labellum II; **prem** – prementum; **lbs** – labial suture. **A** – *Warileya phlebotomanica*; **B** – *Lutzomyia dispar*; **C** – *Psathyromyia shannoni*

apex. The five palpal segments vary in length, all of them present sensorial elements, represented by deciduous hairs that are shed during the maceration process and leave scars; some segments bear tiny hyaline club-shaped sensilla, which for sandflies were first illustrated and denominated modified spines by Newstead (1911), thus since called Newstead's spines, though due to their non-spiniform shape they are more appropriately denominated **Newstead's sensilla.** The arrangement of these sensilla in clusters in the basal half of palpus 3 represents the plesiomorphic state, their dispersion through almost the entire segment, the apomorphic state. Newstead's sensilla are always present on palpus 3, though frequently they may also be present on palpus 2 (Fig. 13-B), occasionally on palpus 4 (Fig. 13-F) and, though rarely, on palpus 5. On palpus 3, these sensilla may be inserted as a well delimited set (cluster) on the basal third, or on the middle of the segment (Figs. 13-C, E), or dispersed

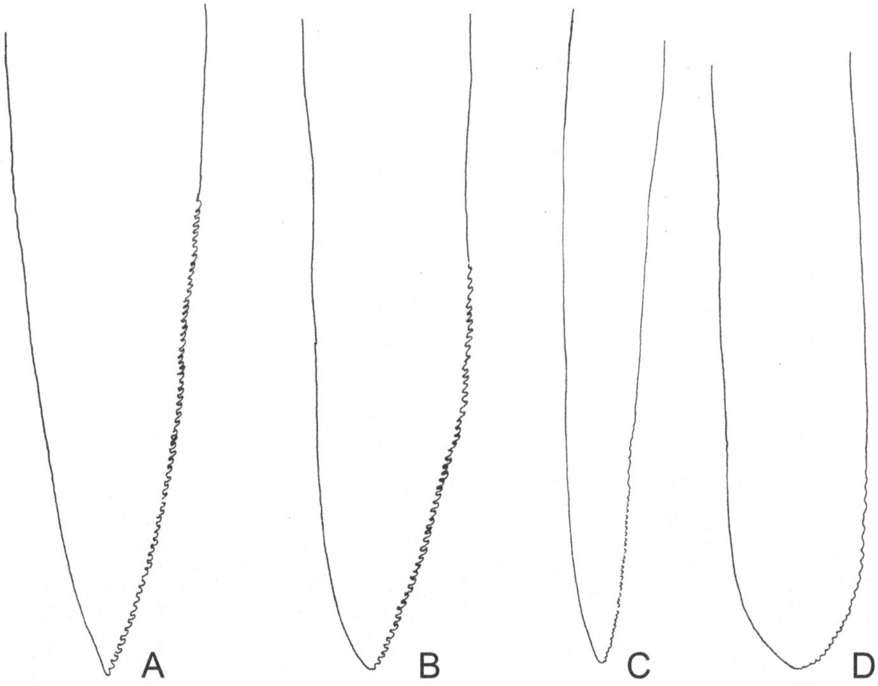

Fig. 11 Apical region of the mandibles of phlebotomine females. **A** – *Warileya phlebotomanica;* **B** – *Lutzomyia longipalpis;* **C** – *Sciopemyia microps;* **D** – *Micropygomyia vexator*

along the segment (Fig. 13-D). This arrangement may help to distinguish different phlebotomine groups. The presence or absence of these sensilla on palpus 2 might also distinguish groups and on the other segments, species. The number of spiniform setae (Figs. 13-C, E) on palpus 4 also permits the distinction of groups within the subgenus *Helcocyrtomyia.* The palpal formula has been used to designate the relationship between the lengths of successive segments of the palpus. Thus, when the first palpal segment is smaller than the second, the second smaller than the third, the third smaller than the fourth and the fourth smaller than the fifth, the palpal formula 1.2.3.4.5 or 1,2,3,4,5 is adopted. When the segments are of equal length, they are put between parentheses, e.g. 1.(2.3).4.5. Among the phlebotomines it seems that the primitive formula is that represented by 1.2.3.4.5, because this is also present in the Bruchomyiinae, the sister group of the Phlebotominae. One first modification seems to have been the shortening of palpus 4 and concomitantly the lengthening of palpus 3, so that the formula then became 1.(2.4).3.5. or 1.2.4.3.5; palpus 4 then continued to become shorter and palpus 2 grew longer, thus giving rise to the formula 1.4.2.3.5 or 1.4.(2.3).5. The palpus 5 that was initially much longer than the sum of 3 and 4 together, underwent shortening, with the consequent inversion of the relationship. The process of the shortening of palpus 5 continued, thus it became shorter than palpus 3 and then shorter than palpus 2, the formula: 1.4.5.2.3 thus arising.

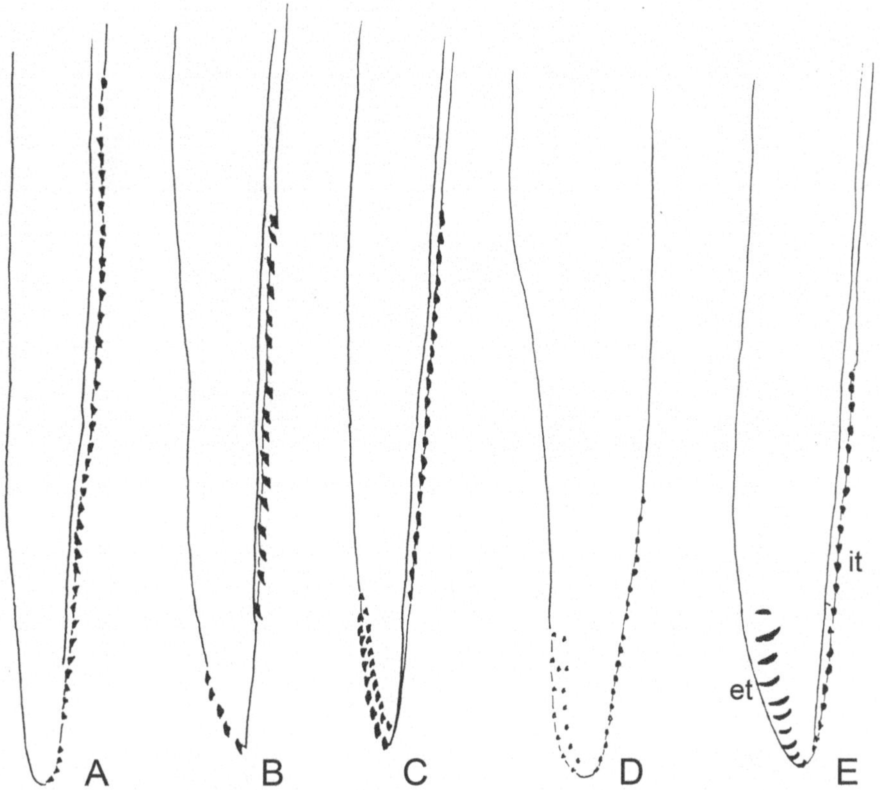

Fig. 12 Lacinia of the maxilla of phlebotomine females: **et** – external teeth; **it** – internal teeth. **A** – *Warileya phlebotomanica;* **B** – *Lutzomyia cruciata;* **C** – *Nyssomyia intermedia;* **D** – *Micropygomyia quinquefer;* **E** – *Micropygomyia longipennis*

Each **antenna**, in accordance with the basic plan of nemathocerous, is formed of three distinct parts or **antennomeres (A)**. The first (AI), called **scape**, is a cylindrical segment inserted in the face; the second (AII) or **pedicel**, a globe-like segment, in which Johnson's organ, a mass of receptor cells for the detection of movements of the flagellum (McAlpine 1981), is present, and the third the **flagellum**, consisting of 14 articles, called the **flagellomeres**. The most basal of these latter (fI) presents about twice the length of fII; the length of the others diminishing gradually to fXIII and sometimes to fXIV.

The length of flagellomere I, whether as an isolated measurement, or expressed as a ratio of the lengths f1/head, or f1/labrum-epipharynx (le), as well as the fXIV/fXIII ratio, permits the distinction between species and groups of species.

The flagellomeres are covered by deciduous hairs, which are shed during the maceration process, and nondeciduous setae. These latter may be **simple setae** (Figs. 15, 16 and 17), more numerous on the apical segments, or with a hyaline blade aspect and called **ascoids** or geniculate spines (the insertion of the latter in the

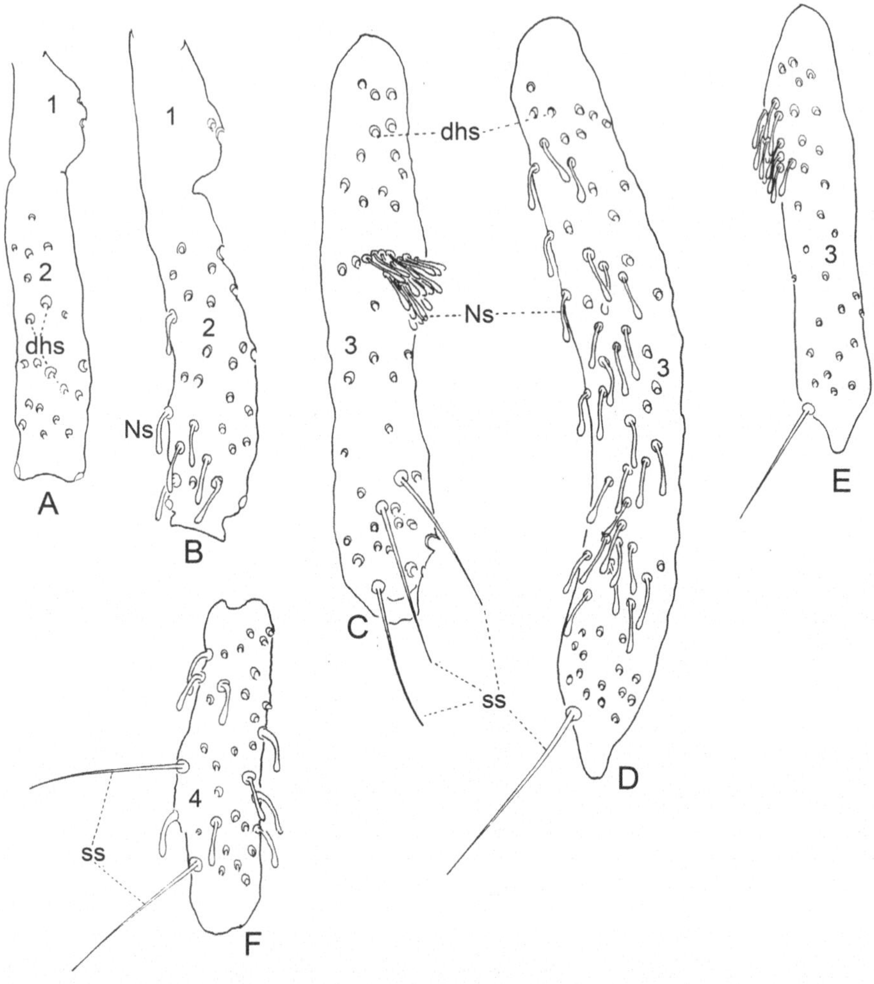

Fig. 13 1st–4th palpal segments of phlebotomine. Ns – Newstead's sensilla; **ss** – simple setae; **dhs** – deciduous hair scars. **A** – *Micropygomyia cayennensis;* **B** – *Psathyromyia aragaoi;* **C** – *Warileya phlebotomanica;* **D** – *Pintomyia fischeri;* **E** – *Mi. cayennensis;* **F** – *Pa. aragaoi*

segment is elbow-like) (Figs. 14, 15, 16 and 17). These are of highly variable length and may (Fig. 15) or may not (Fig. 16) present a posterior spur extending to the base of the article. The ascoids may be inserted in the article on a well-defined peduncle (Figs. 15 and 16-B) or on an atrophied peduncle or are sessile (Fig. 16-A,C). The level of insertion of the ascoids in fI is an important character. So the external ascoid (as regards the mouthparts) may be inserted relatively nearer to the base than the internal one (Fig. 14-A) or the opposite (Fig. 14-C), or both may be on the same level (Fig. 14-B). The number of ascoids on the articles, especially on the apical ones, may also vary. The loss of the ascoids in the Phlebotominae occurred from the apical to the basal articles. The antennal formula, indicating the number of ascoids on each

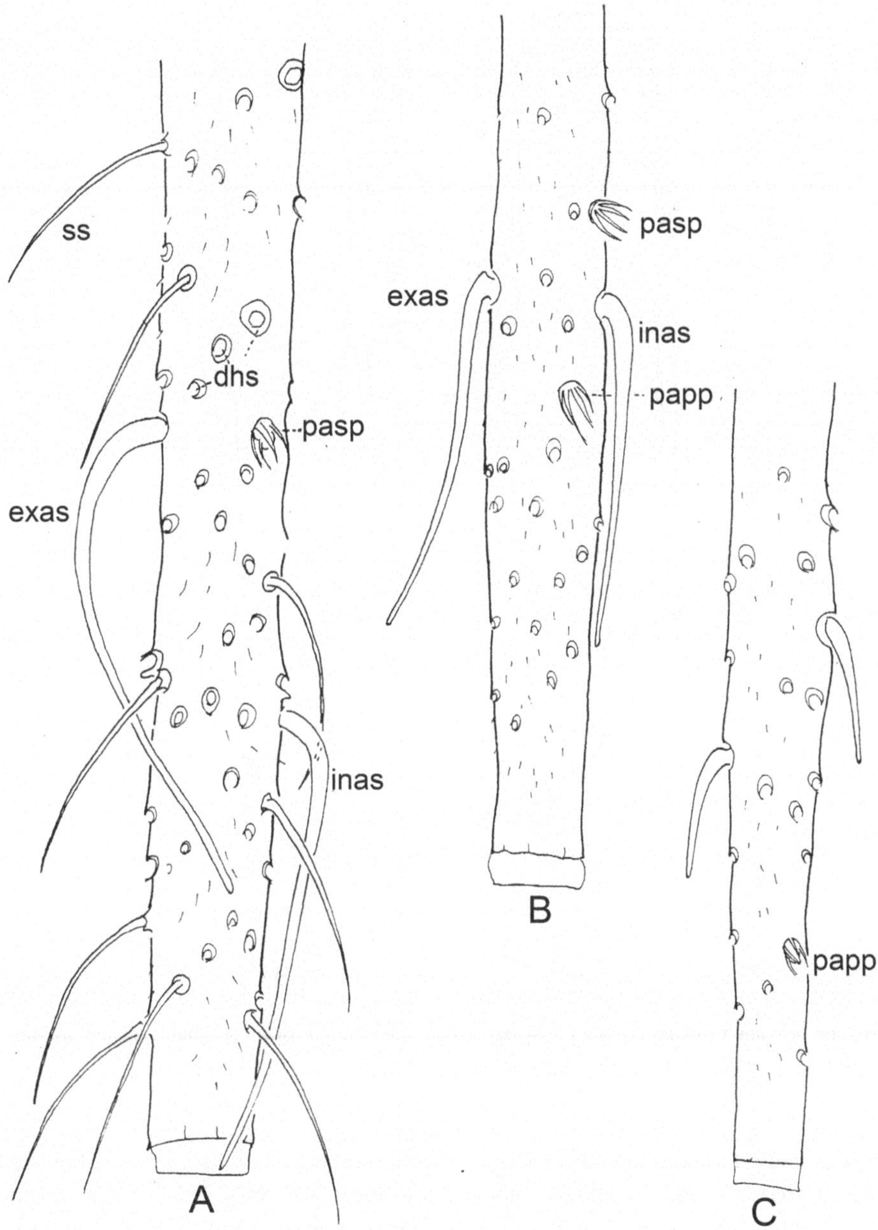

Fig. 14 First flagellomere (fI) of phlebotomine: **dhs** – deciduous hair scar; **exas** – external ascoid; **inas** – internal ascoid; **papp** – preapical papilla; **pasp** – preascoidal papilla; **ss** – simple seta. **A** – *Warileya phlebotomanica;* **B** – *Psychodopygus squamiventris;* **C** – *Pintomyia (Pifanomyia) verrucarum*

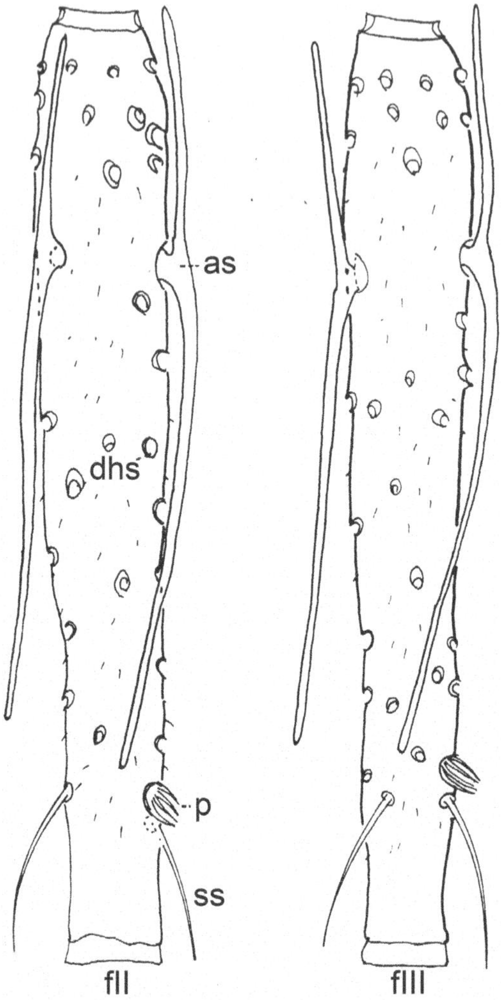

Fig. 15 2nd and 3rd flagellomeres of phlebotomine: **as** – ascoid; **dhs** – deciduous hair scars; **p** – papilla; **ss** – simple setae. *Psathyromyia shannoni*

article, is used to designate this variation. So for both sexes of the species, when the pair of ascoids occurs on all the articles, the formula is fI-fXIV 2; when it is absent only from the fXIV it is fI-fXIII 2, fXIV 0; when it is absent from fXIV and only one ascoid occurs on fXIII, the formula is fI-fXII 2, fXIII 1, fXIV 0, and so on.

Some flagellomeres present papillae, sensorial structures of a rosette shape that may be used for the distinction of groups of species. In 1912, Newstead described these structures for the three last segments and called them "hirsute glands". Parrot (1953) described them as "papilles", also present on fI and fII. On fI, commonly in the preapical area of the internal side of the flagellomere, distally to the insertion of

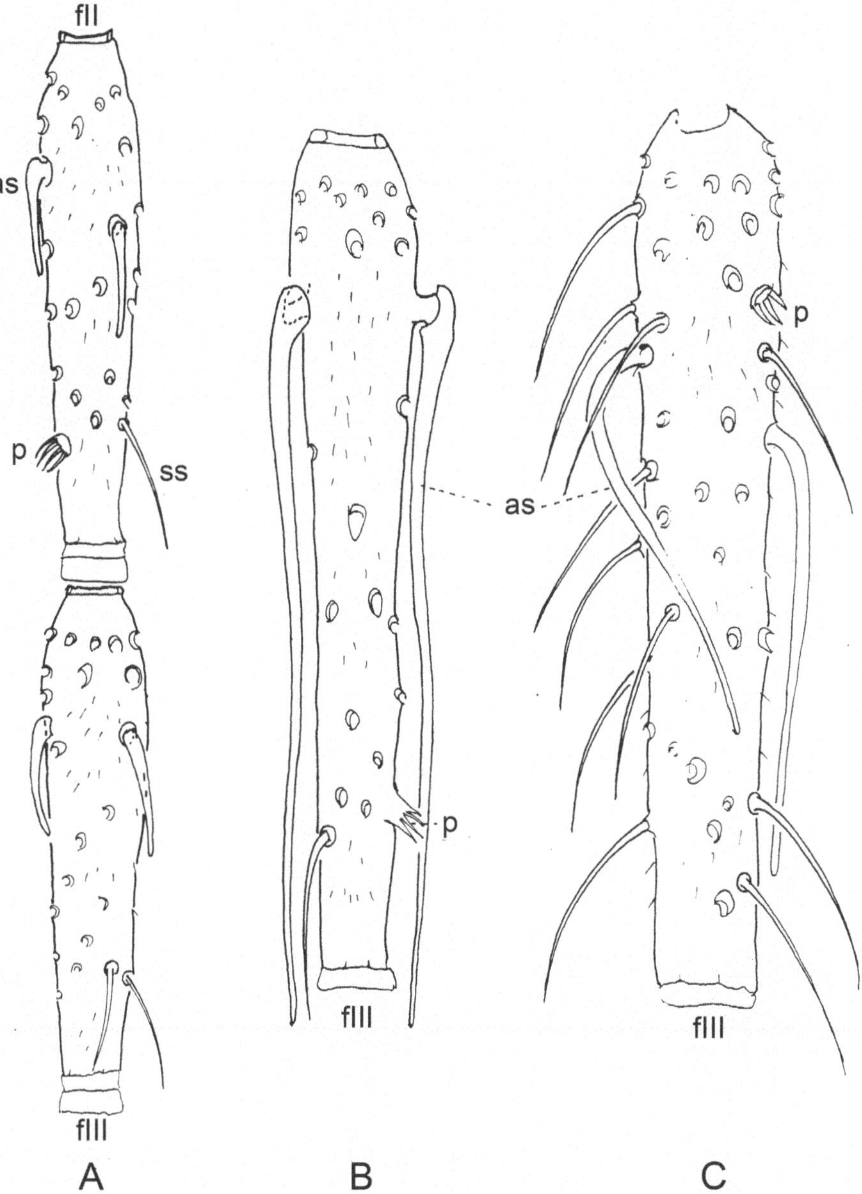

Fig. 16 2nd and 3rd flagellomeres of phlebotomine: **as** – ascoid; **p** – papilla; **ss** – simple seta. **A** – *Pintomyia* (*Pifanomyia*) *ovallesi*; **B** – *Psathyromyia lanei*; **C** – *Warileya rotundipennis*

the internal ascoid, there occurs one papilla, the **pre-apical papilla** (Fig. 14-C). In *Warileya* and *Hertigia* one papilla, called the **pre-ascoidal papilla**, situated nearer the base than the insertion of the internal ascoid, is found (Fig. 14-A). In some species of *Brumptomyia* and *Psychodopygus,* the two papillae may be present (Fig. 14-B), sometimes in this latter more than two papillae may occur – as is also observed in

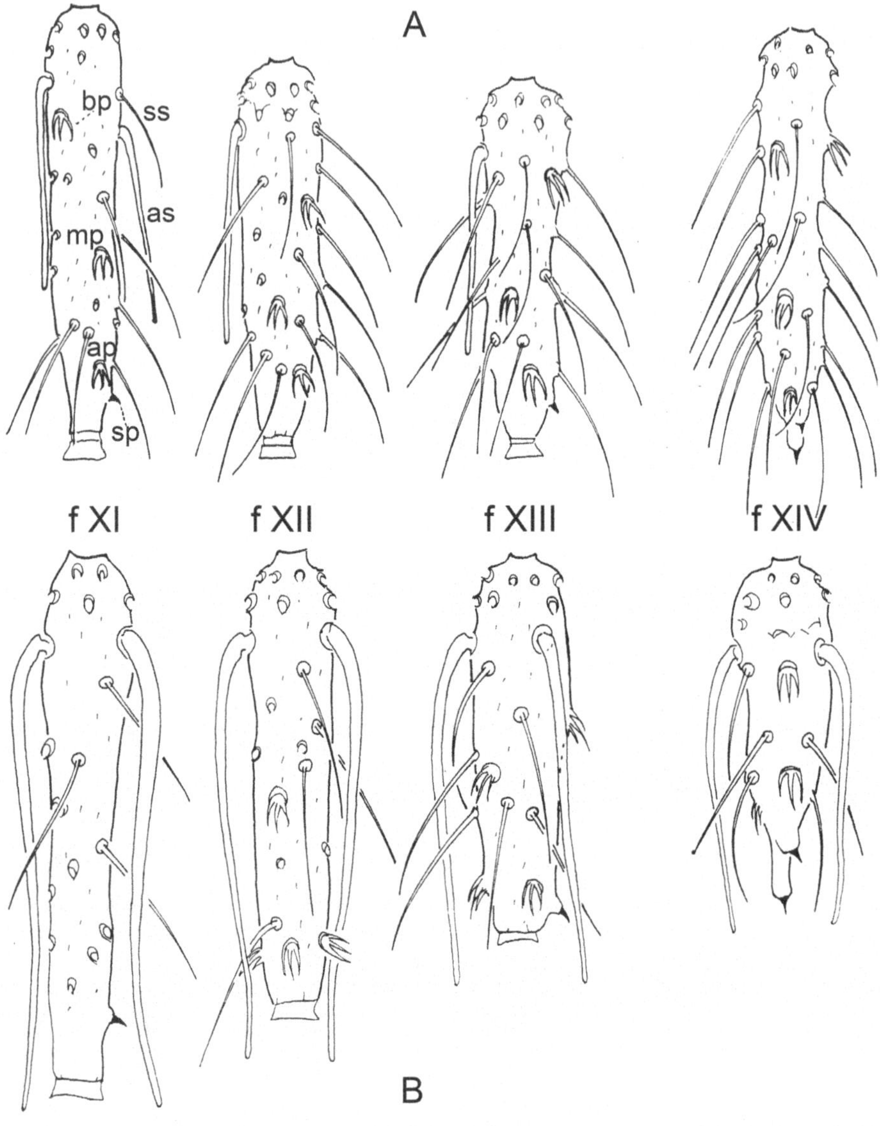

Fig. 17 Apical flagellomeres: fXI – fXIV. **ap** – apical papilla; **as** – ascoid; **bp** – basal papilla; **mp** – median papilla; **ss** – simple seta; **sp** – spiniform papilla. **A** – *Evandromyia tupynambai*; **B** – *Trichophoromyia ubiquitalis*

some Old World species (Parrot 1953). In fII, one papilla is always present, generally in the internal pre-apical area, but in *Warileya* and *Hertigia*, it is situated basally on the article. In fIII many groups have lost the papilla (Fig. 16-A) that, when present, is generally situated in the internal pre-apical area (Fig. 16-B) and, basally in Hertigiina (Fig. 16-C). On the apical segments there are papillae disposed in **apical**, **intermediate** and **basal** areas (Fig. 17-A). In the evolutionary process, the loss of the papillae occurred from fIII towards the apical segments. Some of them are always present on the three last segments of the antenna (fXII-fXIV) (Fig. 17). Some groups also present one or more papillae either on the articles fIX-fXI or fX-fXI, or fXI. A few groups present papilla(e) on fVII-fVIII. There is a **spiniform papilla**, which when magnified up to 400 times, may be confounded with the apical papilla on the apical segments fXIV, fXIII, fXI and fIX (Fig. 17).

The **cibarium** (Figs. 18, 19 and 20) is a cavity formed by plates of dorsoventral sclerites united laterally where they are more sclerotized, thus constituting the **lateral sclerites** (suspender sclerites; Jobling 1987). Dorsally, at the posterior extremity, the lateral sclerites extend transversally and may be united, thus forming the **posterior bridge**. The roof of the cavity may present a posterior protuberance (Theodor 1965) that appears as a curved line, close to the area of junction with the posterior bridge, i.e. when the cibarium is observed dorsally (it seems that this configuration occurs by virtue of a fold formed in the roof, where this and the inferior edge of the posterior bridge join). An area with stronger sclerotization, a **sclerotized area**, in which the posterior clypeal muscles are inserted, may be present in the roof. Various sets of teeth occupying different areas and positions in relation to the lumen and constituting the **bucal armature** may be present on the posterior edge of the ventral surface or mouth floor. When the teeth are implanted close to the lateral sclerites with orientation in a supramedian direction, they are called **lateral teeth**. The teeth implanted in the central area, oriented in the same direction as the lumen, are denominated **horizontal teeth**. In the majority of the phlebotomines, the posterior row of the teeth in the central area has remained in a horizontal position and the **anterior teeth** have assumed a vertical position in relation to the lumen, commonly called **vertical teeth**. Further, on the ventral surface, in an anterior position or above the anterior teeth, the lateral sclerites might present projections that form the sclerotized arch (loral arm, Jobling 1987), in which the muscles of the salivary pump are inserted. The link between the cibarium and the pharynx occurs at the posterior bridge; ventrally the pharynx projects as far as the posterior sclerotized edge of the floor, which is generally situated far from the apex of the cibarium, forming the **cibaral chamber**, less sclerotized, where the posterior and lateral teeth are situated and surrounded by the base of the pharynx. The vertical teeth are implanted in the sclerotized base, which seems to reinforce their role as grinders of particles when in contact with the sclerotized area, while the horizontal teeth act as particle selectors, like a rake (Lewis 1975).

Important modifications have occurred in the cibarium and are useful in the distinction of the groups. The existence of the cibarium in various groups of Phlebotominae suggests that the evolution of the armature obeyed the following sequence; originally, it would have been formed of an ample chamber without teeth or with only small lateral ones, as may be deduced from Hertigiina and Phlebotomina (Fig. 19-A, B). The lateral teeth have become more developed and come to occupy a horizontal position in relation to the lumen, as has occurred with Brumptomyiina (Fig. 19-C) and a few species of Lutzomyiina (Fig. 19-D). At a second point in time, the horizontal and lateral teeth would be maintained, though several of these came to adopt a vertical position in relation to the lumen (Fig. 20-A, B, E, F). At a third time point, the horizontal and vertical teeth remained, the lateral ones disappearing (Fig. 20-C, H, I).

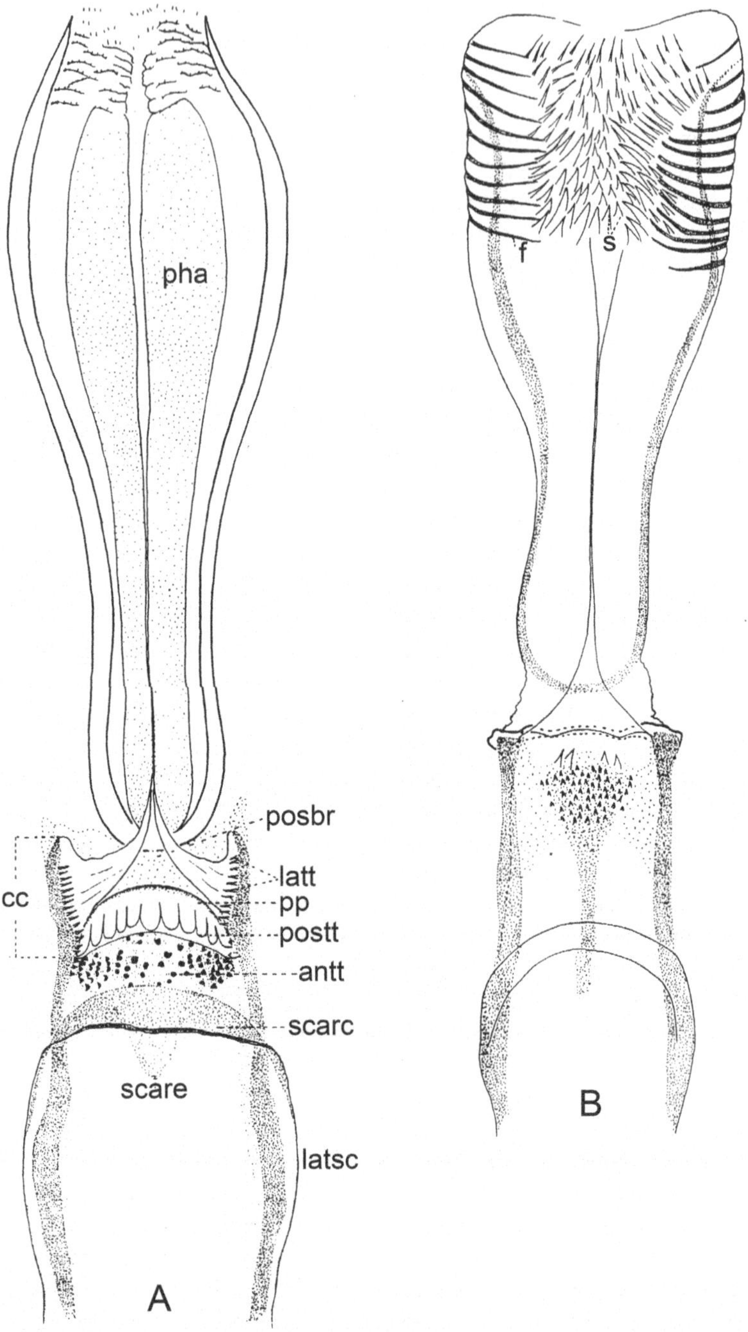

Fig. 18 Cibarium and pharynx of phlebotomine females: **antt** – anterior teeth; **cc** – cibarial chamber; **f** – fold; **latt** – lateral teeth; **pha** – pharynx; **posbr** – posterior bridge; **postt** – posterior teeth; **pp** – posterior protuberance; **latsc** – lateral sclerite; **scare** – sclerotized area; **scarc** – sclerotized arch; **s** – spine. **A** – *Bichromomyia flaviscutellata*; **B** – *Micropygomyia atroclavata*

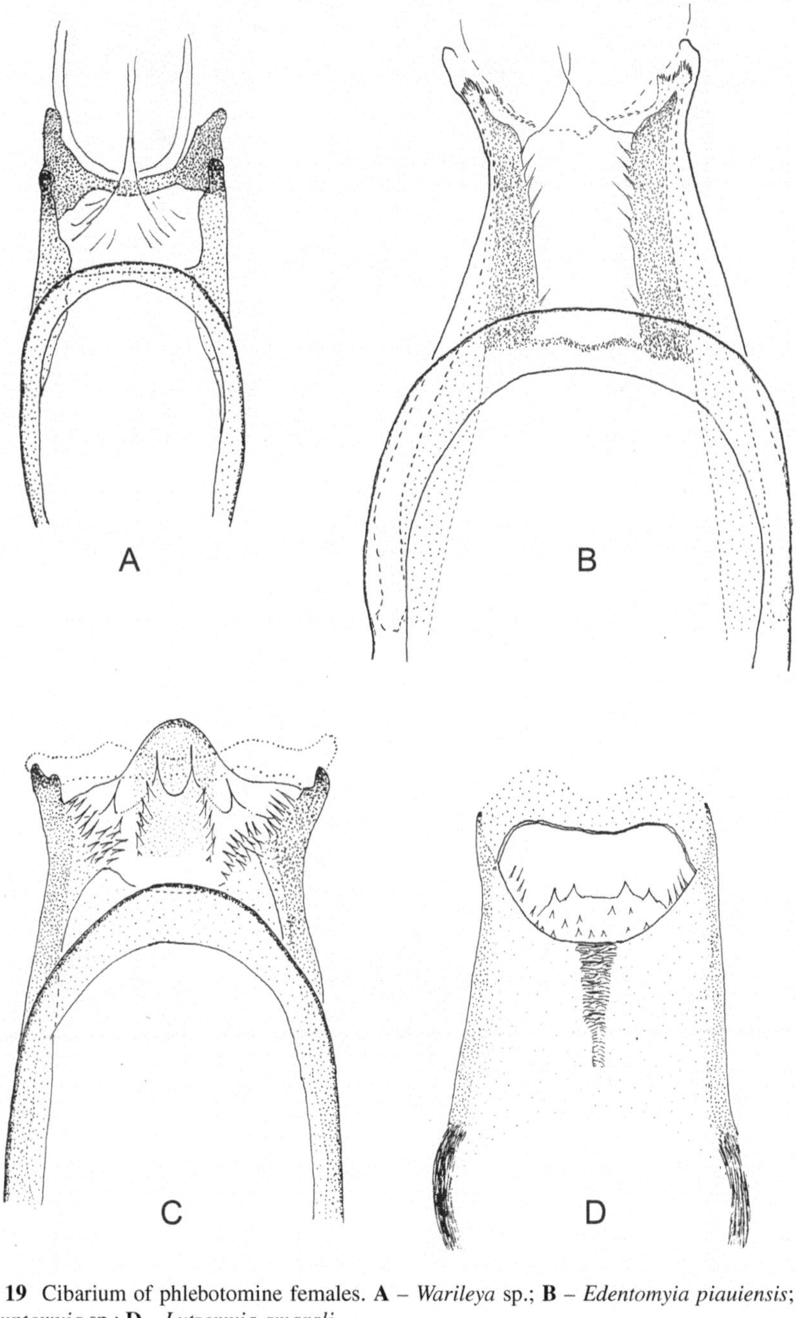

Fig. 19 Cibarium of phlebotomine females. **A** – *Warileya* sp.; **B** – *Edentomyia piauiensis*; **C** – *Brumptomyia* sp.; **D** – *Lutzomyia amarali*

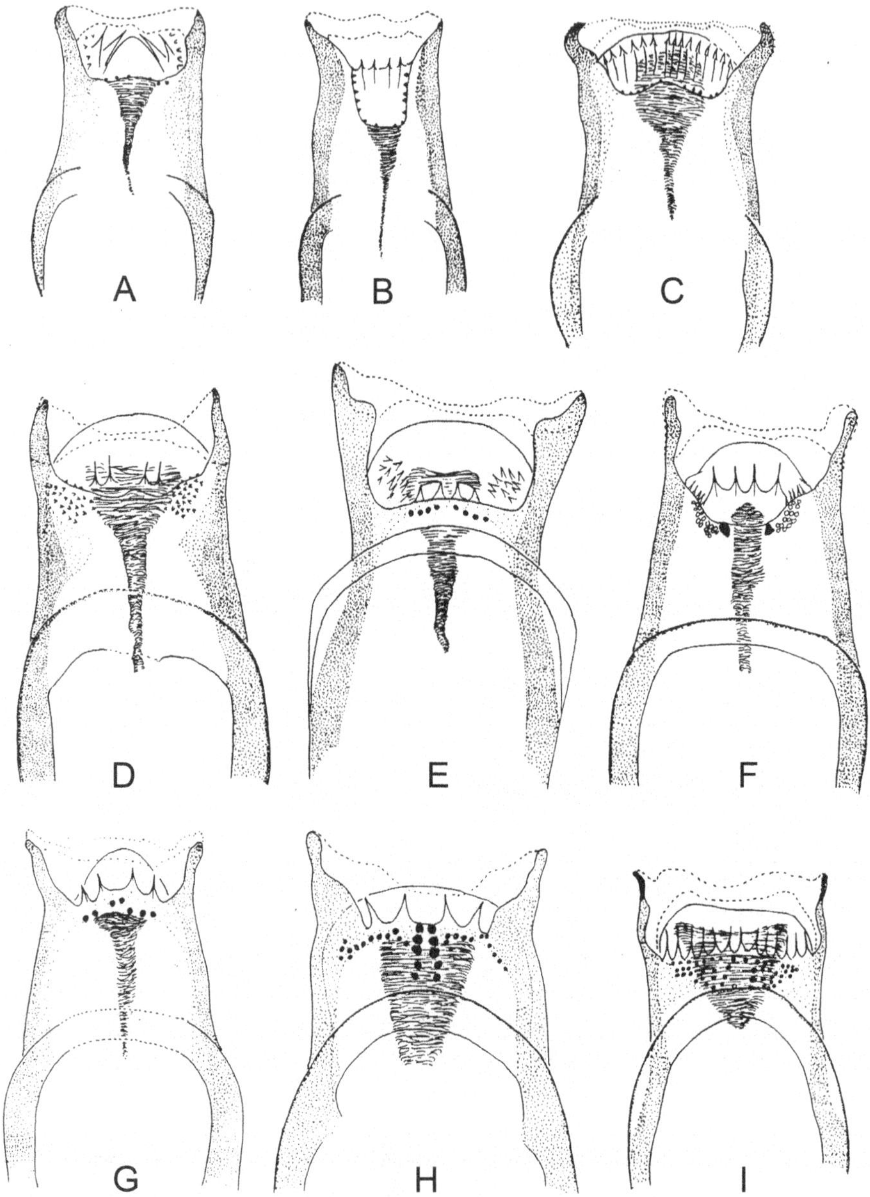

Fig. 20 Cibarium of phlebotomine females. **A** – *Sciopemyia sordellii*; **B** – *Micropygomyia pilosa*; **C** – *Micropygomyia cayennensis cayennensis*; **D** – *Evandromyia walkeri*; **E** – *Dampfomyia anthophora*; **F** – *Pressatia triacantha*; **G** – *Lutzomyia cruciata*; **H** – *Psychodopygus panamensis*; **I** – *Trichophoromyia ubiquitalis*

The horizontal teeth, initially, would have been set in four columns (Fig. 19-C); the most posterior would have remained in this position and the anterior teeth would have tended to adopt a vertical position. The posterior teeth (horizontal) in some groups, would also have increased in number (Fig. 20-C, I) and the base of some of them, originally individual, would have grown together. When this fusion occurred, with the increase in the number of teeth, the set assumed the aspect of a palisade (Fig. 20-C). In another situation the fusion of the apices of some teeth must have occurred (Fig. 20-E). The absence of the lateral teeth would have been accompanied by the shortening of the cibarial chamber (Fig. 20-C, H, I).

The cibarium without the posterior protuberance, sclerotized area and armature or with only small lateral teeth and complete sclerotized arch seems to have been the primitive condition. In accordance with Hennig (1972), who based his argument on the loss of the mandible among males and the concomitant atrophy of the cibarial armature, the cibarium of the females of Phlebotominae, as observed in *Warileya*, *Hertigia* and *Phlebotomus*, has evolved in such a way as to eliminate the cibarial armature.

From what has been observed in the males, it may be deduced that the atrophy actually occurred in the cibarial armature, despite its still maintaining its similarity with the female cibarium, as is shown by the traces in the latter of the horizontal, vertical and even lateral teeth. On the other hand, the presence of only lateral teeth in *Chinius* (an Asiatic genus of Phlebotominae) and in Sycoracinae (another subfamily of Psychodidae) seems to corroborate the sequence suggested.

The **pharynx**, also called the esophageal pump, consists of three plates, one dorsal and two lateral (Forattini 1973), forming a sac-shaped structure, having transversal sclerotized wrinkles and spines on its posterior part. In the majority of the American groups of Phlebotominae the transversal wrinkles are not well sclerotized and the spines are atrophied (Fig. 18-A), but in many species, mainly those of Sergentomyiina, the spines may be well developed and in a few species the transversal wrinkles appear as strongly sclerotized transversal bands (Fig. 18-B). The evolution of the pharynx probably occurred through the atrophying of the spines that is generally accompanied by the further development of the cibarial armature.

Cervix

The cervix is the region that forms the link between the head and the thorax. It is predominantly a membranous structure having on its lateral parts a rectangular sclerite, the **cervical sclerite** (Figs. 21 and 25), into which two or three spiniform sensilla are inserted. The upper anterior part of this sclerite is linked to the head at the junction of the posterior arm of the tentorium and post-occiput, and its basal posterior part is connected to the proepisternum close to its connection with the pronotum. In the ventral area, the cervix may or may not present a pair of spiniform sensilla, the **ventro-cervical sensilla**, very much like those of the cervical

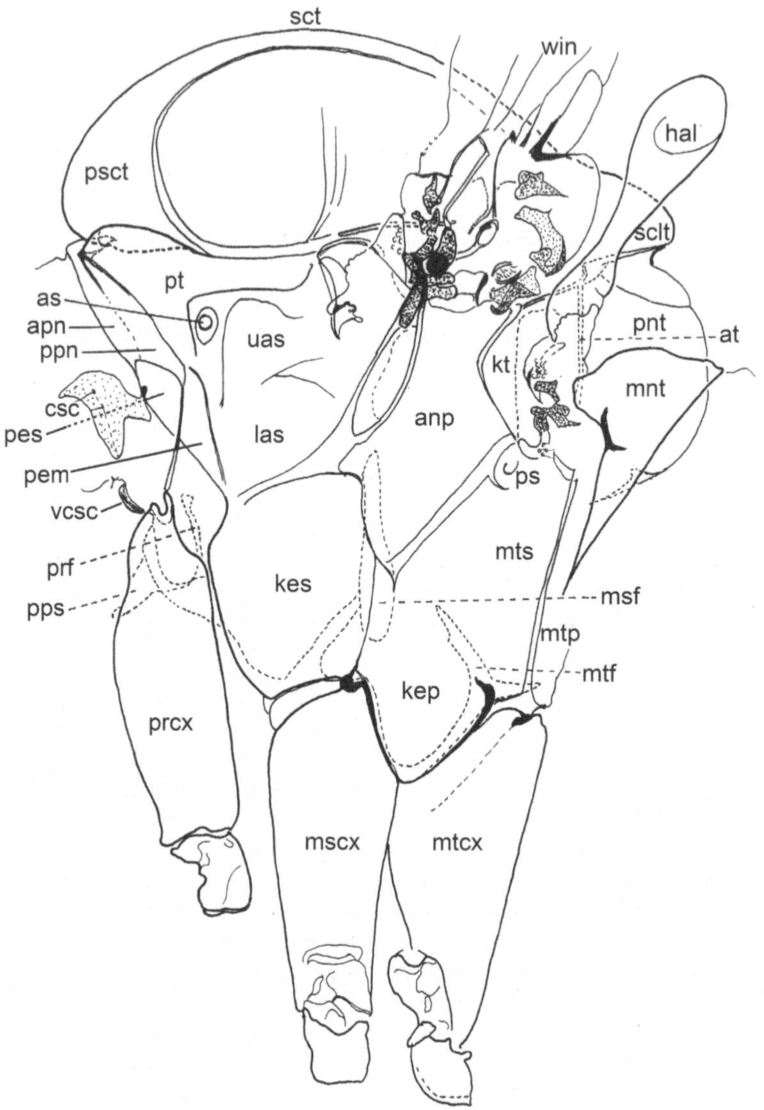

Fig. 21 Cervix and thorax of Phlebotominae. **Cervix**: **csc** – cervical sclerite with a pair of sensilla; **vcsc** – ventrocervical sclerite. **Thorax**: **anp** – anepimeron; **apn** – antepronotum; **as** – anterior spiracle; **at** – anatergite; **hal** – halter; **kep** – katepimeron; **kes** – katepisternum; **kt** – katatergitum; **las** – lower anepisternum; **mnt** – metanotum; **mscx** – mesocoxa; **msf** – mesofurca; **mtcx** – metacoxa; **mtf** – metafurca; **mtp** – metepimeron; **mts** – metepisternum; **pem** – proepimeron; **pnt** – postnotum; **ppn** – postpronotum; **pps** – protuberance of the prosternum; **prcx** – procoxa ; **prf** – profurca; **ps** – posterior spiracle; **psct** – prescutum; **pes** – proepisternum; **pt** – paratergite; **sclt** – scultellum; **sct** – scutum; **uas** – upper anepisternum; **win** – wing. *Lutzomyia peruensis*

sclerites, but inserted in a membranous base located anterior to a small single triangular sclerite, the **ventro-cervical sclerite** (Figs. 21 and 25), situated immediately anterior to the prosternum. The ventro-cervical sclerite, according to Matsuda (1970), was denominated presternum for other groups of Diptera by Crampton (1942), Bonhag (1949) and Sarà and Smerdel (1953), but is probably a secondary cervical sclerite, because the presternum is absent in Mecoptera and in inferior "nematocerous". In Psychodidae, the ventro-cervical sclerite is very close to the basisternum of the prosternum.

Thorax

The thorax consists of three segments and, as in the other Diptera, presents a considerably more developed median segment, because of the insertion of the muscles of the single pair of wings. Dorsally, it is formed by the sclerites known as the **notum**, laterally, the **pleuron** (pl. pleura) and ventrally the **sternum**. As regards each sclerite, the prefixes pro, meso and meta, which define part of the each sclerite, will be used as distinct from pre and post (McAlpine 1981). The segments of the notum may present divisions; those of the pleura are divided by a longitudinal suture, the **pleural suture** in the **episternum** and the **epimeron** and the mesothorax, the most highly developed, is divided transversally, thus forming two more sclerites that are situated in the upper position with the prefix *an* and the lower position *kat*; so they are the **anepisternum**, the **katepisternum** and the **anepimeron** and **katepimeron**. The sternum also presents divisions.

The thoracic appendages are a pair of wings, belonging to the mesothorax, a pair of halters situated in the metathorax, both implanted between the notum and the pleura and a pair of legs inserted in each segment between the pleura and the sternum.

Prothorax

The **pronotum**, dorsally, is represented by a narrow band that is united to the mesonotum in its frontal part and laterally is firmly united to the angle of the paratergite. The **transnotal anterior suture** arises at this point and extends to the proepisternum, separating the **antepronotum** from the **postpronotum**; this suture is not always very distinct throughout its length, only being more evident in its upper part. The postpronotum is separated from the paratergite and from the mesopleura by the tenuous **posterior transnotal suture** that occurs as a continuation of the pleural suture.

The **propleuron** is formed by two sclerites, one in an upper anterior position, called the **proepisternum** and the **proepimeron** situated inferoposteriorly, separated by the **propleural suture** that arises at the articulation of the anterior coxa and extends as far as the posterior transnotal suture. The proepimeron is not always clearly separated from the anepisternum and in view of this it has been called the lower anepisternum by phlebotomine specialists. The proepimeron may or may not present hairs, the **proepimeral setae**.

The **prosternum** (Fig. 25) is divided into two parts; the anterior denominated the **basisternum** and the posterior the **furcasternum**. The basisternum, situated between the anterior coxae, is endowed with a more sclerotized median inflexion that forms the **ridge of the basisternum**. From the basolateral region of the basisternum sprout hairy protuberances, **protuberances of the prosternum** (Abonnenc et al. 1971). Further, on this sclerite is found, basally, a more sclerotized transversal band, the **transversal apodema** (Speight 1969) that is linked, laterally to the **profurca**, to an endosternite in the shape of a cricket bat. Posterior to this apodema occurs the furcasternum, fused with the basisternum of the mesothorax; both these latter are linked by the basal laterals to the pit of the furca. This sclerite is delimited laterally by the pleural aresta and apically by the furcasternum of the mesothorax. In the evolutionary process the protuberance of the prosternum become more slender.

Mesothorax

The **mesonotum** practically covers the whole dorsum of the thorax. It is composed of three basic parts, the **prescutum**, **scutum** and **scutellum** and a fourth element, the **postnotum**. This latter is probably derived from an intersegmental acrotergite (McAlpine 1981). The prescutum is reduced and its separation from the scutum, on the dorsum, is not clear, because the **prescutal suture** continues the **parapsidal suture** with no clear distinction and the prescutal pits are not evident; the posterolateral edge of the prescutal area is crossed by a longitudinal suture [the probable base of the **lateral parapsidal suture** = prescutal suture (Matsuda 1970)] that delimits the lateral sclerite called **paratergite** by McAlpine (1981) or the prescutum by Matsuda (1970). The scutum occupies the greater part of the dorsum of the mesothorax and is incompletely divided by the **transversal suture** [=lateral parapsidal suture (Matsuda 1970)], into the prescutum, the **pre-sutural** and the posterior, **postsutural area**. In the laterodorsal region, the parapsidal sutures that extend into the area of the prescutellum are distinguishable. The **scutoscutellar suture** separates the scutum from the scutellum transversally. The slightly swollen area situated between the posterior base of the wing and the base of the scutellum is called the postalar **callus** and the edge of the postsutural area, immediately above the wing, the **supralar area**.

The setae of the mesonotum (Fig. 22) situated in clearly distinguishable sets in the prescutal area, are known as **prescutal setae**; those in the presutural area, the **presutural setae**; in the supralar area, the **supralar setae**; in the dorsocentral area, delimited by the parapsidal sutures, the **dorsocentral setae**; those situated on the parapsidal suture, the **parapsidal setae** and those localized posterior to the supralar setae and anterior to the scutum-scutellar suture (if present) the **postalar setae**

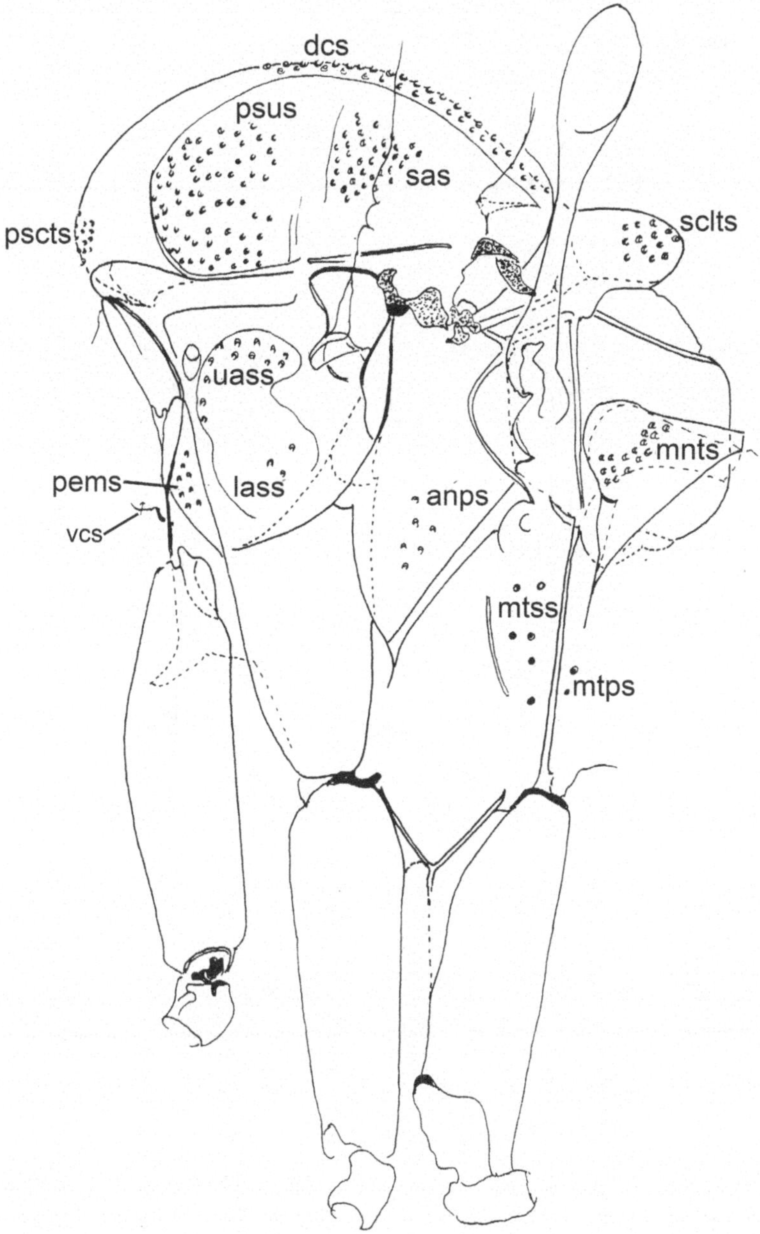

Fig. 22 Setae on the thorax of phlebotomine: **anps** – anepimeral setae; **dcs** – dorsocentral setae; **lass** – lower anepisternal setae; **mnts** – metanotal setae; **mtps** – metepimeral setae; **mtss** – metepsiternal setae; **pems** – proepimeral setae; **pscts** – prescutal setae; **psus** – postsutural setae; **sas** – supralar setae; **sctls** – scutellar setae; **uass** – upper anepisternal setae; **vcs** – ventrocervical sensilla. *Brumptomyia pintoi*

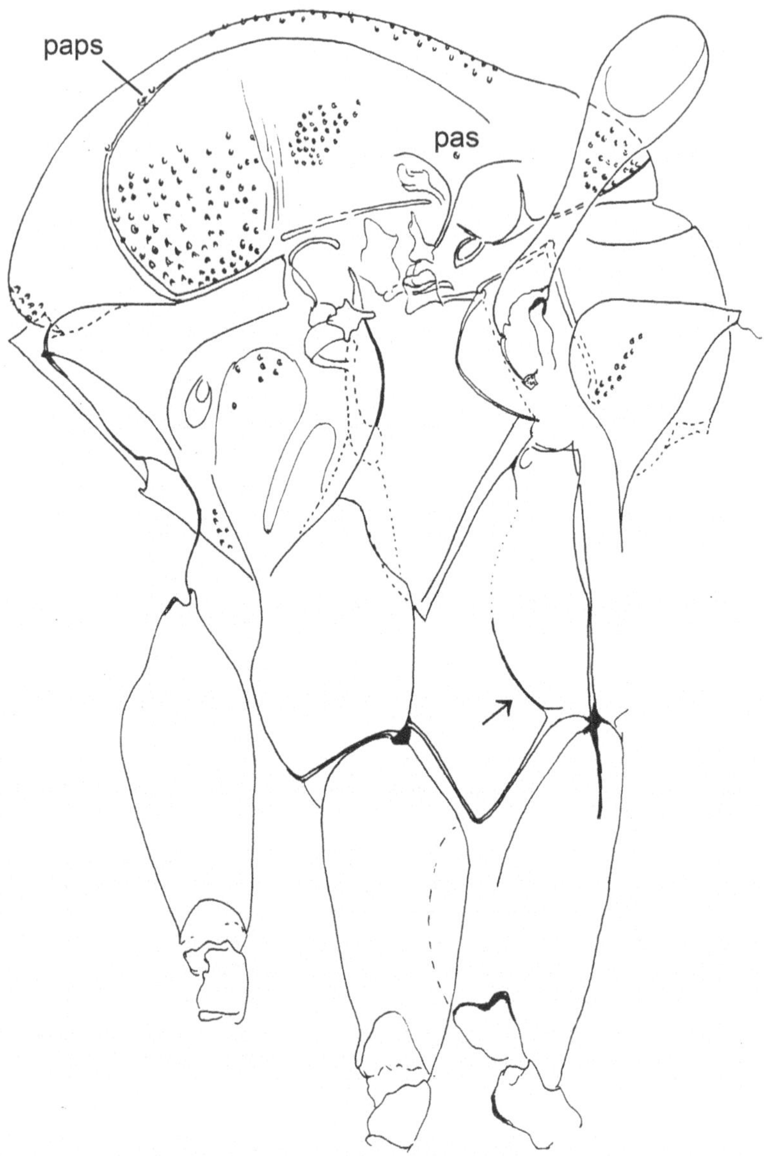

Fig. 23 Thorax in lateral view of Phlebotominae with indication of characteristics in plesiomorphic state: ***arrow*** – indicating the long suture separating the katepimeron and metepisternum; **paps** – parapsidal setae; **pas** – postalar seta. *Oligodontomyia toroensis*

(Figs. 23 and 24). The tendency among the phlebotomines has been to lose these latter setae; in a few groups one or two are still present.

The scutellum, with a rounded edge, presents **scutellar setae** in two lateral sets.

The **Mesopleuron** (Fig. 21) is divided by the **mesopleural suture** into an anterior part, the **mesepisternum** and a posterior, the **mesepimeron**. The former is transver-

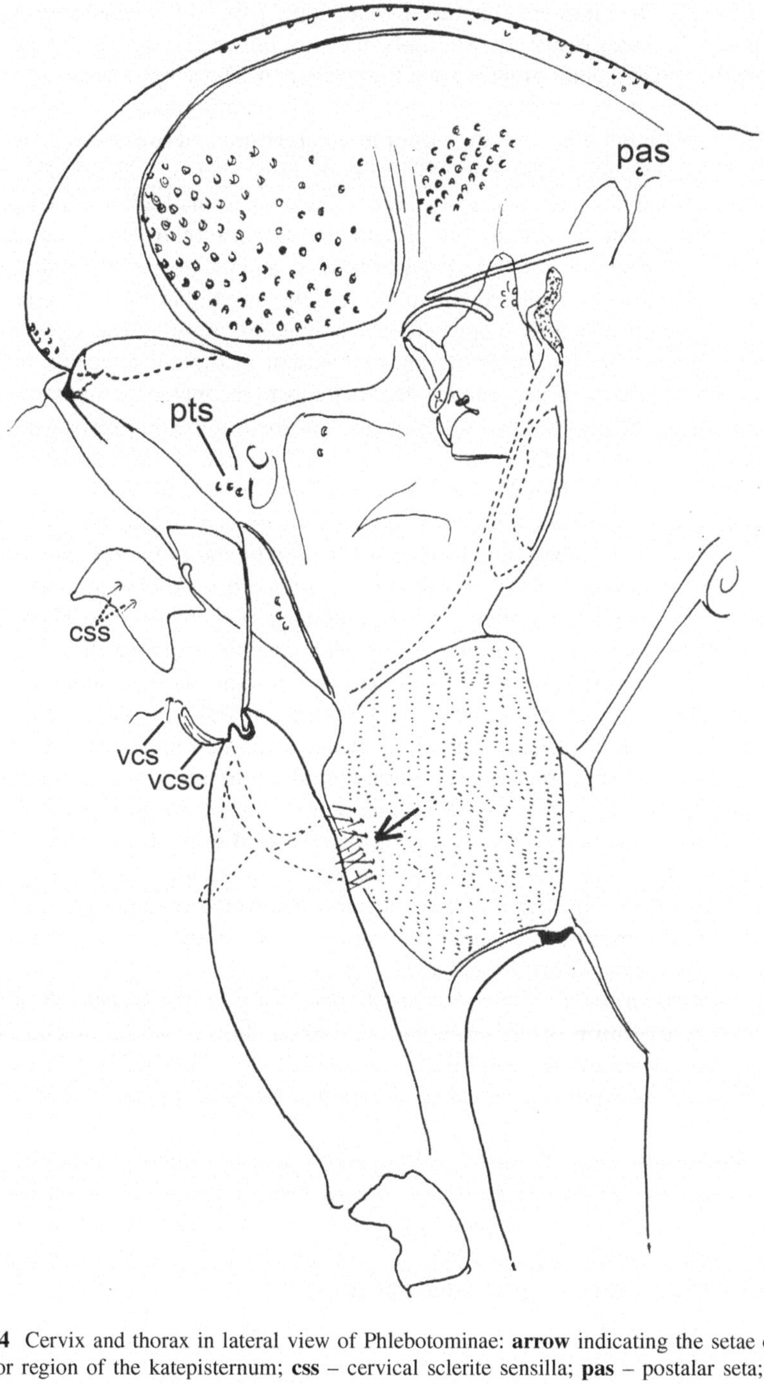

Fig. 24 Cervix and thorax in lateral view of Phlebotominae: **arrow** indicating the setae on the anterior region of the katepisternum; **css** – cervical sclerite sensilla; **pas** – postalar seta; **pts** – paratergital setae; **vcsc** – ventrocervical sclerite; **vcs** – ventrocervical sensilla. *Deanemyia samueli*

sally divided by the **anapleural suture** into an upper part - the **anepisternum** and a lower part – the **katepisternum**. Similarly, the mesepimeron is divided by the **transepimeral suture** into **anepimeron** and **katepimeron**. The anepisternum is not uniformly sclerotized; in its posterior upper part it is membranous and anteriorly is sclerotized and subdivided into the **upper anepisternum** and **lower anepisternum** by a membranous area, inflecting between both, of greater or lesser extent. In the posterior membranous area close to the base of the wing a small sclerite, the **basalar**, occurs, dorsally united by its anterior part, the **basalar anterior**, to the scutum by a transversal suture denominated the **prealar bridge**, and the posterior is linked by its posteroventral part to the anapleural suture. On the upper part of the anepimeron, alar sclerites, with which the wings are articulated, are present. The separation of the katepimeron from the metepisternum is evident in the apical region of these sclerites close to the posterior spiracle, that belongs to the metepisternum; however, the suture that separates the two sclerites has vanished in the majority of the phlebotomines, its traces remaining under the spiracle in some groups and in the basal region next to the tergosternal suture in others (Figs. 21, 22 and 23). In view of this incomplete suture, the katepimeron has been variously denominated by specialists. So Davis (1967), in his drawing, labeled it *katepimero* and in the text, *meropleuro* (*hipopleuro*); Forattini (1973) also called it the *hipopleura*; Abonnenc et al. (1971) and Abonnenc and Leger (1976a, b) denominated it *métépisterne*. These latter authors call the pleurotracantim the *mésocatépimère*. **Pleurotrocantim** is a much reduced sclerite inserted between the katepisternum and the katepimeron, at the junction of the mesopleural and transepimeral sutures. The mesopleuron may present a set of setae on the sclerotized part of the anepisternum in its upper region and rarely in its lower area, the **upper anepisternal setae** and **lower anepisternal setae**, respectively. A few groups may present setae on the katepisternum, the **katepisternal setae** and in the anepimeron, the **anepimeral setae**. In some rare species of Phlebotominae setae may occur above the anterior spiracle, in the paratergite, the **paratergital setae** (Fig. 24). The mesopleuron is covered by short microtrichias, but on the anterior part of the katerpisternum of some groups a set of more developed setae may occur (Figs. 24 and 25).

The **Mesosternum** (Fig. 26) presents the anterior part, the **basisternum** fused with the **furcasternum** of the prothorax, described above, and the much reduced furcasternum linked to the **mesofurca**, as a single highly developed structure. The mesofurca is firmly united to the **pleural arm** that begins at the internal mesopleural suture.

The **Postnotum** (Fig. 27) includes all the posterior parts under the scutellum and is constituted by a **mediotergitum** and two **laterotergites**. Each laterotergite extends down to the posterior spiracle and is divided into a dorsal **anatergite** and a ventral **katatergite** (McAlpine 1981). The mediotergite and the laterotergites are connected internally by the **postnatal bridge**.

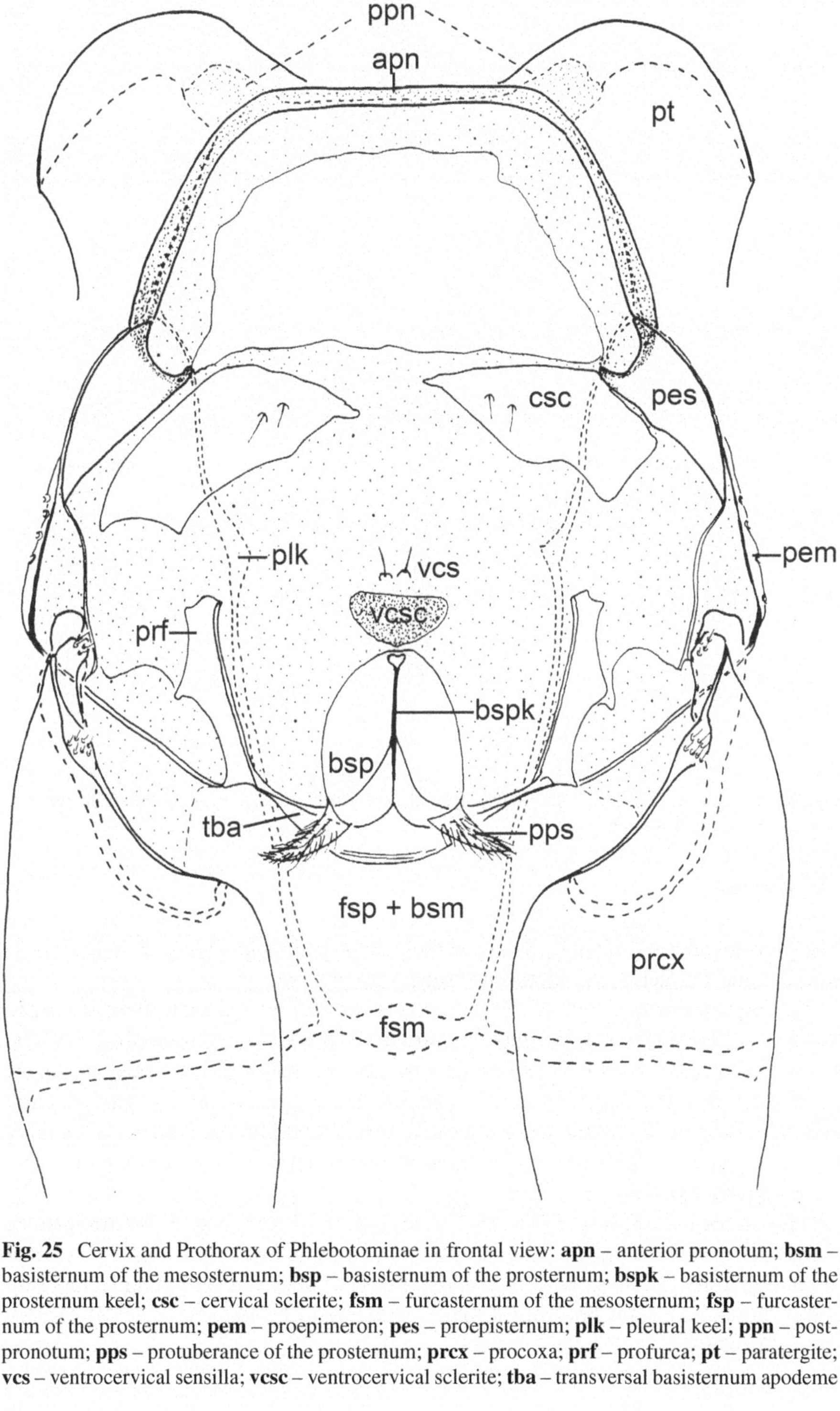

Fig. 25 Cervix and Prothorax of Phlebotominae in frontal view: **apn** – anterior pronotum; **bsm** – basisternum of the mesosternum; **bsp** – basisternum of the prosternum; **bspk** – basisternum of the prosternum keel; **csc** – cervical sclerite; **fsm** – furcasternum of the mesosternum; **fsp** – furcasternum of the prosternum; **pem** – proepimeron; **pes** – proepisternum; **plk** – pleural keel; **ppn** – postpronotum; **pps** – protuberance of the prosternum; **prcx** – procoxa; **prf** – profurca; **pt** – paratergite; **vcs** – ventrocervical sensilla; **vcsc** – ventrocervical sclerite; **tba** – transversal basisternum apodeme

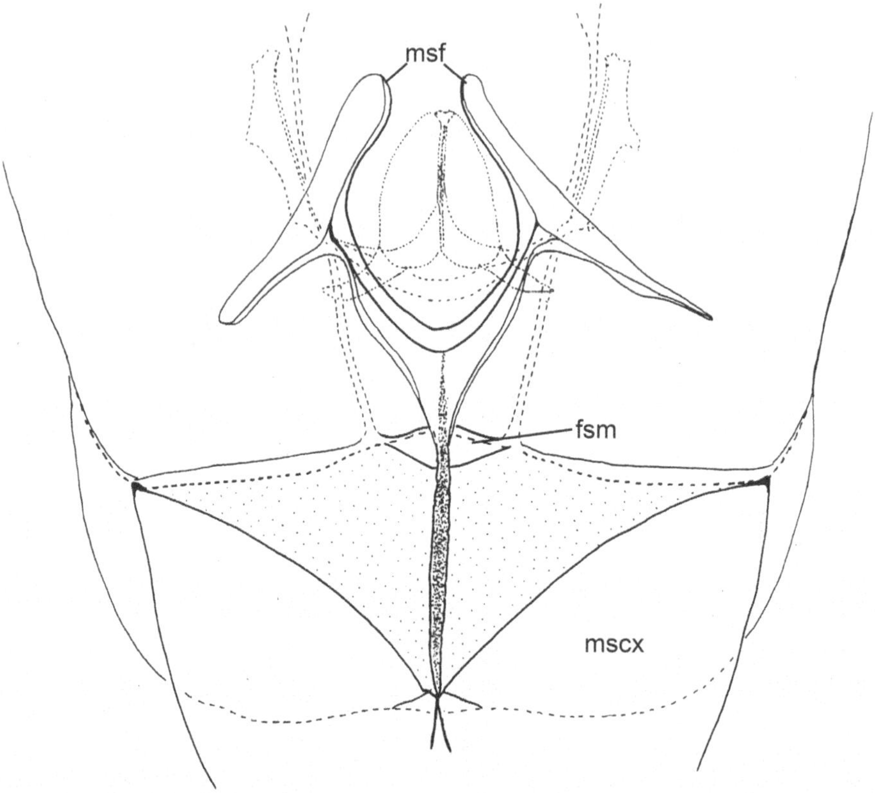

Fig. 26 Mesosternum of Phlebotominae in frontal view: **fsm** – furcasternum of the mesosternum; **msf** – mesofurca; **mscx** – mesocoxa;

Metathorax

The **Metanotum** is formed by a narrow band on the dorsum, widening on each side, where numerous setae, the **metanotal setae**, are present.

The **metapleuron** (Figs. 21, 22 and 23) is divided by the **metapleural suture** into an anterior sclerite, the **metepisternum** and a posterior, the **metepimeron**. The former, as already mentioned, may be fused to the katepimeron, thus forming a mixed sclerite. The latter at its dorsolateral extremities is united with the first abdominal tergite and at the ventral ones, to the first abdominal sternite. Setae may be present on the metepisternum, the **metepisternal setae**, and on the metepimeron, the **metepimeral setae** (Fig. 22).

The metesternum (Fig. 27) is very narrow and well invaginated. The **metafurca** is well developed and may or may not be interlinked by a sclerite along the **vertical arms**. Laterally and basally to these arise extensions with a rhomboid or tapering apex, denominated the **horizontal arms**, which may be long (Fig. 27-B) or very short (Fig. 27-C).

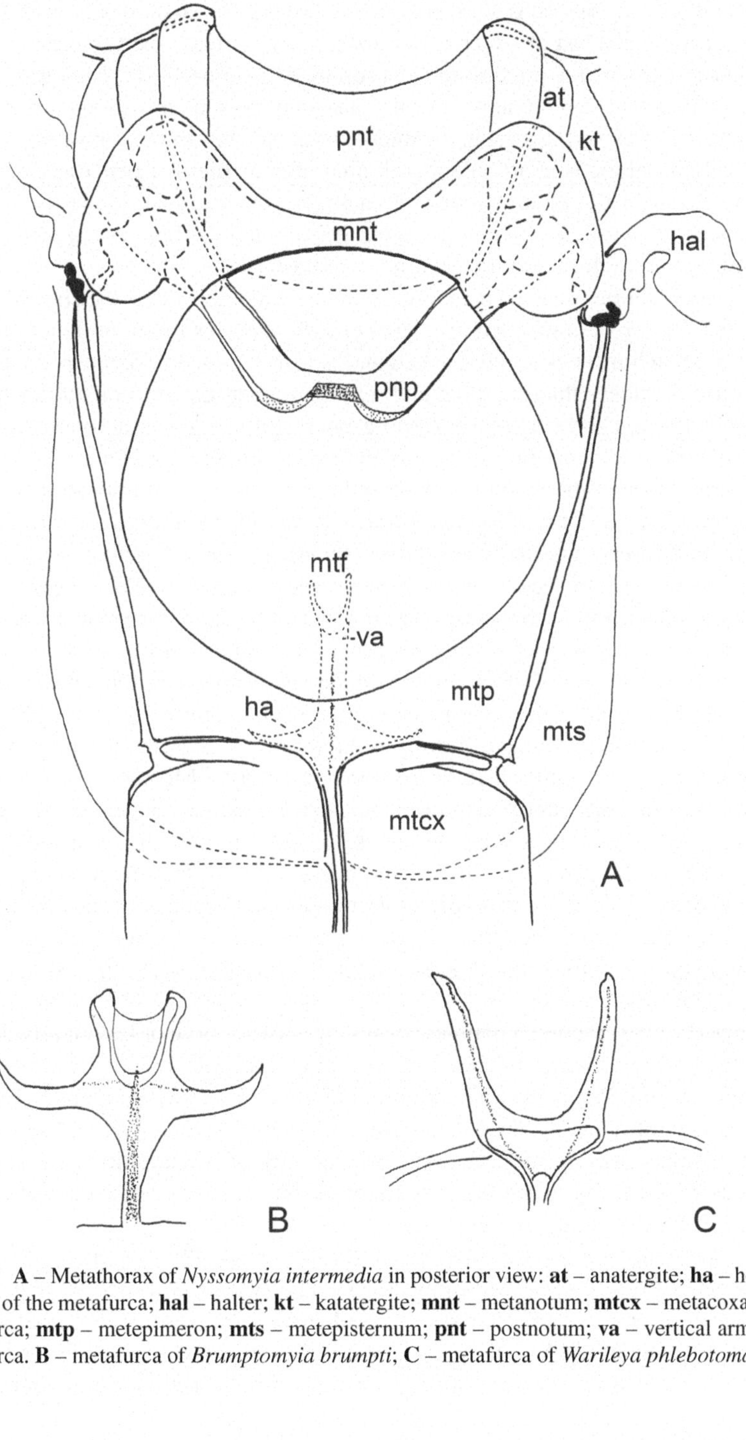

Fig. 27 **A** – Metathorax of *Nyssomyia intermedia* in posterior view: **at** – anatergite; **ha** – horizontal arm of the metafurca; **hal** – halter; **kt** – katatergite; **mnt** – metanotum; **mtcx** – metacoxa; **mtf** – metafurca; **mtp** – metepimeron; **mts** – metepisternum; **pnt** – postnotum; **va** – vertical arm of the metafurca. **B** – metafurca of *Brumptomyia brumpti*; **C** – metafurca of *Warileya phlebotomanica*

The **legs** are long and consist of the following segments: **coxa, trochanter**, **femur**, **tibia** and **tarsus**; the last of which is divided into five articles, the **tarsomeres**, and a small apical segment, the **posttarsus** or **acropod** (McAlpine 1981). The segments are covered with longitudinal bands of scales and may present sets of spines arranged longitudinally or as verticils set in the middle or at the apex of the segments.

Dorsally, all the coxae articulate with a **coxifer** or **pleural process**, situated at the ventral end of the pleural suture of each thoracic segment. Ventrally, the fore coxa articulates with the profurcasternum; the mid and the hind coxae articulate at a common point in the mesofurcasternum + metasternum.

The **coxa** is shorter than the height of the thorax and thicker than the corresponding femur; the fore coxa is inserted at a level clearly superior to the mid and the hind coxae. The **trochanter** is much reduced and is firmly united to the respective coxa. The **femur** is thicker than the tibia and may be longer, equal to or shorter than in each respective leg and may present strong, short spurs on its basal internal surface. The hind femur on its internal surface is endowed with a longitudinal row of spines and in some groups the insertion of these spines presents sclerotization with the appearance of a spur (Fig. 35). The **tibia** is generally the longest segment of the legs. The **tarsomeres** gradually diminish in length, the first of them generally being shorter than the sum of the others, though in some groups it may be longer than this sum. In the separation of the groups, the number of levels of insertion of the spines in tarsomere III may be useful; they are generally arranged as one or more verticils on the intermediate part of the segment with one in its apical region, and sometimes there is only one spine of the intermediate verticils remaining (Figs. 28, 29, 30, 31, 32 and 33). The **acropod** situated distally to tarsomere V presents the **unguitractor plate** localized on its ventral face, as the principal sclerite (McAlpine 1981); basally it is linked to the tarsomere and into its apex two **claws** are inserted. At the apex of the acropod, a more or less membranous sac-like structure, the **arolium**, occurs. At each side of the acropode close to the lateral edge of the unguitractor plate, a small sclerite called the **basipulvillus** (pl. basipulvilli) is present (Figs. 34 and 35).

The ratios between the lengths of some segments of the legs, for example, femur/tibia or tarsomere I/tarsomere II + III + IV + V have been useful in distinguishing species or groups.

Wings (Fig. 36) present a flattened oval shape with a more or less clearly defined point at the extremity and are densely covered by lanceolate scales, with the alula poorly developed. When the phlebotomines are at rest, the wings remain erect and divergent (Forattini 1973) like a delta-wing. The venation is constituted by ten longitudinal veins reaching the alar edge, the majority of the ramifications of which occur basally, with the more distal transverse vein, the r-m, situated between the basal third and the middle of the wing.

The denomination of the alar veins follows basically the interpretation of Redtenbacher (1886) (*apud* McAlpine 1981). The system proposed by that author recognizes six primary veins: **costa (C)**, **subcosta (Sc)**, **radius (R)**, **media (M)**, **cubitus (Cu)** and **anal (A)**. The veins **C** and **Sc** are single. **R**, in its anterior branch, is single and called $\mathbf{R_1}$, the posterior branch, denominated **radial sector (RS)**, gives

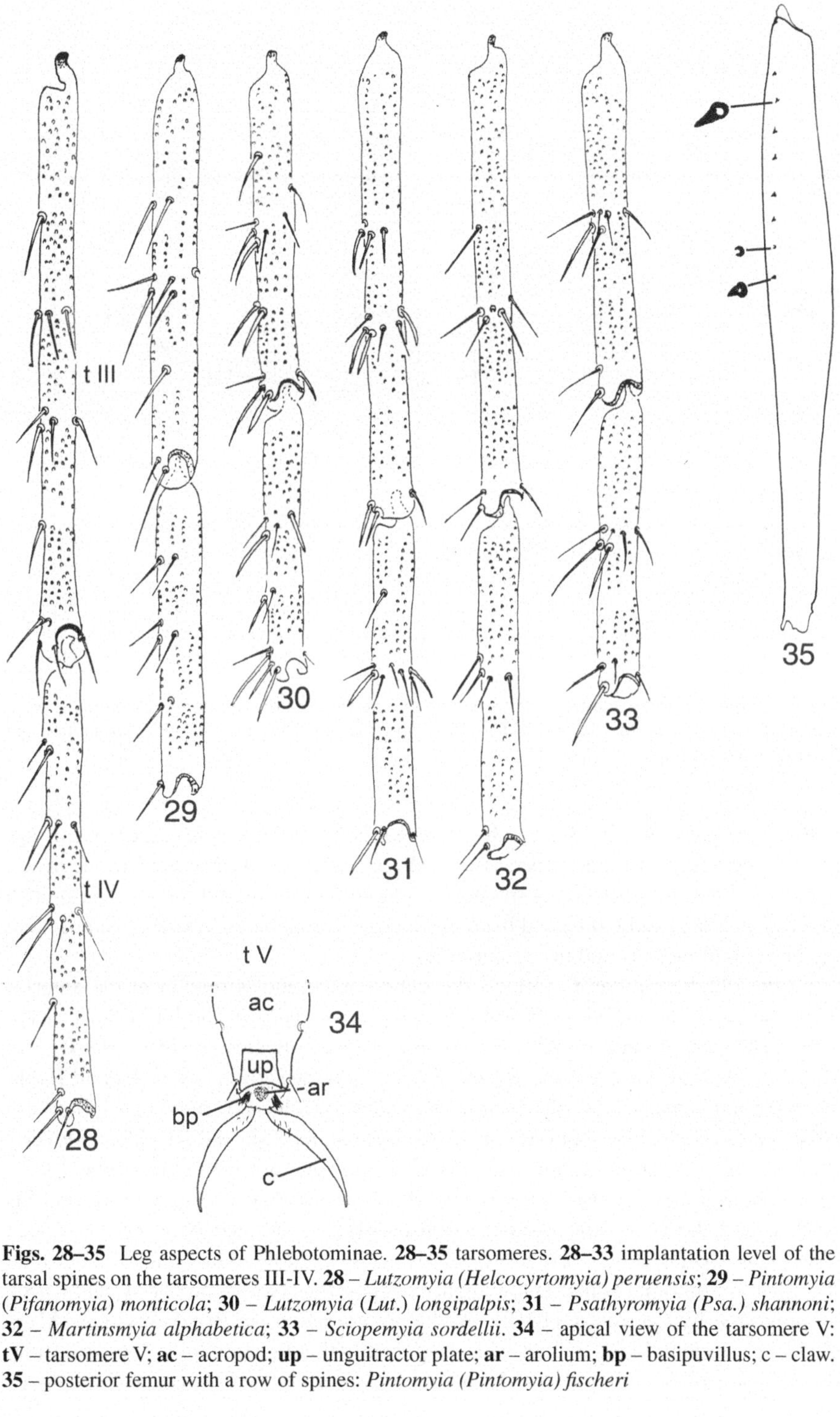

Figs. 28–35 Leg aspects of Phlebotominae. **28–35** tarsomeres. **28–33** implantation level of the tarsal spines on the tarsomeres III-IV. **28** – *Lutzomyia (Helcocyrtomyia) peruensis*; **29** – *Pintomyia (Pifanomyia) monticola*; **30** – *Lutzomyia* (*Lut.*) *longipalpis*; **31** – *Psathyromyia (Psa.) shannoni*; **32** – *Martinsmyia alphabetica*; **33** – *Sciopemyia sordellii*. **34** – apical view of the tarsomere V: **tV** – tarsomere V; **ac** – acropod; **up** – unguitractor plate; **ar** – arolium; **bp** – basipuvillus; c – claw. **35** – posterior femur with a row of spines: *Pintomyia (Pintomyia) fischeri*

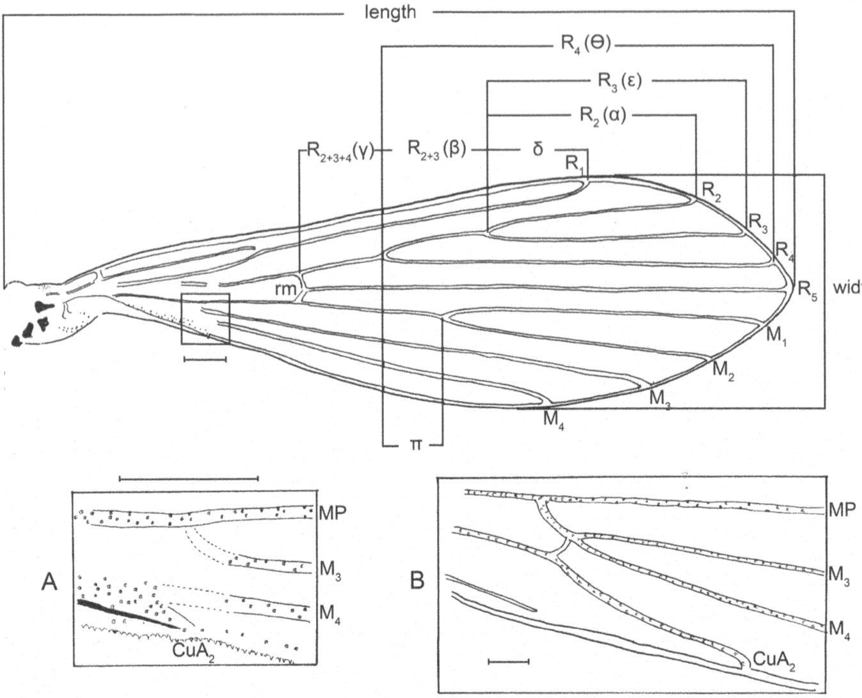

Fig. 36 Wing of Phlebotominae with the mainly alar indices and the square representing the position of the base of the M_3, M_4 and CuA_2 veins which are amplified on **A** – Phlebotominae and **B** – Bruchomyiinae

rise to four veins $\mathbf{R_2}$, $\mathbf{R_3}$, $\mathbf{R_4}$ and $\mathbf{R_5}$. As regards **M**, the anterior branch has disappeared; the posterior branch gives rise to $\mathbf{M_1}$, $\mathbf{M_2}$, $\mathbf{M_3}$ and $\mathbf{M_4}$. For the **Cu**, the designation anterior and posterior is frequently adopted. From the forked anterior branch of Cu stem $\mathbf{CuA_1}$ and $\mathbf{CuA_2}$, and from the posterior branch of Cu, **CuP**. From **A** stem the anterior branch $\mathbf{A_1}$ and the posterior $\mathbf{A_2}$.

In the Phlebotominae **C** follows the wing's edge and is thicker at the anterior edge. **Sc** is single and its apex inflects to $\mathbf{R_1}$, proximally in the basal third of the wing. The **radial sector (RS)**, for the majority of species, divides into $\mathbf{R_5}$ and $\mathbf{R_{2+3+4}}$; this latter divides into $\mathbf{R_4}$ and $\mathbf{R_{2+3}}$ which then gives rise to $\mathbf{R_2}$ and $\mathbf{R_3}$. In *Hertigia* and some species of *Warileya*, $\mathbf{R_5}$ arises from $\mathbf{R_4}$. The media posterior vein **(MP)** gives rise to $\mathbf{M_{1+2}}$ and $\mathbf{M_3}$; from the former stem $\mathbf{M_1}$ and $\mathbf{M_2}$. $\mathbf{M_4}$, following the basic plan of Diptera, never occurs as a free separate vein (McAlpine 1981). In Bruchomyiinae, the sister group of the Phlebotominae, the forking of $\mathbf{M_3}$ and $\mathbf{M_4}$ is clear and the crossvein **m-cu (medial-cubital)** is situated between $\mathbf{M_{3+4}}$ and $\mathbf{CuA_2}$ (Fig. 36-B); posteriorly to **m-cu**, $\mathbf{M_4}$ becomes more evident and thicker than the other longitudinal veins. This evidence suggest the coalescence of $\mathbf{M_4}$ and $\mathbf{CuA_1}$, as suggested by Comstock (1918). In Phlebotominae, as in Bruchomyiinae, $\mathbf{M_4}$ is also more evident than the other veins, suggesting that they are homologous despite the disappearance of the fork $\mathbf{M_3M_4}$ and **m-cu**'s not being clearly evident (Fig. 36-A).

The vein **CuA_2** ends just after the **m-cu**; the **CuP** is greatly reduced and restricted to the stalk of the wing. **A_1** and **A_2** are rudimentary. The crossvein **r-m** (**radial-medial**) is situated between the basal third and the middle of the wing. The areas delimited by veins, **cells**, are named according to the vein that is anterior to them. Thus, the **radial basal (rb)** is anteriorly delimited by the **RS**, and apically by the **r-m**, and the **medial basal (mb)** by the stem of **M** and by the **m-cu**.

Since the beginning of systematic studies of Phlebotominae, various alar measurements, including those of the alar veins, have been suggested, and the most frequently used are:

Width of the wing; measured at its widest point
Length of the wing
Length of **R_5**
alpha (α) – length of **R_2**
beta (β) – length of the branch **R_{2+3}**
gamma (γ) – length of the branch **R_{2+3+4}**
delta (δ) – distance between fork **R_2R_3** and the extremity of **R_1**, positive, when the latter point passes the former; negative, in the reverse situation
pi (π) – distance between the forks **R_{2+3} R_4** and **M_1 M_2**, positive, when the former is situated before the latter; negative, in the reverse situation

França (1919) proposed the use of some alar ratios:

alpha/beta – (α/β) (denominated alar index)
alpha/gamma – (α/γ)
beta/gamma – (β/γ)
delta/alpha – (δ/α)

All these measurements relate to the line perpendicular to R_5.

The wings have undergone successive modifications during the evolutionary process of the phlebotomines, the most evident being the superior portion of the wing has become narrower, so the radial veins have become shorter in relation to the length of the wing; the fork **R_2 R_3** has moved into a more apical position; the posterior portion of the wing has become wider and the medial veins have lengthened, the **radial-basal cell (rb)** has become shorter, with the apical limit (**r-m**) primitively situated in the middle of the wing, having moved to the basal third of the wing. Finally, the whole wing has become narrower.

The halters (Fig. 21) that represent the original posterior wings and have the function of maintaining balance during flight, appear like rackets (Forattini 1973), with the stem being more or less longer than the apical part (knob). In the evolutionary process of the phlebotomines, the stem has become shorter.

Abdomen

The abdomen of the Phlebotominae consists of 11 segments. Each segment is basically formed, dorsally, by the **tergite**, ventrally, by the **sternite** (considerably narrower than the former) with the **pleura** making the connection between the two. The most distal of the segments is denominated **proctiger**, formed by a pair of **cerci** and the **anus** (McAlpine 1981). There are five pairs of spiracles inserted in the pleura close to the basolateral edge of the 3rd–7th tergites.

In both sexes, the tergites of the 1st–7th segments are covered by microtrichias and setae of variable number and aspect. When the setae fall they leave scars that may be distributed as transversal bands (Fig. 92-A.) (primitive state) or randomly (Fig. 92-B).

The males may present, on some tergites, dark points surrounded by clear areas (Fig. 93), sometimes difficult to observe under the optical microscope, or clear points each of which is surrounded by a darker area. These structures may be limited to the central area of the tergites (Fig. 93-A), without the presence of scars of deciduous setae among them, or they may be dispersed over the entire surface of the tergite, mixed with the scars of the deciduous setae (Fig. 93-B). These dark points, visible under the electronic microscope, are **pores** and the surrounding clear area consists of **papillae**, the respective diameters of which on *Lutzomyia longipalpis* are of 0.25 µm and 3.0–3.5 µm (Lane and Ward 1984). The function of these structures is still the subject of speculation. However, these latter authors believe that they are to be associated with the excretion of pheromones. In some species of *Brumptomyia* and of *Nemopalpus* (Bruchomyiinae) each papilla presents an external extension projecting from the pore in the shape of a small candle flame, called **papilla with hair** (Fig. 93-D), while in the other groups of phlebotomines in which the papillae are present, they are invisible under the optical microscope and are called **papillae without hair** (Fig. 93-C, E). In *Sergentomyia*, the papillae are connected internally with structures like well-sclerotized micro-setae (probable glandular ducts). The presence of papillae with hair on the 2nd–7th tergites seems to represent the primitive condition among the phlebotomines. Many groups have lost the papillae on all these tergites, but in others the loss has been but partial. In males, the 8th tergite is very narrow and has generally been deprived of both setae and papillae.

In both sexes, the 1st sternite is much reduced, well sclerotized and presents a pair of small spiniform setae. The 2nd sternite is of a length smaller than, equivalent to or greater than its width and the pigmentation of the central area is sometimes discontinuous and thus presents a bilobed aspect, its base constituted by a well sclerotized narrow concave band linking it to the extremity of the 1st sternite. There are a few deciduous setae in the basal area (primitive state) and these always occur in the apical area. The sternites of the 3rd–6th segments are bell-shaped, with the setae arranged in the shape of a “w”. The 7th sternite is triangular, the scars of the deciduous setae also forming a w-shape, and it presents spiniform setae basolaterally and sometimes, in the central position. The 8th sternite, in the male, is like a tergite though greatly reduced and without setae.

The pleura are membranous and covered with microtrichias. They link the sclerites of the tergite and those of the sternite, as well as laterally the tergite and the sternite. Generally there are no setae on the lateral pleura, but in at least one group (subgenus *Tricholateralis*) they occur in great numbers on the females from the 3rd to the 6th segments (Fig. 38) and in lesser numbers on the males.

The genital opening in both sexes is situated antero-ventrally to the anus. In females, it appears between the 8th and 9th sternites; in the males, the aedeagus ("penis") that contains the opening, arises posteriorly to the 9th sternite. The female terminalia is constituted by the last four segments and in the males, by the last three.

Female Terminalia

The female terminalia are formed by segments 8–11. The internal structures, of ectodermal origin, include a pair of **spermathecae** and the accessory glands (McAlpine 1981), the latter of which are destroyed during the clarification process.

The **spermathecae** (Figs. 37, 42, 43, 44, 45, 46, 47, 48, 49, 50, 51, 52, 53, 54, 55, 56, 57, 58, 59, 60, 61, 62, 63, 64, 65, 66, 67, 68, 69, 70, 71, 72, 73, 74, 75, 76, 77, 78, 79, 80, 81, 82, 83, 84, 85, 86, 87, 88, 89, 90 and 91), sac-like or tubular capsules, are connected by ducts to the **genital chamber**. The opening into the genital chamber may be individually by **individual spermathecal ducts** (Figs. 42, 43, 47, 49 and 50) or these may be joined forming the **common spermathecal duct** of variable length (Figs. 44, 45, 48, 51, 52, 53, 54, 55, 56, 57, 58, 59, 60, 61, 62, 63, 64, 65, 66, 67, 68, 69, 70, 71, 72, 73, 74, 75, 76, 77, 78, 79, 80, 81, 82, 83, 84, 85, 86, 87, 88, 89, 90 and 91). Distally to the spermatheca occur glands and from their cellules arise sclerotized micro-channels that open into a certain apical area (Theodor 1965) called the **terminal knob**. This structure may be formed of a small salience (Figs. 50, 57 and 78) or may stand out from the spermatheca (Figs. 42, 43, 52, 53, 54, 55 and 56). After the clarification process with KOH, the glands disappear and only the micro channels (intracellular ducts) are left on the surface of the terminal knob, giving it a hairy aspect.

The spermatheca is surrounded by a layer of longitudinal muscles. These muscles, when contracted, may provoke segmentation or temporary folds in the walls of the spermatheca when they are membranous or may be flattened when sclerotized. The tube-like spermathecae (Fig. 44), in several groups, suffer permanent segmentation, forming rings (Fig. 37-D), sometimes imbricated (Figs. 60, 61, 62, 63, 64 and 65). A similar process occurs in some groups to the individual or common sperm ducts (Figs. 60, 63 and 65). The sac-like spermatheca may suffer modifications such as segmentation of the apical part, forming simple rings (Figs. 70, 71 and 74) or be of bubble-like shape (Fig. 80). The individual sperm ducts, which are generally tubular and slightly sclerotized (Figs. 44, 45 and 46) may suffer strong dilation, be of sac-like aspect (Fig. 81) or be heavily sclerotized (Figs. 83 and 84).

The 8th segment (Fig. 37-C) is formed by the tergite with a ring aspect; ventrally it is connected by the pleural membrane to the **bilobed sternite**, which is covered by deciduous and permanent setae. Under each lobe of the sternite there occurs a triangular sclerite that is hyaline in its central part and laterally more sclerotized in a narrow band, forming the **ventral branch** and the **dorsal branch**. In this latter the sclerotization is more intense and at its apex a spiniform hair is inserted. A similar structure was described for the Phlebotominae by Christophers and Barraud (1926) and called insula. Smith (1969), in his study of the evolution of the internal mor-

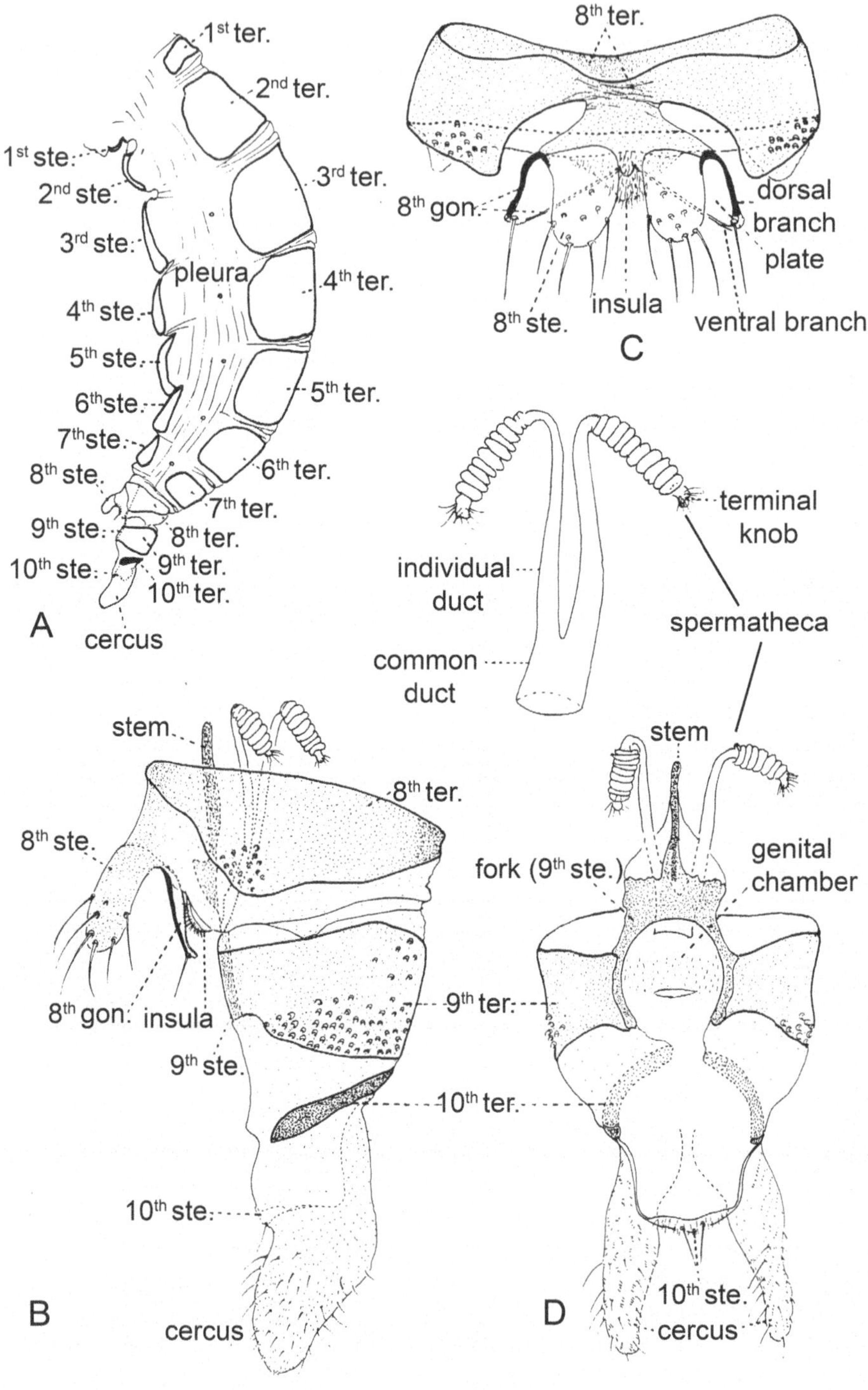

Fig. 37 Abdomen and genitalia of Phlebotomine female. **A** – abdomen in lateral view; **ste** – sternite; **ter** – tergite. **B** – genitalia in lateral view. **C** – 8th segment in ventral view: **gon** – gonopod; **D** – 9th segment, 10th segment and cerci in ventral view. *Nyssomyia neivai*

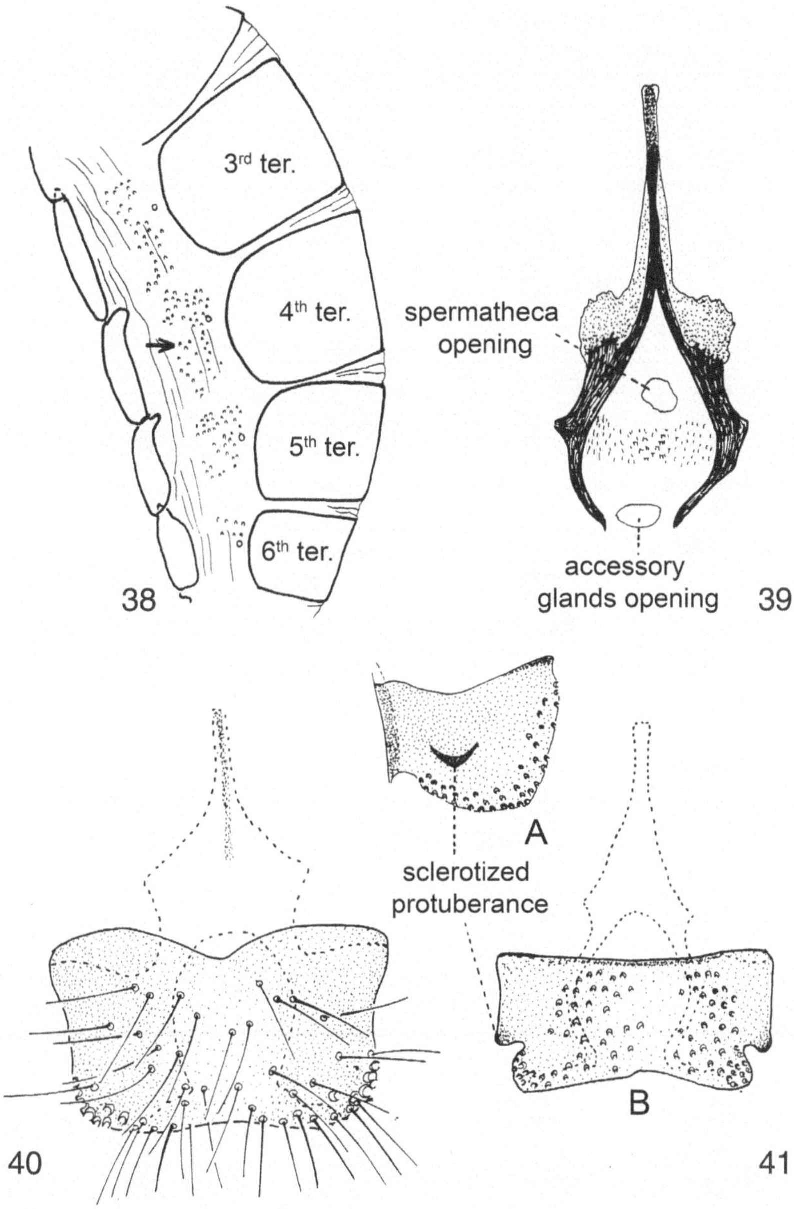

Figs. 38–41 Aspects of the abdomen and the genitalia of phlebotomine females. **38** – 3rd – 6th abdominal segments. **Arrow** indicates the pleural bristles: *Lutzomyia* (*Tricholateralis*) *sherlocki;* **39** – genital fork presenting a notch on the upper part of the genital chamber: *Pintomyia* (*Pifanomyia*) *monticola*; **40** – 9th segment in dorsal view showing non deciduous setae in the median region of the tergite. *Pintomyia* (*Pifanomyia*) sp. (Pia series); **41** – 9th segment showing a sclerotized protuberance on the tergite: **A** – lateral view and **B** – dorsal view: *Migonemyia* (*Migonemyia*) *migonei*

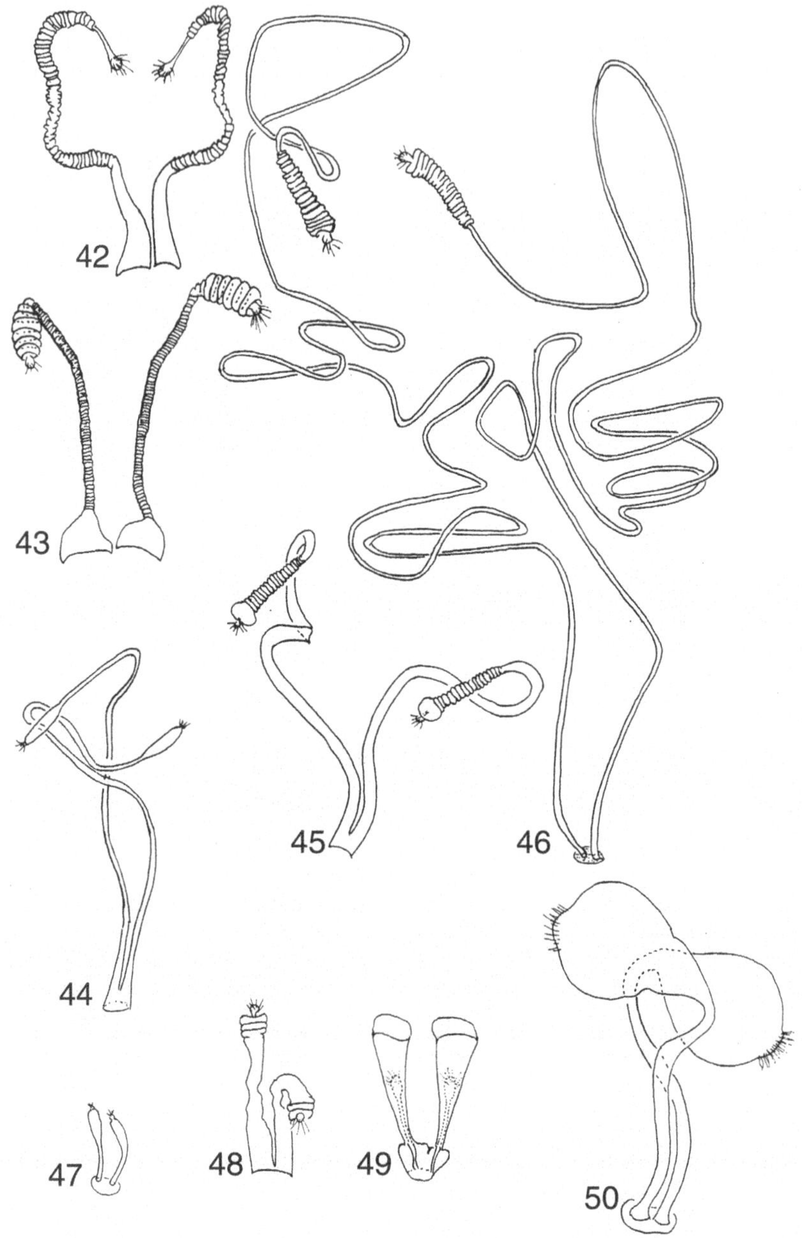

Figs. 42–50 Aspects of spermathecae and their ducts of several groups of Phlebotominae. **42** – *Warileya phlebotomanica*; **43** – *Phlebotomus papatasi*; **44** – *Migonemyia* (*Migonemyia*) *migonei*; **45** – *Lutzomyia* (*Tricholateralis*) *cruciata*; **46** – *Brumptomyia brumpti*; **47** – *Migonemyia* (*Blancasmyia*) *gorbitzi*; **48** – *Micropygomyia pilosa*; **49** – *Viannamyia tuberculata*; **50** – *Psathyromyia* (*Forattiniella*) *aragaoi*

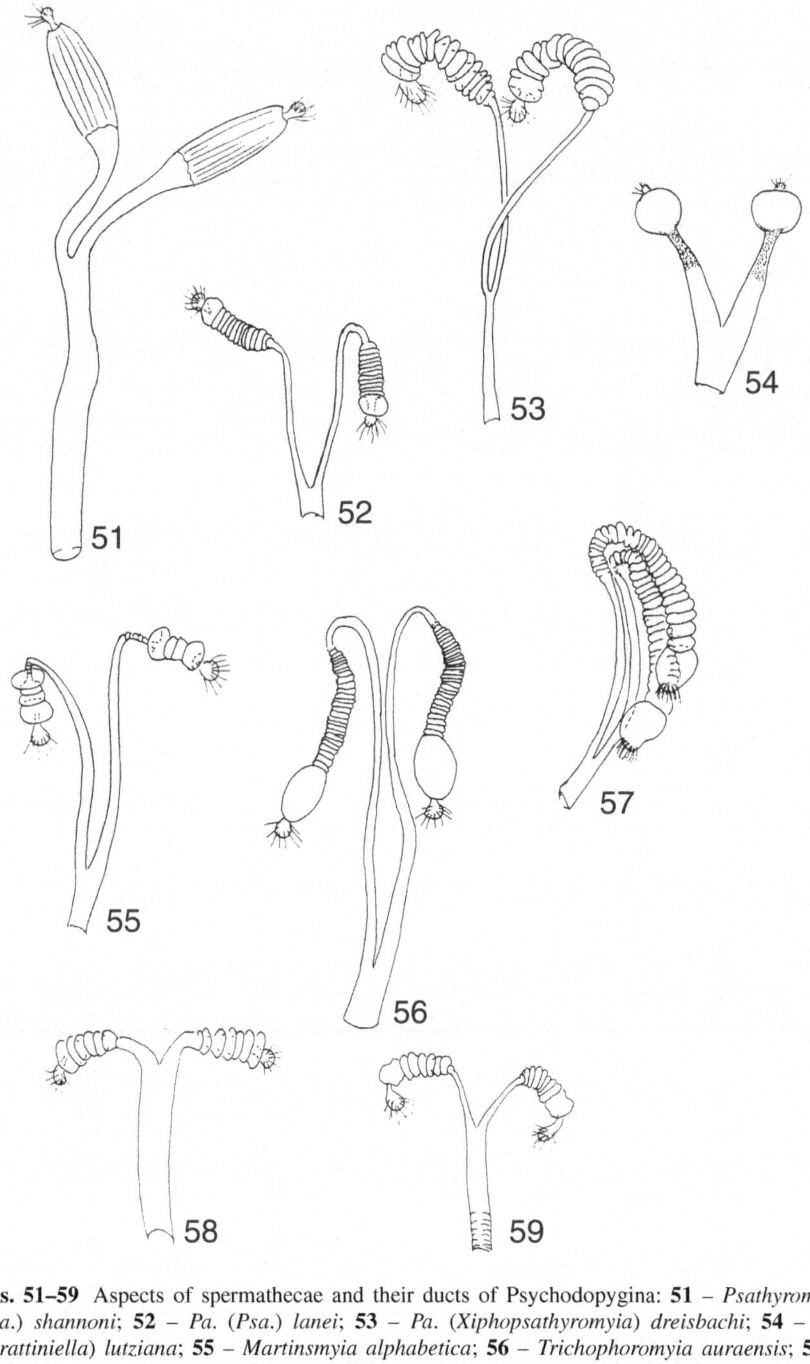

Figs. 51–59 Aspects of spermathecae and their ducts of Psychodopygina: **51** – *Psathyromyia* (*Psa.*) *shannoni*; **52** – *Pa.* (*Psa.*) *lanei*; **53** – *Pa.* (*Xiphopsathyromyia*) *dreisbachi*; **54** – *Pa.* (*Forattiniella*) *lutziana*; **55** – *Martinsmyia alphabetica*; **56** – *Trichophoromyia auraensis*; **57** – *Trichophoromyia ubiquitalis*; **58** – *Martinsmyia gasparviannai*; **59** – *Bichromomyia flaviscutellata*

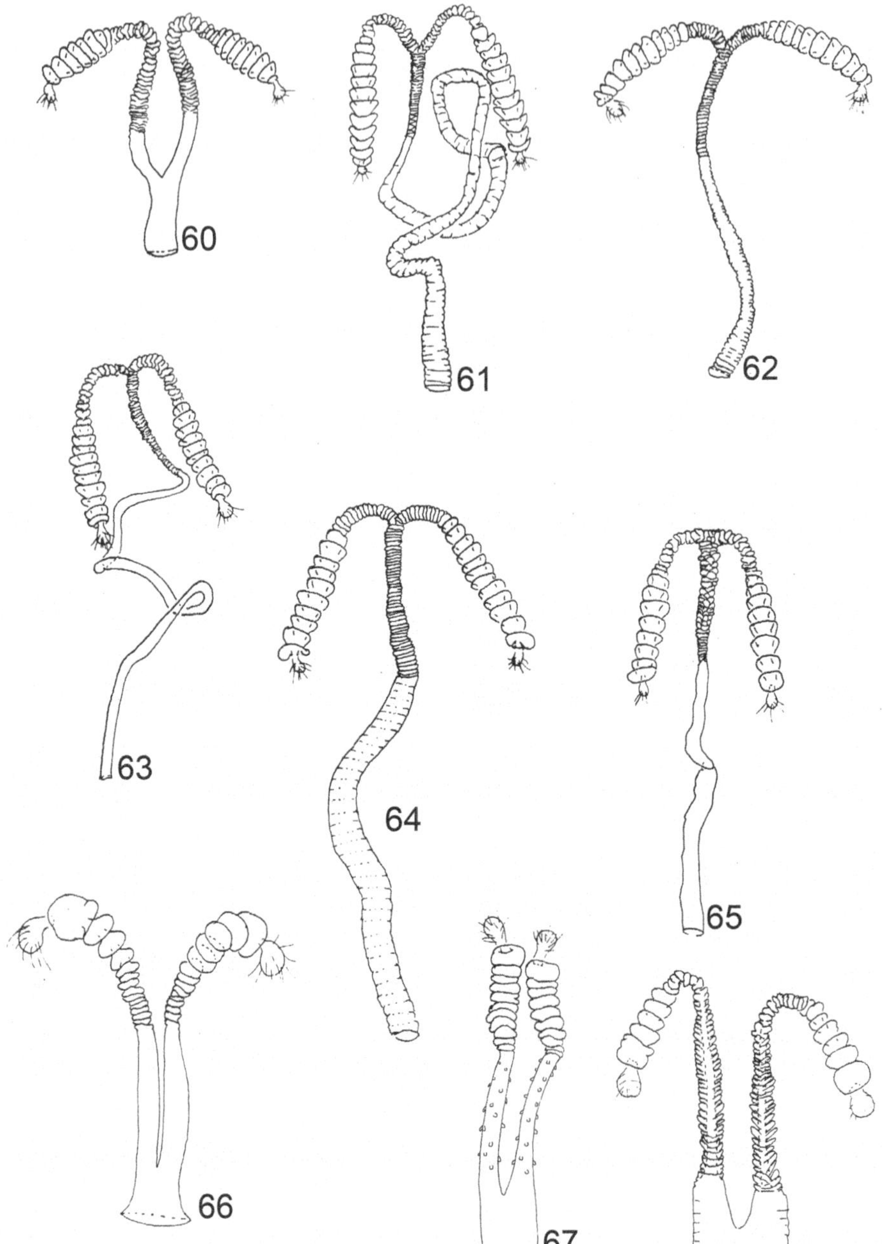

Figs. 60–68 Aspects of spermathecae and their ducts of Psychodopygina: **60** – *Psychodopygus chagasi*; **61** – *Ps. davisi*; **62** – *Ps. panamensis*; **63** – *Ps. ayrozai*; **64** – *Ps. arthuri*; **65** – *Ps. paraensis*; **66** – *Nyssomyia anduzei*; **67** – *Ny. yuilli pajoti*; **68** – *Ny. fraihai*

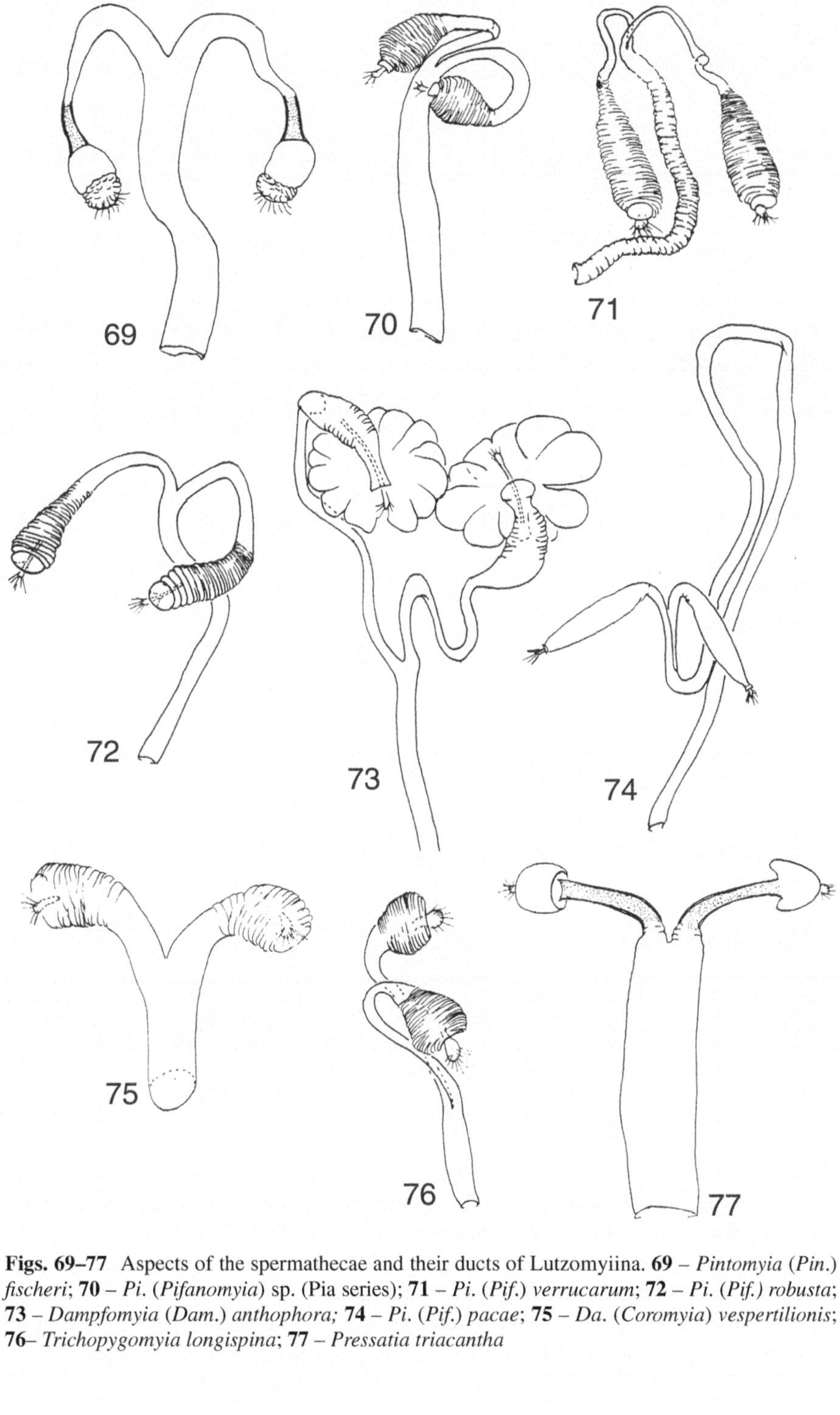

Figs. 69–77 Aspects of the spermathecae and their ducts of Lutzomyiina. **69** – *Pintomyia* (*Pin.*) *fischeri*; **70** – *Pi.* (*Pifanomyia*) sp. (Pia series); **71** – *Pi.* (*Pif.*) *verrucarum*; **72** – *Pi.* (*Pif.*) *robusta*; **73** – *Dampfomyia* (*Dam.*) *anthophora;* **74** – *Pi.* (*Pif.*) *pacae*; **75** – *Da.* (*Coromyia*) *vespertilionis*; **76**– *Trichopygomyia longispina*; **77** – *Pressatia triacantha*

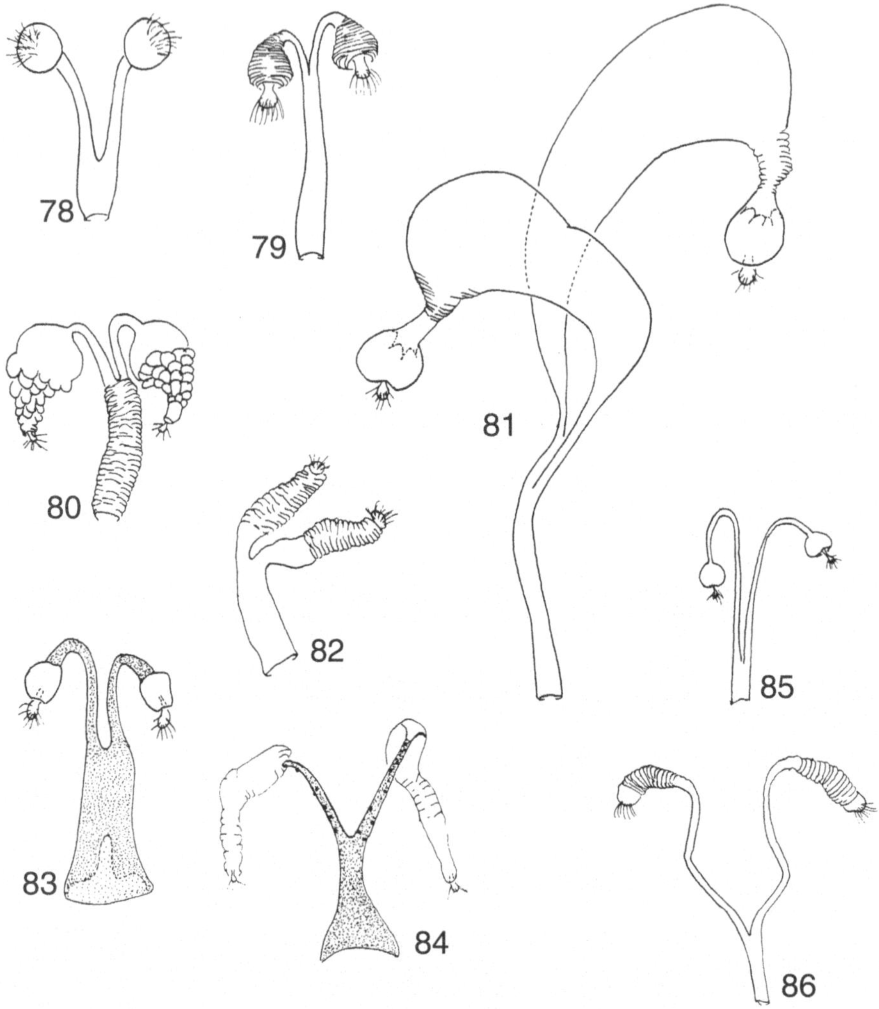

Figs. 78–86 Aspects of the spermathecae and their ducts of Lutzomyiina. **78** – *Expapillata firmatoi*; **79** – *Evandromyia (Eva.) correalimai;* **80** – *Ev. (Eva.) saulensis;* **81** – *Ev. (Barrettomyia) tupynambai*; **82** – *Ev.* (*Eva.*) *infraspinosa*; **83** – *Ev.* (*Aldamyia*) *walkeri*; **84** – *Ev.* (*Bar.*) *monstruosa*; **85** – *Ev.* (*Bar.*) *cortelezzii*; **86**– *Sciopemyia sordellii*

phology of the insects, called this structure the 8th gonapophyse, also known as anterior gonapophyse, first or ventral gonapophyse, hypogynial valve, lancet, saw, stiletto, valve 1 and other denominations. McAlpine (1981) adopted hypogynial valve and Jobling (1987), spatulate stems. Smith's denomination (**8th gonapophyse**) has been adopted here. Internally to this and immediately above the genital chamber there is a membranous projection covered by hairs. No references to the origin of this structure have been found, thus the denomination insula has been adopted for it,

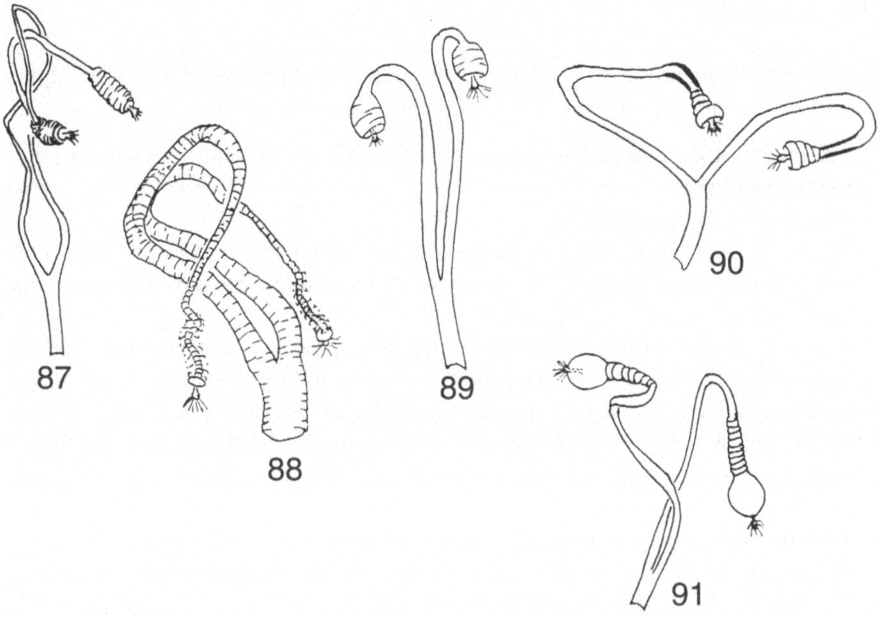

Figs. 87–91 Aspects of the spermathecae and their ducts of Sergentomyiina. **87** – *Deanemyia samueli*; **88** – *Micropygomyia* (*Coquillettimyia*) *vexator*; **89** – *Mi.* (*Sauromyia*) *atroclavata*; **90** – *Mi.* (*Sau.*) *oswaldoi;* **91** – *Mi.* (*Mic.*) *cayennensis*

in accordance with Jobling (1987), as it seems to have the function of closing the entrance to the ovarian channel.

The **9th tergite**, with a semi-anular shape, presents deciduous hairs, more concentrated latero-posteriorly, and in some groups, in its central part, permanent hairs may occur (Fig. 40). In some species occurs a small protuberance situated in the latero-basal (Fig. 41) or latero-apical part.

The **9th sternite** transformed into the branch of the genital fork (Fig. 37).

The **10th tergite** (Fig. 37) occurs in a narrow heavily sclerotized band, obliquely situated between the cercus and the 9**th** tergite.

The **10th sternite** (Fig. 37) closely associated with the cerci.

The **11th segment** (Fig. 37). The tergite and sternite consolidated to form the **proctiger**, forming in its turn, the **cerci** and the **anus**. The cercus may present a pointed or rounded apex.

The genital chamber or vagina (Figs. 37 and 39) is the area between the branches of the genital fork, where the opening of the spermathecae and the accessory glands (collateral or of mucus) are situated. The opening of these latter is situated posterior to the spermathecae (Forattini 1973). The accessory glands are not sclerotized and so they are not observable after clarification with KOH. Their function is to produce adhesive substances for the fixing of the eggs on the substrate. A membrane occurs in the central part of the genital chamber that is united with the branches of the fork. For nuliparous females of *Migonemyia migonei* this membrane has been observed to form folds close to the branches of the furca (Tang and Añez 1995).

Male Terminalia

The male Terminalia follow the basic plan of Diptera, and according to McAlpine (1981) eight elements integrate the terminalia of this group of insects:

- **9th tergite**, the **epandrium** (genital arch) frequently presenting a pair of lobes (**epandrial lobes**)
- **9th sternite** or **hypandrium**
- A pair of gonopods situated posterolaterally to sternite 9 and consisting of a basal **gonocoxite** ('basistylus', 'basimere') and an apical **gonostyle** (gonostylus, 'dististylus', 'distimere', 'clasper')
- A pair of **parameres**, unsegmented paraphallic processes, situated between the posterolateral base of the **parameral sheath** and dorsomedial base of the gonocoxites. In a lateral view, the parameres lie between the gonocoxites and the set formed by tergite 9 and tergite 10; with the parameral sheath between them.
- A central phallic organ, the **aedeagus** ('penis'), lies immediately behind the 9th sternite.
- **10th tergite**, closely associated with the 9th tergite, in those groups in which it is fused (Hertigiini), or with the lateral lobes, as in the majority of phlebotomines in which the 9th tergite is completely separated
- A simple **10th sternite** ('ventral epandrial plate')
- The vestigial tergite and sternite of the 11th segment consolidated to form the **proctiger**, where the cerci and the anus arise

In Phlebotominae the **9th tergite** was, primitively, a single basal band with the epandrial lobes appearing apically (Fig. 95). Posteriorly, the single band disappears completely, only the epandrial lobes remaining (Fig. 94). These latter present a rounded apex covered with deciduous hairs; some groups have permanent setae in this apical region that even resist clarification (Fig. 165). The length of the 9th tergite or epandrial lobes in relation to the proctiger is highly variable. Primitively, both were of equal length in Lutzomyiina, but in the ancestor of *Pressatia*+, the epandrial lobe became longer and underwent an oblique constriction in the pre-apical region (Figs. 157 and 158), and the posterior break and loss of the apical region gave rise to the pointed apex (*Evandromyia*, Figs. 159, 160, 161, 162, 163, 164 and 165). The **9th sternite** is also greatly reduced and bare. The gonostyles (Figs. 114, 115, 116, 117, 118, 119, 120, 121, 122, 123, 124, 125, 126, 127, 128, 129, 130, 131, 132, 133, 134, 135, 136 and 137), except in Hertigiini, are frequently smaller than the gonocoxites; their external surface is generally covered by deciduous hairs though in some groups it may present permanent hairs (Figs. 114 and 115), having spines of varying stages of development and number. The arrangement of these spines tends to occur in the following ways (Fig. 95): one or two apical spines (**apical spine(s)**), when the two are present one of them is situated externally and slightly pre-apically; in several groups this has atrophied and is called the **spiniform pre-apical seta** (Figs. 120, 121, 122, 123, 125, 126, 127, 128 and 129); two are situated on the external surface (**upper external spine** and **lower external spine**) and one on the internal face (**internal spine**). The primitive development and site of these

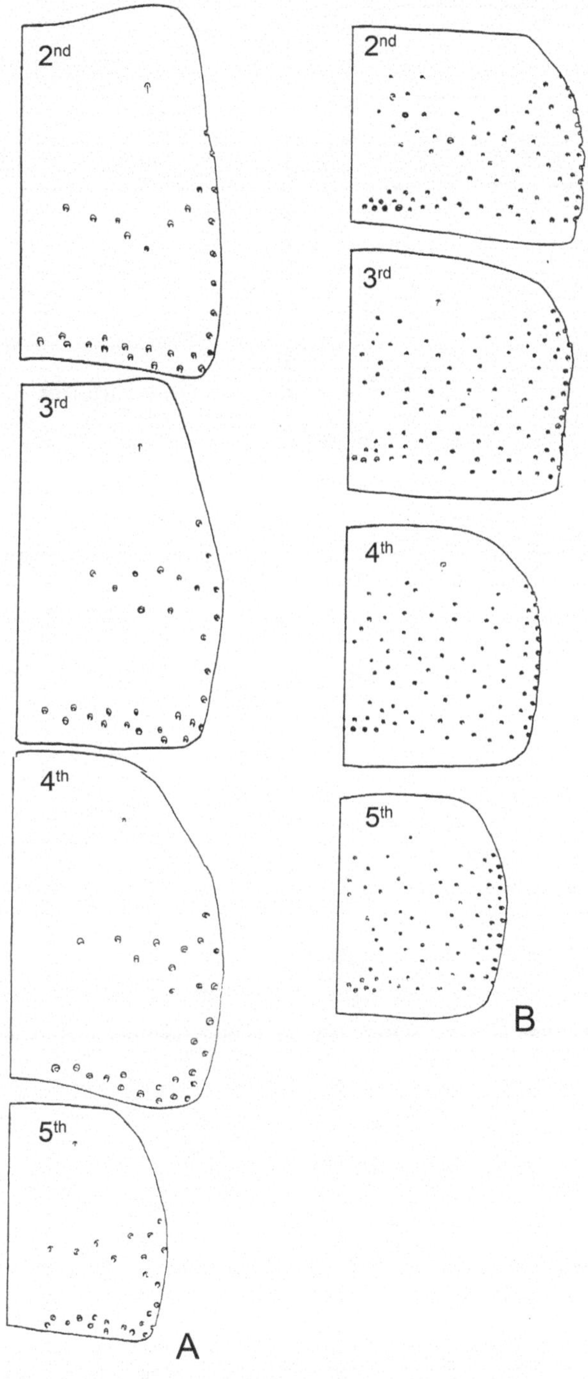

Fig. 92 2nd – 5th male tergites with the arrangement of the deciduous bristles. **A** – in two transversal bands: *Warileya phlebotomanica*. **B** – randomly: *Nyssomyia intermedia*

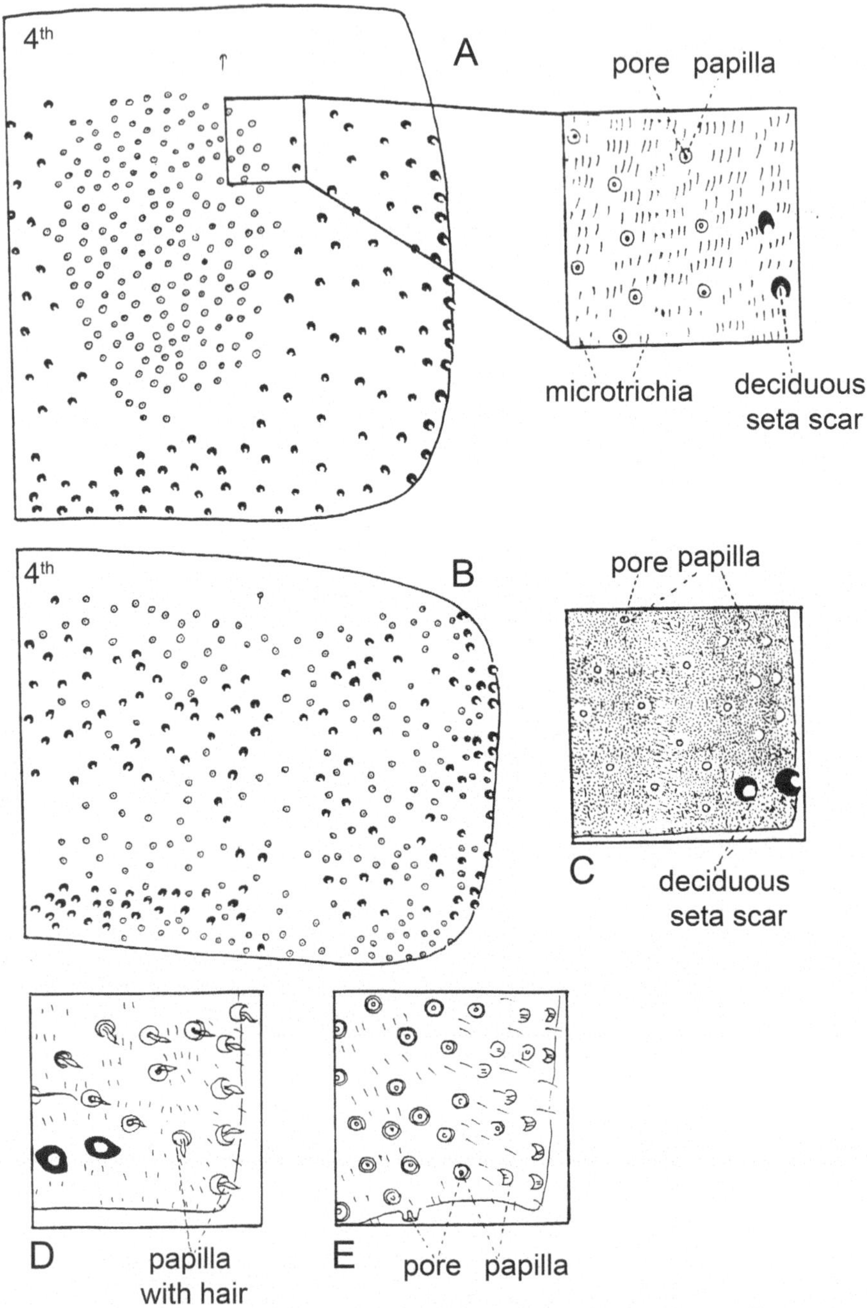

Fig. 93 Distribution and aspects of the tergal papillae on 4th tergite. **A** – restricted to the central area: *Lutzomyia longipalpis*; **B** – dispersed over the entire surface of the tergite, among the deciduous bristle scars: *Pintomyia fischeri*. **C – E** – aspects of the papillae on the 6th tergite: **C** – papillae without hair and without clear demarcation of their borders. **D** – Papillae with hair: *Brumptomyia cardosoi*; **E** – papillae without hair and with a clear demarcation of their borders: *Evandromyia walkeri*

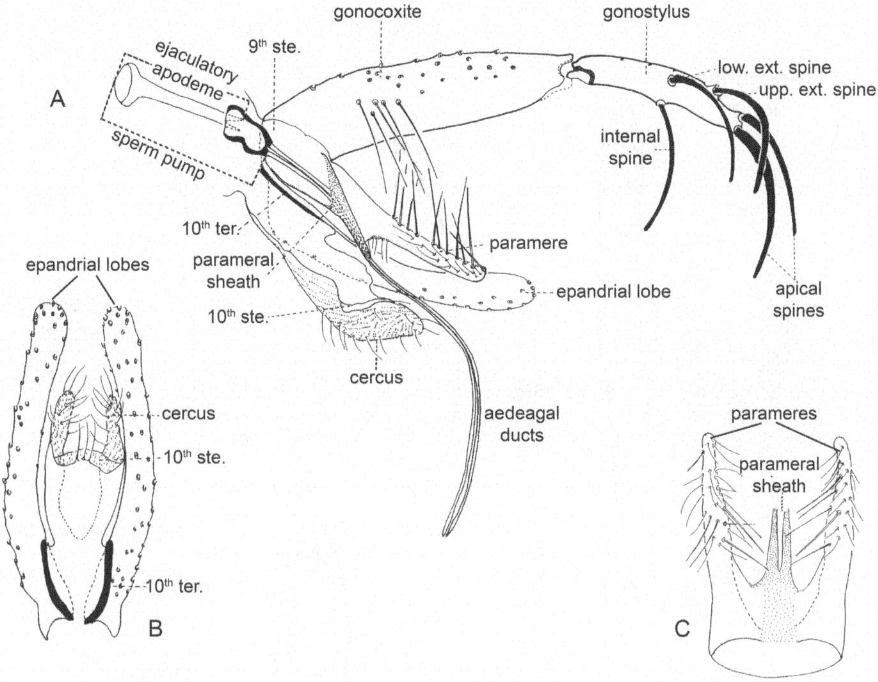

Fig. 94 Male genitalia of Phlebotominae. **A** – lateral view; **B** and **C** – ventral view. *Micropygomyia oswaldoi*. **ter** – tergite; **ste** – sternite; **low. ext.** – lower external; **upp ext.** – upper external

spines are probably represented by five highly developed spines, two of them apical, two external, inserted on the apical half, and the internal one situated on the apical third. The gonocoxites, which are generally cylindrical, have undergone a central constriction (Fig. 153). Their external surface is covered with deciduous hairs and the internal one frequently presents permanent hairs throughout its length (Fig. 149) or forming individualized tufts, apical (Figs. 141, 142 and 143), median (Fig. 145) or basal (Figs. 145, 146, 147 and 148). These hairs may be spiniform (Figs. 146 and 147), semifoliaceous (Fig. 147) or foliaceous (Fig. 144), inserted directly into the surface (Figs. 139, 146 and 149) or in small tubercles (Figs. 141, 142, 143 and 147) or columnar tubercles (Figs. 144 and 148). Various groups of Lutzomyiina present a sclerotized longitudinal band situated on the ventral edge (Figs. 145, 146 and 148). The parameres (Figs. 166, 167, 168, 169, 170, 171, 172, 173, 174, 175, 176, 177, 178, 179, 180 and 181) are those which have undergone the greatest modification both in terms of their shape and their hairy covering. As regards their shape, on a lateral view, it may be simple, digitiform (primitive aspect) (Figs. 94 and 166) or have processes and branches, generally on the dorsal edge (Figs. 169, 170, 177, 178, 179, 180 and 181) and one or more elbows (protuberances) on the ventral edge (Figs. 167 and 168). Their dorsal edge may be straight or concave, with their apex turned towards the gonocoxites, or convex, with their apex turned towards the

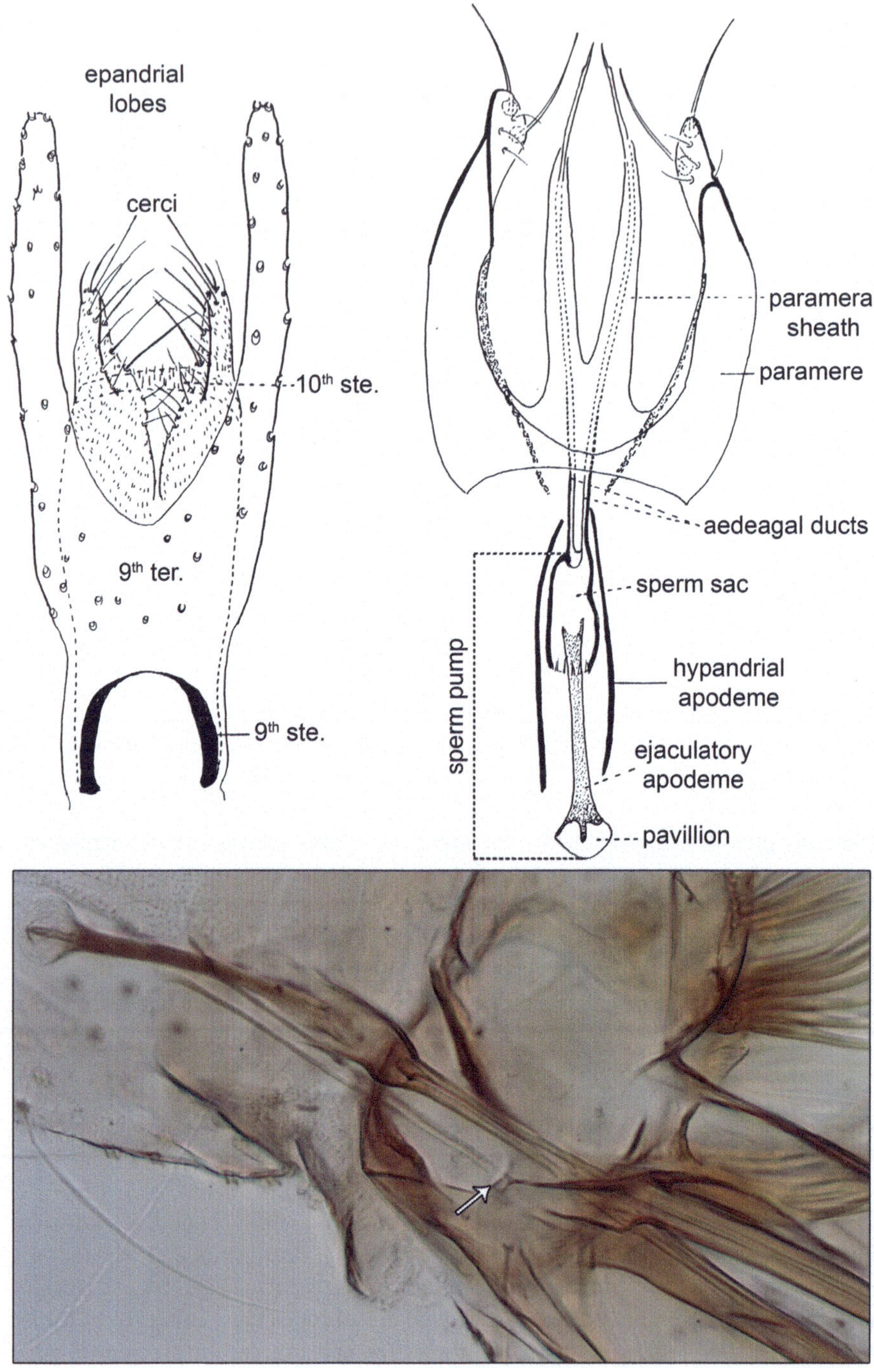

Fig. 95 Ventral view of parts of the male genitalia of Phlebotominae. Photo - arrow indicating the insertion of the hypandrial apodemes in the 9th sternite (hypandrium). *Warileya phlebotomanica*

epandrial lobes (Fig.171); the apical region and the dorsal edge are covered with hairs, generally spiniform, to a greater or lesser extent.

The **aedeagus**, according to Cumming and Wood (2009), consists of an intromitent tubular organ with an external opening, the **phalotrema**. Associated with the aedeagus is the **sperm pump** that is formed of the **ejaculatory apodeme** and the **sperm sac.** In Phlebotominae the aedeagus is bifurcated forming two ducts, the **aedeagal ducts** (Figs. 94 and 95), which have been variously denominated *ejaculatory filaments* (Forattini 1973), *genital filaments* (Jobling 1987; Young & Duncan 1994) and *ejaculatory ducts* (Forattini 1973; Galati 2003), as well as their associated structures, namely, the ejaculatory apodeme [*rod* (Jobling 1987) and *piston* (Galati 2003)], the sperm sac [*chamber* (Galati 2003)], the sperm pump [*pompetta* (Jobling 1987), *genital pump* (Young & Duncan, 1994) and *ejaculatory pump* (Galati, 2003)]. The ejaculatory apodeme, in its anterior part, is called the **pavillion** (Galati 2003a, b) (Fig. 95). The aedeagal ducts can be short (primitive character) or long, thin or thick and are similar in length and width to the length and width of the spermathecal ducts; they may be smooth or transversally striated to a greater or lesser extent; in their preapical region they may be straight, tapered or dilated and their apex may be truncated, oblique, spear-like, hooked, bifurcated, spoon-like or clavate, with or without small teeth, or of some other shape (Figs. 96, 97, 98, 99, 100, 101, 102, 103, 104, 105, 106, 107, 108, 109, 110, 111, 112 and 113). The pair structure, conical (Figs. 94, 166, 167, 168, 169, 170, 171, 172, 173, 174, 175, 176, 177, 178, 179, 180 and 181) or with ramifications (Fig. 178), situated between the parameres and through which the aedeagal ducts can become external has been called the aedeagus by almost all those who deal with phlebotomine taxonomy. However, bearing in mind the definition of this organ presented above and adopted for the majority of the groups of Diptera, we suggest the use of the term **parameral sheath** [= *aedeagal sheath* (Ilango 2004)], which is present in some other subfamilies of Psychodidae. The **ejaculatory duct** forms the connection between the **vesicula seminalis** and the sperm sac (Jobling 1987; Cumming and Wood 2009), both of which are destroyed in the process of clarification of the sand fly specimen. Parallel to the sperm pump and the aedeagal ducts in part of or throughout their length in some genera of Hertigiini, a pair of a sclerotized rods (Fig. 95B) may be present. These structures were interpreted by Ilango (2004) as *the anterior extension of the aedeagal sheath originating between the bases of the gonocoxite, aedeagal sheath and paramere*, named by that author the *gonocoxal/hypandrial apodeme of the aedeagal sheath*. However, in *Warileya* these apodemes seem to originate in the hypandrium at the point of insertion of the parameres as can be seen in Fig. 95-C. Therefore, their denomination as **hypandrial apodemes** is adopted here.

Tergite 10 is formed of sclerotized plates placed longitudinally at the base of tergite 9 (epandrium) or the epandrial lobes. The cerci are well developed, covered with hairs and microtrichia, presenting a rounded or pointed apex. Sternite10 is found in close association with the cerci and tergite 10 (Fig. 94).

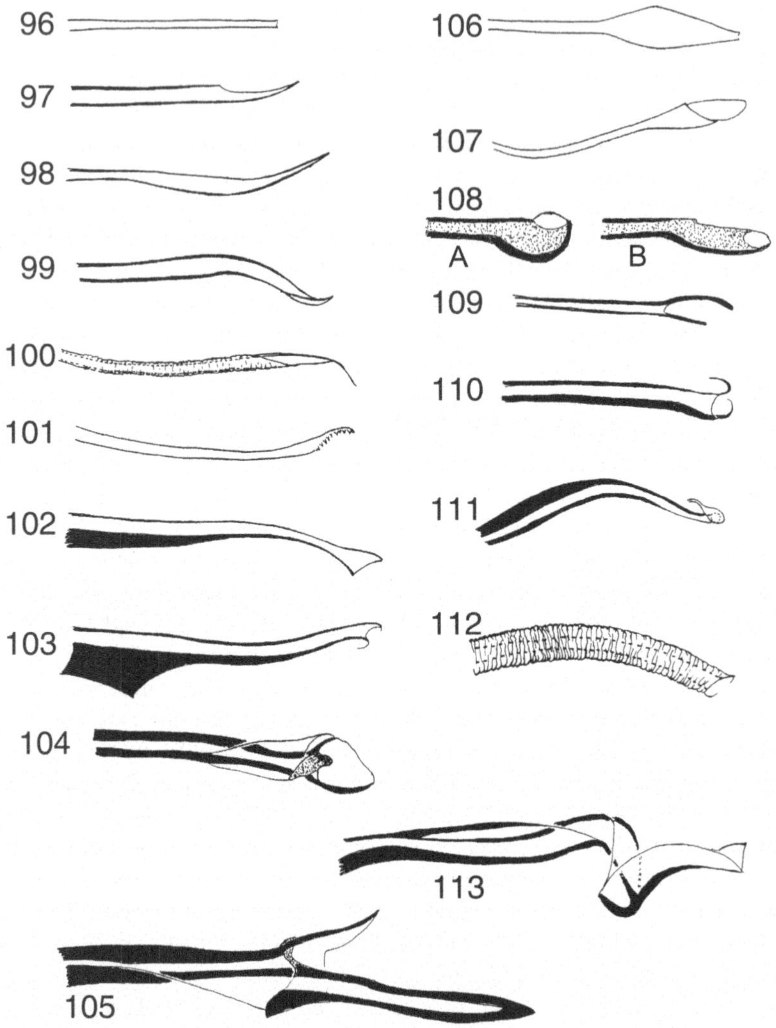

Figs. 96–113 Aspects of the terminal part of the aedeagal ducts: **96** – blunt apex: *Lu. longipalpis*; **97** – beveled apex: *Pa. shannoni*; **98** – apex in the form of semi lance: *Pa. brasiliensis*; **99** – curved in beveled apex: *Mi. longipennis*; **100** – striated duct with curved beveled apex: *Pa. dendrophyla*; **101** – curved and toothed apex: *Ma. gasparviannai*; **102** – funnel-shaped apex: *Ev. teratodes*; **103** – blunt apex with a barb; preapical region of the duct with a sclerotized protuberance: *Ev. termitophila*; **104** – clavate apex: *Ev. walkeri*; **105** – strongly sclerotized bifurcated apex: *Ev. lenti*; **106** – losenge-shaped apex: *Pa. runoides*; **107** – clavate apex: *Th. auraensis*; **108A** – ladle-shaped apex: *Ny. intermedia*; **108B** – spoon-shaped or fish-knife shaped: *Ny. neivai*; **109** – bifurcated apex: *Ny. whitmani*; **110** – apex with barbs: *Ny. yuilli pajoti*; **111** – duct curved in its pre-apical region and apex provided with appendix: *Vi. tuberculata*; **112** – striated duct with blunt apex: *Ev. brachyphala*; **113** – duct with curved preapical region and blunt apex: *Pa. aragaoi*

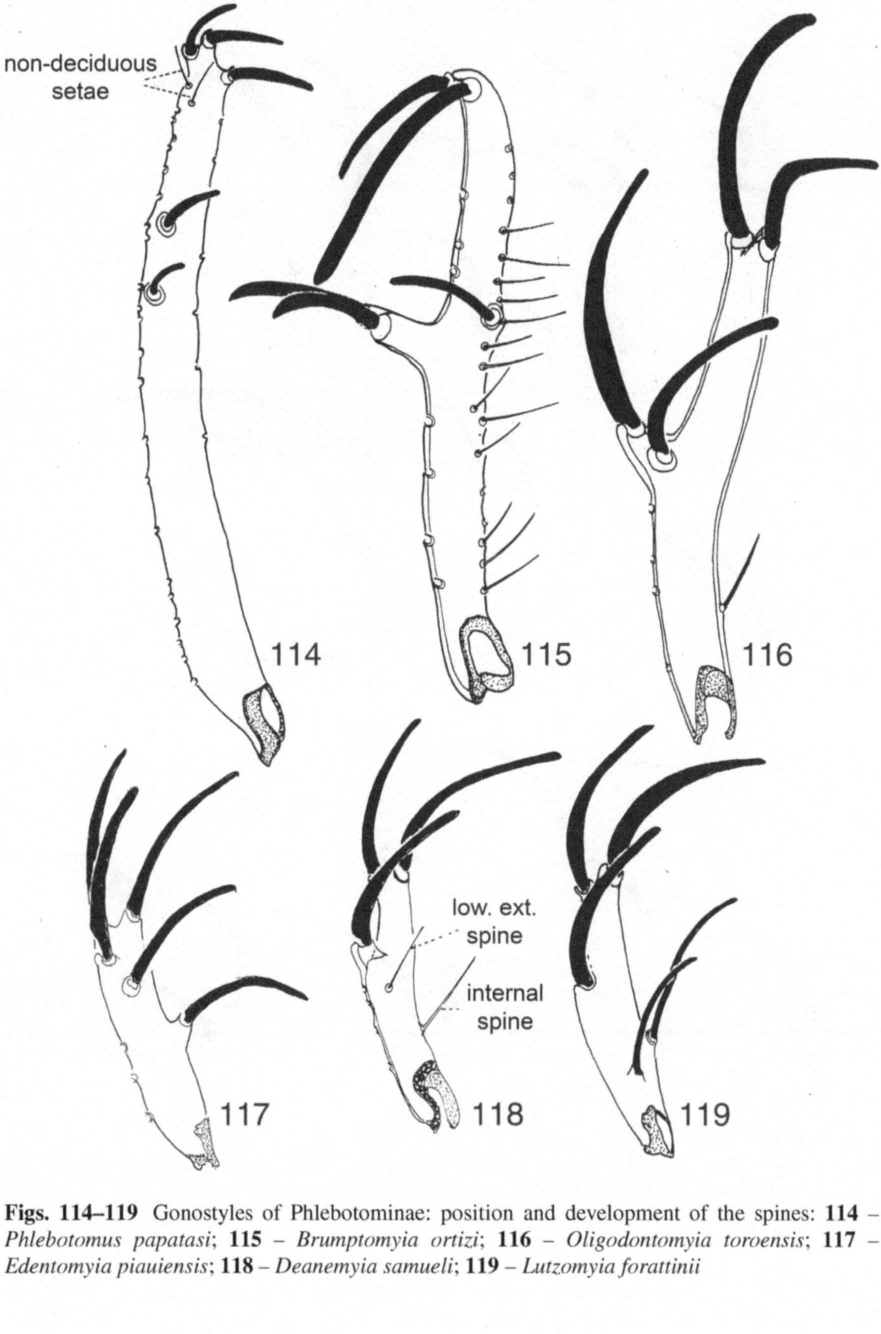

Figs. 114–119 Gonostyles of Phlebotominae: position and development of the spines: **114** – *Phlebotomus papatasi*; **115** – *Brumptomyia ortizi*; **116** – *Oligodontomyia toroensis*; **117** – *Edentomyia piauiensis*; **118** – *Deanemyia samueli*; **119** – *Lutzomyia forattinii*

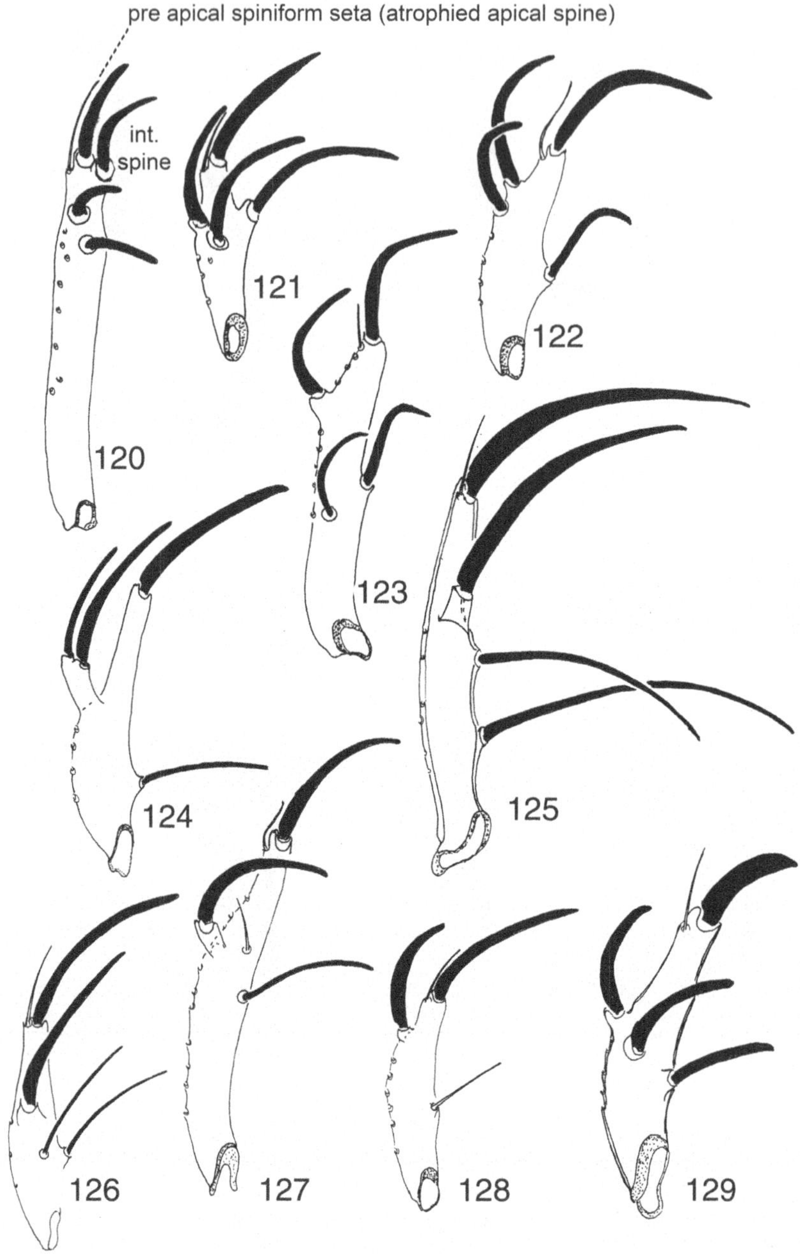

Figs. 120–129 Gonostyles of Phlebotominae: **120** – *Migonemyia gorbitzi*; **121** – *Mg. migonei*; **122** – *Pintomyia fischeri*; **123** – *Lutzomyia longipalpis*; **124** – *Evandromyia correalimai*; **125** – *Trichopygomyia longispina*; **126** – *Evandromyia cortelezzii*; **127** – *Pressatia triacantha*; **128** – *Dampfomyia anthophora*; **129** – *Dampfomyia* sp. de Suchitepequez (Delpozoi group)

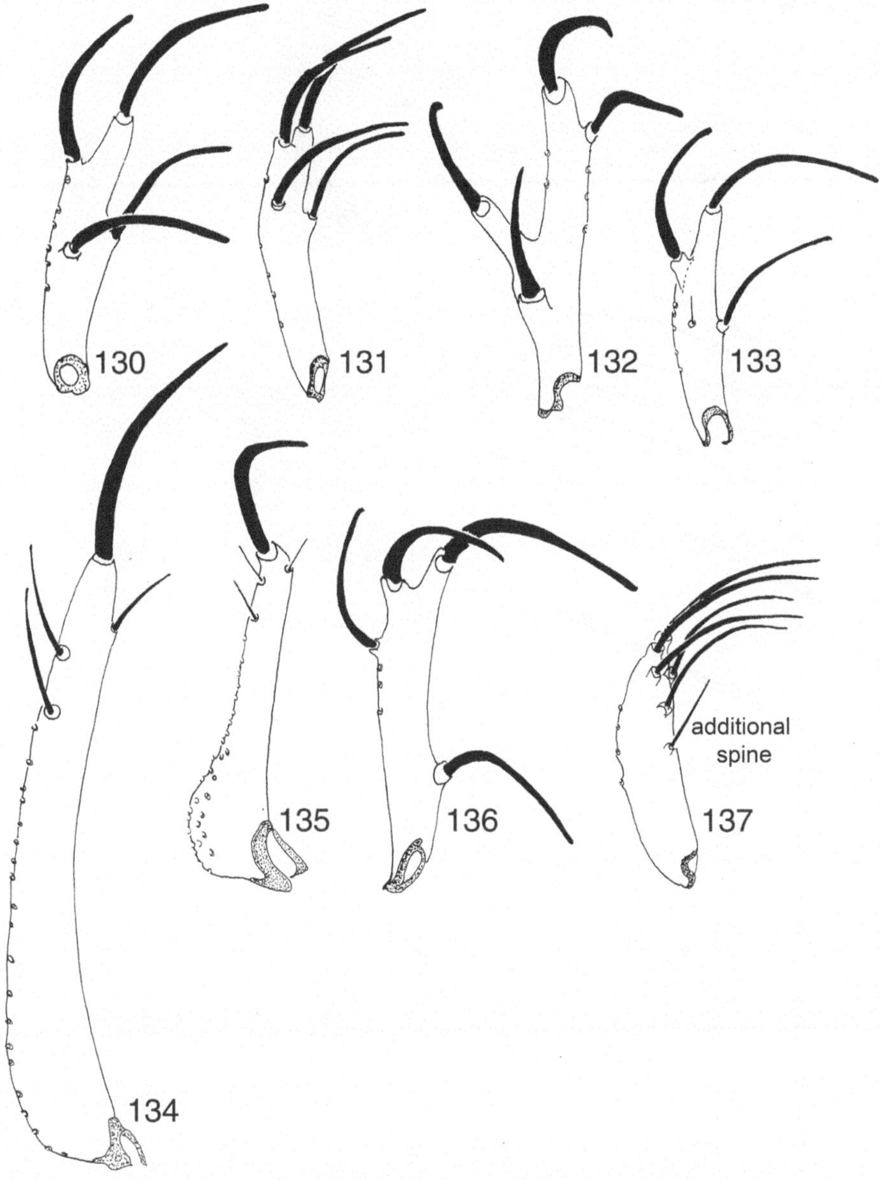

Figs. 130–137 Gonostyles of Phlebotominae. **130** – *Lutzomyia cruciata*; **131** – *Psathyromyia lanei*; **132** – *Viannamyia tuberculata*; **133** – *Micropygomyia pilosa*; **134** – *Psychodopygus geniculatus*; **135** – *Ps. chagasi*; **136** – *Trichophoromyia auraensis*; **137** – *Martinsmyia alphabetica*

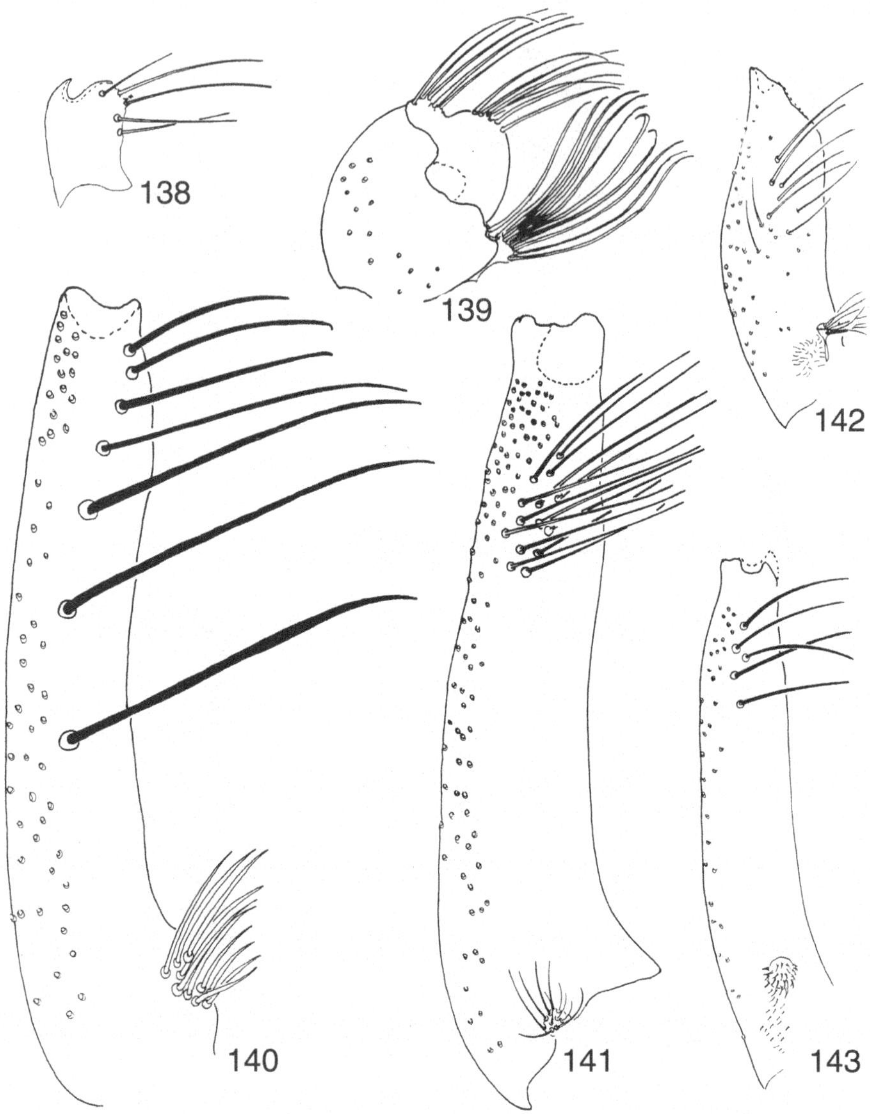

Figs. 138–143 Aspects of the shape and setae of the Phlebotominae gonocoxite. **138** – *Hertigia hertigi*; **139** – *Warileya phlebotomanica*; **140** – *Brumptomyia brumpti*; **141** – *Phlebotomus papatasi*; **142** – *Migonemyia migonei*; **143** – *Mg. gorbitzi*

Identification of the Taxa of the Americas

The identification keys for the supraspecific taxa of American phlebotomine sandflies have been developed according to Galati's (1995) classification presented in the previous section, with the *Edentomyia* genus later described and the series of the

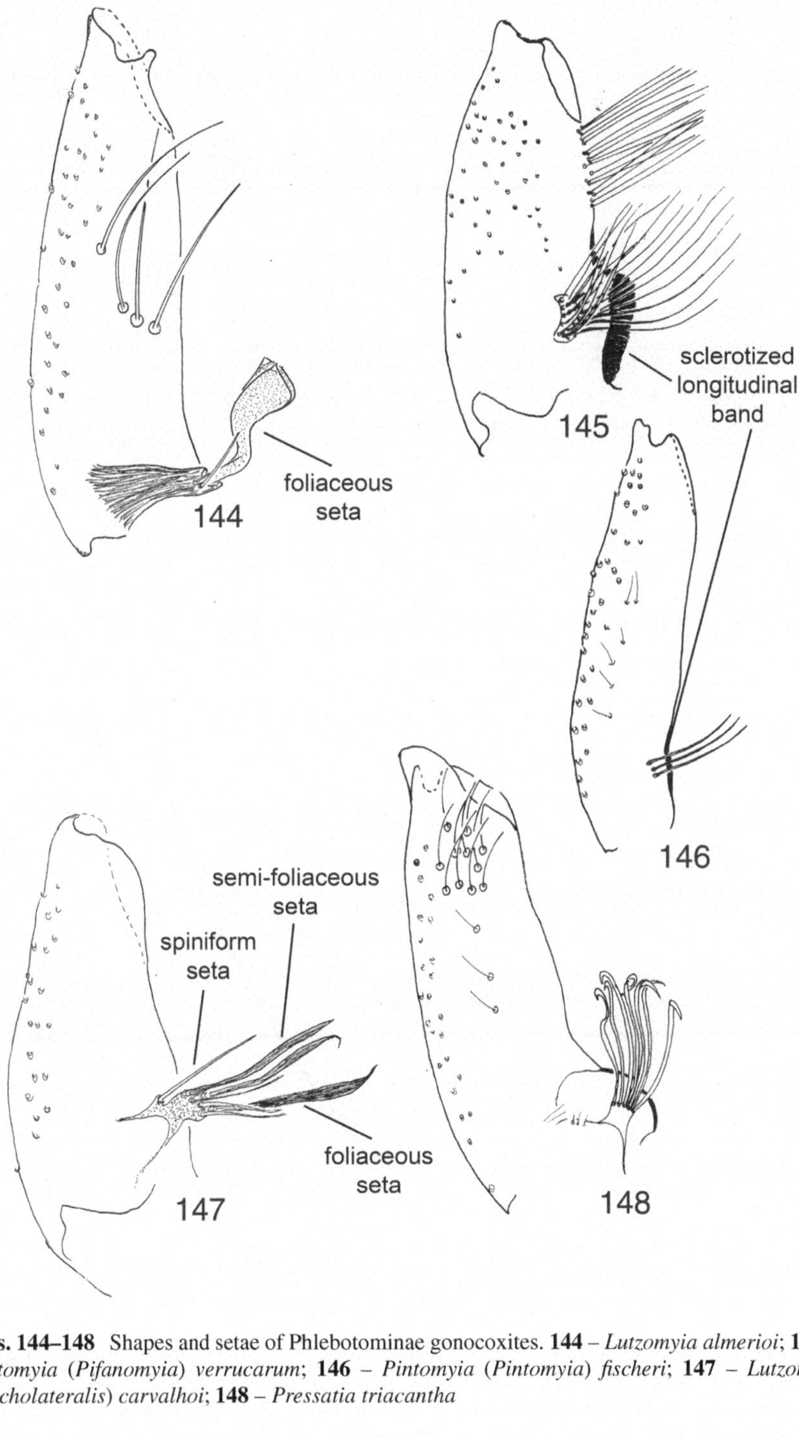

Figs. 144–148 Shapes and setae of Phlebotominae gonocoxites. **144** – *Lutzomyia almerioi*; **145** – *Pintomyia* (*Pifanomyia*) *verrucarum*; **146** – *Pintomyia* (*Pintomyia*) *fischeri*; **147** – *Lutzomyia* (*Tricholateralis*) *carvalhoi*; **148** – *Pressatia triacantha*

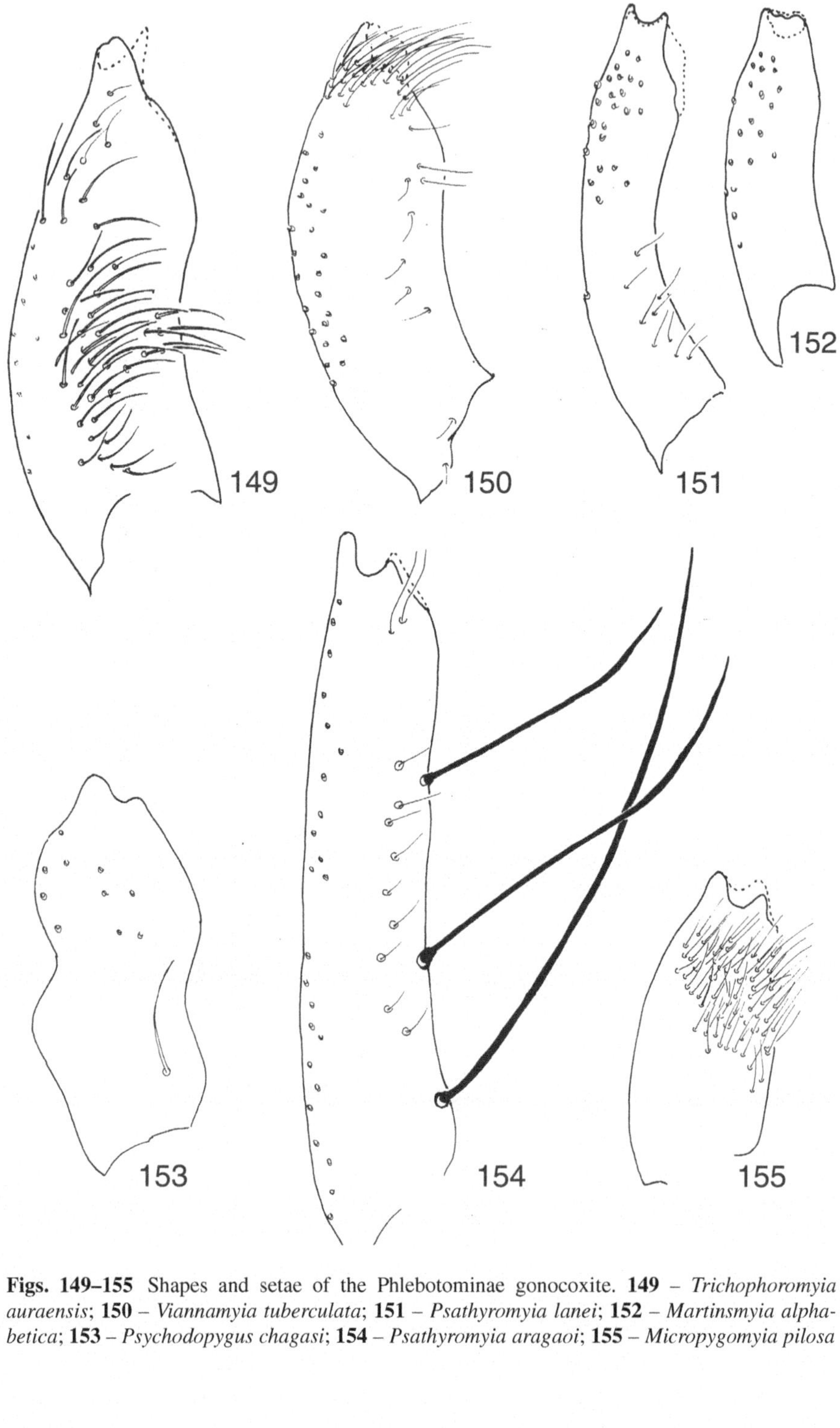

Figs. 149–155 Shapes and setae of the Phlebotominae gonocoxite. **149** – *Trichophoromyia auraensis*; **150** – *Viannamyia tuberculata*; **151** – *Psathyromyia lanei*; **152** – *Martinsmyia alphabetica*; **153** – *Psychodopygus chagasi*; **154** – *Psathyromyia aragaoi*; **155** – *Micropygomyia pilosa*

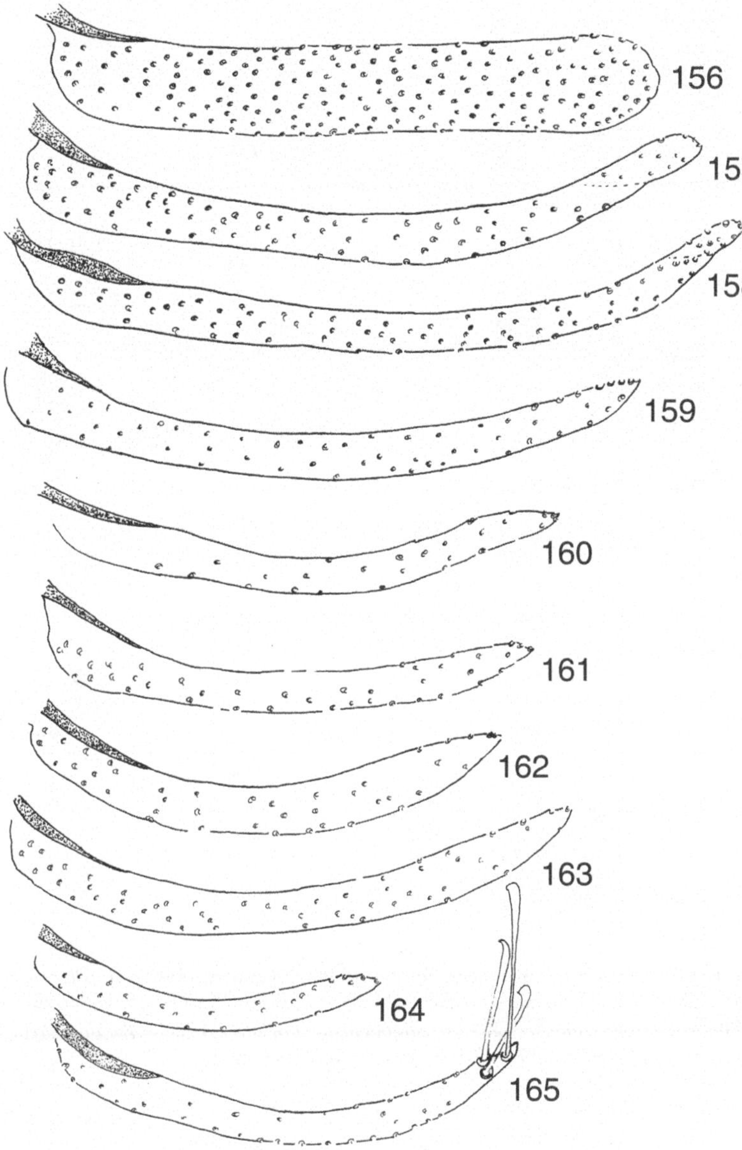

Figs. 156–165 Epandrial lobes. **156** – swollen: *Dampfomyia vespertilionis*; **157, 158** – oblong apex with a preapical notch: **157** – *Pressatia triacantha*; **158** – *Trichopygomyia longispina*; **159 – 165**: tapered beveled: **159** – *Evandromyia walkeri;* **160** – *Ev. monstruosa*; **161** – *Ev. correalimai*; **162** – *Ev. tupynambai*; **163** – *Ev. cortelezzii*; **164** – *Ev. saulensis*; **165** - presence of non-deciduous setae: *Ev. infraspinosa*

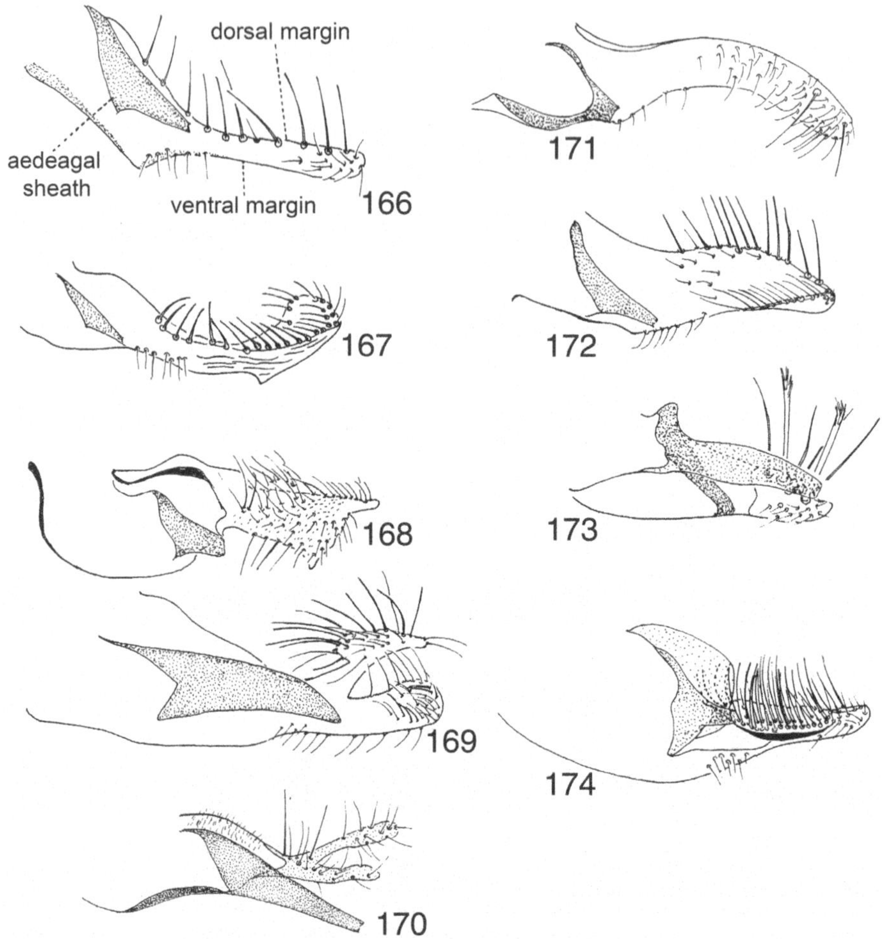

Figs. 166–174 Parameres and parameral sheaths of Phlebotominae. **166** – *Martinsmyia alphabetica*; **167** – *Evandromyia tupynambai*; **168** – *Pressatia triacantha*; **169** – *Trichopygomyia longispina*; **170** – *Ev. infraspinosa*; **171** – *Migonemyia gorbitzi*; **172** – *Psathyromyia lanei*; **173** – *Viannamyia tuberculata*; **174** – *Trichophoromyia auraensis*

subgenus *Helcocyrtomyia* and of the genus *Psychodopygus* added to it. The abbreviation of the names of the genera and subgenera are those of Marcondes (2007).

PSYCHODIDAE: Both sexes: body densely covered by hairs that give a hairy aspect. Antennae with flagellum consisting of articulated flagellomeres. Wing with 9–11 longitudinal veins reaching the wing's edge, the subcosta with the inflection point towards either the costal vein or the R1 before reaching the middle of the wing. The Radial sector has three or more veins, R1 reaching the costa beyond its apical half; the Medial sector with four veins, M4 being united with CuA1; the more distal transversal vein r-m is located in the middle or before the middle of wing.

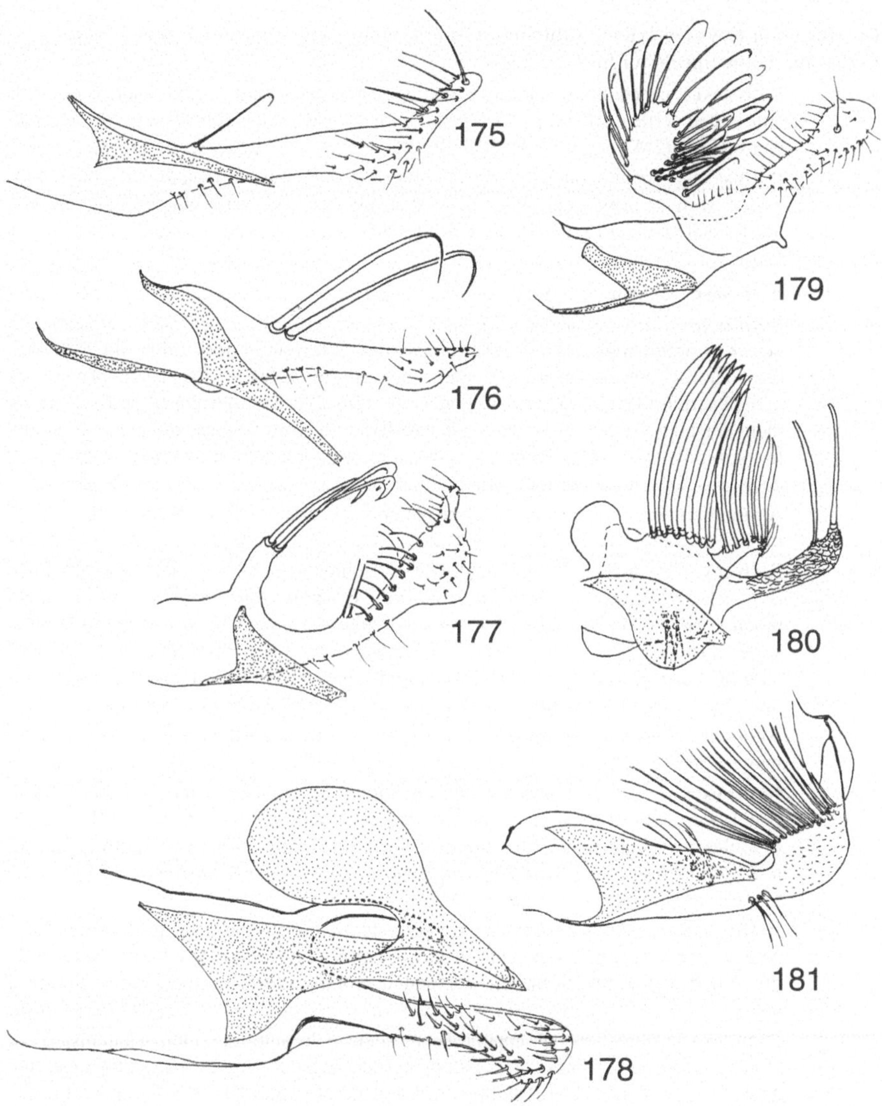

Figs. 175–181 Parameres and parameral sheaths of Phlebotominae. **175** – *Lutzomyia castroi*; **176** – *Lutzomyia longipalpis*; **177** – *Lutzomyia dispar*; **178** – *Trichopygomyia dasypodogeton*; **179** – *Dampfomyia anthophora*; **180** – *Psychodopygus panamensis*; **181** – *Psychodopygus chagasi*

PHLEBOTOMINAE: Both sexes: antennae with flagellum consisting of 14 flagellomeres. Wing: R2, R3, R4, R5 veins isolatedly reaching the wing's edge; the transversal vein r-m is located between the basal third and the middle of the wing and the vein CuA2 is short ending before the beginning of M3 and M4, these latter without any connection between them (Fig. 36). Males: mandibles absent for the great majority of the species, but may be present with an atrophied aspect in some few species. Females: mandibles and a pair of spemathecae present.

Identification Keys for Tribes, Subtribes, Genera, Subgenera, Species Groups, Species Series and Some Species of *Incertae Sedis*

1 **Both sexes**: proepimeral and upper anepisternal setae absent; metathoracic furca with non-united vertical arms (Fig. 27-C); abdominal tergite setae grouped in two transversal bands (Fig. 92-A) ....HERTIGIINI.. 2

Both sexes: proepimeral and upper anepisternal setae present; metathoracic furca with united vertical arms (Fig. 27-A, B); abdominal tergite setae not grouped in two transversal bands (Fig. 92-B)...PHLEBOTOMINI 3

2(1) **Both sexes**: clypeus without setae ....*Hertigia*

Both sexes: clypeus with setae .... *Warileya*

3(1) **Both sexes.** Cibarium: absence of anterior and posterior teeth, sclerotized area and the posterior protuberance (Fig. 19-B); labial sutures forming a fork (Fig. 10-C); thorax: presence of the post-alar seta and long suture between the katepimeron and metapisternum (Fig. 23). **Males**: gonostyle with five well-developed spines, two of them apical, two external implanted isolatedly on the fourth apical and the internal one in its middle (Fig. 117). **Females**: spermathecae segmented, individual spermathecal ducts 3–4 × the length of the common duct.............................. *Edentomyia piauiensis*

Although without the complete set of the above characteristics, one or more of them may be present 4

4(3) **Both sexes**: external ascoid of flagellomere I located more basally than the internal one (Fig. 14-A). **Males**: gonostyle with the two external spines located on one single tubercle (Figs. 115 and 116). **Females**: spermathecae segmented and the individual sperm ducts very long, four or more × the length of the spermatheca (Fig. 46); cibarium with the anterior teeth lying horizontally and sometimes in lateral position (Fig. 19-C) ....BRUMPTOMYIINA.................. 5

Although without the complete set of the above characteristics, one or more of them may be present 6

5(4) **Both sexes**: ascoids with posterior prolongation (Fig. 15), post-alar seta absent. **Males**: gonocoxite with stout, straight setae in the apical region of the inner side (Fig. 140). **Females**: cibarium with the anterior teeth lying horizontally, forming several longitudinal rows, the sclerotized area located posterior to the teeth (Fig. 19-C) .... *Brumptomyia*

Both sexes: ascoids without posterior prolongation, post-alar setae present (Fig. 23). **Males**: gonocoxite without setae in the apical region. **Females**: cibarium with little developed posterior teeth, the anterior teeth do not form rows and there is no sclerotized area .... *Oligodontomyia*

6(4) **Both sexes**: ventro-cervical sensilla absent; setae in the anterior region of katepisternum absent, except for *Bichromomyia*; 4th palpal segment perceptibly smaller than the 2nd. **Males**: external ascoid of flagellomere I located at variable level as compared to the internal one. **Females**: cibarium with the anterior teeth vertically and/or laterally located, well-developed and forming several transversal rows (Figs. 18-A and 20-H, I) ....PSYCHODOPYGINA

Both sexes: generally, ventro-cervical sensilla are present (Figs. 21 and 25), but when absent, setae are present in the anterior region of the katepisternum (Fig. 24). 4th palpal segment longer than, equal to or shorter than the 2nd. **Males**: external ascoid of flagellomere I located more apically than the internal one (Fig. 14-C). **Females**: cibarium with anterior teeth that can be horizontal (Fig. 19-D), lateral (Fig. 20-D, F) or vertical (Fig. 20-A, B, C, E, G), in this latter case, generally, they are located in one or two transversal rows... 7

7(6) **Both sexes**: 2nd palpal segment shorter than or equivalent to the 4th; sets of Newstead's sensilla located in the basal half of the 3rd palpal segment (Fig. 13-E); absence of papilla on flagellomere III (Fig. 16-A); apical, median and basal papillae of flagellomere XI also absent (Fig. 17-B). **Females**: teeth of the hypopharynx generally poorly demarcated or absent (Fig. 9-E, F, G); developed teeth in the pharynx present (Fig. 18-B) or absent (Fig. 18-A)..SERGENTOMYIINA

Both sexes: 2nd palpal segment equivalent to or longer than the 4th; sets of Newstead's sensilla located in the middle of the 3rd palpal segment or scattered, also on the apical half (Fig. 13D); papilla on flagellomere III present (Fig. 16-B) [except for *Sciopemyia, partim*, (*Tricholateralis*), (*Dampfomyia*) and Evansi series]; papillae on flagellomere XI may be present or not. **Females**: hypopharynx bearing well demarcated teeth (except for *Sciopemyia* and *Coromyia*); developed teeth absent from the pharynx (Fig. 18-A) .... LUTZOMYIINA

SERGENTOMYIINA - Genera, Subgenera and Series of Species

1 **Both sexes**: setae present in the anterior region of the katespisternum (Fig. 24); post-alar setae present (Fig. 24); Labial sutures not forming a fork (Fig. 10-B). **Males**: gonostyle with two well developed apical spines and the inner spine atrophied (Fig. 118); paramere with or without appendage on the dorsal margin. **Females**: spermathecae superficially segmented (Fig. 87)..***Deanemyia***

Both sexes: setae absent in the anterior region of the katespisternum (Fig. 24); post-alar setae absent; labial fork present (Fig. 10-C). **Males**: gonostyle with one or two well developed apical spines and the inner spine developed; paramere without appendage on the dorsal margin...***Micropygomyia***.............. 2

2(1) **Both sexes**: well-developed spines present in the pharynx (Fig. 18-B). **Males**: FI longer than head length; gonostyle with one apical spine. **Females**: flagellomere I ca. 0.9 times the head length; cibarium bearing two pairs of posterior (horizontal) teeth...***Mi.* (*Silvamyia*)**

Both sexes: well-developed spines in the pharynx present or absent. **Males**: FI shorter than head length; gonostyle with one or two apical spines. **Females**: flagellomere I shorter than 0.75 times the head length 3

3(2) **Both sexes**: undeveloped spines in the pharynx. **Males**: gonocoxite with basal compact tuft of 4–7 setae; gonostyle with two apical spines and the inner spine located before its middle. **Females**: ascoids of flagellomere II with apex reaching or passing the papilla; cibarium with short cibarial chamber, with two or more pairs of horizontal teeth, the external teeth turned outwards; small and bulbous spermathecae shorter than the terminal knob or elongated, in this case, striated or ringed; visible common sperm duct ....***Mi.* (*Coquillettimyia*)**............ 4

Although without the complete set of the characteristics above, one or more of them may be present 5

4(3) **Males**: epandrial lobe equivalent to or longer than gonocoxite; pre-apical region of aedeagal ducts losange-shaped (Fig. 106). **Females**: cibarium with two or three pairs of horizontal teeth; small and spherical spermathecae, a little wider than the individual sperm ducts that are long and narrow towards the spermatheca (Fig. 88)...***Mi.* (*Col.*)** Vexator series

Males: epandrial lobe shorter than gonocoxite; pre-apical region of aedeagal ducts losange-shaped or not. **Females**: cibarium with two pairs or more of posterior teeth; spermathecae ringed or striated ....***Mi.* (*Col.*)** Chiapanensis series

5(3) **Males**: gonostyle with four spines, the lower external one being atrophied (Fig. 133). **Females**: cibarium with long, narrow chamber, various anterior teeth are laterally placed and the two posterior pairs are united at the base (Fig. 20-B); ascoid of the flagellomere II long, its apex almost reaching that of the article; rudimentary common spermathecal duct (Fig. 48) ....(***Micropygomyia***), *partim* ....***Mi.*** (***Mic.***) Pilosa series

Males: gonostyle with four or five well developed spines (Fig. 94-A). **Females**: cibarium short with wide or narrow chamber; posterior teeth united or not at the base (Fig. 20-B), anterior teeth in vertical or horizontal position; evident common spermathecal duct, even if short (Figs. 89, 90 and 91) 6

6(5) **Males**: gonostyle with four spines and gonocoxite without tuft of setae in the basal and/or median region. **Females**: ascoids of flagellomere II short, generally their apex not reaching the point of insertion of the papilla and their length equivalent to one-third or half that of the article (Fig. 16-A); cibarium with two or more pairs of posterior teeth united at the base (Fig. 20-C); spermathecae smooth or segmented ....(***Micropygomyia***), *partim* .... ***Mi.*** (***Mic.***) Cayennensis series

Males: gonostyle with four or five (Fig.13) spines; in this latter case, the gonocoxite has or not a tuft of setae in the median region. **Females**: cibarium with two pairs of posterior teeth; apex of the ascoids of flagellomere II generally reaches or passes the point of insertion of the papilla or their length is equivalent to more than half that of the article, spermathecae ringed (Fig. 90) or smooth (Fig. 74) ....***Mi.*** (***Sauromyia***) 7

7(6) **Males**: gonostyle with four spines, one apical. **Females**: pharynx with strong sclerotized transversal striae (Fig. 18-B) ....***Mi.*** (***Sau.***) Atroclavata series

Males: gonostyle with five spines, two of which are apical. **Females**: pharynx without sclerotized transversal striae; cibarium with two teeth united or not at the base, may be present and inclined towards the median line and sometimes there is a projection between the inner pairs ....***Mi.*** (***Sau.***) Oswaldoi series

LUTZOMYIINA - Genera, Subgenera, Group and Series of Species

1 **Both sexes**: setae present in the anterior region of the katepisternum (Fig. 24); head and labrum-epipharynx short, so that the sum of both is equivalent to or shorter than the sum of flagellomeres I + II. **Males**: gonostyle with one apical spine and absence of the pre-apical spiniform seta. **Females**: hypopharynx with atrophied teeth (Fig. 9-F); spermathecae segmented (Fig. 86) .... ***Sciopemyia***

Although without the complete set of the characteristics above, one or more of them may be present 2

2(1) **Both sexes**: absence of the papilla on flagellomere III (Fig. 16-A); abdominal pleura with setae (more evident in females) (Fig. 38). **Males**: gonostyle with one apical spine and absence of the pre-apical spiniform seta (Fig. 130). **Females**: spermathecae segmented (Fig. 45) ....***Lutzomyia***, *partim* ....***Lu.*** (***Tricholateralis***)

Although without the complete set of the characteristics above, one or more of them may be present 3

3(2) **Both sexes**: papilla on flagellomere III present (Fig. 15-A); absence of setae on the abdominal pleura. **Males**: gonostyle with two well-developed apical spines; paramere simple, digitiform, without enlarged setae with a hook-shaped apex in the middle region of its dorsal margin. **Females**: presence of setae on the anterior edge of the katepisternum, except in the complex Noguchii which presents small eyes (equal to or smaller than the clypeus), spermathecae segmented...***Lutzom yia***, *partim* ....***Lu.*** (***Helcocyrtomyia***)

Although without the complete set of the characteristics above, one or more of them may be present 4

4(3) **Both sexes**: absence of setae on the anterior region of the katepisternum. **Males**: gonostyle with two or one developed apical spines, in this latter case the pre-apical spiniform seta may be present or not; gonocoxite without a sclerotized longitudinal band in its basal inner area (Fig. 144) and with a tuft of up to seven setae; basal or median region of the dorsal margin of the paramere with one or more setae, different from the apical ones, generally with a hook-shaped apex. **Females**: spermathecae segmented; common spermathecal duct short, the individual spermathecal ducts being at least four times as long as the spermathecae or the common duct .... ***Lutzomyia***, *partim* 5

Males: Although without the complete set of the characteristics above, the gonostyle presents a developed apical spine and generally the pre-apical spiniform seta is present. **Females**: spermathecae not totally segmented [except in *Ev.* (*Ald.*) *termitophila*, but in this case, the individual sperm ducts are as long as the spermatheca], may be elongated (Figs. 44, 74 and 84) or bulbous, with (Figs. 70, 71 and 72) or without transversal striae (Figs. 77, 78, 81 and 83), the length of the individual spermathecal ducts does not exceed 2.5 times the length of the common sperm duct (Figs. 69, 70, 71, 72, 73, 74, 75, 76, 77, 78, 79, 80, 81, 82, 83 and 84) [except in *Ev.* (*Bar.*) Cortelezzii series (Fig. 85)] 6

5(4) **Males**: gonostyle with one apical spine and pre-apical spiniform seta, the external spine located more apically than the internal one, or both at the same level. **Females**: cibarium with incomplete sclerotized arch and horizontal anterior teeth (Fig. 19-D) .... ***Lu.* (*Castromyia*)**

Males: gonostyle with two apical spines or one, in which latter case the pre-apical spiniform seta may or may not be present and the external lower spine located, distinctively, more basally than the inner. **Females**: cibarium with complete sclerotized arch with vertical and/or lateral anterior teeth .... ***Lu.* (*Lutzomyia*)**

6(4) **Both sexes**: flagellomere I short, smaller than or equivalent to half the length of the head. **Males**: inner basal face of gonocoxite without sclerotized longitudinal band (Figs. 142 and 143); gonostyle with the inner spine located very close to the apex (Fig. 120) or in its apical third (Fig. 121), more apically or at the same level as the upper external spine. **Females**: common spermathecal duct very short or absent, spermathecae tubular (Figs. 44 and 47) or bulbous ***Migonemyia*** 7

Both sexes: flagellomere I generally longer than half the length of the head. **Males**: gonostyle with spines otherwise than stated immediately above; inner basal face of gonocoxite with sclerotized longitudinal band (Figs. 145, 146 and 147) (except *Pi.* (*Pif.*) Monticola series and *Trichopygomyia*). **Females**: spermathecae of variable aspect (Figs. 69, 70, 71, 72, 73, 74, 75, 76, 77, 78, 79, 80, 81, 82, 83, 84 and 85) and an evident common duct 8

7(6) **Both sexes**: ventro-cervical sensilla absent. **Males**: gonostyle with the inner spine located in its apical third (Fig. 121). **Females**: small sclerotized protuberance present on 9th tergite (Fig. 41) .... ***Mg.* (*Migonemyia*)**

Both sexes: ventro-cervical sensilla present. **Males**: gonostyle with the inner spine located almost apically (Fig. 120). **Females**: small sclerotized protuberance on 9th tergite absent ....***Mg.* (*Blancasmyia*)**

8(6) **Males**: paramere simple, without jagged protuberance in the pre-apical region of the ventral margin (Figs. 167 and 168); epandrial lobe distinctively narrower than gonocoxite and with rounded apex (Fig. 94). **Females**: apex of the common spermathecal duct reaching or passing the middle of the genital furca stem (Figs. 69, 70, 71, 72 and 74); 8th tergite, generally, with setae (Fig. 37-B) .... ***Pintomyia***............9

Males: paramere simple, generally, with jagged protuberance in the pre-apical region of the ventral margin (Figs. 167, 168 and 179) or with lobe on the dorsal margin (Fig. 169) or branched (Fig. 170); epandrial lobes of variable aspect. **Females**: common spermathecal duct long or short, setae of the 8th tergite absent (except in *Expapillata* and *Trichopygomyia, partim*).. 18

9(8) **Both sexes**: posterior femur with spines (Fig. 35). **Females**: individual spermathecal ducts sclerotized in their distal half (Fig. 69) ....***Pintomyia* (*Pintomyia*)**

Both sexes: posterior femur without spines. **Females**: individual spermathecal ducts not sclerotized in their distal half...***Pintomyia* (*Pifanomyia*)**... 10

10(9) **Both sexes**: presence of the apical papilla on flagellomeres XI and X. **Males**: aedeagal ducts at least four times as long as the sperm pump. **Females**: common spermathecal duct very long, three times or more the length of the genital furca stem.................... 11

Both sexes: absence of the apical papilla on flagellomeres XI and X. **Males**: aedeagal ducts are, at most, five times longer than the sperm pump. **Females**: common spermathecal duct reaches a maximum of twice the length of the genital furca stem 12

11(10) **Males:** aedeagal ducts four times or slightly more the length of the sperm pump. Gonocoxite with a tuft of ca. 12 setae of different lengths. Paramere: ventral margin straight as far as its pre-apical region where it turns towards the gonocoxite almost at a right angle; dorsal margin with accentuated concavity as far as the pre-apical region; spiniform setae restricted to the pre-apical area. **Females**: spermathecae tubular, without clear transition to the individual ducts; common duct smooth in its central region, ca. four × wider than the spermathecae at its widest point................................ ....(*Incertae sedis*)... ***Pi. (Pif.) diamantinensis***

Males: aedeagal ducts long, six or more times the length of the sperm pump. Gonocoxite with a tuft of up to eight setae. Paramere: ventral margin not forming an accentuated angle; dorsal margin pratically straight and with the setae extending along the apical third. **Females:** individual spermathecal ducts with clear transition to the spermatheca; common duct smooth or striated in its central region or narrower than the spermatheca at its greatest width (Fig. 74) ....***Pi.* (*Pif.*)** Pacae series

12(10) **Males**: gonostyle with three external spines. **Females**: 9th tergite frequently has permanent spiniform setae in its central region (Fig. 40); spermathecae vesiculous, though with a clear narrower collar-shaped apical ring (Fig. 70)...***Pi.* (*Pif.*)** Pia series

Males: gonostyle with one or two external spines. **Females**: spermathecae with different aspect from that described above 13

13(12) **Both sexes**: thoracic pleura and coxae dark brown. **Males**: tergal papillae present on the 2nd abdominal tergite. **Females**: medial or basal papillae present on flagellomere XI (Fig. 17-A); genital furca forming an acute angle in the apical area of the chamber (Fig. 39)..***Pi. (Pif.)*** Monticola series

Both sexes: thoracic pleura partially dark brown or completely straw-colored. Brown or straw coxae. **Males**: tergal papillae absent from the 2nd abdominal tergite. **Females**: median or basal papillae on flagellomere XI absent; chamber of genital furca curved in its apical area (Fig. 37-D) 14

14(13) **Both sexes**: papilla on flagellomere III absent; **Males**: gonostyle with four spines, the inner one located in the middle of the gonostyle; gonocoxite without or with just one basal tuft consisting of 1–6 setae ....***Pi.* (*Pif.*)** Evansi series

Both sexes: papilla on flagellomere III present. **Males**: gonostyle with three or four spines, the inner one located at the base or in the middle of the gonostyle; gonocoxite with one or more tufts with variable number of setae.. 15

15(14) **Males**: gonocoxite with apico-central and basal tufts of setae. **Females**: small eyes, in the dorsal view, a little longer than the length of the clypeus and narrower than the interocular distance .... ***Pi.* (*Pif.*)** Verrucarum series

Males: gonocoxite without apico-central tuft of setae, with only the basal one. **Females**: big eyes, in the dorsal view, clearly bigger than the length of the clypeus and equivalent to or wider than the interocular distance.. 16

16(15) **Males**: gonostyle with only one external spine (Fig. 128); tergal papillae present on some abdominal tergites (Fig. 93-C). **Females**: terminal knob of the spermathecae greatly elongated, inserted before the apical ring (Fig. 72)... ***Pi.* (*Pif.*)** Serrana series

Males: gonostyle with two external spines; tergal papillae absent from all abdominal tergites. **Females**: terminal knob of the spermathecae inserted in the apical ring.. 17

17(16) **Males**: gonostyle with the inner spine clearly basal. **Females**: only one ascoid on fXIII; spermatheca with its apical ring narrower than itself ....***Pi.* (*Pif.*)** Townsendi series

Males: gonostyle with the inner spine located in its middle or close to it. **Females**: pair of ascoids present on fXIII; spermatheca with its apical ring as wide as itself ....(*Incertae sedis*)...***Pi.* (*Pif.*) *nuneztovari***

18(8) **Both sexes**: setae absent from the anterior region of the katepisternum; apical papillae on fX either absent or present. **Males**: tergal papillae are absent from all abdominal tergites ....***Dampfomyia***.. 19

Both sexes: setae are generally present in the anterior region of the katepisternum (Fig. 24); apical papillae on flagellomere X present (Fig. 17-A). **Males**: tergal papillae on abdominal tergites present (Fig. 93-A C, E) at least on the 6th tergite (except in *Expapillata*).. 21

19(18) **Males**: epandrial lobe tumid (Fig. 156). **Females**: atrophied teeth in the hypopharynx ***Da.* (*Coromyia*)**

Males: epandrial lobe thinner. **Females**: well-marked teeth in the hypopharynx 20

20(19) **Both sexes**: papilla on flagellomere III absent; **Males**: paramere with hairy appendage (Fig. 179). **Females**: spermathecae with bulbous body, but bearing modifications which look like the petals of a dahlia (probably due to the sclerotization of gland cells) in the apical region (Fig. 73)... ***Da.* (*Dampfomyia*)**

Both sexes: papilla on flagellomere III present. **Males**: paramere without hairy appendage. **Females**: spermathecae simple, bulbous .... ***Da.*** Delpozoi group

21(18) **Males**: tergal papillae absent from all abdominal tergites; epandrial lobe with rounded apex and without a pre-apical constriction. **Females**: flagellomere I shorter than 0.6 times the length of the head; setae present (Fig. 37) or absent from the 8th tergite; set formed by the common and individual sperm ducts, spermatheca and terminal knob very short as this latter reaches only the apex of the genital furca stem (Fig. 78) ....***Expapillata***

Males: epandrial lobe with a pre-apical constriction (Figs. 157 and 158) or with tapered apex (Figs. 159, 160, 161, 162, 163, 164 and 165). **Females**: flagellomere I of variable length; setae absent from the 8th tergite (Fig. 37) [except in *Trichopygomyia*, *partim*], the set formed by the common and individual spermathecal ducts, spermatheca including terminal knob longer than that described immediately above [except in *Ev.* (*Ald.*) *termitophila*]...22

22(21) **Males**: gonostyle with the lower external spine atrophied, seta-like (Fig. 127). **Females**: spermathecae with membranous and long common spermathecal duct, extending beyond the middle of the genital furca stem, the individual spermathecal ducts being sclerotized and much shorter than the common duct (Fig. 77)...........................***Pressatia***

Males: gonostyle with the external lower spine, not seta-like, even though it may be a little less developed than the external upper spine. **Females**: short or long common spermathecal duct, the individual sperm ducts generally being membranous; when they are sclerotized the common duct is generally also sclerotized 23

23(22) **Males**: gonocoxite without compact and basal tuft of setae, but it can present a fringe of setae on the dorsal inner face; paramere bi or tri-branched; lateral lobe not tapered apically. **Females**: ratio between lengths: clypeus/head less than 1/3; eyes/head greater than 1/2; spermatheca striated and the individual spermathecal ducts equivalent to or longer than the common duct (Fig. 76)... ***Trichopygomyia***

Males: gonocoxite with compact and basal tuft of setae; absence of fringe of setae on the dorsal inner face; paramere simple or branched; epandrial lobe with tapered apex (Figs. 159, 160, 161, 162, 163, 164 and 165). **Females**: ratio between lengths: clypeus/head greater than 1/3; eyes/head less than 1/2; spermathecae with variable aspect (Figs. 79, 80, 81, 82, 83, 84 and 85) .... ***Evandromyia*** 24

24(23) **Males:** tergal papillae with evident edges are absent from the 7th tergite and present on the 6th tergite, and sometimes before this latter (Fig. 93-E). Females: cibarium with the anterior teeth arranged in lateral sets (Fig. 20-D); spermathecae, generally apple-shaped (Fig. 83) [except in *Ex. termitophila*, when they are ringed]......***Ev.* (*Aldamyia*)**

Males: tergal papillae present without evident edges on the 7th tergite or before it (Fig. 93-C). **Females**: cibarium with the anterior teeth in lateral sets, some elements present in the central region and in vertical position (Fig. 20-F); spermathecae of variable shape.. 25

25(24) **Males**: gonostyle with external spines located at the apex of a single tubercle (Fig. 124). **Females**: spermathecae with transversal striation and long common spermathecal duct which goes beyond the middle of the genital furca (Fig. 79) .... ***Ev.* (*Evandromyia*)** Rupicola series

Males: gonostyle with the external spines located isolatedly. **Females**: spermathecae not bearing transversal striation, common sperm duct length variable...................... 26

26(25) **Males**: tergal papillae present from the 5th to 7th tergites; paramere with two protuberances on the ventral margin. **Females**: spermatheca presenting in its central part bubble-like evaginations; common sperm duct long (Fig. 80)............. ***Ev.* (*Eva.*)** Saulensis series

Males: tergal papillae present only in the 6th and 7th tergites; paramere with or without one protuberance on the ventral margin, or branched. **Females**: spermatheca without bubble-like evaginations 27

27(26) **Males**: epandrial lobe with non-deciduous and differentiated setae implanted in its apical region (Fig. 165). **Females**: eyes longer than the clypeus; apex of the common spermathecal duct reaches or goes beyond the middle of the genital furca .... ***Ev.* (*Eva.*)** Infraspinosa series

Males: epandrial lobe without non-caducous and differentiated setae. **Females**: eyes shorter than or equivalent to length of clypeus; common spermathecal duct of variable length.. ***Ev.* (*Barrettomyia*)**............................ 28

28(27) **Males**: paramere branched and the parameral sheath with appendage. **Females**: spermatheca greatly elongated, common and individual sperm ducts sclerotized (Fig. 84).. ***Ev.* (*Bar.*)** Monstruosa series

Males: paramere and parameral sheath simple. **Females**: spermathecae globular or elongated, in this latter case not sclerotized.. 29

29(28) **Males**: paramere presenting in its central region of the dorsal margin semifoliaceous setae with apex curved towards the apex of the paramere; aedeagal ducts sinuous in their pre-apical region. **Females**: Spermathecae flattened at the poles; apex of the individual spermathecal ducts does not reach the apex of the genital furca stem............ ....(*Incertae sedis*)................................ ***Ev. (Bar.) edwardsi***

Males: parameres without semifoliaceous and curved setae in their dorsal margin; aedeagal ducts not sinuous in their pre-apical region. **Females**. Spermatecae spherical with the apex of individual spermathecal ducts reaching or going beyond the apex of the genital furca stem.. 30

30(29) **Males**: gonostyle with the external upper spine located in its middle and the internal one on its basal third (Fig. 126). **Females:** spermathecae with tubular individual spermathecal ducts (Fig. 85) ....***Ev.*** **(*Bar.*)** Cortelezzii series

Males: gonostyle with the external upper spine located on its apical third and the inner one in its middle region. **Females**: individual spermathecal ducts in part or completely sac-like (Fig. 81) .... ***Ev.*** **(*Bar.*)** Tupynambai series

PSYCHODOPYGINA: Genera, Subgenera, Groups and Series of Species

1 **Both sexes**: ascoids with evident posterior prolongation (Fig. 15) or if they are rudimentary, their insertion in the article is pedunculated (Fig. 16-B); 1st metatarsomere longer than or equivalent to the total sum of the other tarsomeres ....***Psathyromyia***...................................... 2

Both sexes: ascoids without developed posterior prolongation, if there is a prolongation it is rudimentary; their insertion in the article is not pedunculated (Fig. 16-A); ratio between 1st metatarsomere and the total length of the other tarsomeres variable........ 6

2(1) **Both sexes**: simple setae are absent from flagellomere I and/or II and/or III (Fig. 15). **Males**: frequently, Newstead's sensilla present on the 2nd palpal segment (Fig. 13-B). **Females**: Newstead's sensilla present on the 2nd palpal segment................................ 3

Both sexes: generally, simple setae are present on flagellomere I and/or II and/or III; Newstead's sensilla absent from 2nd palpal segment 4

3(2) **Both sexes**: ascoids with long posterior prolongation, in flagellomere II they reach its base; **Male:** parameral sheath short, its length less than twice the width of its base; gonostyle with all the spines clearly implanted beyond its middle. **Females**: spermathecae with distinct rings; common spermathecal ducts as long as 1.5 × the individual duct ....*Incertae sedis* ....*Pa. pifanoi*

Both sexes: ascoids with posterior prolongation at most reach the middle part of the region between their implantation and the base of flagellomere II (Fig. 15). **Males**: parameral sheath long, its length equivalent to twice or more the width of its base. **Females**: globous, tubular or ringed spermathecae, but in this latter case the rings are lightly demarcated, the individual ducts are very long and the common duct is rudimentary...***Pa.*** **(*Forattiniella*)**

4(2) **Both sexes**: papillae present on flagellomeres X-XI; **Males**: gonocoxite with sclerotized setae located in its apical region. **Females**: 5th palpal segment shorter than or equivalent to the 3rd; spermathecae ringed, some of them being distinctively imbricated (Fig. 53)...***Pa.*** **(*Xiphopsathyromyia*)**

Both sexes: papillae absent from flagellomeres X-XI. **Males**: gonocoxite without sclerotized setae in its apical region. **Females**: 5th palpal segment longer than the 3rd ....***Pa.*** **(*Psathyromyia*)** 5

5(4) **Both sexes**: ascoids with rudimentary spur (Fig. 16-B). **Males**: gonostyle with the inner spine located beyond its middle. **Females:** spermathecae ringed, individual ducts several times longer than the common duct...............................***Pa.*** **(*Psa.*)** Lanei series

Both sexes: generally, ascoids with long posterior prolongation that reaches or almost reaches the base of the article. **Males**: gonostyle with the inner spine located in its middle or before the middle. **Females**: spermathecae smooth (Fig.51) or segmented, frequently the common spermathecal duct is longer than or equivalent to the individual spermathecal ducts .... ***Pa.* (*Psa.*)** Shannoni series

6(1) **Males**: gonostyle with the upper external spine implanted in a conspicuous tubercle located in or before its middle and the lower external one implanted in a smaller tubercle situated more proximally or at the base of the former tubercle (Fig. 132); paramere simple with differentiated setae in the apical region of its dorsal edge (Fig. 173). **Females**: spermathecae bulbous and enclosed in a sclerotized sheath .... ***Viannamyia***

Males: gonostyle with the external upper spine implanted on its apical third and the external inner spine in or before its middle; paramere simple or lobed, with or without differentiated setae. **Females**: spermathecae ringed 7

7(6) **Both sexes:** 5th palpal segment longer than the 3rd; clypeus longer than or equivalent to 2/3 of eye length ....***Martinsmyia*** 8

Males: short clypeus, equivalent to or smaller than ½ eye length. **Females**: 5th palpal segment shorter than or equivalent to the 3rd; clypeus of variable length.................... 9

8(7) **Males**: gonocoxite with basal tuft of setae located on a tubercle. **Females**: common spermathecal duct longer than the individual spermathecal ducts (Fig. 58) ....
.... ***Ma.*** Gasparviannai Group

Males: gonocoxite without basal tuft of setae. Females: common spermathecal duct shorter than the individual spermathecal ducts (Fig. 55)..........***Ma.*** Alphabetica Group

9(7) **Both sexes**: setae in the anterior region of katepisternum (Fig. 24) may be present; mesonotum, generally, two-colored (posterior part of scutum and scutellum straw-colored, in contrast to the rest which is brown). **Males**: gonostyle with four well-developed spines, the inner one situated on its apical third. **Females**: lacinia of the maxilla with a row of external teeth (Fig. 12-B); clypeus very long, equivalent to the length of the eye .... ***Bichromomyia***

Both sexes: thorax without setae in the anterior region of katepisternum; color of mesonotum variable. **Males**: gonostyle with variable numbers of spines and position of the internal spine. **Females**: generally, lacinia of the maxilla with two rows of external teeth (Fig. 12-C); clypeus shorter than eye length 10

10(9) **Both sexes**: flagellomere I presenting two or more papillae (Fig. 14-B). Males: sum of 4th + 5th palpal segment smaller than or equivalent to the 3rd. **Females**: cibarium with two or more pairs of posterior teeth; spermathecae with all rings imbricated (Figs. 60, 61, 62, 63, 64 and 65) .... ***Psychodopygus***

Both sexes: flagellomere I presenting only the pre-apical papilla. Males: sum of the 4th + 5th palpal segments greater than the 3rd. **Females**: cibarium with three or more pairs of posterior teeth; spermathecae with some or all non-imbricated rings (Figs. 37-D, 56, 57, 66, 67 and 68) 11

11(10) **Both sexes**: Newstead's sensilla absent from the 2nd palpal segment. **Males**: terminalia smaller than the length of thorax; gonostyle with long spines, the apical one equivalent to the length of the gonostyle. **Females**: spermathecae with 5–15 rings (except in *Ny. bibinae*), the apical ring equivalent to or slightly longer than the pre-apical one (Figs. 37-D, 66, 67 and 68)..***Nyssomyia***

Both sexes: Newstead's sensilla present on the 2nd palpal segment (Fig. 13-B). **Males**: terminalia equivalent to or longer than the length of thorax; gonostyle with short spines, the apical one usually shorter than the gonostyle. **Females**: spermathecae with 25 or more rings, the apical one frequently three or more times longer than the pre-apical one (Figs. 56 and 57) ....***Trichophoromyia***

KEYS FOR THE IDENTIFICATION OF THE SPECIES

HERTIGIINA

Warileya

MALES

1 Gonocoxite with one or more compact tufts of setae (Fig. 139) 2

Gonocoxite without tuft of setae, but one or more isolated setae may be present........ 4

2(1) Gonostyle with two developed spines; aedeagal ducts/sperm pump ratio $< 2.0:1.0$ .... *Wa. phlebotomanica*

Gonostyle with three developed spines; aedeagal ducts/sperm pump ratio $\geq 2.6:1.0$ 3

3(2) Gonostyle with one apical spine and two others implanted at the same level at the beginning of its 4th apical one; aedeagal ducts $\geq$700 µm, aedeagal ducts/sperm pump ratio ca. 3.0:1.0... *Wa. lumbrerasi*

Gonostyle with two apical spines and another implanted near its 5th apical one; aedeagal ducts <400 µm; aedeagal ducts/ sperm pump ratio ca. 2.7:1.0 ..... *Wa. yungasi*

4(1) 5th palpal segment clearly longer than the sum of 3rd + 4th............... *Wa. nigrosaccula*

5th palpal segment equivalent to or shorter than the sum of 3rd + 4th......................... 5

5(4) Interocular distance ca. 1/5 of the eye width; aedeagal ducts/sperm pump ratio ca. 0.8:1.0 .... *Wa. rotundipennis*

Interocular distance greater than or equivalent to the eye width; aedeagal ducts $\geq$ sperm pump 6

6(5) 5th palpal segment clearly shorter than the 2nd... *Wa. leponti*

5th palpal segment $\geq$2nd.. 7

7(6) Flagellomere I ca. 410 µm and labrum-epipharynx ca. 250 µm .... *Wa. euniceae*

Flagellomere I ca. 260 µm and labrum-epipharynx ca. 140 µm .... *Wa. fourgassiensis*

FEMALES

1 Spermathecae ringed.. 2

Spermathecae not ringed.. 5

2(1) 5th palpal segment longer than 2nd; terminal knob of spermathecae clearly elongated 3

5th palpal segment $\leq$2nd; terminal knob spherical.. 4

3(2) 5th palpal segment ca. 1.8 times longer than 2nd; individual spermathecal ducts less than half of the spermatheca length; spermathecae with several rings clearly imbricated (Fig. 42).. *Wa. phlebotomanica*

5th palpal segment ca. 1.3 times longer than 2nd; individual spermathecal ducts a little longer than the spermatheca; this latter without imbricated rings .... *Wa. lumbrerasi*

4(2) Interocular distance ca. 1/2 of the eye width; 5th palpal segment as long as 2nd .... *Wa. rotundipennis*

Interocular distance slightly greater than the eye width; 5th palpal segment shorter than 2nd... *Wa. euniceae*

5(1) 5th palpal segment clearly longer than the sum of 3rd + 4th; spermathecae bulb-like with their length slightly greater than their width............................... *Wa. nigrosaccula*

5th palpal segment shorter than the sum of 3rd + 4th; spermathecae elongated, their length less than half their width .... *Wa. fourgassiensis*

Hertigia hertigi

Both sexes: pharynx with short spines in the posterior region. **Male**: gonostyle with three spines; gonocoxite with some scattered setae in the pre-apical region, paramere simple and the epandrial lobe shorter than the cercus. **Female**: cibarium with sclerotized arch complete without teeth; sac-like spermatheca, its length equivalent to three times its width, the terminal knob as a narrow prolongation of the apical surface, its length ca. twice its width.

BRUMPTOMYIINA

Brumptomyia

MALES

1 Thorax: presence of anepimeral setae (Fig. 22) 2

Thorax: absence of anepimeral setae 3

2(1) Thorax: presence of lower anepisternal setae. Gonostyle with one apical spine. Gonocoxite: basal tuft with dispersed setae. Aedeagal ducts 3–4 times longer than the sperm pump .... *Br. pintoi*

Thorax: absence of lower anepisternal setae. Gonostyle with two apical spines. Gonocoxite: basal tuft with the setae implanted compactly in a circular area. Aedeagal ducts ca. ten times longer than the sperm pump .... *Br. guimaraesi*

3(1) Gonostyle with the inner spine located on a more basal level than the external spines; gonocoxite with two sclerotized apical setae and the basal tuft consisting of short, scattered setae 4

Without the above set of characteristics 5

4(3) Gonocoxite with its length ca. five times its width, measured in its median region, basal tuft consisting of more than 30 setae; numerous tergal papillae with a hair like a small candle flame from the 2nd to the 7th tergites .... *Br. cardosoi*

Gonocoxite long and narrow, its length ca. eight times its width, measured in its median region, basal tuft consisting of ca. ten setae; tergal papillae with a hair like a small candle flame absent or scarce from the 2nd to 7th tergites .... *Br. bragai*

5(3) Gonocoxite with two basal tufts of setae, one located more internally with ca. 25 setae with apices curved towards the gonocoxite's apex and the other more externally, with ca. 50 setae with apices curved towards the base of the gonocoxite .... *Br. quimperi*

Gonocoxite with one basal tuft of setae 6

6(5) Gonocoxite: basal tuft of dispersed setae (Fig. 149) 7

Gonocoxite: basal tuft with the setae implanted compactly (Fig. 140) 11

7(6) Basal tuft of the gonocoxite located in an approximately circular area and the setae of equivalent length .... *Br. virgensi*

Basal tuft of the gonocoxite located in a clearly elongated area, with the upper setae longer than the lower 8

8(7) Gonocoxite with a small sclerotized tubercle at its base, located before the tuft of setae; paramere abruptly narrowed in its pre-apical region at the expanse of the ventral margin in such a way as to form a digitiform apical process .... *Br. troglodytes*

Gonocoxite without sclerotized tubercle at its base; paramere without the abrupt narrowing in its pre-apical region 9

9(8) The most proximal of the apical setae of the gonocoxite clearly separated from the apical area of the basal tuft; apex of paramere does not reach the middle of the epandrial lobe 10

The most proximal of the apical setae of the gonocoxite located practically at the same level as the setae of the apical area of the basal tuft; apex of paramere goes beyond the middle of the epandrial lobe .... *Br. ortizi*

10(9) Paramere with elbow in the pre-apical region of its ventral margin. Parameral sheath with bifid apex .... *Br. angelae*

Paramere without elbow in the pre-apical region of its ventral margin. Parameral sheath with simple apex .... *Br. nitzulescui*

11(6) Paramere apically bifurcated .... *Br. orlandoi*

Paramere with simple apex 12

12(11)	Gonocoxite: setae of the basal tuft implanted in a tubercle	13
	Gonocoxite setae of the basal tuft implanted directly on its surface	18
13(11)	Gonocoxite: tuft with setae semifoliaceous, the apices of the lower and upper ones converging towards the centre of the tuft	14
	Gonocoxite basal tuft with all setae thin and the apices of the lower ones not turned towards the centre of the tuft	15
14(13)	Gonocoxite: presence of 5–6 isolated setae in its apical region	*Br. mesai*
	Gonocoxite: presence of 7–8 isolated setae in its apical region	*Br. galindoi*
15(13)	Gonostyle with the internal spine implanted at a level between the apical and the tubercle of the external ones; the apex of the paramere does not reach the middle of the epandrial lobe.	*Br. beaupertuyi*
	Gonostyle with the internal spine implanted at the level of the tubercle of the external spines; the apex of the paramere reaches the middle of the epandrial lobe	16
16(15)	Gonocoxite with four or five well-developed apical setae and one thinner one; basal tuft with the lower setae clearly shorter than (less than half) the upper setae of this same tuft	*Br. pentacantha*
	Gonocoxite with three well-developed apical setae and one clearly thinner one; basal tuft with the setae of almost similar length, the central one somewhat longer than the others	17
17(16)	Paramere with the basal width equal to or twice that of the apex	*Br. carvalheiroi*
	Paramere with its basal width less than twice that of the apex	*Br. mangabeirai*
18(12)	Parameral sheath rectangular as far as its middle, where it widens at the expense of the ventral margin and again narrows, ending with a tapered apex	*Br. cunhai*
	Parameral sheath conical	19
19(18)	Gonocoxite with the setae of the basal tuft semifoliaceous (Fig. 140)	20
	Gonocoxite with the setae of the basal tuft thin	21
20(19)	Basal tuft of gonocoxite consisting of ca. 20 setae; paramere with the apical half densely covered by thin setae and stout spines; the more basal setae of the dorsal margin, ten or more in number, are as long as or longer than the width of the paramere in its narrower region	*Br. brumpti*
	Basal tuft of gonocoxite consisting of less than 15 setae; paramere sparsely covered by setae and spines in its apical half; the most basal setae of the dorsal margin are about five, shorter than the width of the paramere in its narrower region	*Br. avellari*
21(19)	Basal tuft of setae of gonocoxite located in a line, the upper setae clearly longer than the lower ones	*Br. travassosi*
	Basal tuft of setae of gonocoxite located in an approximately circular area	22
22(21)	Gonocoxite with one seta in the apical region of its ventral face	*Br. figueiredoi*
	Gonocoxite with three or more setae in the apical region of its ventral face (Fig. 140).	23
23(22)	Setae of the basal tuft of gonocoxite of approximately equal length	*Br. devenanzi*
	Setae of the basal tuft of gonocoxite in an upper position longer than those in the inferior one	24
24(23)	Aedeagal ducts ca. eight times longer than the sperm pump	*Br. hamata*
	Aedeagal ducts ca. five times longer than the sperm pump	*Br. leopoldoi*

FEMALES

The characteristics described for the females of *Brumptomyia* have not been sufficient for the construction of reliable keys. However, the observation of some common characteristics for male and female specimens captured at the same places has permitted the establishment of an association between the two sexes, but undoubtedly this needs confirmation. So the following key is presented with the intention of drawing attention to the need for more complete information concerning future female descriptions.

1	Thorax: presence of anepimeral setae (Fig. 22)	2
	Thorax: absence of anepimeral setae	3
2(1)	Thorax: presence of lower anepisternal setae (Fig. 22)	*Br. pintoi*
	Thorax: absence of lower anepisternal setae	*Br. guimaraesi*
3(1)	Presence of a row of spines on the internal part of the posterior femur	*Br. spinosipes*
	Absence of a row of spines on the internal part of the posterior femur	4
4(2)	9th tergite with a strongly sclerotized band in the upper part of its median region and also in the basal area of the individual spermathecal ducts near their opening into the genital chamber	*Br. troglodytes*
	9th tergite and the base of the individual spermathecal ducts without sclerotized band	5
5(4)	Spermathecae of robust aspect tending to be cylindrical, that is to say, their length is a little more than twice their apical width which is practically the same as their median width	*Br. cunhai*
	Spermathecae funnel-shaped (Fig. 46)	6
6(5)	Individual spermathecal ducts $\leq$ 6 times the spermatheca's length	7
	Individual spermathecal ducts $\geq$ 8 times the spermatheca's length	8
7(6)	Individual spermathecal ducts ca. four times the spermatheca's length	*Br. hamata*
	Individual spermathecal ducts ca. 5–6 times the spermatheca's length; 9th tergite with spiniform permanent setae in its central region (Fig. 40)	*Br. cardosoi* *Br. ortizi*
8(6)	Individual wavy spermathecal ducts sometimes with coils	*Br. nitzulescui*
	Individual spermathecal ducts with coils but not wavy	9
9(8)	Flagellomere I with two papillae	10
	Flagellomere I with one papilla	11
10(9)	Flagellomere I ca. 300 μm; labrum-epipharynx ca. 250 μm	*Br. avellari*
	Flagellomere I ≤270 μm; labrum-epipharynx ≤230 μm	*Br. mesai* *Br. pentacantha*
11(9)	Spermathecae with 25 or more rings	*Br. carvalheiroi*
	Spermathecae with less than 20 rings	12
12(11)	Spermatheca ca. 4.5 times longer than its apical width	*Br. leopoldoi*
	Spermatheca ca. three times longer than its apical width (Fig. 46)	*Br. brumpti*

In the above key for the identification of the females, *Br. cardosoi, Br. cunhai, Br. ortizi* and *Br. pentacantha* are described for the first time and the aedeagal ducts of the males are of corresponding length to the ducts of the spermathecae; the differential characteristics of *Br. pintoi* and *Br. guimaraesi* have been deduced from the respective males. For *Br. beaupertuyi, Br. mesai* and *Br. Travassosi,* the characteristics described are insufficient to differentiate them and no females were available. The females of the other species: *Br. angelae*, *Br. bragai*, *Br. devenanzi*, *Br. figueiredoi*, *Br. mangabeirai*, *Br. orlandoi*, *Br. quimperi* and *Br. virgensi* have not been described.

Oligodontomyia

MALES

1	Flagellomere I ca. 300 μm; epandrial lobe ≤320 μm	*Ol. oligodonta*
	Flagellomere I > 380 μm; epandrial lobe >380 μm	2
2(1)	Epandrial lobe ca. 540 μm; gonostyle with the internal spine slightly more slender and of ca. ½ the length of the external ones	*Ol. isopsi*
	Epandrial lobe ≤475 μm; gonostyle with a thin internal spine like a seta	*Ol. toroensis*

FEMALES

1	Spermathecae partially ringed	*Ol. isopsi*
	Spermathecae completely ringed	2
2(1)	Thorax with paratergites and pleura brown	*Ol. oligodonta*
	Thorax with paratergites and pleura straw	*Ol. toroensis*

SERGENTOMYIINA

Deanemyia

MALES

1	Paramere with dorsal branch	2
	Paramere simple, without dorsal branch	*De. derelicta*
2(1)	Paramere with the dorsal branch like a small appendix bearing a few setae on its apex	*De. appendiculata*
	Paramere with the dorsal branch well-developed and covered with setae	3
3(2)	Dorsal branch of the paramere narrower than the ventral one, both branches covered with simple short setae	*De. ramirezi*
	Dorsal branch of paramere wider than the ventral one, both branches bearing semifoliaceous setae on the dorsal margin	*De. samueli*

FEMALES

1	Ratio between the lengths: labrum-epipharynx / clypeus >2.0	*De. samueli*
	Ratio between the lengths: labrum-epipharynx / clypeus <2.0	2
2(1)	Cibarium with a pair of posterior (horizontal) developed teeth	*De. maruaga*
	Cibarium with two pairs of posterior (horizontal) developed teeth	*De. ramirezi* *De. derelicta*

Micropygomyia (*Silvamyia*)

MALES

1	Gonostyle with the lower external spine implanted in its apical third and the internal in its middle	*Mi.(Sil.) echinatopharynx*
	Gonostyle with the lower external and the internal spines implanted practically at the same level, a little above its middle	*Mi. (Sil.) acanthopharynx*

FEMALES

1	Approximate lengths (μm): head 360; clypeus 120; flagellomere I 300. Wing width 550 μm	*Mi. (Sil.) echinatopharynx*
	Approximate lengths (μm): head 380; clypeus 140; flagellomere I 320. Wing width 590 μm	*Mi. (Sil.) acanthopharynx*

(*Sauromyia*)

Oswaldoi series

MALES

1 Gonostyle with the lower external and the internal spines short (slightly longer than its width); gonocoxite with a cluster of ca. 20 setae distributed throughout almost all its basal, ventral surface .... *Mi.* (*Sau.*) *quechua*

Gonostyle with the lower external and internal spines equal to or longer than twice the gonostyle's width; gonocoxite without cluster of setae, or if present the setae are either fewer or more than 20, presenting a different arrangement from that described above 2

2(1) Wing narrow, R_5/wing width ratio ca. 3.6:1.0........ *Mi.* (*Sau.*) *machupicchu*

Wing wider than that described above, R_5/ wing width ratio $\leq$ 3.3:1.0........ 3

3(2) Wing: delta nil (R_1 ending at the level of R_2,R_3 bifurcation) .... *Mi.* (*Sau.*) *pratti*

Wing: delta clearly positive (R_1 ending beyond the R_2,R_3 bifurcation) 4

4(3) Gonocoxite with a cluster constituted of ca. 30 setae grouped in its median part .... *Mi.* (*Sau.*) *saccai*

Gonocoxite without cluster of setae or if present with less than 20 of them 5

5(4) Clypeus practically as long as the labro-epipharynx and as wide as the eyes .... *Mi.* (*Sau.*) *peresi*

Clypeus shorter than the labro-epipharynx and narrower than the eyes 6

6(5) Gonocoxite with semifoliaceous setae (width $\geq$ twice that of the aedeagal ducts) implanted isolatedly or in a compact tuft in its basal or median region 7

Gonocoxite without setae in its basal or median region or if setae are present they are narrow (width < than that of the aedeagal ducts)........ 10

7(6) Gonocoxite with 5–13 setae arranged in compact tuft. Tergal papillae present only on the VI and VII tergites........ 8

Gonocoxite with 1–6 isolated setae. Tergal papillae present at least from tergites V–VII 9

8(7) Tuft of gonocoxite with 5–6 setae........ *Mi.* (*Sau.*) *dereuri*

Tuft of gonocoxite with 8–13 setae........ *Mi.* (*Sau.*) *quinquefer*

9(7) Tergal papillae present on IV tergite. Gonocoxite with 3–6 setae in line .... *Mi.* (*Sau.*) *ferreirana*

Tergal papillae absent on IV tergite. Gonocoxite with only one seta .... *Mi.* (*Sau.*) *huacalquensis*

10(6) Apex of aedeagal ducts tapered and curved (Fig. 99)........ 11

Apex of aedeagal ducts rounded or tapered, but not as described above........ 12

11(10) Gonocoxite with setae implanted sparsely in its median third .... *Mi.* (*Sau.*) *longipennis*

Gonocoxite with the setae implanted more compactly in its basal third .... *Mi.* (*Sau.*) sp. 2 de Araracuara

12(10) Gonocoxite with tuft constituted of 12 or more long, thin, curved setae, situated between the basal and median quarters........ *Mi.* (*Sau.*) *rorotaensis*

Gonocoxite without tuft of setae or if present with fewer than 12 and in another position........ 13

13(12) Ratio of lengths aedeagal ducts/sperm pump $\geq$3.5: 1.0........ 14

Ratio of lengths aedeagal ducts/sperm pump $\leq$3.0: 1.0........ 16

14(13) Epandrial lobe as long as gonocoxite; clypeus 1/3 the length of the head .... *Mi.* (*Sau.*) *zikani*

Epandrial lobe shorter than gonocoxite; clypeus longer than 1/3 the length of the head ... 15

15(14) Terminalia: length ca. 370 μm; paramere tapering from the base to the apex ... *Mi.* (*Sau.*) *villelai*

Terminalia: length ca. 330 μm; paramere thins abruptly so that its apical half is digitiform ... *Mi.* (*Sau.*) *trinidadensis*

16(13) Tergal papillae present from the III to VII tergites ... 17

Tergal papillae absent on III tergite ... 18

17(16) Pharynx with spines in its apical region ... *Mi.* (*Sau.*) *petari*

Pharynx without spines ... *Mi.* (*Sau.*) *oswaldoi*

18(16) Terminalia $<$ length of head ... *Mi.* (*Sau.*) *pusilla*

Terminalia $\geq$ length of head ... 19

19(18) Terminalia clearly longer than the head ... *Mi.* (*Sau.*) *capixaba*

Terminalia as long as the head ... *Mi.* (*Sau.*) *vonatzingeni*

FEMALES

1 Pharynx with developed spines ... 2

Pharynx with or without atrophied spines ... 8

2(1) Spermathecae long and smooth (banana-shaped) ... *Mi.* (*Sau.*) *trinidadensis*

Spermathecae ringed or striated ... 3

3(2) Spermathecae striated ... *Mi.* (*Sau.*) *saccai*

Spermathecae with distinct rings ... 4

4(3) Spermathecae with 2–3 rings, the apical one almost spherical ... 5

Spermathecae with ca. ten rings, the apical one clearly elongated ... *Mi.* (*Sau.*) *pusilla*

5(4) Cibarial chamber covered with a membrane with many sclerotized points ... *Mi.* (*Sau.*) *vonatzingeni*

Cibarial chamber covered with a membrane without sclerotized points ... 6

6(5) Pharynx with the area covered with spines sclerotized and much wider than that before it ... *Mi.* (*Sau.*) *capixaba*

Pharynx with the area covered with spines unsclerotized and slightly wider than that before it ... 7

7(6) Cibarium with the anterior teeth very small and situated laterally and two considerably more developed ones arranged in the central part; flagellomere I ca. 1.5 × the length of the labrum-epipharynx ... *Mi.* (*Sau.*) *petari*

Cibarium with the anterior teeth vertical and not lateralized, all being practically of the same size; flagellomere I equivalent to the length of the labrum-epipharynx ... *Mi.* (*Sau.*) *ferreirana*

8(1) Spermathecae long and smooth (banana-shaped) ... *Mi.* (*Sau.*) *villelai*

Spermathecae ringed or striated ... 9

9(8) Spermathecae striated, their basal and apical widths practically equal ... *Mi.* (*Sau.*) *peresi*

Spermathecae elongated with the apical ring clearly longer and wider than those before it ... 10

10(9) Spermathecae with 3–6 rings (Fig. 90) ... *Mi.* (*Sau.*) *oswaldoi*

Spermathecae with larger number of rings ... 11

11(10) Lacinia of the maxilla with the external teeth in transversal position (Fig. 12-E) ... *Mi.* (*Sau.*) *longipennis*

Lacinia of the maxilla with the external teeth in longitudinal position (Fig. 12-D) ... 12

12(11) Cibarium: apex of the two pairs of posterior teeth (horizontal) clearly turned towards the central area .... *Mi.* (*Sau.*) *quechua*
Mi. (*Sau.*) *rorotaensis*

Cibarium: apex of the two pairs of posterior teeth (horizontal) clearly turned towards the pharynx 13

13(12) Lacinia of the maxilla with the external teeth in three lines; the more apical teeth being so close to the internal ones that they give the impression of forming four lines .... *Mi.* (*Sau.*) *dereuri*

Lacinia of the maxilla with the external teeth in two lines; the more apical being so close to the internal ones that they give the impression of forming three lines 14

14(13) Lacinia of the maxilla with the most external line consisting of 11–13 teeth .... *Mi.* (*Sau.*) *huacalquensis*

Lacinia of the maxilla with the most external line consisting of 6–9 teeth .... *Mi.* (*Sau.*) *quinquefer*
Mi. (*Sau.*) *zikani*

Atroclavata series

MALES

1 Gonocoxite with ca. four setae with a sclerotized stem and curved apex; paramere with apex curved towards the gonocoxite .... *Mi.* (*Sau.*) *atroclavata*

Gonocoxite with ca. eight simple setae; paramere with straight apex .... *Mi.* (*Sau.*) *venezuelensis*

FEMALES

1 Cibarium with the anterior teeth horizontally arranged in several transversal lines (Fig. 18) .... *Mi.* (*Sau.*) *atroclavata*

Cibarium with the anterior teeth vertically arranged in one transversal line .... *Mi.* (*Sau.*) *venezuelensis*

(*Coquillettimyia*)

Vexator series

MALES

1 Gonocoxite with one basal tuft and another median-apical; gonostyle with the lower external spine implanted before its middle on a level close to that of the internal one 2

Gonocoxite with only one basal tuft, the gonostyle having the lower external spine on a clearly more apical level than the internal one 3

2(1) Gonocoxite with one tuft of ca. seven setae; paramere with a pre-apical elbow on its ventral margin .... *Mi.* (*Col.*) *vindicator*

Gonocoxite with one tuft of ca. five setae; paramere without a pre-apical elbow on its ventral margin .... *Mi.* (*Col.*) *apache*

3(1) Paramere concave in the median region of its dorsal margin and with the setae restricted to its apical third .... *Mi.* (*Col.*) *oppidana*

Paramere straight in the median region of its dorsal margin with the setae implanted in its apical half .... *Mi.* (*Col.*) *vexator*

FEMALES

1 Spermathecae spherical. Individual spermathecal ducts without striation near the spermathecae .... *Mi.* (*Col.*) *vindicator*

Spermathecae of flattened spherical shape. Individual spermathecal ducts striated near the spermathecae 2

2(1) Ascoids on flagellomere II: apex does not reach that of the segment .... *Mi.* (*Col.*) *oppidana*

	Ascoids on flagellomere II: apex reaches or goes beyond that of the segment	3
3(2)	Individual spermathecal ducts ca. two × longer than the stem of genital fork	*Mi.* (*Col.*) *vexator*
	Individual spermathecal ducts ca. four × longer than the stem of genital fork	*Mi.* (*Col.*) *apache*

Chiapanensis series

MALES

1	Aedeagal ducts with the pre-apical region enlarged, lozenge-shaped	2
	Aedeagal ducts without the enlarged, lozenge-shaped pre-apical region	*Mi.* (*Col.*) *chiapanensis*
2(1)	Gonocoxite with tuft consisting of ca. four setae	*Mi.* (*Col.*) *stewarti*
	Gonocoxite with tuft consisting of ca. ten setae	*Mi.* (*Col.*) *californica*

FEMALES

1	Cibarium with two pairs of posterior (horizontal) teeth	*Mi.* (*Col.*) *stewarti*
	Cibarium with eight or more pairs of posterior (horizontal) teeth	2
2(1)	Individual spermathecal ducts ca. twice as long as the common spermathecal duct	*Mi.* (*Col.*) *chiapanensis*
	Individual spermathecal ducts ca. five times longer than the common spermathecal duct	*Mi.* (*Col.*) *californica*

(*Micropygomyia*)

Cayennensis series

MALES

1	Aedeagal ducts with the apical region enlarged	2
	Aedeagal ducts without the enlarged apical region	4
2(1)	Aedeagal ducts ca. five times longer than the sperm pump	*Mi.* (*Mic.*) *lewisi*
	Aedeagal ducts no longer than three times the sperm pump's length	3
3(2)	Aedeagal ducts with the apical region abruptly tapered, with a small terminal barb	*Mi.* (*Mic.*) *durani*
	Aedeagal ducts with a bulging apical region	*Mi.* (*Mic.*) *schreiberi*
4(1)	Aedeagal ducts >3 times the sperm pump	5
	Aedeagal ducts ≤3 times the sperm pump	6
5(4)	Clypeus completely covered with setae; labella with strong sclerotized pseudotracheae, thorax with mesonotum and pleura straw	*Mi.* (*Mic.*) *ancashensis*
	Clypeus with setae on its apical half; labella without strong sclerotized pseudotracheae; thorax with mesonotum, anepisternum and katepisternum well pigmented, almost black	*Mi.* (*Mic.*) *yencanensis*
6(4)	Flagellomere I ≥ the sum of clypeus + labrum-epipharynx	*Mi.* (*Mic.*) *duppyorum*
	Flagellomere I < the sum of clypeus + labrum-epipharynx	7
7(6)	Mesonotum and paratergite brown	8
	Mesonotum brown and paratergite straw	complex *Mi.* (*Mic.*) *cayennensis* *Mi.*(*Mic.*) *cubensis* *Mi.* (*Mic.*) *hardisoni*
8(7)	Thorax: pleura completely dark brown, clearly contrasting with the straw coxae	*Mi.* (*Mic.*) *micropyga*
	Thorax: pleura light brown, not contrasting with the straw coxae	*Mi.* (*Mic.*) *absonodonta*

FEMALES

1 Spermathecae ringed 2
Spermathecae smooth, elongated (banana-shaped) .... *Mi.* (*Mic.*) *lewisi*

2(1) Pharynx with developed spines 3
Pharynx without developed spines 16

3(2) Cibarium with two pairs of posterior teeth .... *Mi.* (*Mic.*) *cubensis*
Cibarium with more than two pairs of posterior teeth 4

4(3) Pharynx: spiny area twice as wide as its anterior region .... *Mi.* (*Mic.*) *duppyorum*
Pharynx: spiny area a little wider than its anterior region 5

5(4) Cibarium with 15 or more pairs of posterior (horizontal) teeth, in comb-like formation .... *Mi.* (*Mic.*) *ctenidophora*
Cibarium with few posterior (horizontal) teeth 6

6(5) Spermathecae with the terminal knob inserted in a depression .... *Mi.* (*Mic.*) *wirthi*
Spermathecae with the terminal knob not inserted in a depression 7

7(6) Wing: *gamma* ≥ *alpha* 8
Wing: *gamma* < *alpha* 9

8(7) Individual spermathecal ducts ca. twice as long as the common spermathecal duct .... *Mi.* (*Mic.*) *durani*
Individual spermathecal ducts ≥4 common spermathecal duct .... *Mi.* (*Mic.*) *yencanensis*

9(7) Cibarium with the posterior teeth in palisade (Fig. 20-C) 10
Cibarium with the posterior teeth not in palisade .... *Mi.* (*Mic.*) *absonodonta*

10(9) Cibarium with six or more anterior vertical teeth; head integument brown 13
Cibarium with four anterior vertical teeth; head integument straw 11

11(10) Posterior central teeth of the cibarium with the united part ca. 5 times higher than the individualized part .... *Mi.* (*Mic.*) *cayennensis braci*
Posterior central teeth of the cibarium with the united part ca. 3 times or less higher than the individualized part 12

12(11) Posterior central teeth of the cibarium with the united part ca. 3 times higher than the individualized part .... *Mi.* (*Mic.*) *cayennensis jamaicensis*
Posterior central teeth of the cibarium with the united part ca. 2 times higher than the individualized part .... *Mi.* (*Mic.*) *cayennensis viequesensis*
Mi. (*Mic.*) *cayennensis cruzi*

13(10) Posterior central teeth of the cibarium with the united part clearly longer than the individualized part 15
Posterior central teeth of the cibarium with the united part equal to or shorter than the individualized part 14

14(13) Anterior (vertical) teeth of cibarium clearly visible, the central ones being more developed than the external ones .... *Mi.* (*Mic.*) *cayennensis maciasi*
Anterior (vertical) teeth of cibarium difficult to discern .... *Mi.* (*Mic.*) *cayennensis puertoricensis*

15(13) Cibarium with 8–9 pairs of posterior teeth .... *Mi.* (*Mic.*) *cayennensis cayennensis*
Mi. (*Mic.*) *farilli*
Mi. (*Mic.*) *hardisoni*
Cibarium with six pairs of posterior teeth .... *Mi.* (*Mic.*) *cayennensis hispaniolae*

16(2) Mesonotum, paratergite and pleura completely dark brown, contrasting with the clear coxae .... *Mi.* (*Mic.*) *micropyga*
Paratergite and anepimeron clear, mesonotum and other pleura parts brown, but with no evident contrast with the coxae .... *Mi.* (*Mic.*) *schreiberi*

Pilosa series

MALES

1 Gonocoxite with tuft of setae.. 2

Gonocoxite without tuft of setae.. *Mi. (Mic.) chassigneti*

2(1) Gonocoxite with open tuft consisting of several scattered setae placed on the apical half (Fig. 155).. *Mi. (Mic.) pilosa*

Gonocoxite with median compact tuft and a basal one consisting of 4–5 setae.. *Mi. (Mic.) mangabeirana*

FEMALES: indistinguishable.

Incertae sedis

Micropygomyia xerophila

Male: palpal formula: 1.2.4.3.5; gonostyle with four spines: one apical, the upper external one located in the apical quarter and the lower external and the inner ones located at the same level, in its middle; gonocoxite with 1–2 setae located at its base on a small tubercle; paramere simple and the aedeagal ducts ca. 1.6 times longer than the sperm pump, the apex bearing a row of small teeth. **Female:** palpal formula: 1.2.4.3.5; cibarium with four pairs of central posterior teeth in horizontal position and four pairs of lateral teeth inclined towards the median line of the posterior teeth, the anterior teeth are vertical and placed in an irregular row; common spermathecal duct and the individual spermathecal ducts short, their apices do not reach the beginning of the genital furca stem, spermathecae spherical, with length equivalent to that of the individual spermathecal ducts.

LUTZOMYIINA

Sciopemyia

MALES

1 Gonocoxite without basal tuft of setae.. 2

Gonocoxite with basal tuft of setae.. 3

2(1) Flagellomere I < 400 µm; aedeagal ducts ≥470 m; epandrial lobe ≥ 167 µm .... *Sc. vattierae*

Flagellomere I ≥ 405 µm; aedeagal ducts ≤440 µm; epandrial lobe ≤157 µm.. *Sc. sordellii*

3(1) Paramere bearing a small tubercle with setae at its apex in the median region of the dorsal margin; gonocoxite with basal tuft located on a highly sclerotized columnar tubercle.. *Sc. pennyi*

Paramere without tubercle in the median region of the dorsal margin; gonocoxite with basal tuft located on a tubercle without columnar aspect 4

4(3) Gonocoxite with tuft consisting of 15 or more setae 5

Gonocoxite with tuft consisting of eight or fewer setae 6

5(4) Paramere: ventral margin without pre-apical protuberance (Fig. 167), the median region of the dorsal margin bearing some setae of curved apex and longer than those on the apical half... *Sc. nematoducta*

Paramere: ventral margin with pre-apical protuberance but with no differentiated setae in the median region of the dorsal margin .... *Sc. preclara*

6(4) Gonostyle: lower external spine more basal than the internal one; dorsal margin of paramere has a set of curved setae separated from the setae on the apical half by a glabrous area .... *Sc. fluviatilis*

Gonostyle: lower external spine more apical than the internal one; dorsal margin of paramere without the glabrous area separating the setae.. 7

7(6) Gonocoxite with basal tuft of fine setae.. *Sc. microps*

Gonocoxite with basal tuft of semifoliaceous setae.............................. *Sc. servulolimai*

FEMALES

1 Papilla on flagellomere III present (Fig. 15); 5th palpal segment shorter than or as long as 3rd 2

Papilla on flagellomere III absent (Fig. 16-A); 5th palpal segment longer than 3rd 3

2(1) Spermathecae tubular, transversally striated with pedunculated terminal knob .... *Sc. vattierae*

Spermathecae also transversally striated, though without the tubular aspect, clearly wider than the individual spermathecal ducts and with sessile terminal knob (Fig. 86) .... *Sc. sordellii*

3(1) Individual spermathecal ducts clearly narrower at the junction with the spermatheca than at the junction with the common spermathecal duct and ca. 15 times longer than the spermatheca, this latter presents intermediate rings clearly wider than the rings at both its extremities .... *Sc. nematoducta*

Individual spermathecal ducts practically of the same width throughout their length and eight times or less longer than the spermatheca; this latter with the intermediate rings perceptibly of the same width as the apical rings 4

4(3) Width of the individual spermathecal ducts at the junction with the spermatheca equivalent to a third of the spermatheca's width, measured at its central part........ *Sc. preclara*

Width of individual spermathecal ducts at the spermatheca's junction equivalent to half or more of the spermatheca's width, measured at its central part 5

5(4) Basal part of the individual spermathecal ducts (junction with the common duct) wider than the spermatheca's width, at its central part .... *Sc. fluviatilis*

Width of the basal part of the individual spermathecal ducts (junction with the common duct) ca. half of the spermaheca's width, measured at its central part........ 6

6(5) 5th palpal segment longer than the sum of 3rd + 4th. Cibarium with the posterior teeth clearly lateralized, the distance between the internal ones twice that between these and the external ones. Individual spermathecal ducts ca. six times longer than the spermatheca........ *Sc. microps*

5th palpal segment clearly shorter than the sum of 3rd + 4th. Cibarium with the posterior teeth not lateralized and the distance between the internal ones slightly greater than that between these latter and the external ones. Individual spermathecal ducts ca. 3.5 times longer than the spermatheca .... *Sc. servulolimai*

Lutzomyia
(*Helcocyrtomyia*)

MALES

1 Gonostyle with five spines 2

Gonostyle with four or six spines, gonocoxite with a basal tuft bearing ca. 20 setae .... *Incertae sedis* 4

2(1) Epandrial lobe equivalent to or longer than the gonocoxite .... Peruensis series

Epandrial lobe shorter than the gonocoxite 3

3(2) Clypeus longer than one-third the head's length; gonocoxite with dense tuft consisting of 9 or more setae in basal position .... Osornoi series

Clypeus shorter than or equivalent to one-third of the head's length; gonocoxite without tuft of setae in basal position, or if present, consisting of up to six setae or scattered tuft, located in the baso-median region .... Sanguinaria series

4(1) Thorax: mesonotum and pleura dark brown. Gonostyle with four spines (the lower external spine absent); aedeagal ducts ca. 2.5 times longer than the sperm pump .... *Lu.*(*Hel.*) *infusca*

Thorax: mesonotum brown and pleura straw. Gonostyle with six spines, the lower external and the internal ones implanted at the same level before the middle of the gonostyle and one additional external spine implanted more basally than the latter two; aedeagal ducts ca. nine times longer than the sperm pump ....*Lu.* (*Hel.*) *vargasi*

FEMALES

1 Completely dark brown; head: width > length.................................... *Lu.* (*Hel.*) *infusca*

Of variable coloring; head: width ≤ length.. 2

2 Palpal segments: 4th ≥ 2/3 of 3rd or with four or more pairs of spiniform setae ....Peruensis series

Palpal segments: 4th ca. 1/2 of 3rd or with two or three pairs of spiniform setae........ 3

3(2) Palpal segments: 5th ≤ 1.25 of 3rd Clypeus ≤1/3 head's length....... Sanguinaria series

Palpal segments: 5th ≥ 1.40 of 3rd. Clypeus >1/3 head's length .... Osornoi series

Sanguinaria series

MALES

1 Gonocoxite with one or more setae at its base.. 5

Gonocoxite without setae at its base.. 2

2(1) Wing: *delta* ca. 0.25 of *alpha*.. *Lu.* (*Hel.*) *tortura*

Wing: *delta* measuring between 0.33 and 0.50 of *alpha* 3

3(2) Wing: *delta* ca. 0.40 of *alpha*... *Lu.* (*Hel.*) *tolimensis*

Wing: *delta* ca. 0.50 of *alpha*.. 4

4(3) Large insect, with approximate lengths (μm): Head 520; flagellomere I 550 and 5th papal segment 350; gonostyle 250 and epandrial lobe 370 ....*Lu.* (*Hel.*) *adamsi*

Smaller insect, with the lengths (μm) ≤: head 480; flagellomere I 460 and 5th palpal segment 300; gonostyle 220 and epandrial lobe 330.....................*Lu.* (*Hel.*) *guderiani*

5(1) Ascoids long on flagellomere II, the internal ascoid reaching the level of the papilla .... *Lu.* (*Hel.*) *velezi*

Ascoids short on flagellomere II, the internal ascoid reaching or slightly exceeds the middle of the article 6

6(5) Gonocoxites with ca. 20 scattered setae in basal-median position; aedeagal ducts ca. five times longer than the sperm pump .... *Lu.* (*Hel.*) *cirrita*

Gonocoxites with basal tuft consisting of 1–6 setae; aedeagal ducts less than four times the length of sperm pump 7

7(6) Sperm pump with the pavilion's diameter ca. four times that of the ejaculatory apodeme .... *Lu.* (*Hel.*) *sanguinaria*

Sperm pump with the pavilion's diameter not more than twice that of the ejaculatory apodeme 8

8(7) Wing: *delta* equivalent to 0.50 of *alpha*.................................... *Lu.* (*Hel.*) *monzonensis*

Lu. (*Hel.*) *guderiani*

Wing: *delta* ≤ 0.40 of *alpha*.. 9

9(8) Gonostyle with the external lower spine and internal spine located at the same level a little above its middle.. 10

Gonostyle with the external lower spine located at a more basal level than the internal spine 11

10(9) Gonocoxite with 1–3 subequal setae implanted in its base........... *Lu.* (*Hel.*) *tolimensis*

Gonocoxite with 1–4 setae, one of them more developed, implanted in its base .... *Lu.* (*Hel.*) *scorzai*

11(9) Gonostyle with the internal spine located in the middle... 12

Gonostyle with the internal spine located above the middle 13

12(11) Tuft of the gonocoxite with three subequal setae; labrum-epipharynx ca. 300 μm .... *Lu.* (*Hel.*) sp. de Pichinde

Tuft of the gonocoxite with 1–4 setae, one of them longer than the others; labrum-epipharynx ca. 280 μm .... *Lu.* (*Hel.*) *hartmanni*

13(11) Gonocoxite having on its base one seta ≤ the width of the gonocoxite .... *Lu.* (*Hel.*) *caceresi*

Gonocoxite having a tuft of 4–6 setae, at least one of them longer than the width of the gonocoxite 14

14(13) Apex of paramere almost reaches the epandrial lobe's apex; paramere with the more basal setae of the dorsal margin perceptibly longer than the width of the area in which they are implanted .... *Lu.* (*Hel.*) *kirigetiensis*

Apex of paramere reaching approximately 2/3 of the epandrial lobe; paramere with the more basal setae of the dorsal margin being shorter than the width of the area in which they are implanted .... *Lu.* (*Hel.*) *gonzaloi*

FEMALES

1 Spermathecae sac-like, transversally striated; the apical ring forming a small collar; individual spermathecal ducts ≤ the spermatheca's length .... *Lu.* (*Hel.*) *botella*

Spermathecae without sac-like shape; individual spermathecal ducts longer than the spermathecal 2

2(1) Spermathecae carrot-shaped, the apical ring being narrower than the pre-apical one, like a collar .... *Lu.* (*Hel.*) *sanguinaria*

Spermathecae slender, the apical ring being equal to or slightly narrower than the pre-apical one 3

3(2) Labrum-epipharynx ≥450 μm 4

Labrum-epipharynx ≤400 μm .... *Lu.* (*Hel.*) *gonzaloi*
Lu. (*Hel.*) *hartmanni*
Lu. (*Hel.*) *kirigetiensis*
Lu. (*Hel.*) *tortura*

4(3) Individual spermathecal ducts ca. 3.5 times longer than the spermatheca .... *Lu.* (*Hel.*) *cirrita*

Individual spermathecal ducts less than three times the spermatheca's length 5

5(4) Wing: *delta* ca. 0.50 of *alpha* 6

Wing: *delta* measuring between 0.30 and 0.45 the value of *alpha* 7

6(5) Labrum-epipharynx ca. 620 μm; flagellomere I 550 μm .... *Lu.* (*Hel.*) *adamsi*

Labrum-epipharynx on average ca. 500 μm (maximum value, 580 μm); flagellomere I, on average 440 μm (maximum value 537 μm) .... *Lu.* (*Hel.*) *guderiani*
Lu. (*Hel.*) *monzonensis*

7(5) Wing: *delta* = *beta* .... *Lu.* (*Hel.*) sp. de Pichinde

Wing: *delta* > *beta* .... *Lu.* (*Hel.*) *scorzai*
Lu. (*Hel.*) *tolimensis*

Osornoi series

MALES

1 Large size, lengths: head ≥500 μm, wing ≥3.2 mm, gonocoxite ≥450 μm 2

Median size, lengths: head ≤430 μm, wing ≤2.7 mm, gonocoxite ≤400 μm 5

2(1) Wing: *delta* ca. 0.50 of *alpha* .... *Lu.* (*Hel.*) *erwindonaldoi*
Lu. (*Hel.*) *larensis*

Wing: *delta* ≤ 0.25 of *alpha* 3

3(2) Wing: *delta* ca. 0.17 of *alpha* ...... *Lu.* (*Hel.*) *wattsi*

Wing: *delta* ca. 0.25 of *alpha* 4

4(3) Paramere with the apical half digitiform, its width being ca. 2/3 that of the epandrial lobe at this latter's narrowest part ...... *Lu.* (*Hel.*) *caballeroi*

Paramere with the apical half digitiform, however thicker, its width being equal to that of the epandrial lobe at this latter's narrowest part ...... *Lu.* (*Hel.*) *osornoi*

5(1) Wing: *delta* ca. 0.33 of *alpha*; *beta* 0.40 of *gamma* 6

Wing: *delta* ca. 0.25 of *alpha*; *beta* 0.60 of *gamma* 8

6(5) Tuft of gonocoxite consisting of ca. 25 fine setae; paramere with straight dorsal margin ...... *Lu.* (*Hel.*) *strictivilla*

Tuft of gonocoxite consisting of up to 21 semifoliaceous setae; paramere with a concavity in the middle of its dorsal margin, so that it curves towards the gonocoxite 7

7(6) Tuft of the gonocoxite of 18–21 setae ...... *Lu.* (*Hel.*) *rispaili*

Tuft of the gonocoxite of 12–15 setae ...... *Lu.* (*Hel.*) *munaypata*

8(5) Tuft of gonocoxite of ca. 30 setae implanted in tubercle ...... *Lu.* (*Hel.*) *quillabamba*

Tuft of gonocoxite of 25 or fewer setae, not implanted in tubercle 9

9(8) Tuft of gonocoxite with 9–14 subequal wide setae 10

Tuft of gonocoxite consisting of ca. 20 setae with the basal ones wider than the apical ...... *Lu.* (*Hel.*) *castanea*

10(9) Tuft of gonocoxite with subequal wide setae implanted in a circular non-compact area ...... *Lu.* (*Hel.*) *ceferinoi*

Tuft of gonocoxite implanted in a semilunar compact area ...... *Lu.* (*Hel.*) *herreri*

FEMALES

1 Large size, head length $\geq$ 580 μm, wing $\geq$5 mm, flagellomere I $\geq$ 570 μm 2

Median size, head length ca. 520 μm, wing ca. 3.2 mm, flagellomere I ca. 530 μm 4

2(1) Wing: *delta* ca. 0.50 of *alpha* ...... *Lu.* (*Hel.*) *larensis*

Wing: *delta* $\leq$ 0.25 of *alpha* 3

3(2) Cibarium: sclerotized area twice as long as its basal width; common spermathecal duct length equivalent to 2/3 of the spermatheca's length ...... *Lu.* (*Hel.*) *caballeroi*

Cibarium: sclerotized area four × longer than its basal width; common spermathecal duct's length equivalent to 1/4 of the spermatheca's length ...... *Lu.* (*Hel.*) *osornoi*

4(1) Wing: *delta*/*alpha* ca. 0.4, *beta*/*gamma* $\leq$ 0.5 5

Wing: *delta*/*alpha* $\leq$ 0.28, *beta*/*gamma* $\geq$ 0.7 7

5(4) Wing: *beta*/*gamma* ca. 0.53 ...... *Lu.* (*Hel.*) *munaypata*

Wing: *beta*/*gamma* ca. 0.4 6

6(5) Flagellomere I longer than labrum-epipharynx and this latter ca. 2 × longer than the clypeus ...... *Lu.* (*Hel.*) *strictivilla*

Flagellomere I shorter than labrum-epipharynx and this latter ca. 2.5 × longer than the clypeus ...... *Lu.* (*Hel.*) *quillabamba*

7(4) Wing: width ca. 1.15 mm, *alpha* 1.17 mm, *delta* ca. 0.34 mm ...... *Lu.* (*Hel.*) *imperatrix*

Wing width $\leq$ 1.0 mm, *alpha*, $\leq$ 0.93 mm, *delta* $\leq$ 0.31 mm 8

8(7) Flagellomere I $\leq$ 370 μm; labrum-epipharynx $\leq$410 μm ...... *Lu.* (*Hel.*) *ceferinoi*

Flagellomere I $\geq$ 400 μm; labrum-epipharynx $\geq$460 μm 9

9(8) Pronotum and paratergite dark brown ...... *Lu.* (*Hel.*) *castanea*

Pronotum and paratergite light brown ...... *Lu.* (*Hel.*) *herreri*

Peruensis series

MALES

1 Presence of setae on the anterior region of the katepisternum; tergal papillae present at least on the 6th and 7th tergites.. 2

Absence of setae on the anterior region of the katepisternum; tergal papillae absent on all tergites .... (complex Noguchii).. 6

2(1) Paramere rectangular at its base and the apical half digitiform, bearing a sharp concavity in the middle of its dorsal margin and setae disposed like a toothbrush, that is, located just on the apical quarter of the digitiform part ....*Lu.* (*Hel.*) *galatiae*

Paramere without the above set of characteristics 3

3(2) Gonostyle with the internal spine implanted beyond its middle.................................. 4

Gonostyle with the internal spine implanted before its middle 5

4(3) Gonostyle with the internal spine located at the beginning of the apical third and the external lower one above the middle; paramere with the apical half laminar and densely covered with thin setae...*Lu.* (*Hel.*) *pescei*

Gonostyle with the external lower spine implanted before its middle; paramere with digitiform apex .... *Lu.* (*Hel.*) *tejadai*

5(3) Paramere with pre-apical protuberance on the ventral margin; gonocoxite with basal tuft consisting of fine setae .... *Lu.* (*Hel.*) *ayacuchensis*

Paramere without pre-apical protuberance on the ventral margin; basal tuft of the gonocoxite with the upper setae fine and the lower ones semifoliaceous .... *Lu.* (*Hel.*) *peruensis*

6(1) Gonostyle with the external lower spine implanted at the same level or slightly more apically than the internal one 7

Gonostyle with the external lower spine implanted more basally than the internal one.. 8

7(6) Digitiform part of the paramere covered by setae only in its apical half....................... .... *Lu.* (*Hel.*) *noguchii*

Digitiform part of the paramere covered by setae throughout.. *Lu.* (*Hel.*) *pallidithorax*

8(6) Gonostyle with the distance between the external spines greater than the external upper spine and the apical ones .... *Lu.* (*Hel.*) *chavinensis*

Gonostyle with the distance between the external spines ≤ the external upper spine and the apical ones... *Lu.* (*Hel.*) *blancasi*

FEMALES

1 Absence of setae on the anterior region of the katepisternum.................................... complex Noguchii.. 2

Presence of setae on the anterior region of the katepisternum..................................... 5

2(1) Wing: *delta* ≥ 0.33 of *alpha;* clypeus shorter than the eyes ......*Lu.* (*Hel.*) *pallidithorax*

Wing*: delta* ≤ 0.25 of *alpha;* clypeus longer than the eyes... 3

3(2) Flagellomere I ca. 400 μm .... *Lu.* (*Hel.*) *chavinensis*

Flagellomere I ca. 300 μm 4

4(3) Labrum-epipharynx ca. 370 μm; 3rd palpal segment with ca. 25 Newstead's sensilla.. *Lu.* (*Hel.*) *blancasi*

Labrum-epipharynx *ca* 300 μm; 3rd palpal segment with ca. 50 Newstead's sensilla..*Lu.* (*Hel.*) *noguchii*

5(1) Diameter of the apical ring of the spermathecae perceptibly smaller than that of the pre-apical ring .... *Lu.* (*Hel.*) *ayacuchensis*

Diameter of apical ring of the spermathecae ≥ that of the pre-apical ring 6

6(5) Wing: *delta* ca. 0.50 of *alpha*; apical ring of the spermathecae perceptibly longer and wider than the pre-apical ring, looking somewhat globular............ *Lu.* (*Hel.*) *peruensis*

Wing: *delta* ca. 0.33 of *alpha*; apical ring of the spermathecae with similar diameter and ca. twice the length of the pre-apical ring 7

7(6) Labrum-epipharynx ca. 500 μm; individual spermathecal ducts ca. 10 × longer than the spermatheca.. *Lu.* (*Hel.*) *pescei*

Labrum-epipharynx ca. 350 μm; individual spermathecal ducts ca. 5 × longer than the spermatheca.. *Lu.* (*Hel.*) *tejadai*
Lu. (*Hel.*) *galatiae*

(*Castromyia*)

MALES

1 Gonocoxite with tuft consisting of six setae; paramere with two setae with hooked apex in the median region of the dorsal margin ....*Lu.* (*Cas.*) *amarali*

Gonocoxite with tuft consisting of 4–5 setae; paramere without setae or presenting one seta with or without hooked apex in the median region of the dorsal margin 2

2(1) Gonocoxite with tuft consisting of 5 setae; paramere sclerotized in its apical half and without setae of hooked apex in the median region of the dorsal margin .... *Lu.* (*Cas.*) *caligata*

Gonocoxite with tuft consisting of 4 setae; paramere not sclerotized in its apical half and with a seta of hooked apex in the median region of the dorsal margin .... *Lu.* (*Cas.*) *castroi*

FEMALES: only *Lu. amarali* has been described. The spermathecae present 15–17 rings and the apical one is ca. 4 times longer than the pre-apical and a little wider than this latter; the common spermathecal duct is shorter than the spermatheca; the individual spermathecal ducts are ca. 4 times longer than the spermatheca and at their basal part half as wide as the spermatheca's width, tapering gradually, so at the apical part it is ca. 0.25 of the basal value.

(*Lutzomyia*)

MALES

1 Labial suture not forming a fork (Fig. 10-B); paramere with a protuberance on the ventral margin before the digitiform apex (Fig. 177) 2

Labial suture forming a fork (Fig. 10-A or C), paramere different from the above 3

2(1) Paramere: tubercle (where the setae with hooked apices are implanted) 2.6 × higher than it is wide; area of the pre-apical bulge with sparse coating of very short setae; dorsal margin with 2–3 setae straight and turned towards the gonocoxite ...... *Lu. (Lut.) dispar*

Paramere: tubercle (where the setae with hooked apices are implanted) twice as high as it is wide; area of the pre-apical bulge presenting a row of longer setae in addition to the sparse coating of very short setae; dorsal margin with 6–7 setae straight and turned towards the gonocoxite. ....*Lu. (Lut.) fonsecai*

3(1) Gonostyle with five well-developed spines.. 4

Gonostyle with four well-developed spines 11

4(3) Gonostyle with one apical spine (Figs. 123 or 131) 5

Gonostyle with two apical spines (Fig. 119) 6

5(4) Gonostyle with pre-apical spiniform seta (Fig. 123), gonocoxite with basal tuft of nine setae, three of them semifoliaceous; the paramere's dorsal margin has a tubercle in which are implanted three long spines curving abruptly to form a right angle .... *Lu.* (*Lut.*) *falquetoi*

Gonostyle without pre-apical spiniform seta (Fig. 131); gonocoxite with a foliaceous seta implanted in a sclerotized basal tubercle; paramere's dorsal margin has a tubercle in which are implanted two semifoliaceous hook-shaped setae .... *Lu.* (*Lut.*) *elizabethrangelae*

6(4) Paramere with dorsal margin bearing 5–6 setae with hooked apices .... *Lu.* (*Lut.*) *battistinii*

Paramere with dorsal margin bearing 1–2 setae with hooked apices 7

7(6) Gonocoxite with the basal tuft implanted in prominent tubercle, having two foliaceous setae (Fig. 144) 8

Gonocoxite with the basal tuft presenting a different aspect from the above 9

8(7) Ventral margin of the paramere with an elongated lobe, but with rounded apex covered with setae .... *Lu.* (*Lut.*) *forattinii*

Ventral margin of the paramere without the above lobe .... *Lu.* (*Lut.*) *almerioi*

9(7) Gonocoxite with the basal tuft implanted in a distinct tubercle presenting four fine setae and one semifoliaceous .... *Lu.* (*Lut.*) *bicornuta*

Gonocoxite with the basal tuft implanted directly in its surface, with four setae, three of them fine and one semifoliaceous 10

10(9) Gonostyle with external lower and internal spines equally developed; ventral margin of the paramere tapers brusquely in its apical third making this area digitiform .... *Lu.* (*Lut.*) *cavernicola*

Gonostyle with external lower spine thinner than the internal one; ventral margin of the paramere straight and simple .... *Lu.* (*Lut.*) *renei*

11(3) Gonocoxite with basal tuft of two foliaceous setae (Fig. 144) 12

Gonocoxite with basal tuft of four filiform setae (of width $\leq$ that of the aedeagal ducts) and/or semifoliaceous ones (wider than the aedeagal ducts) 13

12(11) Paramere with two foliaceous setae .... *Lu.* (*Lut.*) *lichyi*

Paramere with two fine setae .... *Lu.* (*Lut.*) *bifoliata*

13(11) Basal tuft of the gonocoxite with semifoliaceous setae. 14

Basal tuft of the gonocoxite with filiform setae 15

14(13) Ratio: aedeagal ducts/sperm pump 3.0:1.0 .... *Lu.* (*Lut.*) *cruzi*

Ratio: aedeagal ducts/sperm pump $\geq$4.5:1.0 .... *Lu.* (*Lut.*) *souzalopesi*

15(13) Gonostyle without spiniform pre-apical seta .... *Lu.* (*Lut.*) *ischnacantha*

Gonostyle with spiniform pre-apical seta 16

16(15) Isolated setae of the dorsal margin of the paramere implanted in a tubercle 17

Isolated setae of the dorsal margin of the paramere not implanted in a tubercle (Fig. 176) .... *Lu.* (*Lut.*) *longipalpis*
Lu. (*Lut.*) *pseudolongipalpis*

17(16) Gonostyle with the internal spine implanted beyond its middle near the external upper one .... *Lu.* (*Lut.*) *gaminarai*

Gonostyle with the internal spine implanted in its middle 18

18(17) Paramere and gonocoxite robust; paramere with the apical part (after the tubercle in which the curved setae are implanted) less than three times longer than its maximum width; gonocoxite ca. 2.3 times its maximum width .... *Lu.* (*Lut.*) *ischyracantha*

Paramere and gonocoxite thinner; paramere with its apical part 4.0 or more times greater than its maximum width; gonocoxite 2.8 times longer than its maximum width 19

19(18) Ratios: flagellomere I/labrum-epipharynx ca.1.45:1.0 and aedeagal ducts/sperm pump ca. 3.7:1.0; paramere with discreet protuberance in the pre-apical region of its ventral margin .... *Lu.* (*Lut.*) *alencari*
Ratios: flagellomere I/labrum-epipharynx *ca* 1.2:1.0 and aedeagal ducts/sperm pump ca. 3.1: 1.0; paramere without the above protuberance .... *Lu.* (*Lut.*) *matiasi*

FEMALES

1 Labial fork absent (Fig. 10-B) 2
Labial fork present (Fig. 10-A or C) 3

2(1) Presence of Newstead's sensilla on 2nd and 4th palpal segments .... *Lu.* (*Lut.*) *dispar*
2nd and 4th palpal segments without Newstead's sensilla .... *Lu.* (*Lut.*) *fonsecai*

3(2) Cercus length ca. 4.0 × its width .... *Lu.* (*Lut.*) *souzalopesi*
Cercus length ca. 2.0 × its width 4

4(3) Cibarium with two pairs of posterior (horizontal) teeth 5
Cibarium with three or more pairs of posterior (horizontal) teeth 12

5(4) Spermathecae with apical ring clearly longer than the rings that precede it 6
Spermathecae with apical ring as long as the rings that precede it 9

6(5) Thoracic pleura brown 7
Thoracic pleura straw 8

7(6) Antennae: flagellomere XIV clearly longer than flagellomere XIII .... *Lu.* (*Lut.*) *lichyi*
Antennae: flagellomere XIV shorter than flagellomere XIII .... *Lu.* (*Lut.*) *bifoliata*

8(6) Wing: ca. 2.65 mm long, 0.83 mm wide .... *Lu.* (*Lut.*) *cavernicola*
Wing: length ≤ 2.5 mm, width < 0.7 mm .... *Lu.* (*Lut.*) *battistinii*
Lu. (*Lut.*) *bicornuta*
Lu. (*Lut.*) *ischnacantha*

9(5) 9th tergite with protuberance (Fig. 41) .... *Lu.* (*Lut.*) *almerioi*
9th tergite without protuberance (Fig. 40) 10

10(9) Clypeus longer than the eye's width; 10th sternite with two apical setae 11
Clypeus ≤ eye's width; 10th sternite with 3–5 apical setae .... *Lu.* (*Lut.*) *renei*

11(10) 8th tergite generally without setae, though sometimes one is present .... *Lu.* (*Lut.*) *forattinii*
8th tergite with 5–11 setae .... *Lu.* (*Lut.*) *elizabethrangelae*

12(4) Cibarium with the anterior (vertical) teeth well developed clearly larger than the apical external teeth of lacinia; 8th sternite wider than its length .... *Lu.* (*Lut.*) *pseudolongipalpis*
Cibarium with the anterior (vertical) teeth as large as or smaller than the apical external teeth of lacinia; 8th sternite equal to, shorter or longer than its width 13

13(12) Cibarium: posterior (horizontal) teeth with their base as wide as their height, like a saw blade 14
Cibarium with posterior teeth longer than the width of their bases (needle-like) .... *Lu.* (*Lut.*) *longipalpis*
Lu. (*Lut.*) *cruzi*
Lu. (*Lut.*) *gaminarai*
Lu. (*Lut.*) *matiasi*

14(13) Cibarium with five or more pairs of posterior (horizontal) teeth; labrum-epipharynx ca. 270 μm .... *Lu.* (*Lut.*) *ischyracantha*
Cibarium with three pairs of posterior (horizontal) teeth; labrum-epipharynx ca. 320 μm .... *Lu.* (*Lut.*) *alencari*

(*Tricholateralis*)

MALES

1 Median region of the dorsal margin of paramere with one or two setae with hooked apex, located separately from the apical setae (Fig. 175) .... *Lu.* (*Trl.*) *falcata*

Median region of the dorsal margin of paramere without the above setae 2

2(1) Basal tuft of gonocoxite located in a prominent tubercle (Figs. 147 and 148)............ 3

Basal tuft of gonocoxite located on a discreet prominence (Fig. 145) 6

3(2) Basal tuft of gonocoxite bearing up to 9 setae 4

Basal tuft of gonocoxite bearing *ca.* 15 setae .... *Lu.* (*Trl.*) *flabellata*

4(3) External lower spine of gonostyle located more basally than the internal spine; this latter located beyond its middle .... *Lu.* (*Trl.*) *spathotrichia*

External lower spine of gonostyle located at the same level as the internal spine, i.e. before its middle 5

5(4) Basal tuft of gonocoxite bearing semifoliaceous setae (Fig. 147) and other spiniform setae.. *Lu.* (*Trl.*) *carvalhoi*

Basal tuft of gonocoxite bearing only of semifoliaceous setae......................... *Lu.* (*Trl.*) *araracuarensis*

6(2) Basal tuft of gonocoxite bearing of 5–8 setae... 7

Basal tuft of gonocoxite bearing 12 or more setae 10

7(6) Pronotum and paratergite dark brown.. 8

Pronotum and paratergite straw .... *Lu.* (*Trl.*) *evangelistai*

8(7) Thoracic pleura straw... *Lu.* (*Trl.*) *legerae*

Anepisternum brown.. 9

9(8) Flagellomere I ca. 310 µm; flagellomere I/labrum-epipharynx ratio ca.1.22:1.00; wing: length/width ratio ca. 3.7:1.0; aedeagal ducts/sperm pump ratio ca. 3.5:1.0................. .... *Lu.* (*Trl.*) *maesi*

Flagellomere I ≤ 270 µm; flagellomere I/labrum-epipharynx ratio ≤ 1.0:1.0; wing: length/width ratio ≤ 3.0:1.0; aedeagal ducts/sperm pump ratio ≤ 2.8:1.0 .... *Lu.* (*Trl.*) *diabolica*

10(6) 5th palpal segment equal to or slightly longer than 3rd.................. *Lu.* (*Trl.*) *cultellata*

5th palpal segment distinctly longer than 3rd.. 11

11(10) Pronotum and paratergite straw 12

Pronotum and paratergite brown.. 13

12(11) Setae of the basal tuft of the gonocoxite distinctly wider than the aedeagal ducts......... .... *Lu.* (*Trl.*) *sherlocki*

Setae of the basal tuft of the gonocoxite as wide as the aedeagal ducts...... *Lu.* (*Trl.*) *gomezi*

13(11) Paramere with straight apical-dorsal margin .... *Lu.* (*Trl.*) *cruciata*

Paramere with the above margin curved toward the gonocoxite .... *Lu.* (*Trl.*) *marinkellei*

FEMALES

1 Pronotum, paratergite and thoracic pleura straw 2

Pronotum and paratergite brown... 4

2(1) Head totally straw and the mesonotum of straw brown sides .... *Lu.* (*Trl.*) *evangelistai*

Head brown, even if partially, mesonotum dark brown 3

3(2) Spermatheca: individual ducts' width equivalent to 0.33 of the apical ring's width .... *Lu.* (*Trl.*) *gomezi*

Spermatheca: individual ducts' width equivalent to half or more of that of the apical ring... *Lu.* (*Trl.*) *sherlocki*

4(1) Thoracic pleura straw 5
Anepisternum, katepisternum and katepimeron dark brown 6

5(4) 9th tergite with several very short and sclerotized spicules at the apex of the dorso-lateral region .... *Lu.* (*Trl.*) *cruciata*
9th tergite without the above mentioned sclerotized spicules............. *Lu.* (*Trl.*) *legerae*

6(4) Labrum-epipharynx, clearly longer than flagellomere I .... *Lu.* (*Trl.*) *diabolica*
Labrum-epipharynx ≤ flagellomere I .... *Lu.* (*Trl.*) *carvalhoi*
Lu. (*Trl.*) *falcata*
Lu. (*Trl.*) *marinkellei*
Lu. (*Trl.*) *spathotrichia*
Lu. (*Trl.*) *maesi*

The females of *Lu.* (*Trl.*) *cultellata, Lu.* (*Trl.*) *flabellata* and *Lu.* (*Trl.*) *araracuarensis* have not yet been described.

Incertae sedis

Lutzomyia chotensis

Male: ventro-cervical sensilla present. Absence of setae on the anterior region of the katepisternum and of setae on the abdominal pleura. Presence of tergal papillae from V to VII abdominal tergites. Gonostyle with four spines and without pre-apical spiniform seta, external lower spine implanted before the middle of the gonostyle and the internal one on its middle. Gonocoxite with two straight setae at its base. Paramere simple, digitiform with spiniform setae. Epandrial lobes shorter than the gonocoxite. **Female**: not yet described.

Lutzomyia ignacioi

Both sexes: ascoids with rudimentary spur; presence of papilla in flagellomere III; absence of setae on the anterior region of the katepisternum and ventrocervical sensilla present. **Male:** gonocoxite with a sparse basal cluster of ca. ten long fine setae; paramere simple with apex slightly bent towards the gonocoxite; aedeagal ducts ca. twice as long as the sperm pump. **Female**: cibarium with two pairs of posterior (horizontal) teeth; common spermathecal duct and the individual spermathecal ducts striated; spermathecae with 11–12 rings, many of them imbricated; the spermatheca's length is half that of the individual spermathecal ducts, these latter ca. five × longer than the common spermathecal duct.

Lutzomyia manciola

Female: clypeus short, equivalent to 0.25 of the head's length and 0.50 of the eyes' length. Palpal formula: 1.4.2.3.5, with the 5th palpal segment slightly longer than the 3rd; on this latter the Newstead's sensilla are scattered along the apical 2/3; flagellomeres with simple ascoids in fII its apex goes slightly beyond the papilla. Labrum-epipharynx equivalent to 0.53 of the head's length and 0.83 of flagellomere I. Cibarium with four posterior teeth and 11–12 anterior vertical teeth very small; discreet sclerotized area and sclerotized arch incomplete. Thorax dusky throughout. Spermatheca piriform and striated with terminal knob bilobated, individual and common spermathecal ducts striated, the former ca. five times longer than the latter. **Male**: not yet described.

Lutzomyia ponsi

Female: closely similar to that of *Lu. ignacioi*; the differences being the presence of four pairs of posterior (horizontal) teeth and terminal knob of the spermatheca more elongated. **Male**: not described.

Lutzomyia tanyopsis

Female: Clypeus longer than the eyes and ca. 0.4 of the head's length. Palpal formula: 1.4.2.3.5; ascoids with distinctive posterior spur; cibarium with two pairs of posterior teeth, with the inner pair's apex turned towards the pharynx and the external pair turned outward and a row of irregular anterior vertical teeth; spermatheca segmented, the terminal knob distinctly longer and wider than the anterior, common spermathecal duct very short and the individual spermathecal ducts eight times longer than the former. **Male**: not yet described.

Migonemyia (*Migonemyia*)

MALES

1 Gonocoxite with basal tuft of 11–15 setae; aedeagal ducts ≤2.9 times longer than the sperm pump........ *Mg.* (*Mig.*) *rabelloi*

Gonocoxite without basal tuft or with tuft of 1–10 setae; aedeagal ducts ≥3.6 times longer than the sperm pump........ 2

2(1) Paramere digitiform and slightly convex in its dorsal margin; aedeagal ducts and sperm pump measuring respectively ≥640 μm and ≥146 μm........ *Mg.* (*Mig.*) *vaniae*

Paramere strongly convex in its dorsal margin between two concavities; aedeagal ducts and sperm pump measuring respectively ≤590 μm and ≤130 μm........ *Mg.* (*Mig.*) *migonei*

FEMALES

1 Set consisting of common spermathecal duct, individual spermathecal duct and spermatheca relatively short, equivalent to the height of genital furca; individual spermathecal ducts striated in the region of transition to the spermatheca........ *Mg.* (*Mig.*) *rabelloi*

Set consisting of common spermathecal duct, individual spermathecal duct and spermatheca very long, ≥ the height of genital furca; individual spermathecal ducts not striated in the region of transition to the spermatheca........ 2

2(1) Tergite VIII with between zero and two setae; terminal knob of the spermathecae sessile........ *Mg.* (*Mig.*) *vaniae*

Tergite VIII with 5–22 setae; terminal knob of the spermathecae pedunculated (Fig. 44)........ *Mg.* (*Mig.*) *migonei*

(*Blancasmyia*)

MALES

1 Paramere apically bifurcated; epandrial lobe apex with permanent setae........ *Mg.* (*Bla.*) *cerqueirai*

Paramere simple, epandrial lobe apex without permanent setae........ 2

2(1) Gonostyle with the external lower spine located on its apical third; aedeagal ducts slightly longer than the sperm pump........ *Mg.* (*Bla.*) *gorbitzi*

Gonostyle with the external lower spine located in its middle; aedeagal ducts at least twice as long as the sperm pump........ 3

3(2) Gonocoxite with basal tuft consisting of ca. four long setae; aedeagal duct apex tapered and curved (Fig. 99)........ *Mg.* (*Bla.*) *bursiformis*

Gonocoxite without basal tuft of long setae; aedeagal duct apex truncated and flat........ *Mg.* (*Bla.*) *moucheti*

FEMALES

1 Spermatheca of sac-like aspect........ *Mg.* (*Bla.*) *bursiformis*

Spermatheca of tubular aspect........ 2

2(1) The set consisting of individual spermathecal ducts and spermatheca very short so that the apex of the spermatheca comes short of the middle of the genital furca (Fig. 47)........ *Mg.* (*Bla.*) *gorbitzi*

The set consisting of individual spermathecal ducts and spermatheca long, so that the apex of the spermatheca goes beyond the apex of the stem of the genital furca........ 3

3(2) Common spermathecal duct as long as the individual spermathecal ducts........ *Mg.* (*Bla.*) *cerqueirai*

Common spermathecal duct several times shorter than the individual spermathecal ducts........ *Mg.* (*Bla.*) *moucheti*

Pintomyia
(*Pintomyia*)

MALES

1 Metepisternal and/or metepimeral setae present (Fig. 22)... 2

Metepisternal and metepimeral setae absent.. 3

2(1) Gonocoxite with basal tuft consisting of 6–8 setae; paramere digitiform with the setae of the dorsal margin occupying the apical half........................... *Pi.* (*Pin.*) *christenseni*

Gonocoxite with basal tuft consisting of a larger number of setae; paramere thicker, with the setae of the dorsal margin occupying the apical third.... *Pi.* (*Pin.*) *damascenoi*

3(1) Gonocoxite without the basal tuft of setae; anterior and median coxae straw and the posterior coxae light brown.. *Pi.* (*Pin.*) *bianchigalatiae*

Gonocoxite with basal tuft consisting of two or more setae, at least the posterior and median coxae being dark 4

4(3) Gonocoxite with basal tuft consisting of several setae in two rows.....*Pi.* (*Pin.*) *pessoai*

Gonocoxite with basal tuft consisting of 2–5 setae 5

5(4) Aedeagal ducts 3.5–4.0 times longer than the sperm pump ....*Pi.* (*Pin.*) *fischeri*

Aedeagal ducts 2.0–2.5 times longer than the sperm pump .... *Pi.* (*Pin.*) *kuscheli*

FEMALES

1 Metepisternal and/or metepimeral setae present (Fig. 22); flagellomeres X and XI with papillae.. 2

Metepisternal and metepimeral setae absent, flagellomeres X and XI with no papilla... 3

2(1) Spine of the femur highly conspicuous....................................... *Pi.* (*Pin.*) *damascenoi*

.... *Pi.* (*Pin.*) *christenseni*

Spine of the femur greatly reduced, sometimes difficult to see........ *Pi.* (*Pin.*) *mamedei*

3(1) Anterior and median coxae straw, posterior coxae light brown; spermatheca with bulbous non-sclerotized aspect .... *Pi.* (*Pin.*) *bianchigalatiae*

At least the posterior and median coxae dusky, spermatheca sclerotized..................... 4

4(3) Anterior coxae straw; terminal knob's diameter equivalent to or larger than that of the spermatheca, this latter being cylindrical (Fig. 69)............................. *Pi.* (*Pin.*) *fischeri*

.... *Pi.* (*Pin.*) *gibsoni*

All coxae dusky; spermatheca sub-spherical 5

5(4) Labrum-epipharynx ≤270 μm; terminal knob at least as long as the spermatheca .... *Pi.* (*Pin.*) *kuscheli*

Labrum-epipharynx ≥350 μm, terminal knob shorter than the spermatheca .... *Pi.* (*Pin.*) *pessoai*

(*Pifanomyia*)

Pacae Series

MALES

1 Gonocoxite with tuft of 5–8 setae; aedeagal ducts ca. 7 × longer than the sperm pump .... *Pi.* (*Pif.*) *pacae*

Gonocoxite with tuft of 2–3 setae; aedeagal ducts ca. 6 × longer than the sperm pump .... *Pi.* (*Pif.*) *gruta*

FEMALES

1 Spermathecae smooth, banana-shaped (Fig. 74).................................... *Pi.* (*Pif.*) *pacae*

Spermathecae cylindrical and transversally striated .... *Pi.* (*Pif.*) *gruta*

Monticola Series

MALES

1 Gonostyle with pre-apical spiniform seta; gonocoxite with sparse tuft of setae in the basal median region .... *Pi.* (*Pif.*) *monticola*

Gonostyle without the pre-apical spiniform seta; gonocoxite without setae .... *Pi.* (*Pif.*) *misionensis*

FEMALES

1 Spermathecae elongated and transversally striated, bearing an apical inverted tapered collar .... *Pi.* (*Pif.*) *misionensis*

Spermathecae vesicular, not transversally striated and without the apical collar .... *Pi.* (*Pif.*) *monticola*

Pia Series

MALES

1 Presence of a compact cluster of setae in baso-median region of the gonocoxite 2

Presence of a sparse cluster of setae in the middle of the gonocoxite 4

2(1) Labrum-epipharynx ca. 280 μm; flagellomere I ca. 350 μm .... *Pi.* (*Pif.*) *reclusa*

Labrum-epipharynx ca. 220 μm; flagellomere I ca. 320 μm 3

3(2) Tuft of the gonocoxite consisting of ca. ten setae .... *Pi.* (*Pif.*) *torrealbai*

Tuft of the gonocoxite consisting of 17–19 setae .... *Pi.* (*Pif.*) *valderramai*

4(1) Dorsal margin of paramere almost straight bearing setae in its apical third .... *Pi.* (*Pif.*) *pia*

Dorsal margin of paramere curved, bearing setae in its apical fifth .... *Pi.* (*Pif.*) *limafalcaoae*

FEMALES

1 Labrum-epipharynx >350 μm 2

Labrum-epipharynx ≤310 μm 4

2(1) Thoracic pleura dusky throughout 3

Thoracic pleura dusky only in the basal region .... *Pi.* (*Pif.*) *tihuiliensis*

3(2) 5th palpal segment ca. one-third longer than the sum of 3rd + 4th; common and individual spermathecal ducts smooth .... *Pi.* (*Pif.*) *reclusa*

5th palpal segment subequal the sum of 3rd + 4th; common and individual spermathecal ducts striated .... *Pi.* (*Pif.*) *pia*

4(1) Thoracic pleura dusky throughout .... *Pi.* (*Pif.*) *limafalcaoae*

Thoracic pleura partially dusky or completely straw 5

5(4) Thoracic pleura straw .... *Pi.* (*Pif.*) *tocaniensis*

Thoracic pleura dusky in their basal part 6

6(5) 5th palpal segment longer than 3rd. Individual spermathecal ducts clearly shorter than the common spermathecal duct .... *Pi.* (*Pif.*) *suapiensis*

5th palpal segment shorter than 3rd. Individual spermathecal ducts longer than the common spermathecal duct .... *Pi.* (*Pif.*) *emberai*

Verrucarum Series

MALES

1 Gonocoxite with a tuft of setae in basal position 2

Gonocoxite with one tuft of setae in a basal and another in a central position 4

2(1) Parameral sheath with curved apex, recalling the head of a goose .... *Pi.* (*Pif.*) *deorsa*

Parameral sheath simple and conical 3

3(2) Gonostyle with the external lower spine implanted in its middle and the internal one basal; aedeagal ducts ca. twice as long as the sperm pump .... *Pi.* (*Pif.*) *itza*

Gonostyle with the external lower and internal spines implanted in its base; aedeagal ducts ca. 3 times longer than the sperm pump .... *Pi.* (*Pif.*) *columbiana*

4(1) Gonostyle with the internal spine located in a prominent tubercle.............................. 5

Gonostyle with the internal spine located in a discreet tubercle (Fig. 122)................. 8

5(4) Gonocoxite with three tufts of setae 6

Gonocoxite with two tufts of setae 7

6(5) Gonostyle without the pre-apical seta; apical part of paramere with incipient lobe; gonocoxite with basal tuft of setae whose apices do not go beyond that of the paramere .... *Pi.* (*Pif.*) *moralesi*

Gonostyle with pre-apical seta; apical part of paramere with two evident lobes; gonocoxite with basal tuft of setae whose apices go beyond that of the paramere.. *Pi.* (*Pif.*) *antioquiensis*

7(5) Gonostyle: the external lower spine with tapered apex ....*Pi.* (*Pif.*) *aulari*

Gonostyle: the external lower spine with truncated and slightly expanded apex .... *Pi.* (*Pif.*) *andina*

8(4) Gonostyle with columnar aspect, the distance between the external lower and upper spines being twice or more the distance of this latter from the apical spine 9

Gonostyle without columnar aspect, the distance between the external lower and upper spines equal to or less than the distance from this latter to the apical spine ....*Pi.* (*Pif.*) *verrucarum*

9(8) Paramere with dorsal margin being markedly concave with a long apical fringe of setae ....*Pi.* (*Pif.*) *cajamarcensis*

Paramere with slightly concave dorsal margin with no long fringe of setae apically...*Pi.* (*Pif.*) *disiuncta*

FEMALES

1 Spermathecae quasi-spherical...*Pi.* (*Pif.*) *disiuncta*

Spermathecae perceptibly elongated... 2

2(1) Flagellomere I $\geq$ labrum-epipharynx.. 3

Flagellomere I < labrum-epipharynx 4

3(2) Common spermathecal duct ca. 1.5 times longer than the individual spermathecal ducts and these latter perceptibly longer than the spermatheca ....*Pi.* (*Pif.*) *aulari*
Pi. (*Pif.*) *andina*

Common spermathecal duct ca. twice as long as the individual spermathecal ducts, these latter being as long as the spermatheca................................*Pi.* (*Pif.*) *columbiana*

4(2) Common spermathecal duct < individual spermathecal ducts.......................................
....*Pi.* (*Pif.*) *cajamarcensis*

Common spermathecal duct ca. twice as long as the individual spermathecal ducts .... *Pi.* (*Pif.*) *moralesi*
Pi. (*Pif.*) *verrucarum*

The females of *Pi.* (*Pif.*) *antioquiensis, Pi.* (*Pif.*) *deorsa* and *Pi.* (*Pif.*) *itza* have not yet been described.

Evansi Series

MALES

1 Paramere with straight dorsal margin, completely covered by spiniform setae 2

Paramere with dorsal margin slightly curved at the apex towards the gonocoxite, setae restricted to this apical area.. *Pi.* (*Pif.*) *ovallesi*

2(1) Sperm pump ca. 150 μm long, the pavilion's diameter being larger than that of the sperm sac; aedeagal ducts ca. 670 μm; parameres thick, the digitiform part's width slightly smaller than that of the base and as long as the epandrial lobe..
....*Pi.* (*Pif.*) *evansi*

Sperm pump ca. 120 μm long; the pavilion's diameter being smaller or equal to that of the sperm sac; aedeagal ducts ≤570 μm; parameres thinner, with the digitiform part's width perceptibly narrower than its base and smaller than the epandrial lobe ... 3

3(2) Gonocoxite without tuft of setae; thorax with paratergite and pleura straw ... *Pi.* (*Pif.*) *nevesi*

Gonocoxites with tuft of two or more setae; thorax with paratergite brown ... *Pi.* (*Pif.*) *maranonensis*

FEMALES

1 Spermathecae without apical ring ... 2

Spermathecae with apical ring ... 3

2(1) Labrum-epipharynx ≥330 μm; Thorax with paratergite and mesonotum brown contrasting with the straw pleura ... *Pi.* (*Pif.*) *maranonensis*

Labrum-epipharynx ≤270 μm; Thorax with paratergite and pleura straw ... *Pi.* (*Pif.*) *nevesi*

3(1) Spermatheca with a constriction in its apical half ... *Pi.* (*Pif.*) *ovallesi*

Spermatheca without constriction in its apical half ... *Pi.* (*Pif.*) *evansi*

Serrana Series

MALES

1 Paramere simple ... 2

Paramere branched ... 12

2(1) Gonostyle with the internal spine atrophied or perceptibly thinner than the external spine ... 3

Gonostyle with the internal spine of equivalent thickness to the external ... 9

3(2) Gonocoxite with a multibranched seta ... *Pi.* (*Pif.*) *piedraferroi*

Tuft of gonocoxite with three or more simple setae ... 4

4(3) Tuft of gonocoxite with straight setae ... 5

Tuft of gonocoxite with curved setae with apices turned towards the gonocoxite ... 6

5(4) Tuft of gonocoxite with five setae, the longest seta's apex reaching that of the paramere ... *Pi.* (*Pif.*) *oresbia*

Tuft of gonocoxite with 3–4 setae, the longest seta's apex almost reaching the middle of that of the paramere ... *Pi.* (*Pif.*) *ottolinai*

6(4) Tuft of gonocoxite with five setae ... 7

Tuft of gonocoxite with 6–7 setae ... *Pi.* (*Pif.*) *odax*

7(6) Tuft of gonocoxite setae longer than the paramere on its dorsal margin ... *Pi.* (*Pif.*) *guilvardae*

Tuft of gonocoxite setae shorter than the paramere on its dorsal margin ... 8

8(7) Paramere: apical area's length (bearing setae) less than twice its width and with evident concavity preceding the apical area on its dorsal margin ... *Pi.* (*Pif.*) *robusta*

Paramere: apical area's length (bearing setae) greater than or equivalent to 2.5 times its width; dorsal margin, approximately straight ... *Pi.* (*Pif.*) *serrana*

9(2) Mesonotum straw ... 10

Mesonotum brown ... 11

10(9) Gonocoxite with tuft of ca. 20 setae located on a small tubercle ... *Pi.* (*Pif.*) *christophei*

Gonocoxite with tuft of 4–5 setae located on its surface ... *Pi.* (*Pif.*) *orestes*

11(9) Gonocoxite with tuft of 10–12 setae..*Pi.* (*Pif.*) *diazi*

Gonocoxite with two tufts of setae in a median position......................*Pi.* (*Pif.*) *novoae*

12(1) Paramere: ventral lobe as long as the dorsal one; with 3–5 setae in the interlobular region .... *Pi.* (*Pif.*) *torresi*

Paramere: ventral lobe perceptibly shorter than the dorsal one; presence of a tuft with ca. ten setae in the interlobular region 13

13(12) Head length, including clypeus, ca. 360 μm; gonostyle with the preapical seta .... *Pi.* (*Pif.*) *boliviana*

Head length, including clypeus, ca. 310 μm; gonostyle without the preapical seta ....*Pi.* (*Pif.*) *salomoni*

FEMALES

1 Spermatheca with a detached apical ring.. 2

Spermatheca without apical ring... 6

2(1) Spermathecae: marked narrowing separating the apical ring from the rest of the spermatheca.. 3

Spermathecae with no narrowing... 5

3(2) Individual spermathecal ducts as long as the common spermathecal duct.................... ....*Pi.* (*Pif.*) *torresi*

Individual spermathecal ducts ca. twice as long as the common spermathecal duct.... 4

4(3) Head length, including clypeus, ca. 500 μm; labrum-epipharynx ca. 300 μm................. ....*Pi.* (*Pif.*) *boliviana*

Head length, including clypeus ca. 360 μm; labrum-epipharynx ca. 220 μm................. ....*Pi* (*Pif.*) *salomoni*

5(2) Spermatheca with the apical ring's length ca. 0.50 of that of the striated part.................. .... *Pi.* (*Pif.*) *oresbia*

Spermatheca with the apical ring's length 0.33 or less that of the striated part ....*Pi.* (*Pif.*) *ottolinai*
Pi. (*Pif.*) *odax*
Pi. (*Pif.*) *serrana*
Pi. (*Pif.*) *robusta*

6(1) Spermatheca's width at least ca. twice that of the apex of the genital furca.................. .... *Pi.* (*Pif.*) *christophei*

Spermatheca's width equivalent to that of the apex of the genital furca 7

7(6) Mesonotum straw... *Pi.* (*Pif.*) *orestes*

Mesonotum brown 8

8(7) Cibarium with a wide base of the sclerotized area extending throughout the anterior region of the horizontal teeth ....*Pi.* (*Pif.*) *novoae*

Cibarium with a narrow base of the sclerotized area, occupying a narrow band in the anterior median region of the horizontal teeth ....*Pi.* (*Pif.*) *diazi*

Townsendi Series

MALES

1 Gonostyle with thick apical spine, twice the width of the pre-apical one with a concavity on pre-apical part of its superior margin...................................... *Pi.* (*Pif.*) *spinicrassa*

Gonostyle with the apical spine slightly thicker than the pre-apical one without the concavity on its pre-apical part 2

2(1) Paramere without protuberance at the beginning of the apical third of the ventral margin 3

Paramere with protuberance at the beginning of the apical third of the ventral margin 4

3(2) Gonocoxite with one tuft of pratically straight setae, with the more basal ones slightly smaller than the others .... *Pi.* (*Pif.*) *nadiae*

Gonocoxite with the tuft of two distinct groups of setae, one with short, straight setae and the other with long, curved setae .... *Pi.* (*Pif.*) *amilcari*

4(2) Paramere with its apical part bearing a group of setae of length approximately twice the width of the region in which they are located .... *Pi.* (*Pif.*) *torvida*

Paramere with its apical part bearing spines of length equal to or shorter than the width of the region in which they are located 5

5(4) Basal tuft of the gonocoxite with the setae reaching the paramere's apex .... *Pi.* (*Pif.*) *longiflocosa*

Basal tuft of the gonocoxite with the setae reaching, at most, the middle of the paramere 6

6(5) Dorsal margin of the paramere with setae restricted to the apical quarter .... *Pi.* (*Pif.*) *youngi*

Dorsal margin of the paramere with setae occupying its apical third or half 7

7(6) Dorsal margin of the paramere with setae distributed on its apical third .... *Pi.* (*Pif.*) *quasitownsendi*

Dorsal margin of the paramere with setae distributed on its apical half 8

8(7) Paramere's apex reaches or goes beyond the epandrial lobe's apex...... *Pi.* (*Pif.*) *townsendi*

Paramere's apex does not reach that of the epandrial lobe .... *Pi.* (*Pif.*) *sauroida*

FEMALES: indistinguishable among the species.

Incertae sedis

Pintomyia (Pifanomyia) maracayensis

Male: wing with R_1 going slightly beyond the level of the bifurcation R_2R_3, *beta* slightly smaller than *alpha*, *gamma* clearly bigger than *alpha*; gonostyle with four developed spines; epandrial lobe as long as the gonocoxite and perceptibly longer than the paramere. **Female**: not yet described.

Pintomyia (Pifanomyia) rangeliana

Male: clypeus as long as the eyes and ca. one-third of the head's height; gonostyle with spiniform pre-apical seta; the external upper and lower spines located on the apical quarter and third, respectively, and the inner spine (atrophied) at its base; gonocoxite without tuft of setae, paramere thick bearing a concavity on its dorsal margin with the apex slightly curved towards the gonocoxite having its narrower part wider than the epandrial lobe, this latter being longer than the gonocoxite and perceptibly longer than the paramere. **Female**: Apex of the common spermathecal duct reaching the apex of the stem of the genital furca, individual spermathecal ducts practically absent and the spermatheca longer than wider, its width equivalent to twice the common spermathecal duct's, small and spherical terminal knob, perceptibly detached from the spermatheca wall.

***Pintomyia (Pifanomyia)* sp. de Anchicaya**

Male: clypeus shorter than 0.33 of the head's height and ca. 0.50 of the eyes' length. Palpal formula: 1.4.2. (3.5); the posterior femur of the specimen described lacking; gonostyle with the spiniform pre-apical spine, the upper and lower external spines located on the apical quarter and third, respectively, the internal one implanted slightly before its median region; gonocoxite without tuft of setae; paramere simple and digitiform; epandrial lobe slightly longer than the gonocoxite; aedeagal ducts twice as long as the sperm pump. **Female**: not yet described.

Pintomyia (Pifanomyia) naiffi

Male: clypeus as long as 0.33 of the head. Palpal formula 1.4.(2.3).5. Posterior femur of the two specimens described lacking. Gonostyle with pre-apical seta, the apical and the two external spines well developed and the internal one thinner and shorter than the external; the external upper spine implanted in the apical quarter and the external lower spine a little beyond the middle of the gonostyle. Gonocoxite with a tuft consisting of 20 long setae (as long as the width of the gonocoxite) and implanted in a circular area of the median region of the gonocoxite. Paramere with a protuberance in the pre-apical region of the ventral margin; in its dorsal margin there occurs a concavity a little beyond the middle; the apical third being covered with setae. Parameral sheath with its apex almost reaching the middle of the paramere. Epandrial lobes slightly shorter than the gonocoxites and present rounded apex. Aedeagal ducts with slightly striated apex; ca. five times longer than the sperm pump. **Female**: not yet described.

Dampfomyia
(*Dampfomyia*)

MALES

1	Gonostyle with the two external spines	*Da.* (*Dam.*) *insolita*
	Gonostyle with only one external spine	2
2(1)	Paramere with a basal dorsal branch shorter than the width of the base of the paramere, with ca. 3 setae in its apical region; apex of the paramere bifurcated	*Da.* (*Dam.*) *leohidalgoi*
	Paramere with a basal dorsal branch as long as the width of the base of the paramere, with 10 or more setae; apex of the paramere not bifurcated	3
3(2)	Gonocoxite short and thick, its length being twice its width	*Da.* (*Dam.*) *permira*
	Length of gonocoxite three or more times its width	4
4(3)	Dorsal branch of paramere with less than ten setae	*Da.* (*Dam.*) *rosabali*
	Dorsal branch of paramere with more than 20 setae	5
5(4)	Dorsal branch of paramere digitiform and curved	6
	Dorsal branch of paramere with oval extension	*Da.* (*Dam.*) *anthophora*
6(5)	Dorsal branch of paramere bearing setae on its apical half	*Da.* (*Dam.*) *atulapai*
	Dorsal branch of paramere bearing setae on most of its surface, only the basal fifth lacking setae	*Da.* (*Dam.*) *dodgei*

FEMALES

1	Individual spermathecal ducts practically absent; common spermathecal duct more than half the width of the genital chamber	*Da.* (*Dam.*) *permira*
	Individual spermathecal ducts present, though they may be shorter than the spermatheca	2
2(1)	Spermatheca bearing globular projections in its apical region	3
	Spermatheca bearing digitiform projections in its apical region	*Da.* (*Dam.*) *leohidalgoi*
3(2)	Cibarium with just one pair of posterior blade-shaped teeth (Fig. 20-E)	*Da.* (*Dam.*) *anthophora* *Da.* (*Dam.*) *dodgei*
	Cibarium with two or more pairs of posterior teeth	4
4(3)	Cibarium with two pairs of posterior teeth	*Da.* (*Dam.*) *atulapai* *Da.* (*Dam.*) *rosabali*
	Cibarium with three pairs of posterior teeth	*Da.* (*Dam.*) *insolita*

Delpozoi Group

MALES

1 Gonostyle with the inner spine clearly implanted in its base; width of the epandrial lobe greater than that of the gonostyle; paramere without protuberance on the apical third of its ventral margin .... *Da. delpozoi*

Gonostyle with the inner spine implanted near its middle; width of the epandrial lobe smaller or equivalent to that of the gonostyle; presence of protuberance on the apical third of its ventral margin 2

2(1) Paramere thick, its median region wider than the epandrial lobe, however its apical region tapers .... *Da.* sp. de Suchitepequez

Paramere thin, its median region being approximately of the same width as the epandrial lobe, though it presents a small apical extension .... *Da. inusitata*

FEMALES: indistinguishable

(*Coromyia*)

MALES

1 Gonostyle with two well-developed external spines 2

Gonostyle with only one well-developed external spine 3

2(1) Gonocoxite with the basal tuft of less than ten setae; paramere narrow in its distal 2/3 with similar width in the anterior and posterior areas to the protuberance of the ventral margin .... *Da.* (*Cor.*) *deleoni*

Gonocoxite with the basal tuft of more than 12 setae; paramere clearly wider in its distal 1/3, whose width is twice that of the anterior area of the protuberance of the ventral margin .... *Da.* (*Cor.*) *zeledoni*

3(1) Gonostyle with a well-developed inner spine 4

Gonostyle with a seta-like inner spine 5

4(3) Epandrial lobe shorter and narrower or as wide as the gonocoxite .... *Da.* (*Cor.*) *vesicifera*

Epandrial lobe longer and clearly narrower than the gonocoxite .... *Da.* (*Cor.*) *aquilonia*

5(3) Gonostyle with the external spine implanted in its apical quarter or beyond 6

Gonostyle with the external spine implanted in its apical third 8

6(5) Gonostyle short and the lateral lobe thick, so that the former is less than 1.5 times longer than the width of the latter; gonocoxite with tuft of ca. six setae .... *Da.* (*Cor.*) *viriosa*

Gonostyle ca. three times longer than the width of the lateral lobe; gonocoxite with a tuft of 8–10 setae 7

7(6) Setae on the ventral margin of the paramere covering the apical third .... *Da.* (*Cor.*) *isovespertilionis*

Setae on the ventral margin of the paramere covering the apical half .... *Da.* (*Cor.*) *vespertilionis*

8(5) Gonocoxite narrower than the epandrial lobe, both measured in their median region . 9

Gonocoxite wider than the epandrial lobe, both measured in their median region .... *Da.* (*Cor.*) *disneyi*

9(8) Gonostyle with pronounced curve; the external spine implanted a little beyond its middle .... *Da.* (*Cor.*) *steatopyga*

Gonostyle without pronounced curve; the external spine implanted at the beginning of its apical third .... *Da.* (*Cor.*) *beltrani*

FEMALES

1 Cibarium with two pairs of posterior (horizontal) teeth whose base is merged in a small mound; spermathecae with lateral extension and terminal knob elongated .... *Da.* (*Cor.*) *aquilonia*

Cibarium with the posterior (horizontal) teeth with individualized base; spermathecae with or without lateral extension and terminal knob elongated or button-shaped 2

Terminal knob as long as the spermatheca, this latter with apical extension .... *Da.* (*Cor.*) *vesicifera*

Da. (*Cor.*) *deleoni*

Da. (*Cor.*) *zeledoni*

Terminal knob button-shaped, spermatheca without apical extension 3

3(2) Individual spermathecal ducts practically absent; spermathecae communicating directly with the common spermathecal duct .... *Da.* (*Cor.*) *steatopyga*

Da. (*Cor.*) *beltrani*

Da. (*Cor.*) *disneyi*

Individual spermathecal ducts perceptible 4

4(3) Individual spermathecal ducts as long as the common spermathecal duct .... *Da.* (*Cor.*) *viriosa*

Individual spermathecal ducts shorter than the common spermathecal duct (Fig. 75) .... *Da.* (*Cor.*) *vespertilionis*

Incertae sedis

Dampfomyia caminoi

Both sexes. Clypeus longer than the eyes and a little smaller than half the head's length. **Male**: gonocoxite with a longitudinal sclerotized band at its base and a tuft consisting of ca. ten long setae whose length is ca. twice the width of the gonocoxite; paramere simple with a pre-apical protuberance on its ventral margin; the specimen described lacks the gonostyle. **Female**: common spermathecal duct very short ca. 0.25 of the length of the individual spermathecal ducts and pratically of the same width as the genital chamber; the individual spermathecal ducts, at their extremities, are a little narrower than the common duct and in their middle as wide as the common duct, their length being equivalent to ca. twice the height of the stem of the genital furca; spermatheca bulbous with striation, width and length similar to the common spermathecal duct, narrow terminal knob, longer than the spermatheca and inserted in a funnel-shaped depression.

Expapillata

MALES

1 Gonocoxite: basal tuft consisting of thin straight setae, a little longer than the width of the gonocoxite; paramere tapers from the base to the apex .... *Ex. cerradincola*

Gonocoxite: tuft consisting of setae clearly longer than the width of the gonocoxite and apically sinuous; paramere rectangular in the basal 2/3 tapering abruptly in the apical third .... *Ex. firmatoi*

FEMALES

4th palpal segment bearing several Newstead's sensilla; spermatheca tubular with a small apical ring, in which the terminal knob is inserted .... *Ex. cerradincola*

4th palpal segment without the Newstead's sensilla; spermatheca spherical and with the terminal knob undetached from the spermatheca wall (Fig. 78) .... *Ex. firmatoi*

Pressatia

MALES

1 Paramere bearing a protuberance on the ventral margin and a digitiform process on the dorsal margin (Fig. 168) so that the former's width in this area is greater than that of the epandrial lobes; basal tuft of the gonocoxite with spiniform and other flattened setae (Fig. 148).. 2

Paramere with the apex narrower than the width of the epandrial lobes; the basal tuft of the gonocoxite having only flattened setae .... *Pr. dysponeta*

2(1) Paramere with ca. ten flattened setae inserted in the middle of its ventral margin .... *Pr. camposi*

Paramere with only spiniform setae on its ventral margin.. 3

3(2) Gonocoxite with basal, median and apical tufts, the median one with numerous setae.. *Pr. equatorialis*

Median tuft of gonocoxite when present, having few setae... 4

4(3) Basal tuft of gonocoxite of 4–5 very fine, short setae and 6–7 flattened, long setae.. *Pr. triacantha*

Tuft of gonocoxite of at least eight fine, though highly visible, setae, the flattened setae being present in variable number.. 5

5(4) Dorsal margin of paramere concave from its middle, so that the apical half tapers towards the apex, where it widens again at the expanse of a dorsal and ventral protuberance; there are a few setae on the apical half .... *Pr. calcarata*

Dorsal margin of paramere convex or straight, the apical half being hairy 6

6(5) Basal tuft of gonocoxite with 6–7 flattened setae and several fine setae, so short that their apices do not reach those of the flattened setae 7

Basal tuft of gonocoxite with ca. four flattened and several fine setae, the apices of many of them going beyond those of the flattened setae ....*Pr. choti*

7(6) Basal tuft of gonocoxite with ca. ten fine, short and six flattened, long setae; paramere with the dorsal margin of the apical half bearing setae whose length is equivalent to the width of the paramere .... *Pr. duncanae*

Basal tuft of gonocoxite with more than 20 short, fine setae and seven flattened, long setae; the dorsal margin of the apical half of the paramere bearing very short spiniform setae...*Pr. trispinosa*

FEMALES: indistinguishable.

Trichopygomyia

MALES

1 Parameral sheath with apical lobe (Fig.178)... 2

Parameral sheath without apical lobe (Fig. 169)... 5

2(1) Lobe of the parameral sheath curved with tapered apex, its base almost forming a right angle with the parameral sheath; the apex of the parameral sheath when observed in dorsal-ventral position looking like a ship's anchor ....*Ty. depaquiti*

Apical lobe of the parameral sheath with rounded apex... 3

3(2) Apical lobe of the parameral sheath large, its apex turned towards the base of the genitalia and going beyond the middle of the parameral sheath.................................. 4

Parameral sheath's apical lobe rudimentary, its height equivalent to the width of the sheath's apex .... *Ty. conviti*

4(3) Ventral branch of paramere with discreet lobe on the apical third of the dorsal margin; parameral sheath's lobe having an apex narrower than its base.......................*Ty. pinna*

Ventral branch of paramere with straight dorsal margin; parameral sheath's lobe with rounded apex, ca. 3 times wider than its base (Fig. 178)..................*Ty. dasypodogeton*

5(1)	Parameral sheath conical (Fig. 169)	6
	Parameral sheath rectangular, only the apex of the ventral margin being tapered; paramere three-branched	*Ty. martinezi*
6(5)	Paramere with three branches, the central one may be much smaller than the other two	7
	Paramere with two branches	10
7(6)	Central branch of paramere as long as the ventral one, presenting a projection, covered by setae, turned towards the base of the genitalia	*Ty. triramula*
	Central branch of paramere rudimentary or much smaller than the ventral one, with or without setae	8
8(7)	Central branch of paramere with setae on its apical region, the dorsal branch being T-shaped, with the transversal part hairy	*Ty. turelli*
	Central branch of paramere without setae; dorsal branch not T-shaped	9
9(8)	Dorsal branch of paramere with tuft of semifoliaceous setae with straight apices; central arm with tapered apex	*Ty. elegans*
	Dorsal branch of paramere with tuft of fine setae, whose apices curve towards the apex of the paramere, the central branch having a rhomboid apex	*Ty. gantieri*
10(6)	Dorsal branch of paramere with width equivalent to ca. one-third of the ventral one	*Ty. trichopyga*
	Dorsal branch of paramere with width equivalent to or greater than that of the ventral one	11
11(10)	Paramere with a free area between the two branches resembling the shape of the head of a wrench	*Ty. rondoniensis*
	Paramere with a free area between the two branches but of another shape	12
12(11)	Apex of the dorsal branch of paramere reaching the same level as or going beyond the ventral one	13
	Apex of the dorsal branch of paramere does not reach the ventral one	*Ty. witoto*
13(12)	Dorsal branch of paramere much longer than the ventral one, both with a hairy and rounded protuberance	*Ty. ferroae*
	Dorsal branch of paramere as long as the ventral one	14
14(13)	Dorsal branch of paramere with apex wider than its base	*Ty. wagleyi*
	Dorsal branch of paramere with apex narrower than its base	*Ty. longispina* *Ty. ratclifffei**

* In the ten specimens observed from Utinga, Belém, Pará (municipality of the type-locality of *Ty. longispina*), the dorsal branch of the paramere is of varied aspect, including that described for the holotype of *Ty. ratcliffei*. On the other hand, none of them is similar to the drawing of *Ty. longispina* given by the authors of *Ty. ratcliffei* to illustrate the difference between them.

FEMALES: indistinguishable

Evandromyia
(*Aldamyia*)

MALES

1	Paramere with a markedly sclerotized ventral protuberance; aedeagal ducts with bifurcated apex (Fig. 105)	2
	Paramere simple	3
2(1)	Apex of the aedeagal ducts with the two branches of practically equal lengths	*Ev.* (*Ald.*) *carmelinoi*
	Apex of the aedeagal ducts with one of the branches longer than the other (Fig. 105)	*Ev.* (*Ald.*) *lenti*

3(1) Paramere with a hairy projection at the base of the dorsal margin 4
Paramere without the hairy projection at the base of the dorsal margin 5

4(3) Paramere with a marked protuberance in the region of the implantation of the setae on its ventral margin; tuft of gonocoxite with 4–5 short setae and 11–13 longer ones, but shorter than the width of the gonocoxite; apex of the aedeagal ducts tapered like arrows .... *Ev.* (*Ald.*) *orcyi*
Paramere with or without a slight protuberance in the region of the implantation of the setae on its ventral margin; tuft of gonocoxite with four short setae and 3–6 longer ones, but shorter than the width of the gonocoxite; aedeagal ducts with truncated apex .... *Ev.* (*Ald.*) *evandroi*

5(3) Gonostyle with the lower external spine implanted more basally than the internal one .... *Ev.* (*Ald.*) *bacula*
Gonostyle with the lower external spine implanted more apically than the internal one 6

6(5) Paramere with a protuberance in the pre-apical region of the ventral margin 7
Paramere without any protuberance in the pre-apical region of the ventral margin 8

7(6) Parameral sheath bilobed; apices of ejaculatory ducts expanded, cup-shaped, each with a tooth-shaped projection in its concave tip .... *Ev.* (*Ald.*) *williamsi*
Parameral sheath simple, apices of ejaculatory ducts taper abruptly after the distension (Fig. 103) .... *Ev.* (*Ald.*) *termitophila*

8(6) Parameral sheath simple; gonostyle with the distance between the implantation level of the internal and lower external spines practically equal to the distance between the lower and upper external spines 9
Parameral sheath bilobed, one of the lobes may be greatly reduced; gonostyle with the distance between the implantation level of the internal and lower external spines clearly greater than the distance between the lower and upper external spines 10

9(8) Apex of the paramere reaches the apical third of the epandrial lobe; apex of the parameral sheath tapered .... *Ev.* (*Ald.*) *dubitans*
Apex of the paramere reaches only the middle of the epandrial lobe; apex of the parameral sheath tapers abruptly like the nipple of a feeding-bottle .... *Ev.* (*Ald.*) *aldafalcaoae*

10(8) Parameral sheath with a very short dorsal branch like a callus; apex of the aedeagal ducts like ceiling fan blades, each with several small tooth-like projections in its concave tip .... *Ev.* (*Ald.*) *hashiguchii*
Parameral sheath with the dorsal branch not callus-shaped; the apex of the aedeagal ducts wine glass or spoon-shaped, each with (or without) one tooth-like projection in its concave tip 11

11(10) Gonocoxite with a basal tuft of 10–12 setae arranged in two parallel rows ....
.... *Ev.* (*Ald.*) *apurinan*
Gonocoxite with a basal tuft of more than 18 setae arranged in two parallel rows 12

12(11) Parameral sheath with the ventral branch longer then the dorsal; apex of the aedeagal ducts wine glass-shaped, its width being twice or more that of the preceding part of the aedeagal ducts (Fig. 104) .... *Ev.* (*Ald.*) *walkeri*
Parameral sheath with the ventral branch shorter than the dorsal; apex of the aedeagal ducts spoon-shaped and narrower than twice the width of the preceding part of the aedeagal ducts 13

13(12) Parameral sheath with a rudimentary ventral branch; apex of the aedeagal ducts spoon-shaped, each with a tooth-like projection in its middle .... *Ev.* (*Ald.*) *sericea*
Parameral sheath with the ventral branch as long as half the dorsal; apex of the aedeagal ducts without any tooth-like projection 14

14(13) Setae of the tuft of the gonocoxite of practically the same length ....
.... *Ev.* (*Ald.*) *andersoni*
Setae of the tuft of the gonocoxite of varying length .... *Ev.* (*Ald.*) sp. de Baduel

FEMALES

1 Spermathecae ringed .... *Ev.* (*Ald.*) *termitophila*

Spermathecae not ringed 2

2(1) Common spermathecal duct and basal part of the individual spermathecal ducts membranous and highly transparent, it being difficult to distinguish between them .... *Ev.* (*Ald.*) *aldafdalcoae*

Common and individual spermathecal ducts sclerotized, with a clear transition between them 3

3(2) Common and individual spermathecal ducts of similar lengths 4

Common spermathecal duct length $\leq$ 0.50 of that of the individual spermathecal ducts 5

4(3) Individual spermathecal ducts of uniform width along their length; spermatheca wider than its length .... *Ev.* (*Ald.*) *dubitans*

Individual spermathecal ducts widen at the junction with the spermatheca; this latter as long as its width (Fig. 83) .... *Ev.* (*Ald.*) *walkeri*

5(3) External margin of the common spermathecal duct with markedly sclerotized band .. 6

External margin of the common spermathecal duct different from the above 7

6(5) Individual spermathecal ducts ca. four times as long as the common spermathecal duct .... *Ev.* (*Ald.*) *carmelinoi*

Individual spermathecal ducts ca. 2.5 times as long as the common spermathecal duct .... *Ev.* (*Ald.*) *lenti*

7(5) Spermatheca's width at least four times that of the narrowest region of the individual spermathecal duct 8

Spermatheca's width up to two times that of the narrowest region of the individual spermathecal duct 11

8(7) Junction between individual spermathecal duct and spermatheca sclerotized .... *Ev.* (*Ald.*) *williamsi*

Junction between individual spermathecal duct and spermatheca not sclerotized 9

9(8) Individual spermathecal ducts ca. 2.5 times as long as the common spermathecal duct .... *Ev.* (*Ald.*) *apurinan*

Individual spermathecal ducts at least four times as long as the common spermathecal duct 10

10(9) Common spermathecal duct wide, its insertion occupying pratically all the genital chamber, its basal width being twice that of the spermatheca .... *Ev.* (*Ald.*) sp. de Baduel

Common spermathecal duct narrow, its basal width is clearly smaller than that of the genital chamber as also that of the spermatheca .... *Ev.* (*Ald.*) *bacula*

11(7) Bases of the individual spermathecal ducts narrower than the width of the spermatheca 12

Bases of the individual spermathecal ducts wider than the width of the spermatheca.... 13

12(11) Individual spermathecal ducts at least four times longer than the common spermathecal duct .... *Ev.* (*Ald.*) *evandroi*

Individual spermathecal ducts up to three times longer than the common spermathecal duct .... *Ev.* (*Ald.*) *orcyi*

13(11) Base of the common spermathecal duct occupying half the genital chamber's width .... *Ev.* (*Ald.*) *sericea*

Base of the common spermathecal duct wide, occupying practically all the genital chamber's width .... *Ev.* (*Ald.*) *andersoni*

(*Evandromyia*)
Infraspinosa Series

MALES

1 Paramere simple.. 2
Paramere bearing two or three lobes in its apical region 4
2(1) Apex of the epandrial lobe with two spatulate setae............................ *E. (Eva.) sipani*
Apex of the epandrial lobe with three or four spatulate setae..................................... 3
3(2) Apex of the epandrial lobe with three spatulate setae; paramere tapering towards apex *Ev. (Eva.) bourrouli*
Apex of the epandrial lobe with four spatulate setae; paramere with abrupt narrowing in the apical quarter of the ventral margin ....*Ev. (Eva.) pinottii*
4(1) Paramere trilobed in its apical region .... *Ev. (Eva.) begonae*
Paramere bilobed in its apical region... 5
5(4) Gonocoxite with a tuft of ca. four setae; apex of the lateral lobe with two spatulate setae..*E. (Eva.). brachyphalla*
Gonocoxite with a tuft of more than ten setae; apex of the lateral lobe with, generally, three spatulate setae 6
6(5) Paramere with the ventral lobe more intensely sclerotized... 7
Paramere with the two lobes equally sclerotized (Fig. 170)....................................... 8
7(6) Aedeagal ducts heavily striated and sclerotized as far as their pre-apical region; from this area to their apex they taper but are not sclerotized or striated .... *Ev. (Eva.) inpai*
Aedeagal ducts practically uniform throughout their length, but with slight apical expansion, slightly sclerotized and with fine striation in their apical half ....*Ev. (Eva.) tarapacaensis*
8(6) Paramere with the lower branch without protuberance on its ventral margin 9
Paramere with the lower branch with a protuberance on its ventral margin .... *Ev. (Eva.) georgii*
9(8) Paramere with a slight curve just before the ramification so that its width in this area is about ½ that of its base; length of the lower branch ca. 0.75 that of the upper branch ....*Ev. (Eva.) ledezmaae*
Paramere with an accentuated curve on its dorsal magin before the ramification, so that its width in this area is less than one-third of the width of its base; length of the lower branch about 0.65 of the length of the upper branch (Fig. 170)..*Ev. (Eva.) infraspinosa*

FEMALES

1 Mesonotum and thoracic pleura brown; spermathecae, individual and common spermathecal ducts completely striated; spermathecae with a small apical ring ....*Ev. (Eva.) pinottii*
E. (Eva.) bourrouli
Mesonotum light brown or pale; thoracic pleura straw; at least the individual spermathecal duct not striated, spermathecae without apical ring.............................. 2
2(1) Spermathecae ca. twice as long as their width, with clear striation in their central part and with the terminal knob inserted in a funnel-shaped invagination; individual ducts practically absent ....*E. (Eva.) inpai*
Spermathecae at least 3 × as long as their width and without the set of other characteristics described above... 3

3(2) Spermathecae completely ringed .... *Ev (Eva.) brachyphalla*
Spermathecae completely smooth or partially ringed or striated.......... 4
4(3) Flagellomere I longer than the labrum-epipharynx; individual and common ducts of the spermathecae smooth-walled .... *Ev. (Eva.) tarapacaensis*
Flagellomere I shorter than the labrum-epipharynx; spermathecae totally or partially striated.......... 5
5(4) Spermathecae striated in their apical 2/3; common duct smooth (Fig. 82) .... *Ev. (Eva.) infraspinosa*
Spermathecae striated in their central region and apical region smooth; common duct striated.......... 6
6(5) Labrum-epipharynx ca. 290 μm; 5th palpal segment 2.5 × longer than the 4th .... *Ev. (Eva.) begonae*
Labrum-epipharynx ca. 240 μm; 5th palpal segment 3.0 × longer than the 4th .... *Ev. (Eva.) georgii*
The female of *Ev. (Eva.) ledezmaae* has not yet been described.

Saulensis Series

MALES

1 Paramere with the two protuberances of the ventral margin located very close to its middle; dorsal margin bearing a dense covering of hairs in its apical area .... *Ev. (Eva.) wilsoni*
Paramere with one of the protuberances of the ventral margin located in the pre-apical area; dorsal margin with sparse covering of setae in its apical region .... *Ev. (Eva.) saulensis*

FEMALES

Spermathecae with the widest smooth apical part ca. 1/2 of that with bubble-like evaginations.......... *Ev. (Eva.) wilsoni*
Spermatheca with the widest smooth apical part ca. 1/3 of that with bubble-like evaginations (Fig. 80).......... *Ev. (Eva.) saulensis*

Rupicola Series

MALES

1 Gonostyle with the inner spine a little thinner than the external ones; paramere with the apex of the apical appendix turned towards the base of the genitalia; this appendix is covered with setae shorter than itself.......... *Ev. (Eva.) correalimai*
Gonostyle with the inner spine seta-like; paramere with the apex of the apical appendix turned towards the gonocoxites; this appendix is covered with setae longer than itself.......... 2
2(1) Parameral sheath with tubercle on the basal region of dorsal margin.......... *Ev. (Eva.) tylophala*
Parameral sheath without the above tubercle.......... 3
3(2) Aedeagal ducts ca. 370 μm long .... *Ev. (Eva.) rupicola*
Aedeagal ducts ca. 500 μm long .... *Ev. (Eva.) grimaldii*

FEMALES

1 Clypeus clearly longer than the eyes.......... 2
Clypeus as long as or shorter than the eyes 3
2(1) Spermatheca as wide as the terminal knob .... *Ev. (Eva.) gaucha*
Spermatheca clearly wider than the terminal knob (Fig. 79).......... *Ev. (Eva.) correalimai*

3(1) Spermatheca with the width of its apical half ca. one-third that of the basal half .... *Ev. (Eva.) grimaldii*

Spermatheca of uniform width throughout or slightly wider in its median part........... 4

4(3) Labrum-epipharynx ca. 260 μm; spermatheca completely ringed, its width less than twice that of the individual spermathecal duct................................ *Ev. (Eva.) rupicola*

Labrum-epipharynx >300 μm; spermatheca ringed in its basal half and superficially striated in the apical half, its width greater than that of the individual spermathecal duct .... *Ev. (Eva.) tylophalla*

(*Barrettomyia*)

Monstruosa Series

MALES

1 Parameral sheath with the ventral branch shorter than the dorsal one .... *Ev. (Bar.) teratodes*

Parameral sheath with the ventral branch at least twice as long as the dorsal one.. *Ev. (Bar.) monstruosa*

FEMALES

1 Individual spermathecal duct clearly narrower than the spermatheca, both of similar length (Fig. 84).. *Ev. (Bar.) monstruosa*

Individual spermathecal duct slightly narrower than the spermatheca and ca. 1/3 as long as the spermatheca .... *Ev. (Bar.) teratodes*

Cortelezzii Series

MALES

1 Clypeus longer than the eyes 2

Clypeus as long as or shorter than the eyes 3

2(1) Gonocoxite with basal tuft of four small setae implanted in its upper part and six long setae in its lower part .... *Ev. (Bar.) corumbaensis*

Gonocoxite with a tuft two small setae implanted in its upper part and four long setae in its lower part .... *Ev. (Bar.) spelunca*

3(2) Paramere thick, with straight dorsal margin as far as the apical region, where it resembles a hood; gonocoxite with the basal tuft of four long setae .... *Ev. (Bar.) sallesi*

Paramere with a concavity on its dorsal margin and without the hood-shaped apical structure; gonocoxite with the basal tuft of five long setae............ *Ev. (Bar.) cortelezzii*

FEMALES

1 Clypeus ca. 0.66 × as long as the head, excluding the clypeus; individual sperm ducts ca. twice as long as the height of the stem of the genital furca .... *Ev. (Bar.) corumbaensis*

Ev. (Bar.) spelunca

Clypeus ca. 0.60 × or less the length of the head, excluding the clypeus; individual sperm ducts a little longer than the height of the stem of the genital furca .... *Ev. (Bar.) cortelezzii*

Ev. (Bar.) sallesi

Tupynambai Series

MALES

1 Gonocoxite with the basal tuft of 2–3 small setae on its upper part and 3 long setae on the lower part .... *Ev. (Bar.) tupynambai*

Gonocoxite with the basal tuft of five or more long setae 2

2(1) Gonostyle with five or six thick spines, apex of the aedeagal ducts simple .... *Ev. (Bar.) bahiensis*

Gonostyle with four thick spines; apex of the aedeagal ducts enlarged 3

3(2) Gonocoxite with a basal tuft of more than ten long setae................*Ev.* (*Bar.*) *callipyga*
Gonocoxite with a basal tuft of five or six setae.. 4
4(3) Paramere with one spine implanted in a small tubercle in the basal region of its dorsal margin; gonocoxite with a tuft of five setae................................ *Ev.* (*Bar.*) *costalimai*
Paramere with one spine implanted directly in the basal region of the dorsal margin; gonocoxite with a tuft of six setae .... *Ev.* (*Bar.*) *petropolitana*

FEMALES: indistinguishable

PSYCHODOPYGINA

Psathyromyia

(*Forattiniella*)

MALES

1 Aedeagal ducts ≥7 times the length of the sperm pump... 2
Aedeagal ducts ≤6 times the length of the sperm pump... 3
2(1) Aedeagal ducts ≥10 times the length of the sperm pump.............*Pa.* (*For.*) *brasiliensis*
Aedeagal ducts ca. seven times the length of the sperm pump..... *Pa.* (*For.*) *abunaensis*
3(1) Gonostyle with the internal spine implanted at the beginning of its apical quarter ....*Pa.* (*For.*) *carpenteri*
Gonostyle with the internal spine implanted near its middle.. 4
4(3) Aedeagal ducts sinuous in their pre-apical region (Fig. 113) 5
Aedeagal ducts straight 8
5(4) Paramere with semifoliaceous setae implanted just beyond the middle of the ventral margin .... *Pa.* (*For.*) *antezanai*
Paramere with spiniform setae implanted just beyond the middle of the ventral margin 6
6(5) Parameral sheath rectangular almost to the apex, which tapers abruptly at the expanse of the dorsal margin, finishing in a fine point .... *Pa.* (*For.*) *aragaoi*
Parameral sheath conical... 7
7(6) Gonocoxite bearing in its median region a dense cluster of 60–90 setae .... *Pa.* (*For.*) *naftalekatzi*
Gonocoxite without dense cluster of setae..................................... *Pa.* (*For.*) *coutinhoi*
8(4) Aedeagal ducts without lozenge-shaped dilation in their apical or pre-apical region.....9
Aedeagal ducts with lozenge-shaped dilation in their apical or pre-apical region (Fig. 106)... 13
9(8) Gonostyle with the lower external spine implanted at a more basal level than the internal one. Gonocoxite with a median cluster of ca. 35 sparsely distributed setae... *Pa.* (*For.*) *texana*
Gonostyle with the lower external spine implanted at the same level as or more apical than the internal one. Gonocoxite without cluster of setae or at the most ca. ten short, sparse setae.. 10
10(9) Parameral sheath long, its apex reaching the apical quarter of the paramere.................. .... *Pa.* (*For.*) *pascalei*
Parameral sheath short, its apex extending slightly beyond the middle of the paramere... 11
11(10) Aedeagal ducts/ sperm pump ratio ca. 1.5:1.0 and width of the apex of aedeagal duct ca. 8 μm.. *Pa.* (*For.*) *campograndensis*
Aedeagal ducts/ sperm pump ratio ≥ 1.9:1.0 and width of the apex of aedeagal duct ≤4.7 μm.. 12

12(11) Aedeagal ducts/ sperm pump ratio ca. 1.9:1.0 and width of the apex of aedeagal ducts ≥3.6 µm.......... *Pa.* (*For.*) *lutziana*

Aedeagal ducts/ sperm pump ratio ca. 2.3:1.0 and width of the apex of aedeagal duct ca. 2.6 µm.......... *Pa.* (*For.*) *elizabethdorvalae*

13(8) Aedeagal ducts with a lozenge-shaped dilation preceding its apex, then tapering; the length of the posterior filiform part is ca. twice that of the losange-shaped part.......... *Pa.* (*For.*) *inflata*

Aedeagal ducts with apical lozenge or spear-shaped dilation (Fig. 106).......... 14

14(13) Epandrial lobe much longer than the gonocoxite, its apex reaching the level of the implantation of the internal spine of the gonostyle. Paramere tapers from base to apex.......... 17

Epandrial lobe as long as or slightly longer than the gonocoxite, its apex does not reach the level of the implantation of the internal spine of the gonostyle. Paramere with a narrowing at the beginning of the apical third of the ventral margin, where the setae are implanted, but then it widens slightly and narrows again towards the apex.......... 15

15(14) Apex of the aedeagal ducts with lozenge-shaped dilation (Fig. 106).......... 16

Apex of the aedeagal ducts spear-shaped with a long fine tip. Aedeagal ducts ca. five times longer than the sperm pump.......... *Pa.* (*For.*) *castilloi*

16(15) Aedeagal ducts ca. 3 × longer than the sperm pump; width of apical dilation of the aedeagal ducts ca. 3 × that of the anterior filiform part.......... *Pa.* (*For.*) *runoides*

Aedeagal ducts 3.7–5.2 × longer than the sperm pump; width of apical dilation of the aedeagal ducts ca. twice that of the anterior filiform part... *Pa.* (*For.*) *pradobarrientosi*

17(14) Gonocoxite with ca. 20 setae in its apical region. Epandrial lobe >550 µm.......... *Pa.* (*For.*) *barrettoi majuscula*

Gonocoxite with <10 setae in its apical region. Epandrial lobe <500 µm.......... *Pa.* (*For.*) *barrettoi barrettoi*

FEMALES

1 Cibarium with two pairs of posterior (horizontal) teeth (Fig. 20-H).......... 2

Cibarium with more numerous posterior (horizontal) teeth (Fig. 20-I).......... 5

2(1) Spermathecae tubular, individual spermathecal ducts ca. 20 times longer than the spermathecae.......... *Pa.* (*For.*) *brasiliensis*

Spermathecae globose, short individual spermathecal ducts whose apices do not reach the apex of the stem of the genital furca.......... 3

3(2) Width of the spermathecae ca. four × that of the individual spermathecal ducts at the beginning of their sclerotized part.......... *Pa.* (*For.*) *campograndensis*

Width of the spermathecae ca. 2.5 × or less that of the individual spermathecal ducts at the beginning of their sclerotized part.......... 4

4(3) Walls of the individual spermathecal ducts sclerotized almost throughout the ducts' lengths.......... *Pa.* (*For.*) *elizabethdorvalae*

Walls of the individual spermathecal ducts only sclerotized at the ducts' apical third.......... *Pa.* (*For.*) *lutziana*

5(1) Spermathecae tubular.......... 6

Spermathecae vesiculous (Fig. 50).......... 7

6(5) Cibarium with the anterior (vertical) teeth arranged in several irregular rows; individual spermathecal duct tapering towards the spermatheca, so that at the junction with this latter its width is ca. ½ of that of its base.......... *Pa.* (*For.*) *abunaensis*

Cibarium with the anterior (vertical) teeth arranged in a single transversal row; individual spermathecal ducts practically the same width throughout.......... *Pa.* (*For.*) *inflata*
Pa. (*For.*) *runoides*

7(5) Spermathecae globose 8
Spermathecae ovoid 14
8(7) Terminal knob detached from the wall of the spermatheca .... *Pa.* (*For.*) *pascalei*
Terminal knob sessile (Fig. 50) 9
9(8) Individual spermathecal ducts striated .... *Pa.* (*For.*) *carpenteri*
Individual spermathecal ducts smooth 10
10(9) Stem of the genital furca with a great apical expansion .... *Pa.* (*For.*) *antezanai*
Stem of the genital furca with slight pre-apical expansion 11
11(10) 4th palpal segment without Newstead's sensilla .... *Pa.* (*For.*) *texana*
4th palpal segment with Newstead's sensilla 12
12(11) Presence of a pair of ascoids on the last three flagellomeres ...... *Pa.* (*For.*) *naftalekatzi*
Absence of ascoids on the last three flagellomeres 13
13(12) Flagellomere I (FI) ca. 320 µm; labrum-epipharynx (LE) ca. 200µm; FI/LE ratio ca. 1.60:1.00 .... *Pa.* (*For.*) *aragaoi*
Flagellomere I (FI) ca. 210 µm; labrum-epipharynx (LE) ca. µm; 180; FI/LE ratio ca. 1.15:1.00 .... *Pa.* (*For.*) *coutinhoi*
14(7) Cibarium with 5–6 pairs of posterior (horizontal) teeth; individual spermathecal ducts not dilated at the junction of the spermathecae .... *Pa.* (*For.*) *barrettoi barrettoi*
Cibarium with 7–8 pairs of posterior (horizontal) teeth; individual spermathecal ducts dilated at the junction with the spermathecae .... *Pa.* (*For.*) *barrettoi majuscula*

(*Xiphopsathyromyia*)

MALES

1 Paramere without branch on the dorsal margin, median region of the gonocoxite with some sclerotized setae at their bases and flexible, dilated and curved apices .... *Pa.* (*Xip.*) *aclydifera*
Paramere with branch on the dorsal margin; median region of the gonocoxite without modified setae 2
2(1) Paramere with the branch on the dorsal margin covered with setae 3
Paramere with the dorsal branch without setae .... *Pa.* (*Xip.*) *ruparupa*
3(2) Paramere: branch on the dorsal margin thin, with narrow apex presenting a small hairy expansion at its base .... *Pa.* (*Xip.*) *hermanlenti*
Paramere: branch on the dorsal margin with an apex more dilated than its base, with no hairy expansion .... *Pa.* (*Xip.*) *dreisbachi*

FEMALES

1 Labrum-epipharynx ca. 350 µm, > the eye's length .... *Pa.* (*Xip.*) *aclydifera*
Labrum-epipharynx ca. 250 µm, ≤ the eye's length .... *Pa.* (*Xip.*) *dreisbachi*
Pa. (*Xip.*) *hermanlenti*
Pa. (*Xip.*) *ruparupa*

(*Psathyromyia*)

Lanei Series

MALES

1 Paramere finger-shaped in its distal third, dorsal margin bearing ca. 20 semifoliaceous setae, their apices turned in a posterior direction and, more internally, densely covered by ca. 25 spiniform setae, short and turned towards the ventral margin .... *Pa.* (*Psa.*) *pelloni*
Paramere tending to rectangular or triangular shape in its distal third, dorsal margin bearing thin sparse setae and, more internally 15 or fewer spiniform short setae 2

2(1) Paramere with shape tending to rectangular, but with the apex of the dorsal margin rounded, where ca. 10 setae of slightly curved apex are inserted, these being turned towards the base of the genitalia, with the distal setae longer than the median ones; on the dorsal margin a small lobe with straight very short setae; the middle area being covered by a few sparse spiniform setae... *Pa.* (*Psa.*) *digitata*

Paramere with shape tending to triangular, the setae covering the dorsal margin having straight apices; on the ventral margin a small lobe, with setae of curved apex turned towards the base of the genitalia and pointing to the middle of the paramere; the central area being covered by 15 or fewer spiniform setae (Fig. 172)................*Pa.* (*Psa.*) *lanei*

FEMALES of *Pa. lanei* and *Pa. pelloni* indistinguishable; *Pa. digitata* has not yet been described.

Shannoni Series

MALES

1 Paramere with a hairy appendage on the basal region of the dorsal margin................. 2

Paramere different from the above.. 3

2(1) Gonostyle with the external lower spine and the inner spine located on the basal third; appendage of the paramere bearing ca. 20 setae.............................*Pa.* (*Psa.*) *campbelli*

Gonostyle with the external lower and the inner spines located a little before its middle; appendage of paramere bearing ca. ten setae.................................*Pa.* (*Psa.*) *dasymera*

3(1) Posterior prolongation of ascoids long, on flagellomere II, their apices coming close to or going beyond the base of this segment... 4

Posterior prolongation of ascoids short, on flagellomere II, their apices being closer to the point of their insertion.. 18

4(3) Paramere with protuberance on the pre-apical region of the ventral margin.................... .... *Pa.* (*Psa.*) *soccula*

Paramere without protuberance on the pre-apical region of the ventral margin........... 5

5(4) Paramere with sinuous setae on the apical region of its dorsal margin .... *Pa.* (*Psa.*) *undulata*

Paramere without sinuous setae on the apical region of its dorsal margin 6

6(5) Apex of the aedeagal duct dilated as a spoon (Fig. 108-B) .... *Pa.* (*Psa.*) *barretti*

Apex of the aedeagal not dilated.. 7

7(6) Gonostyle with the external lower spine and the internal spine located on its basal third.. 8

Gonostyle with the external lower spine and the internal spine located on or beyond its middle 9

8(7) Paramere: region covered with setae ca. 1.5 × longer than wide, the apical area of the ventral margin having ca. 6 sclerotized striations directed towards the middle of the paramere..*Pa.* (*Psa.*) *cratifer*

Paramere: region covered with setae ca. 3.0 × longer than wide, the apical area of the ventral margin having no sclerotized striations .... *Pa.* (*Psa.*) *lerayi*

9(7) Paramere with the apical half digitiform (Fig. 166).. 10

Paramere with the apical half of almost triangular shape (Fig. 172).......................... 16

10(9) Paramere with a set of spiniform, long setae situated between the dorsal and ventral margins, at the beginning of the digitiform part, and another set of setae on the apical region of the ventral margin... *Pa.* (*Psa.*) *volcanensis*

Paramere without the above sets of setae.. 11

11(10) Paramere: basal region ca. 5 × wider than the digitiform region.................................. .... *Pa.* (*Psa.*) *souzacastroi*

Paramere: basal region ≤3 × wider than the digitiform region.................................. 12

12(11) Pronotum and paratergite clearer than the mesonotum brown 13
Pronotum, paratergite and mesonotum brown 14

13(12) Epandrial lobes almost as long as the gonocoxites. Paramere with the setae of the dorsal margin extending from its apex to the same level as or beyond the apical setae of the set on the ventral margin .... *Pa.* (*Psa.*) *shannoni*
Epandrial lobes shorter than the gonocoxites. Paramere with the setae of the dorsal margin extending from its apex to its middle, not reaching the apical level of the setae of the set on the ventral margin .... *Pa.* (*Psa.*) *baratai*

14(12) Paramere with the setae of the dorsal margin restricted to its apical third .... *Pa.* (*Psa.*) *abonnenci*
Paramere with the setae of the dorsal margin covering the apical half 15

15(14) Thorax with whitish pleura and coxae, clearly contrasting with the brown mesonotum and paratergites .... *Pa. ribeirensis*
Thorax with light brown upper anepisternum and posterior coxae .... *Pa. (Psa.) bigeniculata*

16(9) Thorax with whitish pleura and coxae, clearly contrasting with the brown mesonotum and paratergites .... *Pa.* (*Psa.*) *limai*
Thorax with the pleura light brown, anepisternum and katepisternum darker brown 17

17(16) Paramere with spiniform setae uniformly distributed throughout the apical half .... *Pa.* (*Psa.*) *scaffi*
Paramere with semifoliaceous setae in the apical area of the dorsal margin having a glabrous area between the most apical and pre-apical sets, spiniform setae in a more internal, median position .... *Pa.* (*Psa.*) *dendrophyla*

18(3) Posterior spur of ascoids with pointed apices; paramere digitiform with short setae located on the distal third of the dorsal margin; gonostyle with the external lower spine located more apically than the internal spine .... *Pa.* (*Psa.*) *guatemalensis*
Posterior spur of ascoids of rhomboid apex; paramere triangular with straight setae on the dorsal margin; gonostyle with the external lower spine located at the same level as the internal spine .... *Pa.* (*Psa.*) *punctigeniculata*

FEMALES

1 Common spermathecal duct does not reach the middle of the stem of the genital furca; individual spermathecal ducts striated 2
Common spermathecal duct reaches or goes beyond the middle of the stem of the genital furca; individual spermathecal ducts smooth 3

2(1) Cibarium with 4–5 pairs of posterior (horizontal) teeth and the anterior (vertical) teeth arranged in 6–7 rows in the more central part .... *Pa.* (*Psa.*) *dasymera*
Cibarium with three pairs of posterior (horizontal) teeth and the anterior (vertical) teeth arranged in three transversal rows, in the more central part .... *Pa.* (*Psa.*) *campbelli*

3(1) Spermathecae ringed 4
Spermathecae banana-shaped (Fig. 51) 6

4(3) Cibarium with two pairs of posterior (horizontal) teeth 5
Cibarium with three or more pairs of posterior (horizontal) teeth .... *Pa.* (*Psa.*) *undulata*
Pa. (*Psa.*) *cratifer*

5(4) Posterior spur of the ascoids rudimentary .... *Pa.* (*Psa.*) *guatemalensis*
Posterior spur of the ascoids long, its apex almost reaching the base of the 2nd flagellomere (Fig. 15) .... *Pa.* (*Psa.*) *volcanensis*

6(3) Cibarium with two pairs of posterior (horizontal) teeth 7
Cibarium with more than two pairs of posterior (horizontal) teeth .... *Pa.* (*Psa.*) *punctigeniculata*

7(6) Thorax with paratergite, anepisternum and katepisternum totally brown..................... 8

Thorax with paratergite partially brown, the region below the pre-sutural setae straw; the pleura totally straw or may present light pigmentation on the anepisternum and the katepisternum.. 9

8(7) Thorax with dark brown anepimeron.. *Pa.* (*Psa.*) *scaffi*

Thorax with straw anepimeron .... *Pa.* (*Psa.*) *dendrophyla*

9(7) Individual spermathecal ducts as wide as the spermatheca; terminal knob's length equal to the spermatheca's width .... *Pa.* (*Psa.*) *barretti*

Individual spermathecal ducts narrower than the spermatheca; terminal knob's length perceptibly shorter than the spermatheca's width... 10

10(9) Thorax with whitish pleura and coxae (Fig. 182 A,B)... 11

Thorax: upper anepisternum brown and coxae straw or light brown.......................... 12

11(10) Pronotum and paratergite brown..*Pa.* (*Pas.*) *limai*

Pa. (*Psa.*) *ribeirensis*

Pronotum straw and paratergite off-white..*Pa.* (*Psa.*) *baratai*

12(10) Thorax: pronotum and paratergite clearer than the mesonotum; anterior coxa light brown throughout its length (Fig. 182 C) ....*Pa.* (*Psa.*) *shannoni*

Thorax: pronotum, paratergite and mesonotum brown; anterior coxa straw or brown only at its apex 13

13(12) Thorax: metanotum light brown (Fig. 182 D)...............................*Pa.* (*Psa.*) *abonnenci*

Thorax: metanotum straw (Fig. 182 E)......................................*Pa.* (*Psa.*) *bigeniculata*

Incertae sedis

Psathyromyia maya

Female: ascoids with rudimentary spur; palpal formula 1.4.2.(5.3); cibarium with two pairs of posterior (horizontal) teeth. Ventro-cervical sensilla absent. Thorax dusky throughout. Common and individual spermathecal ducts striated; the length of the common spermathecal duct is ½ that of the individual spermathecal ducts, both together being as long as the height of the stem of the genital furca; spermatheca as long as the common spermathecal duct, striated, widens from the base to the apex, with a constriction in its middle; the terminal knob is longer than wide and presents a small digitiform projection. **Male**: as yet undescribed.

Viannamyia

MALES

1 Gonostyle with two external spines located on the same tubercle; paramere with curved apex turned towards the lateral lobe.. 2

Gonostyle with two external spines not located on the same tubercle; paramere with straight apex 3

2(1) Paramere with one well developed spine at its apex, 2–3 thick setae, sclerotized, with enlarged multi-brunched apices on the pre-apical region of the dorsal margin.... *Vi. furcata*

Paramere without the spine on its apex, the setae on the pre-apical region of the dorsal margin simple.. *Vi. caprina*

3(1) Paramere conical, bearing 2–3 setae of enlarged, hairy apices on the pre-apical region of the dorsal margin (Fig. 173) .... *Vi. tuberculata*

Paramere with digitiform apex; the setae on the pre-apical region of the dorsal margin simple, one apical seta with hooked apex .... *Vi. fariasi*

FEMALES

1 Spermatheca enclosed in a sclerotized chalice-shaped sheath (Fig. 49)....... *Vi. tuberculata*

Spermatheca enclosed in a sclerotized stem-shaped sheath .... *Vi. caprina*

Vi. furcata

Martinsmyia

MALES

1 Gonocoxite bearing a tuft of setae located on a tubercle in its central-basal region....... Grupo Gasparviannai 2

Gonocoxite without the above tuft of setae..................Grupo Alphabetica 3

2(1) Apex of the aedeagal ducts straight .... *Mt. cipoensis*

Apex of the aedeagal ducts lightly curved (Fig. 101) .... *Mt. gasparviannai*

3(1) Gonostyle with five, rarely six spines (Fig. 137) .... *Mt. alphabetica*

Gonostyle with four spines 4

4(3) Ratio: aedeagal ducts/sperm pump = 2.0:1.0 5

Ratio: aedeagal ducts/sperm pump ≥2.5:1.0 6

5(4) 5th palpal segment slightly longer than the 3rd; apex of the aedeagal ducts present row of small teeth .... *Mt. brisolai*

5th palpal segment clearly longer than the 3rd; apex of the aedeagal ducts without teeth .... *Mt. waltoni*

6(5) Paramere straight... 8

Paramere with the apical region curved towards the gonocoxite 7

7(6) Gonostyle with the external lower spine located in its middle, the inner spine located before this latter .... *Mt. oliveirai*

Gonostyle with the external lower spine located in its apical third, the inner spine, located beyond its middle .... *Mt. mollinedoi*

8(6) Epandrial lobe ≤ gonocoxite.. 9

Epandrial lobe > gonocoxite 10

9(8) Gonocoxite with a seta in its submedian region; flagellomere I clearly longer than the head including the clypeus.. *Mt. pisuquia*

Gonocoxite without setae; flagellomere I shorter than the head including the clypeus .... *Mt. minasensis*

10(8) Flagellomere I/labro-epipharynx ratio < 1.2: 1.0.. *Mt. reginae*

Flagellomere I/labro-epipharynx ratio = 1.5: 1.0............................. *Mt. quadrispinosa*

FEMALES

1 Common spermathecal duct longer than the individual spermathecal ducts (Fig. 58). Gasparviannai group 2

Common spermathecal duct shorter than the individual spermathecal ducts (Fig. 55). Alphabetica group 3

2(1) Cibarium with one pair of posterior (horizontal) teeth; individual spermathecal ducts smooth, the spermatheca with ca. eight rings (Fig. 58) .... *Mt. gasparviannai*

Cibarium with two pairs of posterior (horizontal) teeth; individual spermathecal ducts striated, the spermatheca with 12–13 rings... *Mt. cipoensis*

3(1) Spermatheca with 4–5 rings, the central rings of clearly smaller diameter than the basal and apical ones (Fig. 55) .... *Mt. alphabetica*

Spermatheca with six or more rings of subequal diameter 4

4(3) Cibarium with four pairs of posterior (horizontal) teeth.............................. *Mt. waltoni*

Cibarium with one or two pairs of posterior (horizontal) teeth 5

5(4) Cibarium with one pair of posterior (horizontal) teeth; flagellomere I as long as the head's length, including the clypeus .... *Mt. pisuquia*

Cibarium with two pairs of posterior (horizontal) teeth, which may or not be united at their bases; flagellomere I clearly shorter than the head's length, including the clypeus 6

6(5) Cibarium with the posterior (horizontal) teeth united at their bases, in such a way that, sometimes, only one of their apices may be observed 7
Cibarium with individualized posterior teeth 8
7(6) Individual spermathecal ducts: the widest part ca. five times the narrowest part .... *Mt. oliveirai*
Individual spermathecal ducts: the widest part 2.5–3.0 times the narrowest part .... *Mt. minasensis*
8(6) Cibarium with the sclerotized arch approximately rectangular .... *Mt. brisolai*
Cibarium with the sclerotized arch rounded 9
9(8) Common and individual spermathecal ducts smooth; individual spermathecal ducts with the basal width twice that of the apical one (at the junction with the spermathecae) .... *Mt. reginae*
Common and individual spermathecal ducts striated; individual spermathecal ducts with the basal width ca. 5 × that of the apical one (at the junction with the spermathecae) .... *Mt. mollinedoi*

Bichromomyia

MALES

1 Scutellum dark .... *Bi. inornata*
Scutellum straw 2
2(1) Ratio: aedeagal ducts/sperm pump ca. 2.0:1.0 3
Ratio: aedeagal ducts/sperm pump ≥3.0:1.. 4
3(2) Ratio between the lengths: flagellomere I/head <1.0:1.0; epandrial lobe surpasses the paramere by ca. 1/5 of the latter's length .... *Bi. flaviscutellata*
Ratio between the lengths: flagellomere I/head > than 1.0:1.0; epandrial lobe surpasses the paramere by ca. 1/3 of the latter's length .... *Bi. reducta*
4(2) 5th palpal segment equal to the 3rd; epandrial lobe as long as the gonocoxite .... *Bi. olmeca olmeca*
5th palpal segment shorter than 3rd; epandrial lobe shorter than the gonocoxite 5
5(4) Interocular distance ≤1/3 of the eyes' width .... *Bi. olmeca nociva*
Interocular distance >1/2 of the eyes' width .... *Bi. olmeca bicolor*

FEMALES

1 Common spermathecal duct wider than the median region of stem of the genital furca .... *Bi. reducta*
Common spermathecal duct narrower than the median region of the stem of the genital furca 2
2(1) Cibarium with 6–8 posterior (horizontal) teeth; one or more metepisternal setae may be present; spermatheca: apical ring narrower than the pre-apical one; terminal knob with length equivalent to the sum of the three apical rings having its apex evidently curved towards the genital furca's stem .... *Bi. flaviscutellata*
Cibarium with eight or more posterior (horizontal) teeth; metepisternal setae absent; spermatheca: apical ring as long as or longer than the pre-apical one; terminal knob as long as or shorter than the sum of the two apical rings, its apex may or not be curved towards the genital furca's stem 3
3(2) Apical region of the common spermathecal duct and the individual spermathecal ducts lightly striated, these latter one-third as long as the spermatheca; terminal knob clearly curved towards the genital furca's stem .... *Bi. olmeca bicolor*
Common and individual spermathecal ducts smooth, the latter as long as the spermatheca, the terminal knob being slightly curved towards the genital furca's stem 4

4(3) Interocular distance smaller than half the clypeus' width; flagellomere/labrum-epipharynx ca. 1.1:1.0 .... *Bi. olmeca nociva*

Interocular distance equivalent to the clypeus' width; flagellomere/labrum-epipharynx ca. 1.4:1.0 .... *Bi. olmeca olmeca*

Psychodopygus

MALES

1 Anepimeron (Fig. 21) with setae on its upper region .... Chagasi Series

Anepimeron without setae 2

2(1) Gonostyle with five well developed spines .... Davisi Series

Gonostyle with four or fewer spines, developed or not 3

3(2) Gonostyle with only two spines .... *Ps. bispinosus*

Gonostyle with four spines, developed or some atrophied 4

4(3) Gonostyle with the apical spine developed, the other three being atrophied (Fig. 134) .... Guyanensis Series

Gonostyle with a greater number of developed spines 5

5(4) Paramere with hairy lobe on the dorsal margin; gonostyle with the internal spine atrophied .... Panamensis Series

Paramere simple; gonostyle with the four developed spines, the internal may be thinner .... Arthuri Series

Arthuri Series

1 Pronotum and paratergite brown; pleura and coxae straw; epandrial lobe clearly longer than the gonocoxite .... *Ps. lloydi*

Pronotum and paratergite straw; katepisternum, katepimerom and anterior coxae brown; epandrial lobe $\leq$ the gonocoxite 2

2(1) Paramere with tuft of long and sinuous setae located on the basal third of the dorsal margin; gonostyle with the internal spine located on its distal third .... *Ps. matosi*

Paramere without tuft located on the basal third of the dorsal margin; gonostyle with the internal spine located in its middle .... *Ps. arthuri*

Chagasi Series

1 Paramere with branch that originates on the subapical region of the ventral margin ... 2

Paramere without branch on the ventral margin 6

2(1) Dorsal margin of paramere with tuft of long and sinuous setae located at the same level as the base of the ventral branch .... *Ps. bernalei*

Dorsal margin of paramere without the above tuft, though short and straight setae may be present 3

3(2) Apex of paramere truncated, with semicircular aspect .... *Ps. fairtigi*

Apex of paramere different from the above 4

4(3) Aedeagal ducts ca. 3.5 × longer than the sperm pump .... *Ps. leonidasdeanei*

Aedeagal ducts ca. 2.5 × longer than the sperm pump 5

5(4) Ventral branch of paramere cylindrical up to its middle part, where it is enlarged, to become narrow again, in such a way that its apex is narrower than its base and presents the spiniform setae turned towards the other branch .... *Ps. complexus*

Ventral branch of paramere cylindrical from the base up to its pre-apical region, where it is enlarged, becoming triangular shaped, with the apex represented by the longer side of the triangle; absence of the above mentioned setae .... *Ps. wellcomei*

6(1) Paramere pipe-shaped and with a dense tuft of setae located at its apex (Fig. 181) .... *Ps. chagasi*

Paramere different from above 7

7(6) Apical region of the paramere turned towards the gonocoxite, in such a way as to form a concavity on the dorsal margin; without setae 8

Dorsal margin of paramere with concavity in its median region and apical third densely covered by setae .... *Ps. killicki*

8(7) Apico-dorsal margin of paramere bearing a small tongue-shaped projection turned towards the base of the genitalia and extending to the more concave part .... *Ps. douradoi*

Apico-dorsal margin of paramere without the above projection 9

9(8) Pronotum straw; apex of paramere quadrangular with a small and foliaceous seta on the subapical area of the dorsal margin........ *Ps. squamiventris maripaensis*

Pronotum dark brown; apex of the dorsal margin of paramere rounded, the subapical seta not being foliaceous .... *Ps. squamiventris squamiventris*

Davisi Series

1 Aedeagal ducts 5.0 × longer than the sperm pump; tergites I-III straw or light brown .... *Ps. davisi*

Aedeagal ducts 4.5 × longer than the sperm pump; tergites I-III brown 2

2(1) Pronotum and paratergite brown; ventral branch of paramere ca. 3 × longer than its greatest width .... *Ps. claustrei*

Pronotum and paratergite clear; ventral branch of paramere ca. 5 × longer than its greatest width .... *Ps. amazonensis*

Guyanensis Series

1 Epandrial lobe longer than the gonocoxite; paramere lobed .... *Ps. lainsoni*

Epandrial lobe shorter than the gonocoxite; paramere simple, without lobe........ 2

2(1) Paramere: dorsal margin immediately preceding the narrowing of paramere with a set of setae twice or more as long as the width of the area where they are implanted; aedeagal ducts with a slight apical dilation in its ventral margin .... *Ps. luisleoni*

Paramere: dorsal margin immediately preceding the narrowing of paramere with a set of setae shorter or slightly longer than the width of the area where they are implanted; aedeagal ducts without apical dilatation in its ventral margin 3

3(2) Paramere curved in its median region towards the gonocoxite; apex dilated, its width being ca. three times that of its narrowest part .... *Ps. corossoniensis*

Paramere in its apical third curved towards the gonocoxite; apex tapered or slight dilated, its maximum width being twice that of its narrowest part 4

4(3) Paramere bent forming almost a right angle; aedeagal with apical half pigmented 5

Paramere bent at an obtuse angle (nearly 130°), straight apical region; aedeagal ducts slightly pigmented........ *Ps. dorlinsis*

5(4) Gonocoxite: ratio length/width $\geq$ 3.5:1.0; paramere with the apical region, post-curvature with uniform width .... *Ps. geniculatus*

Gonocoxite: ratio length/width $\leq$ 3.0:1.0; paramere: the apical region with a slight narrowing preceding the apex .... *Ps. francoisleponti*

Panamensis Series

1 Dorsal margin of paramere bearing thin curved branch 2

Dorsal margin of paramere without any branch........ 3

2(1) Scutum and coxae brown; paramere: dorsal branch bearing 4–5 well developed apical setae; ca. seven long setae implanted in a line on the median region of ventral margin .... *Ps. llanosmartinsi*

Scutum darker brown than the coxae; paramere: dorsal branch of four setae, the most developed at the beginning of the distal third, the others smaller and situated between this latter and the apex; two well developed setae implanted in a tubercle in the median region of ventral margin........ *Ps. recurvus*

3(1) Dorsal lobe of paramere with the setae turned towards the gonocoxite and arranged in two sets on the dorsal margin, the more basal of which having longer setae than the apical .... *Ps. panamensis*

Dorsal lobe of paramere with the setae turned towards the apex of the genitalia and arranged in the apical region 4

4(3) Setae of the dorsal lobe of paramere located in an apical line 5

Setae of the dorsal lobe of paramere located in several distinct lines 6

5(4) Dorsal lobe of paramere bearing a seta implanted in a small tubercle in the median region of the dorsal margin .... *Ps. hirsutus nicaraguensis*

Dorsal lobe of paramere without isolated setae in the median region of the dorsal face .... *Ps. hirsutus hirsutus*

6(4) Ventral branch of paramere as long as ca. 0.50 the dorsal lobe (the latter measured from the base of paramere) 7

Ventral branch of paramere as long as the dorsal one (the latter measured from the base of paramere) 8

7(6) Katepisternum and anterior coxa brown; aedeagal ducts/sperm pump ratio < 3.2:1.0 .... *Ps. ayrozai*

Pleura whitish; aedeagal ducts/ sperm pump ratio ca. 4.1:1.0 .... *Ps. joliveti*

8(6) Scutum straw 9

Scutum brown 12

9(8) Epandrial lobe shorter than the gonocoxite; apex of the aedeagal ducts hook-like .... *Ps. fairchildi*

Epandrial lobe longer than the gonocoxite; apex of the aedeagal ducts simple 10

10(9) Aedeagal ducts *ca.* 2.8 × longer than the sperm pump .... *Ps. carrerai thula*

Aedeagal ducts *ca.* 3.0 × longer than the sperm pump 11

11(10) Paramere: ventral lobe ca. 8.0 times as long as its apical width; the dorsal lobe practically rectangular, its apical-dorsal area forming an acute - almost a right - angle, being slightly more dilated than the pre-apical region of the lobe, due to a weak convexity in this area; the apical setae of the dorsal lobe are semifoliaceous and those of the apico-dorsal angle present the apical third curved towards the ventral lobe .... *Ps. carrerai carrerai*

Paramere: ventral lobe ca. 5.5 times as long as its apical width; the dorsal lobe with its apical region clearly rounded, i.e. the apical dorsal area forming an angle >90°; the apical setae are spiniform and divergent .... *Ps. yasuniensis*

12(8) Katepisternum and coxae straw 13

Katepisternum and coxae brown .... *Ps. paraensis*

13(12) Paramere with dorsal lobe as wide as the lateral lobes .... *Ps. nocticolus*

Paramere with dorsal lobe twice as wide as the lateral lobes .... *Ps. yucumensis*

FEMALES

1 Anepimeron (Fig. 21) with setae on its upper region .... Chagasi series 2

Anepimeron without the above setae 4

2(1) Common spermathecal duct longer than the individual spermathecal ducts .... *Ps. leonidasdeanei*

Common spermathecal duct shorter than or of equal length to the individual spermathecal ducts 3

3(2) Common spermathecal duct striated; the individual spermathecal ducts unlike fish-bone pattern .... *Ps. bernalei*

Common spermathecal duct smooth, the individual spermathecal ducts of fish-bone pattern (Fig. 60) .... *Ps. chagasi*

Ps. complexus

Ps. douradoi

Ps. fairtigi

Ps. killicki

Ps. squamiventris maripaensis

Ps. squamiventris squamiventris

Ps. wellcomei

4(1) Individual spermathecal ducts 2 × as long as the spermatheca Guyanensis Series 5

Individual spermathecal ducts <2 × the spermatheca's length 6

5(4) Mesonotum straw .... *Ps. lainsoni*

Mesonotum brown .... *Ps. corossoniensis*

Ps. geniculatus

Ps. guyanensis

Ps. francoisleponti

Ps. sp. de Trés Esquinas

6(4) Individual spermathecal ducts forming an angle ≥180° at their junction (Fig. 65)...... 7

Individual spermathecal ducts forming an angle <180° at their junction (Figs. 61, 62, 63 and 64) 8

7(6) Thoracic pleura straw.............. Panamensis Series .... *Ps. nocticolus*

Thoracic pleura brown............. Panamensis Series........ *Ps. paraensis*

8(6) Cibarium: three or four pairs of posterior (horizontal) teeth, the more external ones not inclined to the midline 9

Cibarium: two pairs of posterior (horizontal) teeth; if three pairs, they are inclined to the midline 10

9(8) Pronotum and paratergite brown; pleura and coxae totally straw .... Arthuri Series........ *Ps. lloydi*

Pronotum and paratergite straw; katepisternum, katepimeron and fore coxae brown.... Arthuri Series........ *Ps. arthuri*

10(8) Individual spermathecal ducts as long as or longer than the spermatheca........ 11

Individual spermathecal ducts shorter than the spermatheca........ 12

11(10) Junction of the individual spermathecal ducts forming an acute angle (90°); apex of the common spermathecal duct perceptibly wrinkled........ Panamensis Series .... *Ps. recurvus*

Junction of the individual spermathecal ducts forming an obtuse angle (> 90°); apex of the common spermathecal duct slightly wrinkled.............Arthuri Series......*Ps. matosi*

12(10) Common sperm ducts completely striated, except in its apical region.... *Ps. bispinosus*

Common spermathecal ducts striated or smooth, but with wrinkles in its apical region (Figs. 61 and 65) 13

13(12) Common spermathecal duct striated along the region that precedes the apical wrinkles........ 14

Common spermathecal duct smooth along the region that precedes the apical wrinkles........ 18

14(13) Scutum straw.............................Panamensis Series 15

Scutum brown 16

15(14) Labrum-epipharynx shorter than flagellomere I; LE/fI ratio ca. 0.9:1.0 .... *Ps. fairchildi*

Labrum epipharynx longer than flagellomere I; LE/fI ration ca. 1.1:1.0.. *Ps. carrerai carrerai*

16(14) Junction of the individual spermathecal ducts forming an evident furca, with an acute angle (< 90°) (Fig. 61).......................Davisi Series 26

Junction of the individual spermathecal ducts forming an obtuse angle (> 90°)........ 17

17(16) Fore coxae clear brown.......... Panamensis Series .... *Ps. hirsutus nicaraguensis*

Fore coxae straw.......................... Panamensis Series ....*Ps. hirsutus hirsutus*

18(13) Mesonotum straw 19

Mesonotum brown 20

19(18) Labrum-epipharynx >300 μm, subequal to the length of the head ....Panamensis Series.. *Ps. carrerai thula*

Labrum-epipharynx <300 μm, shorter than the length of the head.............Panamensis Series.. *Ps. carrerai carrerai*

20(18) Katepisternum, katepimeron and all coxae brown 21

Katepimeron, mid and hind coxae straw.. 22

21(20) Cibarium with the anterior (vertical) teeth arranged in a central set; space between the internal posterior (horizontal) space filled with many denticles.......................... Panamensis Series .... *Ps. paraensis*

Cibarium with the anterior (vertical) teeth dispersed; space between the internal posterior (horizontal) teeth without denticles.............................. Panamensis Series.. *Ps. llanosmartinsi*

22(20) Katepisternum and fore coxa brown................Panamensis Series.............. *Ps. ayrozai*

Katepisternum and fore coxa straw.. 23

23(22) Pronotum and paratergites brown 24

Pronotum and paratergites straw 25

24(23) Spermathecae with the apical ring asymmetrical, the terminal knob implanted laterally in the pre-apical ring......................Panamensis Series......................... *Ps. panamensis*

Spermathecae with the terminal knob implanted at the centre of the apical ring... Davisi Series.. *Ps. claustrei*

25(23) Cibarium with the anterior (vertical teeth) arranged in an irregular transversal row, not forming a central longitudinal set.................. Davisi Series ....*Ps. amazonensis*

Cibarium with the anterior (vertical) teeth forming a central longitudinal set.................... Panamensis Series..*Ps. yucumensis*

26(16) Cibarium with the anterior (vertical) teeth arranged in an irregular transversal row, not forming a central longitudinal set; paratergite straw, scutellum brown..............Davisi Series..*Ps. amazonensis*

Cibarium with the anterior (vertical) teeth arranged in a median set; paratergite brown, scutellum straw................ Davisi Series...*Ps. davisi*

The female of *Ps. parimaensis* cannot be identified by the characteristics presented in its description.

Nyssomyia

MALES

1 Gonostyle with the internal spine implanted in its basal region 2

Gonostyle with the internal spine implanted in its middle or slightly beyond this....... 3

2(1) Tegument of the head, pronotum and paratergite straw, contrasting with the mesonotum brown .... *Ny. richardwardi*

Tegument of the head, pronotum, paratergite and mesonotum brown.............*Ny. shawi*

3(1) Gonocoxite with a cluster of setae in its basal-median region ... 4
Gonocoxite without cluster of setae, though some isolated setae may occur in its basal-median region ... 6

4(3) Paramere with apex sharply curved toward the gonocoxite; gonocoxite bearing a cluster with ca. 40 setae ... *Ny. antunesi*
Paramere with apex straight or slightly curved toward the gonocoxite; cluster of the gonocoxite bearing ≤25 setae ... 5

5(4) Aedeagal ducts/sperm pump ratio ≤ 3.0:1.0; cluster of the gonocoxite bearing 12–22 setae ... *Ny. delsionatali*
Aedeagal ducts/sperm pump ration >3.2:1.0; cluster of the gonocoxite bearing 3–7 setae ... *Ny. urbinattii*

6(3) Aedeagal ducts/sperm pump ratio ca. 6.0:1.0 ... *Ny. elongata*
Aedeagal ducts/sperm pump ratio ≤ 4.0:1.0 ... 7

7(6) Aedeagal ducts/sperm pump ratio ≥ 2.3:1.0 ... 8
Aedeagal ducts/sperm pump ratio ≤ 2.0:1.0 ... 10

8(7) Alar ratio *alpha/delta* ca. 2.7:1.0; apex of the aedeagal duct spoon-shaped ... *Ny. hernandezi*
Alar ratio *alpha/delta* ca. 2.0:1.0; apex of the aedeagal duct dilated and bifurcated ... 9

9(8) 5th palpal segment shorter than, equal to or slightly longer than 3rd; flagellomere I/labrum-epipharynx ratio ca. ≤ 1.1:1.0 ... *Ny. whitmani*
5th palpal segment ca. 1/5 longer than 3rd; flagellomere I/labrum-epipharynx ratio ca. 1.3:1.0 ... *Ny. sylvicola*

10(7) Paratergite straw ... 11
Paratergite brown ... 17

11(10) Mesonotum brown in the region of the pre-sutural setae ... 16
Mesonotum straw in the region of the pre-sutural setae ... 12

12(11) 5th palpal segment ≤3rd; mesonotum straw in the central region ... 13
5th palpal segment >3rd; mesonotum brown in the central region ... *Ny. trapidoi*

13(12) Apex of aedeagal ducts bifurcated (Fig. 109) ... *Ny. anduzei*
Apex of aedeagal ducts with only one or two barbs (Fig. 110) ... 14

14 (13) Parameral sheath: basal width clearly shorter than the length of the dorsal margin and equal to the length of the ventral margin; paramere: dorsal margin of the digitiform part with a dense coating of setae longer than the width of the area where they are implanted ... 15
Parameral sheath: basal width equal to the length of the dorsal margin and greater than the length of the ventral margin; paramere: dorsal margin of the digitiform part with a sparse coating of setae (excepting the apical ones) which are as long as or shorter than the width of the area in which they are implanted ... *Ny. yuilli pajoti*

15(14) Paramere: apical digitiform part shorter than the basal rectangular part (this latter measured from its base at the ventral margin to the end of the protuberance with setae); the apical/basal ratio ca. 0.6–0.8:1:00 ... *Ny. fraihai*
Paramere: apical digitiform part as long as the basal rectangular part; the apical/basal ratio ca. 0.8–1.0:1:0 ... *Ny. yuilli yuilli*

16(11) Gonostyle with the upper external spine implanted almost apically ... *Ny. ylephiletor*
Gonostyle with the upper external spine implanted on a level clearly anterior to the apical ... *Ny. umbratilis*

17(10) Gonostyle with the upper external spine implanted almost apically ... *Ny. edentula*
Gonostyle with the upper external spine implanted on a level clearly anterior to the apical ... 18

18(17) Apex of the aedeagal duct ladle-shaped (Fig. 108-A) ....*Ny. intermedia*

Apex of the aedeagal duct spoon-shaped (Fig. 108-B)................................. *Ny. neivai*

FEMALES

1 Common spermathecal duct reaches or goes beyond the middle of the genital furca stem; thorax with paratergite straw.. 2

Common spermathecal duct shorter than the above; thorax with paratergite of varying color 3

2(1) Common spermathecal duct reaches the middle of the genital furca stem; individual spermathecal ducts smooth in their basal half and wrinkled in their distal half .... *Ny. delsionatali*

Common spermathecal duct goes beyond the apex of the genital furca stem ....*Ny. antunesi*

Ny. urbinattii

3(1) Individual spermathecal ducts with sclerotized wrinkles (Fig. 68) .... *Ny. fraihai*

Ny. yuilli yuilli

Individual spermathecal ducts without sclerotized wrinkles 4

4(3) Spermathecae with 26–33 rings; individual spermathecal ducts ca. 3 × wider at their bases than their apices.. *Ny. bibinae*

Spermathecae without the set of characteristics above.. 5

5(4) Paratergite straw... 6

Paratergite brown 11

6(5) Apex of the terminal knob on the same longitudinal axis as the spermatheca (Figs. 37-D and 67)... 7

Apex of the terminal knob lateralized in relation to the longitudinal axis of the spermatheca (Fig. 66)... 9

7(6) Terminal knob with its basal width twice that of the individual spermathecal ducts at their junction with the spermatheca; this latter with ca. five well-delimited rings, without rudimentary rings at their bases.. *Ny. trapidoi*

Terminal knob with its basal width ≤ that of the individual spermathecal ducts at their junction with the spermatheca; this latter having seven or more rings........................ 8

8(7) Mesonotum brown; terminal knob with its maximum width subequal to that of the apical ring and the basal width equivalent to that of the apex of the individual spermathecal ducts at their junction with the spermatheca; this latter having 6–7 well-developed rings and some other rudimentary ones at its base ....*Ny. ylephiletor*

Mesonotum predominantly straw. Terminal knob with its greatest width clearly less than that of the apical ring; spermathecae with ca. ten rings (Fig. 67)........... *Ny. yuilli pajoti*

9(6) Mesonotum light brown, contrasting only slightly with the paratergite.......*Ny. anduzei*

Mesonotum dark brown, contrasting with the straw paratergite................................ 10

10(9) Tegument of the head dark brown; basal flagellomeres without simple setae; individual spermathecal ducts clearly striated and as long as the spermathecae ....*Ny. umbratilis*

Tegument of the head straw; basal flagellomeres with simple setae; individual spermathecal ducts smooth and shorter than the spermathecae........... *Ny. richardwardi*

11(5) Individual spermathecal ducts shorter than the spermathecae........................*Ny. shawi*

Individual spermathecal ducts longer than the spermathecae.................................... 12

12(11) Basal diameter of the terminal knob clearly greater than that of the individual ducts at their junction with the spermatheca 13

Basal diameter of the terminal knob ≤ that of the individual spermathecal ducts at their junction with the spermatheca 14

13(12) Thorax with the anepisternal region brown .... *Ny. edentula*

Thorax with the anepisternal region straw..*Ny. intermedia*

14(12) Individual spermathecal ducts long, their apices (junction with the spermathecae) clearly go beyond the apex of the genital furca stem; spermathecae with 12 or more rings........ 15

Individual spermathecal ducts shorter, their apices (junction with the spermathecae) do not go beyond the apex of the genital furca stem; spermathecae with 6–11 rings (Fig. 37-D) .... *Ny. neivai*

15(14) Individual spermathecal ducts 3–4 times longer than the spermathecae .... *Ny. whitmani*

Individual spermathecal ducts ca. 2.5 times longer than the spermathecae........ *Ny. hernandezi*

Ny. singularis: terminal knob and spermathecae similar to those of *Ny. neivai*

Trichophoromyia

MALES

1 Aedeagal ducts ≤3 times the length of the sperm pump........ 2

Aedeagal ducts >4 times the length of the sperm pump........ 6

2(1) Gonostyle with the lower external spine implanted in its middle in such a way that it is nearer than the internal one to the upper external one........ *Th. reburra*

Gonostyle with the lower external spine implanted beyond its middle in such a way that it is closer to the upper external than the internal one........ 3

3(2) Epandrial lobe longer than the gonocoxite........ 4

Epandrial lobe as long as or shorter than the gonocoxite........ 5

4(3) Paramere: dorsal margin with a concavity which extends as far as the pre-apical region; ventral margin with a discreet elbow at the beginning of its apical third which extends as far as the pre-apical area........ *Th. meirai*

Paramere: dorsal margin with a straight or slightly convex apical half in the area preceding the apex; apex of the paramere slightly dilated at the expanse of the ventral margin which becomes concave after the convexity of the preceding area........ *Th. ubiquitalis*

5(3) Paramere: dorsal margin with concavity extending as far as the pre-apical region; ventral margin with a protuberance between the beginning of the apical third and the pre-apical area........ *Th. omagua*

Paramere: dorsal margin slightly convex between its middle and the apical quarter; pre-apical region slightly dilated at the expanse of the concavity of the ventral margin........ *Th. uniniensis*

6(1) Paramere: without any lobe on the dorsal margin; at the beginning of the apical third it presents ca. 10 semifoliaceous setae with curved apices; immediately after these, it turns abruptly towards the gonocoxite, in this area there are tiny spines with rounded apices........ *Th. reinerti*

Paramere different from the above........ 7

7(6) Gonocoxite with its length less than twice its greatest width; gonostyle with a very short upper external spine whose apex never goes beyond the level of the implantation of the apical one........ 8

Gonocoxite twice or three times as long as its greatest width; gonostyle with the apex of the upper external spine going beyond the level of the implantation of the apical one 11

8(7) Paramere with a trilobed apex........ *Th. lopesi*

Paramere without the trilobed apex........ 9

9(8) Paramere practically of the same caliber throughout, the dorsal margin having no concavity after the median lobe; dorsal margin completely covered with short hairs........ *Th. bettini*

Paramere with its apex clearly wider than its middle due to the pronounced concavity of the dorsal margin, after a median lobe; the dorsal margin is only partly covered with short hairs........ 10

10(9) Gonocoxite with a median cluster of ca. 15 setae, five of them sclerotized and clearly thicker than the others; paramere with the dorsal area posterior to the median lobe concave and more than twice as long as its width .... *Th. gibba*

Gonocoxite with a median cluster of ca. 30 setae of the same thickness; paramere with the dorsal area posterior to the median lobe concave and as long as its width .... *Th. eurypyga*

11(7) Gonocoxite with a median cluster of 2–5 setae much thicker than the others 12

Gonocoxite with the setae of the median cluster all of the same thickness 15

12(11) Epandrial lobe as long as the gonocoxite and without permanent setae in its apical region .... *Th. readyi*

Epandrial lobe clearly longer than the gonocoxite and with permanent setae in its apical region 13

13(12) Gonocoxite: cluster of ca. 25 setae, the thickest ones being thinner than the width of the internal spine of the gonostyle .... *Th. adelsonsouzai*

Gonocoxite: cluster of less than 20 setae, the thickest ones being as thick as the internal spine of the gonostyle 14

14(13) Gonocoxite with the median cluster of ca. five setae thicker than the others; gonostyle with the apical spine thick, i.e. its width is equal to that of the area immediately before its implantation; paramere with median lobe of the dorsal margin oblong .... *Th. brachipyga*

Gonocoxite with the median cluster of two setae thicker than the others; gonostyle with the apical spine narrower than the area immediately before its implantation; paramere with the median lobe of the dorsal margin rectangular .... *Th. viannamartinsi*

15(11) Gonocoxite with two well sclerotized setae in its base, one seta in its middle and a sparse set of thin setae in its apical region .... *Th. dunhami*

Gonocoxite with another arrangement and aspect of setae 16

16(15) Gonocoxite without setae in its basal and median areas, only ca. six sclerotized long setae (as long as twice the width of the gonocoxite) implanted in its apical third .... *Th. saltuosa*

Gonocoxite with setae in its base and/or in its middle 17

17(16) Paramere digitiform posterior to the median lobe on the dorsal margin 18

Paramere without median lobe or if it is present, it is not digitiform posterior to this lobe 22

18(17) Paramere with the digitiform area narrow (its length > six × its width); setae of the median dorsal lobe clearly longer than the lobe's width .... *Th. castanheirai*

Paramere with the digitiform area ≤ 4 × as long as its width; setae of the dorsal median lobe as long as the lobe's width 19

19(18) Paramere with the digitiform area ca. twice as long as its maximum width; some setae that cover the digitiform area are also ca. twice as long as its width 20

Paramere with the digitiform area ≥ 3 times as long as its maximum width; the setae that cover the digitiform area are shorter than its width 21

20(19) Paramere: lobe of the dorsal margin with the setae concentrated in its apical region; dorsal margin of the digitiform area with three long setae .... *Th. napoensis*

Paramere: lobe of the dorsal margin completely covered by setae; digitiform area with ca. five long setae implanted close to the ventral margin .... *Th.* sp. 1. de Araracuara

21(19) Gonocoxite with median cluster of ca. 15 setae .... *Th. beniensis*

Gonocoxite with median cluster of ca. 30 setae .... *Th. howardi*

22(17) Gonocoxite with the setae of the basal cluster united with the median cluster without clear transition (Fig. 149) 23

Gonocoxite with only a median cluster of setae or if the basal cluster is also present the transition between them is clear 29

23(22) Paramere with the dorsal lobe very near its apex, so that this latter seems bilobed; the dorsal lobe is covered in its apical area with setae as long as its height and in the dorsal margin of the apex of the paramere there are short, straight setae; these two sets of setae are separated by a hairless area .... *Th. melloi*

Paramere with the apex without bilobed aspect 24

24(23) Paramere with pronounced concavity in its dorsal margin, so that in its apical region it assumes trapezoidal shape, with the smallest side being located in its middle and the largest at its apex; this latter is coated with thin, short, spiniform setae; internally and near to the apex of the ventral margin there arises a small lobe whose apex turns towards the base of the genitalia; this lobe is also coated with spiniform setae but slightly longer than those on the longest side of the trapezium .... *Th. ininii*

Paramere different from the above 25

25(24) Paramere with concavity in the dorsal margin, however the apex of the paramere is rounded and presents a tuft of setae whose apices are curved towards the base of the genitalia; from the point at which the paramere becomes narrow to its peak, there is a dense coating of microtrichia and in the apical third of the inferior part there is a sparse set of short setae .... *Th. pastazaensis*

Paramere different from the above 26

26(25) Paramere: dorsal margin with the lobe arising at its apex and extending towards its middle; apical area of the lobe with setae as long as the width of the lobe, their apices being directed towards the apex of the genitalia (Fig. 174) 27

Paramere tapering from the base towards its middle, from this point it expands slightly and then again narrows slightly, so the dorsal margin assumes a convex shape 28

27(26) Cluster of the gonocoxite with the setae in ventral position practically as long and wide as the dorsal ones, the diameter of their implantation points being slightly smaller than those of the dorsal setae (Fig. 149) .... *Th. auraensis*

Cluster of the gonocoxite with the setae in ventral position clearly shorter and narrower than those in the dorsal position, the diameter of their implantation points being ca. half that of those of the dorsal setae .... *Th. rufreitasi*

28(26) Paramere with the convex area of the dorsal margin with a fringe of setae as long as the width of this area, with those near the median area of the apex curved towards the base of the genitalia and the setae of the apical area straight; at the apex of the ventral margin, there is a narrow lobe that is directed to the base of the paramere and inflects towards the dorsal margin near the median narrowing of the paramere .... *Th. loretonensis*

Paramere with the convex area of the dorsal margin covered internally with setae shorter than the width of the area; absence of lobe on the ventral margin .... *Th. ruii*

29(22) Epandrial lobe ca. one-third longer than the gonocoxite 30

Epandrial lobe shorter than the above 32

30(29) Epandrial lobe slightly more dilated beyond its middle where numerous permanent setae are implanted. Gonocoxite with three clusters of setae, one apical (of ca. ten setae), one slightly beyond the middle (of ca. 35 setae) and another slightly before the middle (of more than 60 setae). Paramere with a marked concavity in the middle of its dorsal margin; protuberance in the pre-apical region of the ventral margin finishing in a digitiform apex .... *Th. arevaloi*

Epandrial lobe, gonocoxite and paramere with aspects different from the above 31

31(30) Gonocoxite with a median cluster of ca. 20 thin setae. Gonostyle: apex of the upper external spine goes slightly beyond the implantation of the apical spine. Paramere of practically the same caliber as from its basal third and presenting a small salience in its apical quarter covered with setae curved towards the base of the genitalia and on the ventral margin there is a small lobe directed towards the dorsal margin on whose apex ca. four setae are implanted .... *Th. acostai*

Gonocoxite with a median cluster of ca. 12 setae, above which are ca. another five setae thinner than the preceding ones. Gonostyle: apex of the upper external spine goes beyond the middle of the apical spine; paramere tapers abruptly as from its middle, being covered with setae on its apical third .... *Th. rostrans*

32(29) Paramere: apical third assumes trapezoidal shape, with the shortest side being located in the middle of the paramere and the longest at its apex; where on its dorsal margin there are four well sclerotized setae longer than the others .... *Th. wilkersoni*

Paramere with apex different from the above 33

33(32) Gonocoxite with a basal tuft of ten or more setae as long as those of the median tuft; paramere with two or three apical lobes 34

Gonocoxite with a few basal setae, shorter than those of the median tuft; paramere different than the above 35

34(33) Gonocoxite with a basal cluster of ten or more setae and without a distinct median cluster, having only a row of ca. eight setae extending from the middle to the apical region. Paramere: dorsal margin with small lobe covered with short setae as long as the height of the lobe; ventral margin with a wide lobe, practically occupying the entire apical area, arising on the apex and extending towards the apical third, being covered with small setae .... *Th. incasica*

Gonocoxite with basal tuft of ca. 20 setae and in the central region a set of long setae separate from the apical ones. Paramere with three apical lobes, the ventral one shorter than the median and covered with spiniform setae whose apices point in all directions; the median one also having spiniform setae but their apices are directed towards the gonocoxite and the dorsal lobe shorter than the ventral one with semifoliaceous setae whose apices are also directed towards the gonocoxite .... *Tr. nautaensis*

35(33) Gonocoxite: median region with two clusters of setae, one of ca. five setae implanted more internally and another of 8–20 longer setae implanted more externally, all of them more apical than the internal ones or with some at the same level as the internal ones 36

Gonocoxite with only one cluster of setae in its median region 38

36(35) Gonocoxite: external cluster of ca. eight setae; external cluster with the setae implanted more apically than the internal one; paramere with dorsal lobe thin at its apex but widening towards the beginning of the apical third; the lobe being covered with short setae, more concentrated on the apex of the paramere .... *Th. cellulana*

Gonocoxite: external cluster of 10–20 setae, some them implanted at the same level as the internal cluster; paramere different from the above 37

37(36) Paramere: dorsal margin with a lobe arising on its apex and extending towards its middle, densely covered with setae, those situated at the apex of the lobe being ca. twice as long as the others and curved towards the apex of the paramere.... *Th. sinuosa*

Paramere: dorsal margin with a lobe arising on its apex and extending towards the beginning of its apical third with a concavity between its extremities; the apical area of the lobe is covered with short setae, the basal area having a few setae, longer than the apical ones .... *Th. velascoi*

38(35) Paramere: dorsal margin with a lobe arising on its apex and extending towards its middle 39

Paramere without a distinct lobe, having only a slight concavity preceding the bristly part situated after its middle 41

39(38) Paramere: dorsal margin with apical lobe presenting a concave, glabrous area between its two extremities .... *Th. clitella*

Paramere: dorsal margin with the apical lobe completely covered with setae 40

40(39) Paramere with digitiform lobe arising on its ventral margin inflecting obliquely towards the dorsal margin at the beginning of the apical third of the paramere; the apical area of the dorsal margin being covered with long setae with curved apices; there being a glabrous area between the apico-dorsal margin and the ventral lobe .... *Th. flochi*

Paramere: dorsal lobe of apical origin, extending towards the middle region of the paramere; lobe completely covered with setae, those on the apex of the lobe being longer than the lobe's width .... *Th. octavioi*

41(38) Paramere with the apical half triangular; dorsal margin with the setae of the median region separated from the apical ones by a glabrous area .... *Th. pabloi*

Paramere with a digitiform apical half, the dorsal margin being covered with setae throughout the length of its apical third .... *Th. nemorosa*

FEMALES

1 Individual spermathecal ducts with outgrowths, terminal knob ca. three times longer than its width .... *Th. reburra*

Individual spermathecal ducts without outgrowths, spherical terminal knob 2

2(1) Spermathecae: apical ring ≤ 3 times as long as the pre-apical ring 3

Spermathecae: apical ring ≥ 6 times as long as the pre-apical ring .... *Th. adelsonzouzai*
Th. auraensis
Th. beniensis
Th. bettinii
Th. brachipyga
Th. castanheirai
Th. eurypyga
Th. howardi
Th. ininii
Th. napoensis
Th. pabloi
Th. pastazaensis
Th. readyi
Th. rostrans
Th. ruii
Th. velascoi
Th. viannamartinsi
Th. wilkersoni

3(2) Spermathecae: apical ring ca. three times as long as the pre-apical ring .... *Th. ubiquitalis*

Spermathecae: apical ring < twice as long as the pre-apical ring 4

4(3) Diameter of the terminal knob greater than that of the individual ducts .... *Th. omagua*

Diameter of the terminal knob smaller than that of the individual ducts .... *Th. cellulana*

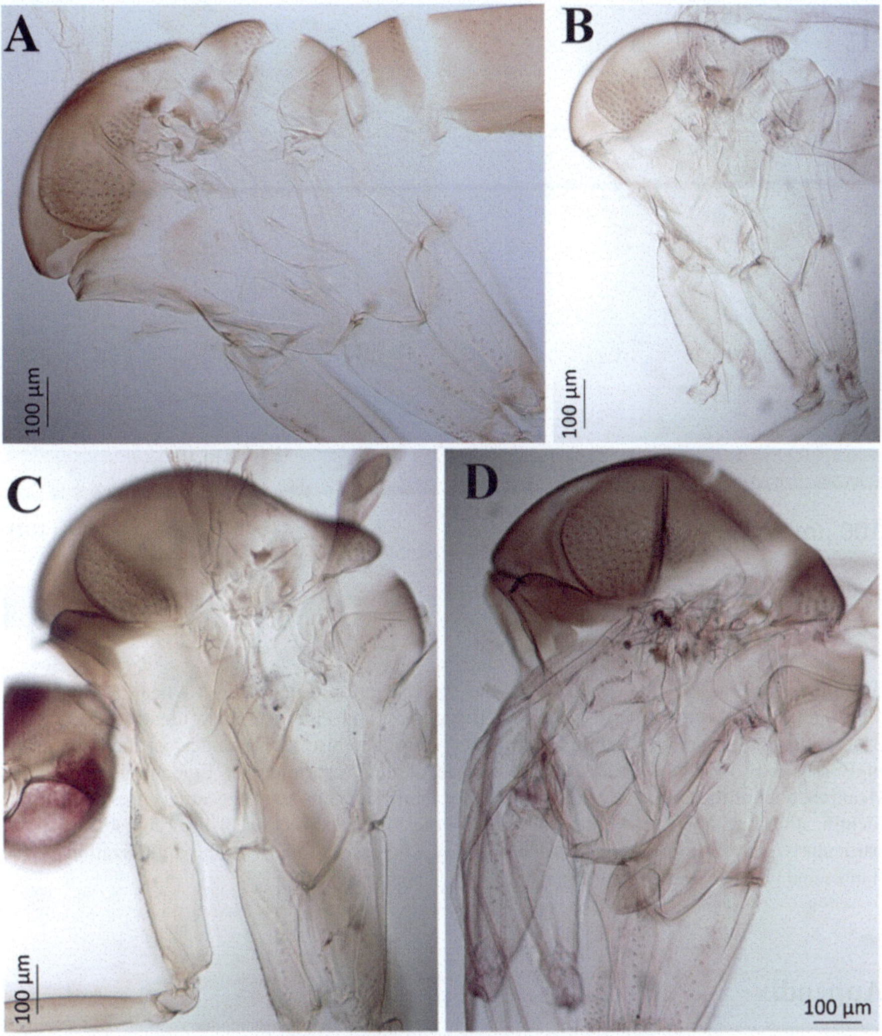

Fig. 182 Coloration of the thoracic sclerites: **A** – *Psathyromyia shannoni;* **B** – *Psathyromyia baratai;* **C** – *Psathyromyia limai* (note: the brown strip under the thorax is the femur)*;* **D** – *Psathyromyia ribeirensis*; **E** – *Psathyromyia bigeniculata*; **F** – *Psathyromyia abonnenci*

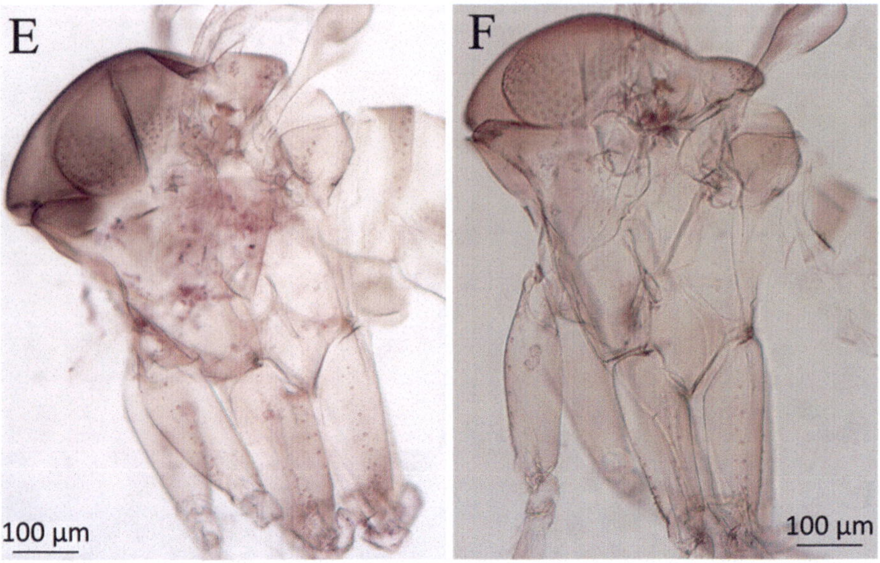

Fig. 182 (continued)

Acknowledgments To the students of the Specialization Course in Medical Entomology, 1995–2001 and those of the discipline of Bioecology and Phlebotominae Identification of the post-graduation program in Public Health 1998–2016, and also to the trainees of the Laboratory of Entomology in Public Health/Phlebotominae of the Department of Epidemiology—Public Health School of the University of São Paulo, for their criticisms and suggestions presented to the identification keys. To Mr. Arthur Anthony Boorne for his correction of the English text. I am particularly grateful to Rodrigo Espíndola Godoy who assisted me with the digitalization of the figures and revision of the manuscript.

Appendix

Index for the tribes, subtribes and genera of Phlebotominae of the world and subgenera, species goups, species series, species and subspecies of the Americas.

Species and subspecies

Bichromomyia	flaviscutellata
Bichromomyia	inornata
Bichromomyia	olmeca bicolor
Bichromomyia	olmeca nociva
Bichromomyia	olmeca olmeca
Bichromomyia	reducta
Brumptomyia	angelae
Brumptomyia	avellari
Brumptomyia	beaupertuyi
Brumptomyia	bragai

Brumptomyia	brumpti
Brumptomyia	cardosoi
Brumptomyia	carvalheiroi
Brumptomyia	cunhai
Brumptomyia	devenanzii
Brumptomyia	figueiredoi
Brumptomyia	galindoi
Brumptomyia	guimaraesi
Brumptomyia	hamata
Brumptomyia	leopoldoi
Brumptomyia	mangabeirai
Brumptomyia	mesai
Brumptomyia	nitzulescui
Brumptomyia	orlandoi
Brumptomyia	ortizi
Brumptomyia	pentacantha
Brumptomyia	pintoi
Brumptomyia	quimperi
Brumptomyia	spinosipes
Brumptomyia	travassosi
Brumptomyia	troglodytes
Brumptomyia	virgensi
Dampfomyia	anthophora
Dampfomyia	aquilonia
Dampfomyia	atulapai
Dampfomyia	beltrani
Dampfomyia	caminoi
Dampfomyia	deleoni
Dampfomyia	delpozoi
Dampfomyia	disneyi
Dampfomyia	dodgei
Dampfomyia	insolita
Dampfomyia	inusitata
Dampfomyia	isovespertilionis
Dampfomyia	leohidalgoi
Dampfomyia	rosabali
Dampfomyia	sp. de Suchitepequez
Dampfomyia	permira
Dampfomyia	steatopyga
Dampfomyia	vesicifera
Dampfomyia	vespertilionis
Dampfomyia	viriosa
Dampfomyia	zeledoni
Deanemyia	appendiculata
Deanemyia	derelicta
Deanemyia	maruaga

Deanemyia	ramirezi
Deanemyia	samueli
Edentomyia	piauiensis
Evandromyia	aldafalcaoae
Evandromyia	andersoni
Evandromyia	apurinan
Evandromyia	bacula
Evandromyia	bahiensis
Evandromyia	begonae
Evandromyia	bourrouli
Evandromyia	brachyphalla
Evandromyia	callipyga
Evandromyia	carmelinoi
Evandromyia	correalimai
Evandromyia	cortelezzii
Evandromyia	corumbaensis
Evandromyia	costalimai
Evandromyia	dubitans
Evandromyia	edwardsi
Evandromyia	evandroi
Evandromyia	gaucha
Evandromyia	georgii
Evandromyia	grimaldii
Evandromyia	hashiguchii
Evandromyia	infraspinosa
Evandromyia	inpai
Evandromyia	ledezmaae
Evandromyia	lenti
Evandromyia	monstruosa
Evandromyia	orcyi
Evandromyia	petropolitana
Evandromyia	pinottii
Evandromyia	rupicola
Evandromyia	sallesi
Evandromyia	saulensis
Evandromyia	sipani
Evandromyia	sericea
Evandromyia	sp. de Baduel
Evandromyia	spelunca
Evandromyia	tarapacaensis
Evandromyia	teratodes
Evandromyia	termitophila
Evandromyia	tupynambai
Evandromyia	tylophalla
Evandromyia	walkeri
Evandromyia	williamsi

Evandromyia	wilsoni
Expapillata	cerradincola
Expapillata	firmatoi
Hertigia	hertigi
Lutzomyia	adamsi
Lutzomyia	alencari
Lutzomyia	almerioi
Lutzomyia	amarali
Lutzomyia	araracuarensis
Lutzomyia	ayacuchensis
Lutzomyia	battistinii
Lutzomyia	bicornuta
Lutzomyia	bifoliata
Lutzomyia	blancasi
Lutzomyia	botella
Lutzomyia	caballeroi
Lutzomyia	caceresi
Lutzomyia	caligata
Lutzomyia	carvalhoi
Lutzomyia	castanea
Lutzomyia	castroi
Lutzomyia	cavernicola
Lutzomyia	ceferinoi
Lutzomyia	chavinensis
Lutzomyia	chotensis
Lutzomyia	cirrita
Lutzomyia	cruciata
Lutzomyia	cruzi
Lutzomyia	cultellata
Lutzomyia	diabolica
Lutzomyia	dispar
Lutzomyia	elizabethrangelae
Lutzomyia	erwindonaldoi
Lutzomyia	evangelistai
Lutzomyia	falcata
Lutzomyia	falquetoi
Lutzomyia	flabellata
Lutzomyia	fonsecai
Lutzomyia	forattinii
Lutzomyia	galatiae
Lutzomyia	gaminarai
Lutzomyia	gomezi
Lutzomyia	gonzaloi
Lutzomyia	guderiani
Lutzomyia	hartmanni
Lutzomyia	herreri

Lutzomyia	ignacioi
Lutzomyia	imperatrix
Lutzomyia	infusca
Lutzomyia	ischnacantha
Lutzomyia	ischyracantha
Lutzomyia	kirigetiensis
Lutzomyia	larensis
Lutzomyia	legerae
Lutzomyia	lichyi
Lutzomyia	longipalpis
Lutzomyia	maesi
Lutzomyia	manciola
Lutzomyia	marinkellei
Lutzomyia	matiasi
Lutzomyia	monzonensis
Lutzomyia	munaypata
Lutzomyia	noguchii
Lutzomyia	osornoi
Lutzomyia	pallidithorax
Lutzomyia	peruensis
Lutzomyia	pescei
Lutzomyia	ponsi
Lutzomyia	pseudolongipalpis
Lutzomyia	quillabamba
Lutzomyia	renei
Lutzomyia	rispaili
Lutzomyia	sanguinaria
Lutzomyia	scorzai
Lutzomyia	sherlocki
Lutzomyia	souzalopesi
Lutzomyia	spathotrichia
Lutzomyia	sp. de Pichinde
Lutzomyia	strictivilla
Lutzomyia	tanyopsis
Lutzomyia	tejadai
Lutzomyia	tolimensis
Lutzomyia	tortura
Lutzomyia	vargasi
Lutzomyia	velezi
Lutzomyia	wattsi
Martinsmyia	alphabetica
Martinsmyia	brisolai
Martinsmyia	cipoensis
Martinsmyia	gaspaviannai
Martinsmyia	minasensis

Martinsmyia	mollinedoi
Martinsmyia	oliveirai
Martinsmyia	pisuquia
Martinsmyia	quadrispinosa
Martinsmyia	reginae
Martinsmyia	waltoni
Micropygomyia	absonodonta
Micropygomyia	acanthopharynx
Micropygomyia	ancashensis
Micropygomyia	apache
Micropygomyia	atroclavata
+Micropygomyia	brandaoi
Micropygomyia	californica
Micropygomyia	capixaba
Micropygomyia	cayennensis braci
Micropygomyia	cayennensis cayennensis
Micropygomyia	cayennensis cruzi
Micropygomyia	cayennensis hispaniolae
Micropygomyia	cayennensis jamaicensis
Micropygomyia	cayennensis maciasi
Micropygomyia	cayennensis puertoricensis
Micropygomyia	cayennensis viequesensis
Micropygomyia	chassigneti
Micropygomyia	chiapanensis
Micropygomyia	ctenidophora
Micropygomyia	cubensis
Micropygomyia	dereuri
+Micropygomyia	dorafeliciangeliae
Micropygomyia	duppyorum
Micropygomyia	durani
Micropygomyia	echinatopharynx
Micropygomyia	farilli
Micropygomyia	ferreirana
Micropygomyia	hardisoni
Micropygomyia	huacalquensis
Micropygomyia	lewisi
Micropygomyia	longipennis
Micropygomyia	machupicchu
Micropygomyia	mangabeirana
Micropygomyia	micropyga
Micropygomyia	oppidana
Micropygomyia	oswaldoi
+Micropygomyia	paterna
Micropygomyia	peresi
Micropygomyia	petari

Micropygomyia	pilosa
Micropygomyia	pratti
Micropygomyia	pusilla
Micropygomyia	quechua
Micropygomyia	quinquefer
Micropygomyia	rorotaensis
Micropygomyia	saccai
Micropygomyia	schreiberi
Micropygomyia	sp. 2 de Araracuara
Micropygomyia	stewarti
Micropygomyia	trinidadensis
Micropygomyia	venezuelensis
Micropygomyia	vexator
Micropygomyia	villelai
Micropygomyia	vindicator
Micropygomyia	vonatzingeni.
Micropygomyia	wirthi
Micropygomyia	xerophila
Micropygomyia	yencanensis
Micropygomyia	zikani
Migonemyia	bursiformis
Migonemyia	cerqueirai
Migonemyia	migonei
Migonemyia	moucheti
Migonemyia	gorbitzi
Migonemyia	rabelloi
Migonemyia	vaniae
Nyssomyia	anduzei
Nyssomyia	antunesi
Nyssomyia	bibinae
Nyssomyia	delsionatali
Nyssomyia	edentula
Nyssomyia	elongata
Nyssomyia	fraihai
Nyssomyia	hernandezi
Nyssomyia	intermedia
Nyssomyia	neivai
Nyssomyia	richardwardi
Nyssomyia	shawi
Nyssomyia	singularis
Nyssomyia	sylvicola
Nyssomyia	trapidoi
Nyssomyia	umbratilis
Nyssomyia	urbinattii
Nyssomyia	whitmani

Nyssomyia	ylephiletor
Nyssomyia	yuilli pajoti
Nyssomyia	yuilli yuilli
Oligodontomyia	isopsi
Oligodontomyia	oligodonta
Oligodontomyia	toroensis
+Pintomyia	adiketis
Pintomyia	amilcari
Pintomyia	andina
Pintomyia	antioquiensis
Pintomyia	aulari
Pintomyia	bianchigalatiae
Pintomyia	boliviana
+Pintomyia	bolontikui
+Pintomyia	brazilorum
Pintomyia	cajamarcensis
Pintomyia	christenseni
Pintomyia	christophei
Pintomyia	columbiana
Pintomyia	damascenoi
Pintomyia	deorsa
Pintomyia	diamantinensis
Pintomyia	diazi
Pintomyia	disiuncta
+Pintomyia	dissimilis
+Pintomyia	dominicana
Pintomyia	emberai
Pintomyia	evansi
+Pintomyia	falcaorum
+Pintomyia	filipalpis
Pintomyia	fischeri
Pintomyia	gibsoni
Pintomyia	gruta
Pintomyia	guilvardae
Pintomyia	itza
+Pintomyia	killickorum
Pintomyia	kuscheli
Pintomyia	limafalcaoae
Pintomyia	longiflocosa
Pintomyia	mamedei
Pintomyia	maracayensis
Pintomyia	maranonensis
+Pintomyia	miocena
Pintomyia	misionensis
Pintomyia	monticola

Pintomyia	moralesi
Pintomyia	nadiae
Pintomyia	naiffi
Pintomyia	nevesi
Pintomyia	novoae
Pintomyia	nuneztovari
Pintomyia	odax
Pintomyia	oresbia
Pintomyia	orestes
Pintomyia	ottolinai
Pintomyia	ovallesi
Pintomyia	pacae
+Pintomyia	paleopestis
+Pintomyia	paleotownsendi
+Pintomyia	paleotrichia
Pintomyia	pessoai
Pintomyia	pia
Pintomyia	piedraferroi
Pintomyia	quasitownsendi
Pintomyia	reclusa
Pintomyia	robusta
Pintomyia	salomoni
Pintomyia	sauroida
Pintomyia	serrana
Pintomyia	sp. de Anchicaya
Pintomyia	spinicrassa
Pintomyia	suapiensis
+Pintomyia	succini
Pintomyia	rangeliana
Pintomyia	tihuiliensis
Pintomyia	tocaniensis
Pintomyia	torrealbai
Pintomyia	torresi
Pintomyia	torvida
Pintomyia	townsendi
Pintomyia	valderramai
Pintomyia	verrucarum
Pintomyia	youngi
Pressatia	calcarata
Pressatia	camposi
Pressatia	choti
Pressatia	duncanae
Pressatia	dysponeta
Pressatia	equatorialis
Pressatia	triacantha
Pressatia	trispinosa

Psathyromyia	abonnenci
Psathyromyia	abunaensis
Psathyromyia	aclydifera
Psathyromyia	antezanai
Psathyromyia	aragaoi
Psathyromyia	baratai
Psathyromyia	barretti
Psathyromyia	barrettoi barrettoi
Psathyromyia	barrettoi majuscula
Psathyromyia	bigeniculata
Psathyromyia	coutinhoi
Psathyromyia	cratifer
Psathyromyia	dasymera
Psathyromyia	dendrophyla
Psathyromyia	digitata
Psathyromyia	dreisbachi
Psathyromyia	guatemalensis
Psathyromyia	brasiliensis
Psathyromyia	campbelli
Psathyromyia	campograndensis
Psathyromyia	carpenteri
Psathyromyia	castilloi
Psathyromyia	dreisbachi
Psathyromyia	hermanlenti
Psathyromyia	inflata
Psathyromyia	lanei
Psathyromyia	lerayi
Psathyromyia	limai
Psathyromyia	lutziana
Psathyromyia	maya
Psathyromyia	naftalekatzi
Psathyromyia	pascalei
Psathyromyia	pelloni
Psathyromyia	pifanoi
Psathyromyia	pradobarrientosi
Psathyromyia	punctigeniculata
Psathyromyia	ribeirensis
Psathyromyia	runoides
Psathyromyia	ruparupa
Psathyromyia	scaffi
+ Psathyromyia	schleei
Psathyromyia	shannoni
Psathyromyia	soccula
Psathyromyia	souzacastroi
Psathyromyia	texana
Psathyromyia	undulata

Psathyromyia	volcanensis
Psychodopygus	amazonensis
Psychodopygus	arthuri
Psychodopygus	ayrozai
Psychodopygus	bernalei
Psychodopygus	fairchildi
Psychodopygus	fairtigi
Psychodopygus	francoisleponti
Psychodopygus	geniculatus
Psychodopygus	guyanensis
Psychodopygus	bispinosus
Psychodopygus	carrerai carrerai
Psychodopygus	carrerai thula
Psychodopygus	chagasi
Psychodopygus	claustrei
Psychodopygus	complexus
Psychodopygus	corossoniensis
Psychodopygus	davisi
Psychodopygus	dorlinsis
Psychodopygus	douradoi
Psychodopygus	hirsutus hirsutus
Psychodopygus	hirsutus nicaraguensis
Psychodopygus	joliveti
Psychodopygus	killicki
Psychodopygus	lainsoni
Psychodopygus	leonidasdeanei
Psychodopygus	llanosmartinsi
Psychodopygus	lloydi
Psychodopygus	luisleoni
Psychodopygus	matosi
Psychodopygus	nocticolus
Psychodopygus	panamensis
Psychododpygus	paraensis
Psychodopygus	parimaensis
Psychodopygus	recurvus
Psychodopygus	sp. de Trés Esquinas
Psychodopygus	squamiventris maripaensis
Psychodopygus	squamiventris squamiventris
Psychodopygus	wellcomei
Psychodopygus	yasuniensis
Psychodopygus	yucumensis
Sciopemyia	fluviatilis
Sciopemyia	microps
Sciopemyia	nematoducta
Sciopemyia	pennyi
Sciopemyia	preclara

Sciopemyia	servulolimai
Sciopemyia	sordellii
Sciopemyia	vattierae
Trichophoromyia	acostai
Trichophoromyia	adelsonsouzai
Trichophoromyia	arevaloi
Trichophoromyia	auraensis
Trichophoromyia	beniensis
Trichophoromyia	bettinii
Trichophoromyia	brachipyga
Trichophoromyia	castanheirai
Trichophoromyia	cellulana
Trichophoromyia	clitella
Trichophoromyia	dunhami
Trichophoromyia	eurypyga
Trichophoromyia	flochi
Trichophoromyia	gibba
Trichophoromyia	howardi
Trichophoromyia	incasica
Trichophoromyia	ininii
Trichophoromyia	lopesi
Trichophoromyia	loretonensis
Trichophoromyia	meirai
Trichophoromyia	melloi
Trichophoromyia	napoensis
Trichophoromyia	natauensis
Trichophoromyia	nemorosa
Trichophoromyia	octavioi
Trichophoromyia	omagua
Trichophoromyia	pabloi
Trichophoromyia	pastazaensis
Trichophoromyia	readyi
Trichophoromyia	reburra
Trichophoromyia	reinerti
Trichophoromyia	rostrans
Trichophoromyia	ruii
Trichophoromyia	ruifreitasi
Trichophoromyia	saltuosa
Trichophoromyia	sinuosa
Trichophoromyia	sp. 1 de Araracuara
Trichophoromyia	ubiquitalis
Trichophoromyia	uniniensis
Trichophoromyia	velascoi
Trichophoromyia	viannamartinsi
Trichophoromyia	wilkersoni
Trichopygomyia	conviti

Trichopygomyia	dasypodogeton
Trichopygomyia	depaquiti
Trichopygomyia	elegans
Trichopygomyia	ferroae
Trichopygomyia	gantieri
Trichopygomyia	longispina
Trichopygomyia	martinezi
Trichopygomyia	pinna
Trichopygomyia	ratcliffei
Trichopygomyia	rondoniensis
Trichopygomyia	trichopyga
Trichopygomyia	triramula
Trichopygomyia	turelli
Trichopygomyia	wagleyi
Trichopygomyia	witoto
Viannamyia	caprina
Viannamyia	fariasi
Vinnamyia	furcata
Viannamyia	tuberculata
Warileya	euniceae
Warileya	fourgassiensis
Warileya	leponti
Warileya	lumbrerasi
Warileya	nigrosaccula
Warileya	phlebotomanica
Warileya	rotundipennis
Warileya	yungasi

Tribes
Hertigiini
Phlebotomini

Subtribes
Australophlebotomina
Brumptomyiina
Hertigiina
Idiophlebotomina
Lutzomyiina
Phlebotomina
Psychodopygina
Sergentomyiina

Genera
Australophlebotomus
Bichromomyia
Brumptomyia

Chinius
Dampfomyia
Deanemyia
Edentomyia
Evandromyia
Expapillata
Hertigia
Idiophlebotomus
+Libanophlebotomites
Lutzomyia
Martinsmyia
+Mesophlebotomites
Micropygomyia
Migonemyia
Nyssomyia
Oligodontomyia
+Palaeomyia
+Phlebotoiella
+Phlebotomiella
+Phlebotomites
Phlebotomus
Pintomyia
Pressatia
Psathyromyia
Psychodopygus
Sciopemyia
Sergentomyia
Spelaeophlebotomus
Trichophoromyia
Trichopygomyia
Viannamyia
Warileya

Subgenera
(Aldamyia)
(Barrettomyia)
(Blancasmyia)
(Castromyia)
(Coquillettimyia)
(Coromyia)
(Dampfomyia)
(Evandromyia)
(Forattiniella)
(Helcocyrtomyia)
(Lutzomyia)
(Micropygomyia)

(Migonemyia)
(Pifanomyia)
(Pintomyia)
(Psathyromyia)
(Sauromyia)
(Silvamyia)
(Tricholateralis)
(Xiphopsathyromyia)
Index of species groups of the Americas
Alphabetica
Delpozoi
Gasparviannai
Index of species series of the Americas
Arthuri
Atroclavata
Bispinosus
Cayennensis
Chagasi
Chiapanensis
Cortelezzii
Davisi
Envansi
Infraspinosa
Guyanensis
Lanei
Monstruosa
Monticola
Osornoi
Oswaldoi
Pacae
Panamensis
Peruensis
Pia
Pilosa
Rupicola
Sanguinaria
Saulensis
Serrana
Shannoni
Townsendi
Tupynambai
Verrucarum
Vexator

Index of synonyms

Species

Phlebotomus	*basispinosus*
Phlebotomus	*cauchensis*
Phlebotomus	*christophersoni*
Phlebotomus	*colasbelcouri*
Phlebotomus	*falciformis*
Phlebotomus	*foliatus*
Phlebotomus	*gasti*
Phlebotomus	*guadeloupensis*
Phlebotomus	*guayasi*
Phlebotomus	*hansoni*
Phlebotomus	*heckenrothi*
Phlebotomus	*humboldti*
Phlebotomus	*intermedius acutus*
Phlebotomus	*intermedius acutus*
Phlebotomus	*intermedius longiductus*
Phlebotomus	*japignyi*
Phlebotomus	*longicornutus*
Phlebotomus	*longiductus*
Phlebotomus	*lutzi*
Phlebotomus	*machicouensis*
Phlebotomus	*marajoensis*
Phlebotomus	*mazzai*
Phlebotomus	*microcephalus*
Phlebotomus	*monticolus incarum*
Phlebotomus	*montoyai*
Phlebotomus	*nordestinus*
Phlebotomus	*otamae*
Phlebotomus	*pestanai*
Phlebotomus	*pifanoi*
Phlebotomus	*pinealis*
Phlebotomus	*pinotti, non* Damasceno & Arouck
Phlebotomus	*rachoui*
Phlebotomus	*rangeli*
Phlebotomus	*rooti*
Phlebotomus	*rubidulus*
Phlebotomus	sp. Floch & Abonnenc
Phlebotomus	sp. 1 de Baduel
Phlebotomus	sp. II de Baduel
Phlebotomus	sp. B du Gallion
Phlebotomus	sp. C Velasco
Phlebotomus	sp. de Cayenne

Phlebotomus	sp. de Crique Anguille
Phlebotomus	sp. de Maripa
Phlebotomus	sp. de Rorota
Phlebotomus	sp. de Saul
Phlebotomus	sp. de Souvenir
Phlebotomus	sp. M
Phlebotomus	sp. n° 768
Phlebotomus	sp. O
Phlebotomus	sp. X
Phlebotomus	*spinosus*
Phlebotomus	*squamiventris, non* Lutz & Neiva
Phlebotomus	*sylvestris, non* Sinton
Phlebotomus	*tejerae*
Phlebotomus	*tikalaensis*
Phlebotomus	*vexator occidentis*
Phlebotomus	*vexillarius*
Phlebotomus	*yucatanensis*
Phlebotomus	*yucatanensis baduelensis*
Phlebotomus	*zulianensis*
Psychodopygus	sp. no. 401.63
Sergentomyia	*pessoana*

Tribe
Australophlebotomini

Genera of the Americas
Françaia
Lutzia
Lutziola
Lutziomyia

Subgenera of the Americas
(Aguayoi)
(Anthophorus)
(Eupsychodopygus)
(Isolutzomyia)
(Oophoromyia)
(Shannonomyia)
(Shannonomyina)
(Xiphomyia) non Townsend

References

Abonnenc E, Leger N (1976a) Sur une classification rationnelle des Diptères Phlebotomidae. Cah ORSTM Entomol Méd Parasit 14:69–78

Abonnenc E, Leger N (1976b) Rectificatif à la note "Sur une classification rationnelle des Diptères Phlebotomidae". Cah ORSTM Entomol Méd Parasit 14:357

Abonnenc E, Pastre J, Quetin RM (1971) Contribution à l'étude de l'exosquelette pleural du thorax des Phlébotomes (Diptera: Psychodidae). Cah ORSTOM Entomol Méd Parasit 9(4):379–385

Abonnenc E, Leger N, Fauran P (1979) Sur deux nouveaux phlébotomes de la Guyane Française: *Lutzomyia yuilli pajoti* n.ssp. et *Lutzomyia claustrei* n. sp. Bull Soc Path Exot 72:75–86

Addis CJ (1945) *Phlebotomus* (*Dampfomyia*) *anthophorus*, n. sp. and *Phlebotomus diabolicus* Hall from Texas (Diptera: Psychodidae). J Parasitol 31:119–127

Adler S, Theodor O (1926) The mouth parts, alimentary tract and salivary apparatus of the females in *Phlebotomus papatasi*. Ann Trop Med Parasit 20:109–142

Adler S, Theodor O (1929) The distribution of sandflies and leishmaniasis in Palestine, Syria and Mesopotamia. Ann Trop Med Parasit 23:269–306

Agassiz L (1846) Nomenclatoris Zoologici Index Universalis. Soloduri, pg. 285

Alexander CP (1920) A new subfamily of Tanyderid flies (Diptera). Ann Entomol Soc Amer 13:402–406

Alexander CP (1928) The Australian species of the genus *Nemopalpus* Psychodidae, Diptera. Proc Linn Soc N S Wales 53:291–294

Alexander CP (1929) A revision of the American two-winged flies of the subfamily Bruchomyiinae. Proc US Nat Mus 75:1–9

Alexander CP (1940) Further observations on the Psychodid subfamily Bruchomyiinae (Diptera). Rev Entomol 11:793–799

Andrade AJ, Shimabukuro PHF, Galati EAB (2013) On the taxonomic status of *Phlebotomus breviductus* Barretto, 1950 (Diptera: Psychodidae: Phlebotominae). Zootaxa 3734(4):477–484

Andrade AJ, Shimabukuro PHF, Galati EAB (2014) The taxonomic status of *Phlebotomus oliverioi* (Diptera, Psychodidae) and a lectotype designation for *Psathyromyia brasiliensis*. Iheringia, Série Zoologia 104(4):426–431

Artemiev MM (1991) A classification of the subfamily Phlebotominae. (*Ist. Intern. Symp. Phleb. Sandflies*). Parassitologia 33(suppl. 1):69–77

Azar D, Nel A, Solignac M, Paichelei JC, Bouchet F (1999) New genera and species of psychodoid flies from the lower cretaceous amber of Lebanon. Paléo 42(6):1101–1136

Barrett TV, Freitas RA, Albuquerque MIC, Guerrero JMH (1996) Report on a collection of *Lutzomyia* sand flies (Diptera: Psychodidae) from the middle Solimões (Amazonas, Brazil). Mem Inst Oswaldo Cruz 91:27–35

Barretto MP (1946a) Sôbre a sinonímia de flebótomos americanos (Diptera, Psychodidae). Primeira nota. Rev Bras Biol 6:527–536

Barretto MP (1946b) Sôbre a sinonímia de flebótomos americanos (Diptera, Psychodidae). Segunda nota. Ann Fac Med Univ S Paulo 22:1–27

Barretto MP (1947) Catálogo dos flebótomos americanos. Arq Zool Est S Paulo 15:177–242

Barretto MP (1950a) Uma questão de nomenclatura: *Phlebotomus* ou *Flebotomus*? Fol Clin Biol 16:199–202

Barretto MP (1950b) Nova contribuição para o estudo da distribuição geográfica dos flebotomíneos americanos (Diptera, Psychodidae). Arch Hig Saúde Pública 15:211–226

Barretto MP (1953) Sôbre a sinonímia de flebótomos americanos (Diptera, Psychodidae). Terceira Nota. Fol Clin Biol 20:209–214

Barretto MP (1955) Sôbre a sistemática da subfamília Phlebotominae Rondani (Diptera, Psychodidae). Revta Bras Ent 3:173–190

Barretto MP (1961) Subfamílias e gêneros neotropicais da família Psychodidae Big., 1854 (Diptera). Pap Avuls Dep Zool Sec Agr S Paulo 14:211–225

Barretto MP (1962) Novos subgêneros de *Lutzomyia* França, 1924 (Psychodidae, Subfamília Phlebotominae). Rev Inst Med Trop S Paulo 4:91–100

Barretto MP, Pessôa SB 1946 Contribuição para o estudo da distribuição geográfica dos flebótomos americanos (Diptera: Psychodidae). *In Livro de Homenagem a Romualdo Ferreira D'Almeida*, São Paulo, Imprensa Oficial do Estado, p 77–102

Bigot M (1854) Essai d· une classification générale en synoptique de l·ordre des insects Diptères. Ann Soc Entomol Fr 3:447–482

Bonhag PF (1949) The thoracic mechanism of the adult horsefly (Diptera: Tabanidae). Mem Cornell Univ Agric Exp Stn 285:3–39

Castro M 1959. Nota sobre Flebotominae Argentinos con la descripcion de Flebotomus misionensis sp. n. (Dipt. Psych.). In Bejarano JFR, Del Ponte E, Orfila RN (eds) Primeras Jornadas Entomoepidemiologicas Argentinas. 2, p 331–334

Christensen HA, Rutledge LC (1973) Two new synonyms in the genus *Lutzomyia* (Diptera, Psychodidae). J Med Entomol 10:314

Christophers SP, Barraud PJ (1926) The development of the male and female hypopygium of *Phlebotomus*. Indian J Res Med Calcuta 13:853–857

Cipa Group (1991) Proposition of a standard description for Phlebotomine sandflies (*Ist. Intern. Symp. Phleb. Sandflies*). Parassitologia 33(suppl.1):127–135

Comstock JH (1918) The wings of insects. Comstock Publishing Co, Ithaca

Coquillett DW 1907 Discovery of Blood-sucking Psychodidae in America. Entomol News Proc 18:101–102, 177

Cordero EH, Vogelsang EG, Cossio V 1928 *Phlebotomus gaminarai* n. sp., nueva especie de flebótomo del Uruguay. *Bol Inst Clin Quirurg* Buenos Aires 4:649–52

Costa Lima A (1932) Sôbre os phlebótomos americanos (Diptera: Psychodidae). Mem Inst Oswaldo Cruz 26:15–69

Crampton GC 1942 The external morphology of the Diptera: In Guide to the insects of Connecticut, fasc. 1. p 10–165

Cumming JM, Wood DM (2009) Adult morphology and terminology. In: Brown BV, Borkent A, Cumming JM, Wood DM, Woodley NE, Zumbado MA (eds) Manual of Central American Diptera, vol 1. National Research Council of Ottawa, Ottawa, pp 9–502. https://doi.org/10.1899/30.3.br.868

Dampf A (1944) Notas sobre flebotómidos mexicanos. Observaciones generales y descripcion de dos especies nuevas (*Phlebotomus oppidanus* y *Phl. vindicator*) encontradas en la Ciudad de Mexico (Ins. Diptera, Fam. Phlebotomidae). Rev Soc Mex Hist Nat 5:237–254

Davis N (1967) Leishmaniasis in the Sudan Republic. 28. Anatomical studies on *Phlebotomus orientalis* Parrot and *P. papatasi* Scopoli (Diptera: Psychodidae). J Med Entomol 4:50–65

Dias ES, Martins AV, Falcão AL, Silva JE (1986) Descrição de uma nova espécie do subgênero *Helcocyrtomyia, Lutzomyia pusilla* sp. n. (Diptera, Psychodidae, Phlebotominae). Mem Inst Oswaldo Cruz 81:395–399

Duckhouse DA (1973) A catalogue of the Diptera of the Americas South of the United States. Pap Avuls Dep Zool Sec Agr S Paulo 6A:1–29

Dyar HG (1929) The present knowledge of the American species of *Phlebotomus* Rondani (Diptera, Psychodidae). Amer J Hyg 10:112–124

Dyar HG, Nuñez-Tovar M (1926/1927) Notes on biting flies from Venezuela. Insect Inscit Menst 14:152–155

Edwards FW (1921) A note in the subfamily Bruchomyiinae (Diptera, Nematocera). Ann Nat Hist London 7:437–439

Fairchild GB (1949) A new fly related to *Phlebotomus* from Panama (Diptera: Psychodidae). Proc Entomol Soc Wash 51:81–84

Fairchild GB (1955) The relationships and classification of the Phlebotominae (Diptera, Psychodidae). Ann Entomol Soc Amer 48:182–196

Fairchild GB, Hertig M (1948) Notes on the *Phlebotomus* of Panama (Diptera, Psychodidae). III. *P. cruciatus, trinidadensis* and *gomezi*. Ann Entomol Soc Amer 41:247–257

Fairchild GB, Hertig M (1950) Notes on the *Phlebotomus* of Panama (Diptera, Psychodidae). VI. *Phlebotomus shannoni* Dyar and related species. Ann Entomol Soc Amer 43:523–533

Fairchild GB, Hertig M (1952) Notes on the *Phlebotomus* of Panama (Diptera, Psychodidae). IX. Description of seven new species. Ann Entomol Soc Amer 45:505–528

Fairchild GB, Hertig M (1957) Notes on the *Phlebotomus* of Panama. XIII. The vexator group, with descriptions of new species from Panama and California. Ann Entomol Soc Amer 50:325–334
Fairchild GB, Hertig M (1958) Notes on the *Phlebotomus* of Panama (Diptera, Psychodidae). XV. Four apparently new synonymies. Proc Entomol Soc Wash 60:203–205
Fairchild GB, Hertig M (1961a) Notes on the *Phlebotomus* of Panama. XVI. (Diptera, Psychodidae). Description of new and little-know species from Panama and Central America. Ann Entomol Soc Amer 54:237–255
Fairchild GB, Hertig M (1961b) Three new species of *Phlebotomus* from Mexico and Nicaragua (Diptera: Psychodidae). Proc Entomol Soc Wash 63:22–28
Feliciangeli MD (1985) Taxonomic review of *Lutzomyia walkeri* (Newstead,1914) [= *Lutzomyia marajoensis* (Damasceno & Causey, 1944)] and the resurrection of *Lutzomyia dubitans* (Sherlock, 1962) (Diptera, Psychodidae). Mem Inst Oswaldo Cruz 80:307–311
Floch H, Abonnenc E 1945. Phlébotomes de la Guyane Française (XV). *P. elongatus, P. syvicolus, P. trichopygus* n. sp., *P. brasiliensis, P. barrettoi* et *P*. sp. de Crique Anguille. *Inst Pasteur Guyane* 106: 16
Floch H, Abonnenc E 1952. Dipteres phlébotomes de la Guiane et des Antilles Françaises. *Faune de L·Union Français*. 14
Floch H, Kramer R 1965 Phlébotomes du Venezuela (IV). *P. vexillarius* Fairchild et Hertig 1952 et *P. foliatus* Mirsa et Ortiz 1953 synonymes de *P. lichyi* Floch et Abonnenc 1950. *Inst Pasteur Guyane* 496: 8
Forattini OP (1971) Sôbre a classificação da subfamília Phlebotominae nas Américas (Diptera, Psychodidae). Pap Avuls Zool 24:93–111
Forattini OP 1973 *Entomologia Médica. IV: Psychodidae. Phlebotominae. Leishmanioses. Bartonelose*. São Paulo, Ed. Edgard Blücher Ltda. p658
Fraiha H, Shaw JJ, Lainson R (1970) *Lutzomyia falciformis* (Floch & Abonnenc, 1944) sinônimo de *Lutzomyia monstruosa* (Floch & Abonnenc, 1944). Rev Saúde Pública 4:215–216
Fraiha H, Shaw JJ, Lainson R (1978) Fauna antropófila de flebótomos da rodovia Transamazônica, Brasil. Bull PAHO 84:134–139
Fraiha H, Ward RD, Ready P (1981) Flebotomíneos brasileiros V. *Lutzomyia* (*Nyssomyia*) *shawi* n. sp., nova espécie antropófila de flebótomo da Amazônia (Diptera, Psychodidae). Rev Brasil Biol 41:699–703
Fraiha H, Ryan L, Ward RD, Lainson R, Shaw JJ (1986) *Psychodopygus leonidasdeanei*, a new species of sand fly (Diptera: Psychodidae) from Pará State, Brazil. Mem Inst Oswaldo Cruz 81:333–339
França C (1919) Observations sur le genre Phlebotomus. Brotéria Ser Zool 17:102–160
França C (1920) Observations sur le genre Phlebotomus. II. Phlébotomes du Nouveau Monde (Phlébotomus du Brésil et du Paraguay). Bull Soc Port Sci Nat 8:215–236
França C (1924) Notes parasitologiques IV. Phlebotomes. J Sciênc Mat Fis Nat 3:22–25
França C, Parrot L (1920) Introducion à l· étude systématique des Diptères du genre *Phlebotomus*. Bull Soc Path Exot 13:695–708
França C, Parrot L (1921) Essai de classification des Phlébotomes. Arch Inst Pasteur Afr N 1:279–284
Galati EAB (1981) Sobre a identificação de *Psychodopygus* (*Psychodopygus*) *arthuri* (Fonseca, 1936) e *P.* (*P.*) *lloydi* (Antunes, 1937) (Diptera, Psychodidae, Phlebotominae). Revta Bras Ent 25:321–322
Galati EAB (1995) Phylogenetic systematics of the Phlebotominae (Diptera, Psychodidae) with emphasis on American groups. (*II Intern. Symp. Phlebotomine Sandflies*). Bol Dir Malariol Saneam Amb 35(supl. 1):133–142
Galati EAB (2003a) Classificação de Phlebotominae. In Rangel EF, Lainson R. Flebotomíneos do Brasil. Rio de Janeiro, Ed. FIOCRUZ, pp 23–51
Galati EAB (2003b) Classificação de Phlebotominae. In Rangel EF, Lainson R. Flebotomíneos do Brasil. Rio de Janeiro, Ed. FIOCRUZ, pp 53–175
Galati EAB, Cáceres AG (1994) Descrição de *Lutzomyia pallidithorax*, sp. n. e de *Lutzomyia castanea*, sp.n. do Peru e análise cladística das séries dos subgênero *Helcocyrtomyia* Barretto (Diptera, Psychodidae) Rev Bras Entomol 38(2):471–488
Galati EAB, Cáceres AG (1999) Descrição de três espécies novas de Phlebotominae (Diptera, Psychodidae) do Departamento de Pasco-Peru. Rev Bras Ent 43(3/4):293–299

Galati EAB, Andrade Filho JD, Falcão AL (2002) *Micropygomyia ferreirana* (Barretto, Martins & Pelegrino) as senior synonym of *Micropygomyia borgmeieri* (Martins, falcão & Silva) and denunciation of an error of identification related to this taxon. Entomol. Vect. *9* (Supl.) ISOPS IV: 156–157

Galati EAB, Andrade Filho JD, Silva ACL, Falcão AL (2003) Description of a new genus and new species of New World Phlebotominae (Diptera, Psychodidae). Rev bras Ent 47(1):63–70

Galati RAB, Le Pont F, Galvis Ovallos F (2011) Fonsecai complex of the genus *Lutzomyia* (Diptera, Psychodidae, Phlebotominae). Abstracts: 7th International Symposium on Phlebotomine Sand Flies. 25–30 Apri, 2011m Turkey

Godoy RS, Galati EAB (2016) Revalidation of *Nyssomyia fraihai* (Martins, Falcão & Silva, 1979) (Diptera: Psychodidae). J Med Entomol:1–9

Gomes AC, Rabello EX, Galati EAB (1978) Flebotomíneos encontrados em galinheiros experimentais nos Estados de São Paulo e Minas Gerais (Brasil) e algumas observações ecológicas. Rev Saúde Pública 12:403–407

Grassi B (1907) Ricerche sui Flebotomi. Mem Mat Fis Soc Ital Sc 14:353–393

Hennig W (1972) Insektenfossilien aus der unteren Kreide. IV. Psychodidae (Phlebotominae), mit einer kritischen Übersicht über das phylogenetische System der Familie und die bisher beschriebenen Fossilien (Diptera). Stutt Beitr Naturkende 241:1–69

Hertig M (1948) A new genus of bloodsuking Psychodids from Peru (Diptera: Psychodidae). Ann Entomol Soc Amer 41:8–16

Ibáñez-Bernal S (1999) Phlebotominae (Diptera: Psychodidae) de México. I. *Brumptomyia* França y Parrot; *Lutzomyia* França, las especies de *L.* (*Lutzomyia*) França y del grupo *Verrucarum*. Fol Entomol Mex 107:61–116

Ibáñez-Bernal S (2001) Notes on the Psychodidae (Diptera) of Belize: Subfamilies Bruchomyiinae and Phlebotominae. Ann Entomol Soc Amer 94:367–385

Ibánez-Bernal S, Marina CF, Muñoz J, Rebollar-Téllez EA, Pech-May A (2015) Phlebotomine sand flies (Diptera: Psychodidae) of Chiapas collected near the Guatemala border, with additions to the fauna of Mexico and a new subgenus name. Zootaxa 3994(2):151–186

Ilango K (2004) Phylogeny of the Old World Phlebotomine sandflies (Diptera, Psychodidae) with special reference to structural diversity of female spermathecae. Orient Insects 38:419–462

International Comission on Zoological Nomenclature (ICZN) (1950) "Flebotomus" Rondani, 1840: (Class Insecta, Order Diptera). Emendation of name to "Phlebotomus". Bull Zool Nomencl 4:358–359

International Comission on Zoological Nomenclature (ICZN) (1954) Opinion 256: Emendation to *Phlebotomus* of the generic name *Flebotomus Rondani*, 1840 (Class Insecta, Order Diptera) under Plenary Powers. Opinions and Declarations rendered by the International Comission on Zoological Nomenclature 5:199–230

International Comission on Zoological Nomenclature (ICZN) (1999). INTERNATIONAL CODE OF ZOOLOGICAL NOMENCLATURE *online*. The International Trust for Zoological Nomenclature. http://www.iczn.org/iczn/index.jsp. Accessed 3 Oct 2017

International Comission on Zoological Nomenclature (ICZN) (2017). Frequently asked questions. What is a nomen nudum? http://iczn.org/faqs#faqs-page-34. Accessed 29 Sept 2017

Jobling B (1987) Anatomical drawings of biting flies. British Museum (Natural History)/Wellcome Trust, London

Lainson R, Ward RD, Young DG, Shaw JJ, Fraiha H (1976) Preliminary entomological and parasitological studies in Humboldt, Aripuanã, Mato Grosso State, Brazil. Acta Amazonica (suplem) 6:55–60

Lane RP, Ward RD (1984) The morphology and possible function of abdominal patches in males of two forms of leishmaniasis vector *Lutzomyia longipalpis* (Diptera: Phlebotominae). Cah ORSTOM Entomol Méd Parasit 22:245–249

Larrousse F (1921) Étude systématique et médicale des Phlébotomes. Thèse médecine. Vigot frères, Paris, 106 p

Le Pont F, Martinez E, Torrez-Espejo JM, Dujardin JP (1998) Phlébotomes de Bolivie: description de cinq nouvelles espèces de *Lutzomyia* de la région subandine (Diptera, Psychodidae). Bull Soc Entomol Fr 103(2):159–173

Lebbe J, Vignes R, Dedet JP 1987 *Identification assistée par ordinateur des phlébotomes de la Guyane Française (Diptera: Psychodidae).* Cayenne, Inst. Pasteur de la Guyane Française.[Ed. l· Institut Pasteur de la Guyane Française à l·occasion du centenaire de l· Institut Pasteur], 165 p
Leger N, Abonnenc E, Pajot FX, Kramer R, Claustre J (1977) Liste comentée de Phlébotomes de la Guyane française. Cah ORSTOM Entomol Méd Parasit 15(3):217–232
Leng YJ (1987) A preliminary survey of phlebotomine sandflies in limestone caves of Sichuan and Guizhou province, south-west China, and description and discussion of a primitive new genus *Chinius*. Ann Trop Med Parasit 81:311–317
Lewis DJ (1975) Functional morphology of mouth parts in New World phlebotomine sandflies (Diptera:Psychodidae). Trans R Entomol Soc London 126:497–532
Lewis DJ, Young DG, Fairchild GB, Minter DM (1977) Proposals for a stable classification of the phlebotomine sandflies (Diptera: Psychodidae). System Entomol 2:319–332
Llanos BZ 1966. Nueva denominación para las especies *Phlebotomus townsendi* n. sp. y *Phlebotomus adleri* n. sp. Llanos, 1964. Rev Brasil Malariol Doen Trop 18: 369
Lutz A, Neiva A (1912) Contribuição para o conhecimento das espécies do gênero *Phlebotomus* existentes no Brasil. Mem Inst Oswaldo Cruz 4:82–95
Malak MA, Salamé Y, Azar D (2013) New phlebotomine flies from Burmese Amber (Diptera: Psychodidae: Phlebotominae). Terrestrial Arthropod Reviews 6:81–101
Mangabeira O (1941a) 3ª contribuição ao estudo dos Flebotomus. *Evandromyia* n. subg. (Diptera; Psychodidae). Mem Inst Oswaldo Cruz 36:215–223
Mangabeira O (1941b) 4ª contribuição ao estudo dos Flebotomus. *Psychodopygus* n. subg. (Diptera: Psychodidae). Mem Inst Oswaldo Cruz 36:237–250
Mangabeira O (1941c) 5ª contribuição ao estudo dos Flebotomus. *Viannamyia* n. subg. (Diptera: Psychodidae). Mem Inst Oswaldo Cruz 36:251–262
Mangabeira O (1942) 7ª contribuição ao estudo dos Flebotomus (Diptera: Psychodidae). Descrição dos machos de 24 novas espécies. Mem Inst Oswaldo Cruz 37:111–218
Marcondes CB (1996) A redescription of *Lutzomyia* (*Nyssomyia*) *intermedia* (Lutz & Neiva, 1912), and resurrection of *L. neivai* (Pinto, 1926) (Diptera, Psychodidae, Phlebotominae). Mem Inst Oswaldo Cruz 91:457–462
Marcondes CB (2007) A proposal of generic and subgeneric abbreviations for phlebotomine sandflies (Diptera: Psychodidae: Phlebotominae) of the world. Entomol News 118:351–356
Martins AV (1970) *Lutzomyia (Psychodopygus) fairtigi* n. sp. from Colombia. (Diptera: Psychodidae): Phlebotominae. Proc Entomol Soc Wash 72:279
Martins AV, Silva JE (1965) *Barretomyia* subg. n., novo subgênero de *Lutzomyia* França, 1924 (Diptera, Psychodidae), com a descrição de uma espécie nova. Ci Cult 17:235–236
Martins AV, Silva JE (1968) *Barrettomyia* subg. n., novo subgênero de *Lutzomyia* França, 1924 (Diptera, Psychodidae), com a descrição de uma espécie nova. *Bol Mus Hist Nat Univ Fed M Gerais Zool 3*: 3
Martins AV, Maciel CS, Silva JE (1968) Nota sobre os flebotomus do grupo *squamiventris* do subgênero *Psychodopygus* Mangabeira, 1941 (Diptera, Psychodidae). Bol Mus Hist Nat Univ Fed M Gerais Zool 1:1–33
Martins AV, Williams P, Falcão AL (1978) American sand flies. Academia Brasileira de Ciências, Rio de Janeiro, 195 p
Matsuda R (1970) Morphology and evolution of the insect thorax. Mem Entomol Soc Canada 76:1–431
McAlpine JF (1981) Morphology and terminology – adults. In: McAlpine JF, Peterson BV, Shewell GE, Teskey HJ, Vockeroth JR, Wood DM (eds) Manual of Neartic Diptera, vol 1. Research Branch Agric. Monograph N° 27, Canada, pp 9–63
Newman E (1834) Attempted division of British insects into natural orders. Ent Mag 2:379–431
Newstead R (1911) The papataci flies (*Phlebotomus*) of the Maltese Islands. Bull Entomol Res 2:27–78
Newstead R (1912) Notes on *Phlebotomus*, with descriptions of new species. Part I. Bul Entomol Res 3:361–367
Newstead R (1914) Notes on *Phlebotomus*, with descriptions of new species. Part. II. Bull Entomol Res 5:179–192
Nitzulescu V (1931) Essai de classification des phlébotomes. Ann Parasit Hum Comp 9:271–275
Ortiz I, Scorza JV (1963) Notas biologicas y taxonomicas sobre algunos Phlebotominae (Diptera, Psychodidae) de Rancho Grande, Venezuela. Acta Biol Venez 3:341–361

Parrot L (1951) Notes sur les Phlébotomes. LXI: a propos de classification. Arch Inst Pasteur Algér 29:28–45

Parrot L (1953) Notes sur les Phlébotomes. LXVII. - "Les papilles" des antennes. Arc Inst Pasteur d'Algerie 31(2):110–116

Pifano F, Álvarez A, Ortiz I (1962) Investigaciones sobre la leishmaniasis tegumentaria en los Valles del Tuy. Venezuela. Arch Venez Med Trop Parasit 4:149–155

Poinar (2004) *Palaeomyia burmitis* (Diptera: Phlebotominae) a new genus and species of cretaceous sandflies with evidence of blood-sucking habits. Proc Entomol Wash 106(3):598–605

Pratt HD (1947) *Shannonomyiina*, new name for *Shannonomyia* Dyar (not Alexander) (Diptera, Psychodidae). Proc Entomol Soc Wash 49:86

Ready P, Fraiha H (1981) Brazilian phlebotomines. VI. *Lutzomyia richardwardi* sp. n., a new species of *Nyssomyia* from Amazonia with keys for this subgenus (Diptera, Psychodidae). Rev Brasil Biol 41:705–712

Ready P, Fraiha H, Lainson R, Shaw JJ (1980) *Psychodopygus* as a genus: reasons for a flexible classification of the phlebotomine sand flies (Diptera: Psychodidae). J Med Entomol 17:75–88

Redtenbacher J (1886) Vergleichende Studien über das Flügelgeäder der Insekten. Ann Naturh Mus Win 1:153–232

Rohdendorf BB (1964) The historical development of Diptera. *Trudy Paleont Inst 100*: 1–312. Tradução inglesa: University of Alberta Press, 1974

Rondani C (1840) *Sopra una specie di insetto dittero. Memoria prima per servire alla Ditterologia italiana*. Parma 1.

Rondani C (1843) Species italicae generis *Hebotomi*, Rndn ex insectis dipteris: fragmentum septimum ad inserviendam dipterologiam italicam. Ann Soc Entomol Fr 1:263–267

Rosabal R, Trejos A (1964) *Phlebotomus* de El Salvador (Diptera, Psychodidae). I. Espécies conocidas. Rev Biol Trop 12:167–173

Rozeboom LE (1947) The identity of the Phlebotomus associated with the bartonellosis in Colombia. Ann Entomol Soc Amer 40:705–714

Ryan L (1986) *Flebótomos do Estado do Pará, Brasil. (Diptera, Psychodidae: Phlebotominae)*. Belém. Wellcome Parasitology Unit. Instituto Evandro Chagas, documento técnico n° 1, 154 p

Sábio PB, Andrade AJ, Galati EAB (2014) Assessment of the taxonomic status of some species included in the Shannoni Complex, with the description of a new species of *Psathyromyia* (Diptera: Psychodidae: Phlebotominae). J Med Entomol 51(2):331–341

Sabio PB, Brilhante AF, Quintana MG, Andrade AJ, Galati EAB (2016) On the synonyms of *Psathyromyia* (*Psathyromyia*) *shannoni* (Dyar, 1929) and *Pa. bigeniculata* (Floch & Abonnenc, 1941) and the resuscitation of *Pa. pifanoi* (Ortiz, 1972) with the description of its female (Diptera: Psychodidae: Phlebotominae). J Med Entomol:1–8

Sarà M, Smerdel A (1953) Morfologia del dermascheletro del torace di *Asilus crabroniformis* L (Dip. Asilidae). Ann Inst Mus Zool Univ Napoli 5(3):1–31

Scopoli JA (1786). *Deliciae florae et faune Insubricae*. Pavia, 85p

Sinton JA (1928) The synonymy of the Asiatic species of *Phlebotomus*. Indian J Med Res 16:297–324

Smith EL (1969) Evolutionary of external insect genitalia. 1. Origin and relationships to other appendages. Ann Entomol Soc Amer 62:1051–1079

Snodgrass RE (1944) The feeding apparatus of biting and sucking insects affecting man and animals. Smithson Misc Colls 104(7):1–113

Solórzano Kremer MM, Wagner R (2009) The first psychodid (Diptera: Psychodidae: Phlebotominae) species from the Lower Eocene amber of Vastan, Gujarat, India. Zootaxa 2152:63–68

Speight MC (1969) The prothoracic morphology of acalyptrates (Diptera) and use in systematics. Trans Entomol Soc London 121:325–421

Strand E (1932) Miscellanea nomenclatorica zoologica et paleontologica. IV. Diptera. Fol Zool Hidrobiol 4:195

Tang Y, Añez N (1995) The use of genital atrium to distinguishing parous and nulliparous neotropical sandflies. Bol Dir Malariol San Amb 35(Supl. 1):337–342

Theodor O (1932) Über eine neue Methode zur Klassifizierung von Phlebotomen und ihre Bedeutung für das studium der durch Phlebotomen übertragenen Krankheiten. Arch Zool Torino 16:1340–1349

Theodor O (1948) Classification of the world species of the subfamily Phlebotominae (Diptera, Psychodidae). Bull Entomol Res 39:85–115

Theodor O (1965) On the classification of American Phlebotominae. J Med Entomol 2:171–197

Tonnoir AL (1922) Notes sur le genre *Nemopalpus* (Dipt., Psychodidae) et description d·une espèce nouvelle. Ann Soc Entomol Belgique 62:125–136

Tonnoir AL (1933) Description of remarkable Indian Psychodidae and their early stages, with a theory of the evolution of the ventral suckers of dipterous larvae. Rec Indian Mus 35:53–75

Vargas L (1949). *Phlebotomus octavioi* n.n. Rev Inst Salub Enf Trop 10:267–268

Vargas L (1978) Los Lutzomyiinae como transmissores de enfermedades. *Forattiniella* n. subgen. y *Aguayoi* n. subgen. ambos del género *Lutzomyia* (Diptera: Phlebotominae). Rev Med México 58:85–93

Walker F (1851) Insecta Britannica Diptera, 1. Reeve & Benham, London

Ward RD, Ready P (1975) Chorionic sculpturing of some sandfly eggs (Diptera: Psychodidae). J Entomol (A) 50:127–134

Ward RD, Shaw JJ, Lainson R (1973) Leishmaniasis in Brazil. VIII. Observations on the phlebotomine fauna of an area highly endemic for cutaneous leishmaniasis, in the Serra dos Carajas, Pará State. Trans R Soc Trop Med Hyg 67:174–183

Williams P (1987) Description of *Lutzomyia* (*Coromyia*) *disneyi*, n. sp. (Diptera: Psychodidae) from Belize, Central America. Mem Inst Oswaldo Cruz 82:525–529

Williams P (1993) Relationships of Phlebotomine sand flies (Diptera). Mem Inst Oswaldo Cruz 88:177–183

Young DG (1979) A review of the bloodsucking Psychodid flies of Colombia (Diptera: Phlebotominae and Sycoracinae). Tech. Bull. 806, Agric. Exp. Station IFAS, Univ. Florida, Gainesville, p 226

Young DG, Arias JR (1977) *Lutzomyia* sand flies of the subgenus *Evandromyia* Mangabeira with description of a new species from Brazil (Diptera: Psychodidae). Acta Amazon 7:59–70

Young DG, Arias JR (1984) The *microps* group of *Lutzomyia* with descriptions of two species from South America (Diptera: Psychodidae: Phlebotominae). Mem Inst Oswaldo Cruz 79:425–431

Young DG, Duncan MA (1994) Guide to the identification and geographic distribution of *Lutzomyia* sand flies in Mexico, the West Indies, Central and South America (Diptera: Psychodidae). *Mem.* Amer Entomol Inst 54:1–881

Young DG, Fairchild GB (1974) Studies of Phlebotomine sand flies. Annual Report, Gainesville: Department of Entomology and Nematology, University of Florida, p 152

Young DG, Morales A (1987) A new species and records of phlebotomine sand flies from Colombia (Diptera, Psychodidae). J Med Entomol 24:651–665

Young DG, Perkins PV (1984) Phlebotomine sand flies of North America (Diptera: Psychodidae). Mosq News 44:263–304

Young DG, Porter H (1974) *Lutzomyia cirrita* n. sp. from Colombia with a new synonym in the genus (Diptera: Psychodidae: Phlebotominae). Fla Entomol 57:321–325

Molecular and Biochemical Markers for Investigating the Vectorial Roles of Brazilian Sand Flies

Paul D. Ready, Felipe M. Vigoder, and Elizabeth F. Rangel

Introduction

The aim of this chapter is to illustrate how molecular and biochemical methods have been developed in recent decades in order to incriminate sand flies as vectors of human diseases in Brazil. The focus is on sand fly species that are incriminated vectors of *Leishmania* species causing human leishmaniasis, and most references are to the vectors of *Leishmania* (*Leishmania*) *infantum chagasi* causing zoonotic visceral leishmaniasis (VL), *Leishmania* (*Leishmania*) *amazonensis* causing zoonotic cutaneous leishmaniasis (CL), and *Leishmania* (*Viannia*) *braziliensis* and other *Leishmania* (*Viannia*) species also causing zoonotic CL (Rangel and Lainson 2009; Ready 2013). There are brief references to the transmission of Carrion's disease (*Bartonella bacilliformis*) elsewhere in South America (Cohnstaedt et al. 2011), but non-vectors and sand flies from other countries and continents are mentioned only if they were investigated by novel approaches not fully applied to Brazilian vectors.

Depaquit (2014) reviewed the molecular-systematics literature on sand flies worldwide. This did not include all the major molecular reports on Brazilian sand flies, owing to the exclusion of population-genetics research when it did not inform taxonomy and perhaps also to the difficulty of finding pre-1990 literature in online databases such as PubMed. The current chapter does consider systematics, but only when it is relevant to vector incrimination. Vectorial traits are not always shared by

P. D. Ready (✉)
London School of Hygiene and Tropical Medicine, and Natural History Museum, London, UK
e-mail: paul.ready@lshtm.ac.uk

F. M. Vigoder
Instituto de Biologia, Universidade Federal do Rio de Janeiro, Rio de Janeiro, Brazil

E. F. Rangel
Instituto Oswaldo Cruz, Fundação Oswaldo Cruz, Rio de Janeiro, Brazil

E. F. Rangel, J. J. Shaw (eds.), *Brazilian Sand Flies*,
https://doi.org/10.1007/978-3-319-75544-1_3

a species in a single monophyletic lineage, as illustrated by the incriminated or suspected vectors of *Le.* (*Vi.*) *braziliensis* in South America being classified in several subgenera of the genus *Lutzomyia sensu lato* including *Helcocyrtomyia, Migonei, Nyssomyia, Pifanomyia*, and *Psychodopygus* (*sensu* Ready 2013). Therefore, a preference for a classification that places these and other vectors in many genera (Galati 2003), rather than the single genus *Lutzomyia* (Young and Duncan 1994), might impair information storage and retrieval (Lewis et al. 1977), and it will not help public-health interventions if the new phylogeny is inaccurate.

Consequently, the current chapter focuses on the complementary molecular and biochemical approaches necessary for identifying the interbreeding populations of a biological species, which are those more likely to share genotypes associated with phenotypes (or traits) of biomedical importance, such as vector competence and insecticide resistance (Ready 2011, 2013). Among these biological species, many will be distinctive morphological species; some will consist of geographical populations characterized by distinctive phylogenetic lineages; and a few might be sympatric with sibling species that are often indistinguishable morphologically but reproductively isolated by courtship behaviour. Molecular and biochemical markers are powerful tools for unravelling these differences, which must be recognized when planning the integrated control of leishmaniasis (Ready 2013; Bates et al. 2015).

Multilocus Enzyme Electrophoresis for Species Identification and Population Genetics

Miles and Ward (1978) were the first to publish methods for the isoenzymatic characterization of a sand fly, *Lutzomyia* (*Nyssomyia*) *flaviscutellata* from Brazil, after which multilocus enzyme electrophoresis (MLEE) was widely used for about 20 years (Ready and Rangel 2003). This can be explained by the universality of most of the targeted biochemical pathways, the widespread development of the techniques by medical entomologists and parasitologists, and the relatively low costs compared with molecular biology. Later, MLEE was much less frequently used, and so this section includes methodological details and some older references that risk being missed by online searches.

MLEE involves an indirect characterization of polymorphism in enzyme-coding genes by electrophoretic fractionation of the proteins, which are extracted from specimens soon after death or after cryopreservation (– 80° C to – 196° C) to avoid protein denaturation. For most sand fly investigations, coarse homogenates of whole fly bodies (or thoraces) in ionic buffers were used to separate variant enzymes (= isoenzymes or isozymes) migrating through gels of starch (Miles and Ward 1978; Lanzaro et al. 1993), cellulose acetate (Ready and Silva 1984), or polyacrylamide (Munsterman et al. 1998), with velocities differing according to protein size and ionic charge. Among all of the proteins in each fractionated homogenate, isoenzymes (or electromorphs) specified by one locus are revealed when they react with a specific substrate in the presence of co-factors, usually producing one or more chromatic bands (Ready and Rangel 2003). Each of these isoenzyme bands has a characteristic migration distance, when compared with standards on the same gel, and is designated either qualitatively by a letter code (in order of mobility) or quantitatively as a percentage of the distance migrated by the most frequently occurring band in that species or sample (100) (Richardson et al. 1990). Bands are recognized as alleles at the same locus (= alloenzymes or allozymes) when Mendelian inheritance is demonstrated by laboratory crosses or inferred in field populations by a lack of deviations between observed genotype frequencies and those expected from Hardy-Weinberg equilibria (Ready and Silva 1984). Heterozygous individuals have two bands for monomeric enzymes but multiple bands for multimeric enzymes (Richardson et al. 1990). Isoelectric focusing provides higher resolution of electromorphs, but it has only been usefully applied for sand flies from the Eastern Hemisphere (Pesson et al. 2004).

Most MLEE investigations on Brazilian sand flies refer to widespread vectors such as *Lutzomyia* (*Lutzomyia*) *longipalpis* (Mukhopadhyay et al. 1998b; Arrivillaga et al. 2003) and *Lutzomyia* (*Nyssomyia*) *intermedia sensu lato* (Meneses et al. 2005), and the findings will be integrated with those from other biochemical and molecular research in the section Major Vectors. More generally, the number of enzyme loci to be characterized should depend on the objectives. Only one to three loci might be sufficient to distinguish sibling species, although investigations in Brazil failed to discover diagnostic loci for sibling or sister species of *Lutzomyia* (*Psychodopygus*) *wellcomei* (Ready & Silva 1984), *Lu.* (*Ny.*) *intermedia s.l.* (Meneses et al. 2005), and *Lu.* (*Lu.*) *longipalpis* (Maingon et al. 2008). Far more loci (up to 21) (Kreutzer et al. 1990) should be characterized if the objective is to infer phylogenetic relationships or to investigate population differentiation in different habitats or geographical regions by using either genetic distances based on the presence or absence of alloenzyme bands or on population-genetics approaches (Meneses et al. 2005).

In general, uncorrected genetic distance (p) increases with phylogenetic distance, with sibling species ($p = 0.194–0.236$) being more differentiated than geographical populations ($p = 0.004$) but less differentiated than species from different subgeneric groups ($p = 0.612–1.250$) (Lanzaro and Warburg 1995). However, genetic distance does not provide proof of species status, and few reports have demonstrated phylogenetic relationships based on isoenzyme data. An exception is the subgenus *Pifanomyia*, found mostly in Andean and nearby neotropical regions, which includes the vector of Carrion's disease (Kreutzer et al. 1990).

Population genetics provides a better approach for discovering intraspecific populations. However, MLEE analysis faces many of the same challenges encountered using polymorphic DNA markers, such as microsatellites, including the presence of null alleles and haploid loci (Hartl and Clark 2007; Richardson et al. 1990). Unfortunately, MLEE studies alone did not demonstrate the presence of sibling species of *Lu.* (*Lu.*) *longipalpis* (Maingon et al. 2008) in Brazil or ecotope populations of *Lutzomyia* (*Nyssomyia*) *umbratilis* (Tibayrenc et al. 1980) in nearby French Guiana, although some new species were differentiated from morphologically similar ones in other countries (Caillard et al. 1986; Kreutzer et al. 1990; Dujardin et al. 1996).

For Brazilian sand flies, MLEE often failed to provide diagnostic markers for sibling species and to reveal intraspecific population differentiation. Together with the constraints of having to use fresh or cryopreserved material, this lack of success led many researchers to focus on more polymorphic molecular and biochemical markers. However, population genetics can provide important insights if based on mixed DNA and isoenzyme data sets, especially when maternally inherited mitochondrial DNA haplotype data is complemented by nuclear alloenzyme frequencies, for example, for *Lu.* (*Lu.*) *longipalpis s.l.* in Latin America (Arrivillaga et al. 2003).

Characterizations of Nucleic Acids for Species Identification, Population Genetics, Phylogenetics and Vector Competence

Development of Techniques

Molecular research on sand flies developed more slowly than it did for mosquitoes and some other arthropods of biomedical importance (Ready and Rangel 2003). The impacts of early techniques for analysing DNA were limited owing to long processing times, lack of sensitivity, a paucity of commercial kits and specialist training, and the targeting of diverse non-coding and coding loci. At first, following the approaches for mosquitoes, repetitive DNAs of non-coding loci or multi-copy nuclear ribosomal RNA genes (rDNA) were targeted because of their abundance in whole genomic DNA. Abundance provided higher sensitivity, and this made it relatively easy to construct DNA probes using bacteriophage and plasmid cloning (Ready et al. 1988, 1991), to analyse restriction fragment length polymorphisms (RFLPs) by using Southern hybridization (Fig. 1) (Ready et al. 1988 1991), and to detect the presence or abundance of diagnostic DNA molecules by dot-blot or squash-blot hybridizations (Ready et al. 1988, 1991; Rangel et al. 1996). These techniques relied on labelling cloned diagnostic DNA molecules with radioisotopes or other labels, followed by their hybridization as single-stranded DNA probes to whole genomic single-stranded DNA bound to hybridization membranes, and the revelation of the hybridized molecules by autoradiography or chemiluminescence. The whole genomic DNA was either extracted and known concentrations pipetted onto membranes for Southern and dot-blot hybridizations, or it was released chemically by cell lysis from body parts crushed (under coverslips or plastic sheets) onto squash-blot hybridization membranes. Such techniques permitted the identification of individual sand flies and any *Leishmania* species infecting them (Ready et al. 1988). However, they were less successful at distinguishing or relating sibling species (Ready et al. 1991; Rangel et al. 1996), often because only one locus was targeted.

The introduction of the polymerase chain reaction (PCR) with locus-specific primers permitted the targeted amplification of DNA fragments at specific loci (Booth et al. 1994; Warburg et al. 1994). One of the early applications of PCR was random amplified polymorphic DNA (RAPD) analysis, which uses one short, non-specific primer to amplify alleles that differ in size according to the genomic

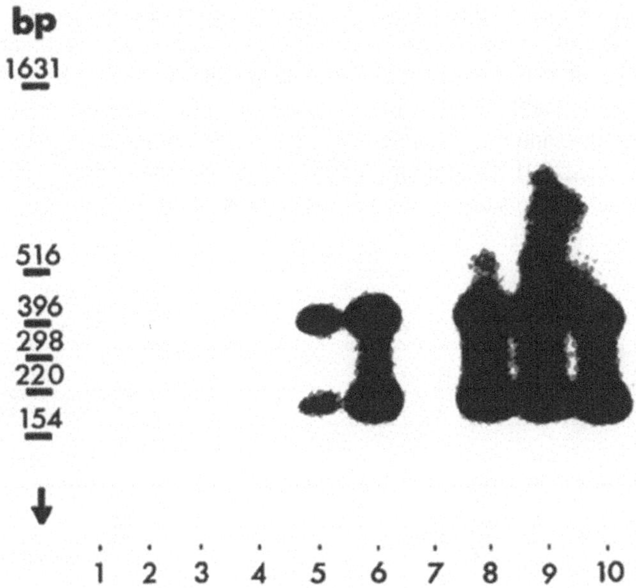

Fig. 1 Autoradiogram of Southern-blotted genomic DNA extracted from individual sand flies collected in Brazil, after digestion with restriction enzyme *Sau3A*, size fractionation in a 1.2% agarose gel, capillary blotting, hybridization with the oligo-labelled insert of the DNA probe (pPsw.105) diagnostic for *Lu. wellcomei*, low-stringency washes, and autoradiography for 2 days at –80° C. Size markers (bp = base pairs) are restriction fragments of plasmid pAT 153 digested with *Hinf I*. Lanes: 1, 2 = male *Lutzomyia complexa* from Marajó island, Pará State; 3, 4 = male *Lu. complexa* from Paranapanema, Serra dos Carajás, Pará State; 5 = male and 6, 7 = female *Lu. wellcomei/complexa* from Parauapebas, Serra dos Carajás; 8 = male and 9, 10 = female *Lu. wellcomei/complexa* from N2, Serra dos Carajás. (Source: Ready et al. 1991)

distance between inverted pairs of primer-like DNA sequences. Using a suite of such primers, RAPD-PCR showed promise for distinguishing sibling species (Adamson et al. 1993) and later for intraspecific population genetics (Dias et al. 1998), but it is no longer widely used because of the difficulties of standardizing experimental conditions and identifying the amplified alleles.

All these techniques detect molecular polymorphism caused by point mutations of nucleotide bases involving transitions [purine adenosine (A) ↔ purine guanine (G); pyrimidine cytosine (C) ↔ pyrimidine thymine (T)]; and transversions [A ↔ C; A ↔ T; G ↔ C; G ↔ T]. However, they do not detect all the mutations in the targeted fragment of genomic DNA, which became routinely possible only after the development of semi-automated Sanger sequencing of DNA fragments obtained by PCR, amplified either directly from genomic DNA or indirectly from complementary (c)DNA synthesised by reverse transcription PCR of messenger (m)RNA. The complete DNA nucleotide sequences generated by such techniques could then be aligned for comparative sequence analyses [now often using Geneious software (www.geneious.com)] in order to identify species-specific polymorphisms, regional populations or lineages (Fig. 2) and phylogenetic relationships, by using a range of often free software for analyses based on the principles of population genetics, cladistics, and Bayesian inference (e.g., Ready et al. 1997; Testa et al. 2002; Mahamdallie et al. 2011; Araki et al. 2013; Depaquit 2014). Further developments of PCR-based approaches have mostly focused on typing *Leishmania* species infecting wild sand flies, and they include field-deployable real-time or quantitative PCR (e.g., McAvin et al. 2012) and multiplex PCR (e.g., Neitzke-Abreu et al. 2014). The epidemiological findings are covered in the section Major Vectors.

LINEAGE and whit no. haplotypes (Some former ones share cyt b seq.)	No.males, No.females sequenced / capture location (see Fig. 1 for location codes)	Position of the 26 variant characters in last 294 bp of cyt b (positions 3-288) 111111111222222222222 24667901138999013457778888 34509892458002510149038678
NORTH-SOUTH		
whit06*	1M /TS	GAGACAGCAACGGTTTACAGTAATCA
whit07*	1M /TS	AAGACAGCAACGGTTTACAGTGATCA
whit08/09*	2F/TS	AAGACAGCAACGGTTTACAGTAATCA
whit10*	2M /MA	AAAACAACAACAGTTTACAGTAATCA
whit11/13*	1M /MA; 1F/VI	AAAACAACAACAGTTTACAGTAGTCA
whit12*	1M /MA	AAAACAACGACAGTTTACAGTAGTCA
		± +
E AMAZÔNIA		
whit14*	1F/DE	AGAATAACAACAGCTTACAGTAGGCT
whit15/16**	2M,1F/DE; 1M,3F/AP	AGAATAACAACAGTTTACAGTAGGCT
whit17**	2F/CA; 1M /AP	AGAATAATAACAGTTTACAGTAGGCT
whit19	1M /AP	AAAATAGCAACAGTTTACAATAGGCT
whit20	1M /AP	AGAATAACAACAGTTTTCAGTAGGCT
whit21	1F/AP	AAAATAACAACAGTTTACAGTAGGCT
whit22	1F/AP	AGAATAACAGCAGTTTACAGTAGGCT
whit23	4M /NAV;1M /AP	AAAATAGCAACAGTTTACAGTAGGCT
whit24	1M /NAV	AAAATAGCAACTGTTTACAGTAGGCT
whit18*	2M,1F/JA	AAAATAGCAACAGTTTACAGTAGGAT
		±
RONDÔNIA		
whit25	3M /BRO;2M,1F/CAC;2M /MON	AGAATAACAACAGTTCATAGTAAGCA
whit26	1M /BRO; 1F/CAC;1M /MON	AGAATGACAACAGTTCATAGTAAGCA
whit27	1M /CAC	AGAATAATAATGGTTCACCGTAAGCA
whit28	2F/CAC	AAAATAACAACAGTTCATAGTAAGCA
whit29	1M /CAC	AGAATAACAACGGTTCATAGTAAGCA
whit30	1M,1F/CAC	AAAATAACAACAGTTCATAGTAAGCT
whit31	1M /MON	AAAATAGCAACAGTTCACAGTAAGCA
		± ±
NORTHEAST		
whit01*	1M /IL	AAAGCAACAACGGCCCACAGTAGGCA
whit02*	1M /IL	AAAGCAACAACGACCCACAGTAGGCA
whit03*	2M /CP; 1M /RE	AAAGTAACAACGGCCCACAGTAGGCA
whit04*	2M /CP	AAAGTAACAACGGCCCACAGCAGGCA
whit05*	1M /RE	AAAATAACAACGGCCCACAGTAGGCA
		± ±+±
Nucleotide position in codon		33333333333331333331133212 3
Type of nucleotide substitution†		iiiiiiiiiiibiiiiviviiiiivvv
Effect on amino acid††		ssssssssNssNsssssNNssNNNs

Fig. 2 (Upper) *Cytochrome b* haplotypes from individual males (M) or females (F) of *Lu. whitmani* from Brazil grouped by geographical lineage and variant 3′ nucleotides. Lineage synapomorphies are diagnostic (+) or informative (±). Nucleotide substitutions † are transitions (i), transversions (v), or both (b), and substitutions †† are synonymous (s) or non-synonymous (N). (Lower) Single shortest phylogenetic tree (cladogram) relating the haplotypes by maximum parsimony analysis (rescaled consistency index = 0.62, 100 heuristic searches). Above each branch are % bootstrap support (bold) and Bremer index. Lineages are North–South (NS), Eastern Amazonia (EA), Northeast (NE), and Rondonia (RO). (Source: Ishikawa et al. 1999)

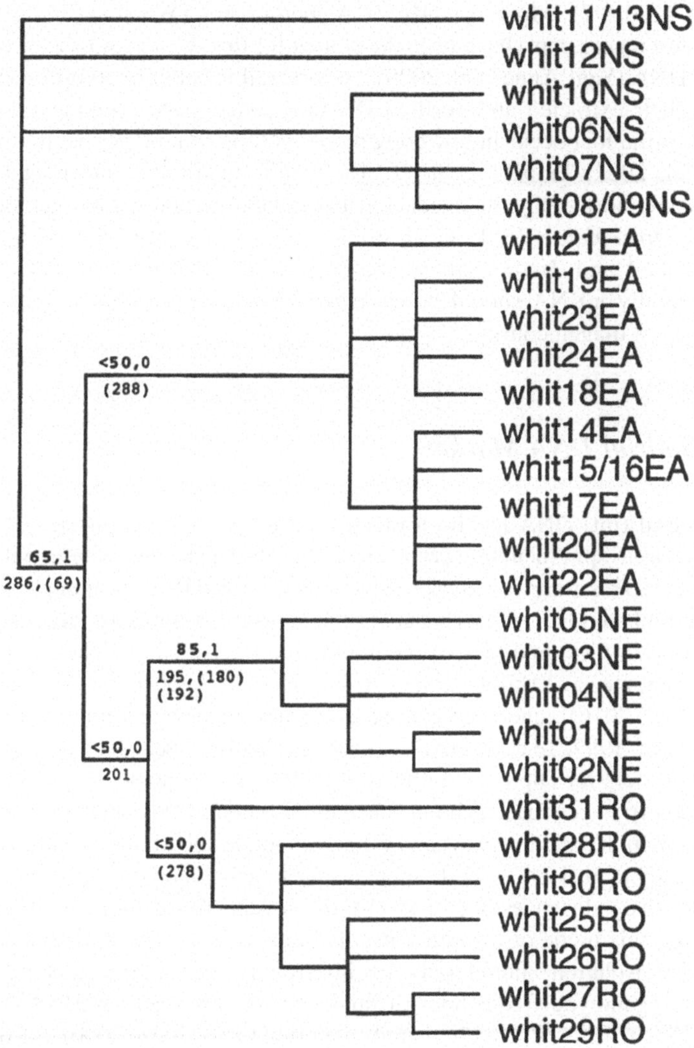

Fig. 2 (continued)

DNA molecules denature more slowly than most proteins, and so, unlike MLEE, specimens do not always need to be cryopreserved. Sand flies should be killed quickly, to avoid biochemical degradation, ideally by freezing (−20°C for 30 min or −80° C to − 196° C for 5 min) or, alternatively, with cigarette smoke. Many of the copies of short, repetitive DNA sequences will survive in specimens stored dry (e.g., over silica gel) or in 70–90% analytical-grade ethanol (or isopropanol), always away from direct sunlight. Such samples were often used for dot- and squash-blot hybridizations (Ready et al. 1988), and they can contain enough short DNA fragments [≤550 base pairs (bp)] for PCR amplification and Sanger sequencing

(Ready et al. 1997). More recently, FTA membranes (Whatman™ FTA Cards: www.gelifesciences.com) have been developed for dry storage of tissues, cells, and extracted DNA (Sant'Anna et al. 2008). However, it is better to cryopreserve specimens (or their extracted and purified DNA) if longer DNA fragments are to be amplified or the loci occur in low copy numbers (one to nine per haploid genome) (Testa et al. 2002; Tabbabi et al. 2014). PCR is facilitated, and readable DNA sequences are more easily obtained, if genomic DNA is purified by methods established for *Drosophila* (Ready et al. 1988, 1991) or by using kits (e.g., Qiagen DNeasy Blood and Tissue Kit: www.qiagen.com; DNAzol: www.thermofisher.com). Normally mRNA should be extracted from fresh specimens (e.g. RNeasy Mini-kit: www.qiagen.com).

Mitochondrial DNA Markers

Mitochondrial (mt) DNA has been much used to identify morphospecies of sand flies as well as to split them into cryptic species, sibling species, or regional lineages (Ready 2013).This follows widespread research on mtDNA by animal-population geneticists and phylogeneticists because of its uniparental mode of inheritance (usually maternal in flies) and lack of recombination (Avise 2000). The circular mitochondrial genome is <20 kilobase pairs in length [15.717 kbp in *Lu.* (*Ny.*) *umbratilis*) (Kocher et al. 2016)], and so it can be easily amplified by cloning or Long PCR (Esseghir et al. 1997; Ye et al. 2015) in order to check the sequence conservation of universal primers for Diptera (Simon et al. 1994). The genome is cytoplasmic and evolves rapidly to produce regional lineages of unique DNA sequences, or haplotypes, that share fixed, mostly synonymous nucleotide substitutions within a lineage (Fig. 2; Ready et al. 1997) and show relatively large genetic distances between lineages [pairwise *p* distances of c. 2.3%/1 million years (Esseghir et al. 2000)].

However, this mode of evolution has its drawbacks for delimiting species, and reproductive isolation should only be inferred by population genetic analyses that include genotype frequencies at nuclear loci, not solely mtDNA haplotype frequencies at what is effectively a single maternal locus (Avise 2000). Introgression of mtDNA is frequent even between good biological species of sand flies (Marcondes et al. 1997; Testa et al. 2002; Mazzoni et al. 2006; Araki et al. 2013), and population bottlenecks commonly produce new mtDNA lineages (Mahamdallie et al. 2011). This means that mtDNA barcoding alone cannot provide proof of species status. It is only indicative of the presence of more than one species in an internally branching group of mtDNA haplotypes (Fig. 3; Ready et al. 1997), especially if such a tree is not a phylogram but only a phenogram (or dendrogram) constructed by clustering sequences by genetic distance using a neighbour-joining or similar algorithm, as often advocated (see Hebert et al. 2003).

According to the narrow definition (www.barcodeoflife.org/), barcoding should be based on the comparative sequence analysis of a specific 648-bp fragment of the mtDNA gene, *COI*, which encodes the protein cytochrome c oxidase 1, but most

```
A            1111112222222222222233
            11678991123456666777827
            94591165762210248368460
sob2s11A    TTCTGGCTCTGTTCATCTATCTT
sob2s13A    ......T......T.G.....C.
sob2s14A    ......T......TC......CA
sob2s15A    ......T......T.G.....C.
sob2s16A    ......T......T.G.....C.
sob2s16B    ......T......TC......C.
sob2s17A    ......T......T.G..G..C.
sob2s17B    .G....T......T.G....TC.
sob2s18A    ......T......T.G....TC.
sob2s18B    ......T......T.G.....C.
sob2s1A     ......T......T.G.....C.
sob2s4A     ......C......T.G....TC.
sob2s4B     ......T......T.G.....C.
sob2s5A     ......T......T.G....CC.
sob2s6A     ......T......T.G....TC.
sob2s7A     ......T......T.G.....C.
sob2s7B     ......T......T.......C.
sob2s8A     ......T......T.G.....C.
sob2s8B     ......T......T.G....TC.
sob2s9A     ......T......TCG.......
sob2s9B     ......T......T.G.....C.
sob1s10A    C.TG..T....CATC......C.
sob1s10B    C.CG..T....CATC......T.
sob1s11A    C..G..T....CATCT.....C.
sob1s12A    C..G..TC...CATC......C.
sob1s12B    C..G..TT...CATC......C.
sob1s13A    C..G..T....CATC......C.
sob1s14A    C..G..T....CATC......C.
sob1s15A    C..G..T....CATC......C.
sob1s15B    C..G..T...ACATC..G...C.
sob1s16A    C..G..T....CATC........
sob1s16B    C..G..T...ACATC..G...C.
sob1s17A    C..G..T..A.CATC.G....C.
sob1s17B    C..G.AT....CATC......C.
sob1s18A    C..GA.T....CATC......C.
sob1s18B    C..G..TT...CATC......C.
sob1s1A     C..G..T.T..CATC..T.T.C.
sob1s1B     C..G..T....CATC..A.G.C.
sob1s2A     C..G..T....CATC......C.
sob1s2B     C..GA.T....CATC......C.
sob1s4A     C..G..T....CATC......C.
sob1s5A     C..G..T....CATC......C.
sob1s5B     C..G..T....CATC......C.
sob1s6A     C..G..T....CATC..A.G.C.
sob1s6B     C..G..T....CATC......C.
sob1s8A     C..G..T..T.CATCT.....C.
sob1s8B     C..G..T..T.CATCT.....C.
sob1s9A     C.....T....CATC......C.
```

Fig. 3 (**A**) Alignment of the variable nucleotide sites of the consensus sequences of the *paralytic* gene fragment from individual *Lu. longipalpis* collected in Sobral, Ceará State, Brazil. Exon sites are in bold. Dots indicate the same nucleotide as in the first aligned sequence. The three fixed differences between the two siblings (1S, sob1s; 2S, sob2s) are highlighted. (**B**) Minimum evolution tree constructed using pairwise genetic distances (*p*) and the *para* sequences from the siblings 1S (black circles) and 2S (white squares). Bootstrap values greater than 50% (1000 permutations) are shown. (Source: Lins et al. 2008)

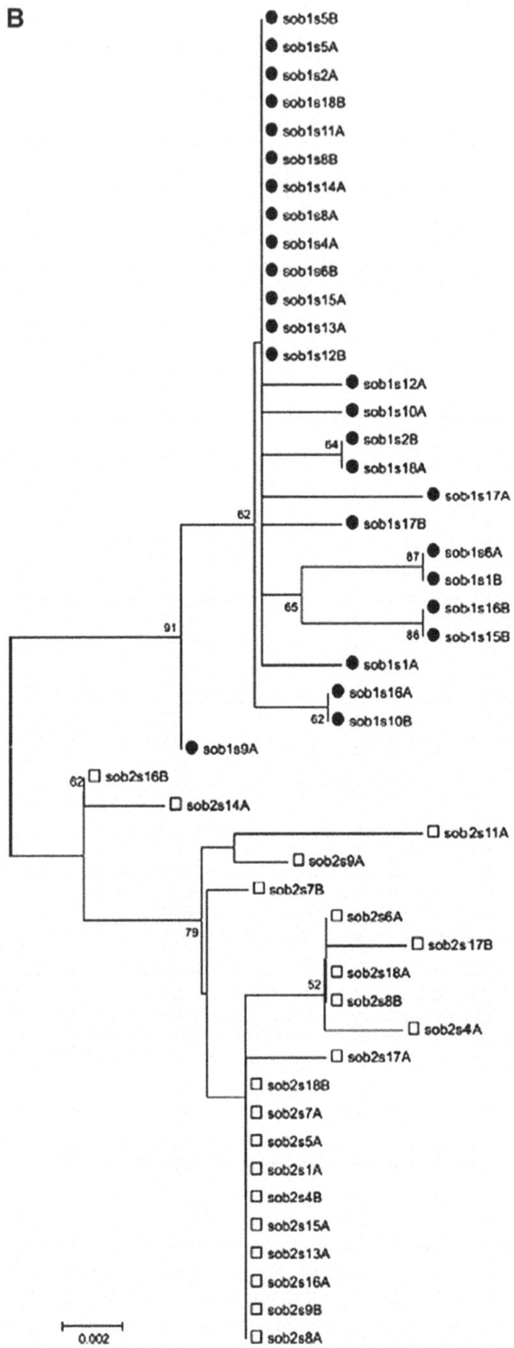

Fig. 3 (continued)

sand fly reports have not targeted *COI*. For Brazilian sand flies, exceptions include research on *Lu.* (*Lu.*) *longipalpis* and its siblings (Arrivillaga et al. 2002, 2003; Maingon et al. 2003), on populations of *Lu.* (*Ny.*) *umbratilis* and *Lutzomyia* (*Ny.*) *anduzei* (Scarpassa and Alencar 2012, 2013; Scarpassa et al. 2015) and on 47 morphospecies (Pinto et al. 2015). DNA barcodes correctly identified 42 of 47 (90%) Brazilian morphospecies and recognized a synonymized species related to *Lutzomyia* (*Psathyromyia*) *shannoni* (Pinto et al. 2015). DNA barcodes also correctly identified all morphospecies surveyed in Colombia (Contreras Gutiérrez et al. 2014; Romero-Ricardo et al. 2016) including four *Brumptomyia* species and *Warileya rotundipennis*, but neither of these studies tackled the recognition of sibling species, in part because of small sample sizes. In contrast, the efficacy of *COI* to recognise sibling species was considered for *Lutzomyia* (*Pifanomyia*) species sampled from Colombia and Peru including populations of *Lu.* (*Pf.*) *verrucarum*, the regional vector of *Bartonella bacilliformis* (Cohnstaedt et al. 2011): Seven morphospecies formed well-supported phylogenetic lineages, but four morphospecies grouped in two paraphyletic lineages, namely, *Lu.* (*Pf.*) *longiflocosa-Lu.* (*Pf.*) *sauroidea* and *Lu.* (*Pf.*) *quasitownsendi-Lu.* (*Pf.*) *torvida*.

Early comparative sequence analyses of c. 490bp of mtDNA (CB3 fragment) containing the 3′ terminal of the gene encoding the protein cytochrome b (C*yt b*) led to a focus on this marker for phylogenetic research on species and populations of *Phlebotomus* (Esseghir et al. 1997) and *Lutzomyia* (Ready et al. 1997; Hodgkinson et al. 2003; Torgerson et al. 2003). However, the CB3 fragment is complex, containing DNA with different rates of mutation, including not only c. 316 bp of C*yt b* but also all c. 82 bp of a transfer (t) RNA gene (*tRNAser*^{TCN}) on the same DNA strand and c. 90 bp of the 3′ terminal of the gene, *Nadh1*, which encodes the protein nicotinamide adenine dinucleotide dehydrogenase on the complementary DNA strand, as well as two short intergenic spacers that can be difficult to align. Consequently, primers were designed for the comparative sequence analysis of c. 717 bp of the more homogeneous 3′ terminal of C*yt b* (>60% of the gene) based on the PCR amplification of two fragments (Esseghir et al. 2000; Testa et al. 2002) or one larger fragment (Tabbabi et al. 2014). The smaller fragments can be easier to amplify from degraded genomic DNA.

Notable research on the mtDNA of *Lutzomyia* species has also been based on the protein-coding gene *Nadh4* (Soto et al. 2001; Arrivillaga et al. 2003; Beati et al. 2004) and the 12S and 16S rRNA genes (Arrivillaga et al. 2003; Cohnstaedt et al. 2011). The alignment of DNA sequences of rRNA genes and tRNA genes should be based on secondary structure models (Esseghir et al. 1997; Vivero et al. 2007).

Conserved Nuclear Markers

Two conserved nuclear loci have been widely used for molecular research on flies, namely, the tandem array of rRNA genes (including 18S rDNA, 28S rDNA and spacers) and the gene encoding the protein elongation factor−1 alpha (*EF-1α*). The slow rate of evolution of much rDNA makes it suitable for the molecular phylogenetics of morphospecies and generic groups (Beati et al. 2004; Depaquit 2014), although the alignment of the sequences depends on imprecise secondary structure models (Aransay et al. 2000). RFLPs of 18S rDNA can be useful for species identification (Terayama et al. 2008), and the more polymorphic intergenic spacers (IGS1, IGS2) of rDNA sometimes provide markers within species complexes (Kuwahara et al. 2009; Zapata et al. 2012).

Comparative sequence analysis of *EF-1α* permitted the delimitation of species and the discovery of mtDNA introgression in non-Brazilian *Lutzomyia* (*Pifanomyia*) species from the Andes and northern Neotropics (Testa et al. 2002), but this gene is not routinely used for sand flies partly because PCR primers must be designed specifically for some subgenera to avoid amplifying more than one locus of multi-copy *EF-1α* or even *EF-2* (Esseghir et al. 2000).

The same lack of locus specificity was met when investigating families of non-LTR retrotransposons in *Lutzomyia* and *Phlebotomus* species (Booth et al. 1994, 1996). Other nuclear genes are discussed in the following sub-sections.

Lovesong Genes: Per, Cacophony *and* Paralytic

Genes associated with behaviour started to be used for the molecular analysis of sand fly species and populations at the turn of the millenium (Oliveira et al. 2001). Genes involved with sexual behaviour are thought to be good molecular markers for phylogenetic studies because they tend to evolve faster due to disruptive selection (Ting et al. 2000), and this also makes them particularly useful for population genetics. The first two behaviour genes investigated in sand flies were *period* (*per*) and *cacophony* (*cac*) (Bauzer et al. 2002a, 2002b; Mazzoni et al. 2002; Lins et al. 2002). The *per* gene is involved in the control of circadian and song rhythms in *Drosophila melanogaster* (Kyriacou and Hall 1980). Circadian rhythm determines a fly's daily activities, such as mating time, which can lead to isolation between populations. In *Drosophila*, *per* also affects rhythms in male song, which are important for reproductive success (Kyriacou and Hall 1980; Ritchie et al. 1999), thus leading to reproductive isolation between closely-related species (Sakai and Ishida 2001). The *cac* gene encodes a subunit of a voltage-gated calcium channel (Smith et al. 1998). Among other phenotypes, *cac* also affects *Drosophila* male songs, with mutants having abnormal inter-pulse intervals and polycyclic pulses. Variations in such characters are frequently species-specific and can be used by females for species recognition (Peixoto and Hall 1998).

Both genes were used for phylogenetic analysis of several Brazilian sand fly species (Bauzer et al. 2002a, 2002b; Lins et al. 2002; Mazzoni et al. 2008), sometimes demonstrating a lack of concordance with the morphological classification (Young and Duncan 1994). However, the greatest contributions from studies using these markers relate to variation among Brazilian populations of *Lu.* (*Lu.*) *longipalpis* (Souza et al. 2017) the taxonomic status of which have been controversial (Brandão-

Filho et al. 2009). Behavioural analysis suggested that *Lu.* (*Lu.*) *longipalpis* was a complex of cryptic species in Brazil (Souza et al. 2004; Hamilton et al. 2005), but molecular investigations often reached conclusions that varied according to the marker. A key step was the molecular analysis of two sympatric populations from Sobral (Ceará State), one with males exhibiting one pair of dorsal spots with pheromone papules (1S) on the fourth abdominal segment and the males of the other population having two pairs of dorsal spots (2S) on the third and fourth segments (Ward et al. 1983; Bauzer et al. 2002b, 2007). Molecular analysis based on *per* shows a high level of genetic structure between the two populations, even though *cac* suggests that some level of introgression still occurs (Bauzer et al. 2002a, 2002b; Bottecchia et al. 2004; Araki et al. 2009, 2013). Such genetic structuring of sympatric populations indicates that the genetic differences among allopatric populations will not always result from geographical variation alone.

More than a dozen Brazilian populations have been analysed using *per*, including Sobral, Estrela de Alagoas (Alagoas State) and Jaíba (Minas Gerais State), with sympatric populations in each of the three localities separated by the abdominal-spot morphotype. Analysis shows a split into two groups associated with the copulatory songs produced by males, named pulse-type and burst-type (Bauzer et al. 2002a, 2002b; Araki et al. 2009, 2013; Vigoder et al. 2015). The burst-type group has similar songs and a low genetic differentiation (F_{st} <0.1) consistent with a single species. On the other hand, the pulse-type group shows more variation in song patterns and a moderate to high F_{st}, which is consistent with the presence of up to five different species. Comparison of *per* in some locations shows fixed polymorphisms diagnostic for the sympatric populations. However, the diagnostic polymorphisms are not consistent among localities, and this is true even when only populations of the same song sub-group are considered (Araki et al. 2009). Another issue is that both *per* and *cac* have shown introgression between populations of *Lu.* (*Lu.*) *longipalpis s.l.* (Araki et al. 2013), which could produce substantial temporal variation in allele frequencies. This could explain, for instance, why Lima Costa et al. (2015) were able to observe fixed polymorphisms between Sobral 1S and Sobral 2S in a segment of *per* that was not diagnostic in the samples of Bauzer et al. (2002b).

Lutzomyia (*Lu.*) *cruzi* is a species closely related to *Lu.* (*Lu.*) *longipalpis* (Young and Duncan 1994), and molecular analysis of *per* showed that the pairwise genetic distances between the two species can be smaller than those between any burst-like and pulse-like pair of populations of *Lu. longipalpis* (Mazzoni et al. 2002). On this evidence, *Lu. cruzi* should be treated as a member of the *Lu. longipalpis* species complex (subsection *Lu. (Lu.) longipalpis* Species Complex as Vectors of *Le. (Le.) infantum chagasi*). Outside this complex, analysis of *per* showed introgression between the closely related *Lu.* (*Ny.*) *intermedia* and *Lutzomyia* (*Nyssomyia*) *whitmani* (Mazzoni et al. 2006) and the presence of a putative sibling species of *Lu.* (*Ny.*) *umbratilis* (Scarpassa and Alencar 2012; subsection *Lu. (Ny.) intermedia*, *Lu. (Ny.) neivai* and *Lu. (Ny.) whitmani* as Vectors of *Leishmania (Viannia)* Species).

Another *Drosophila* song gene studied in sand flies is *paralytic* (*para*) (Peixoto and Hall 1998). This gene encodes a subunit of a voltage-dependent sodium channel and is of wider biomedical interest because some mutations can cause insecticide resistance in arthropod vectors (Pittendrigh et al. 1997). The use of *para* for population genetics research in Brazil again showed a clear separation between the two population groups of *Lu. longipalpis* characterized by copulatory songs (Lins et al. 2008). Comparisons between both sympatric and allopatric populations of the two groups showed higher divergences for *para* than for any other molecular marker ($F_{st} = 0.79$) (Lins et al. 2012). This was the first molecular marker to show fixed-sequence polymorphisms between the two sympatric populations of Sobral 1S and Sobral 2S (Fig. 3), and these polymorphisms are also fixed in several allopatric populations, suggesting that *para* could be a good diagnostic marker within the *Lu. longipalpis* complex (Lins et al. 2008, 2012). The analysis of *para* in *Lu.* (*Lu.*) *cruzi* confirmed it as a member of the *Lu. longipalpis* complex but with a somewhat different result from *per*. Both genes showed *Lu. cruzi* as an independent clade within the complex, but *per* places *Lu. cruzi* closer to pulse-type populations (Mazzoni et al. 2002), while *para* places it closer to burst-type populations (Lins et al. 2008). *Lutzomyia cruzi* has a burst-like copulatory song but produces the male pheromone 9-methyl-germacrene, which is found only in pulse-type populations of *Lu. longipalpis* (Brazil and Hamilton 2002). This suggested to Araki et al. (2013) the possibility that *para* might be differentiating populations in association with copulatory song while *per* might do so in association with the pheromone.

Other Behaviour Genes

The molecular circadian clock determines an organism's daily cycle including mating and feeding times (Meireles-Filho and Kyriacou 2013). The endogenous circadian clock consists of biochemical feedback loops that are self-regulated by a series of complex interactions. Many genes are involved in the core pacemaker in *Drosophila*, and *per* was the first to be described (Mendoza-Viveros et al. 2017). In brief, the proteins formed by the *clock* (*clk*) and *cycle* (*cyc*) genes interact to up-regulate several clock genes including *per* and *timeless* (*tim*). Throughout the day, PER and TIM proteins accumulate and inhibit CLK/CYC activity, thus controlling their own expression (Mendoza-Viveros et al. 2017).

Some circadian-clock genes have been cloned from *Lu.* (*Lu.*) *longipalpis*, but not from other sand flies. Both *clk* and *cyc* were isolated and have amino-acid sequences that indicate a possible change in molecular function compared with *Drosphilia melanogaster* (Meireles-Filho et al. 2006; Gesto et al. 2015). In *Lu. longipalpis*, *cyc* has a transactivation domain at the C-terminal region that is lacking in *Drosophila* but conserved in eukaryotes. In addition, *cyc* shows rhythmic daily expression in males of *Lu. longipalpis*, but it has constitutive expression in *D. melanogaster* and most Diptera (Meireles-Filho et al. 2006). The C-terminal region of *clk* also shows differences in comparison with fruit flies, exhibiting low sequence conservation and a shortened poly-glutamine region, which is usually conserved because of its importance for activating transcription (Gesto et al. 2015).

The last circadian-clock gene to be characterized in *Lu.* (*Lu.*) *longipalpis* was *vrille* (*vri*). Its protein contains a bZIP domain (basic leucine zipper) that mediates DNA binding and is highly conserved in this sand fly (Gesto et al. 2015). Both *clk* and *vri* have been used for phylogenetics to produce the expected relationships among many insect taxa (Gesto et al. 2015).

Salivary Peptide Genes

Peptides are among the salivary-gland molecules that female sand flies pump into their blood-feeding pools in mammalian skin (Ready 2013). Some of the salivary peptides (SPs) are phlebotomine specific, while others have structural homologies to mosquito and tick SPs known to counteract haemostatic and immuno-modulatory responses to blood-vessel damage (Anderson et al. 2006). Many of the SP gene families shared by *Lutzomyia* species were isolated from the model species *Lu.* (*Lu.*) *longipalpis*, and their immuno-modulatory effects were tested in laboratory-bred mammalian hosts (Charlab et al. 1999; Collin et al. 2009).

Only the SP Maxadilan has been intensively characterized to investigate genetic variation in wild sand fly populations, with most of this research being accomplished before the establishment of methods for the routine cloning of potential SP DNA vaccines (Valenzuela et al. 2004; Anderson et al. 2006; Oliveira et al. 2006). This potent vasodilator was the first molecule to be identified in sand fly saliva when it was characterized from *Lu. longipalpis* (Titus and Ribeiro 1988; Lerner et al. 1991). Natural variation in Maxadilan has been investigated by characterizing its multi-copy genes (*max*), antigenic diversity and mRNAs from *Lu. Longipalpis* and sibling species sampled from widespread neotropical locations (Lanzaro et al. 1999; Yin et al. 2000; Milleron et al. 2004), and associations have even been made with different human immune responses to infection by *Le.* (*Le.*) *infantum chagasi* (Warburg et al. 1994). Previous exposure to Maxadilan exacerbates experimental infections, whereas exposure to other SPs can provide protection (see Oliveira et al. 2008). Recently, a *max*-like gene has been was identified in *Lu.* (*Ny.*) *intermedia* (de Moura et al. 2013) and the immunogenicity of the homologue from the sister species, *Lutzomyia* (*Nyssomyia*) *neivai*, studied in patients with cutaneous leishmaniasis (Aires et al. 2017).

Polymorphic Markers for Related Species

RAPD-PCR (subsection Development of Techniques) was soon followed by single-strand conformational polymorphism analysis (SSCP), in which alleles of a single- or multi-copy gene are amplified by PCR and then separated as single-strand DNA molecules by electrophoretic fractionation in polyacrylamide gels. This method was explored for investigating sequence polymorphisms in multi-copy *max* (Warburg et al. 1994) and mtDNA (*Cyt b*, *COI*, *16S*) (Hodgkinson et al. 2002; Arrivillaga et al. 2003) among populations of *Lu.* (*Lu.*) *longipalpis s.l.* Some of the alleles should be sequenced in order to relate electromorphs to point mutations in haplotypes (Lanzaro et al. 1999).

Microsatellite DNA consists of tandem copies of simple repeats (usually two to four nucleotides long) and, after PCR amplification, alleles with different repeat numbers are identified by length in polyacrylamide sequencing gels. This was first investigated for *Lu.* (*Ny.*) *whitmani* and *Lu.* (*Lu.*) *longipalpis* from Brazil (Day and Ready 1999) and proved very informative for relating populations of *Lu. longipalpis* (Maingon et al. 2008). Microsatellite loci sufficiently polymorphic for population genetics can be difficult to clone and isolate from DNA libraries (Aransay et al. 2001; Watts et al. 2001). Therefore, many population ecologists have turned to using single nucleotide polymorphisms (SNPs) identified by next-generation sequencing (NGS) unless they are researching model organisms for which microsatellite DNA sequences and SNPs can be found by online searches of annotated whole-genome sequences (Campo et al. 2013).

Transcriptomics, Whole Genome Sequencing and Vector Competence

Genomics, proteomics, and transcriptomics are progressing rapidly by using ever more advanced high-throughput NGS techniques (Campo et al. 2013). These approaches will increasingly provide the comparative nucleic acid sequences (RNA and DNA) and peptides for developing diagnostic tests to incriminate vectors in natural settings, as well as for understanding the molecular interactions between parasites and vectors that determine vector competence (Ready 2013).

Recently, almost the complete mitochondrial genome of *Lu.* (*Ny.*) *umbratilis* was obtained on a NGS platform (Illumina Hiseq 2500) with a genome-skimming strategy, which also enabled the assembly of a nuclear rDNA repeat (Kocher et al. 2016). However, most whole genome sequencing (WGS) and transcriptomics have been applied to investigate vector competence, including the comparison of sugar-fed, blood-fed and *Leishmania*-infected female sand flies (Jochim et al. 2008), and the identification of the function of sand fly genes (Telleria et al. 2012) and *Leishmania* genes (Doehl et al. 2017) by gene-depletion or gene-replacement experiments. Much of this research is performed with two model sand flies, *Lu.* (*Lu.*) *longipalpis* and *Phlebotomus* (*Phlebotomus*) *papatasi*, and the challenge is to extend the research to other sand flies, especially neotropical vectors of cutaneous leishmaniasis (Ready 2013). EST libraries are available for *Lu.* (*Lu.*) *longipalpis* (Dillon et al. 2006; Azevedo et al. 2012), and regularly updated sequence data for this species are available on Vectorbase (www.vectorbase.org/; Bates et al. 2015).

Cuticular Hydrocarbons

Non-functional Characterization of Cuticular Hydrocarbons

Cuticular hydrocarbons can have various structural or behavioural functions, but the early analyses of medically important flies focused on using gas–liquid chromatography and/or mass spectrophotometry to discover profiles of often unidentified molecules that were diagnostic for closely related species, and *Lu.* (*Ps.*) *wellcomei* was a Brazilian sand fly differentiated by such an approach (Ryan et al. 1986) [subsection Other Vectors of *Leishmania (Viannia)* Species].

Pheromones

Some cuticular hydrocarbons are pheromones, and these have potential as baits in traps for sand fly control (Bray et al. 2014). Pheromones are one type of semiochemical produced by an individual to elicit a physiological and/or behavioural response by another of the same species, and they are frequently used by insects for species recognition (Johansson and Jones 2007). Males of *Lu.* (*Lu.*) *longipalpis s.l.* produce a volatile pheromone that is released by glands located in the papules of the abdominal tergal spots (Lane et al. 1985; Phillips et al. 1986; Ward et al. 1988; Hamilton and Ward 1991). Analysis showed variation in the main chemical component (comprising >90% of the total) and the chemical blends produced by different Brazilian populations (Ward et al. 1988; Hamilton et al. 1994, 2005; Spiegel et al. 2016). These pheromones can attract females across considerable distances (Ward et al. 1988; Morton and Ward 1989), and the main component is the one responsible for female attraction (Hamilton et al. 1994). Behavioural essays showed that the pheromone attracts both sexes in laboratory experiments (Lane et al. 1985; Ward et al. 1988; Spiegel et al. 2016) and in natural environments (Bray et al. 2009), with each sex having chemical receptors in the antennal ascoid sensilla to detect what could well be a sex-aggregation pheromone (Spiegel et al. 2016).

Four distinct chemotypes were found and characterized by their main component (Ward et al. 1988; Hamilton et al. 1996a, 1996b, 2004, 2005; Casanova et al. 2006; Spiegel et al. 2016). Chemotype 1 is composed mainly of the homosesquiterpene (S)-9-methylgermacrene-B (9MGB); it was first found in males from Lapinha cave in Minas Gerais State and later elsewhere in Brazil and in Central and South America. Chemotype 2 is also a homosesquiterpene with the main component being (1S,3S,7R)-3-methyl-α-himachalene (3MαH); it was first described in Jacobina (Bahia State, Brazil). Chemotype 3 is a monocyclic diterpene classified as cembrene-1 (Cemb-1); it is the most widespread pheromone in Brazilian populations and is strongly associated with the burst-type song (see subsection Lovesong Genes: *Per*, *Cacophony* and *Paralytic*). Chemotype 4 is another diterpene called cembrene-2

(Cemb-2); it is found only in males from Jaíba, Minas Gerais State, having one pair of abdominal spots (Jaíba 1S). A fifth chemotype was proposed by Hamilton et al. (2005), based on the proportion of the different chemical components produced by the gland. Found in the Sobral 1S (Céara State) and Montes Claros (Minas Gerais State) Brazilian populations, it also has >90% 9MGB but in a slightly smaller ratio than observed in populations with chemotype 1. It is not clear if this blend should indeed be considered a different chemotype, because it is not known if it causes behavioural or physiological changes.

The role of pheromones in the reproductive isolation of the members of the *Lu. longipalpis* complex requires further clarification. Different populations may vary in their response to potentially interspecific pheromone components. Females from Jacobina with chemotype 2 respond only to their own pheromone, 3MαH, but not to Cemb-1. On the other hand, females from Sobral 2S, where males produce Cemb-1, are attracted to both pheromones when each is presented individually. However, when given a choice, they show a clear preference for their homo-specific Cemb-1 pheromone (Ward and Morton 1991). Males and females from populations with different chemotypes can mate, albeit at a low rate, and their offspring are fertile (Ward et al. 1988; Souza et al. 2008). Males from laboratory crosses of individuals from Lapinha (chemotype 1) and Jacobina (chemotype 2) produce both types of pheromones, 9MGB and 3MαH. In contrast, wild males from sympatric populations do not produce both pheromones, and this indicates effective sexual isolation (Spiegel et al. 2016).

Pheromones have also been characterized for morphologically divergent species of the *Lu. longipalpis* complex. Males of *Lu.* (*Lu.*) *cruzi* also produce chemotype-1 pheromones (9MGB) (Brazil and Hamilton 2002; Hamilton et al. 2005). However, this is the only species in the complex known to produce a burst-type song as well as a pheromone that is not Cemb-1, thus showing the importance of taking an integrative approach based on different markers when dealing with the *Lu. longipalpis* complex (Vigoder et al. 2010a). *Lutzomyia* (*Lu.*) *pseudolongipalpis* is found only in Venezuela, and its males were also found to produce the same 3MαH pheromone found in males of *Lu.* (*Lu.*) *longipalpis* from Jacobina (Arrivillaga and Feliciangeli 2001; Watts et al. 2005).

Sex pheromones are not widely distributed amongst males of other *Lutzomyia* species, some of which have vestigial tergal papules (Hamilton et al. 2002). Males of *Lutzomyia* (*Lu.*) *lenti* from southern Brazil produce a novel diterpene, whereas those from northeastern Brazil do not. A putative diterpene sex pheromone was found in *Lutzomyia* (*Pintomyia*) *pessoai* (Hamilton and Ward 1994).

Major Vectors

Molecular and Biochemical Systematics of Vectors

It is noteworthy that a recent review of the molecular systematics of phlebotomines (Depaquit 2014) did not include a single phylogram. A major project to relate sand fly species at the subgeneric and generic levels, based on comparative analyses of nuclear rDNA and mitochondrial COI sequences, was started by M. D. Bargues and J. Depaquit more than 10 years ago, and the publication of its findings is much anticipated by all in the leishmaniasis community who want to share a single well-supported sand fly classification. Preliminary presentations of the molecular phylogenies at international meetings (Bates et al. 2015) have indicated that not all of the genera proposed by Galati (1995, 2003) are monophyletic. Nevertheless, some regional phylogenies of neotropical species have been published and provide insights of biomedical significance.

In most neotropical foci of VL, *Lu.* (*Lu.*) *longipalpis* or related species are the incriminated vectors of *Le.* (*Le.*) *infantum chagasi* (Rangel and Lainson 2009; Ready 2013), and all are grouped in the monophyletic *Lutzomyia* subgenus (Young and Duncan 1994), or genus (Galati 2003), based on a novel analysis of incomplete concatenated nuclear DNA and mtDNA (Grace-Lema et al. 2015) or *COI* barcoding (Pinto et al. 2015). The same molecular reports, along with others based on mtDNA (Torgerson et al. 2003) and nuclear rDNA (Lins et al. 2002; Mazzoni et al. 2002; Beati et al. 2004), provide some support for the hypothesis that many of the incriminated vectors of CL in the Neotropics are classified in three monophyletic subgenera (Young and Duncan 1994) or genera (Galati 2003), namely, *Nyssomyia*, *Psychodopygus* and *Trichophoromyia*. These can be grouped in a sub-tribe proposed by Galati (2003) (Psychodopygina) separate from that of the VL vectors and the CL vector *Lu.* (*Mi.*) *migonei* (Lutzomyina). However, the different molecular phylogenies do not agree on the relationships of the groups within Psychodopygina or on whether *Lu.* (*Ny.*) *flaviscutellata* and related species should be classified in the genus *Bichromomyia* (Galati 2003).

The report on the *COI* barcoding of Brazilian sand flies (Pinto et al. 2015) included specimens from the eastern and central states of Espirito Santo (15 locations), Bahia (1), Rio de Janeiro (1), Minas Gerais (1), and Mato Grosso (1). A neighbour-joining algorithm was used to group DNA sequences according to pairwise divergences estimated by the Kimura 2-parameter model (K2P), and the phenetic groupings were recognised as species and genera according to thresholds of absolute K2P genetic distances identified by Automatic Barcode Gap Discovery (ABGD) software (Puillandre et al. 2012). This is not a phylogenetic approach (see subsection Mitochondrial DNA Markers) but rather a development of phenetic analyses that tend to group taxa similarly when the genetic distances are calculated from isoenzyme profiles (Torgerson et al. 2003).

Lu. *(Lu.)* longipalpis *Species Complex as Vectors* of Le. (Le.) infantum chagasi

Most molecular and biochemical research on Brazilian sand flies has been on *Lu. longipalpis* and related members of the Longipalpis series of the subgenus *Lutzomyia* (Young and Duncan 1994). The series is treated as the subgenus *Lutzomyia* by Galati (2003), and it contains two species that are often placed in the *Lu. longipalpis* complex because their females and those of *Lu. longipalpis* are isomorphic, namely, *Lu.* (*Lu.*) *cruzi* and *Lu.* (*Lu.*) *pseudolongipalpis* (Arrivillaga and Feliciangeli 2001). There is insufficient space to review all of the research on this complex, although key articles are covered (in this subsection and earlier ones), and additional references can be found in major reviews (Hamilton et al. 2005; Maingon et al. 2008, Araki et al. 2013; Souza et al. 2017) from the research groups of Professor Richard D. Ward and Professor Alexandre A. Peixoto, who together provided the major impetus to this field of research in Brazil.

In summary, based on up to 27 isoenzyme loci and involving 5 mtDNA genes, there are 4 distinctive geographical genetic lineages of *Lu. longipalpis* — in Central America (trans-Andean or species D), northern South America (cis-Andean or species C with sub-lineages in Colombia and Venezuela), Laran in Venezuela (species B = *Lu. pseudolongipalpis*), and Brazil (species A) — and reproductive barriers were demonstrated or inferred between them by producing sterile male hybrids from laboratory crosses, by finding distinctive karyotypes, or by population-genetics analyses (Lanzaro et al. 1993; Yin et al. 1999; Lampo et al. 1999; Arrivillaga et al. 2003). The same tools were applied to demonstrate relatively small genetic distances (increasing with geographical distance) between populations within species C or species D (Munstermann et al. 1998; Mutebi et al. 1998; Soto et al. 2001), but this isolation by distance only held true for some populations of species A in Brazil (Mukhopadhyay et al. 1998a, 1998b; Mutebi et al. 1999; Hodgkinson et al. 2003). Isolation by distance was detected in Brazil among populations that shared a burst-type lovesong and among populations that shared similar pulse-type lovesongs, and this was convincingly demonstrated by population-genetics approaches based on genotyping highly polymorphic microsatellite DNA loci (Watts et al. 2005) and multiple nuclear gene loci (Araki et al. 2009, 2013). Reproductive barriers between Brazilian populations of these two putative *Lu. longipalpis* species were first demonstrated by Ward et al. (1983) and later associated with different aggregation pheromones (see subsection Pheromones; Hamilton et al. 2005), lovesongs (Souza et al. 2004, Vigoder et al. 2010a, 2015) and love-song genes (see subsection Lovesong Genes: *Per*, *Cacophony* and *Paralytic*; Souza et al. 2017).

The pre- and post-copulatory reproductive barriers between these geographical and behavioural variants of *Lu. longipalpis* were considered sufficient for formal species recognition (Brandão-Filho et al. 2009), but until now only *Lu. pseudolongipalpis* has been described as a taxonomic species. Actually, there is no need to name a new species in Brazil, partly because the type material of *Lu. longipalpis* came from Brazil, and partly because it becomes increasingly difficult to split *Lu. longipalpis sensu stricto* from Brazil into two or more well-defined species. It is true that reproductive barriers are strong in at least three locations where there are sympatric populations with burst-type and pulse-type lovesongs, namely, in Sobral (Ceará State, Northeast Brazil), Estrela de Alagoas (Alagoas State, Northeast Brazil), and Jaíba (Minas Gerais State, northern border of Southeast Brazil) (Maingon et al. 2008; Souza et al. 2008; Araki et al. 2009; Vigoder et al. 2015). However, based on the comparative sequence analysis of alleles at 21 nuclear loci, there is increasing evidence of mating between these sympatric populations in the recent past, thus leading to the introgression of allelic sequences of courtship genes (such as *cac* and *per*) and sometimes the loss of associations between such nuclear genetic markers with lovesongs, mating behaviour, aggregation pheromones, and pheromone-gland morphology (1S, 2S, and variants) (Araki et al. 2009, 2013). For example, there is a strong distinction in Sobral between flies characterised by 2S, burst-type song, and Cemb-1 pheromone *versus* flies characterised by 1S, pulse-type song P3, and 9MGB pheromone. In contrast, in Estrela de Alagoas, 1S is characterised by pulse-type song P5 and Cemb-1, whereas in Jaíba, 1S is characterised by pulse-type song P4 and Cemb-2. A population from Mesquita (Rio de Janeiro State, Southeast Brazil) even has a mixed song (Araki et al. 2009).

Araki et al. (2013) discussed the possibility that the process of speciation might be incomplete in Brazil, perhaps after allopatric divergence followed by interbreeding on secondary contact. If true, it would be premature to create one or more taxonomic species. The original intention of Ward et al. (1983) was to investigate any association between regional variants of *Lu. longipalpis* and the endemicity of VL in Northeast Brazil. None of the lovesong lineages has yet been shown to provide such an association. Therefore, future molecular and biomedical investigations should try to link putative sibling species to spatial and temporal variations in phenotypes with genotypic markers that are most likely to affect vector control (Ready 2011, 2013). For example, Lins et al. (2008) discovered amino acid motifs in the *para* gene of *Lu. longipalpis* from Brazil that might cause insecticide resistance, and Souza et al. (2009) reported life-cycle differences among members of this species complex. Certainly, given the regional variation in gene introgression, little is likely to be inferred from molecular investigations of chemotypes alone (Casanova et al. 2015) or of a single gene locus whether mtDNA (Ribolla et al. 2016) or *per* (Lima Costa et al. 2015).

The local and regional incrimination of *Lu. longipalpis s.l.* as a vector of *Le. infantum chagasi* depends on the application of many molecular and biochemical tools (see subsection Transcriptomics, Whole Genome Sequencing and Vector Competence), not least for identifying *Leishmania* infections and bloodmeal sources in wild sand flies (Ranasinghe et al. 2008; Sant'Anna et al. 2008). Salivary peptides are also a major focus for molecular research because some, but not Maxadilan, are potential vaccine candidates and provide a means of estimating exposure to the bites of *Lu. longipalpis* (see subsection Salivary Peptide Genes).

Lu. (Ny.) flaviscutellata *as a Vector of* Le. (Le.) amazonensis

Morphologically, this sand fly belongs to the *Lutzomyia* (*Nyssomyia*) *olmeca* species complex, which contains the incriminated vectors of *Leishmania* (*Leishmania*) *mexicana* in Mexico (*Lu. olmeca*) and Panama [*Lu.* (*Nyssomyia*) *bicolor*] and the potential vectors of this parasite and its sister species *Le. amazonensis* in western Amazonia (*Lu. bicolor*) and of *Le. amazonensis* alone in central and eastern Amazonian Brazil *Lutzomyia* (*Nyssomyia*) *nociva*] (Ready 2013). Throughout tropical Brazil, *Lu. flaviscutellata* is an incriminated or suspected vector of *Le. amazonensis*, and it is the only member of the species complex occurring south of Amazonia (Carvalho et al. 2015). However, molecular markers have not been applied to test for the presence of cryptic sibling species or to assess the status of all members of the species complex. Some could be subspecies as originally described.

Based on a phylogenetic analysis of nuclear 28S rDNA, *Lu. flaviscutellata* from San Martín, Peru, was grouped with *Lu.* (*Ps.*) *geniculata*, *Lu.* (*Trichophoromyia*) *sp.*, *Lu.* (*Ny.*) *yuilli yuilli* and *Lu.*(*Viannamyia*) *tuberculata*, all within the sub-tribe Psychodopygina (Beati et al. 2004); and, based on *COI* mtDNA, *Lu. olmeca* was also grouped with a member of this sub-tribe, *Lu.* (*Psathyromyia*) *aclydifera* (Grace-Lema et al. 2015). However, taxon sampling was inadequate to permit any inferences of epidemiological importance. Of more biomedical interest was the characterization of SPs, including yellow-related proteins, of *Lu. olmeca* from Mexico (Abdeladhim et al. 2016). In this case, molecular phylogenetic analysis was aimed more at understanding the evolution of SP functional differences that might affect epidemiological investigations and leishmaniasis control.

Lu. (Ny.) intermedia, Lu. (Ny.) neivai *and* Lu. (Ny.) whitmani *as Vectors of* Leishmania (Viannia) *Species*

These three members of the subgenus *Nyssomyia* were incriminated as vectors of *Le.* (*Vi.*) *braziliensis* in non-Amazonian regions of Brazil including the Northeast (*Lu. intermedia*, *Lu. whitmani*), the Central-West (*Lu. neivai*, *Lu. whitmani*), the Southeast (all three sand flies), and the South (*Lu. neivai*; also present in adjoining Argentina and Paraguay) (Rangel and Lainson 2009; McIntyre et al. 2017). In addition, *Lu. whitmani* was found infected with *Leishmania* (*Viannia*) *guyanensis* and *Leishmania* (*Viannia*) *shawi* in Amazonia, where its more silvatic populations were considered to be sufficiently distinctive to merit recognition as a cryptic sibling species based on behaviour (Rangel and Lainson 2009) as well as morphometrics and the relative abundance of a repetitive DNA sequence (Rangel et al. 1996). The three sand flies have been a focus for molecular and biochemical research because of their vectorial roles in diverse ecotopes and their morphological similarities: The adults of both sexes of *Lu. whitmani* are distinctive (Young and Duncan 1994), but *Lu. neivai* was long considered a junior synonym of *Lu. intermedia* before Marcondes (1996) resurrected it as a valid morphospecies with distinctive female genitalia.

Molecular or biochemical markers can be useful for the alpha-taxonomy of sand flies (Depaquit 2014), and molecular systematics based on nuclear genes (Lins et al. 2002) or mtDNA barcoding (Pinto et al. 2015) usually groups *Lu. intermedia* with *Lu. whitmani*. However, it is clear that molecular taxonomy must be applied cautiously to these three Brazilian vectors because gene introgression indicates imperfect reproductive barriers, not only between regional populations of *Lu. whitmani* (Ready et al. 1997; Fig. 2) on the southeastern border of Amazonia (Ready et al. 1998) but also between all three species in the Southeast (Marcondes et al. 1997; Mazzoni et al. 2006, 2008). Comparative DNA sequence analyses of the CB3 fragment (containing *Cyt b*) from 94 specimens of *Lu. whitmani* revealed at least two sub-lineages of a species-specific mtDNA lineage with each being associated with populations from distinctive regions (Amazonian with Atlantic Forest biomes; and intervening North–South seasonally drier biomes), and not one *Lu. whitmani* contained a haplotype of the lineage found only in *Lu. intermedia* and *Lu. neivai* (Ready et al. 1997; Marcondes et al. 1997; Ishikawa et al. 1999). Therefore, assuming maternal and non-recombinant mtDNA inheritance (Ishikawa et al. 1999; Avise 2000), the discovery in Espírito Santo State (Viana, near Vitória) of haplotypes of one CB3 lineage of *Lu. whitmani* (North–South) in all seven females with the morphology of *Lu. intermedia* (Marcondes et al. 1997) is most parsimoniously explained by the mating of one or more wild-type females of *Lu. whitmani* with wild-type males of *Lu. intermedia*, followed by the mating of the hybrid females with wild-type males of *Lu. intermedia* alone.

Individuals of both morphospecies were later sampled from a nearby location in Espírito Santo State (Afonso Cláudio) and genotyped at 10 nuclear loci (Mazzoni et al. 2008). An 'isolation with migration' population genetics model provided evidence for introgression from *Lu. intermedia* towards *Lu. whitmani* in three of 10 loci [*cac* (cacophony), *ξCOP* (or zetacop), and the largest non-recombining block of *RpL17A* (ribosomal protein L23)], but there were species-specific fixed polymorphisms and no evidence for gene flow in four loci [*Rp49* (ribosomal protein L32), *RpL36* (ribosomal protein L36), *RpS19a* (ribosomal protein S19a), and the entire *RpL17A*]. Genotype frequencies were not reported for each sex, and mtDNA was not analysed, but all sand flies had morphological characters diagnostic for only one of the two species (Mazzoni et al. 2008), and therefore the introgression could not have been very recent. The direction of introgression is consistent with the mtDNA findings of Marcondes et al. (1997), if hybrid individuals tend to mate with wild-type individuals of one species only. Marcondes et al. (1997) also reported mtDNA introgression from *Lu. neivai* into *Lu. intermedia* in São Paulo State (Caraguatatuba).

The distributions of *per* haplotypes in Afonso Cláudio (Mazzoni et al. 2008) were consistent with no introgression at this locus, which did not match finding evidence of introgression in the other lovesong gene *cac* in the same populations and in *per* in an earlier survey (Mazzoni et al. 2006). The latter characterized 68 *Lu. intermedi*a and 53 *Lu. whitmani*, all F1 individuals bred from flies collected in Afonso Cláudio, two locations in Rio de Janeiro State (Posse, Jacarepaguá) and two locations in Bahia State (Corte de Pedra, Ilhéus). One *Lu. intermedia* from Posse and another from Corte de Pedra had *per* haplotypes placed in the *Lu. whitmani* part of the haplotype network, while two *Lu. whitmani* from Afonso Cláudio and two from Corte de Pedra had *per* haplotypes placed in the *Lu. intermedia* part of the haplotype network. The contradictory findings of the two investigations might result from temporal and spatial changes in the distributions of hybrid individuals (Mazzoni et al. 2008). For *Lu. intermedia*, the failure to detect regional acoustic signals (Vigoder et al. 2011) was matched by the absence of regional *per* lineages (Mazzoni et al. 2006).

Genetic differentiation might produce not only allopatric regional lineages, as for *Lu. whitmani* (Ishikawa et al. 1999; Margonari et al. 2004), but also sympatric cryptic species or incipient species that are adapted to different environments, e.g., the *Anopheles gambiae* complex (Wiebe et al. 2017). The latter possibility was explored in Mesquita municipality (Rio de Janeiro State) where domestic, peri-domestic, and silvatic populations of *Lu. intermedia* were sampled within 800 m of each other in a location 600 m above sea level where *Lu. whitmani* was rare and *Lu. neivai* was absent (Meneses et al. 2005). MLEE with eight loci, SSCP of 3 mtDNA genes, and RAPD-PCR with 6 primers were applied to characterise male flies, but only an analysis of the RAPD-PCR phenotypes provided evidence of non-seasonal population differentiation, with the silvatic population being distinctive despite the MLEE analysis indicating at least 50 flies migrating between the different habitats each generation (Meneses et al. 2005).

Taken together, these molecular and biochemical findings suggest that reproductive barriers do not always prevent substantial gene flow among the three sand fly species, perhaps because courtship gene motifs are shared by introgression (Mazzoni et al. 2006, 2008), and there is little differentiation of local populations of *Lu. intermedia* among habitats (Meneses et al. 2005). However, the three species do maintain distinctive morphologies, different regional and climatic distributions (Ishikawa et al. 1999; McIntyre et al. 2017), specific altitudinal and habitat distributions (Meneses et al. 2005; Mazzoni et al. 2008), and distinguishing behaviours such as anthropophily and endophily (Campbell-Llendrum et al. 1999, 2000). As for *Lu. (Lu.) longipalpis*, this may indicate that we are focusing too much on species boundaries and too little on biological traits that influence vector competence and vectorial capacity (Ready 2013). For these three sand flies species, molecular and biochemical tools are essential for characterising the phenotypes and genotypes associated with, for example, blood-feeding preferences as determined by identifying host kairomones and attractants (Pinto et al. 2001; Andrade et al. 2008) and blood-meal origins as identified by ELISA or comparative DNA-sequence analysis of mammalian *Cyt b* (Marassá et al. 2013); transmission cycles determined by distinguishing specific *Leishmania* species from other trypanosomatids using multiplex PCR and other PCR tests (Pita-Pereira et al. 2009; Ferreira et al. 2015; Neitzke-Abreu et al. 2014); microbiota diversity (Monteiro et al. 2016), which affects the development of *Le. braziliensis* in the sand fly gut (Soares et al. 2014); and sand fly saliva molecules, which could be components of vaccines (de Moura et al. 2013) or assays for estimating exposure to species-specific bites (Aires et al. 2017). Some of these molecular and biochemical findings might not be applicable to all vector populations, and so the challenge is to identify when to investigate natural variation locally and regionally. For example, immunity to whole sand fly saliva can protect against infections of *Le. braziliensis* (*Lu. whitmani*) or exacerbate them (*Lu. intermedia*), while a DNA plasmid vaccine of one vector SP (e.g., Linb11 from *Lu. intermedia*) can provide more carefully defined protection (de Moura et al. 2013). However, this research was based on only one wild population of *Lu. intermedia*, from Corte de Pedra, Bahia State, where there is evidence of interbreeding with *Lu. whitmani* (Mazzoni et al. 2006), and no population genetics was reported.

Other Vectors of **Leishmania (Viannia)** *Species*

Lutzomyia (*Ny.*) *umbratilis* is the incriminated vector of *Le.* (*Vi.*) *guyanensis* in northeastern Amazonia where this parasite frequently infects humans and causes "pian bois" (Rangel and Lainson 2009). The population structure of the vector was investigated because its abundance at ground level varies seasonally in northeastern Amazonian Brazil and French Guiana (Ready et al. 1986); it favours tree-base resting sites only north of the Amazon River; and 'pian bois' is less frequent south of the river (Rangel and Lainson 2009). MLEE did not identify any ecological population differentiation in northeastern Amazonia (Tibayrenc et al. 1980), but molecular research found geographical population differentiation within Brazil.

Based on phylogenetic analysis of 1181 bp of *COI*, Scarpassa and Alencar (2012) found two mtDNA clades of *Lu. umbratilis* in central Amazonian Brazil, with clade I restricted to four populations from north of the Negro and Amazon rivers and clade II being restricted to two populations from an area between the two rivers. However, the pairwise sequence divergence was much less (1%) than that between either clade and the outgroup *Lu.* (*Ny.*) *anduzei* (5.8%). After more widespread sampling of *Lu. anduzei* and phylogenetic analysis of the barcoding region of *COI* (663 bp) from both new and old samples, only the two morphospecies showed pairwise genetic distances consistent with intraspecific variation (average 0.8%; range 0.2–1.4%) and interspecific variation (average 4.4%; range 4.1–4.6%), whereas clades I and II of *Lu. umbratilis* differed by only 0.9–1.0% (Scarpassa and Alencar 2013). A phylogenetic analysis of 74 and 67 sequences of *COI* and *Cyt b* mtDNA haplotypes, respectively, demonstrated no biogeographical variation in the samples of *Lu. anduzei* (Scarpassa et al. 2015).

A phylogenetic analysis of the nuclear lovesong gene *cac* placed *Lu. umbratilis* (from Manacapuru between the Negro and Amazon rivers) as the nearest neighbour to the sister species *Lu. intermedia* and *Lu. whitmani* (from Afonso Cláudio, Espírito Santo State where, however, there is gene introgression) (Lins et al. 2002). A similar analysis of another nuclear lovesong gene, *per,* compared *Lu. umbratilis* from Manacapuru with the same species from one of the populations north of the Amazon River (Rio Preto da Eva) and a population from near Recife, Pernambuco State, in the Northeast region of Brazil and discovered a closer relationship between the latter two despite their distant geographical locations (de Souza Freitas et al. 2016). These results suggest there might be a *Lu. umbratilis* species complex, but this should be confirmed by studying more populations and explaining the contradictory result of the extensive mtDNA barcoding. Even if the populations from the Manacapuru region turn out to represent a cryptic sibling species of *Lu. umbratilis*, there is no evidence that this species will be charactersied by the traits of epidemiological significance associated with many populations south of the Amazon River, for example, from Serra dos Carajás, Pará State (Rangel and Lainson 2009). In fact, the contrary is indicated by the genetic similarities between the population from north of the Amazon River and that from Northeast Brazil.

Serra dos Carajás is also the location where *Lu.* (*Ps.*) *wellcomei* was incriminated as a vector of *Le.* (*Vi.*) *braziliensis* (Rangel and Lainson 2009). Unfortunately, the females of *Lu.* (*Ps.*) *complexa* are morphologically indistinguishable from those of *Lu. wellcomei*, and both species can be abundant in the original disease focus and elsewhere in southeastern Amazonia (Ready et al. 1984). This led to searches for diagnostic molecular markers for use in field studies to resolve the vectorial roles of the two sand flies. Alloenzymes were identified by a population genetics approach, but MLEE provided no diagnostic alleles at 16 loci (Ready and Silva 1984). Individual females of the two species were successfully separated by cuticular hydrocarbon analysis involving the use of gas liquid chromatography and discriminant function analysis (Ryan et al. 1986) and by dot-blot hybridizations with two repetitive DNA probes cloned from males of *Lu. wellcomei* (Ready et al. 1991; Fig. 1). The latter report provided only the second confirmed infection of *Le. braziliensis* in *Lu. wellcomei*, the first arising from rearing male flies from eggs laid by an infected wild-caught female (Ryan et al. 1986), and both infections were found in Serra dos Carajás. Monoclonal antibodies were later used to identify two serodemes of this parasite in four female sand flies of the Squamiventris series of *Lutzomyia* (*Psychodopygus*) from the nearby municipality of Paragominas where *Lu. complexa* predominated at the time (de Souza et al. 1996).

Other *Lutzomyia* (*Psychodopygus*) species have been found infected with *Leishmania* (*Viannia*) species in Brazil including the widespread *Lu.* (*Ps.*) *ayrozai*, *Lu.* (*Ps.*) *paraensis*, *Lu.* (*Ps.*) *davisi*, and *Lu.* (*Ps.*) *hirsuta* (Rangel and Lainson 2009). The latter two, for example, were found infected with *Leishmania* (*Viannia*) *naiffi* in Rondônia (Gil et al. 2003). In a related subgenus, *Lutzomyia* (*Trichophoromyia*) *ubiquitalis* is the incriminated vector of *Leishmania* (*Viannia*) *lainsoni* in Amazonia (Rangel and Lainson 2009). Unfortunately, no insights into the epidemiological roles of these sand flies have been provided by molecular and biochemical research.

Lutzomyia (*Migonemyia*) *migonei* is the only other widespread sand fly in Brazil that has been incriminated as a vector of *Le. braziliensis* (Rangel and Lainson 2009), but the only molecular reports are on phylogenetic analyses (Lins et al. 2002; Mazzoni et al. 2002; Beati et al. 2004), which place this sand fly outside the subtribe Psychodopygina, in which Galati (2003) classifies all the other vectors mentioned in this subsection. It has a copulatory courtship song (Vigoder et al. 2010b) and is morphologically similar to related species of *Migonemyia* (Galati et al. 2007).

New Diagnostic Techniques

Mass Spectrometry of Proteins

Identification of individual sand flies is possible by using the discriminatory power of matrix-assisted laser desorption/ionization time-of-flight mass spectrometry (MALDI-TOF MS), which was first trialled for laboratory-bred *Phlebotomus*

(Dvorak et al. 2014) and then refined by using a reference database that included laboratory-bred *Lu.* (*Lu.*) *longipalpis* of Brazilian origin and wild-caught *Lutzomyia* (*Nyssomyia*) *trapidoi* from Ecuador (Mathias et al. 2015). It can be useful for the identification of specific peptides and species identification by non-specialists, but it has the disadvantage of characterising phenotypes (not genotypes) that can vary because of peptide degradation (e.g., in ethanol), and analysis will often depend on centralized systems administered by commercial enterprises.

Nucleic Acids

Techniques proliferate, and some might well prove useful for sand flies. Checkerboard DNA–DNA hybridization was used to estimate the number of bacteria in the midgut of the mosquito *Aedes aegypti* in Brazil, and one of the infecting species was also detected in the midgut of *Lu.* (*Lu.*) *longipalpis* where microbiota can affect vector competence (Gaio et al. 2011).

Mild-Vectolysis is a non-destructive technique permitting the isolation of genomic DNA without damaging the morphological features of voucher specimens, and Giantsis et al. (2016) demonstrated its ability to provide sufficient DNA for barcoding sand flies.

Kocher et al. (2017) explored the possibility of identifying multiple species from a single bulk sample through high-throughput DNA metabarcoding, and abundant species in French Guiana were successfully detected in ‘vector soup’ by targeting mitochondrial 16S rRNA. The inability to identify less abundant species is a drawback.

Conclusions

Research on species complexes of vectors has provided insights important for targeting control measures, for example, to limit the transmission of malaria by *Anopheles gambiae s.l.* in sub-Saharan Africa (Wiebe et al. 2017), and so it was only natural to investigate speciation among the major Brazilian vectors of leishmaniasis. However, the research in Africa was prompted by knowledge of variation in vector behaviour known to influence control (e.g., differences in the duration of indoor resting by blood-fed *An. gambiae s.s.* and *Anopheles arabiensis*), whereas there was no such knowledge for sand flies (Ready 2011). Molecular and biochemical research carried out on neotropical sand flies in the last 40 years has provided fascinating insights into the speciation of *Lu.* (*Lu.*) *longipalpis* as a vector of VL as well as *Lu.* (*Ny.*) *intermedia*, *Lu.* (*Ny.*) *neivai* and *Lu.* (*Ny.*) *whitmani* as vectors of CL, but this has not been matched by research into polymorphic vector behaviour that could be explained by genetic variation. New national and international goals are required to achieve this.

For example, research on the *Lu. longipalpis* complex was prompted by the aim of explaining why high VL incidence is largely restricted to the Northeast region of Brazil (Ward et al. 1983), and now it is possible to apply the wealth of recent knowledge about population genetics and pheromones to assess if it is regional variations in vector competence, host-seeking, or other behavioural traits that determine high transmission rates, or if human life styles and susceptibility to disease are also important. The VL elimination programme on the Indian sub-continent provides a model for tackling integrated control (Cameron et al. 2016). In Brazil, as on the Indian sub-continent, there is inadequate knowledge about which sand fly resting sites should be targeted for insecticide control and about the extent of insecticide resistance (Alexander et al. 2009). These comments apply also to the vectors of CL, and research on all VL and CL vectors requires multilocus typing because of interspecific gene flow. Single-lineage typing will rarely be sufficiently informative as exemplified by the inadequacy of mtDNA haplotyping for explaining the spatial distribution across Brazil of pheromone chemotypes of *Lu. longipalpis* (Hamilton et al. 2005) and the habitat (peri-domestic or silvatic) and host preferences of *Lu. whitmani* (Ready et al. 1998; Campbell-Llendrum et al. 1999, 2000). A major challenge will be to decide when to apply multilocus population genetics, which is relatively expensive. Its use would be appropriate, for example, to discover the geographical range over which a salivary peptide could be useful for vaccination or estimating the biting rates of a vector on humans and reservoir hosts.

References

Abdeladhim M, V Coutinho-Abreu I, Townsend S, Pasos-Pinto S, Sanchez L, Rasouli M, B Guimaraes-Costa A, Aslan H, Francischetti IM, Oliveira F, Becker I, Kamhawi S, Ribeiro JM, Jochim RC, Valenzuela JG (2016) Molecular diversity between salivary proteins from new world and old world sand flies with emphasis on *Bichromomyia olmeca*, the Sand Fly Vector of *Leishmania mexicana* in Mesoamerica. PLoS Negl Trop Dis 10(7):e0004771

Adamson RE, Ward RD, Feliciangeli MD, Maingon R (1993) The application of random amplified polymorphic DNA for sandfly species identification. Med Vet Entomol 7(3):203–207

Aires J, Casanova C, Vernal S, Nascimento M, Rodrigues S, Lerner EA, Roselino AM (2017) Maxadilan-simile expression in *Nyssomyia neivai*, a sandfly vector in an endemic region of Brazil, and its immunogenicity in patients with American tegumentary leishmaniasis. Mem Inst Oswaldo Cruz 112(2):116–122

Alexander B, Barros VC, Souza SS, Teodoro LP, Soares ZR, Gontijo NF et al (2009) Susceptibility to chemical insecticides of two Brazilian populations of the visceral leishmaniasis vector *Lutzomyia longipalpis* (Diptera: Psychodidae). Trop Med Int Health 14(10):1272–1277

Anderson JM, Oliveira F, Kamhawi S, Mans BJ, Reynoso D, Seitz AE, Lawyer P, Garfield M, Pham M, Valenzuela JG (2006) Comparative salivary gland transcriptomics of sandfly vectors of visceral leishmaniasis. BMC Genomics 7:52

Andrade AJ, Andrade MR, Dias ES, Pinto MC, Eiras AE (2008) Are light traps baited with kairomones effective in the capture of *Lutzomyia longipalpis* and *Lutzomyia intermedia*? An evaluation of synthetic human odor as an attractant for phlebotomine sand flies (Diptera: Psychodidae: Phlebotominae). Mem Inst Oswaldo Cruz 103(4):337–343

Araki AS, Vigoder FM, Bauzer LG, Ferreira GE, Souza NA, Araújo IB et al (2009) Molecular and behavioral differentiation among Brazilian populations of *Lutzomyia longipalpis* (Diptera: Psychodidae: Phlebotominae). PLoS Negl Trop Dis 3(1):e365

Araki AS, Ferreira GE, Mazzoni CJ, Souza NA, Machado RC, Bruno RV et al (2013) Multilocus analysis of divergence and introgression in sympatric and allopatric sibling species of the *Lutzomyia longipalpis* complex in Brazil. PLoS Negl Trop Dis 7(10):e2495

Aransay AM, Scoulica E, Tselentis Y, Ready PD (2000) Phylogenetic relationships of phlebotomine sandflies inferred from small subunit nuclear ribosomal DNA. Insect Mol Biol 9:157–168

Aransay AM, Malarky G, Ready PD (2001) Isolation (with enrichment) and characterization of trinucleotide microsatellites from *Phlebotomus perniciosus*, a vector of *Leishmania infantum*. Mol Ecol Notes 1:176–178

Arrivilaga J, Feliciangeli MD (2001) *Lutzomyia pseudolongipalpis*: the first new species within the longipalpis (Diptera: Psychodidae: Phlebotominae) complex from La Rinconada, Curarigua, Lara state. Venez J Med Entomol 38:783–790

Arrivillaga JC, Norris DE, Feliciangeli MD, Lanzaro GC (2002) Phylogeography of the neotropical sandfly *Lutzomyia longipalpis* inferred from mitochondrial DNA sequences. Infect Genet Evol 2(2):83–95

Arrivillaga J, Mutebi JP, Piñango H, Norris D, Alexander B, Feliciangeli MD et al (2003) The taxonomic status of genetically divergent populations of *Lutzomyia longipalpis* (Diptera: Psychodidae) based on the distribution of mitochondrial and isozyme variation. J Med Entomol 40(5):615–627

Avise JC (2000) Phylogeography: The History and Formation of Species. Harvard University Press, Cambridge, MA, pp 1–447

Azevedo RV, Dias DB, Bretãs JA, Mazzoni CJ, Souza NA, Albano RM, Wagner G, Davila AM, Peixoto AA (2012) The transcriptome of *Lutzomyia longipalpis* (Diptera: Psychodidae) male reproductive organs. PLoS One 7(4):e34495

Bates PA, Depaquit J, Galati EAB, Kamhawi S, Maroli M, McDowell MA et al (2015) Recent advances in phlebotomine sand fly research related to leishmaniasis control. Parasit Vectors 8:131

Bauzer LG, Gesto JS, Souza NA, Ward RD, Hamilton JG, Kyriacou CP et al (2002a) Molecular divergence in the period gene between two putative sympatric species of the *Lutzomyia longipalpis* complex. Mol Biol Evol 19(9):1624–1627

Bauzer LG, Souza NA, Ward RD, Kyriacou CP, Peixoto AA (2002b) The period gene and genetic differentiation between three Brazilian populations of *Lutzomyia longipalpis*. Insect Mol Biol 11(4):315–323

Bauzer LGSR, Souza NA, Maingon RDC, Peixoto AA (2007) *Lutzomyia longipalpis* in Brazil: a complex or a single species? A mini-review. Mem Inst Oswaldo Cruz 102(1):1–12

Beati L, Caceres AG, Lee JA, Munstermann LE (2004) Systematics relationships among *Lutzomyia* sand flies (Diptera: Psychodidae) of Peru and Colombia based on the analysis of 12S and 28S ribosomal DNA sequences. Int J Parasitol 34:225–234

Booth DR, Ready PD, Smith DF (1994) Isolation of non-LTR retrotransposon reverse transcriptase-like sequences from phlebotomine sandflies. Insect Mol Biol 3:89–96

Booth DR, Ready PD, Smith DF (1996) Evolution of multiple families of non-LTR retrotransposons in phlebotomine sandflies. Genet Res 67:227–237

Bottecchia M, Oliveira SG, Bauzer LG, Souza NA, Ward RD, Garner KJ et al (2004) Genetic divergence in the cacophony IVS6 intron among five Brazilian populations of *Lutzomyia longipalpis*. J Mol Evol 58(6):754–761

Brandão-Filho SP, Balbino VQ, Marcondes CB, Brazil RP, Hamilton JG, Shaw JJ (2009) Should reproductively isolated populations of *Lutzomyia longipalpis sensu lato* receive taxonomically valid names? Mem Inst Oswaldo Cruz 104(8):1197–1200

Bray DP, Bandi KK, Brazil RP, Oliveira AG, Hamilton JG (2009) Synthetic sex pheromone attracts the leishmaniasis vector *Lutzomyia longipalpis* (Diptera: Psychodidae) to traps in the field. J Med Entomol 46:428–434

Bray DP, Carter V, Alves GB, Brazil RP, Bandi KK, Hamilton JG (2014) Synthetic sex pheromone in a long-lasting lure attracts the visceral leishmaniasis vector, *Lutzomyia longipalpis*, for up to 12 weeks in Brazil. PLoS Negl Trop Dis 8(3):e2723

Brazil RP, Hamilton JGCH (2002) Isolation and identification of 9-methylgermacrene-B as the putative sex pheromone of *Lutzomyia cruzi* (Mangabeira, 1938) (Diptera: Psychodidae). Mem Inst Oswaldo Cruz 97(3):435–436

Caillard T, Tibayrenc M, Le Pont F, Dujardin JP, Desjeux P, Ayala FJ (1986) Diagnosis by isozyme methods of two cryptic species, *Psychodopygus carrerai and P. yucumensis* (Diptera: Psychodidae). J Med Entomol 23:489–492

Cameron MM, Acosta-Serrano A, Bern C, Boelaert M, den Boer M, Burza S, Chapman LA, Chaskopoulou A, Coleman M, Courtenay O, Croft S, Das P, Dilger E, Foster G, Garlapati R, Haines L, Harris A, Hemingway J, Hollingsworth TD, Jervis S, Medley G, Miles M, Paine M, Picado A, Poché R, Ready P, Rogers M, Rowland M, Sundar S, de Vlas SJ, Weetman D (2016) Understanding the transmission dynamics of *Leishmania donovani* to provide robust evidence for interventions to eliminate visceral leishmaniasis in Bihar, India. Parasit Vectors 9:25

Campbell-Lendrum DH, Pinto MC, Brandão-Filho SP, de Souza AA, Ready PD, Davies CR (1999) Experimental comparison of anthropophily between geographically dispersed populations of *Lutzomyia whitmani* (Diptera: Psychodidae). Med Vet Entomol 13:299–309

Campbell-Lendrum DH, Brandão-Filho SP, Pinto MC, Vexenat A, Ready PD, Davies CR (2000) Domesticity of *Lutzomyia whitmani* (Diptera: Psychodidae) populations: Field experiments indicate behavioural differences. Bull Entomol Res 90:41–48

Campo D, Lehmann K, Fjeldsted C, Souaiaia T, Kao K, Nuzhdin SV (2013) Whole-genome sequencing of two North American *Drosophila melanogaster* populations reveals genetic differentiation and positive selection. Mol Ecol 22:5084–5097

Carvalho BM, Rangel EF, Ready PD, Vale MM (2015) Ecological niche modelling predicts southward expansion of *Lutzomyia* (*Nyssomyia*) *flaviscutellata* (Diptera: Psychodidae: Phlebotominae), vector of *Leishmania* (*Leishmania*) *amazonensis* in South America, under climate change. PLoS One 10(11):e0143282

Casanova C, Hamilton JGC, Trigo JR, Costa AIP (2006) Identification of sex pheromones of *Lutzomyia longipalpis* (Lutz & Neiva, 1912) populations from the state of São Paulo, Brazil. Mem Inst Oswaldo Cruz 101(1):113–115

Casanova C, Colla-Jacques FE, Hamilton JG, Brazil RP, Shaw JJ (2015) Distribution of *Lutzomyia longipalpis* chemotype populations in São Paulo state, Brazil. PLoS Negl Trop Dis 9(3):e0003620

Charlab R, Valenzuela JG, Rowton ED, Ribeiro JM (1999) Toward an understanding of the biochemical and pharmacological complexity of the saliva of a hematophagous sand fly *Lutzomyia longipalpis*. Proc Natl Acad Sci U S A 96(26):15155–15160

Cohnstaedt LW, Beati L, Caceres AG, Ferro C, Munstermann LE (2011) Phylogenetics of the phlebotomine sand fly group *Verrucarum* (Diptera: Psychodidae: *Lutzomyia*). Am J Trop Med Hyg 84(6):913–922

Collin N, Gomes R, Teixeira C, Cheng L, Laughinghouse A et al (2009) Sand fly salivary proteins induce strong cellular immunity in a natural reservoir of visceral leishmaniasis with adverse consequences for *Leishmania*. PLoS Pathog 5:e1000441

Contreras Gutiérrez MA, Vivero RJ, Vélez ID, Porter CH, Uribe S (2014) DNA barcoding for the identification of sand fly species (Diptera, Psychodidae, Phlebotominae) in Colombia. PLoS One 9(1):e85496

Day JC, Ready PD (1999) Relative abundance, isolation and structure of phlebotomine microsatellites. Insect Mol Biol 8(4):575–580

de Moura TR, Oliveira F, Carneiro MW, Miranda JC, Clarêncio J, Barral-Netto M, Brodskyn C, Barral A, Ribeiro JM, Valenzuela JG, de Oliveira CI (2013) Functional transcriptomics of wild-caught *Lutzomyia intermedia* salivary glands: identification of a protective salivary protein against *Leishmania braziliensis* infection. PLoS Negl Trop Dis 7(5):e2242 de Souza Freitas MT, Ríos-Velasquez CM, da Silva LG, Costa CR Jr, Marcelino A, Leal-Balbino TC, Balbino Vde Q, Pessoa FA (2016) Analysis of the genetic structure of allopatric populations of *Lutzomyia umbratilis* using the period clock gene. Acta Trop 154:149–154

de Souza A, Ishikawa E, Braga R, Silveira F, Lainson R, Shaw J (1996) *Psychodopygus complexus*, a new vector of *Leishmania braziliensis* to humans in Pará State, Brazil. Trans R Soc Trop Med Hyg 90(2):112–113

Depaquit J (2014) Molecular systematics applied to Phlebotomine sandflies: review and perspectives. Infect Genet Evol 28:744–756

Dias ES, Fortes-Dias CL, Stiteler JM, Perkins PV, Lawyer PG (1998) Random amplified polymorphic DNA (RAPD) analysis of *Lutzomyia longipalpis* laboratory populations. Rev Inst Med Trop Sao Paulo 40:49–53

Dillon RJ, Ivens AC, Churcher C, Holroyd N, Quail MA et al (2006) Analysis of ESTs from *Lutzomyia longipalpis* sand flies and their contribution toward understanding the insect-parasite relationship. Genomics 88:831–840

Doehl JS, Sádlová J, Aslan H, Pružinová K, Metangmo S, Votýpka J, Kamhawi S, Volf P, Smith DF (2017) *Leishmania* HASP and SHERP genes are required for *in vivo* differentiation, parasite transmission and virulence attenuation in the host. PLoS Pathog 13(1):e1006130

Dujardin JP, Le Pont F, Cruz M, Leon R, Tarrieu LF, Guderian R, Echeverria R, Tibayrenc M (1996) Cryptic speciation in *Lutzomyia* (*Nyssomyia*) *trapidoi* (Fairchild & Hertig) (Diptera: Psychodidae) detected by multilocus enzyme electrophoresis. Am J Trop Med Hyg 54(1):42–45

Dvorak V, Halada P, Hlavackova K, Dokianakis E, Antoniou M, Volf P (2014) Identification of phlebotomine sand flies (Diptera: Psychodidae) by matrix-assisted laser desorption/ionization time of flight mass spectrometry. Parasit Vectors 7:21

Esseghir S, Ready PD, Killick-Kendrick R, Ben-Ismail R (1997) Mitochondrial haplotypes and phylogeography of *Phlebotomus* vectors of *Leishmania major*. Insect Mol Biol 6:211–225

Esseghir S, Ready PD, Ben-Ismail R (2000) Speciation of *Phlebotomus* sandflies of the subgenus *Larroussius* coincided with the late Miocene-Pliocene aridification of the Mediterranean subregion. Biol J Linn Soc 70:189–219

Ferreira TS, Minuzzi-Souza TT, Andrade AJ, Coelho TO, Rocha Dde A, Obara MT, Hecht M, Nitz N, Gurgel-Gonçalves R (2015) Molecular detection of *Trypanosoma* sp. and *Blastocrithidia* sp. (Trypanosomatidae) in phlebotomine sand flies (Psychodidae) in the Federal District of Brazil. Rev Soc Bras Med Trop 48(6):776–779

Gaio Ade O, Rodrigues RC, do Nascimento C, Secundino NF, Lemos FJ, Pimenta PF, Monesi N (2011) Use of the checkerboard DNA-DNA hybridization technique for bacteria detection in *Aedes aegypti* (Diptera:Culicidae) (L.) Parasit Vectors 4:237

Galati EAB (1995) Philogenetic systematics of the Phlebotominae (Diptera, Psychodidae) with emphasis on American groups. Bol Mal Salud Amb 35(Suppl. 1):133–142

Galati EAB (2003) Morfologia e taxonomia: classificação de Phlebotominae. In: Rangel EF, Lainson R (eds) Flebotomíneos do Brasil. Fiocruz, Rio de Janeiro, pp 23–51

Galati EA, Fonseca MB, Marassá AM (2007) The subgenus *Migonemyia* Galati 1995 (Diptera, Psychodidae, Phlebotominae), with description of a new species *Migonemyia vaniae*: a review. Mem Inst Oswaldo Cruz 102(5):605–615

Gesto JSM, Rivas GB, Pavan MG, Meireles-Filho ACA, Amoretty PR, Souza NA, Bruno RV, Peixoto AA (2015) Clocks do not tick in unison: isolation of Clock and vrille shed new light on the clockwork model of the sand fly *Lutzomyia longipalpis*. Parasit Vectors 8:505

Giantsis IA, Chaskopoulou A, Bon MC (2016) Mild-vectolysis: a nondestructive DNA extraction method for vouchering sand flies and mosquitoes. J Med Entomol 53(3):692–695

Gil LH, Basano SA, Souza AA, Silva MG, Barata I, Ishikawa EA, Camargo LM, Shaw JJ (2003) Recent observations on the sand fly (Diptera: Psychodidae) fauna of the State of Rondônia, Western Amazônia, Brazil: the importance of *Psychdopygus davisi* as a vector of zoonotic cutaneous leishmaniasis. Mem Inst Oswaldo Cruz 98(6):751–5

Grace-Lema D, Yared S, Quitadamo A, Janies D, Wheeler WC, Balkew M, Hailu A, Warburg A, Clouse RM (2015) A phylogeny of sand flies (Diptera: Psychodidae: Phlebotominae), using recent Ethiopian collections and a broad selection of publicly available DNA sequence data. Syst Entomol 40(4):733–744

Hamilton JG, Ward RD (1991) Gas-chromatographic analysis of *Lutzomyia longipalpis* tergal pheromone gland extract. Parassitologia 33(Suppl):283–289

Hamilton JG, Ward RD (1994) Chemical analysis of a putative sex pheromone from *Lutzomyia pessoai* (Diptera: Psychodidae). Ann Trop Med Parasitol 88(4):405–412

Hamilton JG, Dougherty MJ, Ward RD (1994) Sex pheromone activity in a single component of tergal gland extract of *Lutzomyia longipalpis* (Diptera: Psychodidae) from Jacobina. J Chem Ecol 20:141–151

Hamilton JGC, Dawson GW, Pickett JA (1996a) 9-Methyl-germacrene B: proposed structure for novel homosesquiterpene from the sex pheromone glands of *Lutzomyia longipalpis* (Diptera: Psychodidae) from Lapinha, Brazil. J Chem Ecol 22:1477–1491

Hamilton JGC, Dawson GW, Pickett JA (1996b) 3-Methyl-a-himachalene: Proposed structure for novel homosesquiterpene sex pheromone of *Lutzomyia longipalpis* (Diptera: Psychodidae) from Jacobina, Brazil. J Chem Ecol 22:2331–2340

Hamilton JG, Brazil RP, Campbell-Lendrum D, Davies CR, Kelly DW, Pessoa FA, de Queiroz RG (2002) Distribution of putative male sex pheromones among *Lutzomyia* sandflies (Diptera: Psychodidae). Ann Trop Med Parasitol 96(1):83–92

Hamilton JG, Brazil RP, Maingon R (2004) A fourth chemotype of *Lutzomyia longipalpis* (Diptera: Psychodidae) from Jaiba, Minas Gerais state, Brazil. J Med Entomol 41(6):1021–1026

Hamilton JG, Maingon RD, Alexander B, Ward RD, Brazil RP (2005) Analysis of the sex pheromone extracts of individual male *Lutzomyia longipalpis* sandflies from six regions in Brazil. Med Vet Entomol 19(4):480–488

Hartl DL, Clark AG (2007) Principles of population genetics, 4th edn. Sunderland, Sinauer Associates

Hebert PD, Ratnasingham S, de Waard JR (2003) Barcoding animal life: cytochrome c oxidase subunit 1 divergences among closely related species. Proc Biol Sci 270(Suppl 1):S96–S99

Hodgkinson VH, Birungi J, Haghpanah M, Joshi S, Munstermann LE (2002) Rapid identification of mitochondrial cytochrome B haplotypes by single strand conformation polymorphism in *Lutzomyia longipalpis* (Diptera: Psychodidae) populations. J Med Entomol 39:689–694

Hodgkinson VH, Birungi J, Quintana M, Dietze R, Munstermann LE (2003) Mitochondrial cytochrome b variation in populations of the visceral leishmaniasis vector *Lutzomyia longipalpis* across eastern Brazil. Am J Trop Med Hyg 69(4):386–392

Ishikawa EA, Ready PD, de Souza AA, Day JC, Rangel EF, Davies CR, Shaw JJ (1999) A mitochondrial DNA phylogeny indicates close relationships between populations of *Lutzomyia whitmani* (Diptera: Psychodidae, Phlebotominae) from the rain-forest regions of Amazônia and northeast Brazil. Mem Inst Oswaldo Cruz 94(3):339–345

Jochim RC, Teixeira CR, Laughinghouse A, Mu J, Oliveira F et al (2008) The midgut transcriptome of *Lutzomyia longipalpis*: comparative analysis of cDNA libraries from sugar-fed, blood-fed, post-digested and *Leishmania infantum chagasi*-infected sand flies. BMC Genomics 9:15

Johansson BG, Jones TM (2007) The role of chemical communication in mate choice. Biol Rev Camb Philos Soc 82(2):265–289

Kocher A, Gantier JC, Holota H, Jeziorski C, Coissac E, Bañuls AL, Girod R, Gaborit P, Murienne J (2016) Complete mitochondrial genome of *Lutzomyia* (*Nyssomyia*) *umbratilis* (Diptera: Psychodidae), the main vector of *Leishmania guyanensis*. Mitochondrial DNA A DNA Mapp Seq Anal 27(6):4219–4221

Kocher A, Gantier JC, Gaborit P, Zinger L, Holota H, Valiere S, Dusfour I, Girod R, Bañuls AL, Murienne J (2017) Vector soup: high-throughput identification of Neotropical phlebotomine sand flies using metabarcoding. Mol Ecol Resour 17(2):172–182

Kreutzer RD, Palau MT, Morales A, Ferro C, Feliciangeli D, Young DG (1990) Genetic relationships among phlebotomine sand flies (Diptera: Psychodidae) in the verrucarum species group. J Med Entomol 27(1):1–8

Kuwahara K, Kato H, Gomez EA, Uezato H, Mimori T, Yamamoto Y, Calvopina M, Caceres AG, Iwata H, Hashiguchi Y (2009) Genetic diversity of ribosomal RNA internal transcribed spacer sequences in *Lutzomyia* species from areas endemic for New World cutaneous leishmaniasis. Acta Trop 112:131–136

Kyriacou CP, Hall JC (1980) Circadian rhythm mutations in *Drosophila* affect short-term fluctuations in the male's court- ship song. Proc Natl Acad Sci U S A 77:6729–6733

Lampo M, Torgerson D, Márquez LM, Rinaldi M, García CZ, Arab A (1999) Occurrence of sibling species of *Lutzomyia longipalpis* (Diptera: Psychodidae) in Venezuela: first evidence from reproductively isolated sympatric populations. Am J Trop Med Hyg 61(6):1004–1009

Lane RP, Phillips A, Molyneux DH, Procter C, Ward RD (1985) Chemical analysis of the abdominal glands of two forms of the *Lutzomyia longipalpis*: site of a possible sex pheromone? Ann Trop Med Parasit 79:225–229

Lanzaro GC, Warburg A (1995) Genetic variability in phlebotomine sandf lies: possible implications for leishmaniasis epidemiology. Parasitol Today 11:151–154

Lanzaro GC, Ostrovska K, Herrero MV, Lawyer PG, Warburg A (1993) *Lutzomyia longipalpis* is a species complex: genetic divergence and interspecific hybrid sterility among three populations. Am J Trop Med Hyg 48(6):839–847

Lanzaro GC, Lopes AH, Ribeiro JM, Shoemaker CB, Warburg A, Soares M et al (1999) Variation in the salivary peptide, maxadilan, from species in the *Lutzomyia longipalpis* complex. Insect Mol Biol 8(2):267–275

Lerner EA, Ribeiro JM, Nelson RJ, Lerner MR (1991) Isolation of maxadilan, a potent vasodilatory peptide from the salivary glands of the sand fly *Lutzomyia longipalpis*. J Biol Chem 266(17):11234–11236

Lewis DJ, Young DG, Fairchild GB, Minter DM (1977) Proposals for a stable classification of the phlebotomine sandflies (Diptera: Psychodidae). Syst Entomol 2:319–332

Lima-Costa CR Jr, Freitas MT, Figueredo CAS Jr, Aragão NC, da Silva LG, Marcondes CB et al (2015) Genetic structuring and fixed polymorphisms in the *period* among natural populations of *Lutzomyia longipalpis* in Brazil. Parasit Vectors 8:193

Lins RM, Oliveira SG, Souza NA, de Queiroz RG, Justiniano SC, Ward RD, Kyriacou CP, Peixoto AA (2002) Molecular evolution of the cacophony IVS6 region in sandflies. Insect Mol Biol 11:117–122

Lins RMMA, Souza NA, Peixoto AA (2008) Genetic divergence between two sympatric species of the *Lutzomyia longipalpis* complex in the *paralytic* gene, a locus associated with insecticide resistance and lovesong production. Mem Inst Oswaldo Cruz 103(7):736–740

Lins RM, Souza NA, Brazil RP, Maingon RD, Peixoto AA (2012) Fixed differences in the *paralytic* gene define two lineages within the *Lutzomyia longipalpis* complex producing different types of courtship songs. PLoS One 7(9):e44323

Mahamdallie SS, Pesson B, Ready PD (2011) Multiple genetic divergences and population expansions of a Mediterranean sandfly, *Phlebotomus ariasi*, in Europe during the Pleistocene glacial cycles. Heredity 106:714–726

Maingon RD, Ward RD, Hamilton JG, Noyes HA, Souza N, Kemp SJ et al (2003) Genetic identification of two sibling species of *Lutzomyia longipalpis* (Diptera: Psychodidae) that produce distinct male sex pheromones in Sobral, Ceará state, Brazil. Mol Ecol 12(7):1879–1894

Maingon RD, Ward RD, Hamilton JG, Bauzer LG, Peixoto AA (2008) The *Lutzomyia longipalpis* species complex: does population sub-structure matter to Leishmania transmission? Trends Parasitol 24(1):12–17

Marassá AM, Galati EA, Bergamaschi DP, Consales CA (2013) Blood feeding patterns of *Nyssomyia intermedia* and *Nyssomyia neivai* (Diptera, Psychodidae) in a cutaneous leishmaniasis endemic area of the Ribeira Valley, State of São Paulo, Brazil. Rev Soc Bras Med Trop 46(5):547–554

Marcondes CB (1996) A redescription of Lutzomyia (Nyssomyia) intermedia (Lutz & Neiva, 1912), and resurrection of L. neivai (Pinto, 1926) (Diptera, Psychodidae, Phlebotominae). Mem Inst Oswaldo Cruz 91(4):457–462

Marcondes C, Day J, Ready PD (1997) Introgression between *Lutzomyia intermedia* and both *Lu. Neivai* and *Lu. whitmani*, and their roles as vectors of *Leishmania braziliensis*. Trans R Soc Trop Med Hyg 91:725–726

Margonari CS, Fortes-Dias CL, Dias ES (2004) Genetic variability in geographical populations of *Lutzomyia whitmani* elucidated by RAPD-PCR. J Med Entomol 41:187–192

Mathis A, Depaquit J, Dvořák V, Tuten H, Bañuls AL, Halada P, Zapata S, Lehrter V, Hlavačková K, Prudhomme J, Volf P, Sereno D, Kaufmann C, Pflüger V, Schaffner F (2015) Identification of phlebotomine sand flies using one MALDI-TOF MS reference database and two mass spectrometer systems. Parasit Vectors 8:266

Mazzoni CJ, Gomes CA, Souza NA, de Queiroz RG, Justiniano SC, Ward RD, Kyriacou CP, Peixoto AA (2002) Molecular evolution of the period gene in sandflies. J Mol Evol 55:553–562

Mazzoni CJ, Souza NA, Andrade-Coelho C, Kyriacou CP, Peixoto AA (2006) Molecular polymorphism, differentiation and introgression in the period gene between *Lutzomyia intermedia* and *Lutzomyia whitmani*. BMC Evol Biol 6:85

Mazzoni CJ, Araki AS, Ferreira GE, Azevedo RV, Barbujani G, Peixoto AA (2008) Multilocus analysis of introgression between two sand fly vectors of leishmaniasis. BMC Evol Biol 8:141

McAvin JC, Swanson KI, Chan AS, Quintana M, Coleman RE (2012) *Leishmania* detection in sand flies using a field-deployable real-time analytic system. Mil Med 177(4):460–466

McIntyre S, Rangel EF, Ready PD, Carvalho BM (2017) Species-specific ecological niche modelling predicts different range contractions for *Lutzomyia intermedia* and a related vector of *Leishmania braziliensis* following climate change in South America. Parasit Vectors 10(1):157

Meireles-Filho ACA, Kyriacou CP (2013) Circadian rhythms in insect disease vectors. Mem Inst Oswaldo Cruz 108(Suppl. 1):48–58

Meireles-Filho ACA, Amoretty PR, Souza NA, Kyriacou CP, Peixoto AA (2006) Rhythmic expression of the cycle gene in a hematophagous insect vector. BMC Mol Biol 7:38

Mendoza-Viveros L, Bouchard-Cannon P, Hegazi S, Cheng AH, Pastore S, Cheng H-YM (2017) Molecular modulators of the circadian clock: lessons from flies and mice. Cell Mol Life Sci 74:1035–1059

Meneses CR, Cupolillo E, Monteiro F, Rangel EF (2005) Micro-geographical variation among male populations of the sandfly, *Lutzomyia* (*Nyssomyia*) *intermedia*, from an endemic area of American cutaneous leishmaniasis in the state of Rio de Janeiro, Brazil. Med Vet Entomol 19:38–47

Miles M, Ward RD (1978) Preliminary isoenzyme studies on phlebotomine sandflies (Diptera: Psychodidae). Ann Trop Med Parasitol 72:398–400

Milleron RS, Ribeiro JM, Elnaime D, Soong L, Lanzaro GC (2004) Negative effect of antibodies against maxadilan on the fitness of the sand fly vector of American visceral leishmaniasis. Am J Trop Med Hyg 70:278–285

Monteiro CC, Villegas LE, Campolina TB, Pires AC, Miranda JC, Pimenta PF, Secundino NF (2016) Bacterial diversity of the American sand fly *Lutzomyia intermedia* using high-throughput metagenomic sequencing. Parasit Vectors 9:480

Morton IE, Ward RD (1989) Laboratory response of female *Lutzomyia longipalpis* sandflies to a host and male pheromone source over distance. Med Vet Entomol 3:219–223

Mukhopadhyay J, Ghosh K, Azevedo AC, Rangel EF, Munstermann LE (1998a) Genetic polymorphism of morphological and biochemical characters in a Natal, Brazil, population of *Lutzomyia longipalpis* (Diptera: Psychodidae). J Am Mosq Control Assoc 14(3):277–282

Mukhopadhyay J, Ghosh K, Rangel EF, Munstermann LE (1998b) Genetic variability in biochemical characters of Brazilian field populations of the *Leishmania* vector, *Lutzomyia longipalpis* (Diptera: Psychodidae). Am J Trop Med Hyg 59:893–901

Munstermann LE, Morrison AC, Ferro C, Pardo R, Torres M (1998) Genetic structure of local populations of *Lutzomyia longipalpis* (Diptera: Psychodidae) in central Colombia. J Med Entomol 35:82–89

Mutebi JP, Rowton E, Herrero MV, Ponce C, Belli A, Valle S et al (1998) Genetic variability among populations of the sand fly *Lutzomyia* (*Lutzomyia*) *longipalpis* (Diptera: Psychodidae) from Central America. J Med Entomol 35(2):169–174

Mutebi JP, Alexander B, Sherlock I, Wellington J, Souza AA, Shaw J et al (1999) Breeding structure of the sandfly *Lutzomyia longipalpis* (Lutz & Neiva) in Brazil. Am J Trop Med Hyg 61:149–157

Neitzke-Abreu HC, Reinhold-Castro KR, Venazzi MS, Scodro RB, Dias Ade C, Silveira TG, Teodoro U, Lonardoni MV (2014) Detection of *Leishmania* (*Viannia*) in *Nyssomyia neivai* and *Nyssomyia whitmani* by multiplex polymerase chain reaction, in Southern Brazil. Rev Inst Med Trop Sao Paulo 56(5):391–395

Oliveira SG, Bottecchia M, Bauzer LGSR, Souza NA, Ward RD, Kyriacou CP et al (2001) Courtship song genes and speciation in sand flies. Mem Inst Oswaldo Cruz 96(3):403–405

Oliveira F, Kamhawi S, Seitz AE, Pham VM, Guigal PM, Fischer L, Ward J, Valenzuela JG (2006) From transcriptome to immunome: identification of DTH inducing proteins from a *Phlebotomus ariasi* salivary gland cDNA library. Vaccine 24(3):374–390

Oliveira F, Lawyer PG, Kamhawi S, Valenzuela JG (2008) Immunity to distinct sand fly salivary proteins primes the anti-*Leishmania* immune response towards protection or exacerbation of disease. PLoS Negl Trop Dis 2(4):e226

Peixoto AA, Hall JC (1998) Analysis of temperature- sensitive mutants reveals new genes involved in the courtship song of *Drosophila*. Genetics 148:827–838

Pesson B, Ready JS, Benabdennbi I, Martin-Sanchez J, Esseghir S et al (2004) Sandflies of the *Phlebotomus perniciosus* complex: mitochondrial introgression and a new sibling species of *P. longicuspis* in the Moroccan Rif. Med Vet Entomol 18:25–37

Phillips A, Ward R, Ryan L (1986) Chemical analysis of compounds extracted from the tergal "spots" of "*Lutzomyia longipalpis*" from Brazil. Acta Trop 43:271–276

Pinto MC, Campbell-Lendrum DH, Lozovei AL, Teodoro U, Davies CR (2001) Phlebotomine sandfly responses to carbon dioxide and human odour in the field. Med Vet Entomol 15(2):132–139

Pinto IS, das Chagas BD, Rodrigues AAF, Ferreira AL, Rezende HR, Bruno RV et al (2015) DNA barcoding of neotropical sand flies (Diptera, Psychodidae, Phlebotominae): species identification and discovery within Brazil. PLoS One 10(10):e0140636

Pita-Pereira D, Souza GD, Zwetsch A, Alves CR, Britto C, Rangel EF (2009) First report of *Lutzomyia* (*Nyssomyia*) *neivai* (Diptera: Psychodidae: Phlebotominae) naturally infected by *Leishmania* (*Viannia*) *braziliensis* in a periurban area of south Brazil using a multiplex polymerase chain reaction assay. Am J Trop Med Hyg 80(4):593–595

Pittendrigh B, Reenan R, ffrench-Constant RH, Ganetzky B (1997) Point mutations in the *Drosophila* sodium channel gene para associated with resistance to DDT and pyrethroid insecticides. Mol Gen Genet 256:602–610

Puillandre N, Lambert A, Brouillet S, Achaz G (2012) ABGD, Automatic Barcode Gap Discovery for primary species delimitation. Mol Ecol 21(8):1864–1877

Ranasinghe S, Rogers ME, Hamilton JG, Bates PA, Maingon RD (2008) A real-time PCR assay to estimate *Leishmania chagasi* load in its natural sand fly vector *Lutzomyia longipalpis*. Trans R Soc Trop Med Hyg 102(9):875–882

Rangel EF, Lainson R (2009) Proven and putative vectors of American cutaneous leishmaniasis in Brazil: aspects of their biology and vectorial competence. Mem Inst Oswaldo Cruz 104:937–954

Rangel EF, Lainson R, Souza AA, Ready P, Azevedo AC (1996) Variation between geographical populations of *Lutzomyia* (*Nyssomyia*) *whitmani* (Antunes & Coutinho, 1939) *sensu lato* (Diptera: Psychodidae: Phlebotominae) in Brazil. Mem Inst Oswaldo Cruz 91(1):43–50

Ready PD (2011) Should sandfly taxonomy predict vectorial and ecological traits? J Vector Ecol 36:S17–S32

Ready PD (2013) Biology of phlebotomine sand flies as vectors of disease agents. Annu Rev Entomol 58:227–250

Ready PD, Rangel R (2003) Caracteres Isoenzimáticos e Moleculares: espécies crípticas. In: Rangel EF, Lainson R (eds) Flebotomíneos do Brasil. Editora Fiocruz, Rio de Janeiro

Ready PD, Silva RMR (1984) An alloenzymic comparison of *Psychodopygus wellcomei* (Diptera: Psychodidae) – an incriminated vector of *Leishmania braziliensis* in Pará State, Brazil – and the sympatric morphospecies *Ps. complexus* (Diptera: Psychodidae). Cahiers Orstom ser Entomol Med Parasitol 22:3–8

Ready PD, Lainson R, Shaw JJ (1984) Habitat and seasonality of *Psychodopygus wellcomei* help incriminate it as a vector of *Leishmania braziliensis* in Amazonia and northeast Brazil. Trans R Soc Trop Med Hyg 78:543–544

Ready PD, Lainson R, Shaw JJ, Ward RD (1986) The ecology of *Lutzomyia umbratilis* Ward & Fraiha (Diptera: Psychodidae), the major vector to man of *Leishmania braziliensis guyanensis* in north-eastern Amazonian Brazil. Bull Entom Res 76:21–40

Ready PD, Smith DF, Killick-Kendrick R (1988) DNA hybridizations on squash-blotted sandflies to identify both *Phlebotomus papatasi* and infecting *Leishmania major*. Med Vet Entomol 2(2):109–116

Ready PD, Lainson R, Shaw JJ, Souza AA (1991) DNA probes for distinguishing *Psychodopygus wellcomei* from *Psychodopygus complexus* (Diptera: Psychodidae). Mem Inst Oswaldo Cruz 86:41–49

Ready PD, Day JC, de Souza AA, Rangel EF, Davies CR (1997) Mitochondrial DNA characterization of populations of *Lutzomyia whitmani* (Diptera: Psychodidae) incriminated in the peri-domestic and silvatic transmission of *Leishmania* species in Brazil. Bull Entomol Res 87:187–195

Ready PD, de Souza AA, Rebelo JM, Day JC, Silveira FT, Campbell-Lendrum D, Davies CR, Costa JM (1998) Phylogenetic species and domesticity of *Lutzomyia whitmani* at the southeast boundary of Amazonian Brazil. Trans R Soc Trop Med Hyg 92(2):159–160

Ribolla PE, Gushi LT, Pires E Cruz Mdo S, Costa CH, Costa DL, Lima Júnior MS, Dorval ME, Gutierrez de Oliveira A, da Cunha Santos MF, Fonseca Camargo-Neves VL, Fortaleza CM, Alonso DP (2016) *Leishmania infantum* genetic diversity and *Lutzomyia longipalpis* Mitochondrial Haplotypes in Brazil. Biomed Res Int 16:9249217

Richardson BJ, Baverstock PR, Adams M (1990) Allozyme electrophoresis. A handbook for animal systematics and population studies. Academic Press, New York

Ritchie MG, Halsey EJ, Gleason JM (1999) *Drosophila* song as a species-specific mating signal and the behavioural importance of Kyriacou & Hall cycles in *D. melanogaster* song. Anim Behav 58:649–657

Romero-Ricardo L, Lastre-Meza N, Pérez-Doria A, Bejarano EE (2016) DNA barcoding to identify species of phlebotomine sand fly (Diptera: Psychodidae) in the mixed leishmaniasis focus of the Colombian Caribbean. Acta Trop 159:125–131

Ryan L, Phillips A, Milligan P, Lainson R, Molyneux DH, Shaw JJ (1986) Separation of female *Psychodopygus wellcomei* and *P. complexus* (Diptera: Psychodidae) by cuticular hydrocarbon analysis. Acta Trop 43:85–89

Sakai T, Ishida N (2001) Circadian rhythms of female mating activity governed by clock genes in *Drosophila*. Proc Natl Acad Sci USA 98:9221–9225

Sant'Anna MR, Jones NG, Hindley JA, Mendes-Sousa AF, Dillon RJ, Cavalcante RR, Alexander B, Bates PA (2008) Blood meal identification and parasite detection in laboratory-fed and field-captured *Lutzomyia longipalpis* by PCR using FTA databasing paper. Acta Trop 107(3):230–237

Scarpassa VM, Alencar RB (2012) *Lutzomyia umbratilis*, the main vector of *Leishmania guyanensis*, represents a novel species complex? PLoS One 7:e37341

Scarpassa VM, Alencar RB (2013) Molecular taxonomy of the two *Leishmania* vectors *Lutzomyia umbratilis* and *Lutzomyia anduzei* (Diptera: Psychodidae) from the Brazilian Amazon. Parasit Vectors 6:258

Scarpassa VM, Figueiredo Ada S, Alencar RB (2015) Genetic diversity and population structure in the *Leishmania guyanensis* vector *Lutzomyia anduzei* (Diptera, Psychodidae) from the Brazilian Amazon. Infect Genet Evol 31:312–320

Simon C, Frati F, Beckenbach A, Crespi B, Liu H, Flook P (1994) Evolution, Weighting, and Phylogenetic Utility of Mitochondrial Gene Sequences and a Compilation of Conserved Polymerase Chain Reaction Primers. Ann Entomol Soc Am 87(6):651–701

Smith LA, Peixoto AA, Kramer EM, Villella A, Hall JC (1998) Courtship and visual defects of *cacophony* mutants reveal functional complexity of a calcium-channel α-1 subunit in *Drosophila*. Genetics 149:1407–1426

Soares VY, Silva JC, Silva KR, Pires e Cruz Mdo S, Santos MP, Ribolla PE, Alonso DP, Coelho LF, Costa DL, Costa CH (2014) Identification of blood meal sources of *Lutzomyia longipalpis* using polymerase chain reaction-restriction fragment length polymorphism analysis of the cytochrome B gene. Mem Inst Oswaldo Cruz 109(3):379–383

Soto SI, Lehmann T, Rowton ED, Vélez BID, Porter CH (2001) Speciation and population structure in the morphospecies *Lutzomyia longipalpis* (Lutz & Neiva) as derived from the mitochondrial ND4 gene. Mol Phylogenet Evol 18(1):84–93

Souza NA, Vigoder FM, Araki AS, Ward RD, Kyriacou CP, Peixoto AA (2004) Analysis of the copulatory courtship songs of *Lutzomyia longipalpis* in six populations from Brazil. J Med Entomol 41:906–913

Souza NA, Andrade-Coelho CA, Vigoder FM, Ward RD, Peixoto AA (2008) Reproductive isolation between sympatric and allopatric Brazilian populations of *Lutzomyia longipalpis s.l.* (Diptera: Psychodidae). Mem Inst Oswaldo Cruz 103(2):216–219

Souza NA, Andrade-Coelho CA, Silva VC, Ward RD, Peixoto AA (2009) Life cycle differences among Brazilian sandflies of the *Lutzomyia longipalpis* sibling species complex. Med Vet Entomol 23(3):287–292

Souza NA, Brazil RP, Araki AS (2017) The current status of the *Lutzomyia longipalpis* (Diptera: Psychodidae: Phlebotominae) species complex. Mem Inst Oswaldo Cruz 112(3):161–174

Spiegel CN, Dias DBS, Araki AS, Hamilton JGC, Brazil RP, Jones TM (2016) The *Lutzomyia longipalpis* complex: a brief natural history of aggregation-sex pheromone communication. Parasit Vectors 9:580

Tabbabi A, Rhim A, Ghrab J, Martin O, Aoun K, Bouratbine A, Ready PD (2014) *Phlebotomus* (*Paraphlebotomus*) *riouxi*: a synonym of *Phlebotomus chabaudi* without any proven vectorial role in Tunisia and Algeria. Med Vet Entomol 28S(1):51–59

Telleria EL, Sant'anna MR, Ortigão-Farias JR, Pitaluga AN, Dillon VM et al (2012) Caspar-like gene depletion reduces *Leishmania* infection in the sand fly host *Lutzomyia longipalpis*. J Biol Chem e:28

Terayama Y, Kato H, Gomez EA, Uezato H, Calvopina M, Iwata H, Hashiguchi Y (2008) Molecular typing of sand fly species (Diptera, Psychodidae, Phlebotominae) from areas endemic for Leishmaniasis in Ecuador by PCR-RFLP of 18S ribosomal RNA gene. J Vet Med Sci 70:907–913

Testa JM, Montoya-Lerma J, Cadena H, Oviedo M, Ready PD (2002) Molecular identification of vectors of *Leishmania* in Colombia: mitochondrial introgression in the *Lutzomyia townsendi* series. Acta Trop 84:205–218

Tibayrenc M, Carriou ML, Corneau B, Pajot FX (1980) Etude allozymique chez *Lutzomyia umbratilis* (Diptera, Psychodidae), vecteur de la leishmaniose en Guyane française. Cahiers Orstom ser Entomol Med Parasitol 18:67–70

Ting CT, Tsaur SC, Wu CI (2000) The phylogeny of closely related species as revealed by the genealogy of a speciation gene, Odysseus. Proc Natl Acad Sci USA 97:5313–5316

Titus RG, Ribeiro JM (1988) Salivary gland lysates from the sand fly *Lutzomyia longipalpis* enhance *Leishmania* infectivity. Science 239(4845):1306–1308

Torgerson DG, Lampo M, Velazquez Y, Woo PT (2003) Genetic relationships among some species groups within the genus *Lutzomyia* (Diptera: Psychodidae). Am J Trop Med Hyg 69:484–493

Valenzuela JG, Garfield M, Rowton ED, Pham VM (2004) Identification of the most abundant secreted proteins from the salivary glands of the sand fly *Lutzomyia longipalpis*, vector of *Leishmania chagasi*. J Exp Biol 207(21):3717–3729

Vigoder FM, Araki AS, Bauzer LG, Souza NA, Brazil RP, Peixoto AA (2010a) Lovesongs and period gene polymorphisms indicate *Lutzomyia cruzi* (Mangabeira, 1938) as a sibling species of the *Lutzomyia longipalpis* (Lutz and Neiva, 1912) complex. Infect Genet Evol 10(6):734–739

Vigoder FM, Souza NA, Peixoto AA (2010b) Copulatory courtship song in *Lutzomyia migonei* (Diptera: psychodidae). Mem Inst Oswaldo Cruz 105(8):1065–1067

Vigoder FM, Souza NA, Peixoto AA (2011) Acoustic signals in the sand fly *Lutzomyia* (*Nyssomyia*) *intermedia* (Diptera: Psychodidae). Parasit Vectors 4:76

Vigoder FM, Souza NA, Brazil RP, Bruno RV, Costa LP, Ritchie MG et al (2015) Phenotypic differentiation in love song traits among sibling species of the *Lutzomyia longipalpis* complex in Brazil. Parasit Vectors 8:290

Vivero RJ, Contreras-Gutierrez MA, Bejarano EE (2007) *Analisis de la estructura primaria y secundaria del ARN de transferencia mitocondrial para serina en siete especies de* Lutzomyia. Biomedica 27:429–438

Warburg A, Saraiva E, Lanzaro GC, Titus RG, Neva F (1994) Saliva of *Lutzomyia longipalpis* sibling species differs in its composition and capacity to enhance leishmaniasis. Philos Trans R Soc Lond Ser B Biol Sci 345(1312):223–230

Ward RD, Morton IE (1991) Pheromones in mate choice and sexual isolation between siblings of *Lutzomyia longipalpis* (Diptera:Psychodidae). Parassitologia 33(Suppl):527–533

Ward RD, Ribeiro AL, Ready PD, Murtagh A (1983) Reproductive isolation between different forms of *Lutzomyia longipalpis* (Lutz & Neiva) (Diptera: Psychodidae), the vector of *Leishmania donovani chagasi* Cunha & Chagas, and its significance to kala-azar distribution in South America. Mem Inst Oswaldo Cruz 78(3):269–280

Ward RD, Phillips A, Burnet B, Marcondes CB (1988) The *Lutzomyia longipalpis* complex: reproduction and distribution. In: Service MW (ed) Biosystematics of haematophagous insects. Systematics Association Special, Clarendon Press, Oxford, UK, pp 257–269

Watts PC, Boyland E, Noyes HA, Maingon RDC, Kemp SJ (2001) Polymorphic dinucleotide microsatellite loci in the sandfly *Lutzomyia longipalpis* (Diptera: Phlebotominae). Mol Ecol Notes 2:60–61

Watts PC, Hamilton JG, Ward RD, Noyes HA, Souza NA, Kemp SJ et al (2005) Male sex pheromones and the phylogeographic structure of the *Lutzomyia longipalpis* species complex (Diptera: Psychodidae) from Brazil and Venezuela. Am J Trop Med Hyg 73(4):734–743

Wiebe A, Longbottom J, Gleave K, Shearer FM, Sinka ME, Massey NC, Cameron E, Bhatt S, Gething PW, Hemingway J, Smith DL, Coleman M, Moyes CL (2017) Geographical distributions of African malaria vector sibling species and evidence for insecticide resistance. Malar J 16(1):85

Ye F, Liu T, King SD, You P (2015) Mitochondrial genomes of two phlebotomine sand flies, *Phlebotomus chinensis* and *Phlebotomus papatasi* (Diptera: Nematocera), the first representatives from the family Psychodidae. Parasit Vectors 8:472

Yin H, Mutebi JP, Marriott S, Lanzaro GC (1999) Metaphase karyotypes and G-banding in sandflies of the *Lutzomyia longipalpis* complex. Med Vet Entomol 13(1):72–77

Yin H, Norris DE, Lanzaro GC (2000) Sibling species in the *Lutzomyia longipalpis* complex differ in levels of mRNA expression for the salivary peptide, maxadilan. Insect Mol Biol 9(3):309–314

Young DG, Duncan MA (1994) Guide to the identification and geographic distribution of *Lutzomyia* sand flies in Mexico, the West Indies, Central and South America (Diptera: Psychodidae). Mem Amer Ent Inst 54:1–881

Zapata S, Leon R, Sauvage F, Augot D, Trueba G, Cruaud C, Couloux A, Teran R, Depaquit J (2012) Morphometric and molecular characterization of the series Guyanensis (Diptera, Psychodidae, *Psychodopygus*) from the Ecuadorian Amazon Basin with description of a new species. Infect Genet Evol 12:966–977

Regional Distribution and Habitats of Brazilian Phlebotomine Species

Gustavo M. de Aguiar and Vanessa R. Vieira

Research on phlebotomine sand-fly fauna (Diptera, Phlebotominae) has been related to the diagnosis of diseases or the identification of their vectors involved in the transmission of pathogens to humans and other animals in certain regions. It is considered that, under natural conditions, these insects are distributed in stable and balanced communities with the variables of the ecosystem as a whole. Thus, it is assumed that the emergence of many epidemic species is directly linked to problems of human ecology caused by the accidental or planned inhabitation of humans into regions where the transmitters are still unknown.

Many studies on the geographic distribution of American sand flies have been published, especially in the last three decades when research on this important group of insects increased. Most studies, however, were limited to locations where these insects had been recorded. Barretto and Pessoa (1946) were the pioneers of systematization studies, and Martins and Morales-Farias (1972), Martins et al. (1978), Young and Duncan (1994), and Shimabukuro et al. (2017) carried out conclusive studies on American sand flies.

A proposal for the classification of Phlebotominae was presented by Galati (2003) who used the cladistics method in her study of American sand flies. Galati reclassified New World sand flies into 22 genera.

From >1000 species of sand flies described worldwide, approximately 10% are involved in the transmission of diseases to human and other mammals and are therefore considered potential vectors of etiological agents of leishmaniasis, arboviruses, and Bartonellosis (Rangel and Lainson 2009). From this proportion, 60% occur in the neotropical region. In Brazil, 279 species have been described so far, accounting for 31% of all species known to occur worldwide.

G. M. de Aguiar (✉) · V. R. Vieira
Instituto Oswaldo Cruz, Fundação Oswaldo Cruz, Rio de Janeiro, Brazil
e-mail: aguiar@ioc.fiocruz.br

E. F. Rangel, J. J. Shaw (eds.), *Brazilian Sand Flies*,
https://doi.org/10.1007/978-3-319-75544-1_4

Twenty-nine species occur in 5 Brazilian regions: North (N), Northeast (NE), Central West (CW), Southeast (SE), and South (S) and therefore are widely distributed.

Species Found in the Five Regions of Brazil

Species	Regions
Brumptomyia avellari (Costa Lima, 1932)	N, NE, CW, SE, S
Br. brumpti (Larrousse, 1920)	N, NE, CW, SE, S
Br. cunhai (Mangabeira, 1942a)	N, NE, CW, SE, S
Br. pintoi (Costa Lima, 1932)	N, NE, CW, SE, S
Evandromyia bacula (Martins, Falcão & Silva, 1965)	N, NE. CW, SE, S
Ev. cortelezzii (Brèthes, 1923)	N, NE, CW, SE, S
Ev. evandroi (Costa Lima & Antunes, 1936)	N, NE, CW, SE, S
Ev. lenti (Mangabeira, 1938)	N, NE, CW, SE, S
Ev. sallesi (Galvão & Coutinho, 1939)	N, NE, CW, SE, S
Lutzomyia longipalpis (Lutz & Neiva, 1912)	N, NE, CW, SE, S
Micropygomyia longipennis (Barretto, 1946)	N, NE, CW, SE, S
Mi. micropyga (Mangabeira, 1942a)	N, NE, CW, SE, S
Mi. oswaldoi (Mangabeira, 1942e)	N, NE, CW, SE, S
Mi. quinquefer (Dyar, 1929)	N, NE, CW, SE, S
Migonomyia migonei (França, 1920)	N, NE, CW, SE, S
Nyssomyia intermedia (Lutz & Neiva, 1912)	N, NE, CW, SE, S
Ny. whitmani (Antunes & Coutinho, 1939)	N, NE, CW, SE, S
Pintomyia christenseni (Young & Duncan, 1994)	N, NE, CW, SE, S
Pi. misionensis (Castro, 1959)	N, NE, CW, SE, S
Pi. monticola (Costa Lima, 1932)	N, NE, CW, SE, S
Psathyromyia aragaoi (Costa Lima, 1932)	N, NE, CW, SE, S
Pa. bigeniculata (Floch & Abonnenc, 1941)	N, NE, CW, SE, S
Pa. lutziana (Costa Lima, 1932)	N, NE, CW, SE, S
Pa. punctigeniculata (Floch & Abonnenc, 1944)	N, NE, CW, SE, S
Psychodopygus ayrozai (Barretto & Coutinho, 1940)	N, NE, CW, SE, S
Ps. geniculatus (Mangabeira, 1941c)	N, NE, CW, SE, S
Ps. hirsutus (Mangabeira, 1942a)	N, NE, CW, SE, S
Sciopemyia microps (Mangabeira, 1942a)	N, NE, CW, SE, S
Sc. sordellii (Shannon & Del Ponte, 1927)	N, NE, CW, SE, S

Another 37 species are also widely distributed and occur in four regions: the CW (36 species), the NE (35 species), the SE (34 species), the N (32 species), and the S (10 species).

Species Found in Four Regions of Brazil

Species	Regions
Bichromomyia flaviscutellata (Mangabeira, 1942a)	N, NE, CW, SE
Brumptomyia mangabeirai (Barretto & Coutinho, 1941a)	N, CW, SE, S
Br. nitzulescui (Costa Lima, 1932)	NE, CW, SE, S
Evandromyia bourrouli (Barretto & Coutinho, 1941b)	N, NE,CW, SE
Ev. carmelinoi (Ryan, Fraiha, Lainson & Shaw, 1986)	N, NE, CW, SE
Ev. corumbaensis (Galati, Nunes, Oshiro & Rego, 1989)	N, NE, CW, SE
Ev. infraspinosa (Mangabeira, 1941b)	N, NE, CW, SE
Ev. sericea (Floch & Abonnenc, 1944)	N, NE, CW, SE
Ev. teratodes (Martins, Falcão & Silva, 1964)	N, NE, CW, SE
Ev. termitophila (Martins, Falcão & Silva, 1964)	N, NE, CW, SE
Ev. walkeri (Newstead, 1914)	N, NE, CW, SE
Martinsmyia oliveirai (Martins, Silva & Falcão, 1970a)	N, NE, CW, SE
Micropygomyia peresi (Mangabeira, 1942a)	N, NE, CW, SE
Mi. trinidadensis (Newstead, 1922)	N, NE, CW, SE
Mi. villelai (Mangabeira, 1942)	N, NE, CW, SE
Migonemyia bursiformis (Floch & Abonnenc, 1944)	N, NE, CW, SE
Nyssomyia fraihai (Martins, Falcão & Silva, 1979)	N, NE, CW, SE
Ny. neivai (Pinto, 1926)	N, CW, SE, S
Ny. umbratilis (Ward & Fraiha, 1977)	N, NE, CW, S
Pintomyia damascenoi (Mangabeira, 1941d)	N, NE, CW, SE
Pi. fischeri (Pinto, 1926)	NE, CW, SE, S
Pi. pessoai (Coutinho & Barretto, 1940)	NE, CW, SE, S
Pi. serrana (Damasceno & Arouck, 1949)	N, NE, CW, SE
Pressatia choti (Floch & Abonnenc, 1941)	N, NE, CW, SE
Psathyromyia abonnenci (Floch & Chassignet, 1947)	N, NE, CW, S
Pa. barrettoi barrettoi (Mangabeira, 1942a)	N, NE, CW, SE
Pa. brasiliensis (Costa Lima, 1932)	N, NE, CW, SE
Pa. dendrophyla (Mangabeira, 1942a)	N, NE, CW, SE
Pa. hermanlenti (Martins, Silva & Falcão, 1970)	N, NE, CW, SE
Pa. lanei (Barretto & Coutinho, 1941)	NE, CW, SE, S
Pa. pascalei (Coutinho & Barretto, 1941c)	N, NE, SE, S
Psychodopygus carrerai (Barretto, 1946)	N, NE, CW, SE
Ps. davisi (Root, 1934)	N, NE, CW, SE
Ps. lloydi (Antunes, 1937)	NE, SE, CW, S
Ps. paraensis (Costa Lima, 1941)	N, NE, CW, SE
Trichophoromyia ubiquitalis (Mangabeira, 1942a)	N, NE, CW, SE
Tr. longispina (Mangabeira, 1942d)	N, NE, CW, SE

In 3 regions there are 41 species, mainly in the CW (35 species), the N (34 species), the NE (31), the SE (16 species), and the S (6 species).

Species Found in Three Regions of Brazil

Species	Regions
Bichromomyia olmeca nociva (Young & Arias, 1982)	N, NE, CW
Brumptomyia galindoi (Fairchild & Hertig, 1947)	CW, SE, S
Br. ortizi Martins, Silva & Falcão, 1971b	N, SE, S
Br. travassosi (Mangabeira, 1942g)	N, NE, SE
Evandromyia edwardsi (Mangabeira, 1941b)	NE, SE, S
Ev. monstruosa (Floch & Abonnenc, 1944)	N, NE, CW
Ev. pinottii (Damasceno & Arouck, 1956)	N, NE, CW
Ev. saulensis (Floch & Abonnenc, 1944)	N, NE, CW
Ev. wilsoni (Damasceno & Causey, 1945)	N, NE, CW
Expapillata cerradincola (Galati, Nunes, Oshiro & Dorval, 1995)	N, CW, SE
Lutzomyia cavernicola (Costa Lima, 1932)	N, CW, SE
Lu. cruzi (Mangabeira, 1938)	NE, CW, SE
Lu. dispar Martins & Silva , 1963	NE, CW, SE
Lu. gomezi (Nitzulescu, 1931)	N, NE, CW
Lu. renei (Martins, Falcão & Silva, 1957)	N, CW, SE
Lu. spathotrichia Martins, Falcão & Silva, 1963	N, NE, CW
Micropygomyia acanthopharynx (Martins, Falcão & Silva, 1962)	N, CW, SE
Mi. ferreirana (Barretto, Martins & Pellegrino, 1956)	CW, SE, S
Mi. pusilla (Dias, Martins, Falcão & Silva, 1986)	N, NE, CW
Mi. schreiberi (Martins, Falcão & Silva, 1955)	NE, SE, S
Mi. vonatzingeni Galati, 2007	N, CW, SE
Nyssomyia anduzei (Rozeboom, 1942)	N, NE, CW
Ny. antunesi (Coutinho, 1939)	N, NE, CW
Ny. richardwardi (Ready & Fraiha, 1981)	N, NE, CW
Pintomyia nevesi (Damasceno & Arouck, 1956)	N, NE, CW
Pressatia triacantha (Mangabeira, 1942c)	N, NE, CW
Pr. trispinosa (Mangabeira, 1942a)	N, NE, SE
Psathyromyia pelloni (Sherlock & Alencar, 1959)	NE, SE, S
Pa. runoides (Fairchild & Hertig, 1953)	N, CW, SE
Pa. scaffi (Damasceno & Arouck, 1956)	N, NE, CW
Psychodopygus amazonensis (Root, 1934)	N, NE, CW
Ps. bispinosus (Fairchild & Hertig, 1951)	N, NE, CW
Ps. chagasi (Costa Lima, 1941)	N, NE, CW
Ps. claustrei (Abonnenc, Léger & Fauran, 1979)	N, NE, CW
Ps. complexus (Mangabeira, 1941c)	N, NE, CW
Ps. guyanensis (Floch & Abonnenc, 1941)	N, CW, SE
Ps. squamiventris squamiventris (Lutz & Neiva, 1912)	N, NE, CW
Ps. wellcomei (Fraiha, Shaw & Lainson, 1971)	N, NE, CW
Sciopemyia servulolimai (Damasceno & Causey, 1945)	N, NE, CW
Viannamyia furcata (Mangabeira, 1941d)	N, NE, CW
Vi. tuberculata (Mangabeira, 1941d)	N, NE, CW

In 2 regions 66 species are listed. The N is more represented (45 species) followed by the CW (33 species), the SE 23 species), the NE (22 species), and the S (9 species).

Species Found in Two Regions of Brazil

Species	Regions
Bichromomyia inornata (Martins, Falcão & Silva, 1965)	N, NE
Bi. olmeca bicolor (Fairchild & Theodor, 1971)	N, CW
Brumptomyia bragai (Mangabeira & Sherlock, 1961)	NE, SE
Br. cardosoi (Barretto & Coutinho, 1941a)	SE, S
Br. figueiredoi Mangabeira & Sherlock, 1961	NE, SE
Br. guimaraesi (Coutinho & Barretto, 1941a)	SE, S
Br. pentacantha (Barretto, 1947)	N, CW
Br. troglodytes (Lutz, 1922)	SE, S
Br. virgensi Mangabeira & Sherlock, 1961	NE, CW
Deanemyia samueli (Deane, 1955)	N, NE
Edentomyia piauienses Galati, Andrade-Filho, Silva & Falcão, 2003	N, NE
Evandromyia andersoni (Le Point & Desjeux, 1988)	N, CW
Ev. begonae (Ortiz & Torres, 1975)	N, CW
Ev. brachyphalla (Mangabeira Fo, 1994)	N, NE
Ev. correalimai (Martins, Coutinho & Luz, 1970)	SE, S
Ev. costalimai (Mangabeira, 1942a)	NE, SE
Ev. dubitans (Sherlock, 1962)	N, CW
Ev. tupynambai (Mangabeira, 1942a)	NE, SE
Expapillata firmatoi (Barretto, Martins & Pellegrino, 1956)	SE, S
Lutzomyia almerioi Galati & Nunes, 1999	CW, SE
Lu. amarali Barretto & Coutinho, 1940	SE, S
Lu. carvalhoi (Damasceno, Causey & Arouck, 1945)	N, NE
Lu. evangelistai Martins & Fraiha, 1971	N, CW
Lu. falcata Young, Morales & Ferro, 1994	N, CW
Lu. flabellata Martins & Silva, 1964	N, CW
Lu. ischnacantha Martins, Souza & Falcão, 1962a	CW, SE
Lu. lichyi (Floch & Abonnenc, 1950)	N, CW
Lu. marinkellei Young, 1979	N, CW
Lu. sherlocki Martins, Silva & Falcão, 1971a	N, CW
Martinsmyia alphabetica (Fonseca, 1936)	SE, S
Mt. gasparviannai (Martins, Godoy & Silva ,1962b)	NE, SE
Mt. minasensis (Mangabeira, 1942a)	N, SE
Micropygomyia capixaba (Dias, Falcão, Silva & Martins, 1987)	NE, SE
Mi. echinatopharynx Andrade Filho, Galati, Andrade & Falcão, 2004	N, CW
Mi. pilosa (Damasceno & Causey, 1944)	N, NE

(continued)

(continued)

Species	Regions
Mi. rorotaensis (Floch & Abonnenc, 1944)	N, NE
Mi. zikani (Barretto, 1950a)	N, SE
Migonomyia cerqueirai (Causey & Damasceno, 1945)	N, NE
Nyssomyia shawi (Fraiha, Ward & Ready, 1973)	N, CW
Ny. urbinattii Galati & Galvis, 2012	N, CW
Pintomyia bianchigalatiae (Andrade Filho, Aguiar, Dias & Falcão, 1999)	SE, S
Pi. mamedei (Oliveira, Afonso, Dias & Brazil, 1994)	CW, SE
Pi. odax (Fairchild & Hertig, 1961)	N, NE
Pi. pacae (Floch & Abonnenc, 1943)	N, NE
Pressatia equatorialis (Mangabeira, 1942a)	N, SE
Psathyromyia campbelli (Damasceno, Causey & Arouck, 1945)	N, CW
Pa. campograndensis (Oliveira, Andrade Filho, Falcão & Brazil, 2001)	N, CW
Pa. coutinhoi (Mangabeira, 1942a)	N, CW
Pa. dasymera (Fairchild & Hertig, 1961)	N, CW
Pa. dreisbachi (Causey & Damasceno, 1945)	N, CW
Pa. inflata (Floch & Abonnenc, 1944)	N, CW
Pa. limai (Fonseca, 1935)	SE, S
Pa. pradobarrientosi (Le Pont, Matias, Martinez & Dujardin, 2004)	N, CW
Psychodopygus arthuri (Fonseca, 1936)	NE, SE
Ps. corossoniensis (Le Pont & Pajot, 1978)	N, NE
Ps. lainsoni (Fraiha & Ward, 1964)	N, CW
Ps. llanosmartinsi (Fraiha & Ward, 1980)	N, CW
Ps. matosi (Barretto & Zago, 1956)	NE, SE
Sciopemyia fluviatilis (Floch & Abonnenc, 1944)	N, NE
Trichophoromyia auraensis (Mangabeira, 1942a)	N, CW
Th. clitella (Young & Perez, 1994)	N, CW
Th. octavioi (Vargas, 1949)	N, CW
Th. ruii (Arias & Young, 1982)	N, CW
Trichopygomyia dasypodogeton (Castro, 1939)	N, CW
Ty. rondonensis (Martins, Falcão & Silva, 1965)	N, CW
Ty. trichopyga (Floch & Abonnenc, 1945)	N, NE

In 1 region, 106 species were found: 64 species in the N, 23 species in the SE, 11 species in the CW, 5 species in the NE, and 3 species in the S.

Species Found in One Region of Brazil

Species	Regions
Bichromomyia reducta (Feliciangeli, Ramirez Pérez & Ramirez, 1988)	N
Brumptomyia angelae Galati, Santos & Silva, 2007	S
Br. carvalheiroi Shimabukuro, Marassa & Galati 2007	SE
Br. orlandoi Fraiha, Shaw & Lainson, 1970	CW
Br. spinosipes Floch & Abonnenc, 1943	N
De. appendiculata (Martins, Falcão & Silva, 1961)	SE
De. derelicta (Freitas & Barrett, 1999)	N
De. maruaga (Alves, Freitas & Barret, 2008)	N
De. ramirezi (Martins, Falcão, Silva & Miranda, 1982)	SE
Evandromyia aldafalcaoae (Santos, Andrade Filho & Honer, 2001)	CW
Ev. apurinan Shimabukuro, Figueira & Silva, 2013	N
Ev. bahiensis (Mangabeira & Sherlock, 1961)	NE
Ev. callipyga (Martins & Silva, 1965)	SE
Ev. gaucha Andrade Filho, Souza & Falcão, 2007	S
Ev. georgii (Freitas & Barrett, 2002)	N
Ev. grimaldii Andrade Filho, Pinto, Santos & Carvalho, 2009	SE
Ev. inpai (Young & Arias, 1977)	N
Ev. orcyi Oliveira, Sanguinette, Almeida & Andrade Filho, 2015	CW
Ev. petropolitana (Martins & Silva, 1968)	SE
Ev. rupicola (Martins, Godoy & Silva, 1962)	SE
Ev. sipani (Fernandez, Carbajal, Alexander & Need, 1994)	N
Ev. spelunca Carvalho, Brazil, Sanguinette & Andrade Filho, 2011	SE
Ev. tarapacaensis (Le Pont, Torres-Espejo & Galati, 1996)	N
Ev. tylophalla Andrade & Galati, 2012	SE
Ev. williamsi (Damasceno, Causey & Arouck, 1945)	N
Lutzomyia alencari Martins, Souza & Falcão, 1962	SE
Lu. araracuarensis (Morales & Minter, 1981)	N
Lu. caligata Martins, Falcão & Silva, 1965	N
Lu. castroi (Barretto & Coutinho, 1941c)	SE
Lu. cruciata (Coquillett, 1907)	CW
Lu. cultellata Barrett, Feitas, Albuquerque & Guerrero, 1996	N
Lu. elizabethrangelae Vilela, Azevedo e Godoy, 2015	N
Lu. falquetoi Pinto & Santos, 2007	SE
Lu. forattinii Galati, Rego, Nunes & Teruya, 1985	CW
Lu. gaminarai (Cordero, Vogelsang & Cossio, 1928)	S
Lu. ischyracantha Martins, Falcão & Silva, 1962b	SE
Lu. souzalopesi Martins, Silva & Falcão, 1970b	SE
Martinsmyia brisolai (Le Pont & Desjeux, 1987)	CW
Mt. cipoensis (Martins, Falcão & Silva, 1964)	SE
Mt. reginae Carvalho, Brazil, Sanguinette & Andrade Filho, 2010	N
Mt. waltoni (Arias, Freitas & Barret, 1984)	N

(continued)

(continued)

Species	Regions
Micropygomyia breviducta (Barretto, 1950)	SE
Mi. cayennensis (Floch & Abonnenc, 1941)	N
Mi. chassigneti (Floch & Abonnenc, 1944)	N
Mi. mangabeirana (Martins, Falcão & Silva, 1963)	N
Mi. petari Galati, Marassá & Andrade, 2003	SE
Migonomyia moucheti (Pajot & Le Pont, 1923)	N
Mg. rabelloi (Galati & Gomes, 1992)	SE
Mg. vaniae Galati, Fonseca & Marassá, 2007	SE
Ny. delsionatali Galati & Galvis, 2012	CW
Ny. pajoti (Abonnenc, Léger & Fauran, 1979)	N
Ny. singularis (Costa Lima, 1932)	SE
Ny. sylvicola (Floch & Abonnenc, 1944)	N
Pi. gruta (Ryan, 1986)	N
Pi. kuscheli (Le Pont, Martinez, Torrez-Espejo, Dujardin, 1998)	CW
Pi. naiffi (Freitas & Oliveira, 2013)	N
Pi. orestes (Fairchild & Trapido, 1950)	NE
Pr. calcarata (Martins & Silva, 1964)	N
*Pr. duncanae (*Le Pont, Martinez, Torrez-Espejo & Dujardin, 1998)	N
Pr. dysponeta (Fairchild & Hertig, 1952)	CW
Pa. abunaensis (Martins, Falcão & Silva, 1965)	N
Pa. baratai Sábio, Andrade & Galati, 2016	SE
Pa. barretti Alves & Freitas, 2015	N
Pa. digitata (Damasceno & Arouck, 1950)	NE
Pa. elizabethdorvalae Brilhante, Sábio & Galati, 2017	N
Pa. naftalekatzi (Falcão, Andrade Filho, Almeida, Brandão-Filho, 2000)	NE
Pa. pifanoi (Ortiz, 1972)	N
Pa. ribeirensis Sábio, Andrade & Galati, 2014	SE
Pa. souzacastroi (Damasceno & Causey, 1944)	N
Ps. bernalei (Osorno-Mesa, Morales & Osorno, 1967)	N
Ps. douradoi (Fé, Freitas & Barrett, 1998)	N
Ps. fairchildi (Barretto, 1966)	SE
Ps. leonidasdeanei (Fraiha, Ryan, Ward, Lainson & Shaw, 1987)	N
Ps. nicaraguensis (Fairchild & Hertig, 1961)	CW
Ps. panamensis (Shannon, 1926)	N
Ps. squamiventris maripaensis (Floch & Abonnenc, 1946)	N
Ps. yucumensis (Le Pont, Caillard, Tibayrenc & Desjeux, 1986)	N
Sc. nematoducta (Young & Arias, 1984)	N
Sc. pennyi (Arias & Freitas, 1981)	N
Sc. preclara (Young & Arias, 1984)	N
Th. adelsonsouzai Santos, Silva, Barata, Andrade & Galati, 2013	N
Th. brachipyga (Mangabeira, 1942a)	N
Th. castanheirai (Damasceno, Causey & Arouck, 1945)	N
Th. dunhami (Causey & Damasceno, 1945)	N

(continued)

(continued)

Species	Regions
Th. eurypyga (Martins, Falcão & Silva, 1963)	N
Th. flochi (Abonnenc & Chassignet, 1948)	N
Th. gibba (Young & Arias, 1994)	N
Th. howardi (Young, 1979)	CW
Th. ininii (Floch & Abonnenc, 1943)	N
Th. lopesi (Damasceno, Causey & Arouck, 1945)	N
Th. loretoensis (Llanos, 1964)	N
Th. meirai (Causey & Damasceno, 1945)	N
Th. melloi (Causey & Damasceno, 1945)	N
Th. readyi (Ryan, 1986)	N
Th. reinerti (Young & Duncan, 1994)	N
Th. rostrans (Summers, 1912)	N
Th. ruifreitasi Oliveira, Teles, Medeiros, Camargo & Pessoa, 2015	N
Th. uniniensis Ladeia-Andrade, Fé, Sanguinette & Andrade Filho, 2014	N
Th. viannamartinsi (Sherlock & Guitton, 1970)	NE
Ty. conviti (Ramírez, Pérez, Martins & Ramírez, 1976)	N
Ty. elegans (Martins, Llanos & Silva, 1976)	N
Ty. pinna (Feliciangeli, Ramirez Pérez & Ramirez, 1989)	N
Ty. ratcliffei (Arias, Ready & Freitas, 1983)	N
Ty. wagleyi (Causey & Damasceno, 1945)	N
Vi. caprina (Osorno-Mesa, Morales e Osorno, 1972)	N
Vi. fariasi (Damasceno, Causey & Arouck, 1945)	N

The N has 204 species, of which 64 are endemic; the CW has 144 species, and 11 are endemic; the SE has 125 species, with 23 of them being endemic; the NE has 122 species, and 5 are endemic; and the S has 57 species, and 3 of them are endemic.

Brazilian sand flies are listed in Table 1 by region and state. The states with the greatest number of species (in parentheses) are listed by region as follows: N: Amazonas (141), Pará (135), Rondônia (125), Acre (95), Roraima (83), Tocantins (73), and Amapá (70); NE: Maranhão (92), Bahia (60), Pernambuco (41), Ceará (28), Piauí (22), Rio Grande do Norte (17), Paraíba (10), Alagoas (9), and Sergipe (7); CW: Mato Grosso (127), Mato Grosso do Sul (65), Goiás (56), and Distrito Federal (31); SE: Minas Gerais (96), São Paulo (77), Rio de Janeiro (65), and Espírito Santo (64); and S: Paraná (54), Rio Grande do Sul (23), and Santa Catarina (17).

Locally distributed species are those known to occupy very restricted areas, sometimes only the type locality. As research progresses, the situation tends to change with the expansion of the area of dispersion. For example, *Lu. cavernicola*, which for several years was restricted to the type locality, a cave of Maquiné in the state of Minas Gerais, was recently captured in other localities of Minas Gerais, Goiás, and the state of Tocantins. However, it was always found in calcareous caves.

Discontinuously distributed species are those recorded in places very distant from each other. For example, *Pa. runoides* occurs in Panamá and Costa Rica

Table 1 Brazilian phlebotomine species by region and state

Region	State	Species	
N	Acre	*Bi. flaviscutellata*	*Pr. calcarata*
		Bi. olmeca bicolor	*Pr. choti*
		Bi. reducta	*Pr. duncanae*
		Br. avellari	*Pr. triacantha*
		Br. pentacantha	*Pa. abonnenci*
		Ev. andersoni	*Pa. abunaensis*
		Ev. bacula	*Pa. aragaoi*
		Ev. begonae	*Pa. bigeniculata*
		Ev. bourrouli	*Pa. brasiliensis*
		Ev. cortelezzii	*Pa. campbelli*
		Ev. evandroi	*Pa. dendrophyla*
		Ev. georgii	*Pa. dreisbachi*
		Ev. infraspinosa	*Pa. elizabethdorvalae*
		Ev. monstruosa	*Pa. lutziana*
		Ev. sallesi	*Pa. pradobarrientosi*
		Ev. saulensis	*Pa. punctigeniculata*
		Ev. sericea	*Pa. runoides*
		Ev. tarapacaensis	*Pa. scaffi*
		Ev. termitophila	*Ps. amazonensis*
		Ev. walkeri	*Ps. ayrozai*
		Ev. williamsi	*Ps. bispinosus*
		Ev. wilsoni	*Ps. carrerai*
		Lu. flabellata	*Ps. chagasi*
		Lu. gomezi	*Ps. corossoniensis*
		Lu. sherlocki	*Ps. davisi*
		Mi. acanthopharynx	*Ps. geniculatus*
		Mi. longipennis	*Ps. guyanensis*
		Mi. micropyga	*Ps. hirsutus*
		Mi. peresi	*Ps. lainsoni*
		Mi. pilosa	*Ps. llanosmartinsi*
		Mi. pusilla	*Ps. paraensis*
		Mi. trinidadensis	*Ps. yucumensis*
		Mi. villelai	*Sc. fluviatilis*
		Mg. migonei	*Sc. preclara*
		Ny. anduzei	*Sc. servulolimai*
		Ny. antunesi	*Sc. sordellii*
		Ny. fraihai	*Th. auraensis*
		Ny. richardwardi	*Th. brachipyga*
		Ny. shawi	*Th. flochi*
		Ny. sylvicola	*Th. ininii*
		Ny. umbratilis	*Th. melloi*
		Ny. whitmani	*Th. ruifreitasi*

(continued)

Table 1 (continued)

Region	State	Species	
		Pi. christenseni	*Th. ubiquitalis*
		Pi. damascenoi	*Ty. dasypodogeton*
		Pi. naiffi	*Ty. elegans*
		Pi. nevesi	*Vi. furcata*
		Pi. odax	*Vi. tuberculata*
		Pi. serrana	
	Amapá	*Bi. flaviscutellata*	*Ev. saulensis*
		Br. cunhai	*Ev. sericea*
		Br. travassosi	*Ev. walkeri*
		Ev. bourrouli	*Ev. williamsi*
		Ev. brachyphalla	*Lu. carvalhoi*
		Ev. evandroi	*Lu. evangelistai*
		Ev. infraspinosa	*Lu. gomezi*
		Ev. inpai	*Lu. longipalpis*
		Ev. monstruosa	*Lu. spathotrichia*
N	Amapá	*Mi. chassigneti*	*Pa. campograndensis*
		Mi. longipennis	*Pa. dendrophyla*
		Mi. micropyga	*Pa. dreisbachi*
		Mi. oswaldoi	*Pa. inflata*
		Mi. pilosa	*Pa. lutziana*
		Mi. pusilla	*Pa. scaffi*
		Mi. rorotaensis	*Ps. amazonensis*
		Mi. trinidadensis	*Ps. ayrozai*
		Mg. migonei	*Ps. bispinosus*
		Ny. anduzei	*Ps. chagasi*
		Ny. antunesi	*Ps. claustrei*
		Ny. fraihai	*Ps. davisi*
		Ny. pajoti	*Ps. geniculatus*
		Ny. richardwardi	*Ps. guyanensis*
		Ny. umbratilis	*Ps. hirsutus*
		Ny. whitmani	*Ps. paraensis*
		Pi. christenseni	*Ps. squamiventris maripaensis*
		Pi. damascenoi	*Ps. squamiventris squamiventris*
		Pi. pacae	*Sc. fluviatilis*
		Pi. serrana	*Sc. sordellii*
		Pr. choti	*Th. brachipyga*
		Pr. trispinosa	*Th. ininii*
		Pa. abonnenci	*Th. ubiquitalis*
		Pa. aragaoi	*Ty. trichopyga*
		Pa. bigeniculata	*Vi. furcata*
		Pa. campbelli	*Vi. tuberculata*

(continued)

Table 1 (continued)

Region	State	Species	
	Amazonas	*Bi. flaviscutellata*	*Mi. peresi*
		Bi. inornata	*Mi. pilosa*
		Bi. olmeca bicolor	*Mi. pusilla*
		Bi. olmeca nociva	*Mi. rorotaensis*
		Bi. reducta	*Mi. trinidadensis*
		Br. brumpti	*Mg. bursiformis*
		Br. pentacantha	*Mg. cerqueirai*
		Br. pintoi	*Mg. migonei*
		De. maruaga	*Mg. moucheti*
		Ev. andersoni	*Ny. anduzei*
		Ev. apurinan	*Ny. antunesi*
		Ev. bacula	*Ny. fraihai*
		Ev. begonae	*Ny. pajoti*
		Ev. bourrouli	*Ny. richardwardi*
		Ev. dubitans	*Ny. shawi*
		Ev. evandroi	*Ny. umbratilis*
		Ev. georgii	*Ny. whitmani*
		Ev. infraspinosa	*Pi. christenseni*
		Ev. inpai	*Pi. damascenoi*
		Ev. monstruosa	*Pi. pacae*
		Ev. pinottii	*Pi. serrana*
		Ev. saulensis	*Pr. choti*
		Ev. sericea	*Pr. triacantha*
		Ev. sipani	*Pr. trispinosa*
		Ev. tarapacaensis	*Pa. abonnenci*
		Ev. walkeri	*Pa. abunaensis*
		Ev. williamsi	*Pa. aragaoi*
		Ev. wilsoni	*Pa. barretti*
		Lu. araracuarensis	*Pa. barrettoi barrettoi*
		Lu. caligata	*Pa. bigeniculata*
		Lu. cultellata	*Pa. brasiliensis*
		Lu. evangelistai	*Pa. campbelli*
		Lu. falcata	*Pa. campograndensis*
		Lu. flabellata	*Pa. coutinhoi*
		Lu. gomezi	*Pa. dasymera*
		Lu. marinkellei	*Pa. dendrophyla*
		Lu. sherlocki	*Pa. dreisbachi*
		Lu. spathotrichia	*Pa. inflata*
		Mt. oliveirai	*Pa. lutziana*
		Mi. cayennensis cayennensis	*Pa. pifanoi*
		Mi. chassigneti	*Pa. punctigeniculata*
		Mi. longipennis	*Pa. runoides*

(continued)

Table 1 (continued)

Region	State	Species	
		Mi. micropyga	*Pa. scaffi*
N	Amazonas	*Pa. souzacastroi*	*Th. castanheirai*
		Ps. amazonensis	*Th. dunhami*
		Ps. ayrozai	*Th. eurypyga*
		Ps. bernalei	*Th. flochi*
		Ps. bispinosus	*Th. gibba*
		Ps. carrerai	*Th. ininii*
		Ps. chagasi	*Th. lopesi*
		Ps. claustrei	*Th. meirai*
		Ps. complexus	*Th. melloi*
		Ps. corossoniensis	*Th. octavioi*
		Ps. davisi	*Th. readyi*
		Ps. douradoi	*Th. rostrans*
		Ps. geniculatus	*Th. ruii*
		Ps. guyanensis	*Th. ubiquitalis*
		Ps. hirsutus	*Th. uniniensis*
		Ps. llanosmartinsi	*Ty. conviti*
		Ps. paraensis	*Ty. dasypodogeton*
		Ps. squamiventris maripaensis	*Ty. longispina*
		Ps. squamiventris squamiventris	*Ty. pinna*
		Ps. wellcomei	*Ty. ratcliffei*
		Sc. fluviatilis	*Ty. rondonensis*
		Sc. nematoducta	*Ty. trichopyga*
		Sc. pennyi	*Ty. wagleyi*
		Sc. preclara	*Vi. caprina*
		Sc. servulolimai	*Vi. fariasi*
		Sc. sordellii	*Vi. furcata*
		Th. auraensis	*Vi. tuberculata*
		Th. brachipyga	
	Pará	*Bi. flaviscutellata*	*Mi. cayennensis cayennensis*
		Bi. olmeca bicolor	*Mi. longipennis*
		Bi. olmeca nociva	*Mi. mangabeirana*
		Br. avellari	*Mi. micropyga*
		Br. brumpti	*Mi. oswaldoi*
		Br. cunhai	*Mi. peresi*
		Br. mangabeirai	*Mi. pilosa*
		Br. ortizi	*Mi. pusilla*
		Br. pentacantha	*Mi. rorotaensis*
		Br. pintoi	*Mi. trinidadensis*
		Br. spinosipes	*Mi. villelai*

(continued)

Table 1 (continued)

Region	State	Species	
		Br. travassosi	*Mi. vonatzingeni*
		De. derelicta	*Mi. zikani*
		De. samueli	*Mg. bursiformis*
		Ed. piauienses	*Mg. cerqueirai*
		Ev. baculaa	*Mg. migonei*
		Ev. begonae	*Ny. anduzei*
		Ev. bourrouli	*Ny. antunesi*
		Ev. brachyphalla	*Ny. intermedia*
		Ev. carmelinoi	*Ny. fraihai*
		Ev. dubitans	*Ny. neivai*
		Ev. evandroi	*Ny. pajoti*
		Ev. georgii	*Ny. richardwardi*
		Ev. infraspinosa	*Ny. shawi*
		Ev. inpai	*Ny. sylvicola*
		Ev. lenti	*Ny. umbratilis*
		Ev. monstruosa	*Ny. urbinattii*
		Ev. pinottii	*Ny. whitmani*
		Ev. saulensis	*Pi. christenseni*
		Ev. sericea	*Pi. damascenoi*
		Ev. tarapacaensis	*Pi. gruta*
		Ev. termitophila	*Pi. monticola*
		Ev. walkeri	*Pi. nevesi*
		Ev. williamsi	*Pi. pacae*
		Lu. carvalhoi	*Pi. serrana*
		Lu. evangelistai	*Pr. choti*
		Lu. gomezi	*Pr. equatorialis*
		Lu. longipalpis	*Pr. triacantha*
		Lu. sherlocki	*Pr. trispinosa*
		Lu. spathotrichia	*Pa. abonnenci*
		Mt. oliveirai	*Pa. aragaoi*
N	Pará	*Pa. barrettoi barrettoi*	*Ps. paraensis*
		Pa. bigeniculata	*Ps. squamiventris maripaensis*
		Pa. brasiliensis	*Ps. squamiventris squamiventris*
		Pa. campbelli	*Ps. wellcomei*
		Pa. coutinhoi	*Sc. fluviatilis*
		Pa. dendrophyla	*Sc. microps*
		Pa. dreisbachi	*Sc. servulolimai*
		Pa. hermanlenti	*Sc. sordellii*
		Pa. inflata	*Th. adelsonsouzai*
		Pa. lutziana	*Th. auraensis*
		Pa. pifanoi	*Th. brachipyga*

(continued)

Table 1 (continued)

Region	State	Species	
		Pa. punctigeniculata	*Th. castanheirai*
		Pa. scaffi	*Th. eurypyga*
		Ps. amazonensis	*Th. melloi*
		Ps. ayrozai	*Th. octavioi*
		Ps. bispinosus	*Th. readyi*
		Ps. carrerai	*Th. reinerti*
		Ps. chagasi	*Th. ruii*
		Ps. claustrei	*Th. ubiquitalis*
		Ps. complexus	*Ty. dasypodogeton*
		Ps. corossoniensis	*Ty. longispina*
		Ps. davisi	*Ty. rondonensis*
		Ps. geniculatus	*Ty. trichopyga*
		Ps. guyanensis	*Vi. fariasi*
		Ps. hirsutus	*Vi. furcata*
		Ps. lainsoni	*Vi. tuberculata*
		Ps. leonidasdeanei	
	Rondônia	*Bi. flaviscutellata*	*Mi. oswaldoi*
		Bi. inornata	*Mi. peresi*
		Bi. olmeca nociva	*Mi. pilosa*
		Bi. reducta	*Mi. pusilla*
		Br. avellari	*Mi. rorotaensis*
		Br. brumpti	*Mi. trinidadensis*
		Br. cunhai	*Mi. villelai*
		Br. pentacantha	*Mg. cerqueirai*
		Br. pintoi	*Mg. migonei*
		Br. travassosi	*Ny. anduzei*
		Ev. bacula	*Ny. antunesi*
		Ev. begonae	*Ny. fraihai*
		Ev. bourrouli	*Ny. pajoti*
		Ev. brachyphalla	*Ny. richardwardi*
		Ev. evandroi	*Ny. shawi*
		Ev. georgii	*Ny. umbratilis*
		Ev. infraspinosa	*Ny. whitmani*
		Ev. inpai	*Pi. christenseni*
		Ev. lenti	*Pi. damascenoi*
		Ev. monstruosa	*Pi. gruta*
		Ev. pinottii	*Pi. nevesi*
		Ev. saulensis	*Pi. odax*
		Ev. sericea	*Pi. pacae*
		Ev. tarapacaensis	*Pi. serrana*
		Ev. termitophila	*Pr. calcarata*
		Ev. walkeri	*Pr. choti*

(continued)

Table 1 (continued)

Region	State	Species	
		Ev. williamsi	*Pr. triacantha*
		Ev. wilsoni	*Pr. trispinosa*
		Lu. caligata	*Pa. abonnenci*
		Lu. carvalhoi	*Pa. abunaensis*
		Lu. evangelistai	*Pa. aragaoi*
		Lu. flabellata	*Pa. barrettoi barrettoi*
		Lu. gomezi	*Pa. bigeniculata*
		Lu. longipalpis	*Pa. brasiliensis*
		Lu. marinkellei	*Pa. campbelli*
		Lu. sherlocki	*Pa. coutinhoi*
		Lu. spathotrichia	*Pa. dasymera*
		Mt. waltoni	*Pa. dendrophyla*
		Mi. acanthopharynx	*Pa. dreisbachi*
		Mi. cayennensis cayennensis	*Pa. inflata*
		Mi. longipennis	*Pa. lutziana*
		Mi. micropyga	*Pa. punctigeniculata*
N	Rondônia	*Pa. runoides*	*Sc. servulolimai*
		Pa. scaffi	*Sc. sordellii*
		Ps. amazonensis	*Th. auraensis*
		Ps. ayrozai	*Th. brachipyga*
		Ps. bispinosus	*Th. castanheirai*
		Ps. carrerai	*Th. clitella*
		Ps. chagasi	*Th. eurypyga*
		Ps. claustrei	*Th. flochi*
		Ps. complexus	*Th. loretoensis*
		Ps. corossoniensis	*Th. melloi*
		Ps. davisi	*Th. octavioi*
		Ps. geniculatus	*Th. readyi*
		Ps. hirsutus	*Th. ruii*
		Ps. lainsoni	*Th. ubiquitalis*
		Ps. leonidasdeanei	*Ty. dasypodogeton*
		Ps. llanosmartinsi	*Ty. longispina*
		Ps. paraensis	*Ty. rondonensis*
		Ps. squamiventris squamiventris	*Ty. trichopyga*
		Ps. wellcomei	*Vi. furcata*
		Ps. yucumensis	*Vi. tuberculata*
		Sc. fluviatilis	
	Roraima	*Bi. flaviscutellata*	*Pi. pacae*
		Bi. olmeca bicolor	*Pi. serrana*
		Br. avellari	*Pr. choti*
		Br. pintoi	*Pr. triacantha*

(continued)

Table 1 (continued)

Region	State	Species	
		Br. spinosipes	*Pr. trispinosa*
		Br. travassosi	*Pa. abonnenci*
		Ev. begonae	*Pa. aragaoi*
		Ev. carmelinoi	*Pa. barrettoi barrettoi*
		Ev. dubitans	*Pa. bigeniculata*
		Ev. evandroi	*Pa. brasiliensis*
		Ev. georgii	*Pa. campbelli*
		Ev. infraspinosa	*Pa. campograndensis*
		Ev. inpai	*Pa. dendrophyla*
		Ev. monstruosa	*Pa. dreisbachi*
		Ev. saulensis	*Pa. lutziana*
		Ev. sericea	*Pa. punctigeniculata*
		Ev. walkeri	*Pa. runoides*
		Ev. williamsi	*Pa. scaffi*
		Lu. gomezi	*Ps. amazonensis*
		Lu. lichyi	*Ps. ayrozai*
		Lu. longipalpis	*Ps. carrerai*
		Lu. sherlocki	*Ps. chagasi*
		Lu. spathotrichia	*Ps. claustrei*
		Mt. oliveirai	*Ps. davisi*
		Mt. waltoni	*Ps. hirsutus*
		Mi. cayennensis cayennensis	*Ps. panamensis*
		Mi. longipennis	*Ps. paraensis*
		Mi. mangabeirana	*Ps. squamiventris maripaensis*
		Mi. micropyga	*Ps. squamiventris squamiventris*
		Mi. peresi	*Sc. sordellii*
		Mi. pusilla	*Th. brachipyga*
		Mi. rorotaensis	*Th. eurypyga*
		Mi. trinidadensis	*Th. octavioi*
		Mg. bursiformis	*Th. ruii*
		Ny. anduzei	*Th. ubiquitalis*
		Ny. antunesi	*Ty. dasypodogeton*
		Ny. fraihai	*Ty. longispina*
		Ny. richardwardi	*Ty. pinna*
		Ny. umbratilis	*Ty. trichopyga*
		Ny. whitmani	*Vi. furcata*
		Pi. christenseni	*Vi. tuberculata*
		Pi. damascenoi	
	Tocantins	*Bi. flaviscutellata*	*Ev. brachyphalla*
		Br. avellari	*Ev. carmelinoi*
		Br. brumpti	*Ev. cortelezzii*

(continued)

Table 1 (continued)

Region	State	Species	
		Ev. bacula	*Ev. corumbaensis*
		Ev. begonae	*Ev. evandroi*
		Ev. bourrouli	*Ev. lenti*
N	Tocantins	*Ev. pinottii*	*Ny. whitmani*
		Ev. sallesi	*Pi. christenseni*
		Ev. saulensis	*Pi. damascenoi*
		Ev. teratodes	*Pi. misionensis*
		Ev. termitophila	*Pr. choti*
		Ev. walkeri	*Pa. abonnenci*
		Ex. cerradincola	*Pa. aragaoi*
		Lu. cavernicola	*Pa. bigeniculata*
		Lu. elizabethrangelae	*Pa. brasiliensis*
		Lu. gomezi	*Pa. campbelli*
		Lu. longipalpis	*Pa. dasymera*
		Lu. renei	*Pa. dendrophyla*
		Lu. sherlocki	*Pa. dreisbachi*
		Mt. minasensis	*Pa. hermanlenti*
		Mt. oliveirai	*Pa. lutziana*
		Mt. reginae	*Pa. pascalei*
		Mi. acanthopharynx	*Pa. punctigeniculata*
		Mi. echinatopharynx	*Pa. runoides*
		Mi. longipennis	*Ps. ayrozai*
		Mi. micropyga	*Ps. claustrei*
		Mi. oswaldoi	*Ps. complexus*
		Mi. peresi	*Ps. davisi*
		Mi. quinquefer	*Ps. hirsutus*
		Mi. rorotaensis	*Ps. llanosmartinsi*
		Mi. trinidadensis	*Ps. paraensis*
		Mi. villelai	*Sc. microps*
		Mi. vonatzingeni	*Sc. sordellii*
		Mg. migonei	*Ty. dasypodogeton,*
		Ny. antunesi	*Vi. furcata*
		Ny. intermedia	*Vi. tuberculata*
		Ny. richardwardi	
NE	Alagoas	*Ev. evandroi*	*Mg. migonei*
		Ev. lenti	*Ny. intermedia*
		Ev. termitophila	*Ny. whitmani*
		Lu. longipalpis	*Pa. brasiliensis*
		Mi. quinquefer	
	Bahia	*Bi. flaviscutellata*	*Ny. intermedia*
		Br. avellari	*Ny. whitmani*
		Br. bragai	*Pi. damascenoi*

(continued)

Table 1 (continued)

Region	State	Species	
		Br. cunhai	*Pi. fischeri*
		Br. figueiredoi	*Pi. misionensis*
		Br. virgensi	*Pi. monticola*
		Ev. bahiensis	*Pi. pessoai*
		Ev. cortelezzii	*Pi. serrana*
		Ev. costalimai	*Pr. choti*
		Ev. edwardsi	*Pa. aragaoi*
		Ev. evandroi	*Pa. barrettoi barrettoi*
		Ev. lenti	*Pa. bigeniculata*
		Ev. sallesi	*Pa. brasiliensis*
		Ev. sericea	*Pa. digitata*
		Ev. termitophila	*Pa. lanei*
		Ev. tupynambai	*Pa. lutziana*
		Lu. gomezi	*Pa. pascalei*
		Lu. longipalpis	*Pa. pelloni*
		Mt. gasparviannai	*Ps. ayrozai*
		Mi. capixaba	*Ps. bispinosus*
		Mi. oswaldoi	*Ps. carrerai*
		Mi. peresi	*Ps. davisi*
		Mi. quinquefer	*Ps. hirsutus*
		Mi. schreiberi	*Ps. matosi*
		Mi. trinidadensis	*Sc. microps*
		Mi. villelai	*Th. viannamartinsi*
		Mg. cerqueirai	*Ty. longispina*
		Mg. migonei	*Ty. trichopyga*
		Ny. anduzei	*Vi. furcata*
		Ny. fraihai	*Vi. tuberculata*
	Ceará	*Bi. flaviscutellata*	*Ev. evandroi*
		De. samueli	*Ev. lenti*
		Ev. cortelezzii	*Ev. sallesi*
NE	Ceará	*Ev. saulensis*	*Mg. bursiformis*
		Ev. sericea	*Mg. migonei*
		Ev. walkeri	*Ny. umbratilis*
		Lu. cruzi	*Ny. whitmani*
		Lu. longipalpis	*Pi. fischeri*
		Mi. oswaldoi	*Pa. bigeniculata*
		Mi. peresi	*Pa. brasiliensis*
		Mi. quinquefer	*Pa. abonnenci*
		Mi. schreiberi	*Ps. wellcomei*
		Mi. trinidadensis	*Sc. servulolimai*
		Mi. villelai	*Sc. sordellii*
	Maranhão	*Bi. flaviscutellata*	*Ny. intermedia*

(continued)

Table 1 (continued)

Region	State	Species	
		Bi. inornata	*Ny. richardwardi*
		Bi. olmeca nociva	*Ny. umbratilis*
		Br. avellari	*Ny. whitmani*
		Br. brumpti	*Pi. christenseni*
		Br. pintoi	*Pi. damascenoi*
		Br. travassosi	*Pi. nevesi*
		De. samueli	*Pi. odax*
		Ev. bacula	*Pi. orestes*
		Ev. bourrouli	*Pi. pacae*
		Ev. brachyphalla	*Pi. serrana*
		Ev. carmelinoi	*Pr. choti*
		Ev. cortelezzii	*Pr. triacantha*
		Ev. corumbaensis	*Pr. trispinosa*
		Ev. edwardsi	*Pa. abonnenci*
		Ev. evandroi	*Pa. aragaoi*
		Ev. infraspinosa	*Pa. bigeniculata*
		Ev. lenti	*Pa. brasiliensis*
		Ev. monstruosa	*Pa. dendrophyla*
		Ev. pinottii	*Pa. hermanlenti*
		Ev. sallesi	*Pa. lutziana*
		Ev. saulensis	*Pa. punctigeniculata*
		Ev. teratodes	*Pa. scaffi*
		Ev. termitophila	*Ps. amazonensis*
		Ev. walkeri	*Ps. arthuri*
		Ev. wilsoni	*Ps. ayrozai*
		Lu. carvalhoi	*Ps. carrerai*
		Lu. dispar	*Ps. chagasi*
		Lu. gomezi	*Ps. claustrei*
		Lu. longipalpis	*Ps. complexus*
		Lu. spathotrichia	*Ps. corossoniensis*
		Mt. oliveirai	*Ps. davisi*
		Mi. longipennis	*Ps. geniculatus*
		Mi. micropyga	*Ps. hirsutus*
		Mi. oswaldoi	*Ps. lloydi*
		Mi. peresi	*Ps. paraensis*
		Mi. pilosa	*Ps. squamiventris squamiventris*
		Mi. pusilla	*Ps. wellcomei*
		Mi. quinquefer	*Sc. fluviatilis*
		Mi. rorotaensis	*Sc. microps*
		Mi. trinidadensis	*Sc. servulolimai*
		Mi. villelai	*Sc. sordellii*

(continued)

Table 1 (continued)

Region	State	Species	
		Mg. bursiformis	*Th. ubiquitalis*
		Mg. migonei	*Th. viannamartinsi*
		Ny. anduzei	*Vi. furcata*
		Ny. antunesi	*Vi. tuberculata*
	Paraíba	*Ev. evandroi*	*Mi. schreiberi*
		Ev. lenti	*Mg. migonei*
		Ev. tupynambai	*Ny. intermedia*
		Ev. walkeri	*Ny. whitmani*
		Lu. longipalpis	*Pa. bigeniculata*
	Pernambuco	*Bi. flaviscutellata*	*Ev. walkeri*
		Br. nitzulescui	*Lu. longipalpis*
		Ev. cortelezzii	*Mi. capixaba*
		Ev. evandroi	*Mi. micropyga*
		Ev. lenti	*Mi. oswaldoi*
		Ev. sallesi	*Mi. peresi*
		Ev. tupynambai	*Mi. quinquefer*
NE	Pernambuco	*Mi. schreiberi*	*Ps. amazonensis*
		Mi. villelai	*Ps. ayrozai*
		Mg. migonei	*Ps. claustrei*
		Ny. intermedia	*Ps. complexus*
		Ny. umbratilis	*Ps. squamiventris squamiventris*
		Ny. whitmani	*Ps. wellcomei*
		Pi. fischeri	*Sc. fluviatilis*
		Pr. choti	*Sc. servulolimai*
		Pa. abonnenci	*Sc. sordellii*
		Pa. aragaoi	*Th. viannamartinsi*
		Pa. barrettoi barrettoi	*Ty. longispina*
		Pa. bigeniculata	*Vi. furcata*
		Pa. brasiliensis	*Vi. tuberculata*
		Pa. naftalekatzi	
	Piauí	*Br. avellari*	*Mt. oliveirai*
		De. samueli	*Mi. oswaldoi*
		Ed. piauienses	*Mi. peresi*
		Ev. carmelinoi	*Mi. quinquefer*
		Ev. evandroi	*Mi. rorotaensis*
		Ev. lenti	*Mi. villelai*
		Ev. saulensis	*Mg. bursiformis*
		Ev. termitophila	*Ny. intermedia*
		Ev. walkeri	*Ny. whitmani*
		Lu. dispar	*Pa. bigeniculata*

(continued)

Table 1 (continued)

Region	State	Species	
		Lu. longipalpis	*Sc. sordellii*
	Rio Grande do Norte	*De. samueli*	*Mi. quinquefer*
		Ev. evandroi	*Mi. trinidadensis*
		Ev. lenti	*Mi. villelai*
		Ev. sallesi	*Mg. migonei*
		Ev. walkeri	*Ny. intermedia*
		Lu. longipalpis	*Ny. whitmani*
		Mi. capixaba	*Ps. wellcomei*
		Mi. oswaldoi	*Sc. sordellii*
		Mi. peresi	
	Sergipe	*Ev. lenti*	*Ny. intermedia*
		Ev. sallesi	*Ny. whitmani*
		Lu. longipalpis	*Pr. choti*
		Mi. villelai	
CW	Distrito Federal	*Bi. flaviscutellata*	*Ny. intermedia*
		Br. avellari	*Ny. neivai*
		Br. pintoi	*Ny. whitmani*
		Ev. bacula	*Pi. christenseni*
		Ev. bourrouli	*Pi. fischeri*
		Ev. corumbaensis	*Pi. kuscheli*
		Ev. evandroi	*Pi. monticola*
		Ev. lenti	*Pa. aragaoi*
		Ev. sallesi	*Pa. bigeniculata*
		Ev. saulensis	*Pa. brasiliensis*
		Ev. teratodes	*Pa. lutziana*
		Ev. termitophila	*Pa. pradobarrientosi*
		Lu. ischnacantha	*Ps. davisi*
		Lu. longipalpis	*Sc. servulolimai*
		Mi. acanthopharynx	*Sc. sordellii*
		Mi. longipennis	
	Goiás	*Bi. flaviscutellata*	*Ev. walkeri*
		Br. avellari	*Ex. firmatoi*
		Br. pintoi	*Lu. cavernicola*
		Ev. bacula	*Lu. cruzi*
		Ev. bourrouli	*Lu. dispar*
		Ev. carmelinoi	*Lu. gomezi*
		Ev. cortelezzii	*Lu. longipalpis*
		Ev. corumbaensis	*Lu. renei*
		Ev. evandroi	*Mt. oliveirai*
		Ev. lenti	*Mi. acanthopharynx*
		Ev. monstruosa	*Mi. longipennis*

(continued)

Table 1 (continued)

Region	State	Species	
		Ev. sallesi	*Mi. micropyga*
		Ev. saulensis	*Mi. oswaldoi*
		Ev. teratodes	*Mi. peresi*
		Ev. termitophila	*Mi. quinquefer*
CW	Goiás	*Mi. trinidadensis*	*Pa. abonnenci*
		Mi. villelai	*Pa. aragaoi*
		Mg. bursiformis	*Pa. barrettoi barrettoi*
		Mg. migonei	*Pa. bigeniculata*
		Ny. intermedia	*Pa. brasiliensis*
		Ny. neivai	*Pa. campbelli*
		Ny. whitmani	*Pa. hermanlenti*
		Pi. christenseni	*Pa. lutziana*
		Pi. damascenoi	*Pa. punctigeniculata*
		Pi. fischeri	*Ps. davisi*
		Pi. misionensis	*Ps. lainsoni*
		Pi. monticola	*Ps. squamiventris squamiventris*
		Pi. pessoai	*Sc. sordellii*
	Mato Grosso	*Bi. flaviscutellata*	*Mi. trinidadensis*
		Bi. olmeca bicolor	*Mi. villelai*
		Bi. olmeca nociva	*Mg. bursiformis*
		Br. avellari	*Mg. migonei*
		Br. brumpti	*Ny. anduzei*
		Br. cunhai	*Ny. antunesi*
		Br. mangabeirai	*Ny. delsionatali*
		Br. nitzulescui	*Ny. fraihai*
		Br. orlandoi	*Ny. intermedia*
		Br. pentacantha	*Ny. richardwardi*
		Br. pintoi	*Ny. shawi*
		Ev. aldafalcaoae	*Ny. umbratilis*
		Ev. andersoni	*Ny. urbinattii*
		Ev. bacula	*Ny. whitmani*
		Ev. begonae	*Pi. christenseni*
		Ev. bourrouli	*Pi. damascenoi*
		Ev. carmelinoi	*Pi. fischeri*
		Ev. cortelezzii	*Pi. nevesi*
		Ev. corumbaensis	*Pi. serrana*
		Ev. dubitans	*Pr. choti*
		Ev. evandroi	*Pr. dysponeta*
		Ev. infraspinosa	*Pr. triacantha*
		Ev. lenti	*Pa. abonnenci*
		Ev. monstruosa	*Pa. abunaensis*

(continued)

Table 1 (continued)

Region	State	Species	
		Ev. pinottii	*Pa. aragaoi*
		Ev. sallesi	*Pa. barrettoi barrettoi*
		Ev. saulensis	*Pa. bigeniculata*
		Ev. sericea	*Pa. brasiliensis*
		Ev. teratodes	*Pa. campbelli*
		Ev. termitophila	*Pa. campograndensis*
		Ev. walkeri	*Pa. coutinhoi*
		Ev. wilsoni	*Pa. dasymera*
		Ex. cerradincola	*Pa. dendrophyla*
		Lu. cruciata	*Pa. dreisbachi*
		Lu. cruzi	*Pa. hermanlenti*
		Lu. dispar	*Pa. inflata*
		Lu. evangelistai	*Pa. lutziana*
		Lu. falcata	*Pa. pradobarrientosi*
		Lu. flabellata	*Pa. punctigeniculata*
		Lu. gomezi	*Pa. runoides*
		Lu. lichyi	*Pa. scaffi*
		Lu. longipalpis	*Ps. amazonensis*
		Lu. marinkellei	*Ps. ayrozai*
		Lu. sherlocki	*Ps. bispinosus*
		Lu. spathotrichia	*Ps. carrerai*
		Mt. brisolai	*Ps. chagasi*
		Mt. oliveirai	*Ps. claustrei*
		Mi. acanthopharynx	*Ps. complexus*
		Mi. echinatopharynx	*Ps. davisi*
		Mi. ferreirana	*Ps. geniculatus*
		Mi. longipennis	*Ps. guyanensis*
		Mi. micropyga	*Ps. hirsutus*
		Mi. oswaldoi	*Ps. lainsoni*
		Mi. peresi	*Ps. llanosmartinsi*
		Mi. pusilla	*Ps. nicaraguensis*
		Mi. quinquefer	*Ps. paraensis*
CW	Mato Grosso	*Ps. wellcomei*	*Th. ruii*
		Sc. microps	*Th. ubiquitalis*
		Sc. servulolimai	*Ty. dasypodogeton*
		Sc. sordellii	*Ty. longispina*
		Th. auraensis	*Ty. rondonensis*
		Th. clitella	*Vi. furcata*
		Th. howardi	*Vi. tuberculata*
		Th. octavioi	
	Mato Grosso do Sul	*Bi. flaviscutellata*	*Mi. pusilla*

(continued)

Table 1 (continued)

Region	State	Species	
		Br. avellari	*Mi. quinquefer*
		Br. brumpti	*Mi. villelai*
		Br. cunhai	*Mi. vonatzingeni*
		Br. galindoi	*Mg. bursiformis*
		Br. nitzulescui	*Mg. migonei*
		Br. pintoi	*Ny. antunesi*
		Ev. aldafalcaoae	*Ny. intermedia*
		Ev. bourrouli	*Ny. neivai*
		Ev. carmelinoi	*Ny. whitmani*
		Ev. cortelezzii	*Pi. christenseni*
		Ev. corumbaensis	*Pi. damascenoi*
		Ev. evandroi	*Pi. fischeri*
		Ev. lenti	*Pi. kuscheli*
		Ev. orcyi	*Pi. mamedei*
		Ev. sallesi	*Pi. misionensis*
		Ev. saulensis	*Pi. monticola*
		Ev. teratodes	*Pi. pessoai*
		Ev. termitophila	*Pa. aragaoi*
		Ev. walkeri	*Pa. bigeniculata*
		Ex. cerradincola	*Pa. brasiliensis*
		Lu. almerioi	*Pa. campbelli*
		Lu. cruzi	*Pa. campograndensis*
		Lu. dispar	*Pa. hermanlenti*
		Lu. forattinii	*Pa. lanei*
		Lu. longipalpis	*Pa. lutziana*
		Lu. renei	*Pa. punctigeniculata*
		Mt. oliveirai	*Ps. claustrei*
		Mi. acanthopharynx	*Ps. davisi*
		Mi. ferreirana	*Ps. lloydi*
		Mi. longipennis	*Sc. sordellii*
		Mi. oswaldoi	*Vi. furcata*
		Mi. peresi	
SE	Espírito Santo	*Bi. flaviscutellata*	*Mi. breviducta*
		Br. avellari	*Mi. capixaba*
		Br. bragai	*Mi. ferreirana*
		Br. cardosoi	*Mi. quinquefer*
		Br. cunhai	*Mi. schreiberi*
		Br. figueiredoi	*Mi. zikani*
		Br. guimaraesi	*Mg. migonei*
		Br. nitzulescui	*Ny. fraihai*
		Br. troglodytes	*Ny. intermedia*
		Ev. callipyga	*Ny. whitmani*

(continued)

Table 1 (continued)

Region	State	Species	
		Ev. cortelezzii	*Pi. bianchigalatiae*
		Ev. costalimai	*Pi. damascenoi*
		Ev. edwardsi	*Pi. fischeri*
		Ev. evandroi	*Pi. mamedei*
		Ev. grimaldii	*Pi. misionensis*
		Ev. lenti	*Pi. monticola*
		Ev. petropolitana	*Pi. pessoai*
		Ev. sallesi	*Pi. serrana*
		Ev. sericea	*Pr. choti*
		Ev. termitophila	*Pr. equatorialis*
		Ev. tupynambai	*Pa. barrettoi barrettoi*
		Ex. firmatoi	*Pa. bigeniculata*
		Lu. alencari	*Pa. lanei*
		Lu. amarali	*Pa. limai*
		Lu. falquetoi	*Pa. lutziana*
		Lu. longipalpis	*Pa. pascalei*
		Lu. souzalopesi	*Pa. pelloni*
		Mt. gasparviannai	*Ps. ayrozai*
SE	Espírito Santo	*Ps. davisi*	*Ps. matosi*
		Ps. fairchildi	*Sc. microps*
		Ps. geniculatus	*Sc. sordellii*
		Ps. hirsutus	*Ty. longispina*
	Minas Gerais	*Bi. flaviscutellata*	*Mi. ferreirana*
		Br. avellari	*Mi. longipennis*
		Br. brumpti	*Mi. micropyga*
		Br. cardosoi	*Mi. oswaldoi*
		Br. cunhai	*Mi. peresi*
		Br. guimaraesi	*Mi. quinquefer*
		Br. mangabeirai	*Mi. schreiberi*
		Br. nitzulescui	*Mi. trinidadensis*
		Br. ortizi	*Mi. villelai*
		Br. pintoi	*Mi. vonatzingeni*
		Br. travassosi	*Mg. migonei*
		Br. troglodytes	*Ny. intermedia*
		De. appendiculata	*Ny. neivai*
		De. ramirezi	*Ny. whitmani*
		Ev. bacula	*Pi. bianchigalatiae*
		Ev. callipyga	*Pi. christenseni*
		Ev. carmelinoi	*Pi. damascenoi*
		Ev. cortelezzii	*Pi. fischeri*
		Ev. corumbaensis	*Pi. mamedei*
		Ev. costalimai	*Pi. misionensis*

(continued)

Table 1 (continued)

Region	State	Species	
		Ev. edwardsi	*Pi. monticola*
		Ev. evandroi	*Pi. pessoai*
		Ev. lenti	*Pi. serrana*
		Ev. petropolitana	*Pr. choti*
		Ev. rupicola	*Pr. equatorialis*
		Ev. sallesi	*Pa. aragaoi*
		Ev. spelunca	*Pa. baratai*
		Ev. teratodes	*Pa. barrettoi barrettoi*
		Ev. termitophila	*Pa. bigeniculata*
		Ev. tupynambai	*Pa. brasiliensis*
		Ev. tylophalla	*Pa. dendrophyla*
		Ev. walkeri	*Pa. hermanlenti*
		Ex. cerradincola	*Pa. lanei*
		Ex. firmatoi	*Pa. limai*
		Lu. alencari	*Pa. lutziana*
		Lu. amarali	*Pa. pascalei*
		Lu. cavernicola	*Pa. pelloni*
		Lu. ischnacantha	*Pa. runoides*
		Lu. ischyracantha	*Ps. ayrozai*
		Lu. longipalpis	*Ps. carrerai*
		Lu. renei	*Ps. davisi*
		Mt. cipoensis	*Ps. fairchildi*
		Mt. gasparviannai	*Ps. hirsutus*
		Mt. minasensis	*Ps. lloydi*
		Mt. oliveirai	*Ps. matosi*
		Mi. acanthopharynx	*Sc. microps*
		Mi. breviducta	*Sc. sordellii*
		Mi. capixaba	*Ty. longispina*
	Rio de Janeiro	*Bi. flaviscutellata*	*Ev. tupynambai*
		Br. avellari	*Ev. walkeri*
		Br. brumpti	*Ex. firmatoi*
		Br. cardosoi	*Lu. alencari*
		Br. cunhai	*Lu. amarali*
		Br. guimaraesi	*Lu. ischyracantha*
		Br. nitzulescui	*Lu. longipalpis*
		Br. pintoi	*Mt. gasparviannai*
		Br. troglodytes	*Mt. minasensis*
		Ev. cortelezzii	*Mi. capixaba*
		Ev. costalimai	*Mi. ferreirana*
		Ev. edwardsi	*Mi. longipennis*
		Ev. lenti	*Mi. micropyga*
		Ev. petropolitana	*Mi. oswaldoi*

(continued)

Table 1 (continued)

Region	State	Species	
		Ev. rupicola	*Mi. peresi*
		Ev. sallesi	*Mi. quinquefer*
		Ev. termitophila	*Mi. schreiberi*
SE	Rio de Janeiro	*Mg. migonei*	*Pa. pascalei*
		Ny. intermedia	*Pa. pelloni*
		Ny. whitmani	*Ps. arthuri*
		Pi. bianchigalatiae	*Ps. ayrozai*
		Pi. fischeri	*Ps. carrerai*
		Pi. mamedei	*Ps. davisi*
		Pi. misionensis	*Ps. fairchildi*
		Pi. monticola	*Ps. geniculatus*
		Pi. pessoai	*Ps. hirsutus*
		Pi. serrana	*Ps. lloydi*
		Pa. aragaoi	*Ps. matosi*
		Pa. barrettoi barrettoi	*Ps. paraensis*
		Pa. bigeniculata	*Sc. microps*
		Pa. brasiliensis	*Sc. sordellii*
		Pa. lanei	*Ty. longispina*
		Pa. lutziana	
	São Paulo	*Bi. flaviscutellata*	*Mg. migonei*
		Br. avellari	*Mg. rabelloi*
		Br. bragai	*Mg. vaniae*
		Br. brumpti	*Ny. fraihai*
		Br. cardosoi	*Ny. intermedia*
		Br. carvalheiroi	*Ny. neivai*
		Br. cunhai	*Ny. singularis*
		Br. galindoi	*Ny. whitmani*
		Br. guimaraesi	*Pi. bianchigalatiae*
		Br. mangabeirai	*Pi. christenseni*
		Br. nitzulescui	*Pi. fischeri*
		Br. ortizi	*Pi. misionensis*
		Br. pintoi	*Pi. monticola*
		Br. troglodytes	*Pi. pessoai*
		Ev. bourrouli	*Pi. serrana*
		Ev. carmelinoi	*Pr. choti*
		Ev. correalimai	*Pr. trispinosa*
		Ev. cortelezzii	*Pa. aragaoi*
		Ev. edwardsi	*Pa. baratai*
		Ev. lenti	*Pa. barrettoi barrettoi*
		Ev. petropolitana	*Pa. bigeniculata*
		Ev. rupicola	*Pa. brasiliensis*
		Ev. sallesi	*Pa. hermanlenti*

(continued)

Table 1 (continued)

Region	State	Species	
		Ev. termitophila	*Pa. lanei*
		Ex. firmatoi	*Pa. limai*
		Lu. alencari	*Pa. lutziana*
		Lu. almerioi	*Pa. pascalei*
		Lu. amarali	*Pa. pelloni*
		Lu. castroi	*Pa. punctigeniculata*
		Lu. dispar	*Pa. ribeirensis*
		Lu. longipalpis	*Ps. arthuri*
		Mt. alphabetica	*Ps. ayrozai*
		Mi. ferreirana	*Ps. geniculatus*
		Mi. longipennis	*Ps. guyanensis*
		Mi. micropyga	*Ps. hirsutus*
		Mi. petari	*Ps. lloydi*
		Mi. quinquefer	*Sc. microps*
		Mi. schreiberi	*Sc. sordellii*
		Mg. bursiformis	
S	Paraná	*Br. angelae*	*Ev. edwardsi*
		Br. avellari	*Ev. evandroi*
		Br. brumpti	*Ev. lenti*
		Br. cardosoi	*Ev. sallesi*
		Br. Cunhai	*Ex. firmatoi*
		Br. galindoi	*Lu. amarali*
		Br. guimaraesi	*Lu. gaminarai*
		Br. mangabeirai	*Lu. longipalpis*
		Br. nitzulescui	*Mt. alphabetica*
		Br. ortizi	*Mi. ferreirana*
		Br. troglodytes	*Mi. longipennis*
		Ev. bacula	*Mi. micropyga*
		Ev. correalimai	*Mi. oswaldoi*
		Ev. cortelezzii	*Mi. quinquefer*
S	Paraná	*Mg. migonei*	*Pa. baratai*
		Ny. intermedia	*Pa. bigeniculata*
		Ny. neivai	*Pa. lanei*
		Ny. umbratilis	*Pa. limai*
		Ny. whitmani	*Pa. lutziana*
		Pi. bianchigalatiae	*Pa. pascalei*
		Pi. christenseni	*Pa. pelloni*
		Pi. fischeri	*Pa. punctigeniculata*
		Pi. misionensis	*Ps. ayrozai*
		Pi. monticola	*Ps. geniculatus*
		Pi. pessoai	*Ps. hirsutus*
		Pa. abonnenci	*Ps. lloydi*

(continued)

Table 1 (continued)

Region	State	Species	
		Pa. aragaoi	*Sc. sordellii*
	Rio Grande do Sul	*Br. cunhai*	*Mg. migonei*
		Br. nitzulescui	*Ny. intermedia*
		Br. pintoi	*Ny. neivai*
		Ev. correalimai	*Ny. whitmani*
		Ev. cortelezzii	*Pi. fischeri*
		Ev. gaucha	*Pi. misionensis*
		Ex. firmatoi	*Pi. monticola*
		Lu. gaminarai	*Pi. pessoai*
		Lu. longipalpis	*Pa. bigeniculata*
		Mt. alphabetica	*Pa. lanei*
		Mi. ferreirana	*Pa. pascalei*
		Mi. schreiberi	
	Santa Catarina	*Br. brumpti*	*Pi. fischeri*
		Br. nitzulescui	*Pi. monticola*
		Br. troglodytes	*Pi. pessoai*
		Ev. edwardsi	*Pa. bigeniculata*
		Ex. firmatoi	*Pa. lanei*
		Mt. alphabetica	*Pa. pelloni*
		Mg. migonei	*Ps. ayrozai*
		Ny. intermedia	*Sc. microps*
		Ny. neivai	

Genus *Bichromomyia* Artemiev, 1991; *Brumptomyia* França & Parrot, 1921; *Deanemyia* Galati, 1995; *Edentomyia* Galati, Andrade Filho, Silva & Falcão, 2003; *Evandromyia* Mangabeira, 1941; *Expapillata* Galati, 1995; *Lutzomyia* França, 1924; *Martinsmyia* Galati, 1995; *Micropygomyia* Barretto, 1962; *Migonemyia* Galati, 1995; *Nyssomyia* Barretto, 1962; *Pintomyia* Costa Lima, 1932; *Pressatia* Mangabeira, 1942; *Psathyromyia* Barretto, 1962; *Psychodopygus* Mangabeira, 1941; *Sciopemyia* Barretto, 1962; *Trichophoromyia* Barretto, 1962; *Trichopygomyia* Barretto, 1962; *Viannamyia* Mangabeira, 1941

Bi. – Bichromomyia, Br. – Brumptomyia, De. – Deanemyia, Ed. – Edentomyia, Ev. – Evandromyia, Ex. – Expapillata, Lu. – Lutzomyia, Mt. – Martinsmyia, Mi. – Micropygomyia, Mg. – Migonemyia, Ny. – Nyssomyia, Pi. – Pintomyia, Pr. – Pressatia, Pa. – Psathyromyia, Ps. – Psychodopygus, Sc. – Sciopemyia, Th. – Trichophoromyia, Ty. – Trichopygomyia, Vi. – Viannamyia

(Fairchild and Hertig 1959) and Rondônia, which is on the border of Brazil with Bolívia (Martins et al. 1965), and in the CW and SE regions. The reason for discontinuity may be associated with the lack of more systematized investigations in the intermediate regions.

Food preferences of sand flies are also predominant factors that directly influence their dispersion. Species of the genus *Brumptomyia*, in their totality, suck the blood of armadillos (Dasypodidae) and are always found in armidillo burrows and only incidentally outside of them. Thus, it can be said that where armidillos do not occur, there are no representatives of the genus.

In short, the more specialized the species, the smaller its range of dispersion and the greater its dependence on habitats that are equally specialized. In contrast, the more generalized the species, the greater its range, particularly by way of its adaptation to less specialized or more diversified habitats.

Habitats of the Brazilian Phlebotomine Species

Sand flies belong to the order Diptera (Psychodidae, Phlebotominae), and, like most Diptera, they are holometabolous insects. The immature forms inhabit terrestrial habitat, developing in places rich in decomposing organic matter, especially that of a vegetal nature. They have cryptozoic habits and are sensitive to changes in temperature and humidity, with a predominantly twilight activity and/or nocturnal period, when they leave their shelters to feed themselves. Males and females feed on carbohydrates, especially nectars and aphid secretions, but only females are hematophagous, which is the precondition for the maturation of their ovaries, with some species being suitable for the spread of pathogens (Brazil and Brazil 2003).

Sand flies occur essentially in wild environments. However, some vector species are adapting to deforested areas and places of human habitation as well as occupying urban environments and peripheries of large cities. The adaptation of these vectors to new habitats, especially those environmentally impacted, indicates a new scenario of transmission of leishmaniasis (Rangel and Vilela 2008; Carvalho et al. 2014).

The types of shelters used by adult sand flies vary according to microhabitat, season, relative humidity, and species. Due to their thin coating, they shelter in places where they can protect themselves from sudden changes in the environment. Thus, they take refuge in places with good moisture content, decomposing organic matter, little or no light, and air movement, that is, in places that protect them from desiccation. Proper refuge spots in particular are fallen leaves in the forest soil, burrows of wild animals, trunks of trees and tabular roots, hollows of trees, crevices in rocks, and caves, etc. In anthropic environments, artificial shelters include shaded and damp areas and shelters for domestic animals such as chicken coops, pigpens, corrals, and others (Aguiar and Vilela 1987; Brazil and Brazil 2003).

To determine the microclimate factors important for sand-fly shelters, Scorza et al. (1968a, 1968b, 1968c) studied the habits of several species of sand flies in the ecotypes of the rainforest zone of Venezuela. The investigators concluded that, during the dry season, the temperature in the shelters is greater than that of the external environment, and it increases during the rainy season. The relative humidity in the shelters, both in the rainy and dry seasons, is always greater than that of the external environment. Another important factor is that, although temperature changes are gradual, the increase in relative humidity is abrupt. Thus, although the temperature remains stable throughout the year, the relative humidity undergoes major changes, which would lead—according to the investigators—to the appearance of sand flies. The relative humidity of the air is therefore the determining factor for the maintenance of these insects in their shelters. The survival of the immature forms in these biotopes during the dry season causes the emergence of adults during the first rains because of the increased humidity. Finally, temperature influences the survival of both the immature and adult forms with the temperature being stable in the shelters, remaining increased during the dry season, and decreasing during the rainy season.

Depending more on the microclimate from places—such as burrows, tree hollows, caves and cracks in rocks—than on the general climatic conditions of the region, certain sand-fly species can overcome efficient barriers to other groups of animals and thereby bind fauna of different zoogeographic regions (Lewis 1965). For example, *Br. pintoi* occurs, like all species of the genus *Brumptomyia*, in armadillo (Dasypodidae) burrows in zones of permanently flooded forest and low altitudes, *e.g.*, the states of Pará and Amazonas. It also occurs in savannas with prolonged droughts and at altitudes of 1000 m, *e.g.*, the state of Minas Gerais.

Most sand-fly species inhabit the Hylian Amazon forests, which are divided into three types: "terra firme," "igapó," and "várzea." These forests occupy about 40% of the Brazilian territorial area representing an area of 3.5 million km^2. In addition to forests of various types, the Hylian Amazon forests are formed of fields and savannas.

In the state of Amapá, on the island of Marajó, and on other islands throughout the Amazonian plain with no forest area, dense herbaceous vegetation occurs, especially tall grasses able to withstand periodic floods.

Savannas in the Amazon forest occur as "islands" of savanoid vegetation distributed in four lines: (1) the first follows the coast and the savannas of Marajó and is considered the most important; (2) the second line is along the lower Amazonas river in the state of Pará; (3) the third is located between the upper courses of the Jari and Trombetas rivers and has the same physiognomy of savannas of Central Brazil but differing in floristic composition; and (4) the fourth line is located in the fields and savannas in the south of the states of Pará and Amazonas, the mid-Tapajós river, and between the Purus and Madeira rivers.

North Eastern "caatinga" vegetation extends throughout the states of Maranhão, Piauí, Ceará, Rio Grande do Norte, Paraíba, Pernambuco, Sergipe, Alagoas, Bahia, and certain parts of northern Minas Gerais and occupies around 11% of the national territory.

The fields of the Brazilian Central Plateau—especially the savannas of Minas Gerais, Tocantins, Goiás, and Mato Grosso, which occupy "islands of vegetation" in the Amazon and "tabuleiro" vegetation in the NE region—cover an area of approximately 23% of the Brazilian territory.

The Atlantic Forests extend along the Brazilian coast, from north to south, in a band that goes inland depending on the location, especially the relief. Great anthropic action decreased the forest to only 6% across its range from Rio Grande do Norte to Rio Grande do Sul. In the South region, fields and woods of various types, including Araucaria of temperate character, predominate. This vegetation covers 14% of the national territory.

The remaining 6% of the Brazilian vegetation includes the palm tree forests of Maranhão, Piauí, and Ceará; the Mato Grosso Pantanal wetland; the mangroves, dunes, and "restingas" of almost all the coast; the rupestrian fields from some the sierras of Minas Gerais and Goiás; the continental aquatic vegetation (fresh water); and the vegetation of the marshes.

The rainforest offers the widest variety of shelters and is therefore the site of the largest number of sand flies. Shelters that become excessively moist are abandoned. However, they can be re-occupied when conditions become favourable again.

The forest soil is home to many species of sand flies including *Psychodopygus ayrozai*, *Bichromomyia flaviscutellata*, *Bi. olmeca bicolor*, and *Bi. olmeca nociva*. These and other species are also found in shrubs and plants, especially when the forest floor becomes too moist.

A great number of sand flies is wild and only infest humans and domestic animals when they penetrate the forest or when houses are near. Some species, however, live in association with humans and their domestic animals; however, contrary to the behaviour of some Culicidae, sand flies cannot yet be considered fully domiciled. Two factors influence the domiciliation process of these insects: the habitat modifications by humans and the adaptive power of the species to the new environment (Aguiar et al. 2014; Vieira et al. 2015).

Anthropic environmental changes have led to substantial alterations in the profile of sand-fly fauna, thus allowing greater adaptation of these insects to the human environment. This has caused some species to disappear, whereas others adapt to the new circumstances as represented by residual forests or cultivated areas, proximity to human homes, in rural areas, or on the outskirts of urban areas with or without rural characteristics (Rangel and Lainson 2009; Aguiar et al. 2014). The use of insecticides, in addition to the proportion of the species in relation to dominance, is also an altering factor in the biodiversity of sand-fly fauna and the size of the population (Teodoro et al. 1999).

Barretto (1943) divided sand-fly species into three categories: (1) wild: living in forests or in non-forest regions but only accidentally found associated with humans and domestic animals; (2) semi-domestic: living outside human and domestic animal habitations and only seeking these to obtain blood repast; and (3) domestic: living in association with humans and domestic animals inside or near dwellings.

All Brazilian sand flies are listed in alphabetical order in Table 2, which also highlights their main habitats and regions of occurrence. Species marked with an asterisk are involved in the transmission of leishmaniasis to humans and animals.

1.	Fallen leaves in forest soil	2.	Armadillo burrows	N -	North
3.	Burrows of other wild animals	4.	Tree trunks and tabular roots	NE-	Northeast
5.	Tree hollows	6.	Treetops	CW-	Central West
7.	Crevices in rocks	8.	Caves	SE-	Southeast
9.	Forest without specified location	10.	Marginal areas	S -	South
11.	Annexes of domestic animals (chicken coops, pigpens, corrals etc.)	12.	Outer and inner walls of human dwellings		

Table 2 Brazilian phlebotomine species, main habitats and regions of occurrence

	Habitats												
Species	1	2	3	4	5	6	7	8	9	10	11	12	Regions
Bi. flaviscutellata (Mangabeira 1942a)*	•	•		•							•	•	N, NE, CW, SE
Bi. inornata (Martins, Falcão & Silva, 1965)									•		•		N, NE
Bi. olmeca bicolor (Fairchild & Theodor, 1971)*	•			•	•								N, CW
Bi. olmeca nociva (Young & Arias, 1982)*	•												N, NE, CW
Bi. reducta (Feliciangeli, Ramirez Pérez & Ramirez, 1988)									•				N
Br. angelae Galati, Santos & Silva, 2007									•				S
Br. avellari (Costa Lima, 1932)	•	•					•	•		•	•		N, NE, CW, SE, S
Br. bragai (Mangabeira & Sherlock, 1961)		•											NE, SE
Br. brumpti (Larrousse, 1920)		•		•			•	•		•	•		N, NE, CW, SE, S
Br. cardosoi (Barretto & Coutinho, 1941a)		•											SE, S
Br. carvalheiroi Shimabukuro, Marassa & Galati 2007									•				SE
Br. cunhai (Mangabeira, 1942a)	•	•								•			N, NE, CW, SE, S
Br. figueiredoi Mangabeira & Sherlock, 1961		•											NE, SE
Br. galindoi (Fairchild & Hertig, 1947)		•		•			•	•					CW, SE, S
Br. guimaraesi (Coutinho & Barretto, 1941a)	•	•		•						•			SE, S
Br. mangabeirai (Barretto & Coutinho, 1941a)		•								•			N, CW, SE, S
Br. nitzulescui (Costa Lima, 1932)	•	•											NE, CW, SE, S
Br. orlandoi Fraiha, Shaw & Lainson, 1970		•											CW
Br. ortizi Martins, Silva & Falcão, 1971b		•											N, SE, S
Br. pentacantha (Barretto, 1947)		•		•									N, CW
Br. pintoi (Costa Lima, 1932)	•	•	•										N, NE, CW, SE, S
Br. spinosipes Floch & Abonnenc, 1943									•				N
Br. travassosi (Mangabeira 1942g)		•	•										N, NE, SE
Br. troglodytes (Lutz, 1922)		•					•	•					SE, S
Br. virgensi Mangabeira & Sherlock, 1961		•											NE, CW
De. appendiculata (Martins, Falcão & Silva, 1961)							•	•					SE

(continued)

Table 2 (continued)

Species	Habitats												Regions
	1	2	3	4	5	6	7	8	9	10	11	12	
De. derelicta (Freitas & Barrett, 1999)									•	•			N
De. maruaga (Alves, Freitas & Barret, 2008)								•					N
De. ramirezi (Martins, Falcão, Silva & Miranda, 1982)							•						SE
De. samueli (Deane, 1955)											•	•	N, NE
Edentomyia piauienses Galati, Andrade-Filho, Silva & Falcão, 2003								•					N, NE
Ev. aldafalcaoae (Santos, Andrade Filho & Honer, 2001)									•				CW
Ev. andersoni (Le Point & Desjeux, 1988)									•				N, CW
Ev. apurinan Shimabukuro, Figueira & Silva, 2013											•		N
Ev. bacula (Martins, Falcão & Silva, 1965)		•	•						•	•			N, NE. CW, SE, S
Ev. bahiensis (Mangabeira & Sherlock, 1961)	•	•	•								•		NE
Ev. begonae (Ortiz & Torres, 1975)				•	•						•		N, CW
Ev. bourrouli (Barretto & Coutinho, 1941b)		•								•			N, NE, CW, SE
Ev. brachyphalla (Mangabeira Fo, 1994)									•				N, NE
Ev. callipyga (Martins & Silva , 1965)		•	•										SE
Ev. carmelinoi (Ryan, Fraiha, Lainson & Shaw, 1986)		•	•						•				N, NE, CW, SE
Ev. correalimai (Martins, Coutinho & Luz , 1970)			•	•									SE, S
Ev. cortelezzii (Brèthes, 1923)				•		•				•	•	•	N, NE, CW, SE, S
Ev. corumbaensis (Galati, Nunes, Oshiro & Rego, 1989)							•	•			•	•	N, NE, CW, SE
Ev. costalimai (Mangabeira, 1942a)		•	•	•			•						NE, SE
Ev. dubitans (Sherlock, 1962)				•	•					•		•	N, CW
Ev. edwardsi (Mangabeira, 1941b)			•							•	•		NE, SE, S
Ev. evandroi (Costa Lima & Antunes, 1936)									•				N, NE, CW, SE, S
Ev. gaucha Andrade Filho, Souza & Falcão, 2007									•				S
Ev. georgii (Freitas & Barrett, 2002)									•				N
Ev. grimaldii Andrade Filho, Pinto, Santos & Carvalho, 2009									•				SE

(continued)

Table 2 (continued)

Species	Habitats												Regions
	1	2	3	4	5	6	7	8	9	10	11	12	
Ev. infraspinosa (Mangabeira, 1941b)		•	•	•							•		N, NE, CW, SE
Ev. inpai (Young & Arias, 1977)									•				N
Ev. lenti (Mangabeira, 1938)			•							•	•	•	N, NE, CW, SE, S
Ev. monstruosa (Floch & Abonnenc, 1944)		•	•	•							•		N, NE, CW
Ev. orcyi Oliveira, Sanguinette, Almeida & Andrade Filho, 2015											•		CW
Ev. petropolitana (Martins & Silva, 1968)				•	•		•						SE
Ev. pinottii (Damasceno & Arouck, 1956)				•	•								N, NE, CW
Ev. rupicola (Martins, Godoy & Silva, 1962)							•						SE
Ev. sallesi (Galvão & Coutinho, 1939)				•	•		•	•			•	•	N, NE, CW, SE, S
Ev. saulensis (Floch & Abonnenc, 1944)				•	•								N, NE, CW
Ev. sericea (Floch & Abonnenc, 1944)									•				N, NE, CW, SE
Ev. sipani (Fernandez, Carbajal, Alexander & Need, 1994)									•	•			N
Ev. spelunca Carvalho, Brazil, Sanguinette & Andrade Filho, 2011								•					SE
Ev. tarapacaensis (Le Pont, Torres-Espejo & Galati, 1996)									•				N
Ev. teratodes (Martins, Falcão & Silva, 1964)			•										N, NE, CW, SE
Ev. termitophila (Martins, Falcão & Silva, 1964)		•	•				•				•		N, NE, CW, SE
Ev. tupynambai (Mangabeira, 1942a)		•	•	•						•			NE, SE
Ev. tylophalla Andrade & Galati, 2012									•				SE
Ev. walkeri (Newstead, 1914)				•	•					•			N, NE, CW, SE
Ev. williamsi (Damasceno, Causey & Arouck, 1945)				•									N
Ev. wilsoni (Damasceno & Causey, 1945)		•		•									N, NE, CW
Ex. cerradincola (Galati, Nunes, Oshiro & Dorval, 1995)									•				N, CW, SE
Ex. firmatoi (Barretto, Martins & Pellegrino, 1956)				•						•	•		SE, S
Lu. alencari Martins, Souza & Falcão, 1962							•						SE
Lu. almerioi Galati & Nunes, 1999				•			•						CW, SE
Lu. amarali Barretto & Coutinho, 1940				•			•						SE, S

(continued)

Table 2 (continued)

Species	Habitats												Regions
	1	2	3	4	5	6	7	8	9	10	11	12	
Lu. araracuarensis (Morales & Minter, 1981)									•				N
Lu. caligata Martins, Falcão & Silva, 1965				•									N
Lu. carvalhoi (Damasceno, Causey & Arouck, 1945)				•									N, NE
Lu. castroi (Barretto & Coutinho, 1941c)									•				SE
Lu. cavernicola (Costa Lima, 1932)							•	•					N, CW, SE
Lu. cruciata (Coquillett, 1907)										•	•		CW
Lu. cruzi (Mangabeira, 1938)*							•	•			•	•	NE, CW, SE
Lu. cultellata Barrett, Feitas, Albuquerque & Guerrero, 1996									•				N
Lu. dispar Martins & Silva , 1963							•	•					NE, CW, SE
Lu. elizabethrangelae Vilela, Azevedo e Godoy, 2015								•					N
Lu. evangelistai Martins & Fraiha, 1971				•									N, CW
Lu. falcata Young, Morales & Ferro, 1994				•	•								N, CW
Lu. falquetoi Pinto & Santos, 2007									•				SE
Lu. flabellata Martins & Silva, 1964				•									N, CW
Lu. forattinii Galati, Rego, Nunes & Teruya, 1985							•	•			•	•	CW
Lu. gaminarai (Cordero, Vogelsang & Cossio, 1928)							•						S
Lu. gomezi (Nitzulescu, 1931)*			•	•	•	•					•	•	N, NE, CW
Lu. ischnacantha Martins, Souza & Falcão, 1962a							•	•					CW, SE
Lu. ischyracantha Martins, Falcão & Silva, 1962b							•						SE
Lu. lichyi (Floch & Abonnenc, 1950)				•	•					•			N, CW
Lu. longipalpis (Lutz & Neiva, 1912)*				•	•		•	•		•	•	•	N, NE, CW, SE, S
Lu. marinkellei Young, 1979				•									N, CW
Lu. renei (Martins, Falcão & Silva, 1957)							•	•					N, CW, SE
Lu. sherlocki Martins, Silva & Falcão, 1971a				•									N, CW
Lu. souzalopesi Martins, Silva & Falcão, 1970b				•									SE
Lu. spathotrichia Martins, Falcão & Silva, 1963				•					•				N, NE, CW
Mt. alphabetica (Fonseca, 1936)									•			•	SE, S
Mt. brisolai (Le Pont & Desjeux, 1987)									•				CW

(continued)

Table 2 (continued)

Species	Habitats 1	2	3	4	5	6	7	8	9	10	11	12	Regions
Mt. Cipoensis (Martins, Falcão & Silva, 1964)							•						SE
Mt. Gasparviannai (Martins, Godoy & Silva ,1962b)									•				NE, SE
Mt. Minasensis (Mangabeira, 1942a)									•				N, SE
Mt. Oliveirai (Martins, Silva & Falcão, 1970a)							•	•				•	N, NE, CW, SE
Mt. Reginae Carvalho, Brazil, Sanguinette & Andrade Filho, 2010								•					N
Mt. Waltoni (Arias, Freitas & Barret, 1984)				•									N
Mi. acanthopharynx (Martins, Falcão & Silva, 1962)				•			•	•					N, CW, SE
Mi. breviducta (Barretto, 1950)									•				SE
Mi. capixaba (Dias, Falcão, Silva & Martins, 1987)									•	•			NE, SE
Mi. cayennensis (Floch & Abonnenc, 1941)			•	•	•								N
Mi. chassigneti (Floch & Abonnenc, 1944)									•				N
Mi. echinatopharynx Andrade Filho, Galati, Andrade & Falcão, 2004									•				N, CW
Mi. ferreirana (Barretto, Martins & Pellegrino, 1956)		•	•	•			•				•		CW, SE, S
Mi. longipennis (Barretto, 1946)				•							•		N, NE, CW, SE, S
Mi. mangabeirana (Martins, Falcão & Silva, 1963)	•	•		•									N
Mi. micropyga (Mangabeira, 1942a)		•	•	•	•						•		N, NE, CW, SE, S
Mi. oswaldoi (Mangabeira, 1942e)				•	•						•	•	N, NE, CW, SE, S
Mi. peresi (Mangabeira, 1942a)		•	•	•			•	•		•	•	•	N, NE, CW, SE
Mi. petari Galati, Marassá & Andrade, 2003									•				SE
Mi. pilosa (Damasceno & Causey, 1944)			•	•	•								N, NE
Mi. pusilla (Dias, Martins, Falcão & Silva, 1986)				•									N, NE, CW
Mi. quinquefer (Dyar, 1929)				•	•		•			•	•		N, NE, CW, SE, S
Mi. rorotaensis (Floch & Abonnenc, 1944)				•									N, NE
Mi. schreiberi (Martins, Falcão & Silva ,1955)				•						•	•	•	NE, SE, S
Mi. trinidadensis (Newstead, 1922)				•	•						•	•	N, NE, CW, SE

(continued)

Table 2 (continued)

Species	Habitats												Regions
	1	2	3	4	5	6	7	8	9	10	11	12	
Mi. villelai (Mangabeira, 1942)				•			•	•	•		•		N, NE, CW, SE
Mi. vonatzingeni Galati, 2007									•				N, CW, SE
Mi. zikani (Barretto, 1950a)									•				N, SE
Mg. bursiformis (Floch & Abonnenc, 1944)		•		•					•				N, NE, CW, SE
Mg. cerqueirai (Causey & Damasceno, 1945)		•	•	•						•			N, NE
Mg. migonei (França, 1920)*				•	•		•			•	•	•	N, NE, CW, SE, S
Mg. moucheti (Pajot & Le Pont, 1923)									•				N
Mg. rabelloi (Galati & Gomes, 1992)									•	•			SE
Mg.vaniae Galati, Fonseca & Marassá, 2007									•	•	•	•	SE
Nyssomyia anduzei (Rozeboom, 1942)*				•						•	•		N, NE, CW
Ny. antunesi (Coutinho, 1939)*		•	•	•		•				•	•		N, NE, CW
Ny. delsionatali Galati & Galvis, 2012										•			CW
Ny. fraihai (Martins, Falcão & Silva, 1979)				•		•				•	•		N, NE, CW, SE
Ny. intermedia (Lutz & Neiva, 1912)*									•	•	•	•	N, NE, CW, SE, S
Ny. neivai (Pinto, 1926)*								•	•	•	•	•	N, CW, SE, S
Ny. pajoti (Abonnenc, Léger & Fauran, 1979)									•				N
Ny. richardwardi (Ready & Fraiha, 1981)				•		•							N, NE, CW
Ny. shawi (Fraiha, Ward & Ready, 1973)									•				N, CW
Ny. singularis (Costa Lima, 1932)									•				SE
Ny. sylvicola (Floch & Abonnenc, 1944)									•				N
Ny. umbratilis (Ward & Fraiha, 1977)*				•		•							N, NE, CW, S
Ny. urbinattii Galati & Galvis, 2012									•	•			N, CW
Ny. whitmani (Antunes & Coutinho, 1939)*				•	•	•				•	•	•	N, NE, CW, SE, S
Pi. bianchigalatiae (Andrade Filho, Aguiar, Dias & Falcão, 1999)									•		•	•	SE, S
Pi. christenseni (Young & Duncan ,1994)				•	•								N, NE, CW, SE, S
Pi. damascenoi (Mangabeira, 1941d)			•	•					•				N, NE, CW, SE

(continued)

Table 2 (continued)

Species	Habitats 1	2	3	4	5	6	7	8	9	10	11	12	Regions
Pi. fischeri (Pinto, 1926)*				•	•	•				•	•	•	NE, CW, SE, S
Pi. gruta (Ryan, 1986)							•	•					N
Pi. kuscheli (Le Pont, Martinez, Torrez-Espejo, Dujardin, 1998)									•				CW
Pi. mamedei (Oliveira, Afonso, Dias & Brazil, 1994)									•				CW, SE
Pi. misionensis (Castro, 1959)				•		•							N, NE, CW, SE, S
Pi. monticola (Costa Lima, 1932)				•	•	•				•	•		N, NE, CW, SE, S
Pi. naiffi (Freitas & Oliveira, 2013)										•			N
Pi. nevesi (Damasceno & Arouck, 1956)				•							•		N, NE, CW
Pi. odax (Fairchild & Hertig, 1961)				•									N, NE
Pi. orestes (Fairchild & Trapido, 1950)									•				NE
Pi. pacae (Floch & Abonnenc, 1943)			•										N, NE
Pi. pessoai (Coutinho & Barretto, 1940)*				•	•	•				•	•	•	NE, CW, SE, S
Pi. serrana (Damasceno & Arouck, 1949)		•	•	•	•					•	•		N, NE, CW, SE
Pressatia calcarata (Martins & Silva, 1964)				•									N
Pr. choti (Floch & Abonnenc, 1941)		•	•	•	•						•		N, NE, CW, SE
Pr. duncanae (Le Pont, Martinez, Torrez-Espejo & Dujardin, 1998)									•				N
Pr. dysponeta (Fairchild & Hertig, 1952)									•				CW
Pr. equatorialis (Mangabeira, 1942a)		•	•	•							•		N, SE
Pr. triacantha (Mangabeira, 1942c)		•	•	•									N, NE, CW
Pr. trispinosa (Mangabeira, 1942a)		•	•	•									N, NE, SE
Pa. abonnenci (Floch & Chassignet, 1947)				•									N, NE, CW, S
Pa. abunaensis (Martins, Falcão & Silva, 1965)	•	•											N
Pa. aragaoi (Costa Lima, 1932)		•		•	•								N, NE, CW, SE, S
Pa. baratai Sábio, Andrade & Galati, 2016									•				SE
Pa. barretti Alves & Freitas, 2015				•									N
Pa. barrettoi barrettoi (Mangabeira, 1942a)		•	•										N, NE, CW, SE

(continued)

Table 2 (continued)

Species	Habitats												Regions
	1	2	3	4	5	6	7	8	9	10	11	12	
Pa. bigeniculata (Floch & Abonnenc, 1941)				•	•	•			•	•	•	•	N, NE, CW, SE, S
Pa. brasiliensis (Costa Lima, 1932)		•		•	•								N, NE, CW, SE
Pa. campbelli (Damasceno, Causey & Arouck, 1945)				•									N, CW
Pa. campograndensis (Oliveira, Andrade Filho, Falcão & Brazil, 2001)									•				N, CW
Pa. coutinhoi (Mangabeira, 1942a)		•	•	•									N, CW
Pa. dasymera (Fairchild & Hertig, 1961)				•	•								N, CW
Pa. dendrophyla (Mangabeira, 1942a)			•	•	•						•		N, NE, CW, SE
Pa. digitata (Damasceno & Arouck, 1950)					•								NE
Pa. dreisbachi (Causey & Damasceno, 1945)		•		•	•					•			N, CW
Pa. elizabethdorvalae Brilhante, Sábio & Galati, 2017									•	•			N
Pa. hermanlenti (Martins, Silva & Falcão, 1970)		•	•	•									N, NE, CW, SE
Pa. inflata (Floch & Abonnenc, 1944)		•	•										N, CW
Pa. lanei (Barretto & Coutinho, 1941)					•	•				•	•		NE, CW, SE, S
Pa. lutziana (Costa Lima, 1932)	•	•	•	•									N, NE, CW, SE, S
Pa. naftalekatzi (Falcão, Andrade Filho, Almeida, Brandão-Filho, 2000)											•		NE
Pa. pascalei (Coutinho & Barretto, 1941c)		•	•										N, NE, SE, S
Pa. pelloni (Sherlock & Alencar, 1959)									•		•		NE, SE, S
Pa. limai (Fonseca, 1935)				•		•							SE, S
Pa. pifanoi (Ortiz, 1972)									•				N
Pa. pradobarrientosi (Le Pont, Matias, Martinez & Dujardin, 2004)									•				N, CW
Pa. punctigeniculata (Floch & Abonnenc, 1944)				•						•	•		N, NE, CW, SE, S
Pa. ribeirensis Sabio, Andrade & Galati, 2014									•				SE
Pa. runoides (Fairchild & Hertig, 1953)		•	•	•									N, CW, SE
Pa. scaffi (Damasceno & Arouck, 1956)				•									N, NE, CW
Pa. souzacastroi (Damasceno & Causey, 1944)				•	•								N

(continued)

Table 2 (continued)

Species	Habitats												Regions
	1	2	3	4	5	6	7	8	9	10	11	12	
Psychodopygus amazonensis (Root, 1934)									•	•		•	N, NE, CW
Ps. arthuri (Fonseca, 1936)									•		•	•	NE, SE
Ps. ayrozai (Barretto & Coutinho, 1940)*	•			•						•			N, NE, CW, SE, S
Ps. bernalei (Osorno-Mesa, Morales & Osorno, 1967)									•				N
Ps. bispinosus (Fairchild & Hertig, 1951)									•				N, NE, CW
Ps. carrerai (Barretto, 1946)	•									•			N, NE, CW, SE
Ps. chagasi (Costa Lima, 1941)									•		•		N, NE, CW
Ps. claustrei (Abonnenc, Léger & Fauran, 1979)									•	•			N, NE, CW
Ps. complexus (Mangabeira, 1941c)*									•		•		N, NE, CW
Ps. corossoniensis (Le Pont & Pajot, 1978)									•				N, NE
Ps. davisi (Root, 1934)	•	•	•	•						•		•	N, NE, CW, SE
Ps. douradoi (Fé, Freitas & Barrett, 1998)									•				N
Ps. fairchildi (Barretto, 1966)							•						SE
Ps. geniculatus (Mangabeira, 1941c)		•							•	•			N, NE, CW, SE, S
Ps. guyanensis (Floch & Abonnenc, 1941)									•		•	•	N, CW, SE
Ps. hirsutus (Mangabeira, 1942a)			•	•		•				•			N, NE, CW, SE, S
Ps. lainsoni (Fraiha & Ward, 1964)									•				N, CW
Ps. leonidasdeanei (Fraiha, Ryan, Ward, Lainson & Shaw, 1987)									•				N
Ps. llanosmartinsi (Fraiha & Ward, 1980)									•				N, CW
Ps. lloydi (Antunes, 1937)									•	•	•	•	NE, SE, CW, S
Ps. matosi (Barretto & Zago, 1956)				•	•					•			NE, SE
Ps. nicaraguensis (Fairchild & Hertig, 1961)									•				CW
Ps. panamensis (Shannon, 1926)									•				N
Ps. paraensis (Costa Lima, 1941)*		•									•		N, NE, CW, SE
Ps. squamiventris maripaensis (Floch & Abonnenc, 1946)									•				N

(continued)

Table 2 (continued)

Species	Habitats												Regions
	1	2	3	4	5	6	7	8	9	10	11	12	
Ps. squamiventris squamiventris (Lutz & Neiva, 1912)*									•		•	•	N, NE, CW
Ps. wellcomei (Fraiha, Shaw & Lainson, 1971)*									•				N, NE, CW
Ps. yucumensis (Le Pont, Caillard, Tibayrenc & Desjeux, 1986)									•				N
Sc. fluviatilis (Floch & Abonnenc, 1944)									•			•	N, NE
Sc. microps (Mangabeira, 1942a)	•		•	•			•						N, NE, CW, SE, S
Sc. nematoducta (Young & Arias, 1984)									•				N
Sc. pennyi (Arias & Freitas, 1981)									•				N
Sc. preclara (Young & Arias, 1984)									•				N
Sc. servulolimai (Damasceno & Causey, 1945)			•	•									N, NE, CW
Sc. sordellii (Shannon & Del Ponte, 1927)			•	•	•		•	•			•	•	N, NE, CW, SE, S
Th. adelsonsouzai Santos, Silva, Barata, Andrade & Galati, 2013									•				N
Th. auraensis (Mangabeira, 1942a)		•	•	•							•		N, CW
Th. brachipyga (Mangabeira, 1942a)		•	•	•							•	•	N
Th. castanheirai (Damasceno, Causey & Arouck, 1945)		•		•		•					•		N
Th. clitella (Young & Perez, 1994)										•			N, CW
Th. dunhami (Causey & Damasceno, 1945)				•									N
Th. eurypyga (Martins, Falcão & Silva, 1963)				•									N
Th. flochi (Abonnenc & Chassignet, 1948)									•				N
Th. gibba (Young & Arias, 1994)									•				N
Th. howardi (Young, 1979)									•				CW
Th. ininii (Floch & Abonnenc, 1943)									•				N
Th. lopesi (Damasceno, Causey & Arouck, 1945)									•	•			N
Th. loretoensis (Llanos, 1964)									•				N
Th. meirai (Causey & Damasceno, 1945)		•		•									N
Th. melloi (Causey & Damasceno, 1945)		•	•										N
Th. octavioi (Vargas, 1949)		•	•										N, CW
Th. readyi (Ryan, 1986)			•										N
Th. reinerti (Young & Duncan, 1994)									•				N
Th. rostrans (Summers, 1912)									•				N
Th. ruifreitasi Oliveira, Teles, Medeiros, Camargo & Pessoa, 2015									•				N
Th. ruii (Arias & Young, 1982)			•										N, CW

(continued)

Table 2 (continued)

Species	Habitats												Regions
	1	2	3	4	5	6	7	8	9	10	11	12	
Th. ubiquitalis (Mangabeira, 1942a)*		•	•	•	•								N, NE, CW, SE
Th. uniniensis Ladeia-Andrade, Fé, Sanguinette & Andrade Filho, 2014									•				N
Th. viannamartinsi (Sherlock & Guitton, 1970)		•		•			•						NE
Ty. conviti (Ramírez, Pérez, Martins & Ramírez, 1976)									•				N
Ty. dasypodogeton (Castro, 1939)		•	•	•									N, CW
Ty. elegans (Martins, Llanos & Silva, 1976)									•	•			N
Ty. longispina (Mangabeira, 1942d)		•	•	•	•								N, NE, CW, SE
Ty. pinna (Feliciangeli, Ramirez Pérez & Ramirez, 1989)									•				N
Ty. ratcliffei (Arias, Ready & Freitas, 1983)									•				N
Ty. rondonensis (Martins, Falcão & Silva, 1965)									•				N, CW
Ty. trichopyga (Floch & Abonnenc, 1945)			•										N, NE
Ty. wagleyi (Causey & Damasceno, 1945)	•	•	•										N
Vi. caprina (Osorno-Mesa, Morales e Osorno, 1972)								•					N
Vi. fariasi (Damasceno, Causey & Arouck, 1945)		•											N
Vi. furcata (Mangabeira, 1941d)	•	•	•	•						•			N, NE, CW
Vi. tuberculata (Mangabeira, 1941d)*		•	•	•	•						•	•	N, NE, CW

*Species found to be vectors of *Leishmania* to human and animals

Brazilian sand flies can be found mainly in 12 different habitats. Following guidance from Barretto (1943), 9 of them (1–9) can be considered wild habitats: (1) fallen leaves in forest soil, (2) armadillo burrows, (3) burrows of other wild animals, (4) tree trunks and tabular roots, (5) tree hollows, (6) treetops, (7) crevices in rocks, (8) caves, and (9) forest without specified location. Three of them (10–12) are semi-domestic and domestic habitats with human influence: (10) marginal areas, (11) annexes of domestic animals, and (12) outer and inner walls of human dwellings. From 279 recorded species, 178 (63.8%) essentially have wild habitats, and contact with humans is restricted to when humans enter the natural environment to perform activities related to agriculture, forestry, mining, and highway and hydroelectric construction, among other professionals who are directly in contact with the forest. The remaining 101 species (36.2%) are distributed among marginal areas (10) represented by cultivated or pastoral areas and residual forests; the annexes of domestic animals (11) such as chicken coops, pigpens, corrals, kennels, etc.; and the outer and inner walls of human dwellings (12). These can be the annexes of domestic animals (pigpens, chicken coops, corrals, etc.) in near-habitat areas (cultivated areas) or even in the residual woodlands close to houses. Of the 22 species found to be vectors of *Leishmania* to human and animals, 17 are present in the human environment, *Bi. flaviscutellata, Lu. cruzi, Lu. gomezi, Lu. longipalpis, Mg. migonei, Ny. anduzei, Ny. antunesi, Ny. intermedia, Ny. neivai, Ny. whitmani, Pi. fischeri, Pi. pessoai, Ps. ayrozai, Ps. complexus, Ps. paraensis, Ps. squamiventris*, and *Vi. tuberculata*. Only 5—*Bi. olmeca bicolor, Bi. olmeca nociva, Ny. umbratilis, Ps. wellcomei*, and *Th. ubiquitalis*—are, to date, essentially wild. Five regions have wide distribution of *Lu. longipalpis*, *Mg. migonei, Ny. intermedia*, *Ny. whitmani*, and *Ps. ayrozai*. In four regions (N, NE, CW, SE), *Bi. flaviscutellata, Ps. paraensis*, and *Th. ubiquitalis* are reported to occur without records in the South region; *Ny. umbratilis* (N, NE, CW, S) without records in the SE region; *Ny. neivai* (N, CW, SE, S) without records in the NE region; and *Pi. fischeri* and *Pi. pessoai* in the North region (NE, CW, SE, S). In 3e regions, *Bi. olmeca nociva, Lu. gomezi, Ny. anduzei, Ny. antunesi, Ps. complexus, Ps. squamiventris squamiventris, Ps. wellcomei*, and *Vi. tuberculata*, (N, NE, CW) and *Lu. cruzi* (NE, CW, SE) occur. In two regions, *Bi. olmeca bicolor* (N, CW) and are known to occur (Table 2).

Leishmaniasis (cutaneous and visceral) is among the diseases that have a major impact on human health. They are considered emerging endemic diseases with a clear territorial expansion. They are included in the list of the Compulsory Notification Diseases System of the Ministry of Health with records in all Brazilian regions. Brazil is the country with the highest prevalence of these diseases in the Americas in both visceral and cutaneous forms (Brasil 2014, 2017; WHO 2010).

The classic profile of leishmaniasis is defined as wild animal zoonoses, which affect humans when they come into contact with the primary cycle in areas of preserved forests, such as in the Amazon, "cerrado" (savannas), and remnants of the Atlantic Forest. They have gradually expanded to rural and deforested areas and to the outskirts of cities. In some Brazilian states, there is a visible process of urbanization (Lainson 1983, 1988; Rangel 1995; Aguiar et al. 2014; Brasil 2017).

Some cycles of American cutaneous leishmaniasis (ACL) transmission deserve to be highlighted, such as the *Leishmania* (*Viannia*) *braziliensis* cycle, which is more widely dispersed in Brazil and is responsible for cutaneous and mucosal forms. Its main vectors are *Mg. migonei, Ny. intermedia, Ny. neivai, Ny. whitmani* and *Ps. wellcomei*. It is worth noting that a severe form of ACL, anergic diffuse leishmaniasis, caused by *Leishmania* (*Leishmania*) *amazonensis* and conveyed by *Bi. flaviscutellata*, was associated for some time with the wetlands of the Amazônia Legal, especially in the states of Pará and Maranhão. Currently, human cases have been registered in the SE, CW, and NE regions, a factor of evident geographical expansion process (Aguiar and Medeiros 2003; Rangel and Lainson 2009). A clinical condition of ACL, restricted to the Amazon, is represented by multiple lesions caused by *Leishmania* (*V.*) *guyanensis* and conveyed by *Ny. umbratilis* (Lainson & Shaw 2005; Rangel & Lainson 2009).

Currently, according to the control and safety strategies of the Ministry of Health, there are three epidemiological patterns of ACL. The first is wild, and transmission occurs in primary vegetation aeas where the disease is characterized as a wild animal zoonosis. The second is wild/occupational and leisure (affected areas) and is associated with disorderly exploitation of the forest and the clearing of forests for different purposes with the approach of humans to the wild cycles. Finally, there is the rural, peri-urban, and areas of ancient colonization related to migratory processes, occupying hillsides and agglomerates in urban centres, which are always linked to secondary or residual forests (Brasil 2017).

American visceral leishmaniasis (AVL) was associated with the rural environment and the outskirts of large cities. However, this epidemiological profile has been changing in some Brazilian regions where the parasitosis is already urbanized. According to the Ministry of Health's AVL Surveillance and Control Manual (2014), two epidemiological standards are defined: (1) the classical pattern associated with the rural environment, outskirts of cities, and low socioeconomic level; and (2) the recent pattern found in the urban environment in medium and large cities.

The most important vector of the etiologic agent of AVL in the Americas is *Lu. longipalpis*, which has been shown to be highly capable of occupying new environments (Brasil 2014).

Thus, leishmaniasis has a close relationship between micro-ecological conditions and links of the epidemiological chain with environmental, natural, or human-induced changes—especially the approximation of humans to wild zoonotic cycles—that directly influence its incidence and expansion. In several Brazilian regions, transmission occurs in areas that are practically deforested. In the peri-urban environment, there are changes in the epidemiological scenario, in which the sand flies, endowed with food eclecticism, would be able to transmit the parasite to humans and other synanthropic and domestic mammals (Lainson and Rangel 2005; WHO 2010). Migratory flow should be considered as an impacting component in the epidemiology of leishmaniasis. Such scenarios—particularly in the NE, CW, SE, and South regions of the country—have been studied (Rangel and Lainson 2009).

The most adapted species to environments that suffered intense anthropic action and that rarely occur in wild environments are *Ny. intermedia* and *Mg. migonei*. The first is the species with a greater capacity for domiciliation and greater adaptation to environments modified by humans. Its presence is always associated with banana crop. In the SE and South regions, its presence in forest areas is practically non-existent (Aguiar et al. 2014).

References

Aguiar GM, Medeiros WM (2003) Distribuição Regional e Habitats das Espécies de Flebotomíneos do Brasil. In: Rangel EF, Lainson R (eds) Flebotomíneos do Brasil. Fiocruz, Rio de Janeiro, pp 207–255

Aguiar GM, Vilela ML (1987) Aspects of the ecology of the sandflies at the Serra dos Órgãos National Park, State of Rio de Janeiro. VI. Shelters and breeding places (Diptera, Psychodidae, Phlebotominae). Mem Inst Oswaldo Cruz 82(4):585–586

Aguiar GM, Azevedo ACR, Medeiros WM, Alves JRC, Rendeiro V (2014) Aspects of the ecology of Phlebotomines (Diptera: Psychodidae: Phlebotominae) in an area of cutaneous Leishmaniasis occurrence. Municipality of Angra dos Reis, Coast of Rio de Janeiro State, Brazil. Rev Inst Med Trop S Paulo 56(2):143–149

Barretto MP, Pessoa SB (1946) Contribuição para o estudo da distribuição geográfica dos Flebótomos americanos. Livro em homenagem RF d'Almeida 7:77–102

Brasil. Ministério da Saúde. Secretaria de Vigilância em Saúde. Departamento de Vigilância Epidemiológica (2014) *Manual de vigilância e controle da leishmaniose visceral*. Ministério da Saúde, Secretaria de Vigilância em Saúde, Departamento de Vigilância Epidemiológica. 1ed 5reimpr. Brasília: Ministério da Saúde 120 p

Brasil. Ministério da Saúde. Secretaria de Vigilância em Saúde. Departamento de Vigilância das Doenças Transmissíveis (2017) *Manual de Vigilância da Leishmaniose Tegumentar*. Ministério da Saúde, Secretaria de Vigilância em Saúde, Departamento de Vigilância das Doenças Transmissíveis. Brasília: Editora do Ministério da Saúde 189 p

Brazil RP, Brazil BG (2003) Biologia de flebotomíneos neotropicais. In: Rangel EF, Lainson R (eds) Flebotomíneos do Brasil. Fiocruz, Rio de Janeiro, pp 257–274

Carvalho BM, Dias CMG, Rangel EF (2014) Phlebotomine sand flies (Diptera, Psychodidae) from Rio de Janeiro State, Brazil: species distribution and potential vectors of leishmaniases. Rev Bras Entomol 58(1):77–87

Fairchild GB, Hertig M (1959) Geographic distribution of the Phlebotomus sandflies of central America (Diptera: Psychodidae). Ann Entomol Soc Am 52:121–124

Galati EAB (2003) Classificação de Phlebotominae e morfologia, terminologia de adultos e identificação dos táxons da América. In: Rangel EF, Lainson R (eds) Flebotomíneos do Brasil. Fiocruz, Rio de Janeiro, pp 23–175

Lainson R (1983) The American leishmaniasis: some observation on their ecology and epidemiology. Trans R Soc Trop Med Hyg 77:569–596

Lainson R (1988) Ecological interaction in the transmission of the leishmaniasis. Phil Trans R Soc London Serie B 321:389–404

Lainson R, Rangel EF (2005) *Lutzomyia longipalpis* and the eco-epidemiology of American visceral leishmaniasis, with particular reference to Brazil: a review. Mem Inst Oswaldo Cruz 100(8):811–827

Lainson R, Shaw JJ (2005) New world Leishmaniasis. In: Collier L, Balows A, Sussman M (eds) Topley & Wilson's microbiology and microbial infections, vol 5, 10th edn. Parasitology. Hodder Arnold, London

Lewis DJ (1965) Internal structural features of some Central American phlebotominae sandflies. Ann Trop Med Parasitol 59:357–385

Martins AV, Morales-Farias EN (1972) Sobre a distribuição geográfica dos flebotomíneos americanos (Diptera, Psychodidae, Phlebotominae). Rev Brasil Biol 32(3):361–371

Martins AV, Falcão AL, Silva JE (1965) Notas sobre os flebótomos do Território de Rondônia, com a descrição de seis espécies novas (Diptera, Psychodidae). Rev Brasil Biol 25(1):1–20

Martins AV, Williams P, Falcão AL (1978) American sand flies. Academia Brasileira de Ciências. Rio de Janeiro – RJ. 195 pp

Rangel EF (1995) Transmission of American cutaneous leishmaniasis in peridomestic foci in Rio de Janeiro State and other similar situations compared to the classical epidemiology in Amazon region. In: Proceedings of the Tropical Disease, Society and the Environment 2:103–110

Rangel EF, Lainson R (2009) Proven and putative vectors of American cutaneous leishmaniasis in Brazil: aspects of their biology and vectorial competence. Mem Inst Oswaldo Cruz 104:937–954

Rangel EF, Vilela ML (2008) Lutzomyia longipalpis (Diptera, Psychodidae, Phlebotominae) and urbanization of visceral leishmaniasis in Brazil. Cad Saúde Pública 24(12):2948–2952

Sangiorgi B, Miranda DN, Oliveira DF, Santos EP, Gomes FR, Santos EO, Barral A & Miranda JC (2012) Natural breeding places for Phlebotomine sand flies (Diptera: Psychodidae) in a semiarid region of Bahia State, Brazil. J Trop Med article ID 124068, 5 pages

Scorza JV, Ortiz I, Gomez I (1968a) Observaciones biológicas sobre algunos flebótomos de "Rancho Grande" (Venezuela). Sobre los factores microclimáticos que determinan la endemicidad de la flebotomofauna de "Rancho Grande". Acta Biol Venez 6:76–83
Scorza JV, Ortiz I, Gomez I, MacLure MT, Ramirez M (1968b) Observaciones biológicas sobre algunos flebótomos de "Rancho Grande" (Venezuela). 2. Microhabitats de Phlebotomus spp (Diptera: Psychodidae). Acta Biol Venez 6:1–27
Scorza JV, Ortiz I, Gomez I (1968c) Observaciones biológicas sobre algunos flebótomos de "Rancho Grande" (Venezuela). 8. Sobre la fluctuación estacional de los microhabitats. Acta Biol Venez 6:97–104
Shimabukuro PHF, Andrade AJ, Galati EAB (2017) Checklist of American sand flies (Diptera, Psychodidae, Phlebotominae): genera, species, and their distribution. ZooKeys 660:67–106
Silva AM, Camargo NJ, Santos DR, Massafera R, Ferreira AC, Postai C, Cristovão EC, Konolsaisen JF, Bisetto A Jr, Perinazo R, Teodoro U, Galati EAB (2008) Diversidade, Distribuição e Abundância de Flebotomíneos (Diptera: Psychodidae) no Paraná. Neotrop Entomol 37(2):209–225
Teodoro U, Kühl JB, Santos DR, Santos ES (1999) Impacto de alterações ambientais na ecologia de flebotomíneos no sul do Brasil. Cad Saúde Pública 15(4):901–906
Vieira VR, Azevedo ACR, Alves JRC, Guimarães AE, Aguiar GM (2015) Ecological aspects of Phlebotomine sand flies (Diptera, Psychodidae, Phlebotominae) in areas of American cutaneous leishmaniasis, in the municipality of Paraty, Rio de Janeiro, Brazil. I-index of abundance by location and type of capture. J Med Entomol 52(5):886–895
WHO, World Health Organization Control of the leishmaniasis: report of a meeting of the WHO Expert Committee on the Control of Leishmaniases. Geneva 22–26 March 2010, report series: 949
Young DG, Duncan MA (1994) Guide to the identification and geographic distribution of *Lutzomyia* sandflies in Mexico, the West Indies Central and South America (Diptera, Psychodidae). Mem Amer Ent Inst 54:1–881

Bionomy: Biology of Neotropical Phlebotomine Sand Flies

Reginaldo P. Brazil and Beatriz G. Brazil

Introduction

Phlebotomine sand flies are included in the order Diptera, a small insect measuring 2 to 3 mm with intense pilosity over its body. Like most Diptera, phlebotomine sand flies are holometabolous (*i.e.*, they present in their life cycle an egg phase; a larval phase that includes four sub-stages; a pupal phase; and, finally, the adult stage). They distinguish themselves, however, from the others of this family by featuring a thinner body and longer and thin legs. In addition, their females need blood to produce eggs, and for this reason they were aggregated by taxonomists in the subfamily Phlebotominae.

What stands out in importance in the neotropical region is that several genera are widely distributed, and many known species participate in the transmission of *Leishmania* species to humans and sylvatic and domestic animals. In agreement with Galati (2016), the number of phlebotomine sand flies has been increasing quickly with the description, each year, of several new species.

Regarding the biology, nevertheless, there is still an important gap in the knowledge of these insects because only some few species have been studied, which makes mistakes likely regarding involving those species whose biology remains unknown. With these considerations taken into account, this chapter provides the reader with an overall view of the subject by presenting some particularities without, however, going into great review.

R. P. Brazil (✉) · B. G. Brazil
Instituto Oswaldo Cruz, Fundação Oswaldo Cruz, Rio de Janeiro, Brazil

E. F. Rangel, J. J. Shaw (eds.), *Brazilian Sand Flies*,
https://doi.org/10.1007/978-3-319-75544-1_5

Immature Forms

Eggs

The eggs of phlebotomine sand flies are oval or ellipsoid in shape and measure, depending on the species, from 0.3 to 0.5 mm in length/0.07 to 0.15 mm in width. Right after they have been laid, the eggs feature a whitened or yellowed color, but within hours they become dark brown. The exochorion presents elevations and depressions, which gives it a characteristic appearance in different species; this does not, however, enable taxonomic importance to be inferred according to present knowledge. The eggs of more than 40 sand-fly species have been described based on the exochorion's structure.

On average, a phlebotomine female in the laboratory lays 40 eggs, although there is great variation according to species. Egg-laying occurs as isolated or small groups of eggs, and the eggs remain adherent to the substrate due to the substance produced by the accessory glands. This substance, rich in fatty acids that coat the eggs, is responsible for their waterproof nature and can also act as oviposition pheromone.

Larvae

The larval forms are small, white, and worm-like. Right after emergence, the larvae feed on the eggshells, bodies of dead adults, and other types of organic matter available.

The larvae are made up of 12 segments; in addition to the head, there are 2 thoracic and 9 abdominal segments. The head is well developed—darker in color than the rest of the body—and covered with small prickles that vary according to the species. The cephalic appendages are the antennae, which vary with the larval stage and the species, and the mouth parts are of the grinder type.

The thorax, made up of three segments, features in the integument a series of bristles, the number and characteristics of which vary according to the stage and the species, thus serving as an element that can be used for specific identification.

The abdomen features nine similar segments, except for the two last ones, which are modified and adapted to the motion function. Each segment's integument presents bristles that vary specifically. The ventral face of the first seven abdominal segments holds false legs that act as locomotive organs. On the dorsal and lateral face of the segments more chitinized areas are noted, which correspond to the tergal plates. The abdominal segments are covered with bristles, the position of which is repeated in all, but they vary according to species and stages. In the eighth segment stands a pair of posterior spiracles, opened on the edge of a small conical protuberance, whereas the ninth segment looks more modified than the eighth: In addition to its importance for locomotion, it is responsible for larva fixation at the change to pupal stage. In this segment tail bristles are noted, the length of which relates to the species' habitat. In Phlebotominae, first-stage larvae feature one pair of tail bristles, and the following stages feature two pairs.

Pupae

The pupa is whitened or yellowed and progressively darkens as adult-emergence approaches. It is made up of 13 segments: The first 4 of them, which are more or less fused, form the cephalothorax; the 9 others, which are well individualized, constitute the abdomen. The abdomen segments are quite uniform, except for the eighth and ninth, which are modified and covered on the exuvia by the fourth larval stage. The pupa does not move: It is fixed to the substrate by means of the larval exuvia and performs just flexion and extension movements of the body. Removal of the larval exuvia allows observation of the last pupal segments, which appear distended and lighter than the former ones. The pupal integument is covered with several types of bristle, and its chaetotaxy is important for specific differentiation. The eyes of the adults can be seen through the pupal wall. Initially they are almost invisible, but they darken as the emergence moment approaches.

Adults

The phlebotomine adults present sexual dimorphism expressed not only in body-shape differences but also in alimentary behavior, which is reflected in the exclusive female's hematophagy.

Males and females morphologically differ in their proboscises, which are shorter in males than in the females, which have long proboscises that are adapted for piercing and sucking. This dimorphism is also expressed in the female by its head, which features inside—in the very ventral region—a set of chitinized structures called "cibarium," which are also associated with hematophagy, although their function is still unknown.

The sexes are also distinguishable by the last abdominal segments, which are modified to constitute the insect's genitalia: In the male stands a set of appendages that are well-developed and ornamented (structure with taxonomic value), whereas in the female they are smaller and more discreet, lying as telescoped structures, which gives a round aspect to the insect's genitalia.

The adults, right after emerging from the pupa, remain little active. The males, which inside the pupa remain with their genitalia appendages folded on the body, need 12 to 24 h so a 180° rotation can occur, after which, it is assumed, they will be ready for the copulation. Other indications of sexual maturation in males are still unknown.

Feeding

The phlebotomine adults, females as much as males, need carbohydrates as energy source for flying and other biological activities. In nature, the major carbohydrate sources are usually obtained from vegetal sap and sugary secretions of aphids or other homopterous. Although the ability of phlebotomine sand flies to use sugars has been known for >70 years, the use of vegetal sap by these insects was suggested in 1963 by Minter and Wijer, who observed some species apparently sucking leaves in Kenya. Such remarks were proved afterward when vegetal fluids were in the diverticulum of wild-caught phlebotomine sand flies (Lewis and Domoney 1966; Willians 1970).

Several natural carbohydrates can take part in the nourishment of phlebotomine sand flies (Table 1). The most common sugars found in different neotropical species are fructose, glucose and saccharose (Lewis and Domoney 1966; Morton et al. 1991). In addition, turanose and melesitose—natural sugars found in the secretions of aphids—have been found in the digestive tract of these insects. Although ingestion of these secretions by phlebotomine sand flies has not been noted, the finding of these two sugars inside the digestive tract of field-caught species is a strong indicator that they use the sweet secretions of aphids because no other natural source of these carbohydrates is known. Moreover, the presence of *Pintomyia verrucarum* was associated with the presence of two species of aphids—*Therioaphis trifoli* and *Acyrthosiphon pisum*—in alfalfa plantations in Peru (Cameron et al. 1995). It is possible that phlebotomine sand flies are directly attracted by the plants contaminated by the secretions of aphids or, indirectly, by the action of pheromones released by these insects, which would act as kairomones, thus attracting the insects toward the location of their secretions. Reinforcing the second hypothesis, Tesh et al. (1992) demonstrated, in the laboratory, that the alarm pheromone of aphids—(E)-b-farnesene—acts as a phagostimulant for *Lutzomyia longipalpis.*

When phlebotomine sand flies ingest a sugary substance, it is stored in the diverticulum, where it is taken slowly from the digestive tract and then absorbed. Experiments with *L. longipalpis* in the laboratory have shown that it takes females

Table 1 Sugars found in wild-caught neotropical phlebotomine sand flies

Species	Sugars[a]	Method(s)	Authors
P.ovalesi, Ny. ylephiletor, Ps. shannoni	Glu/fru/sac	Crom[b]	Lewis & Domoney, 1966
Lu.longipalpis	Fru/sac	Antrona[c]	Magnarelli & Modi, 1988
Lu. longipalpis	Fru/sac	Antronac	Morton et al., 1991
Lu. peruensis	Fru/sac/tur/mel/raf	HPLC[d]	Wallbanks et al., 1991
Ny. intermedia	Fru/Glu/sac/mal/Tre/Mel	HPLC[d]	Souza et al., 1992

[a]Glucose/fructose/saccharose/turanose/melezitose/raffinose/maltose/trehalose
[b]Paper chromatography
[c]Van Handel (1972)
[d]*HPLC* high-performance liquid chromatography

as much as to males, on average, 5 days to digest a saccharose solution at 50% storage in the diverticulum (Brazil, RP, 1997). It is likely that the sugar's passage from the diverticulum to the digestive tract occurs only at the instant when the insect's activities or when the physiological moment demands consumption.

It is also known that sugars play an important role in the development and infectivity of *Leishmania* spp. not only in controlling the intestinal flora and acting as bacteriostatics but also in serving as energy source for parasites, which seem to multiply easier in the digestive tract of phlebotomine sand flies in the presence of sugars.

Blood Feeding

Although only phlebotomine females are hematophagous, hematophagy has been exceptionally noted in *L. longipalpis* males by Coelho (1966), and by Gontijo et al. (1987) in *L. renei* males, both under laboratory conditions. For blood-feeder insects, the saliva plays an essential role in constraining the homeostasis mechanisms of the host, thus inducing vasodilation and inhibiting the blood-clotting actions that facilitate the digestion of food. Therefore, it becomes necessary that the insect, at the moment of feeding, presents developed salivary glands and secretion enough for digestion. It is also important that the mouth parts used to pierce and cut the host's tissues possess the hardness necessary for this function, what occurs just some time after the insect's emergence.

In phlebotomine sand flies, complete digestion might occur 24 h after emergence. However, observations in the laboratory with several species of Phlebotominae indicated that most females begin this activity 48 h after emergence. Such a preference might be associated with salivary-gland maturation and mouth-part hardening.

In *L. longipalpis* females kept under experimental conditions, salivary glands began the process of salivary secretion near the end of the first 24 h after emergence, and they became large, globular, and replete with saliva rich in protein material by the end of 48 h (Gomes Brazil 2000). Such remarks are consistent with the studies of Volf et al. (2000), who detected increased protein content of the salivary glands under the time and temperature conditions the species were kept. This maturation period of the salivary glands certainly influences the beginning and the efficiency of the blood repast. In males, the glands appear minuscule and sometimes cannot be noted in dissections.

Phlebotomine sand flies feature mouth parts that are relatively short (0.2 to 0.4 mm), thus preventing deep penetration of host skin. Laceration of the skin and blood capillaries results in subcutaneous hemorrhage from where the blood is sucked. The time period of suction for complete feeding in species of neotropical Phlebotominae that feed on warm-blooded animals varies considerably, lasting from 1 to 5 min, whereas the process is much slower in species feeding on cold-blooded animals (Chaniotis 1967).

Lewis (1965), studying several species of American phlebotomine sand flies, found well-developed salivary glands that reached the abdomen in species that suck on cold-blooded animals, whereas in species that suck on warm-blooded animals the salivary glands appeared like small sacs fixed to the anterior part of the thorax.

The saliva of phlebotomine sand flies is an important tool in blood feeding. Adler and Theodor (1926) were the first to note, the presence of anti-clotting substances in the saliva of *Phlebotomus papatasi*.

Today great importance is given to the study of salivary constitution in these insects (Ribeiro et al. 1989). The *L. longipalpis* species, from which a peptide has been isolated, identified and denominated maxadilan in allusion to its powerful vasodilating effect, has been the most studied. The presence of anti-clotting and vasodilating substances in the saliva of phlebotomine sand flies enables the insect to neutralize the vertebrate haemostasis, thus facilitating the blood repast, because they allow the formation of small puddles of blood that protrude from the tiny vessels lacerated by the insect's proboscis, which is not long enough to "straw" the vessels like mosquitoes and kissing bugs do (Ribeiro et al. 1989).

The amount of blood ingested varies individually according to the species, but overall it is equal to the insect's own weight, which can range from 0.1 to 0.6 mg (Chaniotis 1967; R. P. Brazil, 1997).

Observations under experimental conditions with *L. longipalpis* have shown that right after feeding, one of the salivary glands is often found to be diminished and to have a thickened wall, whereas the other still appears voluminous with its wall distended, suggesting that the amount of saliva necessary for one repast is found in a sole gland with the other gland serving as a "spare" (Gomes Brazil 2000).

The saliva of phlebotomine sand flies has been highlighted as a factor of importance in establishing infections by *Leishmania* under experimental conditions. The study of this important secretion of phlebotomine sand flies has been intense, both with different species of the parasite and different vertebrate hosts, intending not only to understand the role of vector species in the transmission and maintenance of the disease but also to enable the development of a vaccine that may help control these diseases, the incidence of which increases each year not only in the number of cases but also its regional distribution in regions worldwide.

As in other hematophagous Diptera, in phlebotomine sand flies the blood is necessary to ovary development, with the number of eggs produced being directly proportional to the amount of blood ingested (Ready 1979; Lehane 1991). Despite that being an essential condition for producing eggs, the phenomenon of autogeny (i.e., the production of eggs without previous feeding on blood) has been observed in *Ph. papatasi, L. gomesi*, and *L. cruciata* and has been suggested in *Psychodopygus davisi* (= *L. davisi)* and *Pi. damascenoi* (Forattini, 1973). However, this phenomenon and its meaning remain little studied.

It is known, at least, that two parthegenetic species of Phlebotomine sand fly—one from Rio de Janeiro and *P. mamedei* from the progeny of a laboratory-housed female—did not feed on blood and still produced fertile eggs in the absence of males. The produced eggs originated just females, thus characterizing this sand fly as an autogenic and parthenogenetic species (Oliveira et al. 1994; Brazil and

Oliveira 1999). More recently, a new species from a cave in the Amazon, *Deanemyia maruaga,* showed the same attribute (Alves et al. 2011).

Compared with other hematophagous Diptera, it has been assumed that phlebotomine sand flies usually feature gonotrophic concordance (i.e., there is one blood feeding before each instance of egg-laying). Nevertheless, in some species a second repast before oviposition has been reported (Christensen and Herrer 1980; El-Naien et al. 1992; Brazil et al. 1991a, b). Such is the case of *L. longipalpis* and *N. whitmani* caught in Ceará with human bait and *N. intermedia* caught in a chicken coop in Rio de Janeiro (El-Naien et al. 1992; Brazil et al. 1991a, b). This behavior may vary from species to species or even among populations of the same species. The need for a second blood feeding in pregnant females might be related to water-balance maintenance in view of climatic factors, such as high temperature and low humidity, which would directly affect their capacity to lay eggs. It is not known whether such a characteristic is common among the several phlebotomine species, but this phenomenon takes on great epidemiologic importance because it increases the transmission power of microorganisms by these insects.

In some species of neotropical phlebotomine sand flies, autogenic populations or lineages have been noted sporadically. Autogeny may be not mandatory, like in *Psathyromyia shannoni* sensu *latu*, originally from Florida, and *L. cruciata* and *Danpfomyia beltrani* in Central America (Lewis 1965; Willians 1976) or rather mandatory as noted in a subpopulation of *L. gomezi* in Panamá (Johnson 1961).

Phlebotomine sand flies usually begin their activity period during afternoon or at night, remaining most of the day in their shelters. *Bichromomyia flaviscutellata* shows peak alimentary activity between 10:00 pm and 01:00 am, and it rarely sucks blood during the day (Ward 1977). Hematophagous activity during the day, when it occurs, takes place in low-luminosity environments, especially in caves and forest areas. Among the several phlebotomine sand flies that feature the ability to suck blood during the day, *Ps. wellcomei* is without doubt the most active species. Another species, Nyssomyia *umbratilis*, usually nocturnal, becomes considerably aggressive during the day when dislodged from its shelters.

Many factors stimulate phlebotomine sand flies to suck blood and, like other hematophagous insects, their search for hosts is determined by a series of events intrinsic to the insects but that act along with environmental factors (Lehane 1991). The attraction of the phlebotomine sand flies by the hosts involves, among other stimulus, the host's body temperature and odor.

As for alimentary preference, the species of neotropical phlebotomine sand flies that suck on warm-blooded animals are usually eclectic. Many species of phlebotomine sand flies caught with human bait also suck on other animals; *L. longipalpis* is the best eclecticism example because it has the ability to pierce man, dog, birds, and other animals with considerable avidity in the same biotope. Other species with the same alimentary characteristics include *N. intermedia, Migonemyia migonei* and *Pintomyia fischeri*, which in anthropic environments adapt to the natural shelters of animals and break into residences to suck on humans (Gomes et al. 1982; Rangel et al. 1986; Brazil et al. 1991b). In this case, the break-ins of these locations occur during the beginning of the night and decrease in the first hours of the morning. The alimentary preference may also vary in distinct populations of one same species, such is the case of *N. whitmani* sensu *lato,* which is zoophilic in the Amazon region but is attracted to humans in the northeast and southeast regions of the country (Lainson 1983; Azevedo and Rangel 1991; Brazil et al. 1991a).

Longevity and Dispersion

Adult longevity under natural conditions is practically unknown, but observations in the laboratory indicate that phlebotomine males and females can live for 20 to 30 days, although this may not be the real life span.

Several factors contribute to the longevity of phlebotomine adults. However, decreased humidity seems to be the factor that most affects their survival. The great majority of neotropical phlebotomine sand flies do not survive for long under humidity conditions lower than 50%.

Like their longevity, dispersion of these insects in nature is still little known. Initial studies on dispersion have always been carried out with species of forest habits. Chaniotis et al. (1974) showed that approximately 90% of phlebotomine sand flies were retrieved within a 57-m radius in the Panama forest. Other studies in forest areas have confirmed that the dispersion distance does not exceed 200 m (Chippaux et al. 1984; Alexander 1987). In contrast, studies by Morrison et al. (1993) showed that the dispersion behavior of peri-domiciliary populations of *L. longipalpis* is very similar to the behavior of populations of Phlebotomus, which also presents peri-domiciliary habits in Europe and can disperse within a 500-m distance or further. However, studies in the urban area of Campo Grande, State of Mato Grosso do Sul, Brazil, showed that *L. longipalpis* does not disperse over 241 mts (Oliveira et al. 2012). It is to be admitted that some species (*Pi. fischeri, P. pessoai, M. migonei*, and *N. intermedia*) associated with environments modified by humans feature more dispersion power than forest species (Forattini 1954). This was observed in *Nyssomyia neivai* in Vale da Ribeira, SP, which dispersed over 500 meters (2009).

Shelters

AS insects with crepuscular and nocturnal habits, phlebotomine males and females remain in sheltered places during the day where there are no great environmental changes even if the climatic changes are adverse.

Usually the shelters are dissociated from natural breeding sites, although some of them can serve as places for the development of immature forms In areas of primary forest, a large variety of places can serve as shelter for phlebotomine sand flies. Many species seem to use the space between fallen leaves and the ground, animal burrows, and crevices on tree trunks.

Studies performed in the forest reserve of Ducke, in Central Amazonia, by Cabanillas et al. (1995), showed that sapopema roots, followed by bases of trees and termite mounds, constitute the main shelter for several phlebotomine species. In turn, in the Atlantic forest region (Serra dos Órgãos National Park, State of Rio de Janeiro) the main place used as shelter were animal burrows. In these locations, there has always been the predominance of *Ps. barretoi* followed by *Brumptomyia spp.* This and other findings confirm that the phlebotomine species of the group *P. aragaoi* have an alimentary preference for Dasypodidae and that they use the same environment as their resting place (Young and Duncan 1984; Aguiar and Villela 1987).

Regarding phlebotomine species with occurrence in human residences and their annexes, its shelters are usually within these environmental limits, with pigsties and chicken roosts seeming to be the places most used by *N. intermedia, M. migonei* and *Pi. fischeri* in the region of Southeastern Brazil (Table 2).

Table 2 Shelters of some Brazilian sand-fly species

Species	Places	Author
Ps. aragaoi	Animal burrows in the ground	Aguiar & Villela, 1987
Ps. ayrozai	Animal burrows in the ground	Aguiar & Villela, 1987
Ev. bahiensis	Rodent holes	Mangabeira & Sherlock, 1961
Ps. barretoi	Animal burrows in the ground	Aguiar & Villela, 1987
Ps. dreisbachi	Armadillo holes	Causey & Damaceno, 1945; Ryan, 1986
Pi. fischeri	Trunks of trees, chicken roosts	Aguiar & Villela, 1987; Brazil et al., 1991b
Ps. hirsutus	Rock crevices, under vegetation	Aguiar & Villela, 1987
Ps. inflata	Armadillo holes	Ready et al., 1986
Ny. intermedia	Chicken roosts	Brazil et al., 1991b
Lu. longipalpis	Holes on tree trunks, base of rocks	Mangabeira, 1969
Mi. migonei	Chicken roosts	Brazil et al., 1991b
Mi. oswaldoi	Holes in tree trunks, base of rocks	Mangabeira, 1969
Pi. pessoai	Tree trunks	Aguiar & Vilela, 1987
Mi. quinquefer (= *L. rickardi*)	Holes on tree trunks, base of rocks	Mangabeira, 1969
Ps. shannoni	Holes in tree trunks, base of rocks	Mangabeira, 1969
Ev. termitophila	Termite mounds	Martins et al., 1977
Tr. trichopyga	Armadillo holes	Ready et al., 1986
Ny. umbratilis	Tree trunks, sapopema roots	Ready et al., 1986; Cabanillas et al., 1995

Mating

Regarding mating, little is known about this activity because it occurs in nature. Observations in the laboratory are still not enough to determine whether all phlebotomine species present identical behavior. Some species often mate under laboratory conditions, even in a quite restricted space.

For example, it can be mentioned that *L. longipalpis* copulates very often in cages or in small observation chambers. The nuptial court made by males of this species consists of continuous circular movements and the flapping of wings, which attract susceptible females. After some seconds of intensely vibrating their wings, the males place themselves counter-wise to the female, approaching and using their genital appendages to take the female's posterior edges. During the court, receptive females vibrate their wings in response to the male's call. Nonreceptive females repel males by lowering the abdomen, thus preventing copulation.

Among possible factors before mating, the most important perhaps is the release of the male sexual pheromone to attract females of the same species (Ward et al. 1989). It is admitted that, in nature, copulation occurs during the nocturnal period when phlebotomine sand flies are more active in looking for hosts to suck blood. Observations in the laboratory show that a sole *L. longipalpis* male is capable of copulating many times with different virgin females, even though there is no indication that they are all fertilized. The presence of vertebrate hosts considerably stimulates the copulation activity in this species (Ward et al. 1989).

Other species, such as Ny. *intermedia* and Mg. *migonei,* feature similar behavior; however, it is without the aggressiveness of *L. longipalpis*. According to Ward et al. (1988), the attraction between sexes of *L. longipalpis* is not only olfactory or visual but also auditory, involving the presence of sounds emitted by the male flapping of wings. In this manner, the act of vibrating wings would not only be to release the sexual pheromone existing in the male abdominal glands but also to produce sound vibrations that are responded to by females within a pre-established pattern. This new aspect of mating in *L. longipalpis* has been well studied by the group of Alexandre Peixoto (Souza et al. 2004; Vigoder et al. 2015).

Abdominal glands in phlebotomine sand flies were noted in *L. longipalpis* and *Mi. quinquefer* males first by Barth (1961), who classified them as odorous glands that would have the function of stimulating females before and during copulation. Such remarks were proved afterwards in *L. longipalpis,* by Lane and Ward (1984), because the glands are located in the third and fourth tergites or only in the fourth tergite in different populations. Electron microscopy of the abdominal glands of five Phlebotominae species were observed by Spiegel et al. (2002). Gland extracts of *L. longipalpis* males are able to attract females in laboratory within a distance ≤2.2 m (Morton and Ward 1989; Brazil et al. 1989; Ward et al. 1990). The presence of sexual pheromone has been observed in other phlebotomine species of the genus *Lutzomyia* (Hamilton & Ward, 1994; Hamilton et al. 1999); nevertheless, it has been better studied in *L. longipalpis*.

In extracts of *L. longipalpis* pheromone glands, two compounds were detected: one homosesquiterpene (C_{16} H_{22}) and one diterpene (C_{20} H_{32}). Studies with different original populations demonstrated at least five groups existing that differ for presenting different terpenes and may be cryptic species (Hamilton et al. 1996a, b).

Natural Breeding Sites

Naturally, oviposition in phlebotomine sand flies is made in a substrate rich in organic matter, which serves the development of immature forms. The choice of oviposition place is driven for semiochemicals, which act as aggregation and oviposition pheromones. In *L. longipalpis*, this phenomenon was identified as being attributable to dodecanoic acid (Dougherty and Hamilton 1997), which would hold, along with other organic nature compounds, the function of aggregating pregnant females for laying eggs.

Natural breeding sites are extremely difficult to find. As for neotropical phlebotomine sand flies, the data are scarce and the number of immature forms is quite decreased in places considered breeding sites Ferro et al. 1997). One concrete remarks on natural breeding sites is found in Hanson (1961), the work of whom in Panama identified several breeding sites of forest phlebotomine species.

When considering the places listed in Table 3 as breeding sites of some Brazilian species, the small number of larvae and pupae found in them is not consistent with the high density of adults in these places.

The breeding sites of most of species are still unknown. Only with further studies, all applying several associated techniques, will it be possible to obtain more consistent results.

Table 3 Some breeding sites of phlebotomine sand flies in Brazil

Species	Places	Author
L. longipalpis	Under rocks, in rock crevices, livestock corrals	Deane, 1956; Deane & Deane, 1957
Pi. fischeri	Base of trees	Coutinho & Barreto, 1941
Ny. intermedia	Base of bushes, pigsties	Forattini, 1953, 1954
Ny. whitmani	Banana plantations, pigsties	Forattini, 1960
Ev. bahiensisi	Holes in trees, holes in the ground	Sherlock, 1962
Ev. cortellezzii	Holes in trees, among roots	Sherlock, 1962; Deane, 1956
Mi. oswaldoi	Under rocks, in rock crevices, among roots	Deane, 1956; Deane & Deane, 1957

Biological Cycle

To study the biological cycle of a species, it is necessary to establish colonies of the species in the laboratory. Although it is good to apply a basic methodology to the study of most species, there is a need for adaptations to obtain success in maintaining the colony (Killick-Kendrick 1978; Modi and Tesh 1983).

The life cycle of phlebotomine sand flies and other diptera is characterized by holometaboly (Fig. 1). The larval phase goes through four stages by means of three changes. The term of the different stages, from egg to pupa, varies from one species to another, and for any given species development depends on temperature, humidity, and food availability. Under thermal conditions slightly greater than the temperature considered optimal for most neotropical species (25 to 27 °C), cycle development is faster. As the temperature decreases, the evolution cycle becomes slower. The vital cycle of some Brazilian species under similar conditions in the laboratory is presented in Table 4.

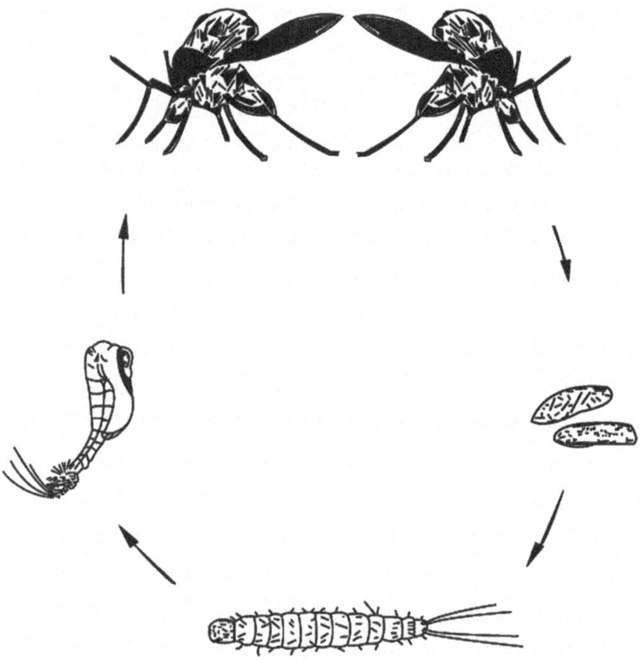

Fig. 1 The life cycle of phlebotomine sand flies

Table 4 Development of three species of Phlebotominae under laboratory conditions (26 to 28 °C/80% UR)[a]

Species	Po	Egg/L1	L1/L2	L2/L3	L3/L4	L4/Pupa	Pupa/ Adult
Evandromyia .Lenti	6.6 + −2.2	5.2 + −1.6	5.8 + −0.7	6.6 + −0.8	6.4 + −0.4	8.2 + −0.7	8.0 + −0.6
Lutzomyia Longipalpis	6.4 + −0.8	2.8 + −0.4	5.0 + −0.6	5.1 + −0.4	4.5 + −0.6	5.7 + −0.4	6.4 + −0.4
Nyssomyia intermedia	7.4 + −0.8	5.6 + −0.4	5.8 + −0.7	5.8 + −0.7	5.4 + −1.0	5.0 + −0.6	5.2 + −0.7

Po = pre-oviposition
L = larva
L1 to L4 = stages 1 to 4
[a]Based on Brazil et al. (1997)

Colonization in Laboratory

The difficulties in finding natural breeding sites of phlebotomine sand flies indicate great importance for the colonization of species so that the vital cycle of these insects, plus the different aspects of their biology and physiology, can be studied. Under the controlled conditions of a laboratory, information on the morphological and behavioral characteristics—as well as their parasite–vector and insect–vertebrate relations—can be obtained, which will enable the establishment of new strategies to control diseases transmitted to humans and other animals.

Few phlebotomine species have been studied under laboratory conditions, so there still exist great gaps in knowledge of the biology of most species, even those of medical importance. This is due to the difficulties encountered in the colonization of these species. Obtaining one or two phlebotomine sand-fly generations in the laboratory is considered by most of researchers because it is very difficult. In turn, establishing a colony requires special care until the insects obtained overtake the eighth generation.

The actual function of the colony is to supply insects in numbers enough for the experimental work without having to insert new individuals brought from the field.

During the last 40 years, many attempts have been made to mass-breed phlebotomine sand flies (Johnson and Hertig 1961; Chaniotis 1975; Killick-Kendrick et al. 1977; Modi and Tesh 1983; Rangel et al. 1985). From approximately 900 species of known Phlebotominae, at least 17 have been colonized in the laboratory for >10 generations (Volf and Volfova 2011).

Colonies of phlebotomine sand flies are usually little productive and demand many hours of daily work for their maintenance. In addition to the adaptation particularities of each species to laboratory conditions, the major difficulties found are related to the high mortality in the first evolution stages, cannibalism among the larvae, fungal proliferation, inappropriate larval diet, and high mortality rate of pregnant females. In our laboratory, with the adaptation of several methodologies, established colonies of phlebotomine sand flies, or only one generation (F1) depending on the need and type of study to be accomplished, have been obtained (Brazil et al. 1997; Brazil and Brazil, 2000 ; Gomes Brazil 2000).

Obtaining Adults

For the initial establishment of a colony in laboratory, adults are usually caught with the aid of light traps, among which the most usual and efficient is the CDC trap (Sudia and Chamberlain 1962). However, the Shannon trap is also used with light or animal bait by employing a mouth aspirator (Castro or similar). Cages employed in catches—or for maintaining phlebotomine sand flies in the laboratory long enough to be used for several procedures—are cubes made of nylon or thin cloth (like that used in curtain manufacture), which provides a sleeve (for the introduction of insects), fixed by strings, sewed on the vertices, to a cuboid metal frame.

Phlebotomine sand flies caught in CDC light traps are taken to the laboratory inside their own textile cages comprising a plastic bag containing a cotton pad or a wet sponge. The cages are kept inside a Styrofoam box during all transportation stages from the field

In the case of manual catch, phlebotomine sand flies can be placed in textile cages to be sent to the laboratory as described earlier or put in pots made of non-toxic plastic with a plaster bottom covered with nylon gauze, which is taken to the laboratory under the same conditions as the cages.

In the case of catches in locations distant from the laboratory, which requires long hours of traveling, it is advisable that the insects should be blood fed while still in the field. The females must be individualized in small pots like those already described. These must be placed in the plastic box, in which females are kept for oviposition, which in turn is placed in the Styrofoam box for the transfer. Before the trip, diluted honey or sugar solution is offered to the females to ensure their survival.

Blood Feeding

Blood feeding is offered to unfed females in laboratory using a cotton pad soaked in diluted honey or 30% to 50% sugar diluted at the ratio of 1:1 with distilled water, which can be directly placed on the upper part of the cage or in a small petri dish placed inside it.

After being returned to the plastic bag under the conditions it was brought from the field, the cage is kept in a BOD incubator with ideal temperature and humidity (the species colonized by our group was kept at a temperature ranging from 25 to 27 °C and a humidity ranging from 70–80%).

For the specimens brought in pots, the diluted honey/sugar solution is directly placed on the netting. These same pots are placed in plastic boxes with a pressure lid, the bottom of which is covered with a layer of filter paper moistened with distilled water. Once the humidity decreases, the paper is moistened again. The appearance of stains may indicate contamination, at which time it is necessary to change the filter paper, taking care to clean the box's bottom with 70% alcohol. Twenty-four hours after arriving at the laboratory, the females caught are fed on blood, which can be done using hamsters (*Mesocriscetus auratus*), which were previously anesthetized with sodium thiopental by intramuscular administration, for 1 to 2 h inside the holding cages kept in the dark.

Most species of sand flies feed well on hamsters or birds (chickens and quails). However, it must be borne in mind that some phlebotomine species feed only on cold-blooded animals, such is the case of *Micropygomyia quinquefer,* thus making it difficult to breed them in laboratory.

After the repast, the females are kept along with the males in their own holding cages. Then they are returned the incubator inside the plastic bag inside the cage and are maintained for 48 h under the same conditions of humidity, temperature, twilight, and carbohydrate feeding as before. This time corresponds to the resting period necessary for the females to produce eggs and to secure fertilization.

Individualization and Identification of Females

Forty-eight hours after the blood meal, the pregnant females are transferred, with the aid of an aspirator, to small plastic pots (4.5-cm diameter × 4.0-cm height) that have a layer of plaster on the bottom covered by a thin textile. The pots are then accommodated in boxes like those used before, in other words, kept in BOD incubators under the previous conditions and monitored to maintain the humidity inside. The boxes are kept in the incubator at 25–27 °C and observed daily to check for oviposition, after which females usually die. The head and the two last abdominal segments of dead females are then dissected and assembled between a slide and a coverslip in Berlese fluid for observation of the cibarium and the respective spermathecae and for specific identification. Eggs belonging to one of the same species are transferred, using a thin paintbrush moistened with distilled water, to the breeding pots (7.0-cm diameter × 8.5-cm height), which are modified like the former ones (i.e., with a plaster bottom), in which all larval stages and pupae are monitored. An average of 350 eggs is transferred to each breeding pot.

Immature Forms

The breeding pots containing the transferred eggs are monitored daily for the removal of fungus and mites and the addition of larval food. During this phase, to minimize the grow of fungus, the food ration must be added only after the emergence of the first larvae.

The emergence period of the eggs varies from species to species and according to incubation temperature. On average, the eggs of phlebotomine sand flies kept at a temperature of 25–27 °C begin their emergence between 5 and 8 days; however, it can take up to 15 days for eggs to emerge.

The food formulae used in the breeding of phlebotomine sand flies may vary considerably. In our laboratory colonies, it is applied in a 1:1 mixture of rabbit dry faeces, vegetal earth, and fern powder, to which 1% of fish food (Vitormonio-RCW/ Brasil) is added. The ration is mixed, sieved, and autoclaved in small glass flasks until its use.

Emergence of Adults

As soon as the first pupae form, the offer of larvae food must be decreased to avoid fungal proliferation and death of the recently emerged adults. When the first adults emerge, they must be transferred to the net cages and kept for 2 or 3 days in the incubator, as described previously, until the first blood feeding.

Maintenance of the colony must be performed taking the same care as its establishment. Essential requirements for success include paying attention to cleaning the environment, rigorously washing the instruments, using alcohol at 70% after washing, and ensuring the absence of strong odors in the insects' maintenance area.

References

Adler S, Theodor O (1926) The mouthparts, alimentary tract end salivary apparatus of the female in Phlebotomus papatasi. Ann Trop Med Hyg 20:109–142

Aguiar GM, Villela ML (1987) Aspects of the ecology of sandflies at the Serra dos Órgãos Park, State of Rio de Janeiro. VI. Shelters and breeding places (Diptera, Psychodidae, Phlebotominae). Mem Inst Oswaldo Cruz 82:585–586

Alexander JB (1987) Dispersal of phlebotomine sand flies (Diptera: Phychodidae) in a Colombian coffee plantation. J Med Entomol 24:552–558

Alves VR, De Freitas RA, Santos FL, Barrett T (2011) Diversity of sandflies (Psychodidae: Phlebotominae) captured in sandstone caves from Central Amazonia, Brazil. Mem Inst Oswaldo Cruz 106:353–359

Azevedo AC, Rangel EF (1991) Study of sandflies species (Diptera: Psychodidae; Phlebotominae) in a focus of cutaneous leishmaniasis in the municipality of Baturité, Ceará, Brazil. Mem Inst Oswaldo Cruz 86:405–410

Barth R (1961) Sobre o aparelho genital interno do macho de *Phlebotomus longipalpis* (LUTZ & NEIVA, 1912) (Diptera: Psychodidae). Mem Inst Oswaldo Cruz 59:23–36

Brazil G, Brazil RP (2000) Sexing sand fly pupae (Diptera: Psychodidae: Phlebotominae). Mem Inst Oswaldo Cruz 95:471–472

Brazil RP, Oliveira SMP (1999) Parthenogenesis in the sandfly *Lutzomyia mamedei* (Diptera: Psychodidae). Med Vet Entomol 13:463–464

Brazil RP, Trumper S, Ward RD (1989) A heated pheromone trap for the sandfly *Lutzomyia longipalpis* (Diptera: Psychodidae). Mem Inst Oswaldo Cruz 84(suppl. II):30

Brazil RP, Morton IE, Ward RD (1991a) Notes of the feeding habits of *Lutzomyia (N) whitmani* (Diptera: Psychodidae) in Ceará state, Northeast Brazil. Mem Inst Oswaldo Cruz 86:497–498

Brazil RP et al. (1991b) Chicken house as a resting site of sandflies in Rio de Janeiro, Brazil. In: MAROLI (Ed.) Proceedings of 1st Symposium on Phlebotomine sandflies, Rome Parassitologia 33(suppl): 113–118

Brazil RP et al (1997) Biology of *Lutzomyia lenti* (Mangabeira) (Diptera: Psychodidae). Anais Soc Entomol Brasil 26:191–193

Cabanillas MRS, Castellon EGB, Alencar MO (1995) Estudo sobre abrigos naturais dos flebotomíneos (Diptera: Psychodidae) na Reserva Florestal Ducke, Manaus, AM, Brasil. Bol/ Direc Malariol San Amb 35:63–75

Cameron MM et al (1995) An association between sandflies and aphids in the Peruvian Andes. Med Vet Entomol 9:127–132

Causey OR, Damaceno RG (1945) Estudo sobre Flebotomus no vale Amazônico. Parte IV. Descrição de *F. cerqueirai, F. dreisbachi, F. meirai,* e *F. ferreirai* (Diptera: Psychodidae). Mem Inst Oswaldo Cruz 42:645–660

Chaniotis BN (1967) The biology of California *Phlebotomus* (Diptera: Psychodidae) under laboratory conditions. J Med Entomol 4:221–233

Chaniotis BN (1975) A new method for rearing *Lutzomyia trapidoi* (Dipetra: Psychodidae), with observations on its development and behavior in the laboratory. J Med Entomol 12:183–188

Chaniotis BN et al (1974) Horizontal and vertical movement of Phlebotomine sandflies in a Panamanian rain forest. J Med Entomol 11:369–375

Chippaux J, Pajot FX, Barbier D (1984) Leishmaniasis in French Guyana. 5. Further data on the ecology of the vector in the cacao forest village. *Chahiers ORSTM*. Série Entomol Medt Parasitol 22:213–218

Christensen HA, Herrer A (1980) Panamanian *Lutzomyia* (Diptera: Psychodidae) host attraction profiles. J Med Entomol 17:522–528

Coelho MV (1966) Desenvolvimento das espécies do gênero *Leishmania* em flebótomos brasileiro do genero Lutzomyia França 1924. Faculdade de Medicina da Universidade de Minas Gerais, Belo Horizonte

Coutinho JO, Barreto MP (1941) Dados bionômicos sobre o *Phlebotomus fischeri* Pinto, 1926 (Diptera: Psychodidae). Rev Bras Biol 1:423–429

Deane LM (1956) Leishmaniose Visceral do Brasil. Serviço Nacional de Educação Sanitária, Rio de Janeiro, 162p

Deane LM, Deane MP (1957) Observações sobre abrigos e criadouros de flebótomos do noroeste do Estado do Ceará. Rev Bras Mal D Trop 9:225–246

Dougherty M, Hamilton JG (1997) Dodecanoic acid is the oviposition pheromone of *Lutzomyia longipalpis*. J Chem Ecol 23:2657–2671

El-Naien DA et al (1992) Field and laboratory evidence for multiple blood feeding by *Lutzomyia longipalpis* Diptera: Psychodidae. Med Vet Entomol 6:13–17

Ferro C et al (1997) Larval microhabitats of *Lutzomyia longipalpis* (Diptera: Psychodidae). J Med Entomol 34:719–728

Forattini OP (1953) Nota sobre criadouros naturais de flebotomos em dependências peridomiciliares no Estado de São Paulo. Arq Fac Hig São Paulo 7:157–168

Forattini OP (1954) Algumas observações sobre biologia de flebotomos (Díptera: Psychodidae) em região da bacia do rio Paraná. (Brasil). Arq Fac Hig São Paulo 8:15–136

Forattini OP (1960) Novas observações sobre biologia de flebotomos (Diptera: Psychodidae) em condições naturais. Arq Fac Hig São Paulo 25:209–215

Forattini OP (1973) Entomologia Médica. São Paulo, Edgar Blucher, 658p

Galati EAB (2016) Phlebotominae (Diptera, Psychodidae) classificação, morfologia, terminologia e identificação de adultos. Departamento de Epidemiologia Faculdade de Saúde

Pública. Apostila Disciplina HEP 5752 Bioecologia e Identificação de Phlebotominae 2016. Universidade de São Paulo

Galati EAB et al (2009) Phlebotomines (Diptera, Psychodidae) in the Speleological Province of the Ribeira Valley: 3. Serra district-area of hostels for tourists who visit the Parque Estadual do Alto Ribeira (PETAR), state of São Paulo, Brazil. Rev Bras Entomol 54:665–676

Gomes Brazil B (2000) Aspectos morfológicos das glândulas salivares de *Lutzomyia* (*Lutzomyia*) *longipalpis* (Lutz & Neiva, 1912) (Diptera: Psychodidae: Phlebotominae) em diferentes tempos pós emergência e pós repasto. Instituto Oswaldo Cruz, Tese de Doutorado, Rio de Janeiro, 101p

Gomes AC et al (1982) Aspectos ecológicos da leishmaniose tegumentar americana. 2. Ecotopo como abrigo de *Psychodopygus intermedius* e observações sobre alimentação e reprodução sob a influência de condições naturais. Rev Saúde Pública 16:149–159

Gontijo CMF et al (1987) The finding of one male specimen of *Lutzomyia renei* (Martins, Falcão & Silva, 1957) experimentally infected by *Leishmania*. Mem Inst Oswaldo Cruz 82:445

Hamilton JGC, Ward RD (1994) Chemical analysis of a putative sex pheromone from *Lutzomyia pessoai* (Diptera: Psychodidae). Ann Trop Med Parasitol 88:405–412

Hamilton JGC, Dawson GW, Pickett JA (1996a) 9-Methylgermagrene-B, a novel homosesquiterpene from sex pheromone glands of *Lutzomyia longipalpis* (Diptera: Psychodidae) from Lapinha, Brazil. J Chem Ecol 20:1477–1491

Hamilton JGC, Dawson GW, Pickett JA (1996b) 3-Methyl-@-himachalene; sex pheromone of, a novel homosesquiterpene from sex pheromone of *Lutzomyia longipalpis* (Diptera: Psychodidae) from Jacobina, Brazil. J Chem Ecol 22:2331–2340

Hamilton JGC et al (1999) Chemical analysis of oxygenated homosesquiterpenes: a putative sex pheromone from *Lutzomyia lichyi* (Diptera: Psychodidae). Bull Entomol Res 89:139–145

Hanson WJ (1961) The breeding places of phlebotomus in Panamá (Diptera: Psychodidae). Ann Entomol Soc America 54:317–322

Johnson PT (1961) Autogeny in Panamanian Phlebotomus sandflies (Diptera: Psychodidae). Ann Entomol Soc America 54:116–118

Johnson PT, Hertig M (1961) The rearing of Phlebotomus sandflies (Diptera: Psychodidae). II. Development and behavior of Panamanian sandflies in laboratory culture. Ann Entomol Soc America 54:764–776

Killick-Kendrick R (1978) Recent advances and outstanding problems in the biology of phlebotomine sandflies. A review Acta Trop 35:297–313

Killick-Kendrick R, Leaney AJ, Ready PD (1977) The establishment, maintenance and productivity of a laboratory colony of *Lutzomyia longipalpis* (Diptera: Psychodidae). Jf Med Entomol 13:429–440

Lainson R (1983) The American leishmaniasis: some observation on their ecology and epidemiology. Trans Roy Soci of Trop Med Hyg 77:569–596

Lane RP, Ward RD (1984) The morphology and possible function of abdominal patches in males of two forms of leishmaniasis vector *Lutzomyia longipalpis*. Cahiers ORSTM, Serie Ent Med Parasitol 22:245–249

Lehane MJ (1991) Biology of blood-sucking insects. Harper-Collins Academic, London, 288p

Lewis DJ (1965) Internal structure of some Central American Phlebotomine sandflies. Ann Trop Med Parasitol 59:375–385

Lewis DJ, Domoney CR (1966) Sugar meals in Phlebotominae and Simuliidae. Proc Roy Entomol Soc Lond 41:175–179

Magnarelli LA, Modi GB (1988) Caloric determination of phlebotomine sandflies (Diptera: Psychodidae). J Med Entomol 20:568–569

Mangabeira O (1969) Sobre a sistemática e a biologia dos phlebotomus do Ceará. Rev Bras Mal D Trop 21:3–25

Mangabeira O, Sherlock I (1961) Descrição de 4 novas espécies de Phlebominae da Bahia (Diptera: Psychodidae). Rev Bras Bio:265–276

Martins AV, Falcão AL, Silva JE (1977) Estudos sobre os flebótomos do Estado de Minas Gerais. XIV. Descrição da fêmea de *Lutzomyia termitophila* Martins, Falcão & Silva, 1964 (Diptera: Psychodidae: Phlebotominae). Rev Bras Biol 37:645–647

Minter DM, Wiijers DJB (1963) Studies on the vector of kala-azar in Kenya IV. Experimental evidence Ann Trop Med Parasitol 57:24–31

Modi GB, Tesh RB (1983) A simple technique for mass rearing *Lutzomyia longipalpis* and *P. papatasi* (Diptera: Psychodidae) in the laboratory. J Med Entomol 20:568–569

Morrison AC et al (1993) Dispersal of the sandfly *L. longipalpis* (Diptera: Psychodidae) at a endemic focus of visceral leishmaniasis in Colômbia. J Medl Entomol 30:427–435

Morton IE, Ward RD (1989) Laboratory response of female *Lutzomyia longipalpis* sandflies to a host and male pheromone source over distance. Med Vet Entomol 3:219–223

Morton IE et al (1991) The presence of fructose in wild-caught *Lutzomyia longipalpis* (Diptera: Psychodidae). Mem Inst Oswaldo Cruz 86:125–116

Oliveira EF et al (2012) Biotic factors and occurrence of Lutzomyia longipalpis in endemic area of visceral leishmaniasis, Mato Grosso do Sul, Brazil. Mem Inst Oswaldo Cruz 107:396–401

Oliveira SMP et al (1994) Description of a new species of sandfly Lutzomyia (P) mamedei (Diptera: Psychodidae) from Rio de Janeiro, Brazil. Mem Inst Oswaldo Cruz 89:319–320

Rangel EF et al (1985) Estabelecimento de colônia em laboratório de *Lutzomyia intermedia* Lutz & Neiva, 1912 (Diptera: Psychodidae). Mem Inst Oswaldo Cruz 80:219–226

Rangel EF et al (1986) Flebotomos de Vargem Grande, foco de leishmaniose tegumentar no Estado do Rio de Janeiro. Mem Inst Oswaldo Cruz 81:347–349

Ready PD (1979) Factors affecting egg production of laboratory-bred *Lutzomyia longipalpis* (Diptera: Psychodidae). J Med Entomol 16:413–423

Ready PD et al (1986) The ecology of *Lutzomyia umbratilis* Ward & Fraiha (Diptera: Psychodidae), the major vector to man of *Leishmania brasiliensis guyanensis* in north-eastern Amazonian Brazil. Bull Entomol Res 76:21–40

Ribeiro JM et al (1989) A novel vasodilatatory peptide from salivary glands of the sandfly *Lutzomyia longipalpis*. Science 243:212–214

Sherlock I (1962) Nota sobre criadouros Naturais de *Phlebotomus* em Salvador, Bahia (Diptera: Psychodidae). Rev Bras Biol 22:125–127

Souza N et al (1992) Carbohydrates in wild caught *Lutzomyia intermedia* (Lutz & Neiva, 1912) (Diptera: Psychodidae: Phlebotominae) as detected by high performance liquid chromatography. Mem Inst Oswaldo Cruz 87(supp. II):208

Souza N et al (2004) Analysis of the copulatory courtship songs of *Lutzomyia longipalpis* in six populations from Brazil. J Med Entomol 41:906–913

Spiegel CN et al (2002) Ultrastructure of male sex pheromone glands in abdominal tergites of five Lutzomyia sandfly species (Diptera: Psychodidae). Arthr Struc Develop 30:219–227

Sudia WD, Chamberlain RW (1962) Battery operated light trap, an improved model. Mosq News 22:126–129

Tesh RB et al (1992) Trans beta farnasene as feeding stimulant for the sand fly *Lutzomyia longipalpis* (Diptera: Psychodidae). J Med Entomol 29:226–231

Vigoder FM et al (2015) Phenotypic differentiation in love song traits among sibling species of the *Lutzomyia longipalpis* complex in Brazil. Parasit Vectors 8:290–295

Volf P, Volfova V (2011) Establishment and maintenance of sand fly colonies. J Vector Ecol 36(suppl 1):S1–S9

Volf P et al (2000) Salivary protein and glycoprotein in sand flies of various species, sex and age. Med Vet Entomol 14:1–6

Wallbanks KR et al (1991) Aphid derived sugar in the neotropical sandfly: Lutzomyia peruensis. Trop Med Parasitol 42:60–62

Ward RD (1977) New world Leishmaniases: a review of the epidemiological changes in the last three decades. In: Proceedings of the XV International Congress of Entomology. pp 505–522

Ward RD et al (1988) The *Lutzomyia longipalpis* complex: reproduction and distribution. In: MW Service (ed) Biosystematics of Haematophagous insects. Oxford University Press, Oxford, pp 257–269

Ward RD et al (1989) Bioassays as indicator of pheromone communication in *Lutzomyia longipalpis* (Diptera: Psychodidae). In: Hart DT (ed) Leishmaniasis. Plenum Publishing Corp. Nato ASI Series, Brazil, p V.163

Ward RD et al (1990) Preliminary laboratory and field trials of a heated pheromone trap for the sandfly *Lutzomyia longipalpis* (Diptera: Psychodidae). Mem Ins Oswaldo Cruz 85:445–452

Willians P (1970) Phlebotomine sandflies and leishmaniasis in British Honduras. Trans Roy Soc Trop Med Hyg 64:317–368

Willians P (1976) The phlebotomine sandflies (Diptera: Psychodidae) of caves in Belize, C. America. Bull Entomol Res 65:601–614

Young DG, Duncan MA (1984) Guide to the identification and geographic distribution of *Lutzomyia* sandflies in Mexico, the West Indies, Central and South America (Diptera: Psychodidae). Mem Amer Entomol Inst 54:1–881

Biology of the *Leishmania*–Sand Fly Interaction

Paulo F. P. Pimenta, Vanessa C. de Freitas, Carolina C. Monteiro, Ana Clara M. A. Pires, and Nágila Francinete C. Secundino

Leishmaniasis is a spectrum of diseases transmitted by sand-fly vector caused by a protozoa parasite from the genus *Leishmania* (Trypanosomatida: Trypanosomatidae) This vector-borne disease is transmitted to humans exclusively through sand-fly bites. The *Leishmania* genus was given to honor Sir William Boog Leishman, an assistant professor of pathology in the British Army Medical School, who discovered the parasite for the first time on a slide spleen smear in 1903.

The *Leishmania* genus is divided into three sub-genera according to parasite development in the vector (Lainson and Shaw 1979). (1) Hypopylarian: Parasites develop within the sand-fly hindgut. This group was previously classified as a subgenus of *Leishmania*, but now it is accepted as Sauroleishmania. (2) Perypylarian: These parasites can establish initial infection in the pyloric region and hindgut of the sand fly, where they are attached to the cuticle (Nieves and Pimenta 2000). This group includes the Vianna subgenus species, such as *L.* (V.) *braziliensis*, and it can be found only in the New World. (3) Suprapylarian: The development of these parasites is restricted to the sand-fly midgut. This *Leishmania* species are classified as *Leishmania* subgenus, and they are distributed in the New World and Old World. Only parasites classified with the perypylarian and suprapylarian patterns have medical importance. The majority of *Leishmania* species that affect humans are from the suprapylarian pattern, and the information presented in this chapter includes a detailed description of this group of *Leishmania* considering its life cycle within its natural vectors.

P. F. P. Pimenta (✉) · V. C. de Freitas · C. C. Monteiro · A. C. M. A. Pires
N. F. C. Secundino
Instituto de Pesquisas René Rachou, Fundação Oswaldo Cruz,
Belo Horizonte, Minas Gerais, Brasil

E. F. Rangel, J. J. Shaw (eds.), *Brazilian Sand Flies*,
https://doi.org/10.1007/978-3-319-75544-1_6

Infective Blood Meal and Promastigote Forms Inside the Sand Fly

The interaction between the parasite and the vector starts at the time when the female sand fly can feed on infective blood to develop its eggs. After choosing an appropriate vertebrate host, the sand-fly mouthparts penetrate the skin, lacerating the epidermis in different spots, forming a blood pool that will be sucked by the insect. During this blood meal, the female occasionally ingests the *Leishmania* parasite present in the infected vertebrate host in a form called "amastigote," an intracellular tissue form found mainly inside macrophages. After the meal is consumed by the sand fly, the recently ingested *Leishmania* amastigotes are stored with the blood in the midgut for digestion. Immediately a differentiation process transforms the amastigote into a flagellated parasite form called "promastigote," the extracellular form that colonizes the vector during its lifetime. Distinct from that of the vertebrate, the life cycle of *Leishmania* in the sand-fly vector is extracellular and occurs inside the digestive tract.

The first promastigotes derived from the differentiation of ingested amastigotes inside the blood meal have high multiplicative capacity, and this form is called "procyclic." The procyclics multiply by binary fission, and subsequently, during the parasite life cycle, they differentiate into morphologically distinct flagellate forms. Different authors have given promastigotes forms distinct designations; however, in this chapter we use the simple nomenclature proposed by Lawyer and colleagues 1990, which has been used in several studies of *Leishmania*–sand fly interaction. Between the second and fifth days after a blood meal, a new promastigote population progressively emerges, the form of which is called "nectomonad." After blood-meal digestion, around the third and fourth days after infection, the nectomonads colonize different parts of the digestive tract according to the sub-genera (as described previously). The nectomonad population decreases with colonization along the midgut, and new promastigote forms emerge, called "haptomonads" and "paramastigotes." Gradually, after total excretion of the blood meal, the "metacyclic," the infective form of vertebrate, appears. The metacyclics migrate to the foregut region of the vector, where they will be ready to be transmitted by the sand-fly bite to a new vertebrate host during the next blood meal.

Natural Barriers Encountered by *Leishmania* Within the Sand-Fly Vector

Development of *Leishmania* within the sand-fly vector is a complex process. Several main events occur during the *Leishmania*–sand fly interaction that constitute barriers to parasite development: (a) *Leishmania* species must be susceptible to the vector when it is present in the vertebrate blood; (b) the action of digestive enzymes on the ingested blood meal; (c) the perithrophic matrix formation, which protects parasites from digestive-enzyme action; (d) the transformation of amastigotes (vertebrate multiplicative form) into promastigotes (sand-fly multiplicative form); (e) escape of the promastigotes from the perithrophic matrix to the foregut region due to parasite production of chitinase; (f) promastigote differentiation and subsequent adhesion to the midgut epithelium, which is dependent on the expression of species-specific molecules (lipophosphoglycan [LPG]) on the parasite surface; (g) parasite migration to the cardia region around the stomodeal valve after complete blood digestion; (h) promastigote differentiation into metacyclic, the infective form able to be injected and infect the vertebrate host at the time of the bite; (i) the successful feeding of an infected sand fly, which inoculates metacyclics into a new vertebrate host during the bite; (j) actions of the saliva produced by the sand-fly vector and the proteophosphoglycan (PPG) produced by promastigotes in the skin-bite site on the vertebrate host, which are able to intensify and modulate the infection; and, finally, (i) recent data suggesting that native microbiota play a role in vector interaction.

In conclusion, due to all the possible barriers to the development of *Leishmania* inside the sand fly, which determine vectorial competence, this phenomenon should be considered a complex dynamic process that is still open to analysis. Further we will discuss the main natural barriers that exist during *Leishmania*–sand fly interaction.

Amastigote Transformation into Promastigote Inside Sand-Fly Midgut

Sand flies, like others blood-sucking insects, digest the blood meal producing amylases, glycosidases, lipases and proteases (mainly trypsin and chymotrypsin), which are secreted by intestinal epithelium cells. Digestive enzymes can be harmful to parasites. The survival and infectivity ability of all microorganisms transmitted by haematophagous insects, including *Leishmania*, is directly associated with resistance to digestive-enzyme action produced by vectors.

The first barrier to *Leishmania* development inside the vector is the transformation of amastigote into promastigote. Studies have indicated that *Phlebotomus papatasi* infected with *Leishmania major* shows about a 50% reduction in parasite numbers during this transformation (Pimenta et al. 1997). Although amastigote differentiation into promastigote is an important step due to their cellular coating, which confers protection against the sand flies' intestinal protease action, many parasites do not complete this process. Thus, the number of parasites can dramatically decrease or even disappear within the vector during blood-meal digestion. Therefore, the sand-fly digestive process causes a drastic decrease in the vector-infection rate in a population (the percentage of infected sand flies) because many insects heal the infection.

After a blood meal and during the digestive process in sand flies, the concentration of digestive enzymes can increase up to 20 times. Some experimental studies have shown that parasite death can be avoided by adding specific proteolytic inhibitors to the infected blood meal (Borosvsky and Schlein 1987; Pimenta et al. 1997). Interestingly it has also been demonstrated that parasite resistance to digestive-enzyme action is stage-specific: Promastigotes are more resistant than amastigotes (Pimenta et al. 1997). This question was explored by comparing amastigotes until they were completely transformed into promastigotes. The transitional form (amastigotes differentiating to promastigotes) is susceptible to death induced by digestive-enzyme action. It was observed that transitional parasite forms are the most susceptible to digestive-enzyme action because this is a critical stage for parasite survival inside the sand fly (Pimenta et al. 1997). After complete transformation, promastigotes acquire greater resistance to digestive-enzyme action in the gut. This resistance is related to surface expression in the promastigote of a molecule not present in amastigotes, a surface lipophosphoglycan (LPG). LPG forms an organized surface coat that covers and protects the promastigote from proteases lysis (Schlein et al. 1990; Pimenta et al. 1991). This observation is consistent with the fact that *Leishmania* promastigote mutants not expressing LPG on their surfaces are killed during their early hours inside the sand-fly midgut. However, when LPG expression is restored, parasites regain the ability to survive in the vector midgut (Pimenta et al. 1994; Sacks et al. 2000; Myskova et al. 2007; Secundino et al. 2010).

Sand-Fly Competence for *Leishmania* Related to Digestive-Enzyme Action

Susceptibility to digestive-enzyme action is species-specific; therefore, different species of *Leishmania* can modulate proteases action differently within their appropriate sand-fly vector. The ability to resist or modulate these enzyme actions is a determining factor for parasite survival; consequently it is related to vector competence. Telleria and colleagues (2010) showed that trypsin expression decreases by about 20% in *Lutzomyia longipalpis* infected with *L. (L.) infantum chagasi*. The enzymatic activity modulation within the vector has been also studied in detail in the *Phlebotomus* genus. Adler and colleagues (1938) noted that *Phlebotomus papatasi* infected by several unsusceptible *Leishmania* species are killed during blood-meal digestion. This matter was taken up in several studies of Schlein and collaborators in 1986, 1987, and 1990, which demonstrated the ability of compatible *L. major* to partially block digestive-enzyme synthesis in its natural vector, *P. papatasi*. This result was not observed with *L. donovani* parasites, which induced a higher enzyme-production level, causing parasite destruction in this unsusceptible vector. Similar studies developed by Pimenta and colleagues (1997) showed that *P. papatasi/L. major* survival is associated with trypsin-enzyme production. Thus, levels of *Leishmania* species and the enzymes secreted by epithelial cells by a specific sand-fly vector in response to a blood meal are determining factors for infection development.

It is interesting to note that *L. major* and *L. donovani* LPG-deficient mutants surviving in their natural vectors while the blood meal is still present (Sacks et al. 2000;

Boulanger et al. 2004; Myskova et al. 2007; Svárovská et al. 2010; Secundino et al. 2010) does not necessarily decrease LPG's contribution to the resistance to initial parasite death in the midgut because in some cases a modest reduction in LPG-mutant survival during initial infection has been noted (Sacks et al. 2000; Myskova et al. 2007; Secundino et al. 2010). Secundino and colleagues (2010) have observed that the addition of PGs derived from LPGs provided to *L. major* LPG-mutants induce resistance in sand fly *P. duboscqi* against lysis due to midgut proteases.

In conclusion, many *Leishmania* ingested along with a blood meal are killed during digestion in sand-fly midgut because of adverse natural conditions. However, a sufficient number of individuals resistant to enzymatic action survive and multiply inside the digestive tract in order to beat the other barriers and maintain infection in the vector.

The Second Blood Meal and Action of Digestive Enzymes in the *Leishmania*–Sand Fly Interaction

Sand flies in nature ingest successive blood meals; however, the interaction studies cited previously were conducted using only one blood meal. Short (1928) demonstrated that a second blood meal shows no deleterious effects to *Leishmania*, and Adler (1964) confirmed that re-feeding sand fly with a normal blood meal does not decrease the *Leishmania* infection rate. However, Elnaiem et al. (1997) showed an increase in the proportion of *L. infantum chagasi* metacyclic in infected *L. longipalpis* after a second normal blood feeding, suggesting that the deleterious effect from the digestive enzymes was not observed in parasites of the established infection. Vivenes and colleagues (2001) also studied the effects of a second normal blood meal in Lu. *evansi* infected with *L. mexicana*, and they concluded that a second meal favors midgut colonization by promastigotes. Furthermore, Nieves and Pimenta (2002), studying *Lu. migonei* (*Migonemia migonei*) infected with *L. braziliensis*, confirmed that re-feeding increases the number of promastigotes in the gut and therefore plays an important role in transmission because a proliferation of the metacyclic population occurs.

Our group has advanced the studies related to second meals in sand-fly vectors. We used the sand-fly model, *L. intermedia* (*Nissomyia intermedia*), infected with *L. braziliensis*, and re-fed the infected sand flies with blood from several animals (Miranda et al. 2008). The sand flies appeared to ingest the same blood amount independent of the animal source, but infection patterns in the vectors changed according to the blood source. There were increased parasite numbers in sand flies that re-fed on chicken and donkey blood but a moderate increase in vectors that re-fed on human, horse, dog, and cattle blood. In contrast, a decreased parasite number was observed in infected vectors that re-fed on pork and lamb blood. As we know, proteolytic-enzyme action is a crucial factor for infection development; it is likely that the second blood meal serves as a new nutrient source; and blood types stimulate *Leishmania* total population growth, thus increasing the number of promastigotes that can modulate enzyme activity. Curiously, the metacyclic proportion did not significantly change in any experimental group of sand flies. The blood-meal influence on *Leishmania* infection is probably related to ingested blood components and hence to the vector digestive process. It is possible that, depending on the specific pair, the re-feeding effect varies depending on each natural pair. Sand flies have a marked ability to opportunistically feed on different vertebrates. It appears that the second blood meal of infected sand-fly vectors from certain domestic animals can increase their ability to transmit *Leishmania* in endemic areas where those animals are found. This may provide a selective advantage to vector competence for some sand-fly species in transmitting the *Leishmania* parasite to vertebrates.

Leishmania Escape from Sand-Fly Perithrophic Matrix

A crucial barrier to *Leishmania* development within sand-fly midgut is the structure called the "perithrophic matrix" (PM). Immediately after blood-meal ingestion, midgut epithelial cells start to synthetize the PM components that will be completely structured in 24 h, and this will persist until the end of the digestive process when the rest of the blood meal is secreted. Similar to other Diptera, sand-fly PM consists mainly of chitin and associated proteins and proteoglycans (Gemetchu 1974; Secundino et al. 2005).

Some classical functions of PM inside insect midguts—such as (a) epithelium protection against damage from food elements (Richards and Richards 1977; Berner et al. 1983); (b) food compartmentalization and digestive-enzyme permeabilization (Terra 1990; Terra and Ferreira 1994); (c) protection of midgut epithelium against ingested pathogens (Peters 1992; Miller and Lehane 1993); (d) flow control of small molecules, such as digestive enzymes and digestion products, toward midgut cells due to PM porosity (Tellam 1996); (e) detoxification by heme sequestering produced by hemoglobin digestion (Pascoa et al. 2002); and, finally, (f) provision of an important physical barrier against pathogen development (Pimenta et al. 1997)—have been considered in the literature.

PM involvement in parasite–vector interaction has been adequately demonstrated in several models (Feng 1951; Billingsley and Rudin 1992; Walters et al. 1992; Shahabuddin et al. 1993; Pimenta et al. 1997). A classic example was developed by Shahabuddin and colleagues (1993) demonstrating that malaria ookinetes secrete a chitinase and partially degrade PM in infected mosquitoes, which allows *Plasmodium* to invade the midgut epithelium to complete its life cycle. Similarly, *Leishmania* species also escape from PM after becoming promastigote, the parasite form that is able to produce chitinase (Schlein et al. 1991; Shakarian and Dwyer 1998; Malta et al. 2016). Schlein and colleagues (1991) found that chitinase and N-acetylglucosaminidase are secreted by cultured *L. major* promastigotes and suggested that these enzymes can catalyze PM degradation, thus allowing parasite escape. Shakarian and Dwyer (2000) identified a chitinase gene in *L. donovani* and demonstrated that the gene locus and the enzymatic activities are conserved in all different species strains.

PM properties—such as thickness, biochemical composition, synthesis and degradation—are related to parasites' ability to survive in sand-fly vectors. Previously we discussed the role of digestive-enzyme action in *Leishmania* drastically decreasing parasite numbers during blood digestion. However, despite the protection conferred by PM to the insect, the parasite uses the PM to protect itself from these digestive enzymes. This fact was well shown in sand flies infected with *L. major* (Pimenta et al. 1997; Pruzinova et al. 2015). The absence of PM possibly arises from feeding sand flies with blood meal containing exogenous chitinase. This lack of PM formation in sand-fly midgut has a deleterious effect on *Leishmania* infection because almost all parasites are killed within the vector. If PM is not synthesized, *Leishmania* is quickly destroyed by digestive-enzyme action. The few parasite survivors are excreted with final digestion products because they did not complete their differentiation into promastigote; therefore, they cannot adhere to the midgut epithelium because such adhesion depends on LPG. This is discussed later in the text.

During the development of infection, a great many free promastigotes remain in the sand-fly midgut lumen after excretion. To explain this fact, Vaidyanathan (2004, 2005) proposed an alternative mechanism to parasite adhesion to the midgut epithelium. A peptide secreted by *L. major* promastigote cells, when added to phlebotomine meal, was demonstrated to be capable of inhibiting *P. papatasi* midgut contraction. Thus, the parasites would not be excreted. This may also be related to PM degradation. In infected sand fly, degradation occurs in the up-front of the PM portion by promastigote chitinase action in contrast to healthy insects, where degradation occurs in the final portion of the abdomen.

In conclusion, the delicate balance between sand-fly vector and *Leishmania* protozoa is PM dependent. Initially the PM protects parasites from the direct action of digestive enzymes at the time of their differentiation into promastigotes. However, the promastigotes must produce a chitinase at the right time to escape from the PM and then to adhere to the midgut epithelium (Pimenta et al. 1992) in order to not be excreted in final step of the digestive process.

Leishmania Specific Adhesion to Sand-Fly Midgut Epithelium

Another important barrier to *Leishmania* within the sand-fly vector is its ability or not to adhere to midgut epithelial cells. Midgut epithelium adhesion is a vital phenomenon for *Leishmania* infection maintenance (Fig. 1). After escaping the PM, *Leishmania* promastigotes must adhere to the midgut epithelium. This adhesion prevents parasite excretion with the rest of the undigested blood and thus allows their proliferation and subsequent differentiation.

The molecule that enables and controls the adhesion of promastigotes is LPG. Importantly, *Leishmania* LPG surface-molecule expression is promastigote specific and is not present on the amastigotes' surface. This observation was impor-

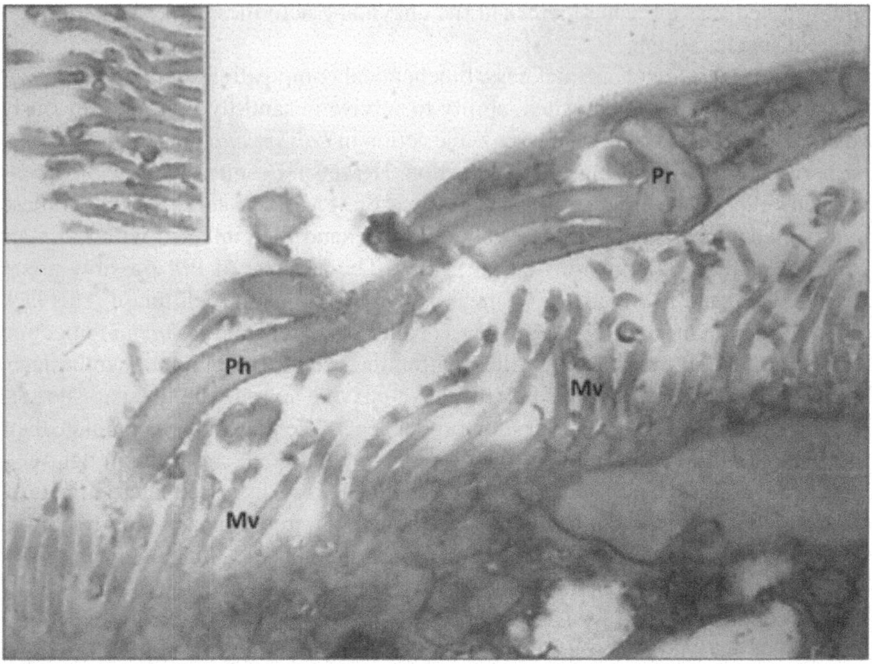

Fig. 1 *L. major* attached to the midgut by the flagellum. The *L. major* promastigote (Pr) is seen attached by the flagellum (Fl) on the microvilli (Mv) of the midgut epithelium of the sand fly *P. papatasi.* Left inset. LPG binding sites over the epithelium microvilli labeled with anti-LPG/gold colloidal particles. Electron micrograph from Paulo Pimenta. (Originally published in Saraiva et al. 1995)

tant and led the first study in 1992 by Pimenta and colleagues about LPG's specific functions in the interaction process with insect vectors. This was the first parasite molecule to be recognized as being active in the interaction process in any pathogen–vector studies. LPG function was first studied in the natural pair *P. papatasi* and *L. major* (Pimenta et al. 1992). LPG was found to expressed specifically in *Leishmania* promastigote, and it covers its entire surface, including the flagellum, being organized as a dense filamentous cellular coat (Pimenta et al. 1989, 1991); since then, it has been found in all *Leishmania* species studied to date (Sacks et al. 2000; Sacks 2001). The characterization of LPG from *L. infantum chagasi* and *L. braziliensis*, two important species circulating in Brazil (Soares et al. 2002, 2005), has also been elucidated.

The LPG molecule consists of a glycan core with a highly conserved lipid anchor and variable oligosaccharide units. LPG's performance in promastigote binding is performed through its saccharide units, which bind specifically to the microvilli of the sand-fly midgut epithelial cells (Pimenta et al. 1992; review Sacks et al. 1995). LPG's role in mediating adhesion was confirmed by studies using promastigote mutants, which are deficient in specific saccharide-unit expression (Butcher et al. 1996) or deficient in total molecule expression (Pimenta et al. 1995; Svárovská et al. 2010; Secundino et al. 2010). In the experiments with LPG-deficient mutants, it was impossible to retain the sand-fly infection after blood-meal excretion because there was no promastigote adhesion to the midgut epithelium. Other studies showed that the vector competence of the sand flies to become infected by certain *Leishmania* species is controlled by an LPG polymorphism on the promastigote surface (Pimenta et al. 1995). The LPG molecule has a large polymorphism on the parasite surface varying its saccharide units in accordance with the *Leishmania* species observed (McConville and Blackwell 1991). Comparative studies conducted with *L. amozonensis*, *L. donovani*, *L. infantum* and *L. major* revealed that in all these species, the ability of promastigotes to adhere to the gut is directly controlled by the LPG structural polymorphism, which is associated with specific receptors present in the microvilli of specific sand-fly species (Pimenta et al. 1995; Kamhawi et al. 2004). In 1994, it was suggested that *Leishmania* LPG should have specific cell receptors (lectin types) in the midgut and that this receptor would be responsible for promastigote adhesion on the epithelium (Pimenta et al. 1994). Only 10 years later, Kamhawi and colleagues (2004) demonstrated for *P. papatasi* and *L. major* the existence of such specific receptors (*P. papatasi* galectin) connecting LPG with sand-fly midgut epithelial microvilli. The results of this study demonstrated that the adhesion mechanism via LPG, which was performed using the natural vector pair *P. papatasi* and *L. major*, plays an important role in the transmission of cutaneous leishmaniasis (Fig. 2). However, it is worth noting thus far that the literature does not clearly demonstrate adherence to the epithelium in other sand-fly species and the possible role of other molecules in this interaction.

Leishmania Metacyclogenesis Inside Sand-Fly Midgut: Promastigote Differentiation into Vertebrate Infective Forms

The parasites' differentiation into metacyclic and subsequent transmission to vertebrate hosts can be considered one of the latest barriers to the process of *Leishmania*–sand fly interaction. After promastigote adhesion to midgut epithelium, which ensures life-cycle continuity within the sand fly, a series of multiplication, changes and differentiation in parasite forms, called "metacyclogenesis," occurs, and this imperative phenomenon warranties competence. Metacyclogenesis is a differentiation process that produces infective metacyclic promastigotes that are able to be transmitted by sand-fly bites and initiate infection in a vertebrate. Metacyclic is considered the only *Leishmania* promastigote form able to initiate the parasite life cycle in a vertebrate. Since the initial studies demonstrating that leishmaniasis are transmitted by sand-fly vectors, there has been suspicion of a specific promastigote form within the vector that adapted to live in vertebrates. Studies by Sacks and colleagues (1984, 1987), comparing promastigotes obtained at subsequent days after sand-fly infective blood feeding, demonstrated conclusively that *Leishmania* parasites are not uniform in transmitting

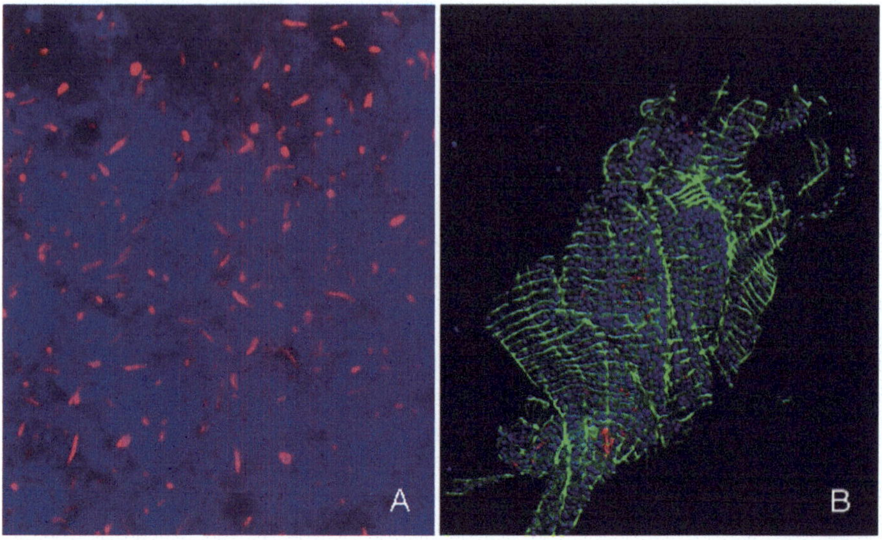

Fig. 2 Scanning confocal microscopy of *P. duboscqi* midguts infected with *L. major*. (**a**) Midgut dissected 9 days after infection. Note in blue (DAPI) the epithelial cells and in red the *Leishmania* promastigote adhered to the midgut epithelium. (**b**) *P. duboscqi* infected midgut dissected 14 days after infection. Note muscle fibers in green (FTC-phalloidin), midgut epithelial cells in blue (DAPI); and red parasites (RFP-expressing strain of *L. major*). (Unpublished photos from Nágila Secundino)

infection to vertebrates. These investigators noted that promastigote populations with the ability to infect vertebrates were gradually obtained only after the fourth day of vector infection. Thus, this experiment demonstrated the existence of a process within the sand flies, i.e., metacyclogenesis, that allowed the development of infective promastigotes being able to initiate infection in a vertebrate host.

After this initial study, many other studies have contributed to our improved knowledge of the metacyclogenesis process because it was possible to obtain metacyclic promastigotes in old axenic cultures and to purify them by agglutination methods with lectins (Sacks et al. 1985). These studies demonstrated that metacyclics are highly differentiated with distinct morphological characteristics, particularly the presence of a very dense cellular coat constituted essentially by LPG (Pimenta et al. 1989, 1991). Subsequently, it was demonstrated that during metacyclogenesis, considerable structural changes occur in the promastigote surface with LPG changing the types and number of saccharide units. The metacyclic LPG molecule becomes about three to five times longer than that in other promastigotes, thus increasing the thickness of the *Leishmania* cell coat (Pimenta et al. 1989). The use of different techniques associated with electron microscopy, immune-staining and cryofracture by Pimenta and collaborators (1989) showed that LPG was present on the *Leishmania* surface as densely filamentous structures distributed exclusively on the metacyclic surface. Other additional studies with different LPG fractions demonstrated that these modifications are crucial for procyclic adhesion and metacyclic subsequent release from midgut epithelium microvilli of sand-fly vectors (Pimenta et al. 1992). The occurrence of these associated phenomena allows *Leishmania* movement to the anterior portion of the sand-fly digestive tract so they can be inoculated into a new vertebrate host. In addition, studies from the same research group with infected sand fly, using a technique that forced infected sand-fly feeding through microcapillaries, demonstrated that only metacyclic inoculated by sand flies and these promastigotes are identical, including their LPG expression, to those obtained *in vitro* (Saraiva et al. 1995).

So, after metacyclogenesis, the metacyclic migrate to the foregut region and can be accumulated in the cardia (a ring coated with cylindrical epithelial cells whose function is control and direct the blood flow by preventing the blood to be regurgitated). Promastigote adhered to this organ can make cuticle damages causing a dysfunction on the blood flow control (Schelein et al. 1992; Volf et al. 2004), than the infected vector regurgitate the metacyclics on the vertebrate skin in the moment of the bite. Moreover, because the insect is not able to ingest enough blood, it repeatedly stings the same or even other vertebrates providing a more efficient parasite transmission.

Finally, studies of metacyclic interaction with mammals (da Silva et al. 1989; Puentes et al. 1990) showed that the metacyclic cell coverage is what permits *Leishmania* survival within the vertebrate. Metacyclics have resistance to complement mediated lysis, which probably promotes their adherence and ingestion by macrophages by way of appropriate receptors. However, it is interesting to note that, after several attempts, the experimental *Leishmania* transmission to vertebrates only occurred in the year 2003 (Kamhawi et al. 2003) with *P. papatasi* infecting mice with *L. major*. This was an important study model to study the roles of sand-fly bite site and saliva in vertebrate *Leishmania* transmission.

In conclusion, metacyclogenesis is a phenomenon occurring inside the sand-fly midgut that makes the parasite pre-adapted to survive within the vertebrate, thus ensuring the continuity of their life cycle.

The Role of the Sand Fly Saliva in *Leishmania* Transmission to the Vertebrate Host

Studies of the role of the vector's saliva *Leishmania* transmission to the vertebrate host were important and affirmed that the insect is not only leading parasites to their target. The initial study in this field was performed by Titus and Ribeiro (1988) and opened up a whole new study field that came to the top with the production of DNA vaccines based on saliva components to combat leishmaniasis. This will presented and discussed next.

The chemical components present in saliva not only modulate the vertebrate's host response, they also assist in locating the blood vessel and determining the parasite's transformation. During the sand-fly bite to obtain the blood meal, the sand fly introduces the mouthparts into the host skin along with its saliva, which assists in the rapid localization of blood vessels. After these vessels are lacerated, the saliva is responsible for blood-flow maintenance, thus promoting rapid blood feeding by the sand fly. Next follows a series of events, i.e. adaptive strategies for the more effective acquisition of blood. The main saliva functions include platelet-aggregation inhibition, anticoagulation, vasodilation and anesthetic function (Kamhawi review in 2000). After blood-vessel laceration, vertebrate host response occurs through hemostasis (platelet activation and coagulation factors, fibrinogen conversion to fibrin) and inflammatory reactions such as erythema, swelling and pain (Ribeiro et al. 1995; Kamhawi 2000).

More than a decade ago, Ribeiro and Titus reported that a small amount of sand-fly saliva could increase infection when inoculated with *Leishmania* promastigotes, thus demonstrating for the first time the role of sand-fly saliva role in disease development (1988). The lesion size and the parasite numbers were increased dramatically when saliva was added to the inoculum during experimental laboratory animal infection. Furthermore, the capacity of the vector's saliva to increase the infectivity of *Leishmania* is restricted to sand-fly saliva because saliva of other arthropods did not produce the same effect (Howard 1986; Ribeiro and Titus 1988). In addition, the increased infectivity due to the saliva was demonstrated with susceptible and resistant mice co-inoculated with *L. major* along with the saliva of *P. papatasi*, both vectors and parasites existing in the Old World (Mbow et al. 1988; Belkaid et al.1998). Until recently, the *L.* salivas of *L. longipalpis* and *P. papatasi* are the most studied species because they are more easily colonized and are recognized as having immunomodulatory activity (see review in Kamawahi 2000).

A pathogenic effect of saliva components was suggested by Warburg et al. (1994). They proposed a possible role in the clinical manifestation related to the variability of *Leishmania* infection inducing or not the *Leishmania* visceralization. They started from the initial observation that *L. infantum chagasi* transmitted by *L. longipalpis* normally causes visceral leishmaniasis in Brazil and Colombia but not in Central America where the infections usually result in skin lesions. The authors found that the molecule maxidilan, a vessel dilator and erythema inducer, found exclusively in sand-fly saliva, has more potent activity in sand fly *L. longipalpis* found in Brazil and Colombia than in Costa Rica. The authors concluded that the maxidilan amount inoculated by the *L. longipalpis* infected *L. infantum chagasi* determines whether the parasites migrate or not to the vertebrate viscera.

In 1998, Belkaid and colleagues developed a natural cutaneous leishmaniasis model in mice to study the effect of saliva in detail. The authors infected mice ears and observed the subsequent lesion development and the acute and chronic inflammation. The authors confirmed that *P. papatasi* saliva enhances infection by *L. major*. The substances present in the saliva causing the increased skin lesion probably are similar to the components found in the saliva of *L. whitmani*, a New World vector species, because the same phenomenon was also shown in mice infected with *L. braziliensis* by this vector (Bezerra and Teixeira 2001). It was also shown that sand-fly saliva antigenic substances are capable of inducing delay-type hypersensibility (DTH) as observed in experimental and human hosts (Belkaid et al. 2000). In Brazil, a significant correlation between the anti-gland immunoglobulin G levels and DTH response in children was found in an endemic area of visceral leishmaniasis (Barral et al. 2000).

Furthermore, protection against *L. major* infection has been shown in mice models when they are pre-exposed or pre-immunized with the saliva of uninfected vectors saliva (Volf et al. 2000; Belkaid et al. 1998). Further studies have shown that saliva components may be candidates for a leishmaniasis vaccine because mice pre-exposed to saliva and inoculated with parasites were protected and associated with a strong DTH reaction (Valenzuela et al. 2002; Kamhawi et al. 2000a, Elnaiem et al. 2005). Exacerbation of the infection was abolished in pre-exposed mice to *P. papatasi* saliva and challenged with *L. major* (Belkaid et al. 1998). Recent data suggest that the response to vector saliva can also be used to monitor human exposure and that of other vertebrate hosts bitten by the sand flies; it can be used as a risk marker for *Leishmania* transmission in endemic areas. In addition to findings from relevant saliva studies, the 15-kDa *P. papatasi* saliva protein was able to protect mice against infection with *L. major*. A vaccine containing the cDNA for this 15-kDa protein led to the same protection in B-cell knockout mice. These results indicated that the DTH response against saliva confers the vaccine protective effect, and salivary components or their cDNAs can be used in vaccine production against leishmaniasis (Valenzuela et al. 2001). Updated literature on this subject can be found in the papers from Valenzuela's group (Anderson et al. 2006; Oliveira et al. 2008; Teixeira et al. 2010; Hosseini-Vasoukolaei et al. 2016; Ferreira et al. 2016).

All studies on the effect of vector saliva related with exacerbation or protection against *Leishmania* infection were performed using laboratory-colonized vectors. However, using the murine model, recent studies have shown significant differences in the exacerbation effect caused by colonized, or wild, *L. longipalpis* salivary gland lysates (Laurenti et al. 2009a, b). The saliva of colonized sand flies co-inoculated with *Leishmania* parasites showed a stronger exacerbation effect, causing injuries twice larger than those caused by the saliva of wild sand-flies inoculated with parasites. Furthermore, it was suggested that the wild-insect saliva does not have the same chemotactic factors encountered in the saliva of colonized sand flies. The protein amount and composition found in the saliva of both groups were also significantly different (Laurenti et al. 2009a). In another study, the same authors suggested that this difference could explain the lesser modulation effect on the *Leishmania* infection observed in mice co-inoculated with parasites and saliva from wild sand fly compared with those co-inoculated with parasites and saliva from colonized sand fly (Laurenti et al. 2009b). Similar studies using the natural pair *P. papatasi* and *L. major* have shown similar results (Ahmed et al. 2010).

Leishmania–Transmission Mechanisms by Infected Sand-Fly Bite

After metacyclogenesis, it is possible to detect metacyclic forms in the foregut region, and the insect is able to transmit *Leishmania*. However, it is believed that simply being infected may influence the vector behavior as we will discuss below.

Different researchers believe in a physical or biological blockage of the *Leishmania*, which is predominant for transmission efficient because fast blood intake is lacking, meaning that the sand fly must bite the host's skin multiple times. Killick-Kendrick and Molyneaux (1981) suggested a transmission mechanism in which metacyclic interferes directly in the mouthpart sensilla, which controls probing and feeding, thus influencing the blood rate and direction in the alimentary canal and promoting the parasite's release on the skin. The stomodeal-valve physical damage was proposed attributed to the *Leishmania* chitinase action (Schlein et al. 1992; Volf 2004). The authors suggested that valve physical damage promotes or facilitates infective metacylic regurgitation into the vertebrate's skin.

Other investigators, however, have described the transmission block to biological parasite masses being embedded in a gel-like matrix in the stomodeal valve (Warburg et al. 1986; Lawyer et al. 1987, 1990; Walters et al. 1987, 1989 Stierhof et al. 1999; Rogers et al. 2002). This gel-like substance secreted by the parasite in the vector midgut is called promastigote secretory gel (PMG), and it is also responsible for mechanical dysfunction on the stomodeal valve formed by plug pressure. This gel was found in *L. longipalpis* infected with *L. mexicana* and *P. papatasi* infected with *L. major*. This obstruction is common in all *Leishmania*–vector combinations studied so far (Stierhof et al. 1999; Rogers et al. 2002, 2004), and it is caused by PPG, a mucin-like substance secreted by parasites inside the midgut. High concentration and limited phlebotomine gut space induce PPGs in the form of PSG (Stierhof et al. 1999). Thereby, it is believed that PSG blocking alters sand fly–probing behavior, which increases the number of attempts to bite and the feeding time (Killick-Kendrick et al. 1977; Beach et al. 1985; Rogers et al. 2002; Rogers and Bates 2007; Bates 2008).

In conclusion, after vessel laceration, saliva is responsible for blood-flow maintenance, thus promoting rapid feeding. These events, both physical and biological, could cause changes in the insect's behavior to facilitate parasite deposition. Actions such as (a) vertebrate tissue injury caused by the bite, (b) PSG, and (c) saliva would help the sand fly retain its meal and facilitate successful parasite entry, thus exacerbating the infection (Schlein et al. 1992; Volf et al. 2004; Rogers et al. 2004; Peters et al. 2008; Oliveira et al. 2008).

After several attempts, *Leishmania* experimental transmission occurred only in the year 2003 (Kamhawi et al. 2003) with the model of *P. papatasi* and *L. major* infecting mice. This model allowed several groups to understand disease transmission from the vector–host point of view, i.e. by bite-like natural transmission. In summary, two aspects were relevant: the bite and the *Leishmania* dose transmitted to the host as follows:

(a) Sand-fly bite: The infected sand-fly female inoculates exclusively infective metacyclic promastigote forms into the host's skin. These are phagocytosed by macrophages, directly or after neutrophil infection, and are rapidly recruited to the bite site (van Zandbergen et al. 2004; Peters et al. 2008). Images of the parasite-transmission process showed a significant role of neutrophils, which are rapidly attracted to the bite site, thus ensuring parasite survival in the early period of infection (Peters et al. 2008).
(b) Sand-fly transmitted dose: In the literature, all knowledge generated about the infection process and the establishment of *Leishmania* in vertebrate hosts has come from studies on subcutaneous or intradermal parasite inoculation.

The traditional, routine experiments use different amounts of parasites (10^2–10^7 parasites/mL) (Belkaid et al. 1998, 2000). The parasite number in the inoculum was appropriate due to the need to establish infection in the natural host without, however, considering the *Leishmania* number inoculated by the sand-fly vector. It has been suggested that the amount of *L. mexicana* regurgitated by *L. longipalpis* varies (10–10,000 parasites). However, this study was conducted using an experimental model, and the parasites were collected through a membrane apparatus (Rogers et al. 2004). Finally, a quantitative polymerase chain reaction study using *P. papatasi* infected with *L. major* demonstrated that the sand fly could inoculate 100–100,000 parasites into the host's tissue. Dose distribution showed a binomial profile: About 75% of the sand flies released ≤600 promastigotes, whereas the other 25% injected >1000 parasites. High doses of infection were strongly associated with sand-fly infection that presented at least 30,000 parasites/midgut (Kimblin et al. 2008). A similar study established a transmission model of *L. infantum chagasi* by the bite of *L. longipalpis*, the vector of American visceral leishmaniasis. The parasites were successfully transmitted by infected sand-fly bites to mice and hamsters (Fig. 3), thus indicating that both animals are good experimental models. The *L. infantum chagasi* dose that was transmitted in each single bite ranged from 10–10,000 parasites, but 75% of the sand flies transmitted <300 parasites (Secundino et al. 2012). These two studies elucidated the *Leishmania*-transmission mechanism by vector bites of both New World and Old World sand flies.

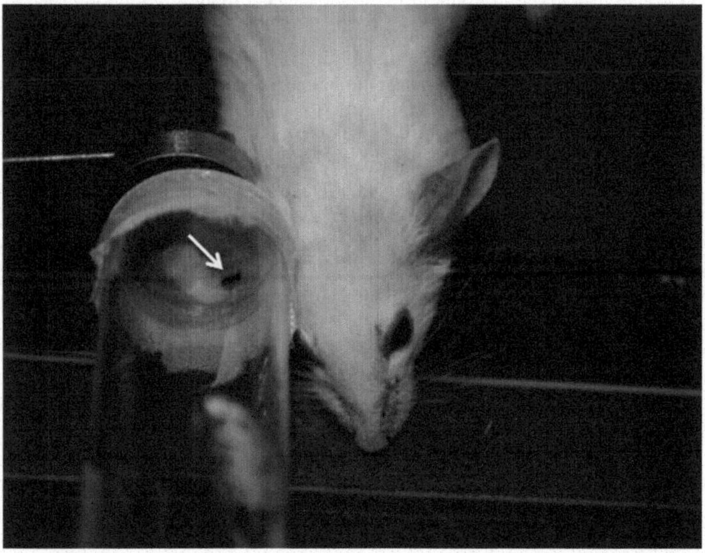

Fig. 3 Experimental transmission of *L. infantum chagasi* by the bite of infected sand fly *L. longipalpis* in an ear of Balb/C mouse. A single sand fly was confined within a vial and allowed to bite the animal's entire ear. Note the blood-meal engorgement of the sand fly (white arrow). (Originally published in Secundino et al. 2012)

Acknowledgements The authors Paulo F. P. Pimenta and Nágila F. C. Secundino received financial support from Fiocruz, CNPq e Fapemig.

References

Adler S (1938) Factors determining the Behaviour of Leishmania sp. in Sandflies. Harefuah 14

Adler S (1964) Leishmania. Adv Parasitol 2:35–96

Ahmed SB, Kaabi B, Chelbi I, Derbali M, Cherni S, Laouini D, Zhioua E (2010) Lack of protection of pre-immunization with saliva of long-term colonized Phlebotomus papatasi against experimental challenge with Leishmania major and saliva of wild-caught P. papatasi. Am J Trop Med Hyg 83:512–514. PM:20810812

Anderson JM, Oliveira F, Kamhawi S, Mans BJ, Reynoso D, Seitz AE, Lawyer P, Garfield M, Pham M, Valenzuela JG (2006) Comparative salivary gland transcriptomics of sandfly vectors of visceral leishmaniasis. BMC Genomics 15:7–52

Barral A, Honda E, Caldas A, Costa J, Vinhas V, Rowton ED, Valenzuela JG, Charlab R, Barral-Netto M, Ribeiro JM (2000) Human immune response to sand fly salivary gland antigens: a useful epidemiological marker? Am J Trop Med Hyg 62:740–745. PM:11304066

Bates PA (2008) Leishmania sand fly interaction: progress and challenges. Curr Opin Microbiol 11:340–344. PM:18625337

Beach R, Kiilu G, Leeuwenburg J (1985) Modifications of sand fly biting behavior by Leishmania leads to increased parasite transmission. Am J Trop Med Hyg 34:278–282

Belkaid Y, Kamhawi S, Modi G, Valenzuela J, Noben-Trauth N, Rowton E, Ribeiro J, Sacks DL (1998) Development of a natural model of cutaneous leishmaniasis: powerful effects of vector saliva and saliva preexposure on the long-term outcome of Leishmania major infection in the mouse ear dermis. J Exp Med 188:1941–1953. PM:9815271

Belkaid Y, Mendez S, Lira R, Kadambi N, Milon G, Sacks D (2000) A natural model of Leishmania major infection reveals a prolonged "silent" phase of parasite amplification in the skin before the onset of lesion formation and immunity. J Immunol 165:969–977. PM:10878373

Berner R, Rudin W, Hecker H (1983) Peritrophic membranes and protease activity in the midgut of the malaria mosquito, Anopheles stephensi (Liston) (Insecta: Diptera) under normal and experimental conditions. J Ultrastruct Res 83:195–204

Bezerra HS, Teixeira MJ (2001) Effect of Lutzomyia whitmani (Diptera: Psychodidae) salivary gland lysates on Leishmania (Viannia) braziliensis infection in BALB/c mice. Mem Inst Oswaldo Cruz 96:349–351. PM:11313642

Billingsley PF, Rudin W (1992) The role of the mosquito peritrophic membrane in blood meal digestion and infectivity of Plasmodium species. J Parasitol 78:430–440. PM:1597785

Borovsky D, Schlein Y (1987) Trypsin and chymotrypsin-like enzymes of the sand fly Phlebotomus papatasi infected with Leishmania and their possible role in vector competence. Med Vet Entomol 1:235–242. PM:2979536

Boulanger N, Lowenberger C, Volf P, Ursic R, Sigutova L, Sabatier L, Svobodova M, Beverley SM, Spath G, Brun R (2004) Characterization of a defensin from the sand fly Phlebotomus duboscqi induced by challenge with bacteria or the protozoan parasite Leishmania major. Infect Immun 72:7140–7146. PM:15557638

Butcher BA, Turco SJ, Hilty BA, Pimenta PF, Panunzio M, Sacks DL (1996) Deficiency in beta1,3-galactosyltransferase of a Leishmania major lipophosphoglycan mutant adversely influences the Leishmania-sand fly interaction. J Biol Chem 271:20573–20579. PM:8702802

da Silva RP, Hall BF, Joiner KA, Sacks DL (1989) CR1, the C3b receptor, mediates binding of infective Leishmania major metacyclic promastigotes to human macrophages. J Immunol 143:617–622. PM:2525590

Elnaiem DA, Hassan HK, Ward RD (1997) Phlebotomine sandflies in a focus of visceral leishmaniasis in a border area of eastern Sudan. Ann Trop Med Parasitol 91:307–318. PM:9229023

Elnaiem DE, Meneses C, Slotman M, Lanzaro GC (2005) Genetic variation in the sand fly salivary protein, SP-15, a potential vaccine candidate against Leishmania major. Insect Mol Biol 14:145–150. PM:15796747

Feng LC (1951) The role of the peritrophic membrane in leishmania and Trypanosome infections of sandflies. Pek Nat Hist Bull 19:327–334

Ferreira VP, Fazito VV, Pangburn MK, Abdeladhim M, Mendes-Sousa AF, Coutinho-Abreu IV, Rasouli M, Brandt EA, Meneses C, Lima KF (2016) SALO, a novel classical pathway complement inhibitor from saliva of the sand fly Lutzomyia longipalpis. Sci Rep 6:19300. PM:26758086

Gemetchu T (1974) The morphology and fine structure of the midgut and peritrophic membrane fo the adult female, Phlebotomus longipes Parrot and Martin (Diptera: Psychodidae). Ann Trop Med Parasitol 68:111–124

Hosseini-Vasoukolaei N, Idali F, Khamesipour A, Yaghoobi-Ershadi MR, Kamhawi S, Valenzuela JG, Edalatkhah H, Arandian MH, Mirhendi H, Emami S (2016) Differential expression profiles of the salivary proteins SP15 and SP44 from Phlebotomus papatasi. Parasit Vectors 9:357. PM:27342811

Howard RF, Ardeshir F, Reese RT (1986) Conservation and antigenicity of N-terminal sequences of GP185 from different Plasmodium falciparum isolates. Gene 46:197–205

Kamhawi S (2000) The biological and immunomodulatory properties of sand fly saliva and its role in the establishment of Leishmania infections. Microbes Infect 2:1765–1773. PM:11137049

Kamhawi S, Belkaid Y, Modi G, Rowton E, Sacks D (2000a) Protection against cutaneous leishmaniasis resulting from bites of uninfected sand flies. Science 290:1351–1354. PM:11082061

Kamhawi S, Modi GB, Pimenta PF, Rowton E, Sacks DL (2000b) The vectorial competence of Phlebotomus sergenti is specific for Leishmania tropica and is controlled by species-specific, lipophosphoglycan-mediated midgut attachment. Parasitology 121(Pt 1):25–33. PM:11085222

Kamhawi S, Ramalho-Ortigao M, Pham VM, Kumar S, Lawyer PG, Turco SJ, Barillas-Mury C, Sacks DL, Valenzuela JG (2003) A role for insect galectins in parasite survival. Cell 119:329–341

Kamhawi S, Ramalho-Ortigao M, Pham VM, Kumar S, Lawyer PG, Turco SJ, Barillas-Mury C, Sacks DL, Valenzuela JG (2004) A role for insect galectins in parasite survival. Cell 119:329–341. PM:15543683

Killick-Kendrick RR, Leaney AJ, Ready PD, Molyneux DH (1977) Leishmania in Phlebotominae sandflies. IV. The transmission of Leishmania mexicana amazonensis to hamster by bite of experimentally infected Lutzomyia longipalpis. Proc R Soc Lond 196:105–115

Killick-Kendrick R, Molyneux DH (1981) Transmission of leishmaniasis by the bite of phlebotomine sandflies: possible mechanisms. Trans R Soc Trop Med Hyg 75:152–154. PM:7268854

Kimblin N, Peters N, Debrabant A, Secundino N, Egen J, Lawyer P, Fay MP, Kamhawi S, Sacks D (2008) Quantification of the infectious dose of Leishmania major transmitted to the skin by single sand flies. Proc Natl Acad Sci U S A 105:10125–10130. PM:18626016

Lainson R, Shaw JJ (1979) The role of animals in the epidemiology of South American leishmaniasis. pp 1–116 In: Lumsden WHR, Evans DA (eds) The Biology of the Kinetoplastida. Academic Press London, 738p

Laurenti MD, da Matta VL, Pernichelli T, Secundino NF, Pinto LC, Corbett CE, Pimenta PP (2009a) Effects of salivary gland homogenate from wild-caught and laboratory-reared Lutzomyia longipalpis on the evolution and immunomodulation of Leishmania (Leishmania) amazonensis infection. Scand J Immunol 70:389–395. PM:19751274

Laurenti MD, Silveira VM, Secundino NF, Corbett CE, Pimenta PP (2009b) Saliva of laboratory-reared Lutzomyia longipalpis exacerbates Leishmania (Leishmania) amazonensis infection more potently than saliva of wild-caught Lutzomyia longipalpis. Parasitol Int 58:220–226. PM:19454323

Lawyer PG, Young DG, Butler JF, Akin DE (1987) Development of Leishmania mexicana in Lutzomyia diabolica and Lutzomyia shannoni (Diptera: Psychodidae). J Med Entomol 24:347–355

Lawyer PG, Ngumbi PM, Anjili CO, Odongo SO, Mebrahtu YB, Githure JI, Koech DK, Roberts CR (1990) Development of Leishmania major in Phlebotomus duboscqi and Sergentomyia schwetzi (Diptera: Psychodidae). Am J Trop Med Hyg 43:31–43

Malta J, Martins GF, Weng JL, Fernandes KM, Munford ML, Ramalho-Ortigao M (2016) Effects of specific antisera targeting peritrophic matrix-associated proteins in the sand fly vector Phlebotomus papatasi. Acta Trop 159:161–169. PM:27012717

Mbow ML, Bleyenberg JA, Hall LR, Titus RG (1998) Phlebotomus papatasi sand fly salivary gland lysate downregulates a Th1, but up-regulates a Th2, response in mice infected with Leishmania major. J Immunol 161(10):5571–5577

McConville MJ, Blackwell JM (1991) Developmental changes in the glycosylated phosphatidylinositols of Leishmania donovani. Characterization of the promastigote and amastigote glycolipids. J Biol Chem 266:15170–15179. PM:1831200

Miller N, Lehane MJ (1993) Peritrophic membranes, cell surface molecules and parasite tropisms within arthropod vectors. Parasitol Today 9:45–50. PM:15463702

Miranda JC, Secundino NF, Nieves E, Souza AP, Bahia AC, Prates DB, Pimenta RN, Pinto LC, Barral A, Pimenta PFP (2008) Studies of the influence of the presence of domestic animals on increasing the transmission probabilities of leishmaniasis. Ann Med Entomol 17:9–15

Myskova J, Svobodova M, Beverley SM, Volf P (2007) A lipophosphoglycan-independent development of Leishmania in permissive sand flies. Microbes Infect 9:317–324. PM:17307009

Nieves E, Pimenta PF (2000) Development of Leishmania (Viannia) braziliensis and Leishmania (Leishmania) amazonensis in the sand fly Lutzomyia migonei (Diptera: Psychodidae). J Med Entomol 37:134–140. PM:15218917

Nieves E, Pimenta PF (2002) Influence of vertebrate blood meals on the development of Leishmania (Viannia) braziliensis and Leishmania (Leishmania) amazonensis in the sand fly Lutzomyia migonei (Diptera: Psychodidae). Am J Trop Med Hyg 67:640–647. PM:12518856

Oliveira F, Lawyer PG, Kamhawi S, Valenzuela JG (2008) Immunity to distinct sand fly salivary proteins primes the anti-Leishmania immune response towards protection or exacerbation of disease. PLoS Negl Trop Dis 2:e226. PM:18414648

Pascoa V, Oliveira PL, Dansa-Petretski M, Silva JR, Alvarenga PH, Jacobs-Lorena M, Lemos FJ (2002) Aedes aegypti peritrophic matrix and its interaction with heme during blood digestion. Insect Biochem Mol Biol 32:517–523. PM:11891128

Peters W (1992) Peritrophic membranes. Springer, New York

Peters NC, Egen JG, Secundino N, Debrabant A, Kimblin N, Kamhawi S, Lawyer P, Fay MP, Germain RN, Sacks D (2008) In vivo imaging reveals an essential role for neutrophils in leishmaniasis transmitted by sand flies. Science 321:970–974. PM:18703742

Pimenta PF, da Silva RP, Sacks DL, da Silva PP (1989) Cell surface nanoanatomy of Leishmania major as revealed by fracture-flip. A surface meshwork of 44 nm fusiform filaments identifies infective developmental stage promastigotes. Eur J Cell Biol 48(2):180–190

Pimenta PF, Saraiva EM, Sacks DL (1991) The comparative fine structure and surface glycoconjugate expression of three life stages of Leishmania major. Exp Parasitol 72(2):191–204

Pimenta PF, Turco SJ, McConville MJ, Lawyer PG, Perkins PV, Sacks DL (1992) Stage-specific adhesion of Leishmania promastigotes to the sand fly midgut. Science 256:1812–1815. PM:1615326

Pimenta PF, Saraiva EM, Rowton E, Modi GB, Garraway LA, Beverley SM, Turco SJ, Sacks DL (1994) Evidence that the vectorial competence of phlebotomine sand flies for different species of Leishmania is controlled by structural polymorphisms in the surface lipophosphoglycan. Proc Natl Acad Sci U S A 91:9155–9159. PM:8090785

Pimenta PF, McConville MJ, Schneider P, Turco SJ (1995) Stage-specific binding of Leishmania donovani to the sand fly vector midgut is regulated by conformational changes in the abundant surface lipophosphoglycan. J Exp Med 181(2):685–697

Pimenta PF, Modi GB, Pereira ST, Shahabuddin M, Sacks DL (1997) A novel role for the peritrophic matrix in protecting Leishmania from the hydrolytic activities of the sand fly midgut. Parasitology 115(Pt 4):359–369. PM:9364562

Pruzinova K, Sadlova J, Seblova V, Homola M, Votypka J, Volf P (2015) Comparison of blood meal digestion and the peritrophic matrix in four sand fly species differing in susceptibility to Leishmania donovani. PLoS One 10:e0128203. PM:26030610

Puentes SM, da Silva RP, Sacks DL, Hammer CH, Joiner KA (1990) Serum resistance of metacyclic stage Leishmania major promastigotes is due to release of C5b-9. J Immunol 145:4311–4316. PM:2147941

Ribeiro JM, Schneider M, Guimaraes JA (1995) Purification and characterization of prolixin S (nitrophorin 2), the salivary anticoagulant of the blood-sucking bug Rhodnius prolixus. Biochem J 308(Pt 1):243–249. PM:7755571

Richards AG, Richards PA (1977) The peritrophic membranes of insects. Annu Rev Entomol 22:219–240. PM:319739

Rogers ME, Bates PA (2007) Leishmania manipulation of sand fly feeding behavior results in enhanced transmission. PLoS Pathog 3:e91. PM:17604451

Rogers ME, Chance ML, Bates PA (2002) The role of promastigote secretory gel in the origin and transmission of the infective stage of Leishmania mexicana by the sand fly Lutzomyia longipalpis. Parasitology 124:495–507. PM:12049412

Rogers ME, Ilg T, Nikolaev AV, Ferguson MA, Bates PA (2004) Transmission of cutaneous leishmaniasis by sand flies is enhanced by regurgitation of fPPG. Nature 430:463–467. PM:15269771

Sacks DL, Kenney RT, Kreutzer RD, Jaffe CL, Gupta AK, Sharma MC, Sinha SP, Neva FA, Saran R (1995) Indian kala-azar caused by Leishmania tropica. Lancet 345(8955):959–961

Sacks DL (2001) Leishmania-sand fly interactions controlling species-specific vector competence. Cell Microbiol 3:189–196. PM:11298643

Sacks DL, da Silva RP (1987) The generation of infective stage Leishmania major promastigotes is associated with the cell-surface expression and release of a developmentally regulated glycolipid. J Immunol 139:3099–3106. PM:3312412

Sacks DL, Perkins PV (1984) Identification of an infective stage of Leishmania promastigotes. Science 223:1417–1419. PM:6701528

Sacks DL, Hieny S, Sher A (1985) Identification of cell surface carbohydrate and antigenic changes between noninfective and infective developmental stages of Leishmania major promastigotes. J Immunol 135:564–569. PM:2582050

Sacks DL, Modi G, Rowton E, Spath G, Epstein L, Turco SJ, Beverley SM (2000) The role of phosphoglycans in Leishmania-sand fly interactions. Proc Natl Acad Sci U S A 97:406–411. PM:10618431

Saraiva EM, Pimenta PF, Brodin TN, Rowton E, Modi GB, Sacks DL (1995) Changes in lipophosphoglycan and gene expression associated with the development of Leishmania major in Phlebotomus papatasi. Parasitology 111(Pt 3):275–287. PM:7567096

Schlein Y, Borut S, Greenblatt CL (1987) Development of sand fly forms of Leishmania major in sucrose solutions. J Parasitol 73:797–805. PM:3625431

Schlein Y, Schnur LF, Jacobson RL (1990) Released glycoconjugate of indigenous Leishmania major enhances survival of a foreign L. major in Phlebotomus papatasi. Trans R Soc Trop Med Hyg 84:353–355. PM:2260168

Schlein Y, Jacobson RL, Shlomai J (1991) Chitinase secreted by Leishmania functions in the sand fly vector. Proc Biol Sci 245:121–126. PM:1682935

Schlein Y, Jacobson RL, Messer G (1992) Leishmania infections damage the feeding mechanism of the sand fly vector and implement parasite transmission by bite. Proc Natl Acad Sci U S A 89:9944–9948. PM:1409724

Secundino NF, Eger-Mangrich I, Braga EM, Santoro MM, Pimenta PF (2005) Lutzomyia longipalpis peritrophic matrix: formation, structure, and chemical composition. J Med Entomol 42:928–938. PM:16465730

Secundino N, Kimblin N, Peters NC, Lawyer P, Capul AA, Beverley SM, Turco SJ, Sacks D (2010) Proteophosphoglycan confers resistance of Leishmania major to midgut digestive enzymes induced by blood feeding in vector sand flies. Cell Microbiol 12:906–918. PM:20088949

Secundino NF, de Freitas VC, Monteiro CC, Pires AC, David BA, Pimenta PF (2012) The transmission of Leishmania infantum chagasi by the bite of the Lutzomyia Longipalpis to two different vertebrates. Parasit Vectors 5:20. PM:22260275

Shahabuddin M, Toyoshima T, Aikawa M, Kaslow DC (1993) Transmission-blocking activity of a chitinase inhibitor and activation of malarial parasite chitinase by mosquito protease. Proc Natl Acad Sci U S A 90:4266–4270. PM:8483942

Shakarian AM, Dwyer DM (1998) The Ld Cht1 gene encodes the secretory chitinase of the human pathogen Leishmania donovani. Gene 208:315–322. PM:9524285

Shakarian AM, Dwyer DM (2000) Pathogenic leishmania secrete antigenically related chitinases which are encoded by a highly conserved gene locus. Exp Parasitol 94:238–242. PM:10831391

Shortt HE, Swaminath CS (1928) The method of feeding of Phlebotomus argentipes with relation to its bearing on the transmission of Kala-azar. Indian J Med Res 15:827–836. https://books.google.com.br/books?id=oK_6OwAACAAJ

Soares RP, Macedo ME, Ropert C, Gontijo NF, Almeida IC, Gazzinelli RT, Pimenta PF, Turco SJ (2002) Leishmania chagasi: lipophosphoglycan characterization and binding to the midgut of the sand fly vector Lutzomyia longipalpis. Mol Biochem Parasitol 121:213–224. PM:12034455

Soares RP, Cardoso TL, Barron T, Araujo MS, Pimenta PF, Turco SJ (2005) Leishmania braziliensis: a novel mechanism in the lipophosphoglycan regulation during metacyclogenesis. Int J Parasitol 35:245–253. PM:15722076

Stierhof YD, Bates PA, Jacobson RL, Rogers ME, Schlein Y, Handman E (1999) Filamentous proteophosphoglycan secreted by Leishmania promastigotes forms gel like three-dimensional networks that obstruct the digestive tract of infected sandfly vectors. Eur J Cell Biol 78:675–689

Svarovska A, Ant TH, Seblova V, Jecna L, Beverley SM, Volf P (2010) Leishmania major glycosylation mutants require phosphoglycans (lpg2-) but not lipophosphoglycan (lpg1-) for survival in permissive sand fly vectors. PLoS Negl Trop Dis 4:e580–PM:20084096

Teixeira C, Gomes R, Collin N, Reynoso D, Jochim R, Oliveira F, Seitz A, Elnaiem DE, Caldas A, de Souza AP, Brodskyn CI, de Oliveira CI, Mendonca I, Costa CH, Volf P, Barral A, Kamhawi S, Valenzuela JG (2010) Discovery of markers of exposure specific to bites of Lutzomyia longipalpis, the vector of Leishmania infantum chagasi in Latin America. PLoS Negl Trop Dis 4(3):e638

Tellam RL (1996) The peritrophic matrix, vol 1. Chapman & Hall, London, pp 86–114

Terra WR (1990) Evolution of digestive systems of insects. Annu Rev Entomol 35:181–200. https://doi.org/10.1146/annurev.en.35.010190.001145

Terra WR, Ferreira C (1994) Insect digestive enzymes: properties, compartmentalization and function. Comp Biochem Physiol B Comp Biochem 109:1–62

Titus RG, Ribeiro JM (1988) Salivary gland lysates from the sand fly Lutzomyia longipalpis enhance Leishmania infectivity. Science 239:1306–1308. PM:3344436

Vaidyanathan R (2004) Leishmania parasites (Kinetoplastida: Trypanosomatidae) reversibly inhibit visceral muscle contractions in hemimetabolous and holometabolous insects. J Invertebr Pathol 87:123–128. PM:15579321

Vaidyanathan R (2005) Isolation of a myoinhibitory peptide from Leishmania major (Kinetoplastida: Trypanosomatidae) and its function in the vector sand fly Phlebotomus papatasi (Diptera: Psychodidae). J Med Entomol 42:142–152. PM:15799523

Valenzuela JG, Belkaid Y, Garfield MK, Mendez S, Kamhawi S, Rowton ED, Sacks DL, Ribeiro JM (2001) Toward a defined anti-Leishmania vaccine targeting vector antigens: characterization of a protective salivary protein. J Exp Med 194:331–342. PM:11489952

Valenzuela JG (2002) High-throughput approaches to study salivary proteins and genes from vectors of disease. Insect Biochem Mol Biol 32(10):1199–1209

van ZG KM, Mueller A, Dannenberg S, Gebert A, Solbach W, Laskay T (2004) Cutting edge: neutrophil granulocyte serves as a vector for Leishmania entry into macrophages. J Immunol 173:6521–6525. PM:15557140

Volf P, Tesarova P, Nohynkova EN (2000) Salivary proteins and glycoproteins in phlebotomine sandflies of various species, sex and age. Med Vet Entomol 14:251–256. PM:11016431

Volf P, Hajmova M, Sadova J, Votypka J (2004) Blocked stomodeal valve of the insect vector: similar mechanism of transmission in two trypanosomatid models. Int J Parasitol 34:1221–1227

Walters LL, Modi GB, Tesh RB, Burrage T (1987) Host-parasite relationship of Leishmania mexicana mexicana and Lutzomyia abonnenci (Diptera: Psychodidae). Am J Trop Med Hyg 36(2):294–314

Walters LL, Modi GB, Chaplin GL, Tesh RB (1989) Ultrastructural development of Leishmania chagasi in its vector Lutzomyia longipalpis (Diptera: Psychodidae). Am J Trop Med Hyg 41(3):295–317

Walters LL, Irons KP, Modi GB, Tesh RB (1992) Refractory barriers in the sand fly Phlebotomus papatasi (Diptera: Psychodidae) to infection with Leishmania panamensis. Am J Trop Med Hyg 46:211–228. PM:1539756

Warburg A, Hamada GS, Schlein Y, Shire D (1986) The effect of post-blood-meal nutrition of Phlebotomus papatasi on the transmission of Leishmania major. Am J Trop Med Hyg 35:926–930

Warburg A, Saraiva E, Lanzaro GC, Titus RG, Neva F (1994) Saliva of Lutzomyia longipalpis sibling species differs in its composition and capacity to enhance leishmaniasis. Philos Trans R Soc Lond Ser B Biol Sci 345(1312):223–230

Sand Fly Vectors of American Cutaneous Leishmaniasis in Brazil

Elizabeth F. Rangel, Ralph Lainson, Bruno M. Carvalho, Simone M. Costa, and Jeffrey J. Shaw

The eco-epidemiology of Brazilian American cutaneous leishmaniasis (ACL) is a complex of epidemiological chains involving different parasites, vectors, and reservoirs. The transmission of the seven *Leishmania* spp. associated with ACL in Brazil involves different phlebotomine species that are closely associated with the parasite's mammalian reservoirs, which range from Xenathra to rodents to primates, resulting in a variety of transmission cycles in the different geographical regions in the country. However, evidence is also accumulating that indicates that there are geographical clades of the different *Leishmania* species that may be associated with different vectors. *Leishmania* species (*L.* (*Leishmania*) *amazonensis*; *L.* (*Viannia*) *braziliensis*; *L.* (*V.*) *guyanensis*; *L.* (*V.*) *lainsoni*; *L.* (*V.*) *shawi*; *L.* (*V.*) *naiffi*; and *L.* (*V.*) *lindenbergi*) are associated with human cutaneous leishmaniasis. However, other Brazilian parasites of the subfamily Leishmaniinae exist, some of which are found in sand flies (*L.* (*V.*) *utingensis*; *Endotrypanum* spp.), whereas others (*L.* (*Mundinia*) *enriettii*; *Porcisia deanei*) have so far not been recorded in them (the parasite nomenclature follows that published in Espinosa et al. 2016).

Over the years well-defined and accepted criteria, data from field studies and—in some cases—experimental results have led to some species being considered as ACL vectors. In our opinion, some are primary vectors, and we will discuss these first. However, an increasing amount of data, principally from molecular studies, suggest that other sand-fly species may be participating in the cycles, and we will discuss these towards the end of this chapter.

E. F. Rangel (✉) · B. M. Carvalho · S. M. Costa
Instituto Oswaldo Cruz, Fundação Oswaldo Cruz, Rio de Janeiro, Brazil
e-mail: efrangel@ioc.fiocruz.br

R. Lainson (deceased)
Instituto Evandro Chagas, Belém, Brazil

J. J. Shaw
Instituto de Ciências Biomédicas, Universidade de São Paulo, São Paulo, Brazil

E. F. Rangel, J. J. Shaw (eds.), *Brazilian Sand Flies*,
https://doi.org/10.1007/978-3-319-75544-1_7

Nyssomyia intermedia (Lutz & Neiva, 1912)

Nyssomyia intermedia was described from specimens collected at Fazenda Ouro Fino in the municipality of Além Paraíba (MG) and also from SP and RJ, where it was found to be abundant in homes. It was one of the first phlebotomine species to be described in the neotropical region (Barretto 1961; Martins et al. 1978; Young and Duncan 1994; Marcondes 1996). The current geographic distribution of *Ny. intermedia* includes Brazil (PA, PI, MA, PE, PR, SC, RS, MG, MS, GO and TO [Fig. 1]), Argentina and Paraguay (Young and Duncan 1994; Marcondes et al. 1997; Rangel and Lainson 2003, 2009).

Recently published projections under climate-change scenarios predict a global decrease in the climatically suitable areas of *Ny. intermedia* in Brazil with a slight expansion in specific areas of the Northeast region for the upcoming decades (McIntyre et al. 2017).

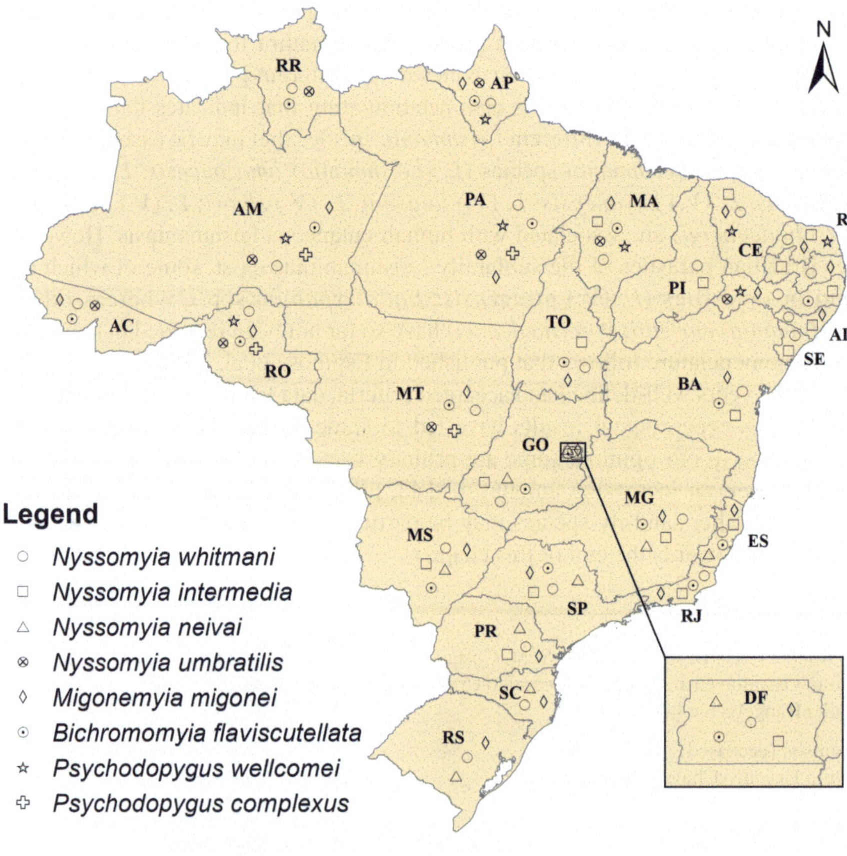

Fig. 1 Distribution of sand-fly vectors of ACL in Brazil

One year after the suggestion of the transmission of cutaneous leishmaniasis by sand flies in the Old World (Sergent et al. 1921), Aragão (1922, 1927) reported in the city of Rio de Janeiro (locality Águas Férreas, currently Cosme Velho and Santa Tereza) the importance of *Ny. intermedia* in the transmission of *L.* (*V.*) *braziliensis* by inoculating a triturated pool of this sand fly in the nose of a dog and thus experimentally reproducing the disease for the first time on the American continent. Subsequently, Costa Lima (1932) recorded the occurrence of *Ny. intermedia* and *Migonemyia migonei* in the neighbourhoods of Copacabana and Laranjeiras, near the slopes of Corcovado and Santa Teresa hills, in Rio de Janeiro. In 1952, Forattini and Santos found promastigotes (referred to as "leptomonads") in *Ny. intermedia*, like those observed by Aragão (1922).

Forattini (1973) pointed out an irregular seasonal behaviour for this species in observing greater population density during the colder months of the year. However, Rangel et al. (1990) reported the occurrence of the vector throughout the year, and Souza et al. (2003) demonstrated that in Rio de Janeiro, *Ny. intermedia* has high density during the hottest times of the year.

Epidemiological evidences suggest *Ny. intermedia* as a vector of *L.* (*V.*) *braziliensis* in endemic areas in the Southeast, considered the main ACL vector in SP and RJ (Forattini 1953, 1973; Forattini et al. 1976; Araújo Filho et al. 1981; Gomes et al. 1986; Rangel et al. 1986, 1990, 1992, 1999; Meneses et al. 2005; Rangel and Lainson 2009). Pita-Pereira et al. (2005) isolated the parasite from *Ny. intermedia*, captured in a focus of ACL in Jacarepaguá (RJ), and identified it as *L.* (*V.*) *braziliensis*. This reinforced all the ecological and epidemiological evidence regarding this vector. In MG and ES, it shares its vector role with *Ny. whitmani* (Barretto 1943; Falqueto 1995; Mayrink et al. 1979; Aguiar & Soucasaux 1984; Souza-Rocha et al. 2007).

Ny. intermedia starts its hematophagous activities at dusk and peaks during the first half of the night. It is abundant inside houses; in peri-domiciliary environments in domestic-animal shelters, such as birds, dogs, equines, and rodents; and in the forest to feed on other wild animals (Forattini 1953, 1973, 1976; Araujo Filho et al. 1981; Rangel et al. 1986; Gomes and Galati 1989; Pirmez et al. 1997; Afonso et al. 2005).

Nyssomia neivai (Pinto, 1926)

This species was described from a sample of males collected in the city of São Paulo at the Butantan Institute. It was considered a synonym of *Ny. intermedia*. Marcondes (1996), however, revalidated *Ny. neivai* as a species after a study of morphological and morphometric characters of specimens collected in Bolivia and the holotypes of both species. The distinction between *Ny. intermedia* and *Ny. neivai* was made possible mainly by morphological differences in the spermathecae (Marcondes 1996; Marcondes and Alexander 2003; Andrade-Filho et al. 2004). The males of the species were not separated by morphology but only by a series of morphometric data

(Marcondes and Borges 2000). According to Andrade-Filho and Brazil (2003), the presence of cryptic species, such as *Ny. intermedia* and *Ny. neivai*, is a result of allopatry, which in the case of these species must have occurred about 250,000 years ago, thus indicating that the separation of species was recent.

Ny. neivai occurs in colder and drier regions of Brazil compared with *Ny. intermedia* (Marcondes et al. 1998, Fig. 1). It was suggested as an important vector mainly in the states of the South and Southeast regions. According to McIntyre et al. (2017), the appropriate climatic range of *Ny. neivai* is currently restricted to the South, Southeast, and Central West regions of Brazil, extending to western Bolivia and Paraguay, like *Ny. intermedia*. The investigators predict that *Ny. neivai* will face changes in its climatic range in the future with the Central West region, in particular, becoming less habitable.

The first record of *Ny. neivai* naturally infected by *L.* (*V.*) spp., possibly *L.* (*V.*) *braziliensis*, was described by Marcondes et al. (1999) using PCR techniques of specimens collected in Piçarras (SC). Later, Silva et al. (2008) found a high incidence of *Ny. neivai* in 37 municipalities of PR where 75.6% of all specimens collected were *Ny. neivai*, thus suggesting their possible participation in the disease cycle. This hypothesis gained strength with the encounter of specimens of *Ny. neivai* infected with *L.* (*V.*) spp. in this state by multiplex PCR techniques (Oliveira et al. 2011). Since 2002, autochthonous cases of ACL by *L.* (*V.*) *braziliensis* have been recorded in the outskirts of the city of Porto Alegre (RS), in vegetation areas near streams (Pita-Pereira et al. 2009). Entomological studies revealed a large predominance of *Ny. neivai* in the area of occurrence of the cases (Gonçalves 2003); later, the natural infection by *L.* (*V.*) *spp.* was found in this species by PCR techniques, thus identifying the phlebotomine as a potential vector of *L.* (*V.*) *braziliensis* (Pita-Pereira et al. 2009).

Recent investigations in the Southeast region suggest that the migration of *Ny. neivai* from forests to residential areas has resulted in the occurrence of ACL cases. However, in a study on the biology of this sand fly in an endemic area in São Paulo, the investigators considered it as having low competence as *Leishmania* vector (Casanova et al. 2009). Studies on the feeding preferences of *Ny. neivai* showed that it is an opportunistic sand fly (feeding on domestic animals such as pigs, dogs, rabbits, and chicken) with few specific host preferences (Dias-Sversutti et al. 2007).

Migonemyia migonei (França, 1920)

The literature suggests that *Mg. migonei* is a sylvatic species found in forests, generally areas of abundant vegetation, occurring less frequently in secondary forests and in capoeiras. However, it is common to observe this phlebotomine inside houses and in domestic animal shelters (Barretto 1943; Forattini 1973; Araújo Filho et al. 1981; Rangel et al. 1986). It is believed that this species has great capacity for adaptation, surviving in degraded areas by man, and approaching impacted environments where it feeds on domestic animals and man, possibly being able to maintain the enzootic cycle of the disease in areas of secondary forest (Ferreira et al. 2001; Queiroz et al. 1994; Rangel and Lainson 2009). It is distributed in Brazil (AP, AC, PA, MA, CE, PB, PE, BA, ES, RJ, SP, PR, SC, RS, MG, MT, Fig. 1), Venezuela, Argentina, Paraguay, Peru, Trinidad, and Tobago (Young and Duncan 1994; Rangel and Lainson 2003).

According to seasonal studies conducted in PE, *Mg. migonei* can be captured during all months of the year; however, other studies have indicated the absence of the vector during the coldest and driest months (Rangel and Lainson 2009; Guimarães et al. 2012).

Its hematophagous activity begins at dusk and advances into the night. According to Nieves and Pimenta (2002), *L.* (*V.*) *braziliensis* infections were observed in *Mg. migonei* females that fed on wild rats, opossums, horses, and males. The first citation of the role of *Mg. migonei* in the ACL-transmission cycle was through the encounter of a species naturally infected by flagellates, probably *Leishmania* promastigotes, in 1941 by Pessoa and Coutinho in São Paulo. Later studies also showed the coincident increase of ACL cases and *Mg. migonei* specimens in SP (Camargo-Neves et al. 2002; Rangel and Lainson 2009). In RJ, the species has been implicated as a probable vector in regions, such as Ilha Grande and Jacarepaguá, where the species has a preference for biting dogs and is associated with the maintenance of canine leishmaniasis in addition to its natural infection with *L.* (*V.*) *braziliensis* in Jacarepaguá (Pita-Pereira et al. 2005; Rangel and Lainson 2009; Gouveia et al. 2012; Carvalho et al. 2014). *Mg. migonei* has also been associated with ACL transmission in MG and ES. In the Northeast region, more precisely in CE, *Mg. migonei* was found to be naturally infected by *L.* (*V.*) *braziliensis* (Azevedo & Rangel 1991, Queiroz et al. 1994). However, this occurred only as a secondary vector of ACL (Rangel and Lainson 2009). Recently molecular studies using PCR techniques identified *Mg. migonei* infected by *L.* (*L.*) *infantum chagasi*, the etiological agent of AVL, in PE and CE (Carvalho et al. 2010; Silva et al. 2014a, b; Rodrigues et al. 2016).

Nyssomyia whitmani (Antunes & Coutinho, 1939)

N. whitmani was described by Antunes and Coutinho in 1939 as *Flebotomus whitmani* in honour of Dr. Whitman of the Rockefeller Foundation, which collaborated with the Brazilian government in the Yellow Fever Campaign. The new species was described based on male and female specimens collected in Ilhéus (BA). Until 1939, it was misidentified as *Ny. intermedia*. In Brazil, this species was registered throughout the country except for SC and RS (Fig. 1) and beyond in French Guiana, Paraguay, Peru, and Argentina (Young and Duncan 1994).

The participation of *Ny. whitmani* in the epidemiological chain of ACL is restricted to the Brazilian territory where its first report as a vector was in SP when it was found infected by flagellates, possibly *Leishmania* (Pessôa & Coutinho 1941). It was considered a sylvatic species, although it could be found inside houses that were located within or near the forest. At dusk, it was found biting man and feeding on dogs and was shown to be present in large numbers in chicken houses. Because the localities Barretto (1943) studied were close to deforested areas, he suggested that the high-density population under these circumstances was simply due to the wide variety of blood sources available to this sand fly.

In 1953, Forattini confirmed the presence of *Ny. whitmani* in both the forest and nearby domestic animal shelters in SP. Later (1960) he commented on the fact that although initially dependent on primary forest, this sand fly could now be captured during several months of the year inside houses, peri-domestic pigsties, and banana plantations and that there was no doubt that it was now thriving in a domestic environment. In the same publication, Forattini (1953) believed that although there were reports indicating that *Ny. whitmani* coexists with *Ny. intermedia* in areas recently occupied by man and may outnumber the latter species as the environmental alterations proceed, more detailed information was needed to confirm this suggestion.

In RJ in general, the *Ny. whitmani* population has for some time remained at a low level (Rangel et al. 1986, 1990; Oliveira et al. 1995). Souza et al. (2001) registered this sand fly's presence in the Atlantic Forest and, more importantly, its frequency in residential areas close to the forest. The same investigators recorded both *Ny. intermedia* and *Ny. whitmani* biting man in a peri-domestic habitat close to the forest and noted that the former prevailed in the residential area, whereas the latter was the most frequent of the two insects in the forest. They observed a higher density of *Ny. whitmani* during winter months, whereas *Ny. intermedia* was most abundant during the hotter months of the year (Souza et al. 2002).

The tendency of *Ny. whitmani* to occupy residential areas in MG was discussed by Mayrink et al. (1979) and Passos et al. (1991). It was found feeding on man and domestic animals and thought to possibly be breeding in this habitat. Conversely, it was found in very small numbers in the neighbouring forest.

In Northeast Brazil, in BA, CE and PE, its behaviour has been shown to be like that seen in the Southeast region, namely, highly attracted to man and well adapted to the domiciliary habitat (Barreto et al. 1982; Vexenat et al. 1986; Brandão-Filho et al. 2003). In BA, it was suggested that *Ny. whitmani* might be breeding on cocoa plantations (França et al. 1991).

Regarding seasonality, differences in behaviour have been noted in different regions of Brazil, probably due to differing climatic conditions. In the Southern region, for example, Barretto (1943) noted its presence year-round, whereas in studies in Petrópolis, RJ, it was found in greater numbers during the months of low temperatures in June through August (Souza et al. 2002).

Although the dusk and nocturnal feeding habit of *Ny. whitmani* follows the usual sand-fly pattern (Barretto 1943), it, too, has been shown to be somewhat variable in different regions of Brazil. In the Northeast, Azevedo and Rangel (1991) showed that it can also be captured during the whole of the dawn period, in chicken houses or feeding on equines, with maximum activity from 1–3 am. Souza et al. (2004) noted that in RJ it could be found feeding on man, in the peri-domestic habitat, between 4 and 6 am, and such behaviour was previously recorded by Teodoro et al. (1993) in PR; studies on host preferences of *Ny. whitmani* among domestic animals, made in the same state, showed that this insect is an opportunistic feeder resulting in a wide choice of hosts in the peri-domestic environments of human colonization.

In the primary forest, in several different regions of PA, Lainson et al. (1979) noted that *Ny. whitmani* has very different habits from those discussed previously in other regions. It was found to be essentially sylvatic and was captured principally from large tree trunks and in the forest canopy. It was disinclined to bite man and attempts to demonstrate its invasion of houses located very near the forest were completely unsuccessful. Subsequent studies confirmed these observations and led to the conclusion that any divergence from this behaviour is likely to occur only under special conditions (Ready et al. 1986; Lainson 1988; Shaw et al. 1991).

This situation led to the suggestion that *Ny. whitmani* might represent a species complex of two or more taxa (Lainson 1988), and this has resulted in several comparative studies on populations of this sand fly from widely different areas of Brazil. Some investigations suggested the existence of at least three different lineages of this sand fly based on biological characters, geographical variations, and morphometric features together with observations on sequences of mitochondrial DNA (Rangel et al. 1996; Ready et al. 1997, 1998). Rangel et al. (1996) made it clear, however, that they did not discard the possibility that the populations they studied—from PA, CE, and BA—could represent a cline. Ishikawa et al. (1999) studied populations from the North, Northeast, Southeast, and the South regions and indicated the existence of a clade from RO within the lineage of forested areas, which included haplotypes of the Amazon and Atlantic forests and Ilhéus (the type locality of *Ny. whitmani*). They suggested that their findings did not sustain the hypothesis of a cryptic species complex but rather the occurrence of a recent crossing-over of populations in forested areas.

Margonari et al. (2004) studied populations of *Ny. whitmani* from the Northeast and the Southeast of Brazil. They confirmed observations on the similar morphometry of these but presented evidence of two biogeographical "clusters." Later, however, they suggested the existence of a genetic flow between the two lineages.

Regarding the incrimination of *Ny. whitmani* as an important vector of ACL, the first suggestion of this was made in 1941, when Pessôa and Coutinho found a specimen from SP infected by flagellates, which were considered possibly to be promastigotes of *Leishmania*: As a result, entomological investigations were intensified in areas of ACL transmission in this region and soon showed that although considered as a sylvatic species, *Ny. whitmani* was a highly anthropophilic sand fly with a particularly dense population (Barretto 1943; Forattini 1954).

In Southeastern Brazil, data in the literature suggest the participation of this sand fly in the transmission of ACL in a focus of the disease in Caratinga, MG, and in the mountainous area of Afonso Cláudio (Mayrink et al. 1979; Falqueto 1995), and Souza et al. (2002) considered that it might be sharing the role of a vector of *L.* (*V.*) *braziliensis*, together with *Ny. intermedia*, in rural RJ. Recently, the finding of a specimen of *Ny. whitmani* infected with a *Leishmania* of the subgenus *Viannia* by PCR analysis, in a region very close to Belo Horizonte, MG, has led to the suggestion that this sand fly could be the vector of cutaneous leishmaniasis in that area (Carvalho et al. 2008).

In the South, *Ny. whitmani* was also considered as a possible vector of ACL in PR, whereas in the northern part of this state a natural infection with *L.* (*V.*) *braziliensis* found in one specimen and the insect's high population density clearly emphasized this sand fly's medical importance (Luz et al. 2000; Teodoro et al. 2003).

In all areas of ACL in Northeast Brazil, this same species of sand fly is considered as an important vector of the disease based on the finding of specimens infected by *L.* (*V.*) *braziliensis* in the area of Três Braços, BA (Hoch et al. 1986; Ryan et al. 1990). The sand fly's predominance in houses and the peri-domestic habitats in general prompted these investigators to suggest the development of a purely domestic cycle of transmission by *Ny. whitmani*. In Ilhéus, BA, the type locality of this sand fly, its fondness for human blood, and its high population density in the domestic habitat led to the same conclusion (Azevedo and Rangel 1991). In the Serra de Baturité, CE, a parasite of the subgenus *Viannia* and others positively identified as *L.* (*V.*) *braziliensis* were found in dissected *Ny. whitmani* (Azevedo et al. 1990; Queiróz et al. 1994) and, once again, these findings—together with population density in the peri-domestic habitat—indicated this fly as the local vector of ACL due to this parasite. Gil et al. (2003) registered *Ny. whitmani* as the second most prolific sand fly in captures made in the central area of RO, stressed its preferential arboreal habits, and recorded the presence of unidentified trypanosomatid parasites in some specimens.

In the municipalities of Rio Branco, Bujari, and Xapuri, AC, a study of sand-fly fauna and the potential vectors of ACL showed that *Ny. whitmani* was the most abundant species with its spatial distribution coinciding with proven transmission sites of *L.* (*V.*) *braziliensis*: it was therefore suggested that this sand fly was a probable vector of this parasite in that region (Azevedo 2008).

The state of Tocantins (TO) has suffered environmental impacts resulting in ecological changes due to the construction of hydroelectric plants, agricultural activities, and the establishment of new settlement areas, and the increasingly high incidence of ACL in this region has probably been due to these activities. *Ny. whitmani* is found in most of the endemic municipalities especially in areas that have been degraded by man (Vilela et al. 2008). In the Central West region, some studies conducted in areas that suffered environmental changes due to human activities have suggested *Ny. whitmani* as an important vector of *L.* (*V.*) *braziliensis* (Galati et al. 1996; Dorval et al. 2009).

This sand fly is one of the principal ACL vectors in Brazil having been recorded in large numbers of endemic areas (Costa et al. 2007) and in association with a wide vegetation diversity. Environmental and climatic changes most probably account for the spread of ACL in Brazil in recent years (Shaw 2007), and *Ny. whitmani* adapts readily to new environments, such as degraded areas, in association with domestic animals and man in rural and peri-urban areas (Costa et al. 2007; Shaw 2008). Peterson and Shaw's (2003) ecological niche modelling of ACL vectors predicted that climate warming would favour the adaptation of *Ny. whitmani* to new areas as well as its geographical expansion within Brazil.

The very different behaviour of *Ny. whitmani* in the primary forest in PA, North Brazil, has already been discussed, and until now this sand fly has not been associated with ACL due to *L.* (*V.*) *braziliensis* in this region. The suggestion was made, however, that promastigotes of a member of the subgenus *Viannia* found in this sand fly in Monte Dourado, PA—an area of ACL due to *L.* (*V.*) *guyanensis*— were probably those of this parasite and that *Ny. whitmani* was participating in its transmission together with the principal vector, *Ny. umbratilis* (Lainson et al. 1981b). The parasite was not identified at the time, and in view of later isolations of *L.* (*V.*) *shawi* from *Ny. whitmani* in another area of primary forest in PA (Lainson et al. 1989), it was suggested that the parasite of *Ny. whitmani* in Monte Dourado, PA, was also *L.* (*V.*) *shawi* (Rangel et al. 1996, Lainson & Shaw 1998). However, the Monte Dourado *Ny. whitmani* infections were recently typed (de Souza et al. 2017) and proved to be *L.* (*V.*) *guyanensis*. Therefore, we conclude that in the Brazilian Guiana Shield *Ny. whitmani* participates in the transmission of *L.* (*V.*) *guyanensis* as previously suggested (Lainson et al. 1981b).

The comment by Lainson (1988), based on years of entomological observations in areas of primary forest in PA by workers in the Instituto Evandro Chagas (i.e. that in Amazonia *Ny. whitmani* is "seldom observed biting man, and never in large numbers") conflicts with the frequency of human *L.* (*V.*) *shawi* infections. One explanation could be that other anthropophilic sand flies are involved in the transmission of *L.* (*V.*) *shawi* with *Ny. whitmani* merely maintaining the enzootic in wild animals. However, Campbell-Lendrum et al. (1999) observed no significant difference in the anthropophily of *Ny. whitmani* from North Brazil and *Ny. whitmani* sensu stricto from other areas in Brazil. The Instituto Evandro Chagas team's observations were made in primary forest biomes, in which it was either absent (de Souza et al. 1996; Ward et al. 1973b) or present in small numbers (de Souza et al. 2016, 2017). Those of Campbell-Lendrum et al. (1999) were performed in a "patch of degraded primary forest" in an area of extensive deforestation. Another study (Donalisio et al. 2012) clearly showed a highly significant difference between the sizes of populations in forested and peri-domiciliary areas of PE. A simple explanation for the apparently conflicting opinions in relation to the anthropophily of Amazonian *Ny. whitmani* populations is variations in population densities related to environmental conditions. The fact that fewer flies are attracted to men in different areas is perhaps not because of difference in their anthropophily rather but differences in populations sizes.

Bichromomyia flaviscutellata (Mangabeira, 1942)

In their various field trips in Brazil, Mangabeira and collaborators captured approximately 17,000 specimens of sand flies representing 57 different species and with 35 of them being new to science. Among the latter was Mangabeira's description of *Bi. flaviscutellata*, which was based only on male insects and collected in the locality of Aurá close to Belém, PA. Later, Sherlock and Carneiro (1962) described the female of the species after the establishment of a laboratory colony of this sand fly from BA. It must be stressed, however, that the taxonomic status of the material from BA has been questioned (Young and Duncan 1994). The specific name of this sand fly was probably chosen in view of the double colouration of the shield where the scutelum is clear and the remainder of the structure a dark brown (Latin *flavus* = golden; yellow + *scutu* = shield).

Bi. flaviscutellata has an extensive geographical distribution (Fig. 1) and can be found in very different habitats such as primary forest, secondary or copse-like vegetation, and lowland *várzea* forest, which during half of the year is subject to various degrees of flooding. Ready et al. (1983) showed that together with the various rodents and marsupials on which it feeds, it rapidly adapts to plantations of introduced trees, such as *Pinus* and *Gmelina*, and it is occasionally captured in the peridomestic habitat of houses located near forest (Lainson et al. 1994). The distribution and population ecology of *Bi. flaviscutellata* are also influenced by climate, particularly by seasonal precipitation (Shaw and Lainson 1972; Ready et al. 1983). Projections from climate-change scenarios suggest an expansion of climatically suitable areas for *Bi. flaviscutellata* in the Southeast and South regions of Brazil in the future (Carvalho et al. 2015). Given continuous environmental and climatic changes, there are modifications in the behaviour of some sand-fly vectors of leishmaniasis, and in the Brazilian *Cerrado* of Central Brazil (extensive, flat areas of low, fire-resistant trees, small palms, and thorny bushes), it is possible to note the spread of *Bi. flaviscutellata* found in association with domestic-animal shelters and the presence of new cases of anergid diffuse cutaneous leishmaniasis (ADCL) (Vilela et al. 2008, 2011; Shaw 2008; Nunes et al. 2008; Queiroz et al. 2012; Brito et al. 2014).

Bi. flaviscutellata is a low-flying sand fly that is essentially nocturnal in its biting habits and highly attracted to rodents but not greatly attracted to man (Lainson and Shaw 1968; Shaw and Lainson 1968; Shaw et al. 1972; Gomes 1994; Vilela et al. 2006, 2007). This is fortunate because it is the proven vector of *Leishmania* (*L.*) *amazonensis*, which, in addition to being an agent of single-lesion cutaneous leishmaniasis, is also the cause of ADCL in individuals with a faulty immunological system. ADCL is highly disfiguring and cured with difficulty. In Brazil, human cases of ADCL were notified in North, Northeast, Central West, and Southeast regions (Costa et al. 2009). In 2007 the first autochthonous human case of ADCL from RJ was notified in the municipality of Paraty (Azeredo-Coutinho et al. 2007). Despite the records of *Bi. flaviscutellata* in the neighbouring municipality of Angra dos Reis (Araújo Filho et al. 1981; Carvalho et al. 2013), the vector remains to be detected in Paraty even after 3 years of monthly sand-fly captures in the region (Vieira et al. 2015). *Bi. flaviscutellata* was not detected in Paraty probably because animal-baited Disney traps were not yet used. This has proven to be the best capture method for zoophilic sand flies such as *Bi. flaviscutellata* (Shaw & Lainson 1968; Dorval et al. 2007, 2009, 2010).

In 1963, Lainson paid a visit to the Instituto Evandro Chagas in Belém, PA, and during a demonstration of the animal-trapping programme of the Rockefeller Virus Laboratory discussed the unique opportunity this held for the examination of these animals for evidence of *Leishmania* infections: ACL was a considerable public health problem in the Amazon Region of Brazil. The director of the programme, the late Dr. Otis Causey, was impressed with the similarity of cutaneous lesions he had seen on the tails of wild rodents and those caused by *Leishmania (L.) mexicana* on the tails of forest rodents in Belize, Central America (Lainson and Strangways-Dixon 1964). He promised to look more closely at the next ones he saw and within a few days presented Lainson with a stained smear of a lesion on the tail of the rodent *Oryzomys capito*, which was rich in amastigotes. At first it was thought that the parasite was *L.* (*V.*) *braziliensis* (Guimarães & Costa 1966), but after subsequent study of the parasite it was given the name of *Leishmania mexicana amazonensis* (Lainson & Shaw 1972) and later amended to *Leishmania (L.) amazonensis* (Lainson & Shaw 1987).

With the knowledge that rodents were important reservoir hosts of the parasite, rodent-baited Disney traps were used to capture sand flies attracted to them. By far the greatest number trapped were *Bi. flaviscutellata*, and dissection of these revealed 8 of 2706 to be heavily infected with promastigotes, which proved to be those of *L.* (*L.*) *amazonensis* (Lainson & Shaw 1968): During this and continuing studies, a total of 45 heavily infected *Bi. flaviscutellata* were recorded in 7498 females dissected, and on no occasion was the parasite encountered in other species of sand flies from the same area.

Finally, *L.* (*L.*) *amazonensis* was experimentally transmitted from hamster to hamster by the bite of *Bi. flaviscutellata* (Ward et al. 1977). This species was replaced in the upper reaches of the Amazon River in the Rondônia and Amazonia states by *Bi. olmeca nociva* and *Bi. reducta*. Infections of *L.* (*L.*) *amazonensis* have been found in both, and it seems likely that these two species are its vector in these areas.

Nyssomyia umbratilis (Ward & Fraiha, 1977)

During a study of the epidemiology of cutaneous leishmaniasis in Surinam in 1966, Wijers and Linger recorded flagellate infections in a tree-trunk inhabiting sand fly, which they referred to as *Phlebotomus anduzei* (syn. *Ny. anduzei*). It was thought to be the most likely vector of "bosch yaws," or *pian-bois*, due to *L.* (*V.*) *guyanensis*, but their attempts to infect hamsters with the flagellates failed, and the parasite remained unidentified.

Lainson et al. (1976) worked in the primary forest of Monte Dourado (Jari), PA, Brazil, north of the Amazon River, where approximately 300 cases of ACL due to *L.* (*V.*) *guyanensis* were recorded, in 1 year, in men working on deforestation. They recorded massive infections with *L.* (*V.*) *guyanensis* in 4 of 55 specimens of a sand fly considered, at the time, to be *Ny. anduzei* and isolated the parasite after the intradermal inoculation of hamsters. Suspicions were aroused during these studies, however, that the vector was not in fact *Ny. anduzei*, and subsequent morphological studies showed that it was a closely related and morphologically very similar sand fly that was new to science.

Ward and Fraiha (1977) described the new sand fly as *Ny. umbratilis* from 10 females collected during the work in Monte Dourado, PA, and an intense study of its behaviour was initiated in the same area (Lainson et al. 1979). It was found that although sand-fly species of the subgenus *Psychodopygus* predominated at ground level, *Ny. umbratilis* was extremely abundant in the forest canopy but descended to ground level, presumably to oviposit, by way of the tree-trunks, on which it could be collected in great numbers in the early morning. In studies conducted in RO, in the area of Samuel Ecological Station; however, it was noted that *Ny. umbratilis* predominated in the canopy (Azevedo et al. 1993). In Monte Dourado, PA, it was noted that *Ny. umbratilis* flies off the tree-trunks when disturbed by man's activities and attacks the nearest person. In the same study, *L.* (*V.*) *guyanensis* was isolated from 16 more specimens of *Ny. umbratilis* and, of 77 sand flies attacking 2 men collecting from the tree trunks, 72 (92.5%) proved to be *Ny. umbratilis*. Some idea of the efficiency of this sand fly in the transmission of ACL in the Monte Dourado, PA, area may be gained by the fact that the 2 men developed a total of 13 leishmanial lesions due to *L.* (*V.*) *guyanensis* on their arms, probably providing the most conclusive incrimination of a vector of ACL ever obtained.

The explanation of this great number of infected sand flies on tree trunks came with the detection of *L.* (*V.*) *guyanensis* in 27 of 59 specimens of the sloth *Choloepus didactylus* in the Monte Dourado area, PA (Lainson et al. 1981a, b). This animal spends most of its time in the forest canopy and has thus become the principal mammalian reservoir host of the parasite. Because the animal may remain in the same tree for a considerable time, there is a gradual build-up of infected *Ny. umbratilis* on a given tree. This sand fly's similar role as a vector of *L.* (*V.*) *guyanensis*, as well as its common presence in the forest canopy and on large tree-trunks at ground level, has been recorded in some other areas of the Amazon region of Brazil (Arias and Freitas 1977a, 1978) and in French Guiana (Le Pont and Pajot 1980). Infection of an undoubted specimen of *Ny. anduzei* with a parasite having development consistent with that of members of the subgenus *Viannia* has been reported in Manaus (AM) (Arias and Freitas 1977b). However, this sand fly can, at most, now be considered only as a possible secondary vector of *L.* (*V.*) *guyanensis* and is probably of low importance with regard to the transmission of ACL to man.

Ready et al. (1986) performed a detailed study of the ecology of *Ny. umbratilis* in the region of Monte Dourado, PA. It is highly anthropophilic and presumably becomes infected after feeding at night, particularly on the two-toed sloth *Choloepus didactylus*, but also on other arboreal animals such as the ant-eater *Tamandua tetradactyla*. In Manaus, AM, precipitin tests on blood in naturally fed *Ny. umbratilis* showed that 66% of them had fed on sloths (Christensen et al. 1982). In a later evaluation with the same method, *Ny. umbratilis* females captured in a non-flooded upland forest in Manaus fed predominantly on rodents (34%) followed by dogs (19%), sloths (18%), humans (16%) and chickens (13%) (Nery et al. 2004). In addition to its nocturnal feeding habits, however, this sand fly clearly will feed in the early daylight hours if disturbed from its resting place on tree trunks. It is recorded biting man in the dry season and, particularly, directly after the rainy season.

Areas of high ACL prevalence due to *L.* (*V.*) *guyanensis* may be found in communities located in or very close to primary forest, and this has led to the erroneous impression that *Ny. umbratilis* is undergoing the process of adapting to a peri-domestic habitat. However, no consistent data exist proving that this is true, and any transmission in this environment is almost certainly due to sand flies that have been attracted to a residential area, from nearby primary forest, by the lights of the houses. Esterre et al. (1986) discussed the acquisition of ACL due to *L.* (*V.*) *guyanensis* in persons living in a small village within forest in French Guyana and came to the same conclusion: When the forest was cleared to about 400–500 m around the village, all peri-domestic transmission ceased. Guerra et al. (2007) discussed this situation in Manaus, AM, and were clearly of the opinion that the eco-epidemiology of ACL there is the same as that recorded in Monte Dourado, PA. In other forested areas on the outskirts of Manaus, however, *Ny. umbratilis* was considered to be present in equal numbers in both the forest and in the peri-domestic habitat (Barbosa et al. 2008).

Observations exist suggesting that *Ny. umbratilis* is a vector of *L.* (*V.*) *guyanensis* in the state of Bolivar, Venezuela (Feliciangeli et al. 1985), possibly indicating an expansion of the Brazilian zoonotic cycle.

Rangel et al. (1998) isolated *L.* (*V.*) *braziliensis* from patients with ACL in Peixoto de Azevedo (MT), and Azevedo et al. (2002) noted that one of the most abundant and highly anthropophilic sand fly in the same area was, morphologically, *Ny. umbratilis*. In addition, they confirmed observations made by workers in the Instituto Evandro Chagas, Belém, PA (Ward et al., 1976) that the population of this sand fly, south of the Amazonas River, behaved very differently from that studied north of the river (Monte Dourado, PA). Although abundant in the forest canopy, it was not found to accumulate on tree trunks at ground level. It was this marked behavioural difference that led Lainson (1988) to suggest that perhaps the populations of *Ny. umbratilis* north and south of the Amazonas River were not identical and, since that time, the taxonomic status of *Ny. umbratilis* started to attract special attention.

Azevedo et al. (2002) studied the morphology and the morphometric characters of the head, thorax, and abdomen of populations of the insect from Brazil (in AP, PA, AM, and MT) and Venezuela (state of Bolivar). They found that analysis of the morphological characters could not separate the populations but that the quantitative characters (morphometry) showed that 77% of these separated the Venezuelan population from the Brazilian ones. The analysis did not, however, supply evidence of heterogeneity among the populations from Brazil, but later studies on *Ny. umbratilis* populations from Brazil and Venezuela suggest the existence of three different populations, which are separated by the geographical barriers of the *planalto* of RR and the two rivers, Negro and Amazon; One is in Venezuela and the other two in Brazil (north and south of the Amazon River (Azevedo 2008)).

The same investigator recorded 52 different species of sand flies in the municipalities of Rio Branco, Xapuri and Bujari, AC (17 being a new record for that state); *Ny. umbratilis* was abundant in the forest canopy in close association with the major reservoir of *L.* (*V.*) *guyanensis*, the sloth *C. didactylus*; and Tojal da Silva et al. (2006) recorded the presence of ACL due to *L.* (*V.*) *guyanensis* in the municipality of Rio Branco. These observations lead Azevedo et al. (2005, 2008) to conclude there is, in fact, a transmission cycle of this parasite south of the Amazonas River involving *Ny. umbratilis*. More recent sand-fly captures in urban and peri-urban Rio Branco did not detect *Ny. umbratilis*, but the other ACL vectors—*Ny. whitmani*, *Ny. antunesi*, and *Bi. flaviscutellata*—were present (Araújo-Pereira et al. 2014).

A biological analysis under laboratory conditions compared *Ny. umbratilis* populations from Manaus and Manacapuru (left and right sides of the Negro River, respectively) and showed differences in their life cycle, fecundity, fertility, adult longevity, and emergence. These differences suggested that some divergence of intrinsic biological features evolved because of their geographical isolation by the Negro River (Justiniano et al. 2004). Further phylogenetic analyses based on mitochondrial DNA detected two distinct lineages in *Ny. umbratilis* populations of opposite sides of the Amazon and Negro rivers, thus reinforcing the thought that these rivers may be acting as effective barriers, preventing gene flow between them (Scarpassa and Alencar 2012).

In PE, where most ACL cases are caused by *L.* (*V.*) *braziliensis* and transmitted by *Ny. whitmani* (Brandão-Filho et al. 1999), studies conducted at a forest reserve in Recife detected *Ny. umbratilis* at very high frequencies (96.5%) and biting rates (≤333.3 flies/person-hour) (Balbino et al. 2001, 2005). Phylogenetic analysis based on wing morphometry and the period clock gene concluded that the Recife population of *Ny. umbratilis* is significantly closer to the Rio Preto da Eva population (north of the Amazon River, AM) and that both populations are genetically distant from Manacapuru (south of Amazon River, AM) (Souza Freitas et al. 2015, 2016).

Molecular taxonomy studies based on a barcode region of mitochondrial DNA of *Ny. umbratilis* and *Ny. anduzei* from different regions of the Amazon clearly separated both species. However, the barcode region did not have enough power to separate the two lineages of *Ny. umbratilis* from opposite sides of the Amazon River, likely reflecting incipient species that have not yet reached the status of distinct species (Scarpassa and Alencar 2013).

Ny. umbratilis has so far not been associated with the transmission of ACL south of the Amazon River, but its behaviour is markedly different from that of the populations from regions located north of the river. One key behavioural difference is the failure of the southern *Ny. umbratilis* populations to concentrate at the base of trees. A parasite isolated from *Ny. umbratilis* captured in Peixoto de Azevedo, MT, proved to be *L.* (*V.*) *braziliensis* and not *L.* (*V.*) *guyanensis* (Azevedo et al. 2002). This raises the question as to its possible participation in the transmission *L.* (*V.*) *braziliensis*. In addition, *L.* (*V.*) *guyanensis* is replaced in the Amazonian forest south of the river by its sister species, *L.* (*V.*) *shawi*, where it is transmitted by *Ny. whitmani*.

Psychodopygus wellcomei (Fraiha, Shaw & Lainson, 1971)

In 1968, the Meridional Mining Company, undertaking mineral exploration in PA, requested the Instituto Evandro Chagas to investigate an alarming number of men acquiring ACL due to *L.* (*V.*) *braziliensis* whilst working on road construction through primary forest in the Serra dos Carajás. It required only a few days for one particular sand fly to become highly suspected as the vector due to its avid feeding on man.

It proved to be a previously undescribed sand fly, which was named *Ps. wellcomei* in honour of Sir Henry Wellcome, founder of the Wellcome Trust, London, who was to sponsor the Institute's leishmaniasis programme for nearly 40 further years.

Ward et al. (1973a) made a study of sand flies captured during a 2-month period (December and January) using human bait, rodent-baited Disney traps, and aspiration from tree trunks, all at ground level, and captures with CDC light traps on platforms built in the trees at 5 and 11 m above the forest floor. A total of 23 different species were caught, and approximately 65% of all the sand flies captured while biting man were *Ps. wellcomei*. Heavy promastigote infections were encountered in three specimens of this sand fly, and the parasite was isolated in culture and the skin of hamsters; subsequent studies showed it to be *L.* (*V.*) *braziliensis*. Finally, Ryan et al. (1987a) performed experimental transmission of the parasite to hamsters by placing the animals in cages with large numbers of newly caught sand flies. All fed flies were separately maintained in glass vials until they had oviposited, at which time they were dissected to detect promastigotes and the eggs of all infected specimens maintained in order to rear males for positive identification. This was necessary because the females of *Ps. wellcomei* are morphologically indistinguishable from those of a sympatric species, *Ps. complexus,* whereas the males have distinctly different morphology.

Ps. wellcomei is an essentially sylvatic and highly anthropophilic species (Ward et al. 1973a; Wilkes et al. 1984). In addition, Ward et al. (1973b) found that 25.5% of all sand flies attracted to rodent-baited traps were of this species: This, and the fact that this sand fly has a vertical flight-range of only 1–2 m above ground level, led to their suggestion that the sylvatic hosts of *Ps. wellcomei* are terrestrial animals, the most highly suspected being rodents and marsupials (Lainson et al. 1973). The isolation of parasites with the biological characters of *L.* (*V.*) *braziliensis* from the rodents *Oryzomys concolor*, *O. capito*, *O. nigripes*, *Akodon arviculoides*, *Proechimys* spp., *Rattus*, and *Rhipidomys leucodactylus*—and the opossum *Didelphis marsupialis* in Brazil—tended to support this view (Lainson and Shaw 1970, 1979; Forattini et al. 1972; Forattini 1973; Lainson et al. 1981b; Rocha et al. 1988). Finally, a more definitive identification of this parasite from the Brazilian rodents *Bolomys lasiurus* and *R. rattus* was obtained by multi-locus enzyme electrophoresis (Brandão-Filho et al. 2003).

Regarding its behaviour and seasonality, *Ps. wellcomei* is most abundant during the rainy season (November–April) and enters into diapause during the dryer months when it is rarely encountered. The same seasonal pattern was observed in more recent studies in areas out of the Amazon Region, in RN, where *Ps. wellcomei* only occurs in months with greater rainfall and lower temperatures (Pinheiro et al. 2013, 2016a, b). Limiting forest work to the dryer months can therefore greatly reduce the risk of acquiring ACL in areas where this sand fly is found. The great importance of *Ps. wellcomei* as a vector of *L.* (*V.*) *braziliensis* is due to its tendency to not only feed at night but also during broad daylight, particularly in cloudy weather. The number of infected females captured during the day was, in fact, found to be greater than that obtained during the night suggesting that transmission is actually most frequent during the day (Wilkes et al. 1984).

The presence of *Ps. wellcomei* has been recorded in other areas out of the Amazon Region such as in forest of the Serra de Baturité, CE (Ready et al. 1983; Azevedo and Rangel 1991). The former investigators suggested that sand flies recorded as *Ps. squamiventris* by Lucena (1953) in Guaramiranga, CE, were possibly *Ps. wellcomei* because the females of the two species are morphologically very similar.

In the Serra de Baturité area, CE, Queiroz et al. (1994) detected flagellates in *Ps. wellcomei* (infection rate 0.05%), but unfortunately the parasites were not identified. *Ps. wellcomei* was also captured in MA, but it was classified as an accessory species (Pereira Filho et al. 2015) because the local main vector of *L.* (*V.*) *braziliensis* is probably *Ny. whitmani* (Rebêlo et al. 2010; Azevedo et al. 2011; Campos et al. 2013).

Other areas of *Ps. wellcomei* recorded outside the Amazon include RN and PE. Despite the recent additional evidence of its high anthropophily in Nísia Floresta, RN (Pinheiro et al. 2016a), the importance of *Ps. wellcomei* as a vector of ACL in Northeast Brazil still must be confirmed. Although *Ps. wellcomei* has been found in the Atlantic Rainforest region of PE, again there is so far no association of this species with local ACL in that region (Andrade et al. 2005; Silva and Vasconcelos 2005).

Psychodopygus complexus (Mangabeira, 1941)

This sand fly was described by Mangabeira from a single male, captured in the municipality of Abaetetuba, PA, in 1938, by members of the Commission of Studies of American Visceral Leishmaniasis. Like *Ps. wellcomei*, the females are highly anthropophilic, although they seem not to share that sand-fly's daytime biting habits.

The females of *Ps. complexus* and *Ps. wellcomei* are morphologically indistinguishable, although the males are easily identified by the structure of the external genitalia. Ready et al. (1991), however, used DNA probes to distinguish the two species and showed that a fragment of DNA highly repetitive for *Ps. wellcomei* was not detected in either sex of *Ps. complexus*.

In Serra dos Carajás, PA, the two species share the same forest habitat, which at first created difficulties in pinpointing the principal vector of *L.* (*V.*) *braziliensis* in that area and required the rearing of males from the eggs of infected females to obtain the all-important males. In a transect running from high up on the range of hills down to the lowland forest, Ready et al. (1984) showed that the predominant species at the higher altitude (≥700 m above sea level) was *Ps. wellcomei* and that this predominance was slowly reversed with decreasing altitude until *Ps. complexus* predominated, in large numbers, in the forest at the foot of the hills (200 m) and *Ps. wellcomei* was completely absent at ≤150 m. Because ACL due to *L.* (*V.*) *braziliensis* is commonly found in the latter lowland forest in various regions of PA, this is a clear indication that vectors other than *Ps. wellcomei* are involved (Shaw et al. 1987). In later studies in Paragominas, where *Ps. wellcomei* is uncommon, several infected females of the *squamiventris* group were found and, because all the males captured proved to be *Ps. complexus*, it was considered sufficient evidence to incriminate this sand fly as the vector of *L.* (*V.*) *braziliensis* in that region (de Souza et al., 1996).

Azevedo et al. (2002) showed that *Ps. complexus* represented 8.2% of all captured sand flies in an area of ACL transmission in Peixoto de Azevedo, MT, although the participation of this species as a vector in this region has yet to be established.

In a military-training area of the Atlantic Forest in Pernambuco, Andrade et al. (2005) found flagellates characteristic of *Leishmania* during dissections of *Ps. complexus* females, but species typing could not be done due to contamination of the cultures. Because *Ps. complexus* predominated (87%) during periods of military activities that were followed by records of human cases of ACL, the investigators considered *Ps. complexus* as the principal suspected vector involved in the local transmission of ACL (Andrade et al. 2005).

In the municipality of Guaraí, Tocantins, *Ps. complexus* was the prevalent sand-fly species in the rural environment associated with human settlements and in captures with Shannon traps, thus confirming its anthropophilic behaviour (Vilela et al. 2013). Additionally, a multiplex PCR analysis of pooled dissected females detected natural infections by *L.* (*V.*) *braziliensis*, which lead the investigators to conclude that although *Ny. whitmani* is thought of as the most important ACL vector in TO, *Ps. complexus* may also play an important role in the transmission cycle of ACL in rural settlement areas of Guaraí (Vilela et al. 2013). Recent studies in the same municipality found positive correlations between *Ps. complexus* abundance and precipitation, which further supports its potential role as a *L.* (*V.*) *braziliensis* vector during the rainy season (Godoy et al. 2017).

Psychodopygus ayrozai (Barretto & Coutinho, 1940)

This species has an extensive geographical distribution in Brazil that encompasses the North, Northeast, Central, Southeast, and South regions (AM, RO, RR, PA, BA, PE, MT, MG, RJ) (Aguiar and Medeiros 2003). However, its level of anthropophily appears to vary in different regions.

Psychodopygus ayrozai is anthropophilic in the more mountainous area in Atlantic Forest of Southeast Brazil (Aguiar and Soucasaux 1984), and its seasonality is associated with the hot and humid months decreasing in frequency during the colder and dryer months of the year. Studies in the Serra dos Órgãos, Rio de Janeiro state, showed that its feeding activity begins at dusk, extending until 12 pm, and that feeding occurred preferentially at ground level (Aguiar and Soucasaux 1984). In the Atlantic Forest of Paraná state, studies indicated it as one of dominant species whose population density fluctuated with temperature and rainfall indices (Marcondes et al. 2001).

Psychodopygus ayrozai has been implicated as a vector of *L.* (*V.*) *naiffi* in the Amazon Region, especially in PA (Lainson and Shaw 1998; Rangel and Lainson 2009). In fact, human *L.* (*V.*) *naiffi* cases are infrequent, probably because *Ps. ayrozai* does not reveal itself as anthropophilic sand-fly species in this region (Lainson and Shaw 1998; Rangel and Lainson 2009). Specimens of this phlebotomine have also been found in *L.* (*V.*) *naiffi* in AP and RO (de Souza et al. 2017; Arias et al. 1985).

In recent studies carried out in TO, in the Cerrado *biome*, *Ps. ayrozai*, which was first recorded in TO, was found with natural infection by *L.* (*V.*) *braziliensis*. These infections occurred in settlements in rural areas in the municipality of Guaraí, an endemic area for ACL with a local transmission profile related to environmental impacts by different purposes. However, the species was not among the most frequent in the study, and the investigators suggest that it may not play a secondary role in local epidemiology (Vilela et al. 2013).

Pintomyia fischeri (Pinto, 1926)

This species was described based on specimens from SP with its occurrence in secondary forested areas from many municipalities (Barretto 1943). Because it could be found close to domestic-animal shelters, it was suggested that it was adapting to a domiciliary environment (Barretto 1943). Currently, the species has its distribution mainly in the states of the South and Southeast regions of Brazil (SC, RS, PR, SP, RJ, MG, ES, MS, MT, and GO) (Aguiar and Medeiros 2003).

Discussion of the epidemiological importance of *Pi. fischeri* began when it was recorded in peri-domestic habitats of São Paulo state where ACL occurred (Forattini 1953). In addition, in MG and SC it was found in endemic ACL areas (Alexander et al. 2002; Marcondes et al. 2005). A study conducted from 1986 to 1995 again found this sand fly in the domiciliary habitat in areas with ACL of SP cases (Camargo-Neves et al. 2002).

Even though it had not found naturally infected with a *Leishmania* sp., there were strong grounds for considering it to be a potential vector. It is highly anthropophilic, and its spatial distribution coincides with reports of human ACL cases in deforested areas, Lainson (1983) suggested that this sand fly could be maintaining transmission of *L.* (*V.*) *braziliensis* among wild animals in forest fragments.

Recently its importance as a vector of *L.* (*V.*) *braziliensis* was reinforced with records of natural infections in females captured in endemic ACL areas of ES

(Rocha et al. 2010). In another study (Pita-Pereira et al. 2011), in the periphery of Porto Alegre (RS), where human cases of *L.* (*V.*) *braziliensis* have occurred, *L.* (*Viannia)* sp. was found infected. This result led the investigators to suggest that it was participating as an ACL vector in the region. In the metropolitan area of Greater Sao Paulo, cases of ACL are sporadic and are associated with fragments of the Atlantic rain forest. In the latter, both within the forest and outside in peri-domiciliary ecotopes *Pi. fischeri* was the dominant species (Moschin et al. 2013). It is interesting to note that neither *Ny. intermedia* nor *Ny. neivai* were found in this habitat but that *Mg. migonei* was present in smaller numbers.

All the previously cited literature reinforces the importance of *Pi. fischeri* in the eco-epidemiology of ACL in Southeast and Southern Brazil, particularly in forested habitats.

Lutzomyia gomezi (Nitzulescu 1931)

This species was described from female sand flies captured in San Cristobal, Tachira state, Venezuela. The male of this species was described from Panama by Rozeboom in 1940 as *Phlebotomus suis*, which was synonymized by Fairchild and Hertig (1948).

In Brazil, this sand fly has been recorded mainly in Northern regions, but it has also been recorded in the Northeast and Central regions (AC, AP, AM, RO, RR, PA, MA, GO, MT, and BA) (Young and Duncan 1994; Aguiar and Medeiros 2003).

Although in northern Brazil, specimens of *Lu. gomezi* were found infected with promastigotes, suggested as being a *Leishmania* sp., belonging to the subgenus *L.* (*Viannia)*, this was not confirmed (Rangel and Lainson 2009).

Historically, this phlebotomine has been associated with *L.* (*V.*) *panamensis* transmission in some South American countries without any evidence of transmission of *Leishmania* spp. in Brazil. However, recently, a natural infection of *L.* (*V.*) *shawi* was found in *Lu. gomezi* captured in Amazonian forest of PA (de Souza et al. 2016). The investigators suggest that this phlebotomine may participate in ACL eco-epidemiology, especially because of its arboreal habits, which is where the mammalian reservoirs of *L.* (*V.*) *shawi* occur.

Other vectors

We consider that the species discussed previously (*Bichromomyia flaviscutellata, Lutzomyia gomezi, Ny. intermedia, Ny. whitmani, Ny. neivai, Ny. umbratilis, Migonemyia migonei, Pintomyia fischeri, Psychodopygus wellcomei, Ps. complexus,* and *Ps. ayrozai*) are primary ACL vectors. However, others exist based on either epidemiological or parasitological evidence, or both, that may be playing roles in ACL transmission.

Trichophoromyia ubiquitalis, the only known vector of *L.* (*V.*) *lainsoni* (Silveira et al. 1991) but an *L.* (*L.*) *amazonensis* infection, was recorded by molecular methods in flies from Lábrea (AM) (Silva et al. 2014b). This sand fly is found in the Brazilian Amazonian forests and is anthropophilic being taken off man in larger numbers where the population is higher. Shannon-trap catches are considered to reflect anthropophily because a man catches the flies as they alight on the traps' surface. In forests, south of the Amazon River, PA, it ranked 16th of 68 species of the females in Shannon-trap catches, but it ranked 3rd in abundance when light-trap catches were included in the calculations (de Souza et al. 2016). In the Brazilian Guiana Shield forest of AP in Shannon-trap catches, it ranked sixth and also sixth when light-trap catches were included in the calculation (de Souza et al. 2017). These figures reflect moderate levels of anthropophily. It is interesting to note that in both places the proportion of males to females in Shannon-trap catches was almost equal (e.g. 21 of 23 in the PA catches and 31 of 36 in those from AP), thus suggesting no great differences in population sizes despite considerable ecological differences reflected by the dominant anthropophilic species in PA being *Ps. complexus/wellcomei* and in AP *Ny umbratilis*. In CDC catches in Lábrea (AM), it was the second most common species (Silva et al. 2014b). Its constant presence in relatively high numbers is consistent with the number of *L.* (*V.*) *lainsoni* ACL cases in forested regions and it being considered this parasite's principal vector.

Evidence exists that *Ps. davisi* participates in the transmission of *L.* (*V.*) *braziliensis* (Grimaldi et al. 1991), but there is stronger evidence for that of *L.* (*V.*) *naiffi* (Gil et al. 2003; de Souza et al. 2016). This sand fly has an extensive distribution throughout Amazonia and in pockets of the Atlantic rainforest. Infections of *L.* (*V.*) *naiffi* have been found in *Ps. davisi* in RO (Gil et al. 2003) and PA (de Souza et al. 2016). It was the dominant species in RO and ranked fifth in the PA study, but consecrated anthropophilic species—such as the *complexus/wellcomei* group and *Ny. umbratilis*—were present in large numbers. However, in the RO study area, *L.* (*V.*) *braziliensis* was the most common parasite, and no cases of human cases of *L.* (*V.*) *naiffi* were detected (Shaw et al. 2007). A possible explanation is that *L.*(*V.*) *naiffi* infections in man are mild and thus go unnoticed.

Another highly anthropophilic species is *Ps. squamiventris*, which occurs in the AM, AP, RR, and regions of PA north of the Amazon River (Ready et al. 1982). Its level of anthropophily is reflected by the fact that 4 times the number of females were captured in Shannon traps than in ground-level CDC traps (de Souza et al. 2017). Infections of *L.* (*V.*) *naiffi* have been found in specimens captured in AM, AP, and PA (Grimaldi et al. 1991; de Souza et al. 2017; Naiff et al. 1991). It has also been found infected with *L.* (*V.*) *braziliensis*, and a natural infection was transmitted experimentally to a hamster (Ryan et al. 1987b). Given its avidity for man, as well as the fact that it has been found infected with two *Leishmania* (*Vianna*) must be considered a highly probable ACL vector.

Records exist in the literature of infections in wild-caught Brazilian phlebotomines that were not identified to the species level. The first was that of Pessôa and Pestana (1940), who found flagellates in *Mg. migonei* and suggested that they were probably *L.* (*V.*) *braziliensis*. Such findings should not be ignored nor forgotten because they are strong circumstantial evidence for the possible role of a species in ACL transmission that needs confirmation. In some cases, subsequent identifications indicate what they most probably were. An example of this are studies (Ryan et al. 1987a) performed more than 30 years ago in the Serra das Carajás. Flagellates were found in 114 of 11,586 phlebotomines, and many identified as "*Leishmania braziliensis* subspecies" were found in *Lu. gomezi, Ny. richardwardi, Ny. shawi, Ny. whitmani, Th. ubiquitalis, Ps. hirsutus,* and *Ps. "wellcomei."* Worthy of mention is 11 infections occurring in *Th. ubiquitalis*. They were almost certainly *L.* (*V.*) *lainsoni* because this species was later identified in this same sand-fly species (Silveira et al. 1991), and this adds weight to the importance of this species as the primary vector of this parasite. *Ps. hirsutus* had also been found infected with *Leishmania* (*Vianna*) in Rio de Janeiro (Rangel et al. 1985), but we do not know what the parasites were. *L.* (*V.*) *shawi* was described in arboreal mammals captured in the Carajás, and it seems quite likely that the infections in the three *Nyssomyia* species belong to this species, but this needs confirmation. In addition to the above-mentioned infections, others have been recorded by different investigators in *Lu. renei, Ny. umbratilis, Pintomyia pessoai, Psathyromyia aragaoi, Pa. dendrophyla, Psychodopygus amazonensis, Ps. claustrei, Ps. davisi,* and *Ps. paraensis*. The *Leishmania* species were not identified; however, based on epidemiological and molecular data they probably belonged to an ACL *Leishmania* species.

Table 1 lists infections of *L.* (*L.*) *amazonensis* in species of *Nyssomyia* and *Trichophoromyia* captured in forests and in *Lutzomyia, Martinsmyia,* and *Nyssomyia* species captured in peri-domestic habitats. *L.* (*V.*) *braziliensis* has been documented in *Evandromyia, Martinsmyia, Micropygomyia,* and *Psychodopygus* species obtained from sylvatic habitats as well as a smaller number near human dwellings. Similarly, infections of *L.* (*V.*) *guyanensis* have been recorded in species of *Martinsmyia* and *Micropygomyia* from forests. The question is this: What do these infections mean in relation to ACL transmission? They may or may not be participating in enzootic or zoonotic ACL cycles, but future studies are needed to answer these questions. The finding of infections using molecular methods in pools of flies must be viewed with caution. It does not mean that the species in question should immediately be considered as a vector. Was blood present? Where were the parasites located? How many were there? Were metacyclic forms present? These are just a few questions, some of which are only answered by viewing the dissected insect's gut. A technique used extensively in the past that can lead to the parasite's isolation. For many years, epidemiological data favored the one parasite/vector hypothesis. However, depending on the *Leishmania* species, recent parasitological results now suggest a more complex situation where one species may be the dominant vector with other species being involved in enzootic and zoonotic ACL transmission.

Table 1 A list of sand fly species implicated in the transmission of American cutaneous leishmaniasis found in sylvatic and peridomestic environments in Brazil

Leishmania species	Sand flies implicated as vectors* or potential vectors in the sylvatic habitat	Sand flies implicated as vectors* or potential vectors in the peridomestic habitat
L. (L.) amazonensis	*Bichromomyia flaviscutellata* #*	*Bichromomyia flaviscutellata*
	Bichromomyia olmeca nociva #*	
	Bichromomyia reducta #*	
		Lutzomyia longipalpis
		Martinsmyia minasensis
		Nyssomyia intermedia
	Nyssomyia umbratilis	
	Nyssomyia yuilli	
	Trichophoromyia ubiquitalis	
L. (V.) braziliensis	*Evandromyia apurinan*	
	Evandromyia edwardsi	
	Evandromyia lenti	*Evandromyia lenti*
	Martinsmyia minasensis	
	Micropygomyia capixaba	
		Micropygomyia ferreirana
	Micropygomyia peresi	
	Migonemyia migonei #*	*Migonemyia migonei**
	Nyssomyia intermedia #*	*Nyssomyia intermedia**
	Nyssomyia neivai #*	*Nyssomyia neivai**
	Nyssomyia whitmani #*	*Nyssomyia whitmani**
	Pintomyia fischeri #*	*Pintomyia fischeri**
	Psychodopygus carrerai	
	Psychodopygus complexus #*	
	Psychodopygus davisi	
	Psychodopygus squamiventris #	
	Psychodopygus wellcomei #*	
L. (V.) guyanensis	*Martinsmyia minasensis*	
	Micropygomyia goiana	
	Nyssomyia anduzei #	
	Nyssomyia umbratilis #*	
L. (V.) naiffi	*Nyssomyia anduzei* #	
	Psychodopygus ayrozai #*	
	Psychodopygus davisi #*	
	Psychodopygus hirsutus #	
	Psychodopygus paraensis #	
	Psychodopygus squamiventris #	
L. (V.) lainsoni	*Trichophoromyia ubiquitalis* #*	
L. (V.) shawi	*Lutzomyia gomezi* #	
	Nyssomyia whitmani #*	*Nyssomyia whitmani**
L. (V.) utingensis	*Viannamyia tuberculata* #*	

Species marked with a hash (#) denotes identifications based on isolated parasites. Species not marked with an asterisk (*) are only potential vectors as they have been found infected by molecular methods with the respective *Leishmania* species

Impacts of Environmental and Climatic Changes

Global human population is facing the impacts of centuries of constant changes in natural environments. Climate change is happening now and impacts in the dynamics of infectious diseases are not only expected but can already be noticed (IPCC 2014; Woodward et al. 2014). Vector-borne diseases are particularly susceptible to environmental and climatic changes because their occurrence depends on the ecological balance between different species in complex transmission cycles (Walsh et al. 1993; Patz et al. 2000; McMichael 2004). Leishmaniases are among the vector-borne diseases most affected by this *ecological chaos* driven by human actions (Shaw 2008), and one of the expected impacts is the expansion of its geographical distribution (Ashford 2000; Dujardin 2006; WHO 2010).

Sand flies are affected by climate, especially by precipitation, humidity, and temperature. These variables influence their distribution, metabolism, and interactions with *Leishmania* (Ready 2008; WHO 2010; Hlavacova et al. 2013). One of the expected impacts of climate change in the eco-epidemiology of leishmaniasis is the expansion of the geographical distribution of its vectors (Peterson and Shaw 2003; González et al. 2010; Moo-Llanes et al. 2013; Carvalho et al. 2015; McIntyre et al. 2017). Given the wide latitudinal range of Brazil, regional climates play a major role in delimiting the distribution of species. Most projections under climate-change scenarios agree that disease vectors should find climatic conditions favourable to their geographic expansions towards higher latitudes in the upcoming decades (Carvalho et al. 2017).

In Brazil, the concept of leishmaniases as a sylvatic zoonosis is restricted to the Amazon Forest, Atlantic Forest fragments, and parts of Cerrado. A new transmission profile has emerged driven mostly by human-made environmental changes. In past decades, human migration of different origins and purposes resulted in major deforestation and unplanned settlements. These changes favour the dispersion of sylvatic animals (some *Leishmania* reservoir hosts) and sand flies (especially those species with eclectic feeding habits) to peri-domestic areas where new transmission cycles may establish close to human dwellings (Rangel 1995; Rangel and Lainson 2009; Costa et al. 2007).

Brazil is currently facing an increasing geographical expansion of ACL, which can probably be explained by the growing environmental changes, which in turn affect vector behaviour (Rangel et al. 2014). Some ACL-vector species have been showing evidences of adaptation to man-modified environments by establishing in peri-domestic areas, even in outskirts of large cities (Brasil 2007; Rangel and Lainson 2009). In this case, two sand-fly species are particularly good examples in different eco-epidemiological situations: *Ny. whitmani* and *Bi. flaviscutellata*. However, there are records of other species (see Table 1) that have been found in or near human dwellings that may be playing secondary or even primary roles in ACL-transmission cycles.

Because of its extensive geographical distribution and its association with two ACL parasites (*L.* (*V.*) *braziliensis* and *L.* (*V.*) *shawi*), *Ny. whitmani* is currently considered the most important ACL vector in Brazil, especially in impacted areas. This sand-fly species was found in several localities associated with the exploitation of natural environments and deforestation caused by the construction of roads, hydroelectric power plants, human settlements, wood extraction, agricultural activities, military training, and ecotourism. These epidemiological patterns occur throughout Brazil and together are considered to be responsible for the geographical expansion of ACL in the country.

Peterson and Shaw (2003) published the first projections of future potential distributions of Brazilian leishmaniasis vectors under climate-change scenarios. The investigators concluded that the ACL vectors *Ny. whitmani*, *Ny. intermedia*, and *Mg. migonei* should expand their distributions by the middle of the twenty-first century in different directions, most notably southwards, with *Ny. whitmani* showing the most dramatic range changes (Peterson and Shaw 2003). More recent projections of the potential distribution of *Ny. whitmani* reinforce the trends described by Peterson and Shaw (2003) and indicate a greater area of expansion of climate suitability in the North region (Costa et al., 2018, Fig. 2). Although climate-change scenarios show

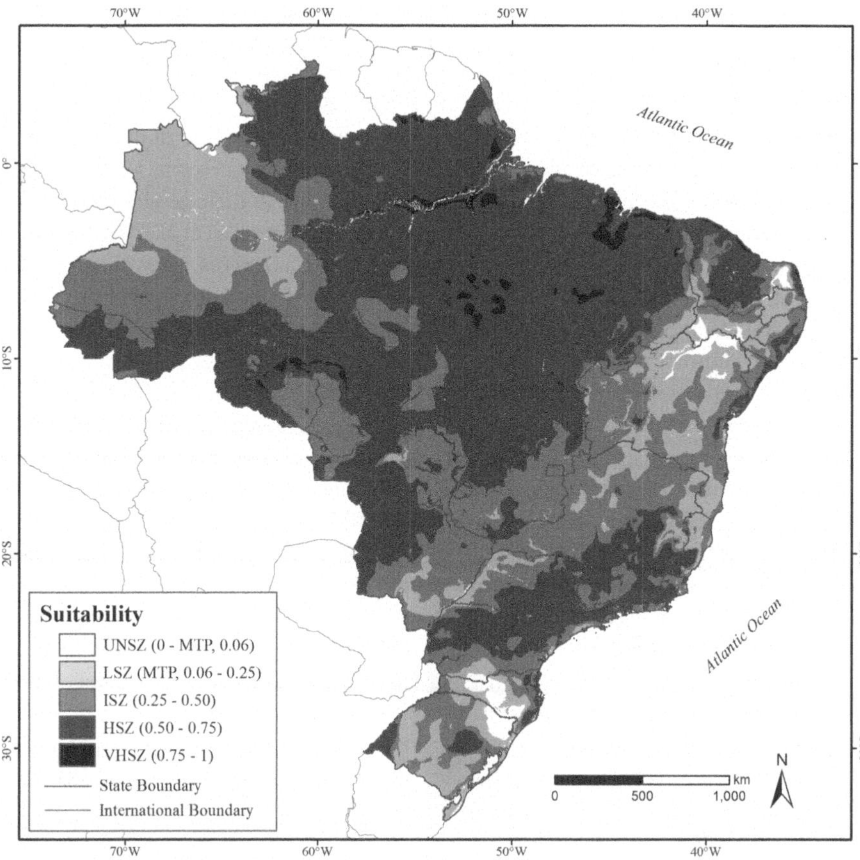

Fig. 2 Climatic suitability of *Ny. whitmani* under a "business as usual" climate-change scenario (average for years 2041–2060). UNSZ unsuitable zone, LSZ low-suitability zone, ISZ intermediate-suitability zone, HSZ high-suitability zone, VHSZ very high-suitability zone

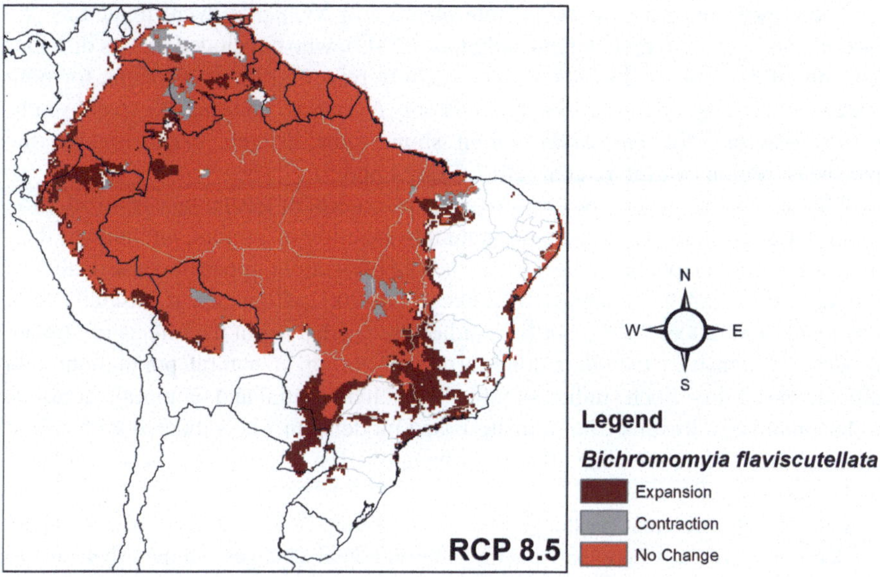

Fig. 3 Climatic suitability of *Bi. flaviscutellata* under a "business as usual" climate-change scenario (average for years 2041–2060). Dark red represent areas that will only become suitable in the future; light red areas are currently suitable and will remain suitable in the future; grey areas are currently suitable but will become unsuitable in the future. RCP representative concentration pathway

that the Amazon Region will become progressively drier (Joetzjer et al. 2013), the updated results state that *Ny. whitmani* will remain present in the region and should expand its area of climate suitability in the future (Costa et al. 2018).

The presence of *Bi. flaviscutellata* in peri-domestic areas, especially in the Cerrado biome, confirms the process of ruralisation of an *L.* (*L.*) *amazonensis* transmission cycle that was previously considered to be strictly sylvatic. Future projections under climate-change scenarios indicate that *Bi. flaviscutellata* might also expand its distribution beyond its current range limits in the Amazon and the Cerrado southwards into the Southeast and South regions (Carvalho et al. 2015, Fig. 3). Human cases of ADCL in Southeast Brazil are currently rare (Costa et al. 2009; Azeredo-Coutinho et al. 2007), although the disease seems to be gradually expanding its occurrence southwards. If the vector reaches these climatically suitable areas and its dispersion is followed by competent hosts and parasites, these can become ADCL-risk areas, especially because these are the most populated areas within the species' range. The possibility of this enzootic cycle to be maintained in secondary forests and even become peri-domestic was previously discussed (Lainson et al. 1994). This could be happening, in part, because of the adaptation process of the vector to man-modified environments. At first, it would be logical to think that a strictly sylvatic cycle would disappear with the deforestation of primary forests (Campbell-Lendrum et al. 2001), but the *L.* (*L.*) *amazonensis* cycle shows evidences of occurrence in secondary forests and peri-domestic areas, where the vector could be dispersing to domestic-animal shelters (Rangel and Lainson 2009).

The closely related species *Ny. intermedia* and *Ny. neivai* were treated as *Ny. intermedia* sensu lato by Peterson and Shaw (2003), who concluded that its distribution might expand southwards. A recent study reviewed the projections for both species separately, demonstrating that it is only *Ny. neivai* that should expand southwards, whereas *Ny. intermedia* might show some discrete expansions in the Northeast region (McIntyre et al. 2017, Figs. 4 and 5).

Climate change poses new challenges to the control of leishmaniasis. In addition to the long-term effects on the geographic distribution of vectors, interannual fluctuations of climate phenomena, such as the El Niño, might impact the seasonality of the sand flies and leishmaniasis (Franke et al. 2002; Chaves and Pascual 2006; Cardenas et al. 2006, 2008). Further studies are needed about the effects of climate in sand-fly densities including long-term monitoring of natural populations and climate variability. Such studies should also include spatial and temporal variations in leishmaniasis. Results from climate-based models must be validated with robust external data before they can effectively be applied in programs of the surveillance and control of leishmaniasis.

Considering the great challenge that is controlling ACL, a disease with complex epidemiology directly associated with environmental changes, studies that aim to characterize and monitor its spatial and temporal trends can support the epidemiological and entomological surveillance actions of health departments. These studies can help to identify receptive areas for new ACL outbreaks and population groups at higher risk of infection so that control actions can be better planned and more effective.

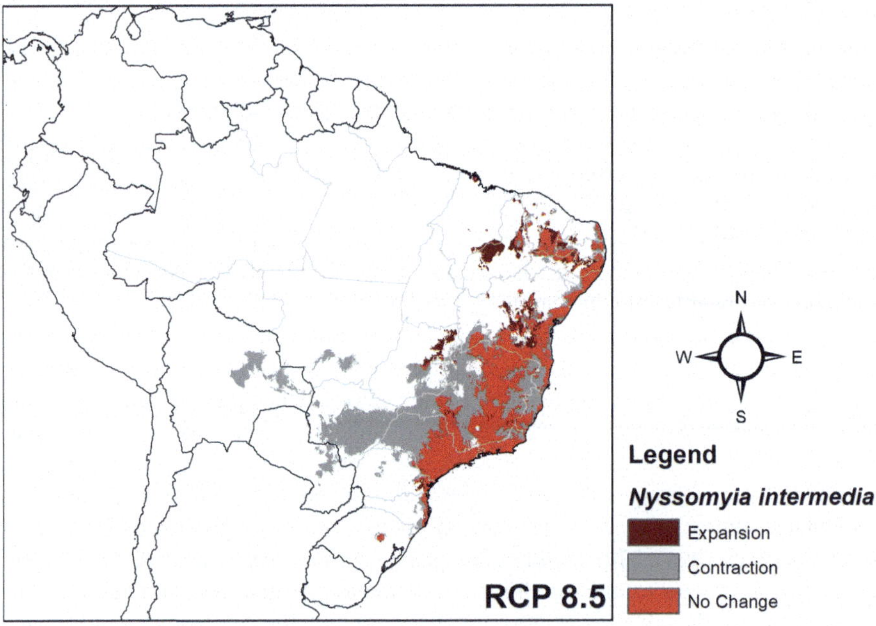

Fig. 4 Climatic suitability of *Ny. intermedia* under a "business as usual" climate-change scenario (average for years 2041–2060). Dark red represent areas that will only become suitable in the future; light red areas are currently suitable and will remain suitable in the future; grey areas are currently suitable but will become unsuitable in the future. RCP representative concentration pathway

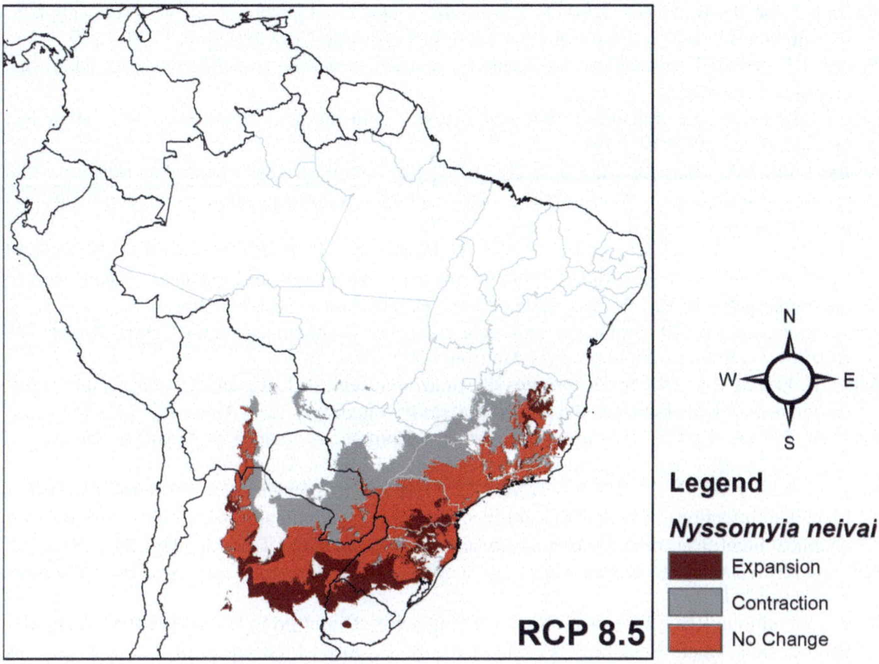

Fig. 5 Climatic suitability of *Ny. neivai* under a "business as usual" climate-change scenario (average for years 2041–2060). Dark red represent areas that will only become suitable in the future; light red areas are currently suitable and will remain suitable in the future; grey areas are currently suitable but will become unsuitable in the future. RCP: representative concentration pathway

References

Afonso MMS, Gomes AC, Meneses CRV, Rangel EF (2005) Studies on the feeding habits of *Lutzomyia* (*N.*) *intermedia* (Diptera: Psychodidae), vector of cutaneous leishmaniasis in Brazil. Cad Saude Publica 21:1816–1820

Aguiar GM, Medeiros WM (2003) Distribuição regional e habitats das espécies de flebotomíneos do Brasil. In: Elizabeth Rangel & Ralph Lainson (org.). Flebotomíneos do Brasil. Editora Fiocruz, Rio de Janeiro. pp 291–310

Aguiar GM, Soucasaux T (1984) Ecological aspects of Phlebotomus of Parque Nacional da Serra dos Órgãos, Rio de Janeiro. I. Monthly frequency in human baits (Diptera: Psychodidae: Phlebotominae). Mem Inst Oswaldo Cruz 79:197–209

Alexander B, Oliveira EB, Haigh E, Almeida LL (2002) Transmission of *Leishmania* in coffee plantations of Minas Gerais, Brazil. Mem Inst Oswaldo Cruz 97:627–630

Andrade MS, Valença HF, da Silva AL, Almeida EL, de Brito ME, Brandão-Filho SP (2005) Sand fly fauna in a military training area endemic for American tegumentary leishmaniasis in the Atlantic Rain Forest region of Pernambuco, Brazil. Cad Saude Publica 21:1761–1767

Andrade-Filho JD, Galati EA, Falcão AL (2004) Biology of the first generation of a laboratory colony of *Nyssomyia intermedia* (Lutz & Neiva, 1912) and *Nyssomyia neivai* (Pinto, 1926) (Diptera: Psychididae). Mem Inst Oswaldo Cruz 99:597–601

Andrade Filho JD, Brazil RP (2003) Relationships of new word phlebotomine sand flies (Diptera: Psychodidae) based on fossil evidence. Mem Inst Oswaldo Cruz 98(Suppl. 1):145–149

Aragão HB (1922) Leishmaniose tegumentar e a sua transmissão pelo *Phlebotomus*. Mem Inst Oswaldo Cruz 82:143

Aragão HR (1927) Leishmaniose Tegumentar e sua Transmissão pelos Flebótomos. Mem Inst Oswaldo Cruz 20(2):177–185

Araújo Filho NA, Sherlock IA, Coura JR (1981) Leishmaniose Tegumentar Americana na Ilha Grande, Rio de Janeiro. V. Observações sobre a biologia dos transmissores em condições naturais. Rev Soc Bras Med Trop 14(4–6):171–183

Araujo-Pereira T, Fuzari AA, Andrade Filho JD, Pita-Pereira D, Britto C, Brazil RP (2014) Sand fly fauna (Diptera: Psychodidae: Phlebotominae) in an area of leishmaniasis transmission in the municipality of Rio Branco, State of Acre, Brazil. Parasit Vect 7(1):360

Arias JR, Freitas RA (1977a) On the vectors of cutaneous leishmaniasis in the Central Amazon of Brazil. I. Preliminary findings. Acta Amazon 7:293–294

Arias JR, Freitas RA (1977b) Flebótomos da Amazônia Central. 1. Resultados obtidos das capturas feitas com iscas humanas e eqüina (Diptera: Psychodidae). Acta Amazon 7:507–527

Arias JR, Freitas RA (1978) Sobre os vetores de leishmaniose cutânea na Amazônia Central do Brasil. 2. Incidência de flagelados em flebótomos selváticos. Acta Amazon 8:387–396

Arias JR, Miles MA, Naiff RD, Povoa MM, de Freitas RA, Biancardi CB, Castellon EG (1985) Flagellate infections of Brazilian sand flies (Diptera: Psychodidae): isolation in vitro and biochemical identification of *Endotrypanum* and *Leishmania*. Am J Trop Med Hyg 34:1098–1108

Ashford RW (2000) The leishmaniasis as emerging and reemerging zoonoses. Int J Parasitol 30:1269–1281

Azeredo-Coutinho RBG, Conceição-Silva F, Schubach A, Cupolillo E, Quintella LP, Madeira MF, Pacheco RS, Valete-Rosalino CM, Mendonça SCF (2007) First report of diffuse cutaneous leishmaniasis and *Leishmania amazonensis* infection in Rio de Janeiro State, Brazil. Trans Roy Soc Trop Med Hyg 101:735–737

Azevedo ACR (2008) Contribuição ao conhecimento da fauna de flebotomíneos do Estado do Acre com ênfase em *Lutzomyia (Nyssomyia) umbratilis* Ward & Fraiha, 1977, importante transmissor de leishmaniose tegumentar americana na Amazônia: biologia, morfologia, morfometria e discussão da identidade taxonômica, PnD Thesis, Instituto Oswaldo Cruz, Rio de Janeiro, 144 pp

Azevedo ACR, Bessa-Luz S, Vilela ML, Rangel EF (1993) Studies on the sand fly fauna of Samuel Ecological Station, Porto Velho municipality, Rondônia State, Brazil. Mem Inst Oswaldo Cruz 88:509–512

Azevedo ACR, Costa SM, Pinto MC, Souza JL, Cruz HC, Vidal J, Rangel EF (2008) Studies on the sand fly fauna (Diptera: Psychodidae: Phlebotominae) from transmission areas of American cutaneous leishmaniasis in State of Acre, Brazil. Mem Inst Oswaldo Cruz 103:760–767

Azevedo ACR, Lainson R, Afonso MMS, Rangel EF (2005) Estudio sobre *Lutzomyia* (*Nyssomyia) umbratilis* Ward & Frailha, 1977 (Psychodidae: Phlebotominae), vector de la leishmaniasis tegumentar Americana, en el Amazonas. Salud Ciencia 14:30–33

Azevedo ACR, Rangel EF (1991) A study of sand fly species (Diptera: Psychodidae: Phlebotominae) in a focus of cutaneous leishmaniasis in the Municipality of Baturité, Ceará, Brazil. Mem Inst Oswaldo Cruz 85:251

Azevedo ACR, Rangel EF, Costa EM, David J, Vasconcelos AW, Lopes UG (1990) Natural infection of *Lutzomyia* (*Nysomyia*) *whitmani* (Antunes & Coutinho, 1939) by *Leishmania* of the *braziliensis* complex in Baturité, Ceará State, Northeast Brazil. Mem Inst Oswaldo Cruz 85:251

Azevedo ACR, Souza NA, Meneses CRV, Costa WA, Costa SM, Lima JB, Rangel EF (2002) Ecology of sand flies (Diptera: Psychodidae: Phlebotominae) in the North of Mato Grosso, Brazil. Mem Inst Oswaldo Cruz 97(4):459–464

Azevedo PCB, Lopes GN, Fonteles RS, Vasconcelos GC, Moraes JLP, Rebêlo JMM (2011) The effect of fragmentation on phlebotomine communities (Diptera: Psychodidae) in areas of ombrophilous forest in São Luís, state of Maranhão, Brazil. Neotrop Entomol 40(2):271–277

Balbino VQ, Coutinho-Abreu IV, Sonoda IV, Marques da Silva W, Marcondes CB (2005) Phlebotomine sandflies (Diptera: Psychodidae) of the Atlantic forest in Recife, Pernambuco state, Brazil: the species coming to human bait, and their seasonal and monthly variations over a 2-year period. Ann Trop Med Parasit 99(7):683–693

Balbino VQ, Marcondes CB, Alexander B, Luna LK, Lucena MM, Mendes A, Andrade PP (2001) First report of *Lutzomyia (Nyssomyia) umbratilis* Ward & Frahia, 1977 outside of Amazonian Region, in Recife, State of Pernambuco, Brazil (Diptera: Psychodidae: Phlebotominae). Mem Inst Oswaldo Cruz 96(3):315–317

Barbosa MG, Fé NF, Marcião AH, Silva AP, Monteiro WM, Guerra JA (2008) Sand fly fauna (Diptera: Psychodidae) in a focus of American cutaneous leishmaniasis on the urban periphery of Manaus, state of Amazonas. Rev Soc Bras Med Trop 41:485–491

Barreto AC, Vexenat JA, Cuba-Cuba CA, Marsden PD (1982) Fauna flebotomínica de uma região endêmica de leishmaniose cutâneo-mucosa no Estado da Bahia. IX Reunião Anual sobre Pesquisa Básica em Doenças de Chagas, 147 pp

Barretto MP (1943) Observações sobre a biologia em condições naturais dos flebótomos do estado de São Paulo (Ditera: Psychodidae). Tese de Livre-Docência, Faculdade de Medicina da USP, São Paulo, p 162

Barretto MP (1961) Subfamilias e gêneros neotropicais da família Psychodidae Big, 1854 (Diptera). Pap Av Dep Zool S Paulo 14:211–225

Brandão-Filho SP, Brito ME, Carvalho FG, Ishikawa EA, Cupolillo E, Floeter-Winter L, Shaw JJ (2003) Wild and synanothropic hosts of *Leishmania (Viannia) braziliensis* in the endemic cutaneous leishmaniasis locality of Amaraji, Pernambuco state, Brazil. Trans R Soc Trop Med Hyg 97:291–296

Brandão-Filho SP, Campbell-Lendrum D, Brito ME, CR SJJD (1999) Epidemiological surveys confirm an increasing burden of cutaneous leishmaniasis in North–East Brazil. Trans Roy Soc Trop Med Hyg 93:488–494

Brasil (2007) Ministério da Saúde. Secretaria de Vigilância em Saúde. Departamento de Vigilância Epidemiológica. Manual de Vigilância da Leishmaniose Tegumentar Americana 2a. ed. Editora do Ministério da Saúde, Brasília

Brito VND, Almeida ADBPF, Nakazato L, Duarte R, Souza CDO, Sousa VRF (2014) Phlebotomine fauna, natural infection rate and feeding habits of *Lutzomyia cruzi* in Jaciara, state of Mato Grosso, Brazil. Mem Inst Oswaldo Cruz 109(7):899–904

Camargo-Neves VLF, Gomes AC, Antunes JLF (2002) Correlação da presença de espécies de flebotomíneos (Diptera: Psychodidae) com registros de casos da leishmaniose tegumentar americana no Estado de São Paulo, Brasil. Rev Soc Bras Med Trop 35(4):1–11

Campbell-Lendrum D, Dujardin JP, Martinez E, Feliciangeli MD, Perez JE, Silans LNMP, Desjeux P (2001) Domestic and peridomestic transmission of American cutaneous leishmaniasis: changing epidemiological patterns present new control opportunities. Mem Inst Oswaldo Cruz 96(2):159–162

Campbell-Lendrum DH, Brandão-Filho SP, Ready PD, Davies CR (1999) Host and/or site loyalty of *Lutzomyia whitmani* (Diptera: Psychodidae) in Brazil. Med Vet Entomol 13:209–211

Campos AM, Matavelli R, Santos CD, Moraes LS, Rebêlo JMM (2013) Ecology of phlebotomines (Diptera: Psychodidae) in a transitional area between the Amazon and the Cerrado in the state of Maranhão, Brazil. J Med Entomol 50(1):52–58

Cardenas R, Sandoval CM, Rodríguez-Morales AJ, Franco-Paredes C (2006) Impact of climate variability in the occurrence of leishmaniasis in northeastern Colombia. Am J Trop Med Hyg 75(2):273–277

Cardenas R, Sandoval CM, Rodríguez-Morales AJ, Vivas P (2008) Zoonoses and climate variability. The example of leishmaniasis in southern departments of Colombia. Ann New York Acad Sci 1149:326–330

Carvalho BM, Dias CMG, Rangel EF (2014) Phlebotomine sand flies (Diptera, Psychodidae) from Rio de Janeiro State, Brazil: species distribution and potential vectors of leishmaniases. Rev Bras Entomol 58(1):77–87

Carvalho BM, Rangel EF, Ready PD, Vale MM (2015) Ecological niche modelling predicts southward expansion of *Lutzomyia (Nyssomyia) flaviscutellata* (Diptera: Psychodidae: Phlebotominae), vector of *Leishmania (Leishmania) amazonensis* in South America, under climate change. PLoS One 10(11):e0143282

Carvalho BM, Rangel EF, Vale MM (2017) Evaluation of the impacts of climate change on disease vectors through ecological niche modelling. Bull Ent Res 107:419–430

Carvalho BM, Maximo M, Costa WA, de Santana ALF, da Costa SM, da Costa Rego TAN, Pita-Pereira D, Rangel EF (2013) Leishmaniasis transmission in an ecotourism area: potential vectors in Ilha Grande, Rio de Janeiro State, Brazil. Parasit Vect 6:325

Carvalho RM, Valença HF, Silva FJ, Pita-Pereira D, Pereira TA, Britto C, Brazil RP, Brandão Filho SP (2010) Natural *Leishmania infantum* in *Migonemyia migonei* (França, 1920) (Diptera:Psychodidae: Phlebotominae) the putative vector of visceral leishmaniasis in Pernambuco. State, Brazil. Acta Trop 116:108–110

Carvalho GM, Andrade Filho JD, Falcão AL, Rocha Lima AC, Contijo CM (2008) Naturally infected *Lutzomyia* sand flies in a *Leishmania*-endemic area of Brazil. Vector Borne Zoonotic Dis 8:407–414

Casanova C, Natal D, Santos FA (2009) Survival, population size and gonotrophic cycle duration of *Nyssomyia neivai* (Diptera: Psychodidae) at an endemic area of American cutaneous leishmaniasis in Southeastern Brazil. J Med Entomol 46:42–50

Chaves LF, Pascual M (2006) Climate cycles and forecasts of cutaneous leishmaniasis, a nonstationary vector-borne disease. PLoS Med 3(8):e295

Christensen HA, Arias JR, de Vasques AM, de Freitas R (1982) Hosts of sand fly vectors of *Leishmania braziliensis guyanensis* in the central Amazon of Brazil. Am J Trop Med Hyg 31:239–242

Costa JML, Costa AAUML, Elkhoury NA, Bezerril ACR, Barral A, Saldanha ACR (2009) Leishmaniose cutânea difusa (LCD) no Brasil após 60 anos de sua primeira descrição. Gazeta Médica da Bahia 79(Supl.3):16–24

Costa SM, Cechinel M, Bandeira V, Zannuncio JC, Lainson R, Rangel EF (2007) *Lutzomyia (Nyssomyia) whitmani s.l.* (Antunes & Coutinho, 1939) (Diptera: Psychodidae: Phlebotominae) and the epidemiology of American cutaneous leishmaniasis in Brazil. Mem Inst Oswaldo Cruz 102:149–153

Costa Lima AM (1932) Sobre os *Phlebotomus* Americanos (Diptera, Psychodidae). Mem Inst Oswaldo Cruz 26(1):15–69

da Costa SM, Luís Passos Cordeiro J, Rangel EF (2018) Environmental suitability for Lutzomyia (Nyssomyia) whitmani (Diptera: Psychodidae: Phlebotominae) and the occurrence of American cutaneous leishmaniasis in Brazil. Parasites & Vectors 11(1)

Dias-Sversutti A, Scodro RB, Reinhold-Castro KR, Neitzke HC, Teodoro U (2007) Preliminary study on feeding preference of *Nyssomyia neivai* (Pinto) and *Nyssomyia whitmani* (Antunes & Coutinho) (Diptera: Psychodidae) in a rural area of the state of Paraná, South Brazil. Neotrop Entomol 36:953–959

de Souza AAA, Santos TV, Jennings YLL, Ishikawa EAY, Barata IR, Silva MGS, Lima JAN, Shaw J, Lainson R, Silveira FT (2016) Natural *Leishmania* (*Viannia)* spp. infections in phlebotomine sand flies (Diptera: Psychodidae) from the Brazilian Amazon region reveal new putative transmission cycles of American cutaneous leishmaniasis. Parasite 23:22

de Souza A, Ishikawa E, Braga R, Silveira F, Lainson R, Shaw J (1996) *Psychodopygus complexus*, a new vector of *Leishmania braziliensis* to humans in Pará state, Brazil. Trans R Soc Trop Med Hyg 90:112–113

de Souza AAA, da Rocha BI, das Gracas Soares Silva M, JAN L, YLL J, EAY I, Prevot G, Ginouves M, Silveira FT, Shaw J, Dos Santos TV (2017) Natural *Leishmania* (*Viannia*) infections of phlebotomines (Diptera: Psychodidae) indicate classical and alternative transmission cycles of American cutaneous leishmaniasis in the Guiana Shield, Brazil. Parasite 24:13

Donalisio MR, Peterson AT, Costa PL, da Silva FJ, Valença HF, Shaw JJ, Brandão Filho SP (2012) Microspatial distributional patterns of vectors of cutaneous leishmaniasis in Pernambuco, northeastern Brazil. J Trop Med 2012:642910

Dorval MEC, Alves TP, Oliveira AGD, Brazil RP, Galati EAB, Cunha RVD (2007) Modification of Disney trap for capture of sand flies (Diptera: Psychodidae: Phlebotominae). Mem Inst Oswaldo Cruz 102(7):877–878

Dorval MEC, Alves TP, Cristaldo G, Rocha HCD, Alves MA, Oshiro ET, Oliveira AG, Brazil RP, Galati EAB, Cunha RVD (2010) Sand fly captures with Disney traps in area of occurrence of *Leishmania* (*Leishmania*) *amazonensis* in the State of Mato Grosso do Sul, mid-western Brazil. Rev Soc Bras Med Trop 43(5):491–495

Dorval ME, Cristaldo G, Rocha HC, Alves TP, Alves MA, Oshiro ET, Brazil RP, Galati EA, Cunha RV (2009) Phlebotomine fauna (Diptera: Psychodidade) of an endemic area in the state of Mato Grosso do Sul, Brazil. Mem Inst Oswaldo Cruz 104:695–702

Dujardin JC (2006) Risk factors in the spread of leishmaniases: towards integrated monitoring? Trends Parasitol 22(1):4–6

Espinosa OA, Serrano MG, Camargo EP, Teixeira MM, Shaw JJ (2016) An appraisal of the taxonomy and nomenclature of trypanosomatids presently classified as *Leishmania* and *Endotrypanum*. Parasitology, ePub, 1–13. https://doi.org/10.1017/S0031182016002092

Esterre P, Chippaux JP, Lefait JF, Dedet JP (1986) Evaluation of a cutaneous leishmaniasis control program in a forest village of French Guyana. Bull World Health Organ 64:559–565

Fairchild GB, Hertig M (1948) An improved method for mounting small insects. Science 108(2792):20

Falqueto A (1995) Especificidade alimentar de flebotomíneos em duas áreas endêmicas de leishmaniose tegumentar no estado do Espírito Santo, PhD Thesis, Fiocruz, Rio de Janeiro, 84 pp

Feliciangeli MD, Ramirez Perez J, Ramirez A (1985) First Venezuelan record of *Lutzomyia umbratilis* Ward & Fraiha, 1977 (Diptera: Psychodidae), a proven vector of *Leishmania braziliensis guyanensis*. Trans R Soc Trop Med Hyg 79:878

Ferreira AL, Sessa PA, Varejão JB, Falqueto A (2001) Distribution of sand flies (Diptera: Psychodidae) at different altitudes in an endemic region of American cutaneous leishmaniasis in the state of Espirito Santo, Brazil. Mem Inst Oswaldo Cruz 96:1061–1067

Forattini OP, Santos MR (1952) Nota sobre infecção natural de *Phlebotomus intermedius* Lutz & Neiva, 1912, por formas leptomonas, em um foco ativo de leishmaniose tegumentar americana. Arq Hig S Paulo 17:171–174

Forattini OP (1953) Nota sobre criadouros naturais de flebótomos em dependências peridomiciliares no estado de São Paulo. Arq Fac Hig Saude Publica Univ Sao Paulo 7:157–168

Forattini OP (1954) Nota sobre a biologia de *Phlebotomus* (Diptera: Psychodidae) em região da Bacia do Rio Paraná (Brasil). Arq Fac Hig Saude Publica Univ Sao Paulo 8:15–136

Forattini OP (1973) Entomologia Médica, vol 4. Editora Edgard Blücher & Editora da Universidade de São Paulo, São Paulo, 658 pp

Forattini OP, Pattoli DBG, Rabello EX, Ferreira OA (1972) Infecção natural de flebotomíneos em foco enzoótico de leishmaniose tegumentar no estado de São Paulo, Brasil. Rev Saude Publica 6:431–433

Forattini OP, Rabello EX, Serra OP, Cotrim MD, Galati EAB, Barata JMS (1976) Observações sobre a transmissão da leishmaniose tegumentar no estado de São Paulo, Brasil. Rev Saude Publica 10:31–43

França F, Lago EL, Tada S, Costa JM, Vale K, Oliveira J, Costa MA, Osaki M, Cheever L, Netto EM, Barretto AC, Johnson WD, Marsden PD (1991) An outbreak of human *Leishmania* (*Viannia*) *braziliensis* infection. Mem Inst Oswaldo Cruz 86:169–174

Franke CR, Ziller M, Staubach C, Latif M (2002) Impact of the El Niño/Southern Oscillation on visceral leishmaniasis, Brazil. Emerg Infect Dis 8(9):914–917

Galati EAB, Nunes VLB, Dorval MEC, Oshiro ET, Cristaldo G, Espínola MA, Rocha HC, Garcia WB (1996) Study of the phlebotomines (Diptera: Psychodidade) in area of cutaneous leishmaniasis in the Mato Grosso state, Brazil. Rev Saude Publica 30:115–128

Gil LH, Basano AS, Souza AA, Silva MG, Barata I, Ishikawa EA, Camargo LM, Shaw JJ (2003) Recent observations on the sand fly (Diptera: Psychodidae) fauna of the state of Rondônia, western Amazonia, Brazil: the importance of *Psychodopygus davisi* as a vector of zoonotic cutaneous leishmaniasis. Mem Inst Oswaldo Cruz 98:751–755

Godoy RE, de Santana ALF, Graser C, Rangel EF, Vilela ML (2017) Aspects on the ecology of phlebotomine sand flies (Diptera: Psychodidae) from Guaraí, state of Tocantins, Brazil, endemic area for American cutaneous leishmaniasis. J Med Entomol 54(1):229–235

Gomes AC (1994) Sand fly vectorial ecology in the state of São Paulo. Mem Inst Oswaldo Cruz 89:457–460

Gomes AC, Rabello EX, Santos JFL, Galati EAB (1986) Ecological aspects of American cutaneous leishmaniasis. 4. Observation on the endophilic behavior of the sand fly and the vectorial role of *Psychodopygus intermedius* in the Ribeira Valley region of the S. Paulo state, Brazil. Rev Saude Publica 20:280–287

Gomes AC, Galati EA (1989) Aspectos ecológicos da leishmaniose tegumentar americana. 7. Capacidade vetorial flebotominea em ambiente florestal primário do sistema da Serra do Mar, região do Vale do Ribeira, estado de São Paulo, Brasil. Rev Saude Publica 23:136–142

Gonçalves BRD (2003) Identificação da Fauna de Flebotomíneos em Função de Casos Autóctones de LTA. Boletim Epidemiológico. Secretaria Municipal de Saúde de Porto Alegre 21(5):5

González C, Wang O, Strutz SE, González-Salazar C, Sánchez-Cordero V, Sarkar S (2010) Climate change and risk of leishmaniasis in North America: predictions from ecological niche models of vector and reservoir species. PLoS Negl Trop Dis 4(1):e585

Gouveia C, Oliveira RM, Zwetsch A, Motta-Silva D, Carvalho BM, Santana AF, Rangel EF (2012) Integrated tools for American cutaneous leishmaniasis surveillance and control: intervention in an endemic area in Rio de Janeiro, RJ, Brazil. Interdis Perspec Infect Dis 2012:1–9

Grimaldi G Jr, Momen H, Naiff RD, McMahon-Pratt D, Barrett TV (1991) Characterization and classification of leishmanial parasites from humans, wild mammals, and sand flies in the Amazon region of Brazil. Am J Trop Med Hyg 44:645–661

Guerra JA, Barbosa MG, Loureiro AC, Coelho CP, Rosa GG, Coelho LI (2007) American tegumentary leishmaniasis in children: epidemiological aspects of cases treated in Manaus, Amazonas, Brazil. Cad Saude Publica 23:2215–2223

Guimarães FN, Costa O (1966) Novas observações sobre a *Leishmania* isolada de "Oryzomys goeldi" na Amazônia (4a nota). Hospital RJ 69:161–168

Guimarães VCFV, Costa PL, Silva FJ, Silva KT, Silva KG, Araújo AI, Rodrigues EHF, Brandão-Filho SP (2012) Phlebotomine sand flies (Diptera: Psychodidae) in São Vicente Férrer, a sympatric area to cutaneous and visceral leishmaniasis in Pernambuco, Brazil. Rev Soc Bras Med Trop 45:66–70

Hlavacova J, Votypka J, Volf P (2013) The effect of temperature on *Leishmania* (Kinetoplastida: Trypanosomatidae) development in sand flies. J Med Entomol 50(5):955–958

Hoch A, Ryan L, Vexenet JA, Rosa AC, Barretto AC (1986) Isolation of *Leishmania braziliensis braziliensis* and other trypanosomatids from phlebotomines in mucocutaneous leishmaniasis endemic area, Bahia, Brazil. Mem Inst Oswaldo Cruz *81*(Suppl):BI 44

IPCC. Core Writing Team, Pachauri RK, Meyer LA (eds) (2014) Climate Change 2014: Synthesis Report. Contribution of Working Groups I, II and III to the Fifth Assessment Report of the Intergovernmental Panel on Climate Change. IPCC, Geneva

Ishikawa EAY, Ready PD, de Souza AA, Day JC, Rangel EF, Davies CR, Shaw JJ (1999) A mitochondrial DNA phylogeny indicates close relationship between populations of *Lutzomyia whitmani* (Diptera: Psychodidae: Phlebotominae) from the rain-forest region of Amazônia and Northeast Brazil. Mem Inst Oswaldo Cruz 94:339–345

Joetzjer E, Douville H, Delire C, Ciais P (2013) Present-day and future Amazonian precipitation in global climate models: CMIP5 versus CMIP3. Clim Dyn 41(11–12):2921–2936

Justiniano SCB, Chagas AC, Pessoa FAC, Queiroz RG (2004) Comparative biology of two populations of *Lutzomyia umbratilis* (Diptera: Psychodidae) of central Amazonia, Brazil, under laboratory conditions. Braz J Biol 64:227–235

Lainson R (1983) The American leishmaniases: some observation on their ecology and epidemiology. Trans R Soc Trop Med Hyg 77:569–596

Lainson R (1988) Ecological interactions in the transmission of the leishmaniases. Philos Trans R Soc Lond Series B 321:389–404

Lainson R, Braga RR, Souza AA, Povoa MM, Ishihawa EAY, Silveira FT (1989) *Leishmania (Viannia) shawi* sp.n, a parasite of monkeys, sloths and procyonids in Amazonian Brazil. Ann Parasit Hum Comp 64:200–207

Lainson R, Shaw JJ (1968) Leishmaniasis in Brazil I. Observations on enzootic rodent leishmaniasis – incrimination of *Lutzomyia flaviscutellata* (Mangabeira) as the vector in the lower Amazonian Basin. Trans R Soc Trop Med Hyg 62:385–395

Lainson R, Shaw JJ (1970) Leishmaniasis in Brazil. V: studies on the epidemiology of cutaneous leishmaniasis in Mato Grosso state and observations on two distinct strains of *Leishmania* isolates from man and forest animals. Trans R Soc Trop Med Hyg 64:654–667

Lainson R, Shaw JJ (1972) Leishmaniasis of the new world: taxonomic problems. Brit Med Bull 28:44–48

Lainson R, Shaw JJ (1979) The role of animals in the epidemiology of South American leishmaniasis. In: Lumsden WHR, Evans DA (eds) Biology of the Kinetoplastida, vol 2. Academic Press, London, pp 1–116

Lainson R, Shaw JJ (1987) Evolution, classification and geographic distribution. In: Peters W, Killick-Kendrick R (eds) The leishmaniases in biology and medicine, vol 1. Academic Press, London, pp 1–20

Lainson R, Shaw JJ (1998) New World leishmaniasis. The neotropical *Leishmania* species. In: Cox FEG, Kreier JP, Wakelin D (eds), Topley & Wilson's microbiology & microbiol infections, 9th edn. vol 5. Parasitology, Arnold, London, pp 242–266

Lainson R, Shaw JJ, Póvoa M (1981a) The importance of edentates (sloths and anteaters) as primary reservoirs of *Leishmania brazi-liensis guyanensis*, causative agent of "pian-bois" in North Brazil. Trans R Soc Trop Med Hyg 75:611–612

Lainson R, Shaw JJ, Póvoa M (1981b) Leishmaniasis in Brazil: XVI. Isolation and identification of *Leishmania* species from sand flies, wild mammals and man in north Pará state, with particular reference to *L. braziliensis guyanensis* causative agent of "pian-bois". Trans R Soc Trop Med Hyg 75:530–536

Lainson R, Shaw JJ, Silveira FT, de Souza AA, Braga RR, Ishikawa EA (1994) The dermal leishmaniases of Brazil, with special reference to the eco-epidemiology of the disease in Amazônia. Mem Inst Oswaldo Cruz 89:435–443

Lainson R, Shaw JJ, Ward RD, Fraiha H (1973) Leishmaniasis in Brazil. IX. Considerations on the *Leishmania braziliensis* complex. Importance of sand flies of the genus *Psychodopygus* (Mangabeira) in the transmission of *Leishmania braziliensis braziliensis* in North Brazil. Trans R Soc Trop Med Hyg 67:184–196

Lainson R, Shaw JJ, Ward RD, Ready PD, Naiff RD (1979) Leishmaniasis in Brazil: XIII. Isolation of *Leishmania* from armadillos (*Dasypus novemcinctus*) and observation on the epidemiology of cutaneous leishmaniasis in north Pará state. Trans R Soc Trop Med Hyg 73:239–242

Lainson R, Strangways-Dixon J (1964) The epidemiology of dermal leishmaniasis in British Honduras. Parte III. Reservoir-host of *Leishmania mexicana* among forest rodents. Trans R Soc Trop Med Hyg 58:136–153

Lainson R, Ward RD, Shaw JJ (1976) Cutaneous leishmaniasis in North Brazil: *Lutzomyia anduzei* as a major vector. Trans R Soc Trop Med Hyg 70:171–172

Le Pont F, Pajot FX (1980) La leishmaniose en Guyane Française. I. Etude de l'ecologie et du taux d'infection naturelle d'un vecteur *Lutzomyia (Nyssomyia) umbratilis* Ward et Fraiha, 1977 en saison sèche. Considerations epidémiologiques. Cah ORSTOM Ser Ent Med Parasitol 18:359–383

Lucena DT (1953) Flebótomos do Nordeste. Morfologia de algumas espécies e sua distribuição. Papéis Avulsos de Zoologia 11:89–107

Luz E, Membrive N, Castro EA, Dereure J, Pratlong F, Dedet JA, Pandey A, Thomaz-Soccol V (2000) *Lutzomyia whitmani* (Diptera: Psychodidae) as vector of *Leishmania (V.) braziliensis* in Paraná state, Southern Brazil. Ann Trop Med Parasitol 94:623–631

Marcondes CB (1996) A redescription of *Lutzomyia (Nyssomyia) intermedia* (Lutz & Neiva, 1912) and resurrection of *L. neivai* (Pinto, 1926) (Diptera: Psychodidae: Phlebotominae). Mem Inst Oswaldo Cruz 91:457–462

Marcondes CB, Day JC, Ready PD (1997) Introgression between *Lutzomyia intermedia* and both *Lu. neivai* and *Lu. whitmani* and their roles as vectors of *Leishmania braziliensis*. Trans R Soc Trop Med Hyg 91:725–726

Marcondes CB, Alexander B (2003) Correlation of male genital filaments and female spermathecal ducts in new world sand flies of the *Lutzomyia intermedia* species complex (Diptera: Psychodidae, Phlebotominae). Mem Inst Oswaldo Cruz 98:611–613

Marcondes CB, Borges PSS (2000) Distinction of males of *Lutzomyia intermedia* (Lutz & Neiva, 1912) complex by ratios between dimensions and by an artificial neural network (Diptera, Psychodidae, Phlebotominae). Mem Inst Oswaldo Cruz 95:685–688

Marcondes CB, Conceição MB, Portes MG, Simão BP (2005) Phlebotomine sand flies in a focus of dermal leishmaniasis in the eastern region of the Brazilian state of Santa Catarina: preliminary results (Diptera: Psychodidae). Rev Soc Bras Med Trop 38:353–355

Marcondes CB, Lozovei AL, Falqueto A, Brazil RP, Galati EAB, Aguiar GM, Souza NA (1999) Influence of altitude, latitude and season of collection (Bergmann's rule) on the dimension of *Lutzomyia intermedia* (Lutz & Neiva, 1912) (Diptera, Psychodidae, Phlebotominae). Mem Inst Oswaldo Cruz 94:693–700

Marcondes CB, Lozovei L, Vilela JH (1998) Distribuição geográfica de flebotomíneos do complexo *Lutzomyia intermedia* (Lutz & Neiva, 1912) (Diptera, Psychodidae) geographic distribution of phlebotomine sandflies of the *Lutzomyia intermedia* (Lutz & Neiva, 1912) complex (Diptera, Psychodidae). Rev Soc Bras Med Trop 31(1):51–58

Marcondes CB, Santos-Neto LG, Lozovei AL (2001) Ecology of phlebotomine sand flies (Diptera: Psychodidae) in Brazilian Atlantic Forest. Rev Soc Bras Med Trop 34:255–260

Margonari CS, Dias-Forte CL, Dias ES (2004) Genetic variability in geographical populations of *Lutzomyia whitmani* elucidated by RAPD-PCR. J Med Entomol 41:187–192

Martins AV, Williams P, Falcão A (1978) American sand flies. Academia Brasileira de Ciências, Rio de Janeiro

Mayrink W, Williams P, Coelho MV, Dias M, Martins AV, Magalhães PA, Costa CA, Falcão AR, Melo MN, Falcão AL (1979) Epidemiology of dermal leishmaniasis in the Rio Doce Valley, state of Minas Gerais, Brazil. Ann Trop Med Parasitol 73:123–137

McIntyre S, Rangel EF, Ready PR, Carvalho BM (2017) Species-specific ecological niche modelling predicts different range contractions for *Lutzomyia intermedia* and a related vector of *Leishmania braziliensis* following climate change in South America. Parasit Vect 10:157

McMichael AJ (2004) Environmental and social influences on emerging infectious diseases: past, present and future. Phil Trans Roy Soc London B 359:1049–1058

Meneses CR, Cupolillo E, Monteiro F, Rangel EF (2005) Micro-geographical variation among male populations of the sand fly, *Lutzomyia* (*Nyssomyia*) *intermedia*, from an endemic area of American cutaneous leishmaniasis in the state of Rio de Janeiro, Brazil. Med Vet Entomol 19:38–47

Moo-Llanes D, Ibarra-Cerdeña CN, Rebollar-Téllez EA, Ibáñez-Bernal S, González C, Ramsey JM (2013) Current and future niche of North and Central American sand flies (Diptera: Psychodidae) in climate change scenarios. PLoS Negl Trop Dis 7(9):e2421

Moschin JC, Ovallos FG, Sei IA, Galati EA (2013) Ecological aspects of phlebotomine fauna (Diptera, Psychodidae) of Serra da Cantareira, Greater São Paulo Metropolitan region, state of São Paulo, Brazil. Rev Bras Epi 16(1):190–201

Naiff RD, Freitas RA, Naiff MF, Arias JR, Barrett TV, Momen H, Grimaldi Junior G (1991) Epidemiological and nosological aspects of *Leishmania naiffi* Lainson & Shaw, 1989. Mem Inst Oswaldo Cruz 86:317–321

Nery LCR, Lorosa ES, Franco AMR (2004) Feeding preference of the sand flies *Lutzomyia umbratilis* and *L. spathotrichia* (Diptera: Psychodidae, Phlebotominae) in an urban forest patch in the city of Manaus, Amazonas, Brazil. Mem Inst Oswaldo Cruz 99(6):571–574

Nieves E, Pimenta PF (2002) Influence of vertebrate blood meals on the development *Leishmania* (*Viannia*) *braziliensis* and *Leishmania* (*Leishmania*) *amazonensis* in the sand fly *Lutzomyia migonei* (Diptera: Psychodidae). Am J Trop Med Hyg 67(6):640–647

Nunes VLB, Galati EAB, Cardozo C, Rocca MEG, Andrade AROD, Santos MFDC, Aquino RB, Rosa DD (2008) Estudo de flebotomíneos (Diptera, Psychodidae) em área urbana do município de Bonito, Mato Grosso do Sul, Brasil. Rev Bras Entomol 52(3):446–451

Oliveira DM, Reinhold-Castro KR, Bernal MVZ, Legriffon CMO, Lonardoni MVC, Teodoro U, Silveira TGV (2011) Natural infection of *Nyssomyia neivai* by *Leishmania* (*Viannia*) spp in the state of Paraná, southern Brazil, detected by multiplex polymerase chain reaction. Vector Borne Zoonotic Dis 11:137–143

Oliveira MR, Macedo VO, de Carvalho EM, Barral A, Marotti JG, Bittencourt A, de Abreu MV, Orge M de La G, Lessa Hde A, Marsden PD (1995) An evolutionary study of mucosal leishmaniasis (a 7 to 17-year follow-up) due to *Leishmania* (*Viannia) braziliensis* in Três Braços, Bahia. Rev Soc Bras Med Trop 28:325–332

Passos VNA, Silva RE, Falcão AL (1991) Fauna flebotomínica de municípios da região metropolitana de Belo Horizonte. Rev Soc Bras Med Trop 24(Suppl. 11):107

Patz JA, Thaddeus KG, Geller N, Vittor AY (2000) Effects of environmental change on emerging parasitic diseases. Int J Parasitol 30:1395–1405

Pereira Filho AA, Bandeira MDCA, Fonteles RS, Moraes JLP, Lopes CRG, Melo MN, Rebêlo JMM (2015) An ecological study of sand flies (Diptera: Psychodidae) in the vicinity of Lençóis Maranhenses National Park, Maranhão, Brazil. Parasit Vect 8(1):442

Pessôa SB, Pestana BR (1940) Infecção natural de "Phlebotomus migonei" por formas em leptomonas provavelmente da "Leishmania brasiliensis". Acta Méd Rio de Janeiro 5:106–111

Pessôa SB, Coutinho JO (1941) Infecção natural e experimental dos flebótomos pela *Leishmania braziliensis* no estado de São Paulo. Hospital 20:25–35

Peterson AT, Shaw J (2003) *Lutzomyia* vectors for cutaneous leishmaniasis in Southern Brazil: ecological niche models, predicted geographic distributions and climate change effects. Int J Parasitol 33:919–931

Pinheiro MPG, da Silva JHT, Inacio CLS, de Melo MDFF (2016a) Anthropophily of *Lutzomyia wellcomei* (Diptera: Psychodidae) in an Atlantic Forest Conservation Unit in Northeast Brazil. J Med Entomol 53(6):1444–1448

Pinheiro MPG, Silva JHT, Cavalcanti KB, de Azevedo PRM (2013) Ecological interactions among phlebotomines (Diptera: Psychodidae) in an agroforestry environment of northeast Brazil. J Vect Ecol 38(2):307–316

Pinheiro MPG, de Medeiros Silva MM, Júnior JBS, da Silva JHT, de Lima AM, De Melo MDFF (2016b) Sand flies (Diptera, Psychodidae, Phlebotominae), vectors of *Leishmania* protozoa, at an Atlantic Forest Conservation Unit in the municipality of Nísia Floresta, Rio Grande do Norte state, Brazil. Parasit Vect 9(1):83

Pirmez C, Oliveira-Neto MP, Franco A, Meneses C, Rangel E, Mayrink A, Gonçalves AJ, Fernandes O, Grimaldi G (1997) Edentates as a possible reservoir of *L.* (*V.*) *braziliensis* in an endemic area of Rio de Janeiro. Mem Inst Oswaldo Cruz *92*(Suppl. I):119

Pita-Pereira D, Alves CR, Souza MB, Brazil RP, Bertho AL, Figueiredo AB, Britto CC (2005) Identification of naturally infected *Lutzomyia intermedia* and *Lutzomyia migonei* with *Leishmania* (*Viannia) braziliensis* in Rio de Janeiro (Brazil) revealed by a PCR multiplex non-isotopic hybridisation assay. Trans R Soc Trop Med Hyg 99:905–913

Pita-Pereira D, Souza GD, Araújo Pereira T, Zwetsch A, Britto C, Rangel EF (2011) *Lutzomyia* (*Pintomyia) fischeri* (Diptera: Psychodidae: Phlebotominae), a probable vector of American cutaneous leishmaniasis: detection of natural infection by *Leishmania* (*Viannia)* DNA in specimens from the municipality of Porto Alegre (RS), Brazil, using multiplex PCR assay. Acta Trop 120(3):273–275

Pita-Pereira D, Souza GD, Zwetsch A, Alves CR, Britto C, Rangel EF (2009) First report of *Lutzomyia* (*Nyssomyia) neivai* (Diptera: Psychodidae: Phlebotominae) naturally infected by *Leishmania* (*Viannia) braziliensis* in a periurban area of South Brazil using a multiplex polymerase chain reaction assay. Am J Trop Med Hyg 80:593–595

Queiroz MFM, Varjão JR, Moraes SCD, Salcedo GE (2012) Analysis of sandflies (Diptera: Psychodidae) in Barra do Garças, State of Mato Grosso, Brazil, and the influence of environmental variables on the vector density of *Lutzomyia longipalpis* (Lutz & Neiva, 1912). Rev Soc Bras Med Trop 45(3):313–317

Queiroz RG, Vasconcelos IA, Vasconcelos AW, Pessoa FA, Souza RN, David JR (1994) Cutaneous leishmaniasis in Ceará state in Northeasten Brazil: incrimination of *Lutzomyia whitmani* (Diptera: Psychodidae) as a vector of *Leishmania braziliensis* in Baturité municipality. Ann J Trop Med Hyg 50:693–698

Rangel EF (1995) Tropical Diseases, Society and the Environment. SAREC Documentation/TDR, pp 103–110

Rangel EF, Costa SM, Carvalho BM (2014) Environmental changes and the geographic spreading of American cutaneous leishmaniasis in Brazil. In: Claborn D (ed) Leishmaniasis–trends in epidemiology, diagnosis and treatment. InTech, Rijeka

Rangel EF, Lainson R (2003) Transmissores de Leishmaniose Tegumentar Americana. In: Elizabeth Rangel & Ralph Lainson (org.). Flebotomíneos do Brasil. Editora Fiocruz, Rio de Janeiro, pp 291–310

Rangel EF, Lainson R (2009) Proven and putative vectors of American cutaneous leishmaniasis in Brazil: aspects of their biology and vectorial competence. Mem Inst Oswaldo Cruz 104(7):937–954

Rangel EF, Ryan L, Lainson R, Shaw JJ (1985) Observations on the sandfly (Diptera: Psychodidae) fauna of Além Paraiba, State of Minas Gerais, Brazil, and the isolation of a parasite of the *Leishmania braziliensis* complex from *Psychodopygus hirsuta hirsuta*. Mem Inst Oswaldo Cruz 80:373–374

Rangel EF, Azevedo ACR, Andrade CA, Souza NA, Wermelinger ED (1990) Studies on sand fly fauna (Diptera: Psychodidae) in a foci of cutaneous leishmaniasis in Mesquita, Rio de Janeiro state, Brazil. Mem Inst Oswaldo Cruz 85:39–45

Rangel EF, Azevedo ACR, Orosko S, Lima JB, Souza NA, Pereira T, Meneses CRV, Costa WA, Cupollilo E, Brahim L (1998) Lutzomyia (Nyssomyia) umbratilis (Ward & Fraiha, 1977) and the ecology of American cutaneous leishmaniasis in Mato Grosso state. I Bienal de Pesquisa da Fundação Oswaldo Cruz, Fiocruz, Rio de Janeiro, 122 pp

Rangel EF, Meneses CRV, Cupolillo E, Azevedo ACR, Costa WA, Costa SM (1999) Aspectos da ecologia de *Lutzomyia intermedia* (Lutz & Neiva, 1912) e a fauna flebotomínica (Diptera: Psychodidae) em área de transmissão da *Leishmania* (*V.*) *braziliensis* no Rio de Janeiro. Rev Soc Bras Med Trop *32*(Suppl. 1):115

Rangel EF, Souza NA, Wermelinger ED, Barbosa AF, Andrade CA (1986) Biology of *Lutzomyia intermedia* Lutz & Neiva, 1912 and *Lutzomyia longipalpis* Lutz & Neiva, 1912 (Diptera: Psychodidae) under experimental conditions. I. Feeding aspects of larvae and adults. Mem Inst Oswaldo Cruz 81:431–438

Rangel EF, Lainson R, Souza AA, Ready P, Azevedo ACR (1996) Variation between geographical populations of *Lutzomyia* (*Nyssomyia) whitmani* (Antunes & Coutinho, 1939) sensu lato (Diptera: Psychodidae: Phlebotominae) in Brazil. Mem Inst Oswaldo Cruz 91:43–50

Rangel EF, Barbosa AF, Andrade CA, Souza NA, Wermelinger ED (1992) Development of *Leishmania* (*Viannia*) *braziliensis* Viannia, 1991 in *Lutzomyia intermedia* (Lutz & Neiva, 1912) (Diptera: Psychodidae) under experimental conditions. Mem Inst Oswaldo Cruz 87:235–238

Ready P (2008) Leishmaniasis emergence and climate change. Rev Sci Tech L'Off Int Epiz 27(2):399–412

Ready PD, Fraiha H, Lane RP, Arias JR, Pajot FX (1982) On distinguishing the female of *Psychodopygus wellcomei*, a vector of mucocutaneous leishmaniasis, from other *squamiventris* series females. I. Characterization of *Ps. squamiventris squamiventris* and *Ps. s. maripaensis* stat. nov. (Diptera: psychodidae). Ann Trop Med Parasitol 76:201–214

Ready PD, Lainson R, Shaw JJ (1983) Leishmaniasis in Brazil: XX. Prevalence of "enzootic rodent leishmaniasis" (*Leishmania mexicana amazonensis*) and apparent absence of *pian-bois* (*Le. braziliensis guyanensis*), in plantations of introduced tree species and in other non-climax forests in eastern Amazonia. Trans R Soc Trop Med Hyg 77:775–785

Ready PD, Lainson R, Shaw JJ (1984) Habitat and seasonality of *Psychodopygus wellcomei* help incriminate it as a vector of *Leishmania* in Amazônia and Northeast Brazil. Trans R Soc Trop Hyg 78:543–544

Ready PD, Lainson R, Shaw JJ, Souza AA (1991) DNA probes for distinguishing *Psychodopygus wellcomei* from *Psychodopygus complexus* (Diptera: Psychodidae). Mem Inst Oswaldo Cruz 86:41–49

Ready PD, Lainson R, Shaw JJ, Ward D (1986) The ecology of *Lutzomyia umbratilis* (Ward & Fraiha, 1977) (Diptera: Psychodiade), the major vector to man of *Leishmania braziliensis guyanensis* in north-eastern Amazonian Brazil. Bull Entomol Res 76:21–40

Ready PD, Day JC, Souza AA, Rangel EF, Davies CR (1997) Mitochondrial DNA characterization of populations of *Lutzomyia whitmani* (Diptera: Psychodiade) incriminated in the peridomestic and silvatic transmission of *Leishmania* species in Brazil. Bull Ent Res 87:187–195

Ready PD, Souza AA, Rebêlo JMM, Day JC, Silveira FT, Campbell-Lendum D, Davies CR, Costa JML (1998) Phylogenetic species and domesticity of *Lutzomyia whitmani* at the south-east boundary of Amazonian Brazil. Bull Entomol Res 87:187–195

Rebêlo JMM, Júnior A, Nascimento A, Silva O, Moraes JLP (2010) Occurrence of sand flies (Diptera, Psychodidae) in leishmaniasis foci in an ecotourism area around the Lençóis Maranhenses National Park, Brazil. Cad Saude Publ 26(1):195–198

Rocha LS, Falqueto A, dos Santos CB, Ferreira AL, da Graça GC, Grimaldi G, Cupolillo E (2010) Survey of natural infection by *Leishmania* in sand fly species collected in southeastern Brazil. Trans R Soc Trop Med Hyg 104(7):461–466

Rocha NMM, Melo MN, Babá EH, Dias M, Michalick MSM, Da Costa CA, Williams P, Mayrink W (1988) *Leishmania braziliensis braziliensis* isolated from *Akodon arviculoides* captured in Ca-ratinga, Minas Gerais, Brazil. Trans R Soc Trop Med Hyg 82:68

Rodrigues ACM, Melo LM, Magalhães RD, de Moraes NB, Júnior ADS, Bevilagua CML (2016) Molecular identification of *Lutzomyia migonei* (Diptera: Psychodidae) as a potential vector for *Leishmania infantum* (Kinetoplastida: Trypanosomatidae). Veterinary Parasitogy 220:28–32

Ryan L, Lainson R, Shaw JJ (1987a) Leishmaniasis in Brazil. XXIV. Natural flagellate infections of sandflies (Diptera: Psychodidae) in Pará state, with particular reference to the role of *Psychodopygus wellcomei* as the vector of *Leishmania braziliensis braziliensis* in the Serra dos Carajás. Trans R Soc Trop Med Hyg 81:353–359

Ryan L, Lainson R, Shaw JJ, Braga RR, Ishikawa EA (1987b) Leishmaniasis in Brazil. XXV. Sandfly vectors of *Leishmania* in Pará State, Brazil. Med Vet Entomol 1:383–395

Ryan L, Vexenet A, Marsden PD, Lainson R (1990) The importance of rapid diagnoses of new cases of cutaneous leishmaniasis in pinpointing the sand fly vector. Trans R Soc Trop Med Hyg 84:786

Scarpassa VM, Alencar RB (2012) *Lutzomyia umbratilis*, the main vector of *Leishmania guyanensis*, represents a novel species complex? PLoS One 7(5):e37341

Scarpassa VM, Alencar RB (2013) Molecular taxonomy of the two *Leishmania* vectors *Lutzomyia umbratilis* and *Lutzomyia anduzei* (Diptera: Psychodidae) from the Brazilian Amazon. Parasit Vect 6(1):258

Sergent ED, Sergent ET, Parrot L, Donatieu A, Béguet M (1921) Transmission du clou de Biskra par le phlébotome (*Phlebotomus papatasi* Scop.) C R Acad Sci 173:1030–1032

Shaw J (2007) The leishmaniases – survival and expansion in a changing world. A mini- review. Mem Inst Oswaldo Cruz 102:541–547

Shaw J (2008) How climatic and environmental variations affect the eco-epidemiology of the leishmaniases and their control. III Workshop de Genética e Biologia Molecular de Insetos Vetores de Doenças Tropicais, 13 pp

Shaw JJ, Ishikawa EAY, Lainson R, Braga RR, Silveira FT (1991) Cutaneous leishmaniasis of man due to *Leishmania (Viannia) shawi* Lainson, de Souza, Póvoa, Ishikawa & Silveira in Pará state, Brazil. Ann Parasitol Hum Comp 66:243–246

Shaw JJ, Lainson R (1968) Leishmaniasis in Brazil: II. Observations on enzootic rodent leishmaniasis in the lower Amazon region – the feeding habits of the vector *Lutzomyia flaviscutellata* in reference to man, rodents and other animals. Trans R Soc Trop Med Hyg 62:396–405

Shaw JJ, Lainson R (1972) Leishmaniasis in Brazil: VI. Observations on the seasonal variations of *Lutzomyia flaviscutellata* in different types of forest and its relationship to enzootic rodent leishmaniasis (*Leishmania mexicana amazonensis*). Trans Roy Soc Trop Med Hyg 66(5):709–717

Shaw JJ, Lainson R, Ryan L, Braga RR, Me-Mahon-Pratt D, David JR (1987) Leishmaniasis in Brazil. XXII. The identification of *Leishmania braziliensis braziliensis* in wild-caught neotropical sand flies using monoclonal antibodies. Trans R Soc Trop Med Hyg 81:69–72

Shaw JJ, Lainson R, Ward RD (1972) Leishmaniasis in Brazil. VII. Further observations on the feeding habitats of *Lutzomyia flaviscutellata* (Mangabeira) with particular reference to its biting habits at different heights. Trans R Soc Trop Med Hyg 66:718–723

Shaw JJ, de Faria DL, Basano SA, Corbett CE, Rodrigues CJ, Ishikawa EA, Camargo LM (2007) The aetiological agents of American cutaneous leishmaniasis in the municipality of Monte Negro, Rondônia state, western Amazonia, Brazil. AnnTrop Med Parasitol 101:681–688

Sherlock I, Carneiro M (1962) Algumas fêmeas de *Phlebotomus* do Brasil (Diptera: Psychodidae). Mem Inst Oswaldo Cruz 60:423–435

Silva AM, de Camargo NJ, dos Santos DR, Massafera R, Ferreira AC, Postai C, Cristóvão EC, Konolsaisen JF, Bisetto A Jr, Perinazo R, Teodoro U, Galati EA (2008) Diversity, distribution and abundance of sand flies (Diptera: Psychodidae) in Paraná state, Southern Brazil. Neotrop Entomol 37:209–225

Silva DF, Vasconcelos SD (2005) Phlebotomine sand flies in fragments of rain forest in Recife, Pernambuco state. Rev Soc Bras Med Trop 38:264–266

Silva DT, Starke-Buzetti WA, Alves-Martin MF, Paixão MS, Tenório MS, Lopes MLM (2014a) Comparative evaluation of several methods for Canine Visceral Leishmaniasis diagnosis. Rev Bras Parasitol Vet 23(2):179–186

Silva TR, Assis MD, Freire MP, Rego FD, Gontijo CM, Shimabukuro PH (2014b) Molecular detection of *Leishmania* in sand flies (Diptera: Psychodidae: Phlebotominae) collected in the Caititu Indigenous Reserve of the Municipality of Lábrea, State of Amazonas, Brazil. J Med Entomol 51:1276–1282

Silveira FT, Souza AA, Lainson R, Shaw JJ, Braga RR, Ishikawa EE (1991) Cutaneous leishmaniasis in the Amazon region: natural infection of the sandfly *Lutzomyia ubiquitalis* (Psychodidae: Phlebotominae) by *Leishmania (Viannia) lainsoni* in Pará State, Brazil. Mem Inst Oswaldo Cruz 86:127–130

Souza CM, Pessanha JE, Barata RA, Monteiro EM, Costa DC, Dias ES (2004) Study on phlebotomine sand fly (Diptera: Psychodidae) fauna in Belo Horizonte, state of Minas Gerais, Brazil. Mem Inst Oswaldo Cruz 99:795–803

Souza Freitas MT, Ríos-Velasquez CM, Costa CRL, Figueirêdo CAS, Aragão NC, da Silva LG, Batista MVA, Balbino TCL, Pessoa FAC, Queiroz BV (2015) Phenotypic and genotypic variations among three allopatric populations of *Lutzomyia umbratilis*, main vector of *Leishmania guyanensis*. Parasit Vect 8(1):448

Souza Freitas MT, Ríos-Velasquez CM, da Silva LG, Costa CRL, Marcelino A, Leal-Balbino TC, Balbino VQ, Pessoa FAC (2016) Analysis of the genetic structure of allopatric populations of *Lutzomyia umbratilis* using the period clock gene. Acta Trop 154:149–154

Souza MB, Cardoso PG, Sanavria A, Marzochi MCA, de Carvalho RW, Ribeiro PC, Ponte CS, Meira AM, Meródio JC (2003) Fauna flebotomínica do município de Bom Jardim, Região Serrana do estado do Rio de Janeiro, Brasil. Rev Bras Parasitol Vet 12:150–153

Souza NA, Andrade-Coelho CA, Vilela ML, Peixoto AA, Rangel EF (2002) Seasonality of *Lutzomyia intermedia* and *Lutzomyia whitmani* (Diptera: Psychodidae: Phlebotominae), occurring sympatrically in area of cutaneous leishmaniasis in the state of Rio de Janeiro, Brazil. Mem Inst Oswaldo Cruz 97:759–765

Souza NA, Vilela ML, Andrade-Coelho CA, Rangel EF (2001) The Phlebotominae sand fly (Diptera: Psychodidae) fauna of two Atlantic Rain Forest Reserves in the state of Rio de Janeiro, Brazil. Mem Inst Oswaldo Cruz 96:319–324

Souza-Rocha L, Falqueto A, dos Santos CB, Grimaldi G Jr, Cupolillo E (2007) Genetic structure of *Lutzomyia (Nyssomyia) intermedia* population from two ecologic regions in Brazil where transmission of *Leishmania (Viannia) braziliensis* reflects distinct eco-epidemiologic features. Am J Trop Med Hyg 76:559–565

Teodoro U, Salvia Filho VL, de Lima EM, Spinosa RP, Barbosa OC, Ferreira MEMC, Verzignassi TG (1993) Flebotomíneos em áreas de transmissão de leishmaniose na região norte do estado do Paraná - Brasil: variação sazonal e atividade noturna. Rev Saude Publica 27:190–194

Teodoro U, Silveira TG, dos Santos DR, dos Santos ES, dos Santos AR, de Oliveira O, Kuhl JB, Alberton D (2003) Influence of rearrangement and cleaning of the peridomiciliary area and building disinsectization on sand fly population density in the municipality of Doutor Camargo, Paraná state, Brazil. Cad Saude Publica 19:1801–1813

Tojal da Silva AC, Cupolillo E, Volpini AC, Almeida R, Romero GA (2006) Species diversity causing human cutaneous leishmaniasis in Rio Branco, State of Acre, Brazil. Tropical Med Int Health 11:1388–1398

Vexenat JA, Barretto AC, Rosa AC (1986) Infecção experimental de *Lutzomyia whitmani* em cães infectados com *Leishmania braziliensis braziliensis*. Inst Oswaldo Cruz 81:125–126

Vieira VR, Azevedo ACR, Alves JRC, Guimarães AE, Aguiar GM (2015) Ecological aspects of phlebotomine sand flies (Diptera, Psychodidae, Phlebotominae) in areas of American Cutaneous Leishmaniasis, in the Municipality of Paraty, Rio de Janeiro, Brazil. I-index of abundance by location and type of capture. J Med Entomol 52(5):886–895

Vilela ML, Azevedo ACR, Costa SM, Costa WA, Silva DM, Grajauskas AM, Moreira-de-Carvalho B, Paes LRDNB, Kozlowsky D, Rangel EF (2008) Sand fly survey in the influence área of Peixe Angical Hydroelectric Plant, state of Tocantins, Brazil. In: 6th International Symposium on Phlebotomine Sandflies, Lima, 95 pp

Vilela ML, Azevedo ACR, Afonso MMS, Silva DM, Rangel EF (2006) Estudos dos vetores das leishmanioses na área de influência do Aproveitamento Hidroelétrico Peixe Angical, estado de Tocantins. XLII Congresso da Sociedade Brasileira de Medicina Tropical. Rev Soc Bras Med Trop *39*(Suppl. I):76

Vilela ML, Azevedo ACR, Motta-Silva D, Costa WA, Rangel EF (2007) Estudos dos vetores das leishmanioses em áreas de influência do Aproveitamento Hidroelétrico Peixe Angical, estado de Tocantins. XLIII Congresso da Sociedade Brasileira de Medicina Tropical. Rev Soc Bras Med Trop *40*(Suppl. I):127

Vilela ML, Azevedo CG, Carvalho BM, Rangel EF (2011) Phlebotomine fauna (Diptera: Psychodidae) and putative vectors of leishmaniases in impacted area by hydroelectric plant, state of Tocantins, Brazil. PLoS One 6(12):e27721

Vilela ML, Pita-Pereira D, Azevedo AC, Godoy RE, Britto C, Rangel EF (2013) The phlebotomine fauna (Diptera: Psychodidae) of Guaraí, state of Tocantins, with an emphasis on the putative vectors of American cutaneous leishmaniasis in rural settlement and periurban areas. Mem Inst Oswaldo Cruz 108(5):578–585

Walsh JF, Molyneux DH, Birley MH (1993) Deforestation: effects on vector-borne disease. Parasitology 106(Suppl):55–75

Ward RD, Fraiha H (1977) *Lutzomyia umbratilis,* a new species of sand fly from Brazil (Diptera: Psychodidae). J Med Entomol 14:313–317

Ward RD, Lainson R, Shaw JJ (1973a) Further evidence of the role of *Lutzomyia flaviscutellata* (Mangabeira) as the vector of *Leishmania mexicana amazonensis* in Brazil. Trans R Soc Trop Med Hyg 67:608–609

Ward RD, Lainson R, Shaw JJ (1977) Experimental transmissions of *Leishmania mexicana amazonensis* Lainson & Shaw, between hamsters by the bite of *Lutzomyia flaviscutellata* (Mangabeira). Trans R Soc Trop Med Hyg 71:265–266

Ward RD, Shaw JJ, Lainson R, Fraiha H (1973b) Leishmaniasis in Brazil. VIII. Observations on the phlebotomine fauna of an area highly endemic for cutaneous leishmaniasis in the Serra dos Carajás, Pará state. Trans R Soc Trop Med Hyg 67:174–183

WHO (2010) World Health Organization. Control of the leishmaniases: report of a meeting of the WHO Expert Committee on the Control of Leishmaniases, Geneva, 22–26 March 2010 (WHO Technical Report Series, n° 949). WHO Press, Genebra

Wijers DJB, Linger R (1966) Man-biting sand flies in Surinam (Dutch Guiana): *Phlebotomus anduzei* as a possible vector of *Leishmania braziliensis*. Ann Trop Med Parasitol 60:501–508

Wilkes TJ, Ready PD, Lainson R, Killick-Kendrick R (1984) Biting periodicities of nulliparous and parous females of *Psychodopygus wellcomei*. Trans R Soc Trop Med Hyg 78:846–847
Woodward A, Smith KR, Campbell-Lendrum D, Chadee DD, Honda Y, Liu Q, Olwoch J, Revich B, Sauerborn R, Chafe Z, Confalonieri U, Haines A (2014) Climate change and health: on the latest IPCC report. Lancet 383:1185–1189
Young DC, Duncan NA (1994) Guide to the identification and geographic distribution of *Lutzomyia* sand flies in México, the West Indies, Central and South America (Diptera: Psychodidae). Mem Ann Entomol Institut 54:1–881

Eco-Epidemiology of American Visceral Leishmaniasis with Particular Reference to Brazil

Elizabeth F. Rangel, Ralph Lainson, Margarete M. S. Afonso, and Jeffrey J. Shaw

Early History: Studies in the States of Sergipe, Pará and Ceará

After the first description of the sand fly *Lutzomyia longipalpis* Lutz and Neiva, 1912, in an indeterminate locality in Brazil, interest in this insect remained largely entomological until the mid-1930s. In 1934, however, Henrique Penna used the viscerotome to examine liver samples from persons who were suspected to have died from yellow fever in various rural localities in Brazil (Penna, 1934). In fact, 41 of these deaths were due to visceral leishmaniasis. His results suggested the major foci of the disease to be in the northeastern states, particularly in Ceará. Carlos Chagas, at that time the director of the Instituto Oswaldo Cruz in Rio de Janeiro, sent his son, Evandro Chagas, to investigate the epidemiology. His first study was made in Sergipe where, in addition to giving the first clinical description of a living case of American visceral leishmaniasis (AVL) in Brazil, he made the important observation that the most frequent blood-sucking insect in and around the patient's house was the phlebotomine sand fly *Lu. longipalpis* (Chagas 1936).

Evandro Chagas was appointed head of a commission set up in 1936 to continue his studies, and in view of the higher prevalence of AVL in the Northeast, it was there that he wished to work. Perversely, the only state governor who offered the

E. F. Rangel (✉) · M. M. S. Afonso
Instituto Oswaldo Cruz, Fundação Oswaldo Cruz, Rio de Janeiro, Brazil
e-mail: efrangel@ioc.fiocruz.br

R. Lainson (deceased)
Instituto Evandro Chagas, Belém, Brazil

J. J. Shaw
Instituto de Ciências Biomédicas, Universidade de São Paulo, São Paulo, Brazil

E. F. Rangel, J. J. Shaw (eds.), *Brazilian Sand Flies*,
https://doi.org/10.1007/978-3-319-75544-1_8

necessary financial and logistic back-up was Da Gama Malcher of Pará in the North of Brazil, where the number of recorded cases was low, and a huge old colonial-style mansion was made available for conversion into the commission's laboratories, which received the imposing name of "The Institute of Experimental Pathology for the North" (IPEN).

Working in the rural areas of Abaetetuba and Moju, where cases of AVL had been recorded by Penna, the commission uncovered more cases of the disease in both humans and dogs (Chagas et al. 1938). Once more, *Lu. longipalpis* was shown to be the principal human-biting insect in and around the houses of the infected persons, and this sand fly became the major suspect as the vector. It was concluded that the disease was essentially rural and only occurred in the close vicinity of forest or copses. For this reason, it was suggested that the origin of the causative parasite, named *Leishmania chagasi* by Cunha and Chagas (1937), was in some wild animal. The commission's hope of confirming the role of *Lu. longipalpis* as the vector and indicating the wild-animal reservoir were dashed in 1940, however, when the tragic death of their leader Evandro Chagas in a mid-air plane collision put an abrupt end to their epidemiological studies. Although the IPEN was renamed "Instituto Evandro Chagas" in his honour, his small band of dedicated workers never recovered fully from the loss of their brilliant and colourful leader, and research on the epidemiology of visceral leishmaniasis in Brazil went into a steady decline.

A rude awakening to the real importance of AVL in Brazil did not occur until 1953, when >100 inhabitants of the small town of Sobral, Ceará, died in a severe outbreak, which jolted the health authorities into activity. Another epidemiological enquiry was organized involving three prominent figures in Brazilian tropical medicine—J. E. Alencar and the married couple L. M. Deane and M. P. Deane—who had formed part of the Evandro Chagas team in Pará. In Ceará, they made two vitally important findings: (1) heavy flagellate infections of what they considered to be promastigotes of *L.* (*L.*) *chagasi* in wild-caught specimens of *Lu. longipalpis* (Deane & Deane 1954a); and (2) the natural infection of foxes with that same parasite (Deane and Deane 1954b). The foxes were identified as *Lycalopex vetulus,* but evidence exists that they were more likely to have been *Cerdocyon thous* (Courtenay et al. 1996). Infections in *Lu. longipalpis* were readily obtained when these sand flies were experimentally fed on an infected fox (Deane and Deane 1954c).

By 1955, Alencar and the Deanes had recorded nearly 1000 new cases of human AVL in Ceará and neighbouring northeastern states. They noted that these occurred in the humid, wooded foothill valleys (boqueirões) and not in the dry lowland plains (sertões) nor on the exposed slopes of the hills where the arid conditions and strong winds were unfavourable for *Lu. longipalpis*. Dogs suffered as badly as humans from the infection: That they were the major reservoir of the human disease was clearly indicated by the high rate of canine infection and the ease with which *Lu. longipalpis* could be infected when fed on infected dogs. In contrast, it was found that humans were a somewhat poor source of the parasite for *Lu. longipalpis* when they were fed on infected patients (Deane 1956).

Fig. 1 Areas of AVL. (**a**) Sobral (Ceará, Brazil), rural profile. (**b**) Araguaína (Tocantins, Brazil), urban profile. (Credit: Afonso MMS)

Distribution of American Visceral Leishmaniasis

Human visceral leishmaniasis was soon shown to have a very wide distribution throughout Latin America extending from Mexico in the north to Argentina in the south. However, until 1984 it was estimated that >90% of the recorded cases in the New World were from Brazil, and of a total of 8959 cases registered in this country, 7882 were from the Northeast and 992 from the Southeast (Deane and Grimaldi 1985). Considering the inadequacy of diagnosis and a general reluctance in permitting autopsies in the more remote rural communities, these figures are likely to have been considerably greater. AVL has been recorded in 12 Latin American countries, and 90% of the cases occurred in Brazil. In most cases, the geographical expansion of *Lu. longipalpis* precedes that of AVL in Brazil, and its expansion is responsible for the establishment of the disease in urban areas of medium and large cities (da Costa et al. 1997; Brasil 2014) (Fig. 1). However, there have been outbreaks in cities, such as Porto Alegre, RS, where AVL is transmitted in the absence of *Lu. longipalpis* (Bianchi et al. 2016, SBMT 2017). Autochthonous human cases have been recorded in most Brazilian states except for Acre, Amapá, Amazonas and Rondônia (Brasil 2014; Sinan 2017; BBC Brasil 2017; Ministério da Saúde 2018 [personal communication]), and infected dogs were absent only in Acre and Amazonas (Maziero et al. 2014; Steindel et al. 2013; Ministério da Saúde 2018 [personal communication] (Fig. 2).

Origin and Taxonomic Position of American Visceral Leishmaniasis's Aetiologic Agent

The name "*Leishmania chagasi* Cunha and Chagas, 1937" long remained in use despite considerable debate regarding the origin of the parasite and its taxonomy. Lainson and Shaw (1987, 1998) considered the parasite as being indigenous to the

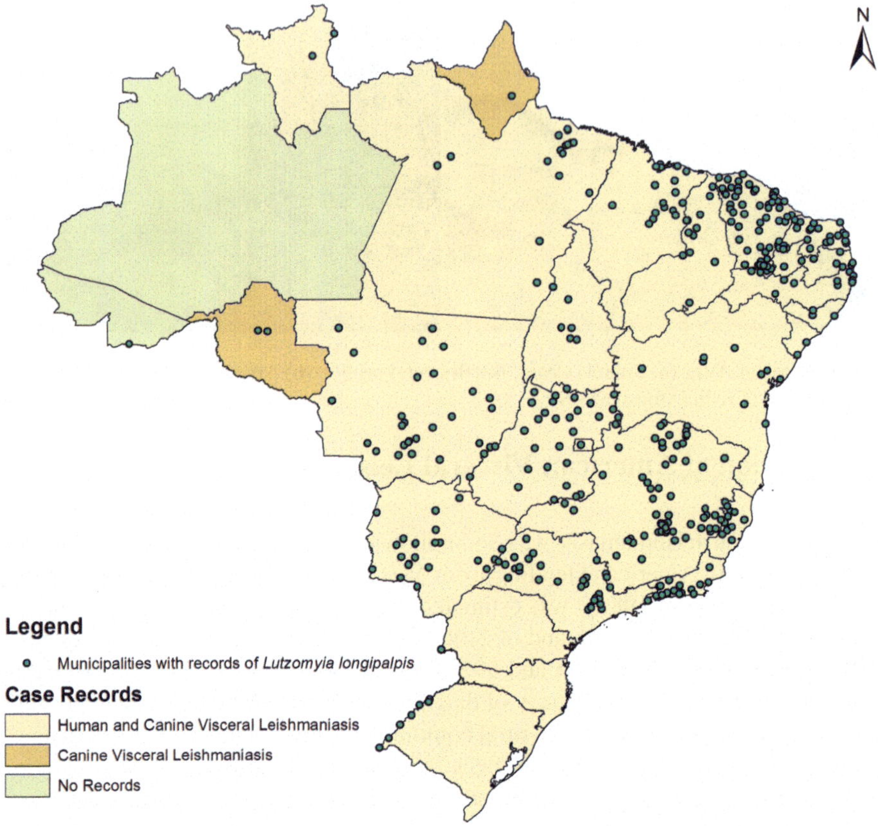

Fig. 2 Geographic distribution of *Lu. longipalpis*: Human and canine visceral cases of AVL in Brazil

Americas. In contrast, Killick-Kendrick (1985) and Rioux et al. (1990) favoured the view that *L.* (*L.*) *infantum* had been imported into Latin America during the Portuguese and Spanish colonization. Another investigator (Lukes et al. 2007), using molecular methods, also considered that AVL had been imported from the Old World. Today there is little support of the parasite's presence in Latin America before human occupation. In a recent paper, Valdivia et al. (2017) estimated that American and Spanish *L.* (*L.*) *infantum* separated around 500 years ago. This coincides with the Iberian colonization of South America. In contrast, *L.* (*L.*) *amazonensis* strains from eastern Brazil and Amazonia had separated some 82,000 years ago, thus indicating their indigenous origin.

The rapid spread of the parasite is almost certainly due to the installation of foci due to the introduction of infected dogs. However, it is possible that in remote newly settled areas, dogs could become infected from sylvatic flies that had become infected after feeding on such wild animals as foxes. In addition, the presence of the parasite in so many different Latin American countries is most likely due to the importation of infected dogs from neighbouring countries or from abroad. It also seems likely that the parasite had been introduced from the Old World on more than one occasion. Evidence for this is the occurrence of infections in many USA states in dogs that have travelled abroad. In some states, a CVL foci is maintained by vertical transmission (Petersen and Barr 2009). Recent studies have shown how rapidly a population of *Lu. longipalpis* can disseminate within a state. After a mere 20 years, the fly expanded some 500 kl in São Paulo State (Casanova et al. 2015).

Similarities of enzyme profiles of stocks identified as *L.* (*L.*) *chagasi* and *L.* (*L.*) *infantum* led Lainson et al. (1981) to suggest that the taxonomic separation of the two parasites "...would best be at subspecific level." Although Mauricio et al. (1999) considered that there were no grounds for any separation, differences have been found between Old and New World *L. infantum* (Jackson et al. 1982, 1984; Decker-Jackson & Tang 1982; Santoro et al. 1986; Martinez et al. 2003; Valdivia et al. 2017). In consideration of these and eco-epidemiological differences, some investigators (Shaw 2002, 2006; Lainson and Rangel 2003a; Lainson and Shaw 2005) adopted the use subspecific names *L.* (*L.*) *infantum* and *L.* (*L.*) *i. chagasi*. We prefer to use the latter name for the parasite circulating in Latin America's endemic visceral leishmaniasis areas but not for the parasite that has recently been imported, such as in the USA.

The *Lutzomyia longipalpis* Complex

Mangabeira (1969) first drew attention to small morphological differences between male examples of *Lu. longipalpis* from Ceará, Northeast Brazil, and others from Pará, North Brazil, and Lainson and Shaw (1979) suggested that the presence of "(...) a *Lu. longipalpis* complex of very similar sand flies (...) may account for certain anomalous situations" and that "a taxonomic revision is needed of (...) *Lutzomyia longipalpis*."

Ward et al. (1983) confirmed Mangabeira's finding, namely, that the male flies from Pará had a pair of white spots on the fourth abdominal tergite, whereas those from Ceará had two pairs of spots on the third and fourth tergites. Furthermore, they showed that the two forms were sexually isolated suggesting that they represented two cryptic species. It was also suggested that this might account for epidemiological differences in AVL in the two geographic areas (Ward et al. 1985).

Using electron microscopy, Lane et al. (1985) showed that the tergal spots were the site of pheromonal glands. Further studies (Ward et al. 1988; Hamilton et al. 1996) showed that the pheromones were specific to a *Lu. longipalpis* population and that the female flies could differentiate the correct one. Successful mating also depends on recognition of the "song" produced by the vibrating wings of the male (de Souza et al. 2002).

More evidence of the existence of a species complex of *Lu. longipalpis* s.l. was offered by Crampton et al. (1989), who prepared a DNA probe specific for a Bolivian population of this sand fly, and Lanzaro et al. (1993), who compared examples from

Costa Rica, Colombia and Brazil by enzyme electrophoresis and cross-breeding experiments. They concluded that these populations were of three distinct species, but they refrained from using any new specific names. Mutebi et al. (2002) added support to this conclusion by demonstrating genetic differentiation of these populations, and this was also demonstrated in populations of the sand fly in Venezuela (Arrivillaga et al. 2000). Further evidence for the presence of cryptic species within a *Lu. longipalpis* complex has been provided by numerous other investigators (Dujardin et al. 1997; Lampo et al. 1999; Uribe 1999; Yin et al. 1999; Arrivillaga and Feliciangeli 2001; Soto et al. 2001; Arrivillaga et al. 2002, 2003).

The existence of such cryptic species in Brazil was disputed by Mukhopadhyay et al. (1998), Mutebi et al. (1999), Azevedo et al. (2000) and Arrivillaga et al. (2002, 2003), all of whom considered that there is only a single species in that country based on a study of several widely separated populations for genetic variability in biochemical characters. They believed that the reasons for any epidemiological variations in AVL should be sought elsewhere. In favour of this view, a recent study of *Lu. longipalpis* populations from six locations in a transect across eastern Brazil by mitochondrial cytochrome-b gene-sequence analysis suggested that sequence divergence also did not adequately indicate cryptic species (Hodgkinson et al. 2003). In contrast, differences in the song patterns between Brazilian populations of *Lu. longipalpis* were consistent with levels of molecular divergence at the *cacophony* locus (Bottecchia et al. 2004; Souza et al. 2004). These groups of investigators suggest that their findings, together with other evidence, supports the existence of a cryptic species-complex under the name of *Lu. longipalpis* in Brazil with as many as four sibling species (Souza et al. 2004). In addition, Jacobina (Bahia), Lapinha (Minas Gerais) and Natal (Rio Grande do Norte) populations had been differentiated on genetic grounds (Bauzer et al. 2002). Maingon et al. (2003) produced genetic evidence of the existence of two sibling species of *Lu. longipalpis* that produce distinct male sex pheromones in Sobral, Ceará, Northeast Brazil. Finally, Watts et al. (2005) correlated microsatellites and the male pheromones of 11 populations from Brazil and Venezuela. They concluded that "Temporal genetic differentiation was mostly not significant at the same site. Spatial genetic differentiation was, however, strong, although there was only a weak relationship between genetic differentiation and the geographic distance separating the samples. (…) Geographic separation explained a much greater (…) percentage of the genetic differences among populations when samples with the same pheromone type were analyzed separately." A cluster analysis showed five groups: *Lu. cruzi* (Brazil) and *Lu. pseudolongipalpis* (Venezuela) as separate species; two Venezuelan and Brazilian groups; and a distinct cluster of Brazilian cembrene populations.

Most investigators have cautiously refrained from giving names to "cryptic species," and it has rightly been asked if these different populations might not simply indicate the initiation of a speciation process rather than the existence of valid species (Bottecchia et al. 2004). It has also been questioned as to whether *Lu. longipalpis* "(…) is a highly polymorphic and geographically variable species, but not a species complex" (Bauzer et al. 2002).

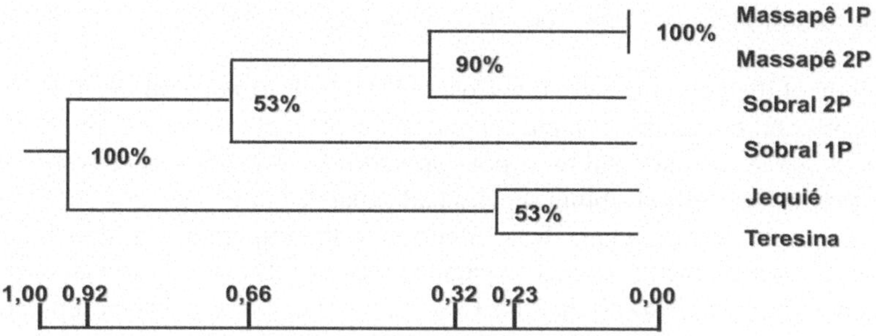

Fig. 3 Dendrogram based on the values of genetic distance inferred from morphometric data of *Lu. longipalpis* populations from different locations of Ceará State. (Source: Afonso MMS 2009)

Considering the observations by Mangabeira (1969) and Azevedo et al. (2000), where morphological differences were clearer, *Lu. longipalpis* males from Jequié (BA), Teresina (PI), Sobral (CE) and Massapê (CE) showed no difference in the spot pattern. In morphometric analyzes, the Jequié population was different from the others by analysis of variance (significant difference in >80% of features) and by Student-Newman-Keuls test (separation based on 37% of analyzed features). Among the four populations from Ceará state, there was a high level of homogeneity; however, the dendogram revealed an isolated branch for the Sobral population with one pair of spots, thus concluding that the populations with one and two pairs of spots would be under speciation, whereas the populations of Massapê are 100% similar (Afonso 2009) (Fig. 3).

Arrivillaga and Feliciangeli (2001), however, gave the name of *Lu. pseudolongipalpis* to a sand fly in Venezuela. The adults are apparently morphologically indistinguishable from those of *Lu. longipalpis*, but the larvae are morphologically distinct. In addition, the adult fly's biting activity was shown to be continuous throughout the night, unlike that of two populations of Venezuelan *Lu. Longipalpis*, which was greatest before 23:00 and steadily decreased from that time onward (Feliciangeli et al. 2004). Arrivillaga et al. (2003) performed phylogenetic analyses of 31 populations of *Lu. longipalpis* s.l. originating throughout this species' geographic range using 7 isozyme loci and mitochondrial genes. The analyses revealed four distinct clades which, it was considered, supported the existence of four species. These had distinct geographic ranges defined as (1) Brazil (*Lu. longipalpis* sensu stricto); (2) Laran (Northwestern Venezuela populations); (3) cis-Andean-Colombia; and (4) *trans*-Andean-Central American populations. The Brazilian clade was represented by 11 populations sampled throughout this country including areas in which *Lu. longipalpis* was originally described: the sand fly of the Laran clade = *Lu. pseudolongipalpis* from Northwest Venezuela; the cis-Andean clade consisted of Colombian populations in Bucaramanga, Palo Gordo, Neiva, Durania and a population from Pacaraima, North Brazil (a mountainous area in Roraima, on the borders of Venezuela and Guyana); and the *trans*-Andean clade included 11 populations from various parts of Central America. The investigators proposed to prepare descriptions and new specific names for the sand flies of the latter two clades.

A study of the period gene (Lima Costa et al. 2015) of populations from Sobral, Ceará state, and Pernambuco again supported the presence of two sibling species in Sobral known as the one- and two-spot populations (1S & 2S, respectively). It was shown that the two Sobral sibling species had expanded into other areas of these two states. These investigators also draw attention to the importance of the abdominal spots in diagnosing the presence of sympatric species. Previously it was suggested (Coutinho-Abreu et al. 2008) that the 1S and 2S populations had been separated by the original course of the San Francisco river. Then its change in course led to a re-encounter of the 1S populations, which could be responsible for their present-day genetic diversity.

Differing opinion will doubtless continue concerning the criteria needed before a considered "cryptic species" is given specific rank, but intrinsic reproductive isolation, as demonstrated by cross-breeding experiments, must surely be high on the list. In this connection, work on the sand-fly's mating "song" is particularly interesting because the vocalization of the males of a number of insects appears to be the most important barrier isolating the different species (Imms 1964; Perdeck 1957).

In a study with five populations of *Lu. longipalpis* from Mato Grosso do Sul State (Campo Grande, Três Lagoas, Aquidauana, Miranda and Bonito), in one from Alagoas State (Estrela de Alagoas) and one population of *Lu. cruzi* from Corumbá (MS), microsatellite analyzes revealed divergence among populations from the two states (due to the large geographical isolation), thus supporting the *Lu. longipalpis* complex. In addition, *Lu. longipalpis* and *Lu. cruzi* showed an introgression between species (Santos et al. 2013).

The high level of the variety of evidence, including chemical, behavioural and molecular characteristics, suggests very recent speciation and complex population structure in the *Lu. longipalpis* species complex. Although significant progress has been achieved to date, the knowledge about recent epidemiological changes, such as urbanization, are essential for the use of effective strategies to control this species (Souza et al. 2017). The finding that the rapid spread of visceral leishmaniasis in São Paulo state was due to the expansion of a population of one chemotype (Casanova et al. 2015) suggests that variations in levels of transmission may be related to specific cryptic species. It also indicates that control measures should concentrate on cryptic species that are epidemiologically important.

The existence and identification of cryptic species, under the name of *Lu. longipalpis* s.l., helps in explaining why different clinical manifestations of AVL exist in Latin America, especially when this is considered in the light of studies on the nature of the saliva of *Lu. longipalpis* s.l. from widely separated geographical areas.

The actual reasons for the drastic changes in the distribution of *Lu. longipalpis* s.l. distribution, especially urbanization, which we will discuss later, are unknown; however, one possibility is climate changes related to global warming. A recent study by Peterson et al. (2017), using ecological-niche models, showed variations that ranged from stability to expansion and even decline. There were indications of expansion in southern Brazil and Argentina and in Amazonia. In this region, the presence of *Lu. longipalpis* was predicted in the northeastern regions Pará state and Amapá state, which are both north of the Amazon River. So far it has only been recorded once in Amapá State (Galardo et al. 2013), and to date no autochthonous cases of human or canine VL have been reported in the region. Data from other areas of Amazonia support the reliability of predictions in this area. However, the models of Peterson et al (2017) failed to predict this species' presence in Uruguay. This may be due to climate changes that occurred after 2000 being too recent to include in the model or due to human introduction.

Influence of *Lutzomyia longipalpis s.l.* Saliva on Infection of Humans with *Leishmania (Leishmania) infantum chagasi*

Although infection with *L. (L.) i. chagasi* is predominately associated with a visceral disease, the same parasite has been shown to produce only non-ulcerative cutaneous lesions in Costa Rica (Zeledón et al. 1989), whereas in Honduras it may cause both visceral and cutaneous leishmaniasis in the same focus (Ponce et al. 1991). The saliva of *Lu. longipalpis* contains a potent vasodilatory peptide, "maxadilan" (Lerner et al. 1991). In experiments investigating the possible influence of the sand fly's saliva on the course of human infection with *L. (L.) i. chagasi*, Warburg et al. (1994) fed *Lu. longipalpis* s.l. of Brazilian, Colombian and Costa Rican origin on the arms of volunteers. They found that the measurements of the resulting erythema at the sites of the bites correlated well with the levels of maxadilan in the sand flies from the three geographical areas. Saliva from the Brazilian colony was the most potent, and that from the Colombian flies was less so. Saliva from the Costa Rican specimens had very little maxadilan, had a very low vasodilatory activity and produced negligible erythema: When mixed with promastigotes of *Leishmania major* and inoculated into the foot-pads of mice, it strongly enhanced proliferation of cutaneous lesions. In contrast, similar inoculations of mixtures of promastigotes and saliva from Colombian and Brazilian *Lu. longipalpis* exacerbated the development of cutaneous lesions to a lesser degree. It was suggested that some of the variability in the clinical presentations of *L. (L.) i. chagasi* infections may be due to the different composition of the saliva of the sand fly, presumably accounting for the manifestation of *L. (L.) i. chagasi* infection in humans as either a visceral or a cutaneous disease. The significance of these findings regarding the nature of infections in wild- or domestic-reservoir hosts in foci of human cutaneous and/or visceral leishmaniasis due to this parasite remains to be studied. A cutaneous lesion due to *L. (L.) i. chagasi* has been reported in a patient from the state of Rio de Janeiro, Brazil (Oliveira et al. 1986). Unlike those described in Costa Rica and Honduras, however, the lesion was ulcerative, and cutaneous manifestations of infection with this parasite in Brazil would appear to be a rarity.

Lutzomyia longipalpis: The Major Vector of American Visceral Leishmaniasis

The overall coincidental distribution of *Lu. longipalpis* and AVL throughout most of Central and South America greatly strengthened the Deanes' conviction that this was the major vector of the disease. Strangely enough, however, although *Lu. longipalpis* is perhaps the most easily colonized of all sand flies in the laboratory, repeated attempts to experimentally transmit the parasite by the bite of this insect failed. Appropriately enough it was in the Instituto Evandro Chagas, where so much of the early history of AVL began, that the chain of evidence incriminating this sand fly was finally completed when five separate transmissions to hamsters were obtained by the bites of experimentally infected laboratory-bred *Lu. longipalpis* (Lainson et al. 1977). The same laboratory (Lainson et al. 1984, 1985) studied a serious outbreak of AVL in the outskirts of Santarém, Pará, where they found this sand fly to be the only species consistently present in and around houses with human and canine infections. Large numbers were captured in the backyard of one house and were fed on clean hamsters, four of which subsequently develop fulminating infections. Dissections of the sand flies used in this experiment indicated an infection-rate of 7%, and 16 isolates were identified as *L.* (*L.*) *chagasi* on enzyme profiles and by monoclonal antibodies. This transmission by the bites of naturally infected *Lu. longipalpis* provided the most conclusive proof possible of the role of this sand fly as a major vector of AVL.

The Ecology of *Lutzomyia longipalpis*: A Sylvatic Origin

Most early studies on AVL in Brazil were conducted in the sparsely forested northeastern states or in other parts of the country that have suffered considerable deforestation: As a result, there developed a tendency to think of the disease only as one that involves a dog and *Lu. longipalpis* in a domestic environment. Observations in the Amazon region of Brazil (Chagas et al. 1938; Lainson et al. 1986; Ryan et al. 1986c), however, indicated that *Lu. longipalpis* is primordially a sylvatic species and that it can still be captured in remote primary forest far from human habitation. In Northern Brazil, this is particularly evident along the length of newly opened roads that pass through forested areas. Primitive houses, inevitably with dogs, chicken houses and other animal shelters, are rapidly thrown up along their length in very close proximity to the forest edge. *Lu. longipalpis* females have catholic feeding habits and quickly invade such habitations; thus, in an epidemiological investigation of cases of AVL along the forest-fringed Igarapé Miri-Tucuruí highway, this sand fly was found in the chicken houses of numerous widely separated houses only 18 months after the road had been opened (Lainson, Shaw, Silveira, & Souza, unpublished observations) (Fig. 4). Finally, even more conclusive evidence came from studies in the municipality of Salvaterra, Island of Marajó, Pará, in a

Fig. 4 Focus of AVL in Pará. Igarapé Miri-Tucuruí highway crossing the primary forest. Dwellings were built close to the road, and within 18 months of their construction they contained *Lu. longipalpis*

focus of AVL (Lainson et al. 1990). Using CDC light-traps variously placed over caged chicken, a fox and sawdust impregnated with the urine and faeces of a fox, attempts were made to capture *Lu. longipalpis* in a pocket of residual primary forest, in the backyard of a house some 500 m distant, and in neighbouring open

Fig. 5 Primary forest in the Serra dos Carajás, Pará State, Brazil, in which very small numbers of *Lu. longipalpis* were captured

savanna. During the dry season, 80 trapping-nights in the forest produced a total of 47 of these sand flies consisting of 22 males and 25 females: None were caught after 14 captures in the savanna, and 2 captures in the backyard of the house provided only 1 male and 4 females. During the wet season, the results were much more impressive: 32 trapping-nights in the forest provided 1161 (463 males and 698 females); 26 captures in the savanna gave a total of 4 (1 male and 3 females); and 24 captures in the backyard of the house produced a total of 1274 (572 males and 702 females). From this and other studies, it was clear that the natural savanna is an unattractive breeding site for *Lu. longipalpis*. In contrast, the large numbers of this sand fly caught in the patch of forest, and the marked association of males and females during both the dry and wet seasons, strongly suggested this to be an important breeding site (Figs. 5 and 6). Galati et al. (2003) reported the capture of *Lu. longipalpis* in the forest environment in the state of Mato Grosso do Sul, Brazil. It remains to be determined, however, if the peri-domestic accumulation of this sand fly is entirely due to their migration from the sylvatic habitat or, at least in part, to the establishment of a secondary peri-domestic breeding site.

The discovery of the latter would be a significant step forward in the control of AVL, but to date all available evidence suggests that the immature stages of *Lu. longipalpis* are thinly dispersed and not concentrated in any particular microhabitat (Deane 1956). In Salvaterra, on the Island of Marajó, the results of an examination of soil removed within and around a small, heavily infested chicken house suggested that the sand flies were not breeding in that microhabitat but had migrated to the chicken house from elsewhere (Dye and Quinnell 1994, personal communication). It has been shown in the laboratory that the male produces a pheromone that

Fig. 6 Eco-epidemiology of AVL in Pará State, Brazilian Amazonia. Full lines = definite transmission lines. Interrupted lines = possible alternative transmission lines, involving other animals as well as humans, as a source of infection for *Lu. longipalpis*. (Source: Lainson and Rangel 2003b)

attracts the female from a substantial distance (Morton and Ward 1989), thus leading these investigators to suggest that the attraction of host odour and male pheromone worked together synergistically at the same time. In contrast, after their observations on the progressive infestation of newly constructed chicken houses by *Lu. longipalpis*, Dye et al. (1991) and Quinnell and Dye (1994a) were led to the conclusion that the females, accompanied by some males, are at first attracted by host odour and later by the pheromone. It was noted, however, that whereas the males tended to remain longer in the chicken houses, most of the females did not rest there during the day.

That *Lu. longipalpis* females feed readily on domestic chicken suggests that wild birds are likely to be among their sylvatic hosts. This sand fly's concentration in chicken houses is of considerable epidemiological importance because it is not customary to spray these with insecticides during antimalarial campaigns, which still remain the principal indirect form of control of AVL.

After experimental studies on the peri-domestic distribution of *Lu. longipalpis* in Salvaterra, Island of Marajó, Quinnell and Dye (1994a, b) concluded that this sand fly tends to congregate at sites outdoors, including animal sheds, where leks can most easily form on abundant, stationary (sleeping) hosts. The flies are much less frequently encountered within houses and, because most dogs sleep outdoors, this probably accounts for a much higher infection rate of AVL in dogs than in humans. It was also suggested that human exposure to the bites of *Lu. longipalpis* was greatest in poorly constructed houses with abundant holes in the walls and the roof.

Urbanization

Visceral leishmaniasis produces major impacts on human health as a consequence of environmental change, mainly through the possible expansion of transmission areas. Continuous environmental change processes—such as deforestation, fires, agriculture, mining, construction of dams and hydroelectric-power plants, migration, unplanned urbanization and lack of urban infrastructure—are examples of situations that have led to an increase in people at risk of infection and fostered the emergence of outbreaks of leishmaniasis in a new eco-epidemiological pattern (Rangel and Vilela 2008; Vilela 2012). AVL shows a persistent scenario in Brazil with most of the factors contributing to its endemicity residing in processes that are external to the health sector. This makes the strengthening of new strategies necessary.

In a study conducted at the state of Tocantins, the results demonstrated a correlation between deforestation and the possible emergence of outbreaks because AVL persists in areas with environmental changes. The increase of deforested areas remained constant in the state and also showed expansion in the record of human cases, especially in the municipality of Araguaína. Some municipalities remained classified as areas of intense transmission according to the methodology proposed by the Brazilian Control Program of Visceral Leishmaniasis (Brasil 2014), especially in the northern and central region, thus showing that the number of human cases remained high and constant throughout the years, principally in areas where there is great environmental impact (Fig. 7). The vector *Lu. longipalpis* was present in all land-use classes because it has adapted to all environments, including impacted areas. This information, coupled with the correlation between the incidence of the disease and urban areas, demonstrate once again the vector's adaptation to anthropic environments (Afonso et al. 2017) (Fig. 8). In the same municipality, a centrifugal dissemination pattern was suggested for AVL, in which the disease showed successive outbreaks in central and peri-urban areas from 2008 onward (Toledo et al. 2017).

The literature contains records of the eclectic behaviour of *Lu. longipalpis*, which feeds on a variety of mammals, including dogs, pigs, equines, bovines and birds (Afonso et al. 2012). Given favourable environmental conditions, the abundance of feeding sources is a determinant factor for the vector's population growth, especially in peri-urban and urban areas, which favours its approximation to man.

In Ceará state (Massapê), specimens that fed on more than one source of animal blood were detected, which illustrates the eclectic feeding behaviour of *Lu. longipalpis* and the general behaviour of sand flies of "tasting" different hosts before completing a blood meal, an important aspect of leishmaniasis transmission (Afonso et al. 2012).

In a study of *Lu. longipalpis* populations from Araguaína (TO), Fortaleza (CE), and Sobral (CE), all of which are areas of intense transmission of AVL, the vector fed mainly on humans (Afonso 2013; Brasil 2014). The anthropophilic behaviour of *Lu. longipalpis* has been demonstrated in many studies (Deane 1956; Lainson and

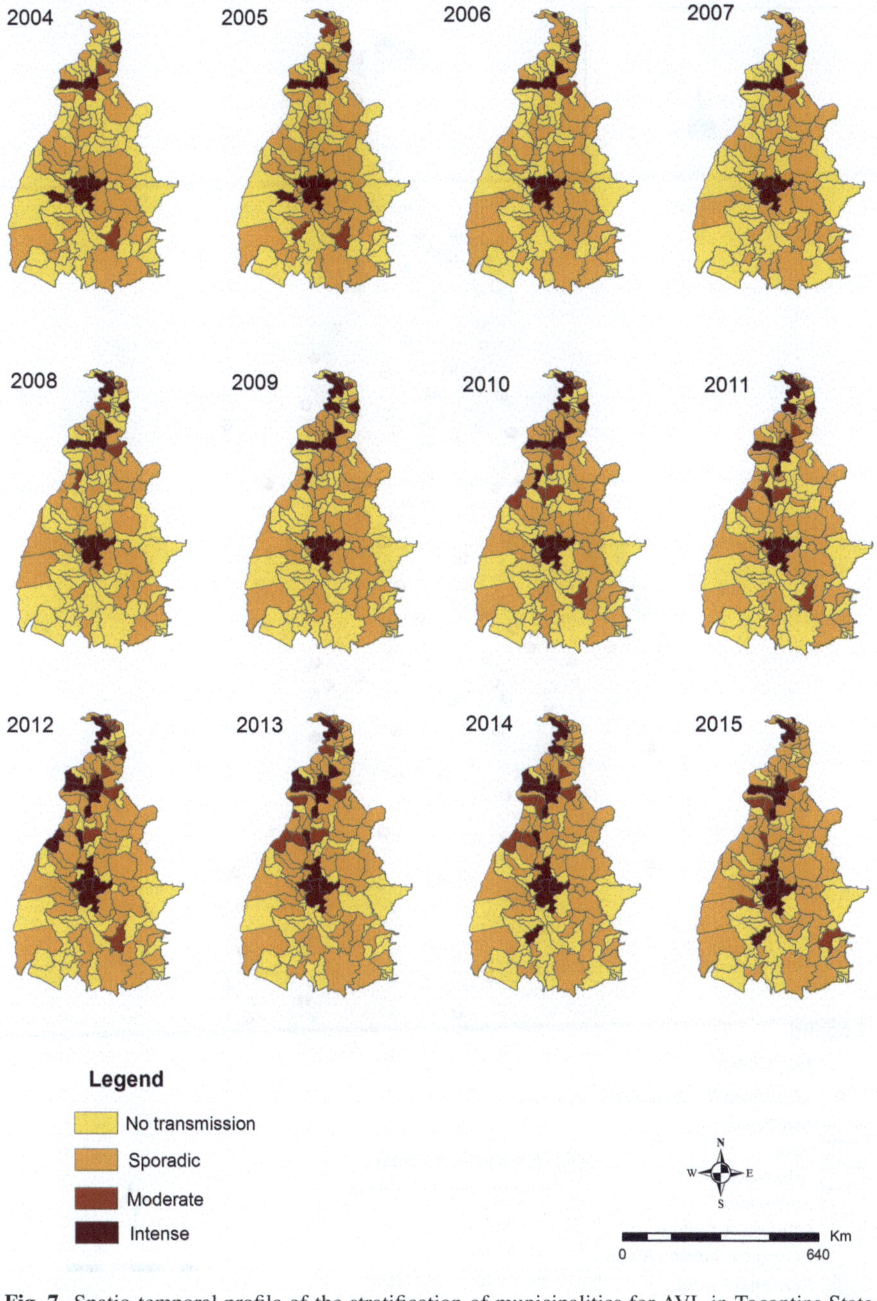

Fig. 7 Spatio-temporal profile of the stratification of municipalities for AVL in Tocantins State, 2004–2015. (Source: Afonso et al. 2017)

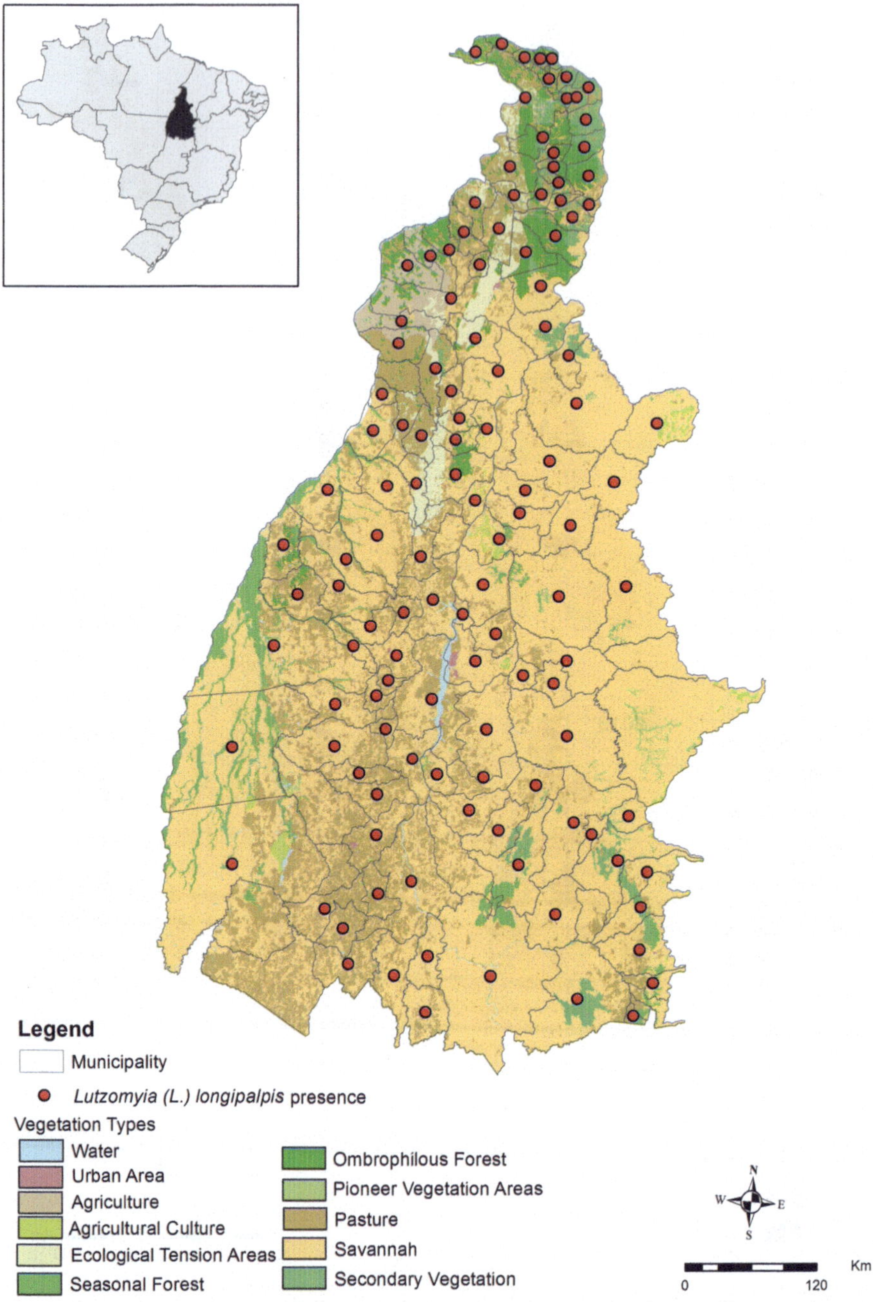

Fig. 8 Map of Tocantins State with land-use classes in association with the presence of *Lu. longipalpis*. (Source: Afonso et al. 2017)

Rangel 2005), which is one of the essential criteria for vector incrimination according to Killick-Kendrick (1990).

The ability to feed on domestic animals (Alexander et al. 2002; Afonso et al. 2012; Brasil 2014) and on synanthropic animals, such as opossums and rodents (Sherlock et al. 1984, 1988; Schalling et al. 2007; Afonso et al. 2012; Lara-Silva et al. 2014), favours the adaptation of *Lu. longipalpis* to human-modified habitats. This allows the maintenance of the transmission cycle of AVL in rural areas and its expansion to urban areas, which characterize the two Brazilian transmission profiles (Brasil 2014).

To understand AVL's urbanization in Brazil, we must use different tools and analyses in entomological surveillance capable of detecting changes in the biology and behaviour of the vector species involved in its transmission. The principal AVL vector in the Americas is *Lu. longipalpis*, which is present in most Brazilian states except for Amazonas and Santa Catarina (Vilela et al. 2014; Borges et al. 2017, Fig. 2). AVL is expanding because of environmental changes and the adaptation of *Lu. longipalpis* to various habitats. To stop this, the planning and implementation of public policies must include measures that minimize the impact of anthropogenic environmental changes favouring AVL transmission. These integrated actions must be incorporated at regional, national and international levels (Salomón et al. 2015).

In the Central Western region of Brazil (Almeida et al. 2015), *Lu. longipalpis* has been shown to be present in all biomes, which corroborates with the results of other studies and demonstratesits adaptability to different environments (Deane and Deane 1962; Lanzaro et al. 1993; Sherlock 1996; Aguiar and Medeiros 2003; Galati 2003; de Almeida et al. 2013). *Lu. cruzi* is predominant in the states of Mato Grosso and Mato Grosso do Sul, thus confirming the findings of Missawa and Lima (2006). *Lu. longipalpis* presented a broader geographic distribution compared with *Lu. cruzi*, although the niche model revealed high climatic suitability for *Lu. cruzi* to the southern region of Goiás.

In the State of São Paulo, a study of spatial and temporal distribution detected that the presence of the vector *Lu. longipalpis* preceded the presence of both canine cases and thus human cases and that the transmission expanded to the central region and then to the north and south regions (Casanova et al. 2015).

Other Possible Vectors of *Leishmania* (*Leishmania*) *infantum chagasi* in Brazil and Neighbouring Countries

The idea that sand flies, other than members of the *Lu. longipalpis* complex, transmit *L.* (*L.*) *infantum* in the Americas originated in epidemiological studies and recently has gained support from molecular studies. However, this must be viewed with caution due to technical limitations. In addition, finding parasite DNA in a sand fly does not relate to transmissibility.

The assumption that *Lu. longipalpis* s.l. was the sole sand-fly vector of *L.* (*L.*) *i. chagasi* throughout the whole geographical range of AVL persisted for >50 years. Suspicions were raised, however, that other species of sand flies might be involved in Venezuela when cases of the disease were recorded in the apparent absence of this sand fly.

Thus, Potenza and Anduze (1942) were unable to find *Lu. longipalpis* in two districts of the state of Bolivar, where two cases of infantile visceral leishmaniasis had been diagnosed, and Pifano and Romero (1964) suggested that *Pintomyia evansi* (Nuñez-Tovar) might be an alternative vector in a focus of AVL in the Turmiquire hills, state of Sucre, Venezuela, where *Lu. longipalpis* was seemingly absent. A further 26 years elapsed, however, before this suspicion was substantiated when Travi et al. (1990) showed that 87% of the sand flies captured in a focus of AVL in the Córdoba Department of Colombia were *Pi. evansi* and that one of these flies was infected with *L.* (*L.*) *i. chagasi* as identified by isolation of the parasite and its characterization by isoenzyme electrophoresis. In further studies in north Colombia, promastigotes were found in nine more specimens of *Pi. evansi* and the parasite again identified as *L.* (*L.*) *i. chagasi* on two occasions (Travi et al. 1996). Thus, the presence of *Pi. evansi* in peri-domestic and intra-domestic habitats throughout the year, as well as the apparent absence of *Lu. longipalpis*, led to the conclusion that *Pi. evansi* is the principal vector of AVL in that region of Colombia, although elsewhere the vector has been shown to be *Lu. longipalpis* (Ferro et al. 1995). Recently, in Carabobo state, Venezuela, Aguilar et al. (1998) recorded the presence of promastigotes in a single specimen of *Pi. evansi* captured in an area endemic for AVL, and—among 1757 sand flies caught in and around houses—72.9% were *Pi. evansi* and only 1.3% were *Lu. longipalpis*.

Finally, Feliciangeli et al. (1999) used k-DNA restriction analysis to show high homologies between the culture forms of the parasite from *Pi. evansi* and a standard stock of *L.* (*L.*) *i. chagasi*. These findings regarding *Pi. evansi* raise two major questions: (1) whether or not it may be an alternative vector of AVL in other parts of this sand fly's geographical range; and (2) whether there exist other alternative vectors. In addition to Colombia and Venezuela, *Pi. evansi* has been recorded in Costa Rica, Honduras, Nicaragua, El Salvador, Guatemala (Young and Duncan 1994) and Mexico (Ibáñez- Bernal et al. 2004). Evidence suggests that adaptation of *L.* (*L.*) *i. chagasi* to *Pi. evansi* is a relatively recent event that is still in progress. Thus, Montoya-Lerma et al. (2003) made a study of the infection-rates and development of *L.* (*L.*) *i. chagasi* in *Lu. longipalpis* and *Pi. evansi* under both natural and experimental conditions. Experimental infection rates and the cycle of *L.* (*L.*) *i. chagasi* in the two flies showed that parasite colonization, differentiation, attachment to the gut epithelium and migration to the foregut were all more frequent and uniform in *Lu. longipalpis* than they were in *Pi. evansi*.

As far as we can ascertain, *Pi. evansi* has not been found in Brazil; however, speculations have been made regarding the possible role of a variety of other sand-fly species in the transmission of *L.* (*L.*) *i. chagasi*. Oliveira et al. (1959) failed to find *Lu. longipalpis* in a village in Minas Gerais where there was a high incidence of AVL, and suspicion fell principally on *Lu. intermedia* and *Lu. whitmani*.

Coelho et al. (1965) were also unable to capture *Lu. longipalpis* in a focus of the disease in southwest Goiás where the most common sand flies were *Lu. intermedia, Lu. whitmani, Lu. shannoni* and *Lu.* (*Psychodopygus*) *davisi*. Ryan et al. (1984) recorded heavy promastigote infections in *Lu. antunesi* captured in a focus of AVL on the Island of Marajó, Pará. Although the organism remained unidentified, its supra-pylarian development in the sand fly raised the question as to whether it was *L.* (*L.*) *i. chagasi*. The same investigators (unpublished observations) found heavy infestations of *Lu. furcata* in pigsties in an area near Belém, Pará, where isolations of *L.* (*L.*) *i. chagasi* had been made from foxes but where *Lu. longipalpis* could not be found. *Lu. furcata* is not anthropophilic but attacks a variety of wild and domestic animals. It could possibly represent, therefore, an alternative vector among such reservoir hosts of *L.* (*L.*) *i. chagasi* as dogs and foxes: Experimentally, it has been shown to can transmit another species of *Leishmania, L.* (*L.*) *amazonensis* (Ryan et al. 1986a).

The female of the sand fly *Lu. cruzi* is morphologically indistinguishable from that of *Lu. longipalpis* (Martins et al. 1984), and the two species can only be reliably separated by small differences when comparing the males. To add to the confusion, a distribution overlap makes it difficult to incriminate either species as the vector of AVL in areas where the two are found together. Santos et al. (1998) dissected many sand flies captured in CDC light-traps around houses in a focus of AVL in Corumbá and Ladário, Mato Grosso do Sul, and found promastigotes in 14 female specimens all with the morphology of *Lu. longipalpis/ Lu. cruzi*. The parasite was identified as *L.* (*L.*) *i. chagasi* by monoclonal antibodies and, in virtue of the apparent absence of males of *Lu. longipalpis* in their captures, these investigators concluded that all of the infected flies were *Lu. cruzi* and that "(...) *Lu. cruzi* is the vector of *Leishmania chagasi* in Mato Grosso do Sul, Brazil." Although the evidence for this supposition is strong, it is not yet conclusive, and Santos et al. (2003) have in fact confirmed the presence of *Lu. longipalpis* in the Corumbá area of study. Until infected females can be conclusively identified as *Lu. cruzi*—by way of biochemical methods (Ryan et al. 1986b), by DNA probes (Ready et al. 1991) or after the production of adults by raising them from the eggs of infected flies (Ryan et al. 1987)—the role of *Lu. cruzi* as a vector of *L.* (*L.*) *i. chagasi* must remain doubtful. Santos et al. (1998) suggested that the epidemiology of AVL in the area of Bolivia bordering Mato Grosso do Sul "(...) certainly should be the same (...)," namely, that *Lu. cruzi* also occurs in Bolivia. The central region of Brazil then presented a differentiated situation regarding the transmission of AVL. As already mentioned, in 1998 it was suggested that *Lu. cruzi* would be involved with the transmission of AVL in Corumbá and Ladário (MS) due to its anthropophily, high density and absence of *Lu. longipalpis* in the transmission areas (Santos et al. 1998). In 2003, the same group reported the finding of natural infection in *Lu. cruzi* (Santos et al. 2003). Subsequently, in 2006, evidence pointed to the participation of *Lu. cruzi* in AVL transmission in the municipality of Jaciara (MT) (Missawa and Lima 2006). Such studies show that in areas where *Lu. longipalpis* is not present, *Lu. cruzi* would be involved in the transmission of AVL.

Among other possible "alternative" vectors, *Lu. intermedia* and *Lu. whitmani* must be included. *Lu. intermedia*, highly suspected as a vector of *L.* (*V.*) *braziliensis* in southeast Brazil, shares a similar habitat to that of *Lu. longipalpis*, is highly anthropophilic and is also known to feed on dogs: In addition, it has been experimentally infected with *L.* (*L.*) *i. chagasi* (Chagas 1940, Paraense & Chagas 1940). In contrast, *Lu. intermedia* has not been recorded further north than Paraíba, in Alagoa Grande and Areia, and part of Pernambuco in Lagoa dos Gatos, Nazaré, Quipapá, Timbaúba and Vitória de Santo Antão (Martins et al. 1978; Young and Duncan 1994). Consequently, it cannot be involved as a secondary vector of AVL in the highly endemic areas in Ceará and Piauí or in the states of Maranhão and Pará. *Lu. whitmani* sensu stricto is a confirmed vector of *L.* (*V.*) *braziliensis* in Northeastern Brazil (Rangel and Lainson 2003) and, as mentioned previously, has been suspected as a vector of AVL in Minas Gerais and Goiás. It is highly anthropophilic and frequently found, together with *Lu. longipalpis*, in chicken houses and human dwelling places. Regarding transmission in the sylvatic habitat in north Brazil, *Lu. whitmani* sensu lato might function as a vector among foxes, but its rarity near houses and its non-anthropophilic habits militates against it being a vector of *L.* (*L.*) *i. chagasi* to humans.

In the Amazon region, suspicion must fall on *Lu. flaviscutellata* as a conceivable alternative vector of *L.* (*L.*) *i. chagasi*. It is better known as the major sylvatic vector of *L.* (*L.*) *amazonensis* among a variety of rodents and marsupials, but this parasite has been isolated from a fox in Pará (Lainson and Shaw 1987), thus indicating that this sand fly does include foxes among its hosts and that these animals are natural hosts of *L.* (*L.*) *i. chagasi*. *Lu. flaviscutellata* is occasionally found invading the peridomestic habitat in areas where isolated cases of Amazonian AVL have been diagnosed (Lainson et al. 1994). It is not greatly attracted to humans, however, so its role as a secondary vector, if indeed it exists, would be of minor importance. Souza et al. (2003) were unable to find *Lu. longipalpis* in 6 of 18 foci of AVL in the municipality of Rio de Janeiro and suggested the participation of other species of sand flies, such as *Lu. migonei* and *Lu. firmatoi*. When considering the apparent absence of *Lu. longipalpis* in such foci of AVL, however, it must be remembered that with the change of rainy to dry season, the population density of this sand fly may fall to such an extent that no examples can be found until the next wet season.

In studies conducted through multiplex polymerase chain reaction (PCR) assays, it was possible to detect two *Mg. migonei* females from an AVL-transmission area in Pernambuco infected by *L.* (*L.*) *infantum*, the first finding of natural infection of this species, thus suggesting that this could be the vector of *L.* (*L.*) *infantum* in areas of AVL where *Lu. longipalpis* is absent (de Carvalho et al. 2010). Guimarães et al. (2016) observed the production of *L.* (*L.*) *infantum* metacyclic forms and the promastigote-secretory gel in experimentally laboratory-bred flies. The infections were similar those seen in experimentally infected *Lu. longipalpis*, which led the investigators to consider that *Mg. migonei* is a permissive vector. In Argentina, there is also epidemiological evidence (Salomón et al. 2009) of a vectorial role of *Mg. migonei* in AVL transmission.

Table 1 lists infections of *L.* (*L.*) *infantum* detected by different molecular methods. The interpretation of the importance of these infections must be viewed with

Table 1 List of sand-fly infections of *Leishmania* (*L.*) *infantum* diagnosed by different molecular methods in sand flies

Genus	Species	Literature source	Country	State
Evandromyia	*cortelezzii*	Carvalho et al. 2008	Brazil	MG
Ev.	*lenti*	Rego et al. 2014	Brazil	MG
Ev.	*salesi*	Saraiva et al. 2008	Brazil	MG
Ev.	*termitophila*	Saraiva et al. 2010	Brazil	MG
Expapillata	*firmatoi*	Donalisio et al. 2017	Brazil	SP
Lutzomyia	*almerio*	Savani et al. 2009	Brazil	MS
Lu.	*cruzi*	de Pita-Pereira et al. 2008	Brazil	MS
Lu.	*forattinii*	de Pita-Pereira et al. 2008	Brazil	MS
Lu.	*ischnacantha*	Rego et al. 2014	Brazil	MG
Micropygomyia	*peresi*	Rego et al. 2014	Brazil	MG
Migonemyia	*migonei*	Moya et al. 2015	Argentina	MI
Nyssomyia	*intermedia*	Rego et al. 2014	Brazil	MG
Ny.	*neivai*	Saraiva et al. 2008	Brazil	MG
Ny.	*whitmani*	Moya et al. 2015	Argentina	MI
Ny.	*whitmani*	Saraiva et al. 2010	Brazil	MG
Pintomyia	*monticola*	Donalisio et al. 2017	Brazil	SP

caution. Under no circumstances can they be considered as vectors but simply as potential vectors that require investigation in greater detail. Savani et al. (2009) observed flagellates in both *Lu. longipalpis* and *Lu. almerio* that were identified by (Small Subunit Ribosomal DNA) SSU rDNA-based PCR as *L.* (*L.*) *infantum*. This raises the intriguing possibility that both were transmitting AVL. Clearly, other sand flies— such as *Lu. cruzi* and *Mg. migonei*—are potential AVL vectors (Santos et al. 1998; Salomón et al. 2010; de Carvalho et al. 2010), but *Lu. longipalpis* is the most important component in AVL transmission and is undoubtedly the principal biological risk factor in the transformation of epidemiological profiles and the increase in the diseases urbanization.

The establishment of *L.* (*L.*) *infantum* in the Americas was due to the parasite finding a permissive vector, in other words, *Lu. longipalpis*. By encountering other permissive vectors that adapt to the peridomestic habitat, it is likely that AVL could become established in areas that are ecologically inhospitable to *Lu. longipalpis* (Peterson et al. 2017).

Other Wild Animal Hosts of *Leishmania* (*Leishmania*) *infantum chagasi*

Sherlock et al. (1984, 1988) isolated *L.* (*L.*) *i. chagasi* from two opossums, *Didelphis albiventris*, captured in a focus of AVL in Jacobina, Bahia, but they considered that it was unlikely that this animal represented an important reservoir of the parasite because of the low infection rate (only 2 of 84 examined).

Investigators in Colombia (Corredor et al. 1989a, b; Travi et al. 1994) registered the isolation of the parasite from the common opossum *Didelphis marsupialis* after the *in vitro* culture of spleen, liver and skin in various media and the intraperitoneal inoculation of hamsters. In one focus of AVL, the infection rate of the opossums was as high as 12 of 37 (32%), and it was concluded that this animal is an important reservoir of *L.* (*L.*) *i. chagasi*. Travi et al. (1998a) followed-up these findings by experimentally infecting *D. marsupialis* with both amastigotes and promastigotes of *L.* (*L.*) *i. chagasi* (dog strain). No parasites could be detected by culture of the opossums' blood, and only very few *Lu. longipalpis* were infected when fed on these animals. They nevertheless considered that xenodiagnosis with the sand fly *Lu. longipalpis* was a more sensitive method for detecting infection than PCR. Travi et al. (1998b) then studied a variety of small mammals captured in both undisturbed and degraded dry forest in northern Colombia using PCR and dot-blot hybridization techniques: They made no attempt to isolate the parasite. Positive PCR/hybridization results for *L.* (*L.*) *i. chagasi* DNA were obtained for 3 of 21 (14.3%) *D. marsupialis* caught in undisturbed forest and 13 of 137 (9.5%) specimens of this animal from the degraded forest. Positive results were also recorded for 3 of 34 specimens of the rodent *Proechimys canicollis* from undisturbed forest and in 2 of 4 specimens from degraded forest. The investigators considered these results to indicate active infections of these rodents with *L.* (*L.*) *i. chagasi*. No foxes were examined in these surveys, although the investigators stated that *C. thous* was present in the study areas and "(…) might contribute to the maintenance of *L. chagasi*." The high percentage of *C. thous* infected in foci of AVL in North Brazil—and the experimental infection of *Lu. longipalpis* fed on an infected fox—together suggest this to be highly likely.

Studies of feeding habits carried out in *Lu. longipalpis* populations from Sobral (CE), Massapê (CE) and Jequié (BA), Brazil and in Colombia (Morrison et al. 1993; Afonso et al. 2012) revealed positivity for opossum blood. These indications, together with reports of natural infection of *D. marsupialis* by *Leishmania* spp., possibly *L.* (*L.*) *infantum chagasi*, add to the discussion about the role of these mammals as potential reservoirs for AVL.

Because of positive PCR hybridization tests for *L.* (*L.*) *i. chagasi* in wild-caught *Pr. canicollis*, the Colombian workers investigated the susceptibility of another spiny rat, *P. semispinosus*, to experimental infection with *L.* (*L.*) *i. chagasi* by the intracardial and intradermal inoculation of promastigotes (Travi et al. 2002). No parasites could be isolated from these spiny rats on periodic culture of liver aspirates, but at autopsy they were isolated in cultures of splenic material from 5 of 10 of the animals. No parasites could be found in stained spleen smears, and repeated xenodiagnosis (*Lu. longipalpis*) failed to reveal parasites. Finally, results of PCR hybridization examination of skin (ears) were all negative.

The investigators concluded that "The inability to infect *P. semispinosus* experimentally with *L. chagasi* indicates that it is not highly susceptible to this *Leishmania* species (…)"; that "(…) *L. chagasi* infection in *Proechimys semispinosus* is contained and compartmentalized"; and that "*Proechimys canicollis*, which is naturally infected with *L. chagasi* in Northern Colombia, may be a more capable reservoir host than *P. Semispinosus*."

During studies on leishmaniasis in the Amazon region of north Brazil by investigators from the Instituto Evandro Chagas, a total of 2637 wild animals— including rodents, marsupials, procyonids, canids and edentates—were examined for leishmanial infection (Lainson et al. 1987): This list included large numbers of the opossum *D. marsupialis* and the spiny rat *P. guyannensis*, many of which were captured near the houses of patients with AVL and—as in the case of opossums—frequently in the backyards of such houses. No infections with *L.* (*L.*) *i. chagasi* were detected in any animal other than the fox *C. thous* after culture of spleen and liver tissue and inoculation of this material intraperitoneally into hamsters. At the time of these studies, the PCR/hybridization technique had not been developed and, in view of the finding of Travi et al. (1998b) (i.e. that tissues of wild-caught *P. canicollis* gave positive PCR results for *L.* (*L.*) *i. chagasi* DNA in Colombia), Lainson et al. (2002) examined the susceptibility of laboratory-bred *P. guyannensis* to experimental infection with a canine strain of *L.* (*L.*) *i. chagasi* from north Brazil. The animal proved to be totally resistant to infection by way of promastigotes and amastigotes after massive intraperitoneal inoculation of the parasite, and the results of subsequent PCR/hybridization tests made on liver and spleen tissue were negative. This failure in attempts to feed laboratory-bred *Lu. longipalpis* on *P. Guyannensis*, or to capture this sand fly in traps baited with the rodent and placed in or near houses infested by *Lu. longipalpis*, led to the conclusion that this species of spiny rat plays no part in the eco-epidemiology of AVL in north Brazil.

At the 3rd World Congress on Leishmaniasis in April 2005, investigators at the Adolfo Lutz Institute, São Paulo, Brazil, presented the results of an examination of wild animals for evidence of *Leishmania* infections in two localities of endemic cutaneous leishmaniasis in the state of São Paulo. Each animal was examined by "(...) one or more of the following methods: detection of rK39 antibody in whole blood; intradermal inoculation of hamsters with skin biopsies from lesions and/or hipocromic spots or culture and/or DNA extraction for PCR and RFLP tests." Among the positive results were "*L.* (*L.*) *chagasi* in 1 *Akodon* sp. and 2 *D. Marsupialis*." Unfortunately, the published abstract of the presentation (Tolezano et al. 2005) does not indicate by which method these results were obtained or, more importantly, if the parasite was isolated from these three animals.

The finding of a benign *L.* (*L.*) *i. chagasi* infection in marsupials and rodents suggests their possible role as source of infection to sand flies. Of greater importance, however, is the fact that it raises the question as to what extent these animals may act as reservoirs of infection for the sand-fly vector and thus play a role in the epidemiology of human AVL.

An effective reservoir host of any parasite is one that can participate in the maintenance and dissemination of that parasite in nature. When parasites are dependent on haematophagous vectors for their transmission, it is clearly necessary to show that these can be infected when fed on the host in question. Until this is done, the infected animal is best referred to as a "potential reservoir." The isolation of *L.* (*L.*) *i chagasi* from numerous specimens of the fox *C. thous* and the opossum *D. marsu-*

pialis—and the experimental infection of *Lu. longipalpis* fed on these animals—places them firmly in the category of natural reservoirs of *L.* (*L.*) *i. chagasi* (Lainson et al. 1990; Travi et al. 1998a). In contrast, although positive results of PCR tests on the tissues of some wild rodents do suggest that these may also represent reservoirs, isolation of the parasite and experimental infection of *Lu. longipalpis* fed on the infected animals is needed to confirm this.

In Minas Gerais, the species *Rattus norvegicus* was found infected by *L. infantum* [= *L.* (*L.*) *i. chagasi*] using the *Leishmania* nested-PCR technique, thus suggesting the possibility of this rodent's participation in the AVL zoonotic cycle (Lara-Silva et al. 2014).

Other Parasites Associated with Visceral Leishmaniasis

L. (*L.*) *amazonensis* has been isolated from AVL patients from Bahia, Brazil (Barral et al. 1991) and *Endotrypanum colombiensis* from cases of AVL in Venezuela (Rodriguez-Bonfante et al. 2003). In 2007, *L.* (*L.*) *amazonensis* was identified in two dogs from Araçatuba, São Paulo state, diagnosed clinically as having canine visceral leishmaniasis (Tolezano et al. 2007). This same parasite has also been identified in dogs from two CVL-endemic locations in Minas Gerais state (Dias et al. 2011; Valdivia et al. 2017).

In the above-mentioned areas, *Lu. longipalpis* is the CVL vector, and Sherlock (1996) transmitted *L.* (*L.*) *amazonensis* with flies of this species from Jacobina, Bahia. In addition, this natural infection of this same parasite has been found using molecular methods in *Lu. longipalpis* captured in an endemic AVL region of Mato Grosso do Sul (Paiva et al. 2006; Savani et al. 2009).

The natural vector of *L.* (*L.*) *amazonensis* is *Bi. flaviscutellata*, which can be found in both primary and secondary forest but not in peri-domestic habitats. Evidence is accumulating that it is being transmitted in this habitat, amongst dogs, by *Lu. longipalpis*. It is possible that this cycle starts by a dog becoming infected by an infected *Bi. flaviscutellata* when it goes into forested areas, and on its return to the peri-domestic area it becomes a source of infection of *Lu. longipalpis*. However, given this fly's presence in forests, it cannot be ruled out that it also takes part in the sylvatic cycle of *L.* (*L.*) *amazonensis*.

Amazonian American Visceral Leishmaniasis: Indigenous or Introduced?

Although there remains little doubt that peri-domestic/intra-domestic infestations by *Lu. longipalpis* and/or *Pi. evansi* originate(d) from sylvatic populations, the origin of *L.* (*L.*) *i. chagasi* in Amazonia has been controversial, particularly in the more remote forested areas. This led Lainson and Shaw (1998) to consider that the parasite was indigenous (i.e., present before humans' presence in the Americas) and not imported by immigrants from the northeast foci. This hypothesis was further motivated by finding infections in healthy wild foxes (Silveira et al. 1982; Lainson et al. 1969, 1987) and records of sporadic, widely separated cases in Pará state that were not significantly associated with immigrant families from the northeastern VL foci. A similar occult infection in *C. thous* was recorded in a focus of AVL in Corumbá, Mato Grosso do Sul, Brazil (Mello et al. 1988). Furthermore, infections registered in men sleeping in lumber camps in or near forest, far from fixed habitations, suggested a feral source of the parasite (Lainson, Shaw, Silveira and Souza, unpublished observations).

If the parasite was indigenous in Amazonian foxes, then where did it come from? Wild canids have been present in South America since the Pleistocene era some 2–3 million years ago (Perini et al. 2010) and are considered to have originated from North America. This was around the same period that *donovani* and *infantum* split from the ancestral visceral line (Barratt et al. 2017). It also marked the beginning of the ice age. which lasted until about 12,000 year ago. Given these scenarios, it seems very unlikely that an African parasite of canids could have reached American canids before the European colonization of the American continent.

Only now are we beginning to appreciate the speed at which vector-borne pathogens can spread. Its introduction into an area where there is a suitable vector results in its rapid expansion as has been witnessed recently with the Zika virus (Bogoch et al. 2016). A more plausible explanation of fox infections in Amazonia is that they became infected from dogs habouring *L .(L.) infantum* that were imported with immigrants from the Iberian Peninsula. Settlers from this region began to arrive in western Amazonia at the end of the seventeenth century. It is inconceivable that during a period of >200 years that infected dogs were not inadvertently imported. Local foxes would have found the farmsteads a source of food and could have become infected while visiting them. In these early settlements, there is also strong evidence that wild *Lu. longipalpis* adapted quickly to a peri-domestic environment. This assumption exists because in the municipality of Igarape Miri, Pará, *Lu. longipalpis* was found in the chicken houses of numerous widely separated houses located along a stretch of road that had only been opened 18 months earlier (Lainson, Shaw, Silveira and Souza, unpublished observations). On returning to their feral habitats, an enzootic was set up with the help of local forest sand flies that probably included *Lu. longipalpis*. A similar situation could have occurred in other regions where foxes are found infected. It is important to remember that the occult nature of *L. (L.) infantum* in wild canids makes them excellent maintenance reservoirs. In the Old World, it has been suggested that the origin of human visceral leishmaniasis, due to parasites of the *L. (L.) donovani* complex, was a rural enzootic of wild canids, such as foxes, jackals and wolves, that later spread to dogs (Lysenko 1971).

Final Comments

We can never know the exact origin of the *L.* (*L.*) *infantum* infections found in American wild animals, but current evidence suggests that they are from infected dogs imported by immigrants during a period of hundreds of years. The presence of sylvatic sand flies, such as *Lu. longipalpis* and *Pi. evansi*, most probably resulted in the maintenance and spread of wild enzootic cycles of *L.* (*L.*) *infantum* in such animals as foxes, which are closely related to Old World foxes (Perini et al. 2010). A similar situation occurred with the yellow fever virus, which was imported from west Africa in the 1600s and spread throughout the Americas where it is now an enzootic of wild animals, such as monkeys, in Amazonia (Bryant et al. 2007).

Within this scenario there are two possible ways in which peri-domestic transmission cycles can be established. One is from peri-domestic flies being infected from a roaming wild animal, such as a fox, or from imported infected dogs. Both situations require established populations of peri-domestic phlebotominae, such as *Lu. longipalpis*. Another feasible source of infection for either a human or a dog is going into a wooded area where there are flies that have become infected from a wild animal. This situation is probably rare.

Transmission of the parasite to dogs or humans in situations where peri-domestic fly populations have become infected is then only a question of time when a single infected dog, often with a vast supply of amastigotes for these sand flies in its skin, sets the scene for a small focus of canine or human visceral leishmaniasis. When human habitations become overcrowded, with conditions of poor hygiene and an abundance of domestic animals, particularly dogs, the concentration of the vector *Lu. longipalpis* (and/or *Pi. evansi* in some parts of Latin America) may reach very high levels, which present the risk of a serious outbreak of canine and human disease. At this stage of events, the dog becomes the major reservoir of infection for humans, and the wild-animal host may in fact no longer be present in the immediate area. Control measures in such foci of infection will certainly decrease the number of human cases of AVL, but—unfortunately—total elimination of the disease is unlikely due to a persistent source of *L.* (*L.*) *i. chagasi* in the wild-animal enzootic, a problem equally difficult to resolve in the matter of controlling cutaneous leishmaniases.

References

Afonso MMS, Duarte R, Miranda JC, Caranha L, Rangel EF (2012) Studies on the feeding habits of populations from endemic areas of American leishmaniasis in northeastern Brazil. J Trop Med 2012:1–5

Afonso MMS (2009) Estudo sobre algumas populações brasileiras de *Lutzomyia* (*L.*) *longipalpis* (Lutz & Neiva, 1912) (Diptera: Psychodidade: Phebotominae): morfologia, morfometria e hábitos alimentares (Dissertation). Minas Gerais: Universidade Federal de Viçosa, 80 pp. Available from: http://locus.ufv.br/bitstream/handle/123456789/3863/texto%20completo.pdf?sequence=1&isAllowed=y

Afonso MMS, Chaves SAM, Magalhães MAM, Garrido R, Azevedo C, Carvalho BM, Rangel EF (2017) Eco-epidemiology of American visceral Leishmaniasis in Tocantins State, Brazil: factors associated with the occurrence and spreading of the vector *Lutzomyia* (*Lutzomyia*) *longipalpis* (Lutz & Neiva, 1912) (Diptera:Psychodidae:Phebotominae). In: David Claborn (org). The Epidemiology and Ecology of Leishmaniasis. Editora Intech, Rikeja, pp 91–115

Afonso MMS (2013) Estudos sobre *Lutzomyia* (*Lutzomyia*) *longipalpis*: hábitos alimentares, infecção natural por *Leishmania* (*Leishmania*) *infantum chagasi* e correlação com a expansão da leishmaniose visceral americana [PhD Thesis]. Escola Nacional de Saúde Pública Sérgio Arouca, Rio de Janeiro, 152 pp

Aguilar CM, Fernández E, Cannova DC, Ferrer E, Cabrera Z, Souza WJS, Coutinho SG (1998) Urban visceral leishmaniasis in Venezuela. Mem Inst Oswaldo Cruz 93:15–16

Aguiar GM, Medeiros WM (2003) Distribuição regional e habitats das espécies de flebotomíneos do Brasil. In: Elizabeth Rangel, Ralph Lainson (org.) Flebotomíneos do Brasil. Editora Fiocruz, Rio de Janeiro, pp 291–310

Alexander B, Carvalho RL, McCallum H, Pereira MH (2002) Role of the domestic chicken (*Gallus gallus*) in the epidemiology of urban visceral leishmaniasis in Brazil. Emerg Infect Dis 8:1480–1485

Almeida PS, Andrade AJ, Sciamarelli A, Raizer J, Menegatti JA, Hermes SCNM, Carvalho MSLC, Gurgel-Gonçalves R (2015) Geographic distribution of phlebotomine sandfly species (Diptera: Psychodidae) in Central-West Brazil. Mem Inst Oswaldo Cruz 110:551–559

Arrivillaga JC, Rangel Y, Oviedo M, Feliciangeli MD (2000) Genetic diversity among Venezuelan populations of *Lutzomyia longipalpis* (Diptera:Psychodidae: Phleboto-minae). J Med Entomol 37:325–330

Arrivillaga JC, Feliciangeli MD (2001) *Lutzomyia pseudolongipalpis*, the first new species within the *longipalpis* (Diptera: Psychodidae: Phlebotominae) complex from Rinconada, Curarigua, Lara State, Venezuela. J Med Entomol 38:783–790

Arrivillaga JC, Norris DE, Feliciangeli MD, Lanzaro GC (2002) Phylogeography of the neotropical sand fly *Lutzomyia longipalpis* inferred from mitochondrial DNA sequences. Infect Genet Evol 2:83–95

Arrivillaga J, Mutebi JP, Pinango H, Norris D, Alexander B, Feliciangeli MD, Lanzaro GC (2003) The taxonomic status of genetically divergent populations of *Lutzomyia longipalpis* (Diptera: Psychodidae) based on the distribution of mitochondrial and isozyme variation. J Med Entomol 40:615–627

Azevedo ACR, Monteiro FA, Cabello PH, Souza NA, Rosa-Freitas MG, Rangel EF (2000) Studies on populations of *Lutzomyia longipalpis* (Lutz & Neiva, 1912) (Diptera: Psychodidae: Phlebotominae) in Brazil. Mem Inst Oswaldo Cruz 95:305–322

Barral A, Pedral-Sampaio D, Grimaldi Junior G, Momen H, McMahon-Pratt D, Ribeiro de Jesus A, Almeida R, Badaro R, Barral-Netto M, Carvalho EM et al (1991) Leishmaniasis in Bahia, Brazil: evidence that *Leishmania amazonensis* produces a wide spectrum of clinical disease. Am J Trop Med Hyg 44:536–546

Barratt J, Kaufer A, Peters B, Craig D, Lawrence A, Roberts T, Lee R, McAuliffe G, Stark D, Ellis J (2017) Isolation of novel trypanosomatid, *Zelonia australiensis* sp. nov. (Kinetoplastida: Trypanosomatidae) provides support for a Gondwanan origin of dixenous parasitism in the Leishmaniinae. PLoS Negl Trop Dis 11:e0005215

Bauzer LGSR, Souza NA, Ward RD, Kyriacou CP, Peixoto AA (2002) The *period* gene and genetic differentiation between three Brazilian populations of *Lutzomyia longipalpis*. Insect Mol Biol 11:315–323

Bianchi MV, Fredo G, Tagliari NJ, Leite-Filho RV, de Lorennzo C, Lupion CG, Driemeier D, Sonne L (2016) Leishmaniose Visceral Canina autóctone na região urbana de Porto Alegre, RS, Brasil. Acta Sci Vet 44(Suppl 1):110

Bogoch II, Brady OJ, Kraemer MUG, German M, Creatore MI, Kulkarni MA, Brownstein JS, Mekaru SR, Hay SI, Groot E, Watts A, Khan K (2016) Anticipating the international spread of Zika virus from Brazil. Lancet 387:335–336

Borges DA, Molina SMG, Pinto MC, Galati EAB, Cesario M, Ortiz DGS (2017) First record of *Lutzomyia* (*Lutzomyia*) *longipalpis* (Diptera: Psychodidae: Phlebotominae) on the Trinational Frontier (Brazil–Peru–Bolivia) of South-Western Amazonia. J Med Entomol 54:1425–1429

Bottecchia M, Oliveira SG, Bauzer LGSR, Souza NA, Ward RD, Garner KJ, Kyriacou CP, Peixoto AA (2004) Genetic divergence in the *cacophony* IVS6 Intron among five Brazilian populations of *Lutzomyia longipalpis*. J Mol Evol 58:754–761

Brasil. Ministério da Saúde (2014) Secretaria de Vigilância em Saúde. Departamento de Vigilância Epidemiológica. Manual de Vigilância e Controle da Leishmaniose Visceral Americana 1a. ed. Editora do Ministério da Saúde, Brasília, 120 pp

British Broadcasting Corporation Brasil [homepage on the Internet] (2017) Available from: http://www.bbc.com/portuguese/brasil-41753612

Bryant JE, Holmes EC, Barrett AD (2007) Out of Africa: a molecular perspective on the introduction of yellow fever virus into the Americas. PLoS Pathog 3:668–673

Carvalho GM, Filho JD, Falcao AL, Lima AC, Gontijo CM (2008) Naturally infected *Lutzomyia* sand flies in a *Leishmania*-endemic area of Brazil. Vector Borne Zoonotic Dis 8:407–414

Casanova C, Colla-Jacques FE, Hamilton JG, Brazil RP, Shaw JJ (2015) Distribution of *Lutzomyia longipalpis* chemotype populations in São Paulo State, Brazil. PLoS Negl Trop Dis 9:1–14

Chagas E (1936) Primeira verificação em individuo vivo, da leishmaniose visceral no Brasil. Bras Méd 50:221–222

Chagas AW (1940) Criação de flebótomos e transmissão experimental da leishmaniose visceral americana. Mem Inst Oswaldo Cruz 35:327–333

Chagas E, Cunha AM, Ferreira LC, Deane L, Deane G, Guimarães FN, Paumgartten MJ, Sá B (1938) Leishmaniose visceral americana (Relatório dos trabalhos realizados pela Comissão Encarregada do Estudo da Leishmaniose Visceral Americana em 1937). Mem Inst Oswaldo Cruz 33:89–229

Coelho MV, Cunha AS, Falcão AR (1965) Notas sobre um foco de calazar no sudoeste do Estado de Goiás. Rev Bras Malariol Doenças Trop 17:143–148

Corredor A, Gallego JF, Tesh RB, Peláez D, Diaz A, Montilla M, Paláu MT (1989a) *Didelphis marsupialis,* an apparent wild reservoir of *Leishmania donovani chagasi* in Colombia, South America. Trans R Soc Trop Med Hyg 83:195

Corredor A, Gallego JF, Tesh RB, Morales A, de Carrasquilla CF, Young DG, Kreutzer RD, Boshell J, Paláu MT, Caceres E, Peláez D (1989b) Epidemiology of visceral leishmaniasis in Colombia. Am J Trop Med Hyg 40:480–486

Courtenay O, Santana EW, Johnson PJ, Vasconcelos IAB, Vasconcelos AW (1996) Visceral leishmaniasis in the hoary zorro *Dusicyon vetulus*: a case of mistaken identity. Trans R Soc Trop Med Hyg 90:498–502

Coutinho-Abreu IV, Sonoda IV, Fonseca JA, Melo MA, Balbino VQ, Ramalho-Ortigao M (2008) *Lutzomyia longipalpis* s.l. in Brazil and the impact of the São Francisco River in the speciation of this sand fly vector. Parasit Vectors 1:16

Crampton J, Knapp T, Ward RD (1989) DNA probes for vector taxonomy. In DDT Hart, *Leishmaniasis*. The Current Status and New Strategies for Control, NATO ASI Series A: Life Sci 163 Plenum Press & NATO Scientific Affairs Division, New York and London, pp 957–964

Cunha AM, Chagas E (1937) Nova espécie de protozoário do gênero *Leishmania* patogênico para o homem. *Leishmania chagasi* n. sp. Nota Prévia. Hospital (Rio de Janeiro) 11:3–9

da Costa AI, Casanova C, Rodas LA, Galati EA (1997) Atualização da distribuição geografica e primeiro encontro de *Lutzomyia longipalpis* em área urbana no Estado de São Paulo, Brasil. Rev Saúde Pública 31:632–633

de Almeida PS, Sciamarelli A, Batista PM, Ferreira AD, Nascimento J, Raizer J, Andrade Filho JD, Gurgel-Gonçalves R (2013) Predicting the geographic distribution of *Lutzomyia longipalpis* (Diptera: Psychodidae) and visceral leishmaniasis in the state of Mato Grosso do Sul, Brazil. Mem Inst Oswaldo Cruz 108:992–996

de Carvalho MR, Valença HF, da Silva FJ, de Pita-Pereira D, de Araújo PT, Britto C, Brazil RP, Brandão Filho SP (2010) Natural *Leishmania infantum* infection in *Migonemyia migonei* (França, 1920) (Diptera: Psychodidae: Phlebotominae) the putative vector of visceral leishmaniasis in Pernambuco State, Brazil. Acta Trop 116:108–110 de Pita-Pereira D, Cardoso MA, Alves CR, Brazil RP, Britto C (2008) Detection of natural infection in *Lutzomyia cruzi* and *Lutzomyia forattinii* (Diptera: Psychodidae: Phlebotominae) by *Leishmania infantum chagasi* in an endemic area of visceral leishmaniasis in Brazil using a PCR multiplex assay. Acta Trop 107:66–69

de Souza NA, Ward RD, Hamilton JG, Kyriacou CP, Peixoto AA (2002) Copulation songs in three siblings of *Lutzomyia longipalpis* (Diptera: Psychodidae). Trans R Soc Trop Med Hyg 96:102–103

Deane LM (1956) Leishmaniose visceral no Brasil. Serviço Nacional de Educação Sanitária, Rio de Janeiro, pp 1–162

Deane MP, Deane LM (1954a) Infecção natural do *Phlebotomus longipalpis* por leptomonas, provavelmente de *Leishmania donovani*, em foco de calazar, no Ceará. Hospital (Rio de Janeiro) 45:697–702

Deane LM, Deane MP (1954b) Encontro de leishmânias nas vísceras e na pele de uma raposa, em zona endêmica de calazar, nos arredores de Sobral, Ceará. Hospital (Rio de Janeiro) 45:419–421

Deane MP, Deane LM (1954c) Infecção experimental do *Phlebotomus longipalpis* em raposa (*Lycalopex vetulus*) naturalmente parasitada pela *Leishmania donovani*. Hospital (Rio de Janeiro) 46:651–653

Deane LM, Deane MP (1962) Visceral leishmaniasis in Brazil: geographical distribution and transmission, São Paulo. Rev Inst Med Trop 4:198–212

Deane LM, Grimaldi G (1985) Leishmaniasis in Brazil. In Chang KP, Bray RS (org.) Leishmaniasis, Elsevier, Amsterdam, pp 247–281

Decker-Jackson FE, Tang DB (1982) Identification of *Leishmania* spp. By radiorespirometry. II. A statistical method of data analysis to evaluate the reproducibility and sensitivity of the technique. In: Chance ML, Walton BC (org.) Biochemical Characterization of Leishmania, proceedings of a workshop held at the PanAmerican Health Organization, Geneva, pp 205–245

Dias ES, Regina-Silva S, Franca-Silva JC, Paz GF, Michalsky EM, Araujo SC, Valadao JL, de Oliveira L-SF, de Oliveira FS, Pacheco RS, Fortes-Dias CL (2011) Eco-epidemiology of visceral leishmaniasis in the urban area of Paracatu, state of Minas Gerais, Brazil. Vet Parasitol 176:101–111

Donalisio MR, Paiz LM, da Silva VG, Richini-Pereira VB, von Zuben APB, Castagna CL, Motoie G, Hiramoto RM, Tolezano JE (2017) Visceral leishmaniasis in an environmentally protected area in southeastern Brazil: epidemiological and laboratory cross-sectional investigation of phlebotomine fauna, wild hosts and canine cases. PLoS Negl Trop Dis 11:e0005666

Dujardin JP, Torrez EM, Le Pont F, Hervas D, Sossa D (1997) Isozymic and metric variation in the *Lutzomyia longipalpis* complex. Med Vet Entomol 11:394–400

Dye C, Davies CR, Lainson R (1991) Communication among phlebotomine sandflies: a field study of domesticated *Lutzomyia longipalpis* populations in Amazonian Brazil. Anim Behav 42:183–192

Feliciangeli MD, Rodriguez N, De Guglielmo Z, Rodriguez A (1999) The reemergence of American visceral leishmaniasis in an old focus in Venezuela. II Vectors and parasites. Parasite 6:113–120

Feliciangeli MD, Arrivillaga JC, Bravo A, Arias F (2004) Activity of *Lutzomyia pseudolongipalpis* and *L. longipalpis* s.l. (Diptera: Psychodidae) in Venezuela. Parasite 11:273–278

Ferro C, Morrison AC, Torres M, Pardo R, Wilson ML, Tesh RB (1995) Age structure, blood feeding behaviour, and *Leishmania chagasi* infection in *Lutzomyia longipalpis* (Diptera: Psychodidae) at an endemic focus of visceral leishmaniasis in Colombia. J Med Ent 32:618–629

Galati EAB (2003) Classificação de Phlebotominae. In: Elizabeth Rangel, Ralph Lainson (org.) Flebotomíneos do Brasil. Editora Fiocruz, Rio de Janeiro, pp 23–51

Galati EAB, Nunes VLB, Cristaldo G, Rocha HC (2003) Aspectos do comportamento da fauna flebotomínea (Diptera:Psychodidae) em foco de leishmaniose visceral e tegumentar na Serra da Bodoquena e área adjacente, Estado de Mato Grosso do Sul, Brasil. Rev Pat Trop 32:235–261

Galardo AKR, Galardo CD, Santana AA, Mendes JCC, de Souza FRA, Duarte JP, Saraiva JF, Pinna LCL, Carvalho RW, Correa APS, Lima ANSF (2013) Primeira ocorrência de *Lutzomyia* (*Lutzomyia*) *longipalpis* Lutz & Neiva, 1912 (Diptera: Psychodidae: Phlebotominae) no Estado do Amapá, Brasil. Biota Amaz 3:197–183

Guimarães VC, Pruzinova K, Sadlova J, Volfova V, Myskova J, Brabdão-Filho SP, Volf P (2016) Lutzomyia migonei is a permissive vector competent for *Leishmania infantum*. Parasit Vectors 9:159

Hamilton JGC, Ward RD, Dougherty R, Maingon C, Ponce E, Noyes H, Zelodón R (1996) Comparison of the sex-pheromone components of *Lutzomyia longipalpis* from areas of visceral and atypical cutaneous leishmaniasis in Honduras and Costa Rica. Ann Trop Med Parasitol 90:533–541

Hodgkinson VH, Birungi J, Quintana M, Deitze R, Munstermann LE (2003) Mitochondrial cytochrome B variation in populations of the visceral leishmaniasis vector *Lutzomyia longipalpis* across eastern Brazil. Am J Trop Med Hyg 69:386–392

Ibáñez-Bernal S, Rodríguez-Domínguez G, Gómez-Hernández CH, Ricardez-Esquinc JR (2004) First record of *Lutzomyia evansi* (Nuñez-Tovar 1924) in México (Diptera: Psychod-idae: Phlebotominae). Mem Inst Oswaldo Cruz 99:127–129

Imms AD (1964) A general textbook of entomology. Methuen & Co Ltd./EP Dutton & Co Inc., London/New York, 886 pp

Jackson PR, Stiteler JM, Wohlhieter JA, Reed SG, Badaró R, Inverso JA, Jackson JE (1984) Characterization of *Leishmania* responsible for visceral disease in Brazil by restriction endonuclease digestion and hybridization of kinetoplast DNA. In: Proceedings of 11th Internat Congr Trop Med Malar Calgary, Canadá, p 68

Jackson PR, Wohlhieter J, Hockmeyer WT (1982) *Leishmania* characterization by restriction endonuclease digestion of kinetoplastic DNA. In: Abstr Vth Internat Congr Parasitol Toronto, Canadá, p 342

Killick-Kendrick R (1985) Some epidemiological consequences of the evolutionary fit between leishmaniase and their phlebotomine vectors. Bull Soc Path Exot 78:747–755

Killick-Kendrick R (1990) Phlebotomine vectors of leishmaniasis: a review. Med Vet Entomol 4:1–24

Lainson R, Shaw JJ, Lins CZ (1969) Leishmaniasis in Brazil, IV. The fox, *Cerdocyon thous* as a reservoir of *Leishmania dovovani* in Pará State, Brazil. Trans R Soc Trop Med Hyg 63:741–745

Lainson R, Ward RD, Shaw JJ (1977) Experimental transmission of *Leishmani chagasi* causative agent of neotropical visceral leishmaniasis, by the sandfly *Lutzomyia longipalpis*. Nature (London) 266:628–630

Lainson R, Shaw JJ (1979) The role of animals in the epidemiology of South American leishmaniasis. In: Lumsden WHR, Evans DA (org.) Biology of the Kinetoplastida, vol 2, Academic Press, London, pp 1–116

Lainson R, Miles MA, Shaw JJ (1981) On the identification of viscerotropic leishmanias. Ann Trop Med Parasitol 75:251–253

Lainson R, Shaw JJ, Ryan L, Ribeiro RSM, Silveira FT (1984) Presente situação da leishmaniose visceral na Amazônia, com especial referência a um novo surto da doença ocorrido em Santarém, Estado do Pará, Brasil. *Bol Epidemiol F SESP,* Rio de Janeiro (número especial), pp 1–8

Lainson R, Shaw JJ, Ryan L, Silveira FT (1985) Leishmaniasis in Brazil XXI. Visceral leishmaniasis in the Amazon Region and further observations on the role of *Lutzomyia longipalpis* (Lutz & Neiva, 1912) as the vector. Trans R Soc Trop Med Hyg 79:223–226

Lainson R, Shaw JJ, Silveira FT, Braga RR, Ryan L, Povoa MM, Ishikawa EAY (1986) A *Leishmania* e as leishmanioses. In: Instituto Evandro Chagas: 50 Anos de Contribuição às Ciências Biológicas e à Medicina Tropical, vol I, Serviços de Saúde Pública, Belém, Pará, pp 83–124

Lainson R, Shaw JJ (1987) Evolution, classification and geo-graphic distribution. In: Peters W, Killick-Kendrick R (org.) The *Leishmaniases* in Biology and Medicine, vol 1, Academic Press, London, pp 1–120

Lainson R, Shaw JJ, Silveira FT, Braga RR (1987) American visceral leishmaniasis: on the origin of *Leishmania* (*Leishmania*) *chagasi*. Trans R Soc Trop Med Hyg 81:517

Lainson R, Dye C, Shaw JJ, MacDonald DW, Courtenay O, Souza AAA, Silveira FT (1990) Amazonian visceral leishmaniasis – distribution of the vector *Lutzomyia longipalpis* (Lutz & Neiva) in relation to the fox *Cerdocyon thous* (Linn) and the efficiency of this reservoir host as a source of infection. Mem Inst Oswaldo Cruz 85:135–137

Lainson R, Shaw JJ, Silveira FT, Souza AAA, Braga RR, Ishikawa EAY (1994) The dermal leishmaniases of Brazil, with special reference to the eco-epidemiology of the disease in Amazonia. Mem Inst Oswaldo Cruz 89:435–443

Lainson R, Shaw JJ (1998) New World leishmaniasis – the neotropical *Leishmania* species. In: Collier L, Balows A, Sussman M (org.) Topley & Wilson's microbiology and microbial infections, vol. 5, Parasitology. Arnold, London, pp 241–266

Lainson R, Ishikawa EAY, Silveira FT (2002) American visceral leishmaniasis: wild animal hosts. Trans R Soc Trop Med Hyg 96:630–631

Lainson R, Rangel EF (2003a) Ecologia das leishmanioses: *Lutzomyia longipalpis* e a eco-epidemiologia da leishmaniose visceral americana (LVA) no Brasil. In: Rangel EF, Lainson R (org.) Flebotomíneos do Brasil. Editora Fiocruz, Rio de Janeiro, pp 311–336

Lainson R, Rangel EF (2003b) Flebotomíneos do Brasil. Editora Fiocruz. Rio de Janeiro, 367 pp

Lainson R, Rangel EF (2005) *Lutzomyia longipalpis* and the eco-epidemiology of American visceral leishmaniasis, with particular reference to Brazil: a review. Mem Inst Oswaldo Cruz 100:811–827

Lainson R, Shaw JJ (2005) Leishmaniasis in the New World. In: Collier L, Balows A, Sussman M (org.) Topley & Wilson's microbiology and microbial infections, 10th edn. vol 5, Parasitology, Arnold, London, pp 313–349

Lampo M, Torgerson D, Márquez LM, Rinaldi M, Garcia CZ, Arab A (1999) Occurrence of sibling species of *Lutzomyia longipalpis* (Diptera: Psychodidae) in Venezuela: first evidence from reproductively isolated sympatric populations. Am J Trop Med Hyg 61:1004–1009

Lane R, Phillips A, Molyneux DH, Procter G, Ward RD (1985) Chemical analysis of the abdominal glands of two forms of *Lutzomyia longipalpis*: site of a possible sex pheromone? Ann Trop Med Parasitol 79:225–229

Lanzaro GC, Ostrovska K, Herrero MV, Lawyer PG, Warburg A (1993) *Lutzomyia longipalpis* is a species complex: genetic divergence and interspecific hybrid sterility among three populations. Am J Trop Med Hyg 48:839–847

Lara-Silva FO, Barata RA, Michalsky EM, Ferreira EC, Lopes MOG, Pinheiro AC, Fortes-Dias CL, Dias ES (2014) *Rattus norvegicus* (Rodentia: Muridae) infected by *Leishmania* (*Leishmania*) *infantum* (syn. *Le. chagasi*) in Brazil. Bio Med Research International 2014:1–7

Lerner EA, Ribeiro JMC, Nelson RJ, Lerner M (1991) Isolation of Maxadilan, a potent vasodilatory peptide from the salivary glands of the sandfly *Lutzomyia longipalpis*. J Biol Chem 266:11234–11236

Lima Costa CR Jr, Freitas MT, Santiago Figueiredo CA Jr, Aragão NC, da Silva LG, Marcondes CB, Dias RV, Leal-Balbino TC, Souza MB, Ramalho-Ortigão M, Balbino Vde Q (2015) Genetic structuring and fixed polymorphisms in the gene period among natural populations of *Lutzomyia longipalpis* in Brazil. Parasit Vectors 8:193

Lukes J, Mauricio IL, Schonian G, Dujardin JC, Soteriadou K, Dedet JP, Kuhls K, Tintaya KW, Jirku M, Chocholova E, Haralambous C, Pratlong F, Obornik M, Horak A, Ayala FJ, Miles MA (2007) Evolutionary and geographical history of the *Leishmania donovani* complex with a revision of current taxonomy. Proc Natl Acad Sci U S A 104:9375–9380

Lutz A, Neiva A (1912) Contribuição para o conhecimento das espécies do gênero *Phlebotomus* existentes no Brasil. Mem Inst Oswaldo Cruz 4:84–95

Lysenko AJ (1971) Distribution of leishmaniasis in the old world. Bull WHO 44:515–520

Maingon RD, Ward RD, Hamilton JG, Noyes HA, Souza N, Kemp SJ, Watts PC (2003) Genetic identification of two sibling species of *Lutzomyia longipalpis* (Diptera: Psychodidae) that produce distinct male sex pheromones in Sobral, Ceará State, Brazil. Mol Ecol 12:1879–1894

Mangabeira O (1969) Sobre a sistemática e biologia dos *Phlebotomus* do Ceará. Rev Bras Malariol Doenças Trop 21:3–26

Martinez E, Alonso V, Quispe A, Thomas MC, Alonso R, Piñero JE, González AC, Ortega A, Valladares B (2003) RAPD method useful for distinguishing *Leishmania* species: design of specific primers for *L. braziliensis*. Parasitology 127:513–517

Martins AV, Williams P, Falcão AL (1978) American sand flies. Acad Bras Ciências, Rio de Janeiro, p 195

Martins AV, Falcão AL, Silva JE, Dias ES (1984) Nota sobre *Lutzomyia* (*Lutzomyia*) *cruzi* (Mangabeira, 1938), com a descrição da fêmea (Diptera, Psychodidae, Phlebotominae). Mem Inst Oswaldo Cruz 79:439–442

Mauricio IL, Howard MK, Stothard JR, Miles MA (1999) Genomic diversity in the *Leishmania donovani* complex. Parasitology 119:237–246

Maziero N, Thomaz-Soccol V, Steindel M, Link JS, Rossini D, Alban SM, Nascimento AJ (2014) Rural-urban focus of canine visceral leishmaniosis in the far western region of Santa Catarina State, Brazil. Vet Parasitol 205:92–95

Mello DA, Rego FA, Oshozo E, Nunes VLB (1988) *Cerdocyon thous* (Carnivora: Canidae) naturally infected with *Leishmania chagasi* (Cunha & Chagas, 1937) in Corumbá (Mata Grosso do Sul State, Brazil). Mem Inst Oswaldo Cruz 83:259

Missawa NA, Lima GB (2006) Spatial distribution of *Lutzomyia longipalpis* (Lutz & Neiva, 1912) and *Lutzomyia cruzi* (Mangabeira, 1938) in the State of Mato Grosso. Rev Soc Bras Med Trop 39:337–340

Montoya-Lerma J, Cadena H, Oveido M, Ready PD, Barazarte R, Travi BL, Lane RP (2003) Comparative vectorial efficiency of *Lutzomyia evansi* and *L. longipalpis* for transmitting *Leishmania chagasi*. Acta Trop 85:19–29

Morrison AC, Ferro C, Tesh RB (1993) Host preference of the sand fly *Lutzomyia longipalpis* at an endemic focus of American visceral Leishmaniasis in Colômbia. Am J Trop Med Hyg 49:68–75

Morton I, Ward RD (1989) Laboratory responses of female *Lutzomyia longipalpis* sandflies to a host and male pheromone source over distance. Med Vet Ent 3:219–223

Moya SL, Giuliani MG, Manteca Acosta M, Salomon OD, Liotta DJ (2015) First description of *Migonemyia migonei* (Franca) and *Nyssomyia whitmani* (Antunes & Coutinho) (Psychodidae: Phlebotominae) natural infected by *Leishmania infantum* in Argentina. Acta Trop 152:181–184

Mukhopadhyay J, Ghosh K, Rangel EF, Munstermann LE (1998) Genetic variability in biochemical characters of Brazilian field populations of the *Leishmania* vector, *Lutzomyia longipalpis* (Diptera: Psychodidae). Am J Trop Med Hyg 59:893–901

Mutebi JP, Alexander B, Sherlock I, Wellington J, Souza AA, Shaw J, Rangel EF, Lanzaro GC (1999) Breeding structure of the sand fly *Lutzomyia longipalpis* (Lutz & Neiva) in Brazil. Am J Trop Med Hyg 61:149–157

Mutebi JP, Tripet F, Alexander JB, Lanzaro GC (2002) Genetic differentiation among populations of *Lutzomyia longipalpis* (Diptera: Psychodidae) in Central and South America. Ann Entomol Soc Amer 95:740–752

Oliveira AC, Batista SM, Falcão AL (1959) Calazar em Minas Gerais. Revisão dos dados epidemiológicos obtidos até 1958. Hospital (Rio de Janeiro) 56:625–643

Oliveira MP, Grimaldi G, Momen H, Pacheco RS, Marzochi MCA, McMahon-Pratt D (1986) Active cutaneous leishmaniasis in Brazil, induced by *Leishmania donovani chagasi*. Mem Inst Oswaldo Cruz 81:303–309

Paiva BR, Secundino NF, Nascimento JC, Pimenta PF, Galati EA, Junior HF, Malafronte RS (2006) Detection and identification of *Leishmania* species in field-captured phlebotomine sandflies based on mini-exon gene PCR. Acta Trop 99:252–259

Paraense L, Chagas AW (1940) Transmissão experimental da leishmaniose visceral americana pelo *Phlebotomus intermedius*. Nota prévia. Bras Med 54:179–180

Penna HA (1934) Leishmaniose visceral no Brasil. Bras Med 48:949–950

Perdeck AC (1957) The isolating value of specific song patterns in two sibling species of grasshoppers (*Chorthippus brunneus* Thumb and *C. biguttulus*). Behaviour 12:1–75

Perini FA, Russo CAM, Schrago CG (2010) The evolution of South American endemic canids: a history of rapid diversification and morphological parallelism. J Evol Biol 23:311–322
Petersen CA, Barr SC (2009) Canine leishmaniasis in North America: emerging or newly recognized? Vet Clin North Am Small Anim Pract 39:1065–1074
Peterson AT, Campbell LP, Moo-Llanes DA, Travi B, Gonzalez C, Ferro MC, Ferreira GEM, Brandao-Filho SP, Cupolillo E, Ramsey J, Leffer AMC, Pech-May A, Shaw JJ (2017) Influences of climate change on the potential distribution of *Lutzomyia longipalpis* sensu lato (Psychodidae: Phlebotominae). Int J Parasitol 47:667–674
Pífano CF, Romero MJ (1964) Comprobación de um nuevo foco de leishmaniose em Venezuela, Valle de Cumanacoa, Edo. Sucre. Gac Méd Caracas 72:473–479
Ponce C, Ponce E, Morrison A, Cruz A, Kreutzer R, McMahon-Pratt D, Neva F (1991) *Leishmania donovani chagasi*: new clinical variant of cutaneous leishmaniasis in Honduras. Lancet 337:67–70
Potenza L, Anduze J (1942) Kalaazar en el Estado de Bolívar, Venezuela. Rev Policlín (Caracas) 11:312–317
Quinnell RJ, Dye C (1994a) An experimental study of the peridomestic distribution of *Lutzomyia longipalpis* (Diptera: Psychodidae). Bull Ent Res 84:379–382
Quinnell RJ, Dye C (1994b) Correlates of the domestic abundance of *Lutzomyia longipalpis* (Diptera: Psychodidae) in Amazonian Brazil. Med Vet Ent 8:219–224
Rangel EF, Vilela ML (2008) *Lutzomyia longipalpis* (Diptera, Psychodidae, Phlebotominae) and urbanization of visceral leishmaniasis in Brazil. Cad Saúde Pública 24:2948–2952
Rangel EF, Lainson R (2003) Ecologia das leishmanioses. Transmissores de leishmaniose tegumentar Americana. In: Rangel EF, Lainson R (org.) Flebotomíneos do Brasil. Editora Fiocruz, Rio de Janeiro, pp 291–309
Ready PD, Lainson R, Shaw JJ, Souza AA (1991) DNA probes for distinguishing *Psychodopygus wellcomei* from *Psychodopygus complexus* (Diptera: Psychodidae). Mem Inst Oswaldo Cruz 86:41–49
Rego FD, Shimabukuro PH, Quaresma PF, Coelho IR, Tonelli GB, Silva KM, Barata RA, Dias ES, Gontijo CM (2014) Ecological aspects of the Phlebotominae fauna (Diptera: Psychodidae) in the Xakriaba Indigenous Reserve, Brazil. Parasit Vectors 7:220
Rioux JA, Lanotte G, Serres E, Pratlong F, Bastien P, Perieres J (1990) Taxonomy of *Leishmania*. Use of isoenzymes. Suggestions for a new classification. Ann Parasitol Hum Comp 65:111–125
Rodriguez-Bonfante C, Bonfante-Garrido R, Grimaldi G Jr, Momen H, Cupolillo E (2003) Genotypically distinct *Leishmania colombiensis* isolates from Venezuela cause both cutaneous and visceral leishmaniasis in humans. Infect Genet Evol 3:119–124
Ryan L, Silveira FT, Lainson R, Shaw JJ (1984) Leishmanial infections in *Lutzomyia longipalpis* and *Lu. Antunesi* (Diptera: Psychodidae) on the island of Marajó, Pará State, Brazil. Trans R Soc Trop Med Hyg 78:547–548
Ryan L, Lainson R, Shaw JJ (1986a) The experimental transmission of *Leishmania mexicana amazonensis* Lainson & Shaw, between hamsters by the bite of *Lutzomyia furcata* (Mangabeira). Trans R Soc Trop Med Hyg 80:164–165
Ryan L, Phillips RA, Milligan P, Lainson R, Molyneux DH, Shaw JJ (1986b) Separation of female *Psychodopygus wellcomei* and *Ps. complexus* (Diptera: Psychodidae) by cuticular hydrocarbon analysis. Acta Trop 43:85–89
Ryan L, Lainson R, Shaw JJ, Fraiha H (1986c) Ecologia de flebotomíneos (Diptera: Psychodidae: Phlebotominae) na Região Amazônica. In: Instituto Evandro Chagas: 50 Anos de Contribuição às Ciências Biológicas e a Medicina Tropical, vol I, Serviços de Saúde Pública, Belém, Pará, pp 307–320
Ryan L, Lainson R, Shaw JJ (1987) Leishmaniasis in Brazil. XXIV. Natural flagellate infections of sandflies (Diptera: Psychodidae) in Pará state, with particular reference to the role of *Psychodopygus wellcomei* as the vector of *Leishmania brazilienisis braziliensis* in the Serra dos Carajás. Trans R Soc Trop Med Hyg 81:353–359

Salomón OD, Quintana MG, Bruno MR, Quiriconi RV, Cabral V (2009) Visceral leishmaniasis in border areas: clustered distribution of phlebotomine sandflies in Clorinda, Argentina. Mem Inst Oswaldo Cruz 104:801–804

Salomón OD, Araki AS, Hamilton JG, Acardi SA, Peixoto AA (2010) Sex pheromone and period gene characterization of *Lutzomyia longipalpis* sensu lato (Lutz & Neiva) (Diptera: Psychodidae) from Posadas, Argentina. Mem Inst Oswaldo Cruz 105:928–930

Salomón OD, Feliciangeli MD, Quintana MG, Afonso MM, Rangel EF (2015) *Lutzomyia longipalpis* urbanisation and control. Mem Inst Oswaldo Cruz 110:831–846

Santoro F, Lemesre JL, Rizvi FS, Afchain D, Sadogirsky M (1986) Spécificité au niveau de protéines de surface des promastigotes de *Leishmania donovani* (Laveran et Mesnil, 1903), *Leishmania infantum* Nicolle, 1908 et *Leishmania chagasi* Cunha et Chagas, 1937. In: Rioux JA (2–6 July, 1984). Leishmania. Taxonomie et Phylogenése. Aplications eco-Épidemiologiques. Montpellier, France: Colloque Internat, IMMEE, pp 71–76

Santos SO, Arias J, Ribeiro AA, Hoffmann MP, Freitas RA, Malacco MAF (1998) Incrimination of *Lutzomyia cruzi* as a vector of American visceral leishmaniasis. Med Vet Ent 12:315–317

Santos SO, Arias JR, Hoffmann MP, Furlan MBG, Ferreira WF, Pereira C, Ferreira L (2003) The presence of *Lutzomyia longipalpis* in a focus of American visceral leishmaniasis where the only proven vector is *Lutzomyia cruzi*, Corumbá, Mato Grosso do Sul State. Rev Soc Bras Med Trop 36:633–634

Santos MFC, Ribolla PEM, Alonso DP, Andrade-Filho JD, Casaril AE, Ferreira AMT, Fernandes CES, Brazil RP, Oliveira AG (2013) Genetic structure of *Lutzomyia longipalpis* populations in Mato Grosso Do Sul, Brazil, based on microsatellite markers. PLoS One 8:1–7

Saraiva L, Carvalho GML, Sanguinette CC, Carvalho DAA, Falcão AL, Andrade Filho JD (2008) Sandflies (Diptera: Psychodidae: Phlebotominae) collected on the banks of the Velhas River in the state of Minas Gerais, Brazil. Mem Inst Oswaldo Cruz 103:843–846

Saraiva L, Andrade Filho JD, Silva S, Andrade AS, Melo MN (2010) The molecular detection of different Leishmania species within sand flies from a cutaneous and visceral leishmaniasis sympatric area in Southeastern Brazil. Mem Inst Oswaldo Cruz 105:1033–1039

Savani ES, Nunes VL, Galati EA, Castilho TM, Zampieri RA, Floeter-Winter LM (2009) The finding of *Lutzomyia almerioi* and *Lutzomyia longipalpis* naturally infected by *Leishmania* spp. in a cutaneous and canine visceral leishmaniases focus in Serra da Bodoquena, Brazil. Vet Parasitol 160:18–24

SBMT (2017) Registrado primeiro óbito por leishmaniose visceral humana em Porto Alegre. Available from: http://www.sbmt.org.br/portal/registrado-primeiro-obito-por-leishmaniose-visceral-humana-em-porto-alegre/

Schalling HD, da Silva ES, Van der Meide WF, Schoone GJ, Gontijo CM (2007) *Didelphis marsupialis* (commom opossum): a potential reservoir host for zoonotic leishmaniasis in the metropolitan region of Belo Horizonte (Minas Gerais, Brazil). Vector Borne Zoonotic Dis 7:387–393

Shaw JJ (2002) In: Farrell J (ed) New world Leishmaniasis: the ecology of leishmaniasis and the diversity of leishmanial species in Central and South America. World Class Parasites: Leishmania/Kluwer Academic Publishers, Boston, Dordrecht, pp 11–31

Shaw JJ (2006) Further thoughts on the use of the name *Leishmania* (*Leishmania*) *infantum chagasi* for the aetiological agent of American visceral leishmaniasis. Mem Inst Oswaldo Cruz 102:541–547

Sherlock IA (1996) Ecological interactions of visceral leishmaniasis in the state of Bahia, Brazil. Mem Inst Oswaldo Cruz 91:671–683

Sherlock IA, Miranda JC, Sadigursky M, Grimaldi G (1984) Natural infection of the opossum *Didelphis albiventris* (Marsupialia: Didelphidae) with *Leishmania donovani* in Brazil. Mem Inst Oswaldo Cruz 79:511

Sherlock IA, Miranda JC, Sadigursky M, Grimaldi G (1988) Observações sobre calazar em Jacobina, Bahia. VI. Investigações sobre reservatórios silvestres e comensais. Rev Soc Bras Med Trop 21:23–27

Silveira FT, Lainson R, Shaw JJ, Póvoa MM (1982) Leishmaniasis in Brazil: XVIII. Further evidence incriminating the fox *Cerdocyon thous* as a reservoir of Amazonian visceral leishmaniasis. Trans R Soc Trop Med Hyg 76:830–832

Sinan: Sistema de Informação de Agravos de Notificação [homepage on the Internet] (2017) Available from: http://dtr2004.saude.gov.br/sinanweb/

Soto SI, Lehmann T, Rowton ED, Velez BID, Porter CH (2001) Speciation and population structure in the morphospecies *Lutzomyia longipalpis* (Lutz and Neiva) as derived from the mitochondrial ND4 gene. Mol Phylogenet Evol 18:84–93

Souza MB, Marzochi MCA, Carvalho RW, Ribeiro PC, Pontes CS, Caetano JM, Meira AM (2003) Ausência da *Lutzomyia longipalpis* em algumas áreas de ocorrência de leishmaniose visceral no Município do Rio de Janeiro. Cad Saúde Púb 19:1881–1885

Souza NA, Vigoder FM, Araki AS, Ward RD, Kyriacou CP, Peixoto AA (2004) Analysis of the copulatory courtship songs of *Lutzomyia longipalpis* in six populations from Brazil. J Med Entomol 41:906–913

Souza NA, Brazil RP, Araki AS (2017) The current status of the *Lutzomyia longipalpis* (Diptera: Psychodidae: Phlebotominae) species complex. Mem Inst Oswaldo Cruz 112:161–174

Steindel M, Menin A, Evangelista T, Stoco PH, Marlow MA, Fleith RC, Pilati C, Grisard EC (2013) Outbreak of autochthonous canine visceral leishmaniasis in Santa Catarina, Brazil. Pesq Vet Bras 33:490–496

Toledo CRS, Almeida AS, Chaves SAM, Sabroza PC, Toledo LM, Caldas JP (2017) Vulnerability to the transmission of human visceral leishmaniasis in a Brazilian urban area. Rev Saúde Pública 51:1–11

Tolezano JE, Taniguchi HH, Araújo MF, Cunha EA, Chioccola VLP PC, Shaw JJ (2005) Evidence of different *Leishmania* enzootic cycles, especially *L.* (*Viannia*) *braziliensis*, in two endemic American cutaneous leishmaniasis (ACL) regions of São Paulo State, Brazil. Abstract book, third world congress on Leishmaniasis,10–15 April, 2005. Palermo-Terrasini, Sicily, Italy, p 213

Tolezano JE, Uliana SR, Taniguchi HH, Araujo MF, Barbosa JA, Barbosa JE, Floeter-Winter LM, Shaw JJ (2007) The first records of *Leishmania* (*Leishmania*) *amazonensis* in dogs (Canis familiaris) diagnosed clinically as having canine visceral leishmaniasis from Aracatuba County, Sao Paulo State, Brazil. Veter Parasitol 149:280–284

Travi BL, Vélez ID, Brutus L, Segura I, Jaramillo C, Montoya J (1990) *Lutzomyia evansi*, an alternative vector of *Leishmania chagasi* in a Colombian focus of visceral leishmaniasis. Trans R Soc Trop Med Hyg 84:676–677

Travi BL, Jaramillo C, Montoya J, Segura I, Zea A, Goncalves A, Vélez ID (1994) *Didelphis marsupialis*, an important reservoir of *Trypanosoma cruzi* and *Leishmania* (*Leishmania*) *chagasi* in Colombia. Am J Trop Med Hyg 50:557–565

Travi BL, Montoya J, Gallego J, Jaramillo C, Llano R, Vélez ID (1996) Bionomics of *Lutzomyia evansi* (Diptera: Psychodidae), vector of visceral leishmaniasis in Northern Colombia. J Med Entomol 33:278–285

Travi BL, Osório Y, Guarin N, Cadena H (1998a) *Leishmania* (*Leishmania*) *chagasi*: clinical and parasitological observations in experimentally infected *Didelphis marsupialis*, reservoir of new world visceral leishmaniasis. Exp Parasitol 88:73–75

Travi BL, Osorio Y, Becerra MT, Adler GH (1998b) Dynamics of *Leishmania chagasi* infection in small mammals of the undisturbed and degraded tropical dry forests of northern Colombia. Trans R Soc Trop Med Hyg 92:275–278

Travi BL, Arteaga LT, Leon AP, Adler GH (2002) Susceptibility of spiny rats (*Proechimys semispinosus*) to *Leishmania* (*Viannia*) *panamensis* and *Leishmania* (*Leishmania*) *chagasi*. Mem Inst Oswaldo Cruz 97:887–892

Uribe S (1999) The status of the *Lutzomyia longipalpis* species complex and possible implications for *Leishmania* transmission. Mem Inst Oswaldo Cruz 94:729–734

Valdivia HO, Almeida LV, Roatt BM, Reis-Cunha JL, Pereira AA, Gontijo C, Fujiwara RT, Reis AB, Sanders MJ, Cotton JA, Bartholomeu DC (2017) Comparative genomics of canine-isolated

Leishmania (*Leishmania*) *amazonensis* from an endemic focus of visceral leishmaniasis in Governador Valadares, southeastern Brazil. Sci Rep 7:40804
Vilela ML (2012) Estudos sobre os flebotomíneos e os potenciais vetores da leishmaniose tegumentar americana no Estado do Tocantins [PhD thesis]. Instituto Oswaldo Cruz, Rio de Janeiro, pp 100
Vilela ML, Afonso MMS, Costa SM, Costa WA, Rangel EF (2014) In: Conceição-Silva F, Alves CR (org.) *Lutzomyia* (*Lutzomyia*) *longipalpis:* Fatores Associados ao Processo de Expansão e Urbanização da Leishmaniose Visceral Americana. Leishmanioses do Continente Americano. Editora Fiocruz, Rio de Janeiro, pp 183–192
Warburg A, Saraiva E, Lanzaro GC, Titus RG, Neva F (1994) Saliva of *Lutzomyia longipalpis* sibling species differs in its composition and capacity to enhance leishmaniasis. Phil Trans R Soc Lond Serie B 345:223–230
Ward RD, Ribeiro AL, Ready PD, Murtagh A (1983) Reproductive isolation between different forms of *Lutzomyia longipalpis* (Lutz & Neiva) (Diptera: Psychodidae), the vector of *Leishmania donovani chagasi* Cunha & Chagas and its significance to kala -azar distribution in South America. Mem Inst Oswaldo Cruz 78:269–280
Ward RD, Ribeiro AL, Ryan L, Falcão AL, Rangel EF (1985) The distribution of two morphological forms of *Lutzomyia longipalpis* (Lutz & Neiva) (Diptera: Psychodidae). Mem Inst Oswaldo Cruz 80:145–148
Ward RD, Phillips A, Burnet B, Marcondes CB (1988) The *Lutzomyia longipalpis* complex: reproduction and distribution. In: MW Service, Biosystematics of Haematophagus Insects. Clarendon Press, Oxford, pp 257–269
Watts PC, Hamilton JG, Ward RD, Noyes HA, Souza NA, Kemp SJ, Feliciangeli MD, Brazil R, Maingon RD (2005) Male sex pheromones and the phylogeographic structure of the *Lutzomyia longipalpis* species complex (Diptera: Psychodidae) from Brazil and Venezuela. Am J Trop Med Hyg 73:734–743
Yin H, Mutebi JP, Marriott S, Lanzaro GC (1999) Metaphase karyotypes and G-banding in sand flies of the *Lutzomyia longipalpis* Complex. Med Vet Entomol 13:72–77
Young DG, Duncan MA (1994) Guide to the identification and geographic distribution of *Lutzomyia* sandflies in Mexico, the West Indies, Central and South America (Diptera: Psychodidae). Mem Am Entomol Inst 54:881
Zeledón R, Hidalgo H, Viquez A, Urbina A (1989) Atypical cutaneous leishmaniasis in a semi-arid region of north-west Costa Rica. Trans R Soc Trop Med Hyg 83:789

Brazilian Phlebotomines as Hosts and Vectors of Viruses, Bacteria, Fungi, Protozoa (Excluding Those Belonging to the Genus *Leishmania*) and Nematodes

Jeffrey J. Shaw, Amélia T. de Rosa, Ana Cecilia Cruz, and Pedro Fernando da C. Vasconcelos

Many microorganisms—ranging from viruses, bacteria, entomophthoralic fungi, trypanosomatids and aseptate gregarines to Tylenchida and Spirurida nematodes (Warburg 1991)—have been isolated from American sand flies. Their importance as vectors of *Leishmania* species has eclipsed their role as vectors or hosts of other organisms except for Bartonellosis. In the case of trypanosomatids, studies on leishmaniasis have generated, directly or indirectly, most of the information that currently exists on these different parasites. In this chapter, we refer to findings of other monoaxenic parasites, including protozoa and bacteria, which can affect leishmanial vector efficiency. Eclectic feeding habits are responsible for epidemiological situations, in which viruses are transmitted between different vertebrates.

It is known that species such as *Bichromomyia flaviscutellata* and *Lutzomyia longipalpis* feed on both birds and mammals. However, there is evidence (Tesh et al. 1971) that some species, such as *M. micropyga*, feed on both warm- and cold-blooded vertebrates. However, so far the participation of sand flies in the transmission between different animal phyla has not been verified. If it does occur, it is most likely to be viruses because there are examples of phyla switching such as that of West Nile Virus passing form birds to man or the overwintering of the western equine encephalitis virus in snakes (Gebhardt et al. 1964).

J. J. Shaw (✉)
Instituto de Ciências Biomédicas, Universidade de São Paulo, São Paulo, Brazil

A. T. de Rosa
University of Texas Medical Branch, Galveston, TX, USA

Instituto Evandro Chagas, Secretaria de Vigilância em Saúde, Ananindeua, Pará, Brazil

A. C. Cruz · P. F. da C. Vasconcelos
Instituto Evandro Chagas, Secretaria de Vigilância em Saúde, Ananindeua, Pará, Brazil

E. F. Rangel, J. J. Shaw (eds.), *Brazilian Sand Flies*,
https://doi.org/10.1007/978-3-319-75544-1_9

Viral Parasites

Vesiculoviruses (Rhabdoviridae – *Vesiculovirus* – vesicular stomatitis virus [VSV] group), phleboviruses (Bunyaviridae – *Phlebovirus* – phlebotomus fever group), and orbiviruses (Reoviridae–*Orbivirus*–Changuinola virus group) are amongst the various virus groups associated with phlebotomines in both the Old World and the New World.

It is important to note that the World Health Organization (WHO 1967) has suggested criteria for recognizing an arthropod as an arbovirus vector, which—more recently—were referenced by the International Committee on Taxonomy of Viruses (ICTV 2016). They include the following:

1. Isolation of the virus from species collected in the field;
2. Demonstration of the ability of the arthropod to become infected by feeding on a viraemic vertebrate host or an artificial suspension of the virus;
3. Demonstration of the ability of the insect to transmit biologically by stinging; and
4. Accumulation of evidence in the field, confirming the association of the arthropod with the vertebrate, in which the disease or infection is occurring.

Based on these criteria, the vectors are distributed as follows:

1. Suspected vectors (those that meet one of the above criteria);
2. Potential vectors (those that satisfy the natural infection and experimental transmission tests); and
3. Confirmed vectors (those that meet all the criteria (WHO 1967)).

From 1961 to 1995, 69 arbovirus serotypes were isolated from sand flies in several areas of the Brazilian Amazon (Fig. 1), distributed among families, genera and groups (Table 1).

Rhabdoviridae – Vesiculovirus – *VSV Group*

This genus currently consists of 16 distinct virus serotypes that have been isolated from a variety of arthropods and mammals in Asia, Africa, Europe and the Americas. Ten of them have been associated with sand flies (Comer and Tesh 1991). Six are known to cause disease in humans and domestic animals (Travassos da Rosa et al. 1984). For this reason, they are of public and veterinary health importance.

In the Brazilian Amazon region, rhabdoviruses belonging to the VSV serogroup have been frequently recovered from phlebotomine sand flies; Carajás and Marabá vesiculoviruses were isolated from pools of *Lutzomyia* spp. captured in Serra Norte, municipality of Marabá, Pará state. In addition, the Carajás virus was isolated from a pool of male *Lutzomyia* spp. This shows that vertical transmission of this agent occurs in nature (Travassos da Rosa et al. 1984a). Both agents replicate and were

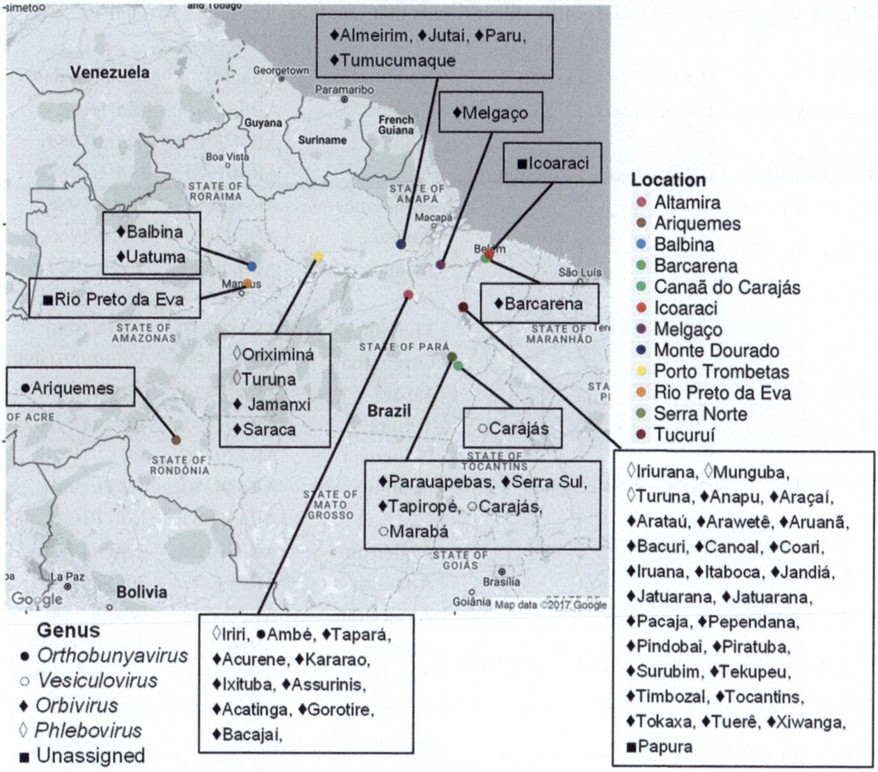

Fig. 1 Map of the Brazilian Amazonia region showing the localization of the capture sites of the sand flies from which viruses were isolated

transmitted vertically by *Lutzomyia longipalpis* after intrathoracic inoculation (Travassos da Rosa et al. 1984a). This species is presumably not the natural vector of these viruses because Carajás and Marabá came from forest areas, which are not the normal habitat of *Lu. longipalpis*. Thus, other sand-fly species are likely to serve as vectors for these viruses (Comer and Tesh 1991).

Neutralizing antibodies to Marabá virus were detected in a single human-serum sample in Brazil (Travassos da Rosa et al. 1984a). However, the source of infection of these viruses for sand flies has not been determined nor has the their potential to cause human disease.

Inhangapi is a member of the Rhabdoviridae, but it does not belong to the genus *Vesiculovirus*. This possible arbovirus was isolated only once from a pool of *L. flaviscutellata* females captured in the Catu Forest, Belém, Pará (Karabatsos 1985). in which it had been tentatively inserted. More recently, the Kairi virus, isolated from a pool of sand-fly females caught in Nova Fronteira Agrovila (km 80), Altamira, Pará, was inserted in this family due to the results obtained by electron microscopy.

Table 1 Arbovíruses isolated from Brazilian Amazonian sand flies between 1961 and 1995

Family	Genus	Antigenic group	Name	Sample	Isolation date
Rhabdoviridae	*Vesiculovirus*	VSV	Carajás	AR4113	28/01/1983
			Marabá	AR 411459	04/02/1983
Bunyaviridae	*Phlebovirus*	Phlehotomus	Ambé	AR407981	14/01/1982
			Ariquemes	AR485678	30/12/1988
			Icoaraci	AN24262	14/10/1960
			Jacundá	AR428329	31/10/1984
			Joa	AR371637	29/03/1979
			Munguba	AR389707	20/09/1980
			Oriximiná	AR385309	26–30/07/1988
			Tapará	AR413570	04/02/1983
			Turuna	AR352492	20–16/07/1978
			Uriurana	AR479776	16/12/1985
Reoviridae	*Orbivirus*	Changuinola	Acatinga	AR428250	06/09/1988
			Acurenê	AR446985	18/09/1985
			Almeirim	AR389709	20/09/1980
			Altamira	AR264277	30/07/1974
			Água Branca	AR505545	08/09/1990
			Anapu	AR496014	28–29/06/1986
			Araçai	AR425269	17–20/02/1984
			Arataú	AR428812	24/08/1984
			Arawete	AR505172	11–14/06/1987
			Aruanã	AR428815	27/08/1984
			Assurinis	Ar482249	03/09/1988
			Bacajaf	AR482267	18/09/1988
			Bacuri	AR496008	06–07/07/1986
			Balbina	AR478620	30/04/1988
			Barcarena	AR511413	18/06/1991
			Canindé	AR54342	05/04/1963
			Canoal	AR433317	01/10/1984
			Catetê	AR495605	31/10/1989
			Coari	AR433343	30/09/1984
			Gorotire	AR482251	06/09/1988
			Gurupi	AR35646	14–20/09/196 1
			Iopaka	AR491065	30/10/1984
			Ipixaia	AR490469	26/06/1986
			Iruana	AR496021	28–29/06/1986
			Itaboca	AR496034	04/07/1986
			Jamanxi	AR243090	04–12/06/1973
			Jandiá	AR440489	20/07/1985
			Jatuarana	AR440497	20/07/1985
			Jutaí	AR397374	28/05/1981
			Kararaô	AR447024	17–18/09/1985

(continued)

Table 1 (continued)

Family	Genus	Antigenic group	Name	Sample	Isolation date
			Melgaço	AR524678	/11/1993
			Ourém	AR41067	02/04/1962
			Pacajá	AR440503	20/07/1985
			Parauapebas	AR415962	20/05/1983
			Parti	AR397370	28/05/1981
			Pependana	AR440504	20/07/1985
			Pindobaí	AR482675	10–21/02/1986
			Piratuba	AR478781	14/12/1985
			Saracá	AR385278	02/07/1980
			Serra Sul	AR498935	11/05/1990
			Surubim	AR440507	20/07/1985
			Tapiropé	AR434080 2	4/10/1984
			Tekupeu	AR505169	28/05–01/06/1987
			Timbozal	AR440541	18–25/07/1985
			Tocantins	AR486776	20/05/1986
			Tocax	AR505170	02–03/06/1987
			Tuerê	AR484704	20/05/1986
			Tumucumaque	AR397956	25/05/1981
			Uatumã	AR478626	29–30/04/1988
			Uxituba	AR452652	04–20/04/1986
			Xaraira	AR490492	26/06/1986
			Xiwanga	AR505172	04–09/06/1987
Not classified	Not classified	Not grouped	Papura	AR450572	14–15/04/1985
			Rio Preto	AR540870	09–13/12/1995
Rhabdoviridae[a]	Not classified	Not grouped	Inhangapi	AR177325	28/11/1969
			Iriri	AR408005	18/02/1982
Bunyaviridae[a]	Not classified	Not grouped	Santarém	AN238758	06/06/1973

[a]Provisional family classification based on morphology and serology

Bunyaviridae – Phlebovirus – *Phlebotomus Fever Group*

Phlebotomines also serve as vectors for an appreciable number of viruses included in the genus *Phlebovirus*. Of the 39 phleboviruses currently known, most (66%) have been associated with phlebotomines. It should be noted that all isolations of Old World sand-fly viruses are from insects belonging to the genus *Phlebotomus*, whereas those from the New World are from insects belonging to American sand-fly genera. Although there are other genera, this virus–insect association undoubtedly reflects the range of vertebrate hosts of these viruses, food preferences and the geographical distribution of vectors. Phleboviruses appear to be limited to mammals, thus reflecting the preferences for mammals of the different phlebotomine genera that are their vectors (Tesh 1988).

Of the 39 recognized serotypes, 25 have been isolated in the New World (Vasconcelos et al. 2001). The prevalence of serotypes in the New World is probably a reflection of the large diversity of sand flies in the Americas. Indeed, approximately 530 different sand-fly species are currently known in the New World, whereas only approximately 370 have been reported from the Old World.

Ten phleboviruses have been associated with disease in humans, who become infected when in contact with the sand-flies' ecological niche. In the New World, this tangential mode of infection results in limited and sporadic numbers of cases, usually in people living near or in forested areas (Guerreiro et al. 1998). In addition, serological evidence of human infection has been reported with other phleboviruses, although its real potential as a disease is unknown. It is noteworthy that the diagnosis of infection by these agents is difficult and that little attention is given to such cases because they are easily confused with malaria, influenza and other viral respiratory diseases or other arbovirus diseases (Guerreiro et al. 1998). Phlebovirus Rift Valley, in addition to being a human pathogen, is also the main problem of veterinary health in Africa where small outbreaks and large epizootics have been reported for decades (Hassan et al. 2017; Peter and Morgan 1981).

In the Brazilian Amazon, 21 (9 not yet registered in the International Catalogue of arboviruses) *Phlebovirus* members have been isolated (Nunes-Neto et al. 2017; Travassos da Rosa et al. 1983). Of these, 9 were obtained from batches of female sand flies collected in various areas of the region. However, Pacui virus was also isolated from a pool of males as well as from rodents (Aitken et al. 1975). Ariquemes virus was isolated only once from a male pool, thus strongly supporting that trans-ovarian transmission of these viruses in nature (Travassos da Rosa et al. 1998).

The other phleboviruses recorded in the region, with the exception of Itaporanga, have been isolated from naturally infected mosquitoes and vertebrates (humans and wild animals) that have not been associated with arthropods until now (Nunes-Neto et al. 2017).

It is important to note that provisionally within the family Bunyaviridae, but not included in any group or genus, is Santarém virus isolated from a rodent *Oryzomys* sp. and from a female pool of *Psychodopygus carrerai* (Travassos da Rosa et al. 1998).

Reoviridae – Orbivirus – *Changuinola Virus Group*

The Changuinola group consists of many antigenically related viruses that have been associated with sand flies, mosquitoes and wild mammals. Such viruses have been found only in tropical America (Travassos da Rosa et al. 1984).

In the Brazilian Amazon, of the 58 serotypes belonging to the Changuinola group, 52 were isolated from sand flies in several areas of the region (Table 1 and Fig. 1). Of these, only 11 are registered in the International Arbovirus Catalogue. Most of these are single isolations, which makes it difficult to understand their natural enzootic cycles. A recent study demonstrated that the orbiviruses isolated in the Brazilian Amazon are closely related with each other and constitute a highly related phylogenetic clade (Silva et al. 2014). In view of their abundance and the well-known ability of orbiviruses to form recombinants (Gorman 1979), new serotypes may continue to emerge.

Unclassified Sand-Fly Viruses

Studies are underway to identify two new possible arboviruses—Papura and Rio Preto—isolated from sand flies in Tucurui Pará and Rio Preto da Eva, Amazonas, respectively (Travassos da Rosa et al. 1998) Rio Preto is a orthobunyavirus, whereas Papura virus has yet to be defined as to taxonomic status and thus remains as new ungrouped and unclassified virus.

Summary

The diversity of virus types indicates the biological complexity of these agents and suggests that the potential vector role as well as that the public-health importance of sand flies has not been sufficiently appreciated. However, viral infection of male sand flies indicates that the transovarian transmission of certain viruses associated with these insects occurs naturally and that the ecology of some of these agents is quite different from those of mosquito-borne viruses.

Four phleboviruses (Alenquer, Candiru, Morumbi and Serra Norte) were found in febrile Brazilian patients; however, to date they have not been recorded in sand flies. Sloths are a known food source for sand flies, and infections of Changuinola and phlebovirus complexes have already been reported from them, but these viruses have not been found in sand flies. This dissociation of the isolates of insects and mammals makes one wonder whether they are in fact transmitted by sand flies.

Bacterial Parasites

In 1964 Brooks (1964) drew attention to the potential importance of intestinal symbionts to insects of medical importance, but phlebotomines were not mentioned. Worldwide there are very few studies on the bacterial fauna of sand-fly intestines and the first of Brazilian sand flies was that of Oliveira et al. in 2000 (Oliveira et al. 2000) who made cultures from the guts of female *Lu. longipalpis* captured in the Lapinha cave in Minas Gerais. This, together with other studies using the same methods (Pereira de Oliveira et al. 2001; Gouveia et al. 2008) and two using metagenomic analyses (McCarthy et al. 2011; Sant'Anna et al. 2012), revealed the presence of 52 different bacteria (Table 2) in the intestinal tracts of Brazilian sand flies of the *Lu. longipalpis* complex. The drawback with culturing is that many bacteria go undetected because they do not grow in culture. In a more recent publication, Kelly et al. (2017) identified 609 microorganism OTUs in a 16S metagenomic analysis of the midgut of laboratory-bred *Lu. longipalpis*. Many of those listed in Table 2, such as Serratia and Enterobacteria, were found; however, of the 121 identified to species only a few coincided with those found previously (Table 2). What is surprising is that species of *Acanthamoeba* and *Wolbachia* were identified. This suggests that other material other than those of the actual midgut contents were inadvertently included in the analysis.

Three important questions exist: Where do the bacteria come from? WHAT is their effect on the fly? How does this effect their vectorial capacity? It seems logical that they are ingested principally in the larval stage and, to a lesser degree, in the adult stage. The larvae feed on decaying material of vegetable and animal origin so they can become infected with both plant and animal pathogens. Some of these may be beneficial to the fly. For instance, the presence of nitrogen-fixing bacteria of the genus *Bradyrhizobium* could help them fix nitrogen. However, pathogenic bacteria could reduce the fly's vigour and longevity. In addition, bacteria

Table 2 Bacteria found in the intestine of Brazilian *Lu. longipalpis* and *Lu. cruzi*

Group/family	Genus/species	Source	*long.*	*cruzi*
Moraxellaceae	*Acinetobacter baumannii*	2,3,4	+	–
Moraxellaceae	*A. bucaniid*	4	+[a]	–
Moraxellaceae	*Acinetobacter* sp.	5	+	–
Moraxellaceae	*A. lwoffi*	1.2	+	–
Bacillaceae	*Anoxybacillus flavithermus*	4	+	–
α Proteobacterium	*Asaia* sp.	6	+	–
Bacillaceae	*Bacillus thuringiensis*	1	+	–
Alphaproteobacteria	*Bradyrhizobium japonicum*	5	+	+
Burkholderiaceae	*Burkholderia cepacia*	2,3	+	–
Alphaproteobacteria	*Caulobacter* sp.	5	+	+
Bacteroidetes	*Chryseobacterium meningosepticum*	5	–	+
Enterobacteriaceae	*Citrobacter freundii*	3	+	–
Firmicutes	*Clostridium disporicum*	5	+	+
Firmicutes	*Clostridium* sp.	5	+	–
Firmicutes	*C. glycolicum*	5	–	+
Enterobacteriaceae	*Enterobacter aerogenes*	2,3	+	–
Enterobacteriaceae	*E. amnigenus*	3	+	–
Enterobacteriaceae	*E. cloacae*	1,2,3	+	–
Enterobacteriaceae	*E. gergoviae*	2,3	+	–
Enterobacteriaceae	*E. sakazakii*	2	+	–

(continued)

Table 2 (continued)

Group/family	Genus/species	Source	*long.*	*cruzi*
Enterobacteriaceae	*E. taylorae*	3	+	–
Gammaproteobacteria	*Erwinia billingiae*	5	+	–
Enterobacteriaceae.	*Escherichia coli*	3	+	–
Pseudomonadaceae	*Flavimonas orizihabitans*	1,3	+	–
Bacillaceae	*Geobacillus kaustophilus*	4	+	–
Enterobacteriaceae	*K. oxytoca*	3	+	–
Enterobacteriaceae	*K. ozaenae*	3	+	–
Enterobacteriaceae	*K. pneumoniae*	5	+	–
Firmicutes	*Lactobacillus zymae*	5	–	+
Microbacteriaceae	*Leifsonia xyli*	4	+[a]	–
Betaproteobacteria	*Leptothrix* sp.	5	+	–
Enterobacteriaceae	*Morganella morganii*	3	+	–
Actinobacteridae	*Nocardioides albus*	5	+	–
Enterobacteriaceae.	*Pantoea agglomerans*	3	+	–
Propionibacteriaceae	*Propionibacterium acnes*	4,5	+	–
Proteobacterium	(uncultured)	4	+[a]	–
Pseudomonadaceae	*Pseudomonas aeruginosa*	2,3	+	–
Pseudomonadaceae	*P. fluorescens*	2	+	–
Pseudomonadaceae	*P. putida*	1,3	+	–
Ralstoniaceae	*Ralstonia* sp.	5	+	+
Ralstoniaceae	*Ralstonia pickettii*	4	+	–
Enterobacteriaceae	*Serratia liquefaciens*	3	+	–
Enterobacteriaceae	*S. marcescens*	2,3	+	–
Enterobacteriaceae	*Serratia* sp.	5	+	–
Bacteroidetes	*Sphingobacterium daejeonense*	5	+	–
Firmicutes; Bacillales	*Staphylococcus* sp.	3	+	–
Firmicutes; Bacillales	*S. xylosus*	5	–	+
Streptomycetaceae	*Streptomyces coelicolor*	4	–	–
Xanthomonadaceae	*Stenotrophomonas maltophilia*	1,2,3	+	–
Veillonellaceae	*Veillonella* sp.	4	+[a]	–
Flavobacteriaceae	*Weeksella virosa*	3	+	–
Enterobacteriaceae	*Yokenella regensburgei*	2	+	–

Source: 1 = Oliveira et al. (2000); 2 = Pereira de Oliveira et al. (2001); 3 = Gouveia et al. (2008); 4 = McCarthy et al. (2011); 5 = Sant'Anna et al. (2012); 6 = Sant'Anna et al. (2014)
[a]Only found in males

of medical importance have been found in sand flies so could they be vectors of these pathogens to man and other animals?

Research on sand fly–gut bacteria is in its infancy. *Bacillus thuringiensis* could be used to control sand-fly populations (Yuval and Warburg 1989), but would this produce unforeseen problems? In addition, it is intriguing and challenging to see if the manipulation of this micro fauna influences vectorial capacity. Adler and Theodor (1927) suggested that bacterial competition could modulate the infections of leishmania in the sand fly. Over the years evidence has been growing to vindicate this opinion. Recent papers (Sant'Anna et al. 2014; Kelly et al. 2017) investigating the interaction between *Leishmania* and the bacterial gut fauna of experimental infections in *Lu. longipalpis* indicated that bacteria can affect the parasite's development positively or negatively. It appears that bacteria facilitate metacyclogenesis. Besides this the level of regurgitated during feeding plays an important role in the immunological response of the host. Their presence in the initial lesion leads to the recruitment of cells that facilitate the parasite's expansion and consequently future pathologies (Kelly et al. 2017). When *Leishmania*-infected flies were treated with antibiotics, parasite multiplication and the development of metacyclic promastigotes was reduced (Wilson et al. 2017). However, not all bacteria are good for *Leishmania*. Sant'Anna et al. (2014) found that *Serratia marcescens* inhibited leishmanial development but that *Lu. longipalpis* infected with *Leishmania* lived longer than those infected with just *Serratia*. Thus, evidence is accumulating suggesting that the beneficial or detrimental effect of gut bacteria on *Leishmania* development is related to the species of bacteria. In the case of infections of pathogenic bacteria, the sand fly is protected by the *Leishmania*.

The epidemiological implications of these recent findings of the pivotal importance of the intestinal microbiome are extremely important. Could the bacterial fauna acquired by the immature stages or the adults feeding on plants increase or decrease vectorial capacity? McCarthy et al. (2011) compared the microbiota of *Lu. longipalpis* captured in an urban area of Argentina with those captured in a wild environment in Minas Gerais. They found that the bacterial microbiota of the two populations was completely different. Only one bacterium was found in the urban flies compared with nine in the wild flies. In a set of experiments, Sant'Anna et al. (2014) showed that when pre-feeding laboratory-reared *Lu. longipalpis* with *Asaia*—or with the yeast *Pseudozyma* together with *Asaia* or *Ochrobactrum* intermedium—both diminished the midgut promastigote population. *O. intermedium* has so far only been found in laboratory *Lu. longipalpis* colonies.

Information of the effects of enteric bacterial entomopathogens in Brazilian sand flies is limited to *Lu. longipalpis* and *Leishmania*; however, it may apply to infections with other trypanosomatids. Sant'Anna et al. (2014) concluded that populations that are susceptible to *Leishmania* infection may survive bacterial enteric pathogens better than those that are resistant to leishmanial infection. This implies that *Leishmania* circulating in an enzootic or endemic environment will favour the expansion of a susceptible vector population.

Fungi

Pathogenic fungi of insects belong to the order Entomophthorales, which consists of five families (Keller and Petrini 2005). Fungi can be a problem in colonies where they may form hyphal mats, in which the larvae become entangled and eventually die. However, they are rarely found in wild-caught flies: The first record is that of Warburg (1991) who noted mycelia in the thoracic muscle of a female *Pintomyia pia* collected in Colombia. The parasite was considered to belong to the genera *Entomophthora* or *Conidiobolus*. The only records we could find of entomophthorales in Brazilian sand flies were those of McCarthy et al. (2011). In their metagenomic analysis they identified *Peronospora conglomerata*, *Cunninghamella bertholletiae*, *Mortierella verticillate* and *Toxicocladosporium irritans* in *Lu. longipalpis* from the Lapinha cave.

Microsporidia

For many years microsporidia were considered protozoa, but they are now accepted as being a sister group to the fungi. They are present in all the major animal groups and are common in insects. Under the light microscope, these infections appear as off white- to white-coloured spots. Phylogenies using ribosomal DNA indicate that morphology is unreliable for genus identification (Brown and Adamson 2006).

Family Pleistophoridae Stempell, 1909

The assignment of the genera *Pleistophora* and *Vavraia* to a family is controversial. They have both been placed in the family Glugeidae, and *Vavraia* has been designated as the type species of the family Thelohaniidae. However, for the moment we follow Sprague et al. (1992) who considered them both to be in the family Pleistophoridae. These controversial issues will undoubtedly be resolved by future molecular studies.

The first infection of a microsporidian in a Brazilian sand fly was registered by Ward and Killick-Kendrick (1974) in the midgut of a female *Ps. lainsoni* collected at km 46 of the Altamira–Itaituba section of the Transmazônica Highway. Subsequently (Lainson et al. 1977), two infections were found in the Malpighian tubules and two in the midgut of *Ny. umbratilis*. Those from the Malpighian tubules were considered to belong to the genus *Pleistophora* due to the morphology of the sporoblasts that appeared in the form of a bunch of grapes. One from the mid-gut was classified as belonging to the family Thelohaniidae due to schizonts with diplokaryon nuclei and octonucleate sporonts. The second midgut parasite was composed of predominantly uninucleate spores often in chains. It was tentatively placed in the genus *Microsporidium*. A *Thelohania*-like parasite has also been recorded in *Ps. maripaensis* (Canning 1977).

Matos et al. (2006) studied the infection of a microsporidian that they found in a *Lu. longipalpis* population close to the city of Teresina, Piaui. The infection was found in the subcuticular tissue of the larval and pupal abdomen and in the

Malpighian tubules and midgut of the adult flies. These investigators described the morphology and ultrastructure of the parasite, considered is to be a new species belonging to the genus *Vavraia*, and named it *Vavraia lutzomyiae*. This is the first named species of a microsporidian from a Brazilian sand fly. It seems very unlikely that the microsporidia recorded from Amazonian sand flies in the 1970s all belong to the same species.

Protozoal Parasites

Family Lecudinidae Mingazzini, 1891 – Aseptate Gregarines

Aseptic gregarines (Eugregarinorida, Aseptatorina and Lecudinidae) of sand flies were initially classified as belonging to the genus *Monocystis*. Later, they were transferred to the genus *Ascocystis* (Scorza & Carnevali 1981). Because this name was preoccupied, the genus *Ascogregarina* was created to replace it. However, analyses of the small-subunit rRNA-gene sequences by Votypka et al. (2009) showed that the sand-fly gregarines are very different from those of mosquitoes, which are classified as *Ascogregarina*. Because of this Votypka et al. (2009) created the genus *Psychodiella* for the sand-fly gregarines. Thus, the valid name for the gregarines of Brazilian sand flies is *Psychodiella chagasi.*

Gregarine infections (*Monocystis mackiei*) were first described in 1927 (Shortt and Swaminath 1927) in Indian specimens of *Phlebotomus argentipes*. However, it was only in 1961 that Adler and Mayrink (1961) found similar parasites in *Lu. longipalpis* from Minas Gerais, Brazil. About 20% of *Lu. longipalpis*, both wild and laboratory-bred, were infected, and the parasite was considered a new species, *Monocystis chagasi* (Adler and Mayrink 1961). Gregarines were subsequently recorded in *Bi. flaviscutellata* (Lewis et al. 1970) *Ev. Evandroi, Ev. Sallesi* and *Ps. complexus.* Due to their morphological similarities to the parasite described by Adler and Mayrink, they were all considered to be *P. chagasi*.

It is debatable as to whether the aseptate gregarines infections found in the different species of Brazilian sand flies are in fact *P. chagasi*. Detailed morphological studies of the sporocysts were not made. In addition, there is also evidence of host specificity. Wu and Tesh (1989) noted that *P. chagasi* infected large numbers of Old and New World sand flies, but its complete cycle only occurred in *Lu. longipalpis*. Molecular studies are needed to elucidate the specific status of the *Psychodiella* infections of different Brazilian sand flies.

The life cycle is simple. Two morphologically similar trophozoites are confined to a cyst. The two become gamonts that sprout and divide to produce gametocytes. These fuse together to form a zygote, the walls of which thicken to give rise to the typical sporocyst. It adheres to the surface of the egg, and the larva becomes contaminated when it hatches. Although infections are frequently severe and result in the enlargement of accessory glands, there is no evidence yet of any effect on fertility or egg production. However, in experimental infections of *P. chagasi* in *Lu. longipalpis*, Wu and Tesh (1989) noted that, although fertility was not affected, there was a significant reduction in adult longevity.

Family Trypanosomatidae Doflein, 1901

Finding trypanosomatids in Brazilian sand flies is mostly a by-product of leishmaniases epidemiological studies. A list of 66 of these infections is given in Table 3 based on the information contained in 26 papers (Arias et al. 1985; Brazil and Ryan 1984; de Souza et al. 2017; Ferreira et al. 2008, 2015; Freitas et al. 2002; Galati et al. 2006; Gil et al. 2003; Hoch et al. 1986; Lainson and Shaw 1979; Lainson et al. 1973; Naiff et al. 1989; Pessoa et al. 2007; Rogers et al. 1988; Ryan et al. 1987a, 1987b; Shaw 1992; Shaw and Lainson 1972; Shaw et al. 1987; Sherlock and Pessôa 1966; Silveira et al. 1991; Souza et al. 1998; Teixeira et al. 2011; Viola et al. 2008; Williams and Coelho 1978; Ryan et al. 1987a). Few have been isolated and characterized. Some have been listed as belonging to the genera *Crithidia, Blastocrithidia, Strigomonas, Endotrypanum* and *Trypanosoma*; however, this characterization should be viewed with reservation because in some cases it is based on morphology. Camargo (1999), when discussing the morphology of trypanosomatids, drew attention to the fact that it is not a reliable character to distinguish genera of trypanosomatids found in plants and insects.

The scarcity of knowledge about dietary preferences of sand flies makes it almost impossible to associate infections with trypanosome species. However, there is strong evidence that Brazilian sand flies are vectors of trypanosomes of mammals, lizards, snakes and frogs. Evidence is slowly accumulating suggesting that certain genera are linked to the transmission of trypanosomes of the different vertebrate phyla, which reflects their host preferences. Potentially there is a danger that fewer flagellates, other than *Leishmania*, will be recorded in the future because present-day molecular methods used to detect *Leishmania* do not detect trypanosomes or monoxenous parasites.

Monoxenous Trypanosomatids

The first confirmed finding of a monoxenous parasite in sand flies is that of Wallace and Hertig (1968). Based on ultrastructural studies of the kinetoplast of a flagellate isolated from a Panamanian phlebotomine, *Lu. sanguinaria*, they considered that the parasite belonged to the genus *Crithidia*. To date there are few records of suspected or confirmed monoxenous flagellates in Brazilian sand flies compared to with those of *Leishmania*.

Table 3 Infections of Brazilian sand flies by different trypanosomatid genera

Sand fly species	Parasite[a]	State	Literature source
Evandromyia edwardsi	*Trypanosoma* sp.	MG	Williams and Coelho (1978)
E. evandroi	*Trypanosoma* sp.	MA	Brazil and Ryan (1984)
E. evandroi	*Trypanosoma* sp.	DF	Ferreira et al. (2015)
E. infraspinosa	*Trypanosoma* sp.	PA	Shaw and Lainson (1972)
E. infraspinosa	*Trypanosoma* sp. (An03)	RO	Gil et al. (2003), Ferreira et al. (2008)
E. inpai	Trypanosomatid[b]	PA	Ryan et al. (1987b)
E. pinottii	*Trypanosoma* sp.	PA	Ryan et al. (1987b)
E. sallesi	*Trypanosoma* sp.	MG	Williams and Coelho (1978)
E. saulensis	*Trypanosoma* sp.	PA	Shaw and Lainson (1972)
Lutzomyia gomezi	*Trypanosoma* sp.	RO	Gil et al. (2003), Ferreira et al. (2008)
L. almerioi	*Strigomonas galati*	MS	Galati et al. (2006), Teixeira et al. (2011)
M. cayennensis	*Trypanosoma* sp.	AM	Ryan et al. (1987a)
M. micropyga	*Trypanosoma* sp.	BA	Sherlock and Pessôa (1966)
M. rorotaensis	*T. thecodactylus*	PA	Lainson and Shaw (1979)
M. rorotaensis	*Trypanosoma* sp.	PA	Arias et al. (1985)
M. rorotaensis	*Trypanosoma* sp.	PA	Arias et al. (1985)
M. rorataensis	Trypanosomatid[b]	PA	Ryan et al. (1987b)
M. rorataensis	Trypanosomatid[b]	RO	Souza et al.(1998)
M. rorataensis	Trypanosomatid[b]	AM	Pessoa et al. (2007)
M. trinidadensis	*Trypanosoma* sp.	PA	Ryan et al. (1987a, 1987b)
Migonemyia migonei	Monoxenous parasite	AP	de Souza et al. (2017)
Nyssomyia anduzei	*Endotiypanum* sp.	PA	Shaw (1992)
Ny. anduzei	*Endotrypanum* sp.[c]	AM	Rogers et al. (1988)
N. antunes	Trypanosomatid[b]	PA	Lainson and Shaw (1979)
N. antunesi	*Trypanosoma* sp.	PA	Silveira et al. (1991)
N. umbratilis	*Endotrypanum* sp.[c]	AM	Rogers et al. (1988)
N. umbratilis	Trypanosomatid[b]	PA	Ryan et al. (1987a)
N. umbratilis	Trypanosomatid[b]	PA	Ryan et al. (1987a)
N. umbratilis	*Trypanosoma*?	AP	Freitas et al. (2002)
N. whitmani	*Trypanosoma* sp.	BA	Hoch et al. (1986)
N. whitmani	*Blastocrithidia*	DF	Ferreira et al. (2015)
*P. damascenoi**	*Trypanosoma* sp	AM	Arias et al. (1985)
P. nevesi	trypanosomatid[b]	PA	Arias et al. (1985)
P. neves	Trypanosomatid[b]	AC	Arias et al. (1985)
P. nevesi	Trypanosomatid[b]	RO	Arias et al. (1985)
P. dendrophila	Trypanosomatid[b]	PA	Lainson and Shaw (1979)
P. dendrophila	Trypanosomatid[b]	PA	Ryan et al. (1987a, 1987b)
P. dendrophila	*Trypanosoma* sp. (An03)	RO	Gil et al. (2003), Ferreira et al. (2008)
P. dendrophila	*Endotrypanum*?	AP	Freitas et al. (2002)
P. (shannoni series) sp	*Endotrypanum* sp.[d]	PA	Arias et al. (1985)

(continued)

Table 3 (continued)

Sand fly species	Parasite[a]	State	Literature source
P. (shannoni series) sp	*E.schaudinni*[d]	RO	Arias et al. (1985)
P. (shannoni series) sp	*Endotrypanum* sp.	RO	Arias et al. (1985)
P. (shannoni series) sp	*T. rangeli*[d]	RO	Arias et al. (1985)
P. (shannoni series) sp	Trypanosomatid[b]	AM	Arias et al. (1985)
P. (shannoni series) sp	Trypanosomatid[b]	AC	Arias et al. (1985)
*P. bigeniculata***	*Endotrypanum* sp.[c]	AM	Rogers et al. (1988)
*P. bigeniculata***	Trypanosomatid[b]	AM	Pessoa et al. (2007)
Sciopemyia fluviatalis	Trypanosomatid[b]	PA	Ryan et al. (1987b)
S. servulolimae	*Trypanosoma* sp.(An03)	RO	Gil et al. (2003), Ferreira et al. (2008)
*S. sordellii****	*Trypanosoma* sp	PA	Ryan et al. (1987a)
*S. sordellii****	*Trypanosoma* sp.	PA	Silveira et al. (1991)
S. sordellii	*Trypanosoma* sp.	PA	de Souza et al. (2016)
S. sordellii	*Trypanosoma* sp.(An03)	RO	Gil et al. (2003), Ferreira et al. (2008)
S. sp.	*Trypanosoma* sp.(An03)	RO	Gil et al. (2003), Ferreira et al. (2008)
Psychodopygus amazonensis	Typanosomatid[b]	PA	Lainson et al. (1973)
P. amazonensis	Trypanosomatid[b]	PA	Ryan et al. (1987b)
P. ayrozai	Trypanosomatid[b]	BA	Hoch et al. (1986)
P. clausteri	trypanosomatid[b]	PA	Ryan et al. (1987a)
P. clausteri	*T. freitasi*	AM	Naiff et al. (1989)
P. "complexus"	Trypanosomatid[b]	PA	Ryan et al. (1987b)
P. davisi	*Trypanosoma* sp.	PA	Shaw and Lainson (1972)
P. davisi	Trypanosomatid[b]	PA	Ryan et al. (1987a, 1987b)
P. davisi	*Trypanosoma* sp.	BA	Hoch et al. (1986)
P. paraensis	Trypanosomatid[b]	PA	Ryan et al. (1987a)
P. "wellcomei"	Monoxenous parasite[e]	PA	Ryan et al. (1987a), Shaw et al. (1987)
Viannamyia tuberculata	*Trypanosoma* sp.[f]	RO	Viola et al. (2008)

Names that are in the original paper: **Lutzomyia spinose*, ***Lutzomyia shannoni*, and ****Lutzomyia nordestina*

Species names in quotes indicate the most likely identification based on males captured at the same time

An03 = An anuran trypanosome clade

[a]Identifications are based mainly on morphology

[b]The authors do not refer to genus except that it is not a *Leishmania*

[c]Based on kinetoplast DNA probes

[d]Based on isoenzyme analysis

[e]Based on the presence of choanomastigotes

[f]Based on V7V8 SSU rDNA the isolate was similar to snake trypanosomes

[g]Unpublished observation quoted in Williams and Coelho (1978)

Flagellates that grew well in culture were found in female *Psychodopygus wellcomei* from Pará State (Ryan et al. 1987a; Shaw et al. 1987) and a female *Migonemyia migonei* from Amapá State (de Souza et al. 2017). They did not react with any of the leishmania-specific monoclonal antibodies and morphologically resembled insect parasites. Because of this and their morphology, they were tentatively considered to be Crithidia. Using primers that amplified the D7 region of the 24Sα rRNA gene Ferreira et al. (2015, no. 5358) identified a *Blastocrithidia* species in a female *Ny. whitmani* captured in a rural peri-domiciliary location of the Federal District.

During epidemiological studies of the phlebotomine fauna of forested areas of the Serra da Bodoquena, Mato Grosso do Sul State, Galati et al. (2006) isolated a flagellate from female *Lu. almerioi* that again did not react with any leishmania-specific monoclonal antibodies. This same isolate was studied by Teixeira et al. (2011) in their review of the monoxenous genera *Angomonas* and *Strigomonas*. They concluded that it was a new species and named it *Strigomonas galati* together with its symbiont, which received the name *Kinetoplastibacterium galatii*. This isolate is the first monoxenous parasite of Brazilian sand flies to receive a specific name.

The negative reaction with leishmanial-specific monoclonal antibodies seen with parasites of the Amazonian sand flies was the same as that seen with *S. galati,* which adds weight to the conclusion that they were monoxenous flagellates. However, it is debatable as to whether they belonged to the genus *Crithidia* or to one of the symbiont-bearing genera such as *Angomonas* or *Strigomonas.* Such questions can best be answered using molecular phylogenetic markers.

Thus far, there are few records of monoxenous trypanosomatids in sand flies. Wallace and Hertig (1968) suggested that the parasite they found may not be specific for sand flies and could represent an accidental infection. Another possibility is that some monoxenous parasites of dipterans are opportunistic and infect a wide variety of different hosts, which would ensure their survival.

Trypanosomes of Amphibia

The first association of sand flies with anuran trypanosomes in the Americas was that of Anderson and Ayala (1968) who experimentally infected *M. vexator* with *Trypanosoma bufophlebotomi* that had fed on California Toads (*Anaxyrus boreas halophilus*). Ferreira et al. (2008) showed that trypanosomes of Brazilian frogs and toads fell into three clades. Clade An01 were trypanosomes of hylids, and clades An02 and An03 harboured trypanosomes of toads and frogs. A fourth clade, An04, was composed of trypanosomes from African, European and North American anurans.

An03 appears to be a lineage confined to Amazonian nocturnal frogs and toads that occupy the same terrestrial habitats and aquatic breeding sites. Trypanosomes were identified in *Evandromyia infraspinosa* from Pará and Rondônia (Shaw and Lainson 1972; Gil et al. 2003); *Psathyromyia dendrophila* from Pará and Rondônia (Gil et al. 2003; Lainson and Shaw 1979; Ryan et al. 1987a, 1987b), *Sciopemyia* sp. (Gil et al. 2003), *Sc. fluviatalis* (Ryan et al. 1987b), *Sc. servulolimae* from Rondônia

(Gil et al. 2003) and *Sc. sordellii* from Pará and Rondônia (de Souza et al. 2016; Ryan et al. 1987a; Silveira et al. 1991). The anuran clade An03 was identified in the following species from Rondônia: *Ev. infraspinosa, Ps. dendrophila, Sciopemyia sp., Sc. Servulolimae* and *Sc. sordellii*. These results strongly suggest that the evandromyias and sciopemyias are important vectors of anuran trypanosomes in Amazonia. It is difficult to know the importance of *Ps. dendrophila* as a vector of anuran trypanosomes. This species appears to be a catholic feeder and has been found infected with mammalian flagellates such as *Endotrypanum*. However, the *Psathyromyia* (Shannoni series), to which this species belongs, contains 19 species that are morphologically similar and can be confused. For instance, both *Ps. dendrophyla* and *Ps. shannoni* have been recorded from the same regions, and their feeding habits are not well defined.

Trypanosomes of Lizards

Christensen and Telford (1972) found a trypanosome, *Trypanosoma thecadactyli*, in a Panamanian forest gecko (*Thecadactylus rapicaudus*) that was experimentally infected *M. trinidadensis*. Both the lizard and sand fly are found from Mexico to Brazil. In the Jari river region of northern Pará, Lainson and Shaw (1979) found flagellates in *Mi. rorotaensis* that had fed on *T. rapicaudus*, which strongly suggested that they belonged to *T. thecadactyli*. Sherlock sand Pessôa (1966) suggested that flagellates found in *Mi. micropyga* captured in animal holes or on tree trunks belonged to a lizard *Leishmania*. This statement seems incorrect because the parasites in their illustrations are like trypanosomes. The close association between sand flies that inhabit tree trunks and lizards suggests that they are the natural vectors of this reptilian trypanosome. In an analysis of three *Mi. micropyga*, two reacted with cold-blooded animal antisera (reptiles and amphibians) and one with mammalian antisera (Tesh et al. 1971). This adds weight to the hypothesis that *M.* species are the vectors of lizard trypanosomes in Brazil.

Trypanosomes of Mammals

There is good circumstantial evidence suggesting that neotropical phlebotomines are the vectors of some *T. (Megatrypanum)* trypanosomes of bats and rodents. In Peru, Herrer (1942) observed that *T. phyllotis* of cricetid rodents of the genus *Phyllotis* developed in the intestine of *Lu. noguchii* and that specimens of this same sand fly collected in rodent nests were infected. In Costa Rica, Zeledon and Rosabal (1969) observed that the trypanosome *T. leonidasdeanei* of insectivorous bats developed in *Dampfomyia vespertilionis*. This phlebotomine is found in tree hollows inhabited by bats. Christensen and Herrer (1975) considered it as a vector suggesting, as had Hoare (1972), that transmission would occur by the ingestion of infected sand flies.

Naiff et al. (1989), working in Amazonas State, found a natural infection of *T. freitasi* in *Psychodopygus clausteri* suggesting it as its vector. This marsupial trypanosome is morphologically similar to *T. phylloti*. Previously, Professor Leonidas Deane (1964) had tried unsuccessfully to experimentally infect triatomines with this same trypanosome, adding weight to the hypothesis that these megatrypanums are transmitted by sand flies.

The infection by *T. rangeli* in sand flies of the Shannoni series from Rondônia is a good example of the difficulties in interpreting the exact meaning of an infection (Arias et al. 1985). Trypanosomes of this group are transmitted by reduviids, and there is no evidence that phlebotomines could be the vectors. This single record was based on a culture made from a fly that had consumed a residual blood meal. This underscores the fact that merely finding a trypanosome in any hematophagous insect should be interpreted with caution, especially if blood is present in the intestine.

Parasites of the Genus *Endotrypanum*

Most of these parasites were found while searching for *Leishmania* infections. It is quite possible that some of infections attributed as being *Leishmania* were in fact *Endotrypanum*. The reservoirs of these parasites are principally sloths, especially two-toed sloths of the genus *Choloepus*, which is also an important reservoir of numerous species of Latin American *Leishmania*. In Pará State, of 33 *Endotrypanum* infections 16 were mixed *Endotrypanum*/*Leishmania*. However, up until now no mixed infections of these two parasites have been found in phlebotomines that are known to feed on sloths such as *Ny. anduzei, Ny. umbratilis* and *Psathyromyia bigeniculata* (Lainson et al. 1981a; Lainson et al. 1981b). Such findings may in part be linked to the methods of isolation employed, which act as septic filters. For example, if sand-fly material is inoculated into hamsters, *Leishmania* may be isolated, but *Endotrypanum* will not because this flagellate does not infect laboratory animals.

However, two other factors can influence mixed infections: There may be unknown levels of specificity of both parasites relative to the host, or one parasite could inhibit the development of the other (Pacheco et al. 1987). Both options are open to experimentation, but to date there are no theoretically reasons for believing that phlebotomine cannot be infected by both parasites. In practice, this observation has never been made, but with DNA-specific oligonucleotides and PCR techniques infections from primary isolates can now be examined in the search for mixed infections.

The first indication that *Endotrypanum* developed in sand flies arose from experimental studies in Panama (Shaw 1964). Arias et al. (1985) found that a total of 13.37% of sand flies of the Shannoni group collected at the base of the trees in the states of Acre, Amazonas (north and south of the Amazon River), Rondônia, and Pará had *Endotrypanum* infections. Their isozyme profiles showed that 6.29% of the isolates, all from Rondônia, belonged to 2 distinct groups. One, consisting of 16 isolates, was considered as *E. schaudinni*, and the other, composed of 8 strains, was called *Endotrypanum* species. The first group was classified as *E. schaudinni* because its iso-enzymatic profile was identical to that of a sloth isolate from the Jari region of Pará, except for the mobility of enzyme MPI, which was slightly faster. However, the 2 groups showed differences in 7 enzymatic loci. Some of the infections identified as *E. schaudinni* were associated with blood-meal remnants (Arias and Naiff personal communication).

The results of experimental infections of Brazilian phlebotomines with parasites identified as *Endotrypanum* are controversial. Shaw (1981) successfully infected laboratory-reared *Bi. flaviscutellata*, *Lu. longipalpis* (Marajó), and *Vi. furcata* fed on *Endotrypanum*-infected *Choloepus didactylus* sloths. The infections were like those described by the same investigator in 1964 in Panamanian phlebotomines. However, the investigator did not succeed in infecting another group of *Lu. longipalpis* from the same colony with *Endotrypanum* cultures. The developmental cycle in the pylorus region was like that of *Leishmania* of the subgenus *L.* (*Viannia*), but the invasion of Malpighian tubules was also observed, which does not occur in peripylarian *Leishmania*. This is characteristic of other sand-fly trypanosomatids, excluding *Leishmania*, and clearly distinguishes *Endotrypanum* infections from those caused by *Leishmania* of the subgenus *L.* (*Viannia*), although both develop as promastigotes. However, others (Franco et al. 1997; Barbosa et al. 2006) failed to see this, but their infections were initiated with culture and not blood forms. One possible explanation is that parasites that tolerate or prefer acid pH levels are lost during cultivation in neutral or alkaline media.

Franco et al. (1997) experimentally infected laboratory-bred *Ps. shannoni* originating from the United States, sand *Lu. longipalpis* from Lapinha cave, Brazil, and Colombia with strains of *Endotrypanum* that had been isolated (Arias et al. 1985) from a *Ps. shannoni* and a two-toed sloth captured in Rondônia. Parasites isolated from flies of the Shannoni series infected 100% of *Ps. shannoni* but only 62.3% of *Lu. longipalpis*. The Colombian lineage of *Lu. longipalpis* was significantly less infected with a sloth strain than that of the Shannoni group (18.2% compared with 55.6%). According to the reference number of the strains, it is probable that the parasites of the Shannoni series and those of sloths originate in different regions of the Amazon. In another set of experiments, Barbosa et al. (2006) found that *L. (V.) guyanensis* inhibited infections of *Endotrypanum* in a Lapinha-cave *Lu. longipalpis* colony.

What could be the reason for the contrasting results of the experimental infections of *Lu. longipalpis* with *Endotrypanum* cultures? In one case (Shaw 1981), no infections were obtained; however, in the others (Franco et al. 1997; Barbosa et al. 2006) they were. One explanation could be in the different origins of the *Lu. longipalpis* colonies. The ones that were negative originated from Marajó, Pará, whereas those that were positive came from the Lapinha Cave, Minas Gerais. The Marajó population has a burst-type mating song and cembrine pheromone, and the Lapinha cave population has a pulse-type 2 mating song and a 9-methylgermacrene-B pheromone. They are therefore different sibling species that could have different susceptibilities, which would explain the completely opposite results. What adds weight to this is that the positive infections were in two different laboratories but with Lapinha-cave *Lu. longipalpis* colonies. A problem with this hypothesis is that blood forms did infect Marajó *Lu. longipalpis*. Further evidence for parasite/vector specificity of flagellates identified as *Endotrypanum* is needed to explain these contradicting results.

Another complex question is the identity of parasites considered to be *Endotrypanum.* Arias et al. (1985) appears to have been justified in placing the parasites for sand flies of the Shannoni series in different taxa. However, the question is whether it was correct to use the name *Endotrypanum* for the group of parasites of that had very isoenzymatic profiles from *E. schaudinni*. Franco et al. (1996) identified 12 *Endotrypanum* zymodemes amongst isolates from sloths, but there was no association between their geographical origins. A more recent molecular study (Espinosa et al. 2016) confirmed high levels of genetic variability and two well-supported clades. Again, there was no clear geographical association with the clades and the geographical origin of their strains, but one was composed of strains from the States of Pará and Amazonas.

There is evidence that Rondônia is in a region that constitutes a biogeographic bridge between east and west (Patton and da Silva 1998). Thus, it is possible that two distinct *Endotrypanum* taxa could coexist in this region. It is an attractive hypothesis; however, it does not explain why *E. schaudinni* isolates of sand flies infected a consistently higher percentage of sand flies reaching 100% in the homologous *Ps. shannoni*. The taxonomic position of sand-fly infections considered to be *Endotrypanum* will only be clearer when a greater number of strains are examined and compared with those from mammals.

Concluding Remarks

This review shows that Brazilian sand flies are hosts to hundreds of microorganisms in addition to those belonging to the different *Leishmania* subgenera. These insects clearly represent a potential meeting point for many different parasites. However, more studies are needed before it can be said that they are in fact vectors of these organisms. A crucial question is this: Can one affect the development of the other? Because viruses are so common in sand flies, it is questioned whether or not they would affect the development of trypanosomatids in the vector organism.

The variety of virus found in sand flies is astounding, and there is evidence of transovarian transmission. The surprising thing is that these viruses are only occasionally found in vertebrate hosts. Pacui was isolated 100 times from pools of *Bi. flaviscutellata* captured in the Utinga forest near Belém (Aitken et al. 1975), and 30% of the wild rodents studied (including two from Amapá) and 10% of the marsupials from the same locality were serologically positive. These findings reinforce the thesis that studies carried out during periods of high transmission are more likely to elucidate the vector/reservoir cycle.

Acknowledgements The authors are grateful to Dr. Robert Tesh and Thiago Vasconcelos for their helpful comments and input.

References

Adler S, Mayrink W (1961) Sobre uma gregarina, *Monocystis chagasi* n. sp., de *Phlebotomus longipalpis*. Notas sobre as glândulas acessórias de *P longipalpis*. Rev Inst Med Trop Sao Paulo 3:230–238

Adler S, Theodor O (1927) The transmission of *Leishnma tropica* from artificially infected sandflies to man. Ann Trop Med Parasitol 21:89–110

Aitken TH, Woodall JP, De Andrade AH, Bensabath G, Shope RE (1975) Pacui virus, phlebotomine flies, and small mammals in Brazil: an epidemiological study. Am J Trop Med Hyg 24:358–368

Anderson JR, Ayala SC (1968) Trypanosome transmitted by Phlebotomus: first report from the Americas. Science 161:1023–1025

Arias JR, Miles MA, Naiff RD, Povoa MM, de Freitas RA, Biancardi CB, Castellon EG (1985) Flagellate infections of Brazilian sand flies (Diptera: Psychodidae): isolation in vitro and biochemical identification of *Endotrypanum* and *Leishmania*. Am J Trop Med Hyg 34:1098–1108

Barbosa AF, Oliveira SM, Bertho AL, Franco AM, Rangel EF (2006) Single and concomitant experimental infections by *Endotrypanum* spp. and *Leishmania (Viannia) guyanensis* (Kinetoplastida: Trypanosomatidae) in the neotropical sand fly Lutzomyia longipalpis (Diptera: Psychodidae). Mem Inst Oswaldo Cruz 101:851–856

Brazil RP, Ryan L (1984) Nota sobre a infecção de *Lutzomyia evandroi* (Diptera: Psychodidae) por *Ascocystis chagasi*(Alder & Mayrink, 1961) no Estado do Maranhão. Mem Inst Oswaldo Cruz 79:375–376

Brooks MA (1964) Symbiotes and the nutrition of medically important insects. Bull World Health Organ 31:555–559

Brown AM, Adamson ML (2006) Phylogenetic distance of *Thelohania butleri* Johnston, Vernick, and Sprague, 1978 (Microsporidia; Thelohaniidae), a parasite of the smooth pink shrimp *Pandalus jordani,* from its congeners suggests need for major revision of the genus Thelohania Henneguy, 1892. J Eukaryot Microbiol 53:445–455

Camargo EP (1999) *Phytomonas* and other Trypanosomatids parasites of plants and fruits. Adv Parasitol 42:29–112

Canning EU (1977) New concepts of microsporida and their potential in biological control. In: Fallis AM (ed) Parasites their world and ours, proceedings of the 18th symposium of the Royal Society of Canada, Toronto, pp. 101–140

Christensen HA, Herrer A (1975) *Lutzomyia vespertilionis* (Diptera: Psychodidae): potential vector of chiropteran trypanosomes in Panama. J Med Entomol 12:477–478

Christensen HA, Telford SR (1972) *Trypanosoma thecadactyli* sp. n. From forest geckoes in Panama, and its development in the sand fly *Lutzomyia trinidadensis* (Newstead) (Diptera, Psychodidae). J Protozool 19:403–406

Comer JA, Tesh RB (1991) Phlebotomine sand flies as vectors of vesiculoviruses: a review. Parassitologia 33:143–150

de Souza AA, Dos Santos TV, Jennings YL, Ishikawa EA, Barata ID, Silva MD, Lima JA, Shaw J, Lainson R, Silveira FT (2016) Natural *Leishmania* (*Viannia)* spp. infections in phlebotomine sand flies (Diptera: Psychodidae) from the Brazilian Amazon region reveal new putative transmission cycles of American cutaneous leishmaniasis. Parasite 23:22

de Souza AAA, da Rocha Barata I, das Gracas Soares Silva M, Lima JAN, Jennings YLL, Ishikawa EAY, Prevot G, Ginouves M, Silveira FT, Shaw J, Dos Santos TV (2017) Natural *Leishmania* (*Viannia*) infections of phlebotomines (Diptera: Psychodidae) indicate classical and alternative transmission cycles of American cutaneous leishmaniasis in the Guiana shield, Brazil. Parasite 24:13

Deane LM (1964) Tripanosomídeos de mamíferos de região Amazônica. III. Hemoscopia e xenodiagnóstico de animais silvestres dos arredores de Belém, Pará. Rev Inst Med Trop Sao Paulo 6:25–232

Espinosa OA, Serrano MG, Camargo EP, Teixeira MM, Shaw JJ (2016) An appraisal of the taxonomy and nomenclature of trypanosomatids presently classified as Leishmania and Endotrypanum. Parasitology, ePub, 1–13. https://doi.org/10.1017/S0031182016002092

Ferreira RC, De Souza AA, Freitas RA, Campaner M, Takata CS, Barrett TV, Shaw JJ, Teixeira MM (2008) A phylogenetic lineage of closely related trypanosomes (Trypanosomatidae, Kinetoplastida) of anurans and sand flies (Psychodidae, Diptera) sharing the same ecotopes in Brazilian Amazonia. J Eukaryot Microbiol 55:427–435

Ferreira TS, Minuzzi-Souza TT, Andrade AJ, Coelho TO, Rocha Dde A, Obara MT, Hecht M, Nitz N, Gurgel-Goncalves R (2015) Molecular detection of *Trypanosoma* sp. and *Blastocrithidia* sp. (Trypanosomatidae) in phlebotomine sand flies (Psychodidae) in the Federal District of Brazil. Rev Soc Bras Med Trop 48:776–779

Franco AM, Momen H, Naiff RD, Moreira CF, Deane MP, Grimaldi G Jr (1996) Enzyme polymorphism in *Endotrypanum* and numerical analysis of isoenzyme data. Parasitology 113:39–48

Franco AM, Tesh RB, Guzman H, Deane MP, Grimaldi Junior G (1997) Development of *Endotrypanum* (Kinetoplastida: Trypanosomatidae) in experimentally infected phlebotomine sand flies (Diptera: Psychodidae). J Med Entomol 34:189–192

Freitas RA, Naiff RD, Barrett TV (2002) Species diversity and flagellate infections in the sand fly fauna near Porto Grande, State of Amapa, Brazil (Diptera: Psychodidae. Kinetoplastida: Trypanosomatidae). Mem Inst Oswaldo Cruz 97:53–59

Galati EA, Nunes VL, Boggiani PC, Dorval ME, Cristaldo G, Rocha HC, Oshiro ET, Damasceno-Junior GA (2006) Phlebotomines (Diptera: Psychodidae) in forested areas of the Serra da Bodoquena, state of Mato Grosso do Sul, Brazil. Mem Inst Oswaldo Cruz 101:175–193

Gebhardt LP, Stanton GJ, Hill DW, Collett GC (1964) Natural overwintering hosts of the virus of western equine encephalitis. N Engl J Med 271:172–177

Gil LH, Basano SA, Souza AA, Silva MG, Barata I, Ishikawa EA, Camargo LM, Shaw JJ (2003) Recent observations on the sand fly (Diptera: Psychodidae) fauna of the State of Rondonia, Western Amazonia, Brazil: the importance of Psychdopygus davisi as a vector of zoonotic cutaneous leishmaniasis. Mem Inst Oswaldo Cruz 98:751–755

Gorman BM (1979) Variation in orbiviruses. J Gen Virol 44:1–15

Gouveia C, Asensi MD, Zahner V, Rangel EF, Oliveira SM (2008) Study on the bacterial midgut microbiota associated to different Brazilian populations of *Lutzomyia longipalpis* (Lutz & Neiva) (Diptera: Psychodidae). Neotrop Entomol 37:597–601

Guerreiro SG, Travassos da Rosa APA, Vasconcelos PFC, Travassos da Rosa ES, Tesh RB, Travassos da Rosa JFS (1998) Characterization of two new phleboviruses associated with human illness from the Amazon region of Brazil. In: Travassos da Rosa APA, Vasconcelos PFC, Travassos da Rosa JFS (eds) An overview of arbovirology in Brazil and neighbouring countries. Evandro Chagas Institute, Belém, pp 100–105

Hassan OA, Affognon H, Rocklov J, Mburu P, Sang R, Ahlm C, Evander M (2017) The one health approach to identify knowledge, attitudes and practices that affect community involvement in the control of Rift Valley fever outbreaks. PLoS Negl Trop Dis 11:e0005383

Herrer A (1942) *Trypanosoma phyllotis* n.sp. e infecciones asociadas en uma titira, el *Phlebotomus noguchi*. Rev Peru Med Exp Salud Publica 1:40

Hoare CA (1972) The trypanosomes of mammals. A zoological monograph. Blackwell Scientific Publications, Oxford, UK, p 749

Hoch A, Ryan L, Vexenat JA, Rosa ACOC, Barretto AC (1986) Isolation of *Leishmania braziliensis* and other trypanosomatids from Phlebotomine in a mucocutaneous endemic area, Bahia, Brazil. Mem Inst Oswaldo Cruz 81:62

ICTV (2016) International Committee on Taxonomy of Viruses. https://talk.ictvonline.org/taxonomy/w/ictv-taxonomy

Karabatsos N (1985) International catalogue of arboviruses including certain other viruses of vertebrates. In, 3rd edn. The American Society of Tropical Medicine and Hygiene. US Department of Health, Education and Welfare, San Antonio, p 1147

Keller S, Petrini O (2005) Keys to the identification of the arthropod-pathogenic genera of the families Entomophthoraceae and Neozygitaceae (Zygomycetes), with description of three new subfamilies and a new genus. Sydowia 57:25–53

Kelly PH, Bahr SM, Serafim TD, Ajami NJ, Petrosino JF, Meneses C, Kirby JR, Valenzuela JG, Kamhawi S, Wilson ME (2017) The gut microbiome of the vector *Lutzomyia longipalpis* is essential for survival of *Leishmania infantum*. MBio 8:e01121–e01116

Lainson R, Shaw JJ (1979) The role of animals in the epidemiology of south American Leishmaniasis. In: Lumsden WHR, Evans DA (eds) Biology of the Kinetoplastida. Academic Press, London/New York/San Francisco, pp 1–116

Lainson R, Shaw JJ, Ward RD, Fraiha H (1973) Leishmaniasis in Brazil. IX. Considerations on the *Leishmania braziliensis* complex. importance of sandflies of the genus *Psychodopygus* (Mangabeira) in the transmission of *L. braziliensis braziliensis* in north Brazil. Trans R Soc Trop Med Hyg 67:184–196

Lainson R, Killick-Kendrick R, Canning EU, Shaw JJ, Ward RD, Leaney AJ, Nicholas JP (1977) Microsporidia of Brazilian sandflies. Trans Roy Soc Trop Med Hyg 71:381

Lainson R, Shaw JJ, Povoa M (1981a) The importance of edentates (sloths and anteaters) as primary reservoirs of *Leishmania braziliensis guyanensis,* causative agent of "pianbois" in north Brazil. Trans R Soc Trop Med Hyg 75:611–612

Lainson R, Shaw JJ, Ready PD, Miles MA, Povoa M (1981b) Leishmaniasis in Brazil: XVI. Isolation and identification of *Leishmania* species from sandflies, wild mammals and man in north Para state, with particular reference to *L. braziliensis guyanensis* causative agent of "pian-bois". Trans R Soc Trop Med Hyg 75:530–536

Lewis DJ, Lainson R, Shaw JJ (1970) Determination of parous rates in Phlebotomine sandflies with special reference to Amazonian species. Bull Entomol Res 60:209–219

Matos E, Mendonca I, Azevedo C (2006) *Vavraia lutzomyiae* n. sp. (Phylum Microspora) infecting the sandfly *Lutzomyia longipalpis* (Psychodidae, Phlebotominae), a vector of human visceral leishmaniasis. Eur J Protistol 42:21–28

McCarthy CB, Diambra LA, Rivera Pomar RV (2011) Metagenomic analysis of taxa associated with *Lutzomyia longipalpis*, vector of visceral leishmaniasis, using an unbiased high-throughput approach. PLoS Negl Trop Dis 5:e1304

Naiff RD, Barret TV, Freitas RA (1989) Isolation of *Trypanosoma freitasi* (Kinetoplasitda: Trypanosomatidae) from *Psychodopygus claustrei* (Diptera: Psychodidae). Mem Inst Oswaldo Cruz 84:273–275

Nunes-Neto JP, Souza WM, Acrani GO, Romeiro MF, Fumagalli M, Vieira LC, Medeiros DBA, Lima JA, Lima CPS, Cardoso JF, Figueiredo LTM, Silva SPD, Tesh R, Nunes MRT, Vasconcelos P (2017) Characterization of the Bujaru, frijoles and Tapara antigenic complexes into the sandfly fever group and two unclassified phleboviruses from Brazil. J Gen Virol 98:585–594

Oliveira SMP, Moraes CA, Gonçalves CM, Giordano-Dias JM, D'almeida MD, Asensi RP, Mello RP, Brazil RP (2000) Prevalence of microbiota in the digestive tract of wild females of *Lutzomyia longipalpis* Lutz & Neiva, 1912 (Diptera: Psychodidae). Rev Soc Bras Med Trop 33:319–322

Pacheco RS, Grimaldi G Jr, Morel CM (1987) Inhibition of growth of *Leishmania mexicana* by *Leishmania mexicana amazonensis* during "in vitro" co-cultivation. Mem Inst Oswaldo Cruz 82:537–542

Patton JL, da Silva MNF (1998) Rivers, refuges and ridges: The geography of speciation in Amazonian mammals. In: Berlocher S, Howard DJ, editors. Endless Forms: Species and Speciation. New York: Oxford University Press; p. 202–213.

Pereira de Oliveira SM, de Morais BA, Goncalves CA, Giordano-Dias CM, Vilela ML, Brazil RP, D'Almeida JM, Asensi MD, Mello RP (2001) Digestive tract microbiota in female *Lutzomyia longipalpis* (Lutz & Neiva, 1912) (Diptera: Psychodidae) from colonies feeding on blood meal and sucrose plus blood meal. Cad Saude Publica 17:229–232

Pessoa FA, Medeiros JF, Barrett TV (2007) Effects of timber harvest on phlebotomine sand flies (Diptera: Psychodidae) in a production forest: abundance of species on tree trunks and prevalence of trypanosomatids. Mem Inst Oswaldo Cruz 102:593–599

Peter CJ, Morgan JM (1981) Rift Valley fever. In: Steele JH (ed) Viral zoonosis. CRC Press, Boca Raton, pp 403–420

Rogers WO, Burnheim PF, Wirth DF (1988) Detection of *Leishmania* within sand flies by kinetoplast DNA hybridization. Am J Trop Med Hyg 39:434–439

Ryan L, Lainson R, Shaw JJ (1987a) Leishmaniasis in Brazil. XXIV. Natural flagellate infections of sandflies (Diptera: Psychodidae) in Para state, with particular reference to the role of

Psychodopygus wellcomei as the vector of *Leishmania braziliensis* in the Serra dos Carajas. Trans R Soc Trop Med Hyg 81:353–359

Ryan L, Lainson R, Shaw JJ, Braga RR, Ishikawa EA (1987b) Leishmaniasis in Brazil. XXV. Sandfly vectors of *Leishmania* in Pará State, Brazil. Med Vet Entomol 1:383–395

Sant'Anna MR, Darby AC, Brazil RP, Montoya-Lerma J, Dillon VM, Bates PA, Dillon RJ (2012) Investigation of the bacterial communities associated with females of Lutzomyia sand fly species from South America. PLoS One 7:e42531

Sant'Anna MR, Diaz-Albiter H, Aguiar-Martins K, Al Salem WS, Cavalcante RR, Dillon VM, Bates PA, Genta FA, Dillon RJ (2014) Colonisation resistance in the sand fly gut: *Leishmania* protects *Lutzomyia longipalpis* from bacterial infection. Parasit Vectors 7:329

Scorza JV, Carnevali M (1981) Evidence for allocating Monocystis chagasi Adler & Mayrnk 1961 to the genus Ascocystis Grassé, 1953 (Gregarunia: Lecudinidae). In: Canning EU (ed) Parasitological topics a presentation volume to PCC Garnham FRS on the occasion of his 80th birthday. Society of Protozoologists Special Publication, Allen Press Inc, Lawrence, pp 229–231

Shaw JJ (1964) A possible vector of *Endotrypanum schaudinni* of the Sloth, *Choloepus hoffmanni*, in Panama. Nature 201:417–418

Shaw JJ (1981) The behaviour of Endotrypanum schaudinni (Kinetoplastidae:Trypanosomatidae) in three species of laboratory-bred neotropical sandflies (Diptera:Psychodidae) and its influence on the classification of the genus Leishmania. In: Canning EU (ed) Parasitological topics a presentation volume to PCC Garnham FRS on the occasion of his 80th birthday. Society of Protozoologists Special Publication, Allen Press Inc, Lawrence, pp 232–241

Shaw JJ (1992) *Endotrypanum*, a unique intraerythrocytic flagellate of new world tree sloths. An evolutionary link or an evolutionary backwater? Ciência e Cultura 44:107–116

Shaw JJ, Lainson R (1972) Leishmaniasis in Brazil. VI. Observations on the seasonal variations of *Lutzomyia flaviscutellata* in different types of forest and its relationship to enzootic rodent leishmaniasis (*Leishmania mexicana amazonensis*). Trans R Soc Trop Med Hyg 66:709–717

Shaw JJ, Lainson R, Ryan L, Braga RR, McMahon-Pratt D, David JR (1987) Leishmaniasis in Brazil: XXIII. The identification of Leishmania braziliensis in wild-caught neotropical sandflies using monoclonal antibodies. Trans R Soc Trop Med Hyg 81:69–72

Sherlock I, Pessôa SB (1966) *Leptomonas* infectando naturalmente *Phlebotomus* em Salvador (Bahia, Brasil). Rev Lat Am Microbiol 8:47–50

Shortt HE, Swaminath CS (1927) *Monocystis mackiei* n. sp. parasitic in *Phlebotomus argentipes*. Indian J Med Res 15:539–552

Silva SP, Dilcher M, Weber F, Hufert FT, Weidmann M, Cardoso JF, Carvalho VL, Chiang JO, Martins LC, Lima CP, Da Silva DE, Vianez-Junior JL, Popov VL, Travassos da Rosa AP, Tesh RB, Vasconcelos PF, Nunes MR (2014) Genetic and biological characterization of selected Changuinola viruses (Reoviridae, Orbivirus) from Brazil. J Gen Virol 95:2251–2259

Silveira FT, Souza AA, Lainson R, Shaw JJ, Braga RR, Ishikawa EE (1991) Cutaneous leishmaniasis in the Amazon region: natural infection of the sandfly *Lutzomyia ubiquitalis* (Psychodidae: Phlebotominae) by *Leishmania* (*Viannia*) *lainsoni* in Pará State, Brazil. Mem Inst Oswaldo Cruz 86:127–130

Souza AAA, Camargo LMA, Faria DL, Shaw JJ (1998) Resultados preliminares: Identificação de flebotomíneos em Montenegro – Rondônia (Amazônia Ocidental Brasileira). Rev Soc Bras Med Trop 31:74–75

Sprague V, Becnel JJ, Hazard EI (1992) Taxonomy of phylum microspora. Crit Rev Microbiol 18:285–395

Teixeira MMG, Borghesan TC, Ferreira RC, Santos MA, Takata CSA, Campaner M, Nunes VLB, Milder RV, de Souza W, Camargo EP (2011) Phylogenetic validation of the genera *Angomonas* and *Strigomonas* of Trypanosomatids harboring bacterial Endosymbionts with the description of new species of Trypanosomatids and of Proteobacterial Symbionts. Protist 162:503–504

Tesh RB (1988) The genus Phlebovirus and its vectors. Annu Rev Entomol 33:169–181

Tesh RB, Chaniotis BN, Aronson MD, Johnson KM (1971) Natural host preferences of Panamanian sandflies as determined by precipitin test. Am J Trop Med Hyg 20:150–156

Travassos da Rosa AP, Tesh RB, Pinheiro FP, Travassos da Rosa JF, Peterson NE (1983) Characterization of eight new phlebotomus fever serogroup arboviruses (Bunyaviridae: Phlebovirus) from the Amazon region of Brazil. Am J Trop Med Hyg 32:1164–1171

Travassos da Rosa AP, Tesh RB, Pinheiro FP, Travassos da Rosa JF, Peralta PH, Knudson DL (1984) Characterization of the Changuinola serogroup viruses (Reoviridae: Orbivirus). Intervirology 21:38–49

Travassos da Rosa AP, Tesh RB, Travassos da Rosa JF, Herve JP, Main AJ Jr (1984a) Carajas and Maraba viruses, two new vesiculoviruses isolated from phlebotomine sand flies in Brazil. Am J Trop Med Hyg 33:999–1006

Travassos da Rosa JFS, Travassos da Rosa APA, Vasconcelos PFC, Pinheiro FP, Rodrigues SG, Travassos da Rosa ES, Dias LB, Cruz ACR (1998) Arboviruses isolated in the Evandro Chagas institute, including some described for the first time in the Brazilian Amazon region, their known hosts, and their pathology in man. In: Travassos da Rosa APA, Vasconcelos PFC, Travassos da Rosa JFS (eds) An overview of Arbovirology in Brazil and neighbouring countries. Evandro Chagas Institute, Belém, pp 19–31

Vasconcelos PF, Travassos da Rosa AP, Rodrigues SG, Travassos da Rosa ES, Degallier N, Travassos da Rosa JF (2001) Inadequate management of natural ecosystem in the Brazilian Amazon region results in the emergence and reemergence of arboviruses. Cad Saude Publica 17(Suppl):155–164

Viola LB, Campaner M, Takata CS, Ferreira RC, Rodrigues AC, Freitas RA, Duarte MR, Grego KF, Barrett TV, Camargo EP, Teixeira MM (2008) Phylogeny of snake trypanosomes inferred by SSU rDNA sequences, their possible transmission by phlebotomines, and taxonomic appraisal by molecular, cross-infection and morphological analysis. Parasitology 135:595–605

Votypka J, Lantova L, Ghosh K, Braig H, Volf P (2009) Molecular characterization of gregarines from sand flies (Diptera: Psychodidae) and description of *Psychodiella* n. g. (Apicomplexa: Gregarinida). J Eukaryot Microbiol 56:583–588

Wallace FG, Hertig M (1968) Ultrastructural comparison of promastigote flagellates (Leptomonads) of wild-caught Panamanian Phlebotomus. J Parasitol 54:606–612

Warburg A (1991) Entomopathogens of phlebotomine sand flies: laboratory experiments and natural infections. J Invertebr Pathol 58:189–202

Ward RD, Killick-Kendrick R (1974) Field and laboratory observations on *Psychodopygus lainsoni* Fraiha & Ward and other sandflies (Diptera: Phlebotomidae) from the Transamazônica highway, Pará State, Brazil. Bull Entomol Res 64:213–221

WHO (1967) Arboviruses and human disease. Report of a WHO Scientific Group. World Health Organ Tech Rep Ser 369:1–84

Williams P, Coelho MV (1978) Taxonomy and transmission of *Leishmania*. Adv Parasitol 16:1–42

Wilson AJ, Morgan ER, Booth M, Norman R, Perkins SE, Hauffe HC, Mideo N, Antonovics J, McCallum H, Fenton A (2017) What is a vector? Philos Trans R Soc Lond Ser B Biol Sci 372:20160087

Wu WK, Tesh RB (1989) Experimental infection of Old and New World phlebotomine sand flies (Diptera: Psychodidae) with *Ascogregarina chagasi* (Eugregarinorida: Lecudinidae). J Med Entomol 6:237–242.

Yuval B, Warburg A (1989) Susceptibility of adult phlebotomine sandflies (Diptera: Psychodidae) to *Bacillus thuringiensis* var. israeliensis. Ann Trop Med Parasitol 83:195–196

Zeledon R, Rosabal R (1969) *Trypanosoma leonidasdeanei* sp. n. in insectivorous bats of Costa Rica. Ann Trop Med Parasitol 63:221–223

Methods for Capturing, Processing and Preserving Phlebotominae

Mauricio Luiz Vilela, Adriana Zwetsch, and Júlia dos S. Silva

Entomological Study of Phlebotominae

The relevance of gathering knowledge associated with phlebotomine fauna is related to these insects' role as the main transmitting agent for several pathogens. Among these pathogens, some of the most important are *Leishmania* spp., *Bartonella bacilliformis* and different arboviroses (Maroli et al. 2012), which affect human populations and that of other vertebrates; they also have a significant effect on public health.

Field work with phlebotomine sand flies contributes to generating data on the bioecology of these insects other than making viable the gathering of information, the identification of possible vectors, the performance of taxonomic studies, and the description of new species. Also important is the development of research on phlebotomine sand-fly bionomy, aiming to identify possible behavioral changes in this vector's species and promote the implementation of more effective measures for control programs (WHO 2010; Brasil 2013, 2014).

The way in which field activities are carried out depends entirely on the objectives envisioned for the research along with knowledge about the local fauna and the area to be explored. Sampling strategies preferably focus on winged adults captured with Castro manual aspirators, electric aspirators, light traps or interception traps (made with cloth). Other techniques—such as animal baits, carbon dioxide, pheromones or sticky traps—may be used. The use of emergency traps is limited to environments identified as potential breeding sites (Alexander 2000; Vilela et al. 2003; Alten et al. 2015).

M. L. Vilela (✉) · A. Zwetsch · J. dos S. Silva
Instituto Oswaldo Cruz, Fundação Oswaldo Cruz, Rio de Janeiro, Brazil

E. F. Rangel, J. J. Shaw (eds.), *Brazilian Sand Flies*,
https://doi.org/10.1007/978-3-319-75544-1_10

Regarding their biology, female phlebotomine sand flies lay their eggs on soil that is rich in decomposing organic matter in terrestrial microenvironments with low lighting and high humidity. Natural breeding sites are hard to locate, and they remain unknown for most species. Finding immature forms in their natural environment is an laborious task because the size, color, means of locomotion and mimicry abilities hinder their identification. Among the possible breeding sites for phlebotomine sand flies in their natural environment include tree trunks and roots, animal holes, spaces on or under rocks, forest soil and the surface of fallen leaves and domestic surroundings, such as pigsties, hen houses, kennels and barns (Forattini 1973). Despite contributions made in the identification of breeding sites, information on this matter remains lacking (Feliciangeli 2004; Alencar et al. 2011; Moncaz et al. 2012, 2014; Sangiorgi et al. 2012; Souza et al. 2014; Vivero et al. 2015).

Due to the biology of phlebotomine sand flies, control of their immature forms has generally not been considered a priority (Brasil 2013, 2014). However, some methods are still available to obtain the larvae and pupae of this insect. One of the most frequent is the daily and long-term observation of substrate samples collected in the field in enameled containers, covered with a glass funnel, and examined until the possible hatching of adults. Another possible technique is the sifting and analysis of substrate samples through flotation in sugar solution to research larvae and pupae. The advantage of this method is that it preserves the larvae alive, thus making it possible to evaluate their development until adulthood (Vilela et al. 2003). Yet another option is the Berlese-Tullgren funnel method (Lincoln and Sheals 1979), which is frequently used to obtain fauna samples of small arthropods. It was proposed by Berlese in 1905 and modified by Tullgren in 1917 (Garay 1989; Rodrigues 2008). This method uses a lamp on top of a funnel, into which samples are placed. Containers with alcohol are placed on the base of the equipment. The larvae then migrate, by thermic and luminous stimulation, to the bottom of the equipment and subsequently fall inside the containers.

Adult emergency traps are another strategy used for potential biotypes. In general, they are used where sand flies can be captured while they are static. The traps are inspected at regular intervals, and sand flies are captured when they emerge. Depending on the type of microhabitat in question, different kinds of emergency traps can be employed (Casanova 2001; Moncaz et al. 2012).

A study performed in Panama used the techniques of sifting and sugar-water flotation for samples found on tree bases, and the technique was able to identify the presence of larvae from 15 different species from a variety of natural habitats (Hanson 1961). In another example, the use of emergency traps for adults in the Valley of the Mogi-Guaçu river, in São Paulo, made it possible to identify potential breeding sites for sand flies and resulted in the capture of 73 individuals from 3 different species (Casanova 2001). Research carried out in natural breeding sites in different microhabitats of the "terra firme" forest in the state of Amazonas (Alencar et al. 2011) highlighted that the sifting and sugar solution–floatation method, combined with the direct examination of samples, presented the best results. The same technique was used in the state of Rio de Janeiro, where 42 individuals from 6 different species were captured (Souza et al. 2014). In Colombia, the direct examination of soil samples, followed by the incubation of substrates and the use of emergency traps, allowed the identification of immature forms in 160 soil samples. The largest number of them was found in tabular roots and on tree bases (Vivero et al. 2015).

Regarding their habits, phlebotomine are insects usually observable from afternoon to morning dusk. Due to their thin cuticle and sensitivity to macroclimate variations, they live sheltered in environments with low lighting and high humidity where the microclimate protects them against desiccation. Natural shelters might be dissociated from natural breeding sites by the hematophagic activity of females. However, in some cases, shelters can also be sites for the development of immature forms. Different environments— such as animal holes and nests, tree trunks or hollow trees, cracks and spaces between or on top of rocks, the surface of fallen leaves, termite mounds or holes dug in the soil—may be considered potential natural shelters for phlebotominae (Forattini 1973; Aguiar et al. 1985a). Despite that, it is important to consider that some species are adapted to home environments, which leads to these diptera being found both inside and on the surrounding area of households and in domestic animal shelters, such as hen houses, pigsties and cannels. In these cases, the presence of phlebotominae depends on the distance of these households from the forest area.

To capture phlebotomine sand flies in their natural environment, a range of methods—from the Castro manual aspirator to the use of techniques that depend on attractive features—can be implemented. These can further be combined with other methods, including interception (cloth) traps, light traps or traps that use animal baits.

Capture Methods

Manual Aspirators

In the field, phlebotominae can be collected using manual aspirators. This is the case when they land in surfaces, such as cloth traps, in-house walls, areas surrounding domiciles and animal shelters. In other situations, aspirators can also be used in insectaries specialized in breeding phlebotominae.

The Castro manual aspirator (Castro 1937) is made with a rubber tube connected to a glass, acrylic or plastic tube with fine cloth, voile or thin nylon mesh cloth placed between the two components used for sealing (Alexander 2000) (Fig. 1a). Plebotominae are aspirated and blown into the interior of a cage (collecting cage or transportation cage), which is also made with fine cloth, voile or thin nylon mesh and sustained by a stainless steel or iron cube-shaped frame (Barraud 1929). The use of an acrylic or plastic tube for this technique is advisable as a measure of biosecurity.

One significant change has been made to the original Castro manual aspirator: A small motorcycle fuel filter was adapted into the acrylic tube and attached to the rubber tube. This filter must be changed periodically. This adaptation avoids the inhalation of insect bristles, small infected soil particles, dust and residues of animal

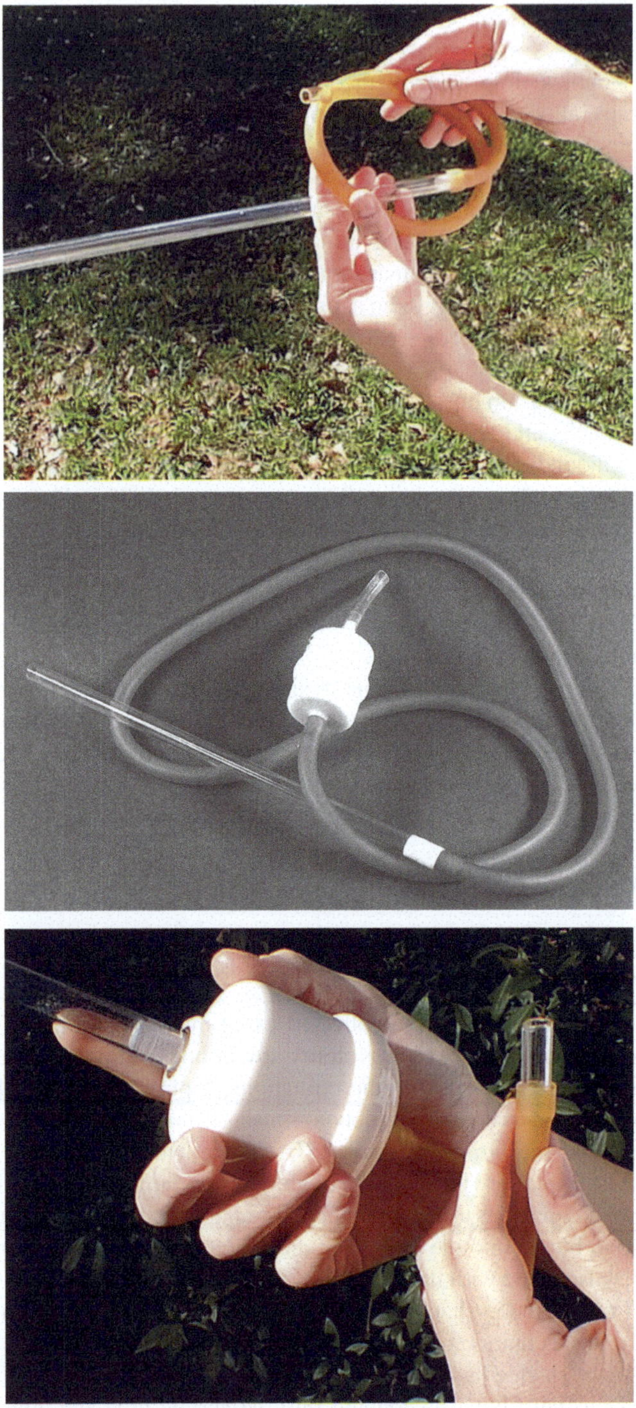

Fig. 1 (**a**) Standard Mouth Aspirator (Model 412). (Photo credit: http://johnwhock.com/products/aspirators/mouth-aspirators/). (**b**) Castro manual aspirator. (Photo credit: Maurício Vilela). (**c**) Mouth Aspirator with HEPA Filter (Model 612). (Photo credit: http://johnwhock.com/products/aspirators/mouth-aspirators/)

feces, which could happen due to the diversity of environments from which phlebotominae are collected. This change minimizes potential risks to human health and was initially idealized in the context of an insectary specialized in phlebotomine breeding (Fig. 1b).

This method can be presented in other forms, such as the simple manual aspirator tube (model 412) and the manual aspirator with filter (model 612). The latter contains a 0.3-μm HEPA (high-efficiency particulate arrestance) filter attached to a 12-inch polycarbonate tube, thus offering protection to the user (Fig. 1c) (http://johnwhock.com/products/aspirators/mouth-aspirators/).

Electrical Aspirators

An additional method used to capture small insects is mechanic suction equipment (Trips 1968). This equipment comes in some different forms, produced by a range of manufacturers, and it is basically composed of an acrylic tube and an electric motor. In this case, the aspirated phlebotominae are kept inside a small tube or collecting chamber, and in case other insects also get sucked in, a laboratory screening is performed.

One example of this method is the vacuum manual aspirator (BioQuip®), which is powered by an 18V battery, built with an acrylic transparent tube, and containing a 12.7-cm long, 5.1-cm wide caliber removable collecting chamber. It also contains an aluminum net on one end for air flow and a valve with a hinge on the other end. A supplementary lid on one of the ends avoids the escape of samples when the chamber is removed. The suction can be adjusted using a cloth piece on the bottom of the chamber to decrease damage to the captured insects (http://www.bioquip.com/search/DispProduct.asp?pid=2820GA) (Fig. 2).

On the field, Castro or electric manual aspirators can be used for in-house capture (on the internal walls of houses) and capture in the surrounding areas of households (external walls of houses, internal or external walls of animal shelters. Capture

Fig. 2 Heavy Duty Hand-Held DC Vac/Aspirator. (Photo credit: http://www.bioquip.com/html/view_prodpics.asp?CatalogNum=2820GA&P=3)

by way of Shannon traps can be performed for wooded areas (tree-trunk capture can employ both Shannon and Damascene traps). These traps can also be placed on platforms built near large trees, in which case data on the acrodendrophilic inclination of the species can be obtained (Aguiar et al. 1985b).

Light Traps

A diversity of capture methods using traps are employed in entomologic surveillance and vector-control actions. Lately, some models of automatic traps based on attraction by a light source and suction by small fans have been developed to assist in capturing small insects in field studies (Lumsden 1958; Minter 1961; Chaniots and Anderson 1968; Falcão 1981; Pugedo et al. 2005).

Center for Disease Control Light Traps

The Center for Disease Control (CDC) light trap (Sudia and Chamberlain 1962) is used both in scientific research activities and in monitoring activities for control programs. Considering its original description, this standard tool is one of the most commonly used in capturing small diptera. The advantages of this trap are its light weight, easy portability, easy transportability to the field, and it's foldable collecting chamber. Furthermore, this technique decreases the exposure time of professionals to insects in the environment.

In general, this method is a first choice for studies of phlebotomine and mosquito vectors. The trap is composed of an acrylic tube with an opening on the upper part, meant to fit a support for the aluminum protection lid. This feature protects the trap from wind and rain and helps with the process of insect suction. The motor triggers a small propeller and the light source, and it is powered by a 6V battery or four 1.5V D-type long-duration batteries. On the bottom end of the trap's body, the acrylic tube is surrounded by a cage of two attached segments: a fine cloth, voile or thin nylon mesh tube and a collecting chamber (a plastic container). This chamber's function is to store the captured insects, which were attracted by light and sucked into its interior, therefore allowing their transportation to the laboratory (Fig. 3a).

A modification on the CDC trap has replaced the collecting chamber of the original model with a new one made with a metal screen and a diameter equal to that of the trap's body. A polyvinyl chloride (PVC) tube is placed around the collecting chamber and the fine cloth, voile or thin nylon mesh sleeve. They are then attached to the lid's support, aiming to protect the trap from the rain (Gomes et al. 1985).

Fig. 3 (**a**) CDC Miniature Light Trap. (Photo credit: Maurício Vilela). (**b**) CDC Miniature Light Trap – HP model. (Photo credit: Maurício Vilela)

*CDC Light Traps: HP (*Hoover Pugedo*) Model*

This trap is a prototype created by modifying the original CDC model. It has a cylindrical body made with white thermoplastic material, PVC or high-resistance resin. The small motor that triggers the propeller and the light source is powered by either four D-type 1.5 V long-duration batteries or one 6 V battery. A portable cage is attached to the body of the trap (foldable cage or collecting foldable cage). This cage is made with fine cloth or voile cloth and sustained by parallel circular hoops, forming a retractable structure (Pugedo et al. 2005). Another cage (collecting cage or transportation cage)—made with fine cloth, voile or thin nylon mesh and sustained by a cubic stainless steel or iron structure—may be used for this trap (Barraud 1929) with the function of storing the phlebotominae attracted by the light source. The choice of the type of cage employed in this method is defined according to the objectives of the activity (Fig. 3b).

Although attraction traps with phototropic stimulation are one of the preferred methods for capturing these dipteran, in the case of anthropophilic species other strategies will be better suited for capture, such as ones with olfactory stimulation. These might be used when there are ethical restrictions to the participation of humans in the research or when the use of light traps shows limitations.

One example of such olfactory stimulation is the use of carbon dioxide, which acts like a long-reach attractive and can be obtained through a compressed gas recipient or a bottle of dry ice suspended over the CDC trap (Kasili et al. 2009; Kline et al. 2011). Another option is using octanol ($C_8H_{16}O$), an alcohol produced by plants and fungi that attracts certain insects through properties like those of pheromones. Octanol is present in human sweat and breath. Finally, a third possibility is to apply human synthetic BG-Mesh Lure® Kayromones (BGML = lactic acid, caproic acid and ammonia), a combination of non-toxic substances found in human skin, such as lactic acid, ammonia and fatty acids. These substances are found in the BG-Sentinela® trap, which is frequently used for monitoring culicidae.

Experimental studies carried out with *Nyssomyia intermedia* and *Lutzomyia longipalpis* showed that octanol was more effective in attracting the former, whereas BGML was more successful with the latter. These tests reinforce the possibility of using these substances as attractive features to capture certain species of phlebotominae (Andrade et al. 2008).

Recently, a trap that does not require the use of batteries, with a low cost and efficiency comparable to the CDC (Leishtrap = HP trap with colored neon bracelets), was developed and used to evaluate *Lu. longipalpis*, and *Ny. whitmani*. The results were promising (Andrade 2010).

The use of different lamps—incandescent, UV and LED with different wavelengths and colors—might also be an alternative for the CDC light trap. With the intent of raising efficiency in terms of sampling and prolonging the battery life, these strategies was evaluated (Bishop et al. 2004; Jenkins and Young 2010). The results show that cold-cathode lamps present some advantages, including their low cost, high durability and rare need for replacement. Some studies, in which LED lamps were used to attract vectors, revealed the preference of some species for specific light wavelengths (Bishop et al. 2004; Hoel et al. 2007). Traps with blue and green LED lights were compared with incandescent lights, but there were no significant differences between the two light wavelengths observed. Still, the trap with a green LED did attract more phlebotominae, indicating that this strategy seems to be more efficient (Silva et al. 2015). In a study developed in Egypt, *Phlebotomus papatasi* showed a strong attraction to red light (Hoel et al. 2007). The data known so far only reinforces the need for more investigation aiming to identify the more adequate type of lamps to attract phlebotominae.

The choice of cage types varies according to the ends of the procedure being undertaken. For phlebotominae meant for taxonomy, it is common to use cages with two attached segments: a fine cloth, voile or thin nylon mesh tube and a collecting

Fig. 4 Collecting cage or a transporting cage for phlebotomine sand flies. (Photo credit: Maurício Vilela)

chamber (plastic container). Alternatively, portable cages (foldable cage or collecting foldable cage) can be used. These are also made with fine cloth or voile and sustained by parallel circular hoops, forming a retractable structure (Pugedo et al. 2005). To store the insects in an insectary, a collecting cage or a transporting cage are used. These are made in fine cloth, voile or thin nylon mesh and sustained by a stainless steel or iron structure (Barraud 1929) (Fig. 4).

Light traps are used in houses to assess the frequency and degree of adaptation of certain species to the environment of human residences—both to the inside of houses and areas close to domestic animal shelters. In areas near woods or forests, the presence of wild species is observable. In these areas, it is possible to identify species with acrodendrophilic habits by putting light traps in tree tops (Aguiar et al. 1985b). Areas destined for monoculture are also likely environments to find phlebotominae.

Cleaning and maintaining the light traps after capture activities is crucial to keep the equipment in proper conditions and to avoid cross-contamination. Acrylic or PVC parts should be cleaned using a moist towel with the neutral detergent solution used in laboratories (with 1% concentration). Attention should be paid not to damage the motor, the lamp support, the batteries or the battery terminal (Souza personal communication, 1998). It is also key to certify that the battery contacts are not oxidized; in case they are, it is recommended to use an iron sponge for maintenance. The battery contacts and the trap's motor should not get wet. Traps should be kept in a dry, sheltered place, disassembled and have usage instructions that are visible to future users. Any specific maintenance, such as changing lamps or pieces, should be required and done before field excursions.

Another indispensable activity is cleaning light-trap cages that have been used in fieldwork. This should be done immediately after the end of capture activities to avoid cross-contamination in further uses, which could lead to the presence of phlebotominae of one geographic area being attributed to another. The washing of cages is done by soaking them in neutral 1% or 5% detergent solution overnight and rinsing it in running water. The final step is drying. According to technical recommendations, cages used in insectaries can also be soaked overnight with a 1% or 10% sodium hypochlorite solution (Souza personal communication, 1997).

New Jersey Traps

The New Jersey light trap (Mulhern 1934) uses a fan for phlebotomine suction, leading them to a funnel with a piece of mesh adapted to the size of the insect, and further leading them to a killing bottle. In case the specimens need to be kept alive, the flask is replaced by a cage. The initial function idealized for this trap was the capture of mosquitoes. Nonetheless, this has been one of the most frequently used pieces of equipment for phlebotomine capture. This trap can be placed at ground level or over it and, for it to be more efficient, it can be set next to phlebotomine shelters or breeding sites (Alten et al. 2015).

Interception Traps

Shannon Trap

The Shannon trap is made with a rectangular cotton cloth structure, with two identical lateral surfaces, measuring approximately 1.6 × 2.5 × 1.6 m. Its assembling for the field can be done in the surroundings of the domestic environment or in the woods by using four pieces of rope tied to the top end of the trap. The ropes should then be tied to trees or wooden pieces so that the trap can be completely extended (Fig. 5) (Shannon 1939). Inside the trap, a light source or an animal is placed as bait for phlebotominae. For the animal model, the most commonly used animal is the *Mesocricetus auratus* hamster.

Phlebotominae landed on internal and external walls of the trap are captured with a Castro manual aspirator and/or an electric aspirator. Later, they are put in cages or in polypropylene tubes. The choice of the type of deposit for phlebotominae (cage or tube) will vary according to the aim of the capture.

The Shannon trap can help gather data on the species of phlebotominae attracted by light source, animal baits, or even anthropophilic species that might be drawn by

Fig. 5 Shannon trap. (Photo credit: Margarete Afonso)

the human individual collecting the samples. An additional interesting aspect of this method is the possibility of establishing a correlation between the captured species and their frequency schedules (Aguiar et al. 1985a, 1985b, 1985c).

Damasceno Trap

Among the strategies deployed for capturing phlebotominae in their natural environment, this has been the most frequent one. The Damasceno trap is built with a cotton cloth structure, open on one of its sides and suspended over a place identified as a possible natural shelter of phlebotominae. One professional must stand in the internal part of the trap with a Castro aspirator and/or an electric aspirator and a lantern to light the place to be investigated. The professional should be careful to surround an area as large as possible with the trap. Outside of the trap, another member of the team assists by keeping the cloth structure extended. In this setting, the professional standing inside the trap will slowly unblock the entrance of the biotype (e.g. with the use of a stick). This will allow the phlebotomine sand flies to leave the inside of the trap and land on the internal wall of the trap where they will be captured (Aguiar and Vilela 1987). The lantern is not only good for lighting; it also serves as a disturbance to insects, provoking them to will move up and down the walls of the trap and thus be captured more easily (Fig. 6).

One other type of trap used in the same situations in made with polyethylene funnels of different sizes, which are attached to a tube with a cork and placed over animal holes. This strategy is primarily used to capture adult phlebotominae.

Fig. 6 Damasceno trap. (Photo credit: Margarete Afonso)

Malaise Trap

The Malaise trap (Townes 1962) was developed to capture insects during flight, and it consists of a nylon structure, similar to a tent, held by trees or bushes that are along likely flight routes for insects. When they enter the trap, insects are led by the inclined roof to two plastic or transparent glass cylinders in each end of the structure.

These cylinders contain a fixating agent, such as ethyl acetate, carbon tetrachloride or potassium cyanide. The Malaise trap and its varietals are not a preferred equipment for phlebotomine specialists (Young and Duncan 1994). Few tests have been conducted using this trap, and the results obtained were unsatisfactory (Alexander et al. 2001; Andrade-Filho et al. 2008; Shimabukuro et al. 2016).

Sticky Traps

Sticky traps are another example of phlebotomine capture by interception. They are made of standard white sheets or paper or cards that are embedded in castor oil. Their use is advisable in environments close to natural shelters, during the night, for two or three consecutive nights. The number of sheets used for each sampling must be constant (Alten et al. 2015). For this method to show significant quantitative results, one of the sides of the sheet, or both sides, must be completely exposed. In Brazil, in the context of control activities for AVL, the standard of use established was five sheets of legal size II or A4 sulfite paper (Brasil 2014). This is a low-cost strategy for phlebotomine capture, and it is easy to produce on a large scale (Alten et al. 2015). However, due to the castor oil's viscosity, this technique is inefficient for environments with high relative air humidity (Young and Duncan 1994).

Traps with Animal Baits

Disney Trap

The Disney trap (Disney 1966) was developed to capture phlebotominae with zoophilic behavior, as well as low-flight species, using animal baits. The trap is composed of a flat surface—in general, an aluminum, zinc or stainless-steel tray—with an opening in the center where a cage with the animal bait (hamster) is placed. This cage is sustained by a smaller tray placed under it, and the surrounding area is embedded in castor oil. In general, this trap is placed on the ground. To avoid ant attacks, it is advisable to place the trap over four wooden stumps or branches, which are also embedded in castor oil. The insects, attracted by the animal bait, become jumpy and stick to the oil-covered surface. Furthermore, they can be easily removed with a brush. It is important to highlight that all the equipment can be covered with the same material of the trap, plastic or similar, to protect it from excessive exposure to sunlight or rain.

One relevant change was made to the Disney trap by Thatcher (1968), by using two trays simultaneously—one on top and another one under the trap. The bait is placed on the upper tray. The castor oil can be used for either one or both trays.

Dorval et al. (2007) also implemented a change to the original model of this trap, aiming to increase security and well-being of the animal bait (hamster) and avoid

Fig. 7 Disney trap. (Photo credit: Dorval MEC et al. 2007)

the trap being damaged by predators. The animal trap was assembled between two pieces of galvanized zinc that had been sprayed with castor oil so they could slide and be removed with minimum effort. This facilitates the cleaning and identification of the phlebotominae captured. The corners of the cage, the roof and the bottom are made of stainless steel, and the legs are removable to facilitate assembly. The animal baits were stored in plastic boxes, 30cm above the ground, individually covered with iron mesh, appropriately sized per the ILAR (Institute of Laboratory Animal Resources) (ILAR 1996) population-density regulations, and placed on wooden beds. These innovations allowed the animal bait to drink water and eat during the entire capture procedure. In addition, these traps offer greater levels of protection and hygiene by avoiding excrements fall on the zinc pieces, which could damage the captured phlebotomine sample (Fig. 7).

In captures performed with the Disney trap, it is prudent to wash the insects in a solution of type II water and neutral (commercial) soap to remove the castor oil used on the surface of the trap. Importantly, the use of an animal bait (hamster) requires previous authorization from the *Ethics Commission on Animal Use*, which is responsible for analyzing and assessing experiment protocols for research using animals. Other authorizations are also necessary to develop research with animal baits.

Processing Methods

Phlebotomine Transportation

This procedure starts as soon as field activities are completed and ends in the laboratory. To transport phlebotominae, some of the materials used include polypropylene flasks or similar containers, black or dark plastic bags, Styrofoam boxes or isothermal transportation boxes, paper, pencil, labels and 70% alcohol solution.

When the capture method includes the use of light traps, the sample will contain not only phlebotominae but other arthropods. The material captured can be anesthetized and killed on the field before transportation or in the laboratory. The following step will be the screening of the material.

The transportation of live insects aimed at maintaining an insectary is done using cages (collecting cages or transportation cages) made of fine cloth, voile or thin nylon mesh and sustained by a cube-shaped stainless steel or iron structure (Barraud 1929) (Fig. 4). The cages are stored in a chilled Styrofoam box with moist flannel pieces to maintain the humidity and temperature for transportation (Rangel et al. 1985). The cage used in the field for phlebotomine transportation is surrounded by a dark plastic bag and placed inside a labelled Styrofoam box. It is worth mentioning that all the material captured should be transported to the laboratory respecting the rules of transportation of biologic material currently in place. No disposals should be made on the field.

Phlebotomine Anesthesia and Killing Through Cooling

Anesthesia and euthanasia are techniques in which the adult phlebotomine is killed by going through a period of anesthesia through cooling. This procedure can be performed on the field (in Styrofoam boxes or in isothermal boxes with recyclable ice) or in the laboratory (with equipment such as a refrigerator with a freezer).

After capture, phlebotomine sand flies are removed from the cage with the help of a Castro manual aspirator or an electric aspirator that has its extremity closed with a piece of cotton. Alternatively, phlebotominae can be kept inside the cage and brought along with the rest of the material from the field. Both the aspirator and the cage must be wrapped with a black or dark plastic bag to avoid, which goes through the structure of the cooling units and compromises the quality of the phlebotominae. The procedure of anesthesia and killing is carried out by inserting the aspirator or the cage, wrapped in a plastic bag, inside the Styrofoam or isothermal box (in the field)—which should contain recycled ice—and maintaining it for 30–40 min under those conditions. The material can also be kept inside a freezer for 5–10 min.

Screening of Adult Phlebotominae

The screening of phlebotominae is the technique through which adults are separated from other insects (Forattini 1973; Vilela, Rangel and Lainson 2003). For this procedure, it is mandatory to use personal protective equipment (PPE), such as coats, nitrile or latex gloves and PFF2 masks with a frontal breathing valve.

The necessary material for this procedure includes the following: absorbent paper, lining paper for the stand or any type of white paper, a polypropylene transparent tray or similar, a petri dish, watchmaker tweezers, a brush and/or stylet (made with insulin syringes or similar), a polypropylene tube or similar, a manual magnifying glass or a stereoscopic microscope, and type II water and 70% alcohol solution (ASTM 1991).

Dry Screening

In this method, all the insects collected in the cage, propylene tube or Castro manual aspirator are slowly placed on a surface covered with white paper. This paper sheet can be accommodated on a polypropylene transparent tray, which will help to better confine the insects during the screening. The screening of phlebotominae and their separation from other insects by their phenotypic characteristics is done with the help of watchmaker tweezers, a brush and/or stylet lightly moistened with type II water or 70% alcohol solution. A manual magnifying glass or a stereoscopic microscope are also needed. By the end of the screening, the other insects are disposed of in accordance with the biosecurity guidelines for biologic material or material used for scientific collection (Brasil 2010).

It is usually suggested that the screening procedure is carried out by two professionals in order to have a proper revision of the material analyzed. By the end of each analysis, phlebotominae should be transferred from the paper sheet to a polypropylene tube that has been labelled on the inside with the information of the capture written in pencil. This tube may contain a 70% alcohol solution depending on the objectives of the study.

Wet Screening

This is the most commonly used screening method for phlebotominae. The specimens contained in the cage, polypropylene tube or Castro manual aspirator are slowly placed onto a polypropylene tray covered with a sheet of white sheet. After all the insects are on the paper sheet, they are then poured into a polypropylene tube with a 70% alcohol solution (when the screening is not immediate) or into a petri dish with 70% alcohol solution (when the screening is immediate). From then onward, the procedure is the same as the one for dry screening (Barreto and Coutinho, 1940).

Clarification and Diaphanization

Clarification and Diaphanization for Routine Diagnostics

These are procedures entirely performed in the laboratory using a chapel for the exhaust gases (collective protection equipment [CPE]). The material employed in this technique includes cell-culture plate in different sizes and numbers – 6-, 24- or 96-well style (made out of insulin syringes or similar), disposable Pasteur pipettes (5 or 10 ml or similar), absorbent paper, paper for lining the stand, a glass funnel, a paper funnel and a timer. The solutions and chemical substances used in this method are: 10% potassium hydroxide (KOH 10%), type II water (ASTM 1991), acetic acid P.A. and lactophenol solution (Barreto and Coutinho 1940).

Each identified polypropylene tube containing samples from the capture procedure is placed on the board and inside the first well. The label of the tube should be transferred to the board's lid to guarantee that the samples can be identified and traced back afterward. The phlebotominae are slowly transferred to the well with the help of a stylet. The choice of board type for this procedure depends on the amount of phlebotominae to be processed.

The first well should receive the initial sample, whether it is dry or wet, and the process of clarification and diaphanization is initiated from the second well onward.

Step 1: KOH 10% is poured into the well with the phlebotominae, submerging them for 3 h. This step promotes the softening of the chitin.

Step 2: The phlebotominae are transferred, with the help of a stylet, to another well containing acetic acid P.A. They soak for approximately 20 min. This is the step to neutralize the potassium KOH.

Step 3: The phlebotominae are transferred to the well containing type II water and remain there for 20 min for cleaning purposes.

Step 4: Finally, the insects are transferred to a lactophenol solution and are kept there for 24 h to finalize the diaphanization process.

After this period, the material is ready for the next phase, the preparation of the slides.

Phlebotominae may remain in the lactophenol solution for prolonged periods of time without any damage to their quality. Nevertheless, this will make the insects lighter, and their structures will likely become excessively transparent, which leads to difficulties in diagnosis. In that sense, it is important to carefully evaluate the time the sample will remain on the solution, especially considering that some species of phlebotominae already have a lighter color.

For disposal, chemical substances are transferred to plastic bottles (labelled canisters) using Pasteur pipettes. Lactophenol solution is the only substance or solution involved in this process that can be reused for up to 5 years. To keep this material, it must be filtered into a labelled amber flask using glass and paper funnels. This is a fast and efficient process that allows the reuse of the lactophenol solution, although only for temporary collection. The slides prepared in lactophenol are meant for temporary analyses. Due to that fact, these slides can be maintained for a period of 2–3 years. This procedure should not be applied to slides meant for permanent scientific collection.

Clarification and Diaphanization for Scientific Collection

Proposed by Forattini (1973), the preparation of phlebotominae for scientific reference was modified by Galati (1990) and revised by the team of the Laboratory of National Reference in Entomologic Vigilance, Taxonomy and Ecology of Leishmaniosis Vectors at the Oswaldo Cruz Institute/FIOCRUZ. This revision was related to adjustments in biosecurity (Brasil 2010).

This procedure is performed in the laboratory using a in a chapel for the exhaust gases. The use of CPE and appropriate PPE (coat and nitrile gloves) is mandatory (Brasil 2010). The material used in this process is the same as that described in the previous item with the addition of small glass petri dishes replacing the board of wells. The chemical solutions used for this procedure are KOH 15%; 10% acetic acid; 70%, 80% and 95% alcohol; 1% acid fuchsin and the products absolute ethyl alcohol P.A., liquid phenol P.A. and eugenol P.A.

Steps:

Step 1: The phlebotominae stay immersed in phenol for 24 h (optional step).
Step 2: The phlebotominae are transferred to 15% KOH for 12 h.
Step 3: The insects are further transferred to 10% acetic alcohol (quick wash). If necessary, acetic acid P.A. and 1% fuchsine solution are added for coloring.
Step 4: The phlebotominae are immersed for 10 min in 70% alcohol solution.
Step 5: The phlebotominae are immersed for 10 min in 80% alcohol solution.
Step 6: The phlebotominae are immersed for 10 min in 95% alcohol solution.
Step 7: The phlebotominae are immersed for 10 min in absolute alcohol.
Step 8: The phlebotominae are immersed in eugenol for at least 3 h. The phlebotominae can remain submerged in eugenol for long periods of time without compromising their quality or the procedure. After this step, the material is ready for the subsequent phase, preparing the slide for scientific collection.

Preparing the Slide

Slide Preparation for Routine Diagnosis

This procedure is performed in a laboratory using a chapel, for the exhaust gases, with a stereoscopic microscope within it and CPE. It is also mandatory to use PPE during the entire procedure (Brasil 2010).

For this phase, the necessary materials include trays in medium-density fiberboard or medium-intensity wooden fiberboard that can carry 24 slides, a disposable Pasteur pipette (5–10 mL or similar), a stylet (made with insulin syringes or similar), absorbent paper, lining paper for the stand, labels, slides and coverslips and Berlese solution (Barreto and Coutinho 1940). A small dose of Berlese solution is poured into a 10-mL dropper bottle and identified. This recipient should be used in the preparation phase to guarantee the quality of the product and of the technique.

Each processed well (sample) is identified with a label on the lid of its board. New labels with the exact same identification are made and glued to the tray on which slides are placed after being prepared.

The inside of the chapel for the exhaust gases (CPE) is covered with the lining paper, and the preparation is carried out with a stereoscopic microscope. A small drop of Berlese solution is put under the slide, followed by the slow displacement (using a stylet) of a phlebotomine individual. The male specimen can be prepared without any cuts and in a lateral position or with the head cut and in dorsal view (with the eyes turned upward). Traditionally, the head is placed on the left side of the slide and the thorax-abdomen on the right side. For the female specimen, the head is cut and placed in a ventral position, the thorax is laterally exposed, and the abdomen is put in aligned ventral position. It is recommended that the preparation of the female structures is presented in a triangle, instead of in line, which facilitates viewing during the identification process. Traditionally the head of the female is placed on the left side of the slide, the thorax in the middle and the abdomen on the right side. After the material is positioned, it should be covered with a coverslip.

Because the phlebotomine is a small insect, the coverslip is not used in its commercial size (24 × 32 mm among other sizes available on the market). Instead, the coverslip is always divided into four pieces using a disposable 0.1-mm dental drill with a diamond coating. This tool is inserted into a needleless insulin syringe. The procedure is done using protection glasses and nitrile gloves, on an MDF tray or on a piece of wooden fiber board, using a ruler.

The prepared slides remain under a drying process on the trays for approximately 24 h. After this period, the material is ready for taxonomic identification. If there is interest in keeping the slide for collection purposes, it is advisable to use commercial clear polish to seal the coverslip with the slide on the sides. This process is called "caulking," and it stops the passage of liquids or air onto the slides. For slides that will not be kept for collection, there is no need to do this procedure, and they can be disposed of. However, it is recommended to reuse the slides. To do so, they must be soaked overnight in 5% laboratory detergent and then washed under running water with 1% laboratory detergent.

Preparation of a Slide for Scientific Collection

Originally proposed by Forattini (1973) and modified by Galati (1990), this procedure consists of the preparation of a slide for a permanent collection. It was revised by the team of the Laboratory of National Reference in Entomologic Vigilance, Taxonomy and Ecology of Leishmaniosis Vectors at the Oswaldo Cruz Institute/FIOCRUZ aiming to make adjustments in terms of biosecurity (Brasil 2010).

The procedure is performed in a laboratory using a chapel for the exhaust gases (CPE) having a stereoscopic microscope within it. The use of PPE (coat, nitrile gloves and PFF2 masks with frontal breathing valves) is mandatory throughout the procedure (Brasil 2010).

The material used for this procedure is the same as described previously, except for the coverslip, which is round in this case. This is done because this material is assembled for collection. The chemical substance used for preparation in this procedure is a solution of Canada balsam (natural or synthetic) + xylene (Brasil 2003). The preparation is undertaken between the slide and the coverslip with a small drop of Canada balsam plus xylene or Enecê solution.

Phlebotomine Preservation

This is a procedure done in a laboratory, in which adult phlebotominae are maintained for the screening and diaphanization and clarification procedures. It is also applicable for the period before screening. For this procedure, the use of PPE—coat and nitrile gloves—is mandatory.

The preservation of phlebotominae on a dry environment is adequate for short periods of time (i.e. ≤3 days). It is done by keeping the insects in the polypropylene tube in a laboratory at a room temperature of 22°C ± 2°C. The bottles with many insects deserve special attention because they may decompose faster. In this case, the deadline for maintenance is lowered to a few hours or a day.

In the method for phlebotomine preservation in dry environments, it is possible to extend the period of maintenance by adding silica to the tube with the insects and keeping them in laboratory at room temperature (approximately 22°C ± 2°C). There is still an option of preservation in low temperatures where the polypropylene tubes containing the phlebotominae are stored in a refrigerator with a freezer or in liquid nitrogen.

The preservation of phlebotominae in humid environments (in a 70% alcohol solution) is the most usual method for this purpose. In it, properly identified polypropylene flasks are filled with a 70% alcohol solution so the phlebotominae can be maintained in a laboratory at room temperature (approximately 22°C ± 2°C) for longer periods of time. To keep them for periods >2 months, it is necessary to change the alcohol in the flask. Flasks with many phlebotominae (at >40% capacity) need a monthly change of the alcohol solution (Barreto and Coutinho 1940).

PPE

Just as in activities performed inside a laboratory, in field activities it is essential to use PPE. In that sense, it is relevant to highlight that the clothes used by capturers must be adequate and safe to decrease, at the most, the area of exposure to phlebotominae and/or other arthropod vectors. The use of gaiters to avoid snake and domestic animal bites, raincoats, gloves and hats—along with hand and/or head lanterns—are crucial elements for fieldwork security. Another point to be considered is to be current with immunization, paying special attention to vaccines against hepatitis A and B, yellow fever, DTP (diphtheria, tetanus and pertussis), rabies and other vaccines specifically recommended for the area where activities will be performed (Brasil 2010).

Notes made on the field records, aimed at evaluating the environmental conditions found during the capture, consist of fundamental data for future analyses. These documents may present the location of the capture; geographical coordinates and information on altitude obtained through a Global Positioning System (GPS); date and starting and finishing times of capture; temperature; humidity as measured by a hygrometer; and information on whether or not it was raining or not, wind speed and lunar phase. It is also important to write down the method of capture employed and the names of the professionals responsible for the activity. These documents still reserve a space for comments where professionals are free to register any relevant additional information. It is advisable that the fieldwork records are filled out with a pencil to avoid the loss of information in the event of exposure to water and/or alcohol during any step of the work. The cages used on the field must have their information (place of usage, method and date) registered in pencil on a label. The same is true for material that is stored in tubes.

Concluding Notes

It is important to highlight that the choice of appropriate methodology to capture phlebotomine sand flies is directly related to the objectives envisioned for the study in question. The successful development of fieldwork depends on a range of factors—such as the organization, logistics and choice of adequate equipment for the activities to be performed—to be managed by the professionals involved, Still, natural aspects may have significant influence, thus hindering the possibility to attain the final results expected for the research. Within the scope of epidemiologic surveillance of phlebotominae and considering the current context of potential global climatic changes, it is reasonable to consider that this might interfere with the biocenosis of these vectors. Therefore, this might result in changes in behavior, distribution, abundance, frequency and seasonality of each species. However, despite of these possible issues, the data obtained in fieldwork can help support more effective planning for control strategies other than providing further knowledge on the bioecology of these diptera and their natural environment.

References

Aguiar GM, Vilela ML (1987) Aspects of the ecology of sandflies at the Serra dos Órgãos National Park, State of Rio de Janeiro. VI. Shelters and breeding places (Diptera, Psychodidae, Phlebotominae). Mem Inst Oswaldo Cruz 82:585–586

Aguiar GM, Vilela ML, Schuback PD, Soucasaux T (1985a) Aspectos da Ecologia dos Flebótomos do Parque Nacional da Serra dosa Órgãos, Rio de Janeiro. VI – Abrigos e Criadouros Naturais. Anais III Jornada Científica da Fundação Oswaldo Cruz, p. 146

Aguiar GM, Schuback PD, Vilela ML, Azevedo AC (1985b) Aspectos da ecologia dos flebótomos do Parque Nacional da Serra dos Órgãos, Rio de Janeiro: II. Distribuição vertical (Diptera, Psychodidae, Phlebotominae). Mem Inst Oswaldo Cruz 80(2):187–194

Aguiar GM, Vilela ML, Schuback P, Soucasaux T, Azevedo ACR (1985c) Aspectos da ecologia dos flebótomos do Parque Nacional da Serra dos Órgãos, Rio de Janeiro. IV. Freqüência mensal em armadilhas luminosas (Diptera, Psychodidae, Phlebotominae). Mem Inst Oswaldo Cruz 80:465–482

Alencar RB, Queiroz RG, Barret TV (2011) Breeding sites of phlebotomine sand flies (Diptera: Psychodidae) and efficiency of extraction techniques for immature stages in terra-firme forest in Amazonas State, Brazil. Acta Trop 118:204–208

Alexander B (2000) Sampling methods for phlebotomine sand flies. Med Vet Entomol 14:109–122

Alexander B, Freitas JM, Quate LW (2001) Some Psychodidae (Diptera) from Atlantic Forest in South-Eastern Brazil, with descriptions of *Trichomyia dolichopogon* sp. nov. and *Trichomyia riodocensis* sp. nov. Braz J Biol 61(03):467–474

Alten B, Ozbel Y, Ergunay K, Kasap OE, Cull B, Antoniou M, Velo E, Prudhomme J, Molina R, Bañuls AL, Schaffner F, Hendrickx G, Van Bortel W, Medlock JM (2015) Sampling strategies for phlebotomine sand flies (Diptera: Psychodidae) in Europe. Bull Entomol Res 150:1–15

Andrade AJ (2010) Ecologia química de flebotomíneos (Diptera: Psychodidae: Phlebotominae): Desenvolvimento de uma armadilhas e análise dos hidrocarbonetos cuticulares das espécies [PhD Thesis]. Universidade Federal de Minas Gerais, Belo Horizonte, p 167

Andrade AJ, Andrade MR, Dias ES, Pinto MC, Eiras AE (2008) Are light traps baited with kairomones effective in the capture of *Lutzomyia longipalpis* and *Lutzomyia intermedia*? Na evaluation of synthetic human odor as na attractant for phlebotomine sand flies (Diptera: Psychodidae: Phlebotominae). Mem Inst Oswaldo Cruz 103:337–343

Andrade-Filho JD, Oliveira RC, Fonseca AR (2008) Flebotomíneos (Diptera: Psychodidae) coletados com armadilha Malaise no Centro-Oeste de Minas Gerais. Neotrop Entomol 37(01):104–106

ASTM (1991) American Society for Testing and Materials Standart Specificacion for Reagents Water. Document D 1193–91 An American National Standart Federal Test Method Standard No. 7916

Barraud PJ (1929) A simple method for carriage of living mosquitoes over long distance in the tropics. Ind J Med Res 28:281–285

Barreto MP, Coutinho JO (1940) Processos de captura, transporte, dissecção e montagem de flebótomos. Arq Fac Med Univ de São Paulo XVI: 173–191

Bishop AL, Worrall R, Spohr LJ, McKenzie HJ, Barchia IM (2004) Improving light-trap efficiency for *Culicoides* spp. with light-emitting diodes. Vet Ital 40:266–269

Brasil (2003) Ministério da Justiça e Segurança Pública. Portaria MJ N 1274, de 25 de agosto de 2003. Listas dos produtos químicos sujeitos a controle e fiscalização, tendo em vista o potencial emprego no processamento ilícito de drogas. Diário Oficial da União, Brasília, 25 de agosto de 2003

Brasil (2010) Ministério da Saúde. Portaria Nº 3.204, de 20 de outubro de 2010. Dispõe sobre Norma Técnica de Biossegurança para Laboratórios de Saúde Pública. Diário Oficial da União, Brasília, p 50, 21 out. 2010. Seção 1.

Brasil (2013) Ministério da Saúde. Secretaria de Vigilância em Saúde. Departamento de Vigilância Epidemiológica. Manual de Vigilância da Leishmaniose Tegumentar Americana. 2nd ed. Brasília-DF, p 182

Brasil (2014) Ministério da Saúde. Secretaria de Vigilância em Saúde. Departamento de Vigilância Epidemiológica. Manual de Vigilância e Controle da Leishmaniose Visceral Americana. 2nd ed. Brasília-DF, p 122

Casanova C (2001) A soil trap for collections of phlebotomine sandflies. Mem Inst Oswaldo Cruz 96(02):273–275

Castro GO (1937) Sobre o processo de cultura de flebótomos Nota prévia. Soc Biol. Sessão de 8 de outubro. Rio de Janeiro

Chaniots BN, Anderson JR (1968) Age structure, population dynamics and vector potential of *Phlebotomus* in Northern California. II. Field population dynamics and natural flagellate infections in parous females. J Med Entomol 5:273–292

Disney RHL (1966) A trap for phebotomine sandflies attracted to rats. Bull Entomol Res 56:445–451

Dorval MEC, Alves TP, Oliveira AG, Brazil RP, Galati EAB, Cunha RV (2007) Modification of Disney trap for capture of sand flies (Diptera: Psychodidae: Phlebotominae). Mem Inst Oswaldo Cruz 102:877–878

Falcão AR (1981) Um novo modelo de armadilha luminosa de sucção para pequenos insetos. Mem Inst Oswaldo Cruz 76:303–305

Feliciangeli MD (2004) Natural breeding places of phlebotomine sandflies. Med Vet Entomol 18(4):453

Forattini OP (1973) Entomologia Médica. IV Psychodidae. Leishmaniose. Bartonelose. Edgard Blucher, São Paulo, p 658

Galati EAB (1990) Sistemática dos Phlebotominae (Diptera, Psychodidae) das Américas. (PhD Thesis). Universidade de São Paulo, São Paulo, p 311

Garay I (1989) Relations entre l'hetérogéinéité des litières et l'organization des peuplements d'arthrpodes édaphiques. École Normale Supérieure, Publicaciones du Laboratoire de Zoologie, Paris, p 192

Gomes AC, Rabello EX, Natal D (1985) Uma nova câmara coletora para armadilha CDC-miniatura. Rev Saúde Públ 19(2):190–191

Hanson WJ (1961) The breeding places of Phlebotominae in Panama (Diptera, Psychodidae). Ann Entomol Soc Am 54:317–322

Hoel DF, Butler JF, Fawas EY, Watany N, El-Hossary SS, Villinski J (2007) Response of phlebotomine sand flies to light-emitting diodemodified light traps in southern Egypt. J Vector Ecol 32:302–308

ILAR-Institute of Laboratory Animal Resources. National Research Council (1996) Guide for care and use of laboratory animals. The National Academic Press, Washington, DC, p 126

Jenkins AB, Young MB (2010) Investigation of Culicoides spp. preference for colour and source using light emitting diode and fluorescent light. S Afr J Anim Sci 40:514–518

Kasili S, Kutimaa H, Mwandawiro C, Ngumbi PM, Anjili CO (2009) Comparative attractiveness of CO2-baited CDC light traps and animal baits to *Phlebotomus duboscqi* sandflies. J Vector Borne Dis 46:191–196

Kline DK, Hogsette JÁ, Müller GC (2011) Comparison of various configurations of CDC-type traps for the collection of *Phlebotomus papatasi* Scopoli in southern Israel. J Vector Ecol 36(01):212–218

Lincoln RJ, Sheals JG (1979) Invertebrate animals – collection & preservation. British Museum (Natural History), London

Lumsden WHR (1958) A trap for insects biting small vertebrates. Nature 181:819–820

Maroli M, Feliciangeli MC, Bichaud L, Charrel RN, Gradoni L (2012) Phlebotomine sandflies and the spreading of leishmaniases and other diseases of public health concern. Med Vet Entomol 27:123–147

Minter DM (1961) A modified Lumsden suction trap for biting insects. Bull Entomol Res 52:238–240

Moncaz A, Faiman R, Kirstein O, Warburg A (2012) Breeding sites of Phlebotominae Sergenti, the sand fly. Vector of cutaneous Leishmaniasis in the Judean Desert. PLoS Negl Trop Dis 6(7):e1725

Moncaz A, Kirstein O, Gebresellassie A, Lemma W, Gebre-Michael T, Balkew M, Belay S, Hailu A, Warburg A (2014) *Sergentomyia* spp.: breeding sites in vertisols and peri-domestic habitats in North West Ethiopia. Acta Trop 137:88–94

Mulhern TD (1934) A new development in mosquito traps. Proc N. J. Mosq Exterm Assoc, 20: 137

Pugedo H, Barata RA, França-Silva JC, Silva JC, Dias ES (2005) HP: um modelo aprimorado de armadilha luminosa de sucção para captura de pequenos insetos. Rev Soc Bras Med Trop 38(1):70–72

Rangel EF, Souza NA, Wermelinger ED, Barbosa AF (1985) Estabelecimento de colônia, em laboratório, de *Lutzomyia intermedia* Lutz & Neiva, 1912 (Diptera, Psychodidae, Phlebotominae). Mem Inst Oswaldo Cruz 80(2):219–226

Rodrigues KM, Correia MEF, Alves LB, Aquino AM (2008) Funis de Berlese-Tülgren modificados utilizados para amostragem de macroartrópodes de solo. Pesq Agropec Bras 1:6

Sangiorgi B, Miranda DN, Oliveira DF, Santos EP, Gomes FR, Santos EO, Barral A, Miranda JC (2012) Natural Breeding Places for Phlebotomine Sand Flies (Diptera: Psychodidae) in a Semiarid Region of Bahia State, Brazil. J Trop Med 124068:1–5

Shannon RC (1939) Methods for collecting and feeding mosquitoes in jungle yellow fever studies. Am J Trop Med Hyg 19:131–148
Shimabukuro PHF, Moreira JAC, Costa TS (2016) Sand flies (Diptera: Psychodidae) from the Brazilian Atlantic Forest domain collected with malaise traps. J Med Entomol 53(6):1488–1491
Silva FS, Brito JM, Costa-Neta BM, Lobo SEPD (2015) Evaluation of light-emitting diodes as attractant for sandflies (Diptera: Psychodidade: Phlebotominae) in northeastern Brazil. Mem Inst Oswaldo Cruz 110(6):801–803
Souza TL, Figueiredo FB, Almeida AB, Benigno CV, Pontes CS, Souza MB (2014) Natural breeding sites of phlebotomine sand flies (Diptera: Psychodidae) on Marambaia Island, Rio de Janeiro State, Brazil. Acta Trop 136:104–107
Sudia WD, Chamberlain RW (1962) Battery operated light trap, an improved model. Mosq News 22:126–129
Thatcher VE (1968) Studies of Phebotomine sandflies using castor oil traps baited with Panamanian animals. J Med Entomol 5:293–297
Townes H (1962) Design for a Malaise Trap. P Entomol Soc Wash 64:253–262
Trips MA (1968) Sucction apparatus for collecting mosquitoes and other insects. Mosq News 28:647–648
Vilela ML, Rangel EF, Lainson R (2003) Métodos de Coleta e Preservação de Flebotomíneos. In: Rangel EF, Lainson R (eds) Flebotomíneos do Brasil. Editora Fiocruz, Rio de Janeiro, pp 353–367
Vivero RJ, Torres-Gutierrez C, Bejarano EE, Peña HC, Estrada LG, Florez F, Ortega E, Aparicio Y, Muskus CE (2015) Study on natural breeding sites of sand flies (Diptera: Phlebotominae) in areas of *Leishmania* transmission in Colombia. Parasite and Vectors 8:116
WHO (2010) Control of the Leishmaniasis. WHO Tech Rep Ser 949:1–186
Young DC, Duncann NA (1994) Guide to the identification and geographic distribution of *Lutzomyia* sandflies in México, the West Indies, Central and South America (Diptera: Psychodidae). Mem Am Entomol Inst 54:1–881

Leishmaniasis-Vector Surveillance and Control in Brazil: A Challenge to Control Programs

Ana Nilce S. Maia-Elkhoury, Rafaella Albuquerque, and Oscar D. Salomón

Epidemiological Status

Introduction

In Brazil, leishmaniasis are zoonotic and vector-borne diseases, and they are a challenge to control programs due to the clinical complexity and diversity of *Leishmania* species, the vectors, and the domestic and wild reservoirs involved in different transmission cycles (Desjeux 2004; World Health Organization [WHO] 2010). The presence of this disease is associated with poverty, and the social, economic, environmental, and climatic determinants have a direct influence on leishmaniasis epidemiology (Alvar 2006; Franke 2002; Maia-Elkhoury 2016).

In 2012, the WHO developed guidelines for a group of 10 neglected tropical diseases (NTDs), including leishmaniasis, and in the same year the London Declaration reinforced the commitment to support this initiative by focusing on the control and elimination of NTDs. In 2016, this agreement was ratified by the Member States through the CD55.R09 Resolution. Despite technical and scientific advances, the proposed measures for leishmaniasis control are concentrated on the detection and treatment of human cases along with the use of new alternatives, which—at times—must be combined with other control and surveillance preventive measures based on the characteristics and clinical–epidemiological profile of the disease in the Americas (WHO 2012; London Declaration 2012; PAHO 2016;

A. N. S. Maia-Elkhoury (✉)
Communicable Diseases and Environmental Determinants of Health (CDE), VT, Pan American Health Organization (PAHO), Duque de Caxias, Rio de Janeiro, Brazil
e-mail: aelkhoury@paho.org

R. Albuquerque
Communicable Diseases General Office, Ministry of Health, Brasília, Brazil

O. D. Salomón
National Institute of Tropical Medicine, Ministry of Health, Puerto Iguazú, Argentina

E. F. Rangel, J. J. Shaw (eds.), *Brazilian Sand Flies*,
https://doi.org/10.1007/978-3-319-75544-1_11

Gramiccia and Gradoni 2005; de Vlas 2016). Nevertheless, the challenge of controlling these diseases persists because no tools are available to decrease and effectively manage transmission by vectors.

Notification and analysis of consolidated data and epidemiological indicators, spatially disaggregated at the lowest geographical level, have shown their importance in the identification of leishmaniasis-priority areas and in the guidance of surveillance and control actions by contributing to a better technical, operative and financial management (Brazil 2006, 2007; Belo 2013; PAHO 2016).

The authors agree with the taxonomic proposal by Galati (2003) for American phlebotominae and its use in scientific literature. However, because this chapter is mainly for technical agents of the programs of control, in addition to the interested researchers and students, we retained the binomial names given by Young and Duncan (1994). The broader use of the genus *Lutzomyia* is still the regular denomination among programmatic field-based activities. Thus, to introduce new genus here could introduce doubts about the actual identity of the species. Once the taxonomy by Galati is more generalized among the technicians, it could be made uniform.

Tegumentary Leishmaniasis

Tegumentary leishmaniasis (TL) is distributed over the 27 federative units of Brazil and presented an average of 21,089 cases from 2010 to 2014. A total of 20,296 cases were reported in 2014 with an incidence rate of 10 cases/100,000 population. The North region registered 51.1% of cases (10,387) followed by the Northeast (4696) and Central West (3038) regions. The incidence rate in the North region was 6 times higher (60.3 cases/100,000 population) than the average incidence in the country; however, states such as Acre, Amapá, and Roraima presented the following incidence rates: 130.4, 127.6, and 93.2 cases/100,000 population, respectively (Brazil 2016).

From the total, 74% of the cases occurred in males and 7.5% in children younger than 10 years of age, and 4.9% of the reported cases were mucosal leishmaniasis. The spatial analysis of the data, at a municipality level, shows the wide distribution of this disease in the country and displays the case-occurrence areas, which are stratified into five categories: low, moderate, high, intense, and very intense record of cases (Fig. 1a).

Brazil is bordered by 10 countries of South America, and 11 federative units of the 27 share borders with 9 countries. Twenty percent of the TL autochthonous cases were registered in 181 border municipalities (Fig. 1b and Table 1).

In the Americas, among the 15 circulating *Leishmania* species that present as cutaneous, mucosal, and mucocutaneous forms, 7 are present in Brazil. The main species are *Leishmania braziliensis*, *L. guyanensis*, and *L. amazonensis*. The remaining identified species are *L. lainsoni, L. naiffi, L. shawi*, and *L. lindenbergi.*

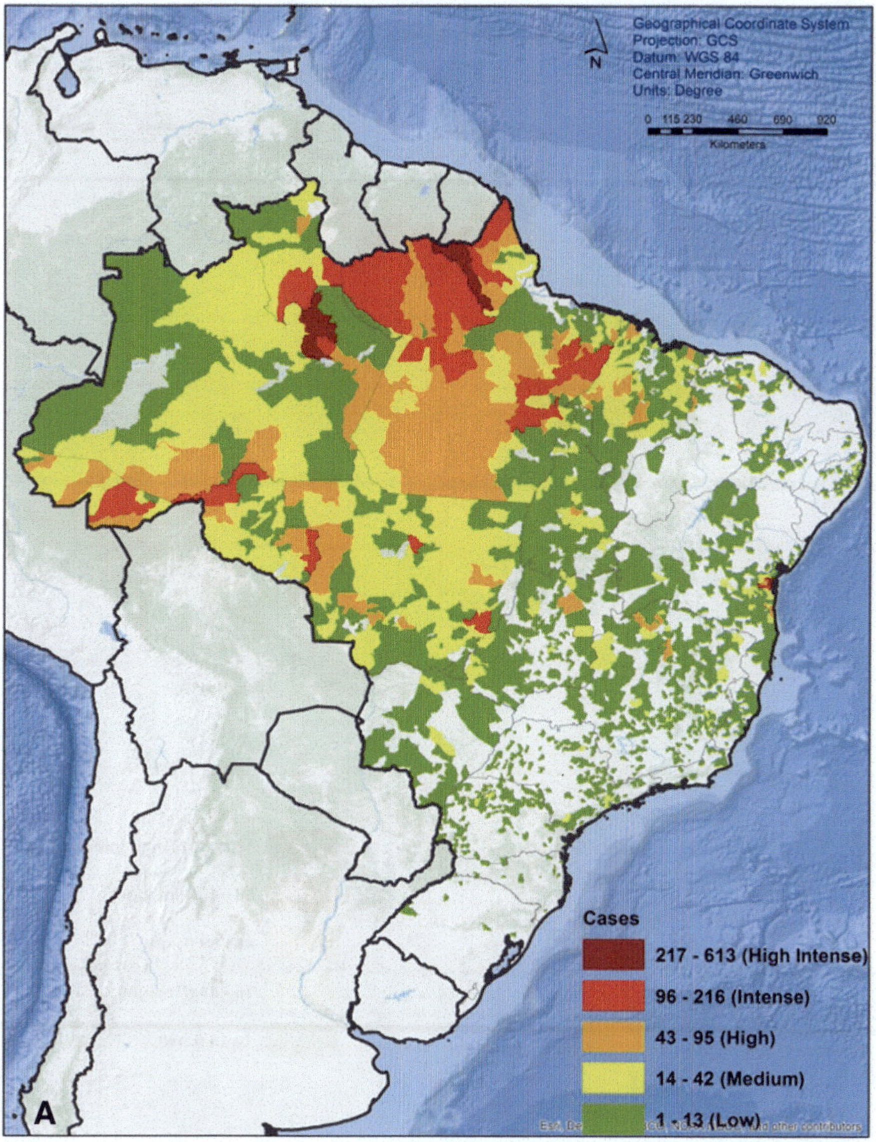

Fig. 1 (**a**) Cases of TL cases by municipality, Brazil, 2014. (**b**) Cases of TL by border municipality, Brazil, and border countries, 2014. (Source: SisLeish-PAHO/WHO: Data reported by National Leishmaniasis Surveillance Programs/Surveillance Services)

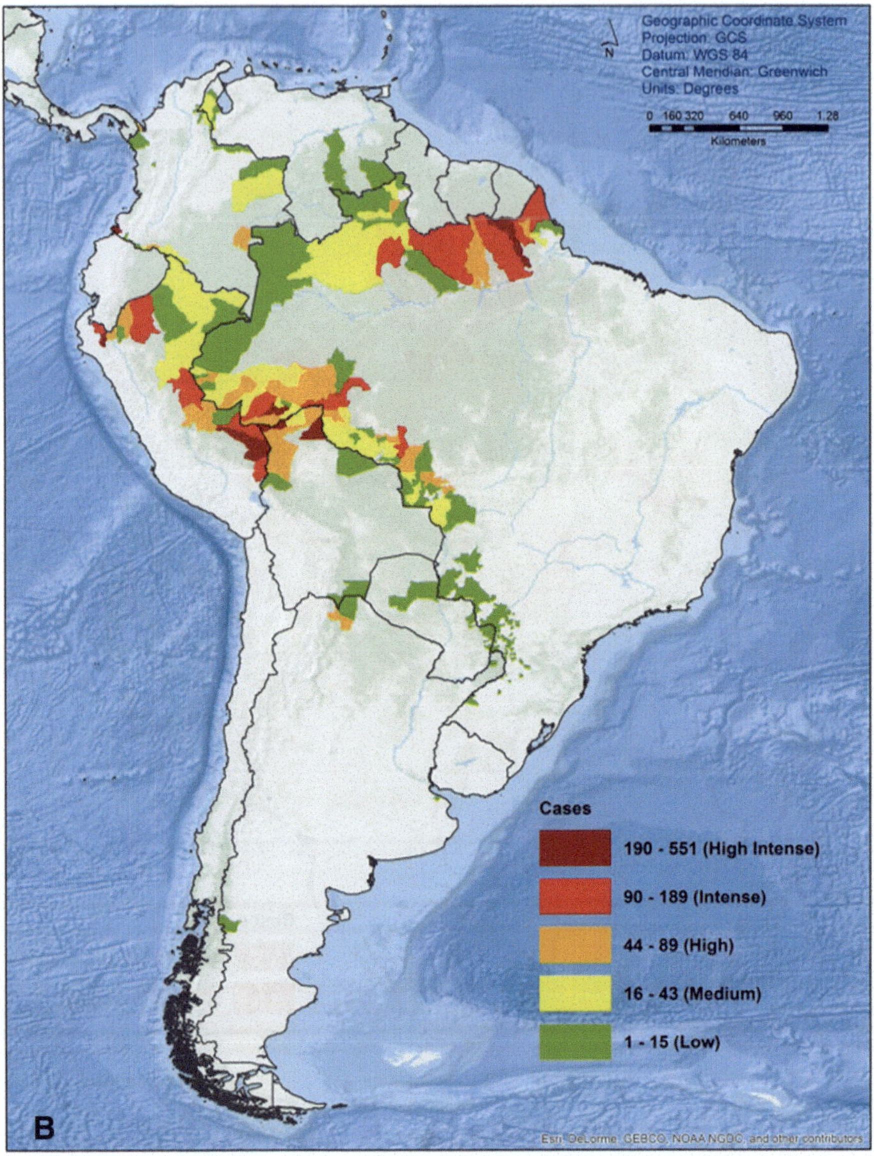

Fig. 1 (continued)

Table 1 Case distribution of TL by federative units, municipalities, and countries sharing borders with Brazil, 2014

Federative units	Total TL cases	TL cases in border municipalities		Border countries with autochthonous cases of TL
		No.	%	
Acre	1044	1044	100	Bolivia & Peru
Amapá	985	699	70.9	French Guyana
Amazonas	1799	284	15.8	Venezuela, Colombia & Peru
Mato Grosso	2329	308	13.2	Bolivia
Mato Grosso do Sul	144	58	40.3	Bolivia & Paraguay
Pará	4385	390	8.9	Suriname & Guyana
Paraná	358	75	20.9	Argentina & Paraguay
Rio Grande do Sul	8	4	50	Argentina
Rondônia	1159	773	66.7	Bolivia
Roraima	463	461	99.5	Guyana & Venezuela
Santa Catarina	15	3	20	Argentina

Source: SisLeish-PAHO/WHO: Data reported by National Leishmaniasis Surveillance Programs/Surveillance Services

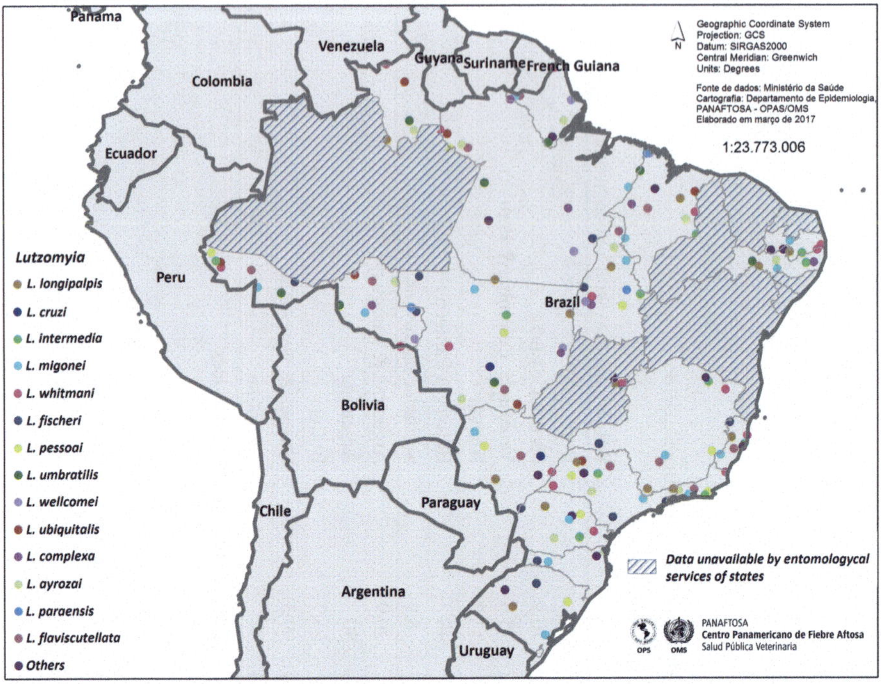

Fig. 2 Distribution according to presence of TL vector by federative unit, Brazil, 2014. (Source: PAHO/WHO: Data reported by National Leishmaniasis Surveillance Programs/Surveillance Services)

Regarding TL vectors, the most important identified species in Brazil are *Lu. (nyssomyia) intermedia Lutzomyia, Lu. (Ny.) whitmani, Lu. (migonemyia) migonei, Lu. (Ny.) flaviscutellata, Lu. (pintomyia) fischeri* and *Lu. (Ny.) neivai.* Data available by the entomology state service show the vector distribution by species and federative units (Fig. 2).

Visceral Leishmaniasis

During the period of 2010–2014, Brazil recorded an annual average of 3450 cases of visceral leishmaniasis (VL) and is listed among the 6 countries with 90% of global case occurrence, which corresponds to approximately 95% of the cases in the Americas (Alvar 2012; PAHO 2016). VL is currently under geographic expansion: It is present in 21 federative units from 5 Brazilian regions. In 2014, 3453 new VL cases were recorded, and 58.8% were from the Northeast region of the country followed by the Southeast (13.2%) and North (11.7%) regions. Maranhão, Ceará, Bahia and Minas Gerais together have reported 50.7% (1753) of the cases in Brazil (Brazil 2017).

From the total, 42.6% of cases occurred in children younger than 10 years of age, 64.9% in males, with a fatality rate of 6.9%. *L. infantum*/HIV coinfection occurred in 6.8% of cases showing growth during the last decades due to expansion, VL urbanization, and concurrent movement of the HIV/AIDS infection to rural areas (Maia-Elkhoury 2008; Lindoso et al. 2014).

In the late 1990s, 85% of VL cases were concentrated in the Northeast region of the country; however, due to geographic expansion and VL urbanization in the Brazilian South Central states, the disease reached Paraguay and subsequently Argentina (Oliveira et al. 2006; Canese 1998; Cousiño 2006; Maia-Elkhoury 2008; Salomón and Orellano 2005; Mestre and Fontes 2007; Salomón et al. 2008). In 2010, the presence of *Lu. longipalpis* was detected in two states from Uruguay, at the border with Argentina and Brazil; and in 2015 and 2016 the first two autochthonous cases of canine VL were confirmed at the border municipalities with Argentina and Brazil, respectively (Salomón et al. 2011; Satragno et al. 2017).

In Brazil, VL is caused by *Leishmania infantum*, which is the only circulating specie in the Americas, and is transmitted by two phlebotomine species, *Lu. longipalpis* and *Lu. cruzi*. *Lu. longipalpis* has a wide geographic distribution and is present in nearly 100% of the municipalities with VL transmission; nevertheless, entomology studies conducted in Brazil and Argentina have shown the presence of only other species of TL-transmitting plebotomines in some areas with autochthonous VL transmission (Salomón et al. 2010; Carvalho et al. 2010; Galvis-Ovallos 2017).

The wide geographic dispersion of *Lu. longipalpis* within Brazilian municipalities and bordering countries has been observed since 2000 because it has accompanied the occurrence of human and canine VL cases (Figs. 3a, b and 4).

Among the areas with the highest density of VL cases in South America, 13 areas are from the Northeast region followed by 2 areas in Tocantins state (North region), 3 distributed in Minas Gerais and São Paulo states (Southeast region), 1 in Mato Grosso do Sul state (Mid-West), and 1 in Paraguay, which corresponds to data from the Central Department. Still, it is also possible to observe a geographic expansion of the disease to the South of the country reaching the tri-border area between Brazil, Argentina and Paraguay (Fig. 4).

Entomological Surveillance and Phlebotomine Control

As of 1999, when the epidemiology and disease control services started to decentralize from the national level to the state and municipality levels, entomological surveillance and control actions were implemented in the country. The state entomology services were strengthened, thus enhancing the technical capability of professionals for capture, identification, taxonomy and vector control. Moreover, these services were reinforced with more supplies for sampling, processing, vector identification and vehicles. On this occasion, the methodologies and VL-entomology techniques were reviewed and agreed upon to improve and guide the surveillance

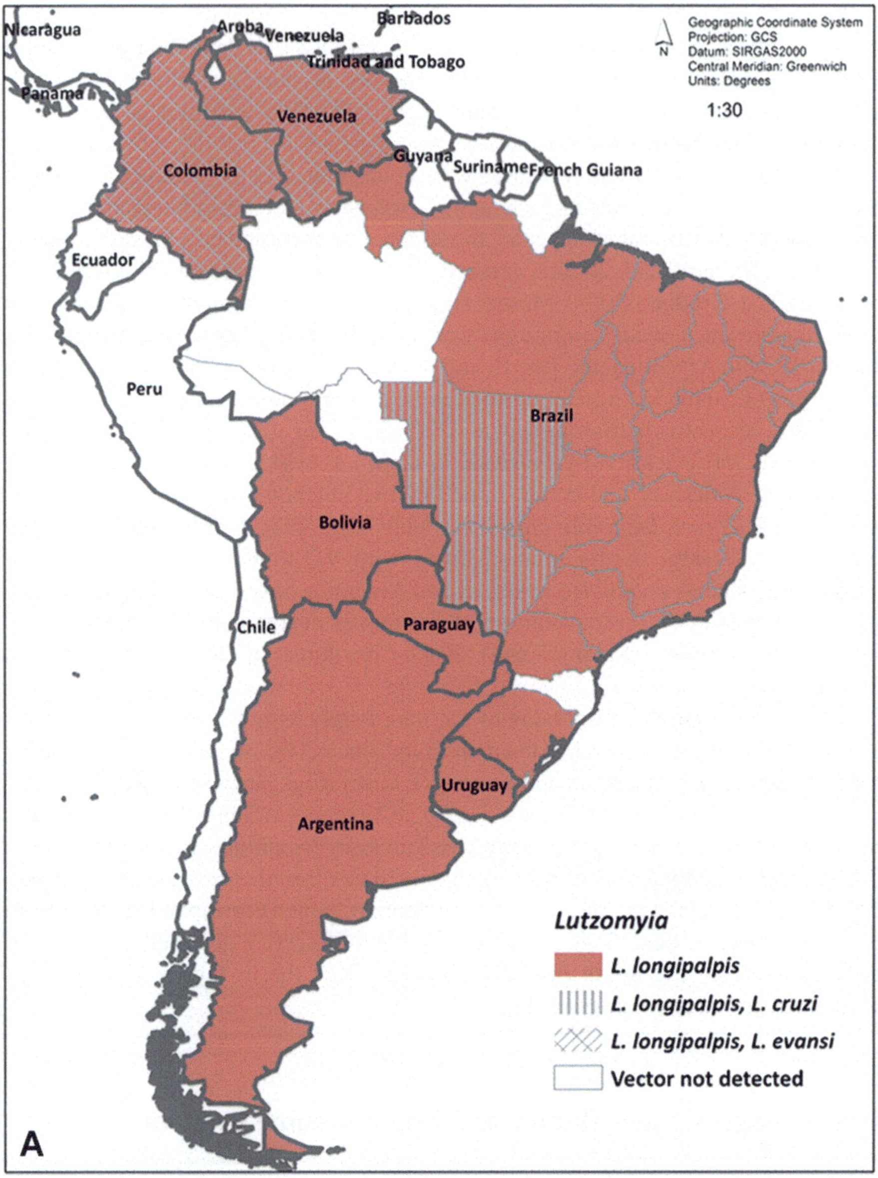

Fig. 3 (**a**) *Lu. longipalpis* distribution by Brazilian Federative Units and bordering countries, 2014. (**b**) *Lu. longipalpis* distribution by municipality, Brazil, 2014. (Source: PAHO/WHO: Data reported by National Leishmaniasis Surveillance Programs/Surveillance Services)

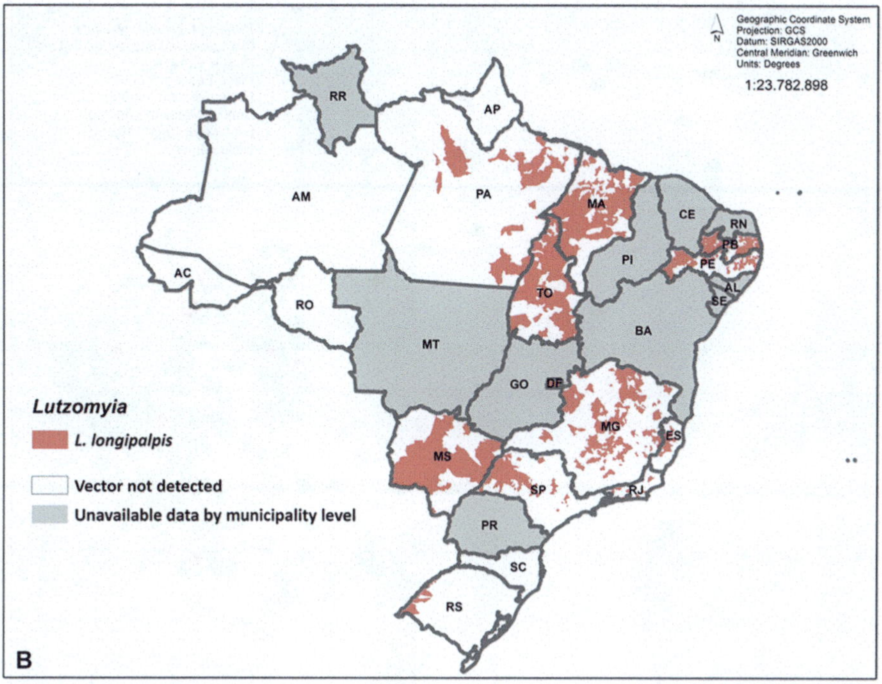

Fig. 3 (continued)

and control actions based on risk stratification. Afterward, the entomological surveillance and control actions for TL were also reviewed and implemented, although some entomology state services and leishmaniasis-control programs still need to improve and systematize the actions.

Leishmaniasis Epidemiological Classification

The leishmaniasis-prevention, -surveillance, and -control activities recommended by the Brazilian Ministry of Health are based in actions aimed at areas with presence of cases and preventative measures in areas with no cases.

For VL, municipalities with no transmission are defined as areas with no record of autochthonous cases of human or canine VL. These areas are classified according to their vulnerability and receptivity because the objective is to identify the risk of introduction or circulation of *L. infantum* and, consequently, to guide measures to avoid the occurrence of human cases of this disease. Vulnerable areas are those with no record of autochthonous cases of human and/or canine VL but meet one or more

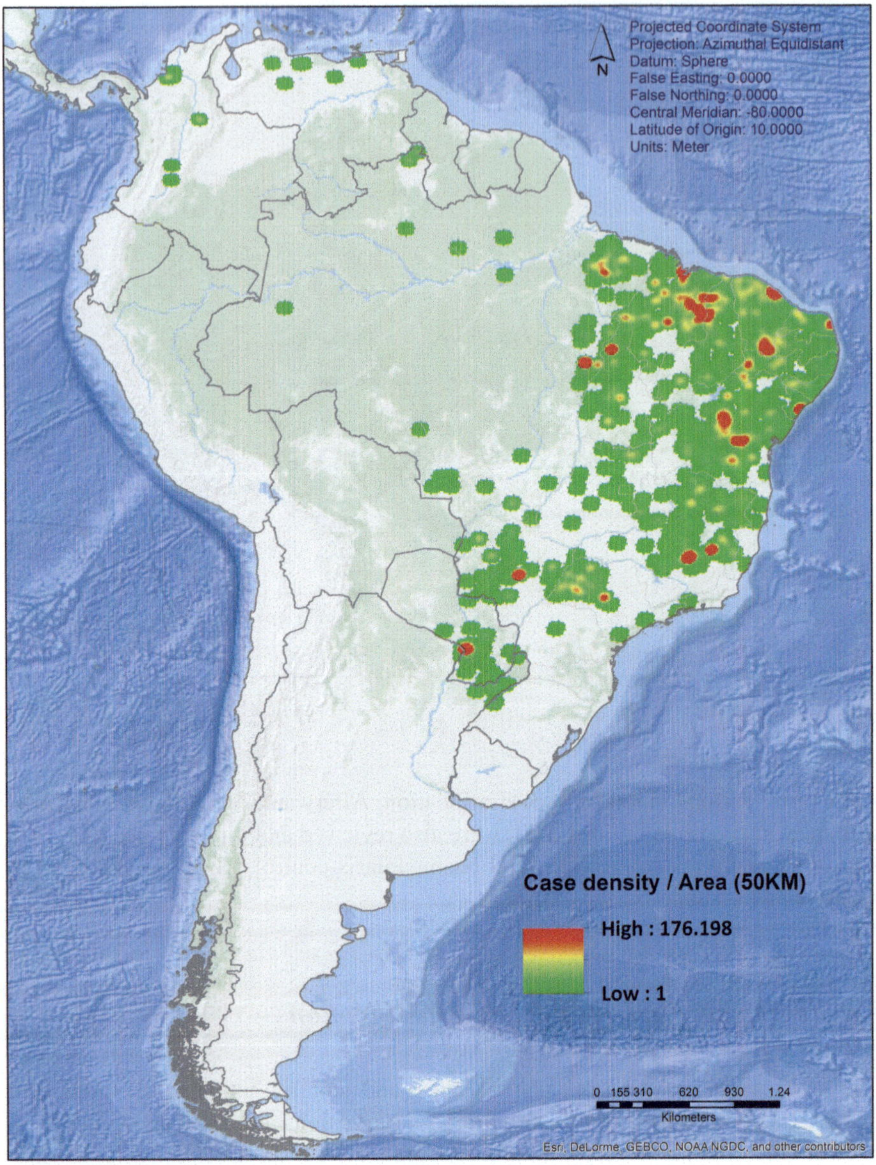

Fig. 4 Distribution density of VL cases in a 50-km radius by municipalities from Brazil and bordering countries, 2014. (Source: SisLeish-PAHO/WHO: Data reported by National Leishmaniasis Surveillance Programs/Surveillance Services)

of the following criteria: contiguous to areas with VL cases, intense migratory flow or shared road connections with municipalities having registered autochthonous cases. Receptive areas are those classified as vulnerable, according to the criteria mentioned previously, with the verified presence of *Lu. longipalpis* or *Lu. cruzi* after an entomological survey (Brazil 2006).

The vulnerability and receptivity criteria are strengthened with the surveillance of existing cases in endemic municipalities and new cases in non-endemic municipalities where the expansion of the disease is justified by the adaptation of the vectors to anthropic areas, mainly those associated with construction works of great environmental impact—such as hydroelectric power plants, settlements and road openings—as well as population changes mostly due to rural exodus (Antonialli et al. 2007; Mestre and Fontes 2007; Cardim et al. 2013, 2016). In São Paulo state, from 1999 to 2011, the disease expanded from the Northeast region to the Southeast region in tandem with the Marechal Rondon highway and the Bolivia–Brazil gas pipeline. As for the Mato Grosso state, the disease has progressed to municipalities of the South Central and Southeast regions of the state in areas with higher population mobility, which is facilitated by the roads network in this region (Mestre and Fontes 2007; Cardim et al. 2013).

The municipalities with human and/or canine VL are those with *L. infantum* transmission. These municipalities can be classified as areas with the first record of a VL autochthonous case, areas with outbreaks or areas with sporadic, moderate or intense transmission. The average number of cases from the last 3 years is used for this classification, and two cut-off points (2.4 and 4.4) were established to determine the magnitude of the disease. Nonetheless, this methodology is under review, and there is an undergoing analysis of the need of its replacement for a composite indicator, which will include readily available variables such as "number of cases" and "incidence rate." Currently, under the present classification, Brazil has 1062 municipalities with sporadic, 127 with moderate and 150 with intense transmission (Fig. 5).

Despite the full expansion of VL and its presence in large areas of the country, it is possible to observe heterogeneity in its distribution and transmission in urban areas. The diversity of the transmission in these areas may be associated with several intrinsic factors related to human population (number of susceptible individuals as well as nutritional, social and economic status), vector and reservoir population (size, susceptibility, vector competence between sibling species and reservoir competence to maintain the parasite), and environmental circumstances (microclimate, vegetation, altitude and relief, among others) (Koopman et al. 2005; Werneck 2008; Almeida and Werneck 2014). It is fundamental to consider these factors to choose adequate intervention and adapt the methodology recommended by the Ministry of Health to the reality of the location.

For TL, municipalities with no transmission are defined as those with no record of autochthonous TL cases in the last 3 years. The criteria of vulnerability and receptivity for TL are like the previously mentioned criteria for areas with no VL cases. TL-vulnerable areas are those with no transmission or having silent transmission with a biome favorable to vector presence meeting the criteria of contiguity to

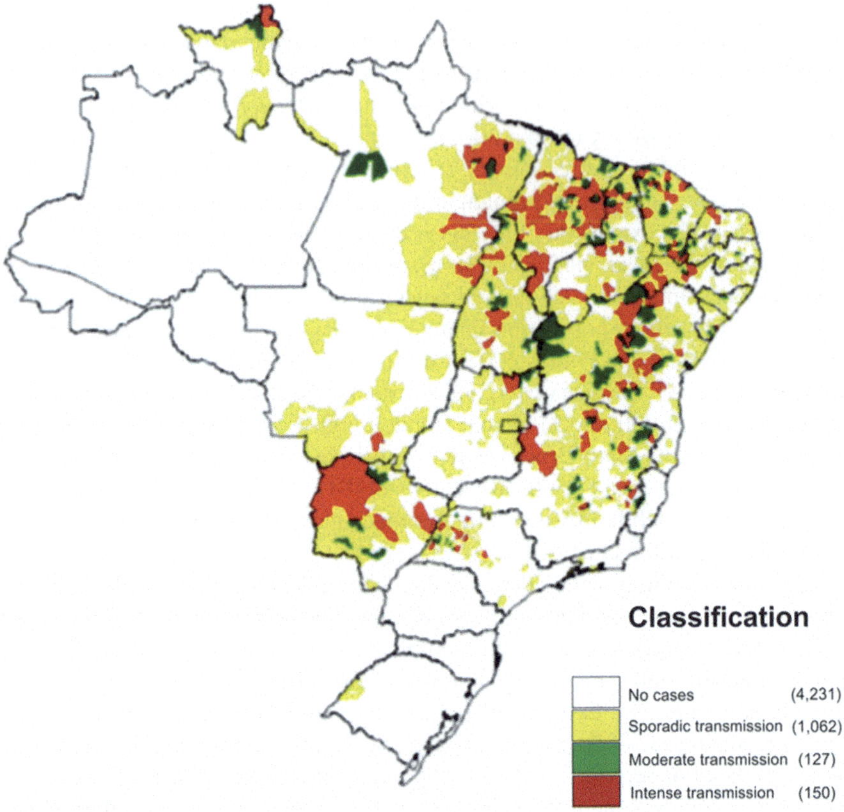

Fig. 5 Classification of municipalities with visceral leishmaniasis cases. (Source: SVS/MS Ministry of Health, Brazil)

areas with transmission or environmental modification (deforestation, settlements development plans, etc.). Receptivity is related to the presence of phlebotomine species with proven or suspected vector competence in the area after recent entomological survey (Brazil 2016).

Municipalities with TL cases are classified through the composite indicator, which was proposed and validated by the Pan American Health Organization (PAHO/WHO), comprising experts and representatives from endemic countries. It is composed of three individual indicators: "number of new cases," "density rate," and "incidence rate." Based on the composite indicator, municipalities with transmission are classified into five strata using the natural-break method: sporadic, moderate, high, intense, and very intense (Fig. 6) (Brazil 2016).

Specific entomological surveillance and vector-control measures are adopted in each of these classifications for either VL or TL as described in the next section.

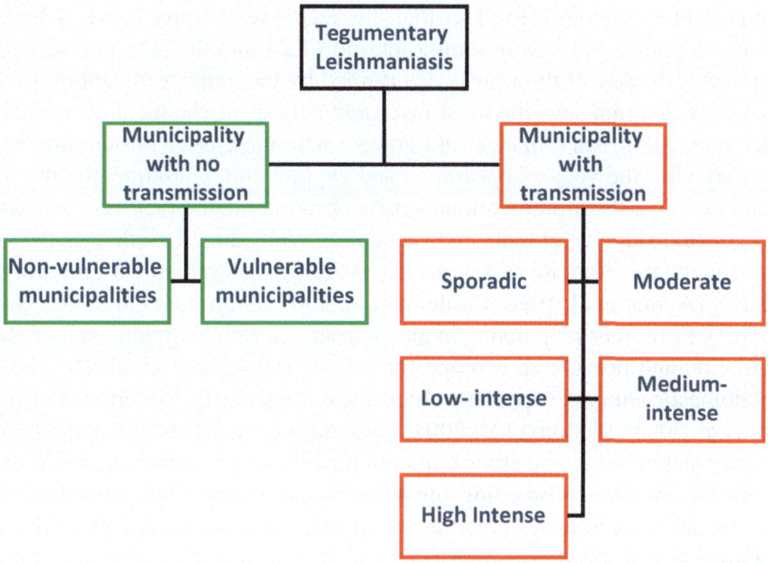

Fig. 6 Epidemiological classification for surveillance and control of TL. (Source: SVS/MS Ministry of Health, Brazil)

Entomological Surveillance

Entomological surveillance is a continuous vector-observation tool to evaluate biological and ecological characteristics—plus their interaction levels with humans, the reservoir and the environment—for the timely detection of changes in their behavior that may affect the disease-transmission pattern (Gomes 2002).

To confirm the presence of VL vectors (*Lu. longipalpis* and/or *Lu. cruzi*), thus confirming autochthony of human or canine VL cases), and the behavior of the vectors over the years, the Ministry of Health advocates three entomology activities: investigation, survey, and entomological monitoring.

Entomological investigation is a qualitative activity based on the capture of phlebotominae on 3 consecutive nights with the main objective of confirming autochthonous transmission in an area. The traps must be displayed at probable infection sites (PIS), with a minimum of two traps per PIS, one for the intra-domicile are and one for the peri-domicile area (Brazil 2006).

Entomological survey is a qualitative activity conducted in areas with no operational conditions for the development of an entomological monitoring with the goal of identifying vector dispersion in an area to guide the location of the canine serological survey and vector-control activities (Brazil 2006).

Unlike other entomological activities, monitoring has both qualitative and quantitative characteristics aiming at the knowledge of vector dispersion and the establishment of the seasonality curve of vectors in an area. It is recommended to conduct this activity in at least 10 sampling points according to the area size (Brazil 2006).

It is important to highlight the fact that the vector-seasonality curve is based on systematic captures within a minimum period of 24 months. The period with the highest vector density of the year is determined by the number of captured vectors per month, which indicates the most favorable period for chemical control as well as other prevention, surveillance and control activities. Notwithstanding, several factors may alter the vector appearance and capture and, consequently, the necessary data to define the aforementioned curve. Environmental factors—such as precipitation, wind velocity, climate and vegetation—may directly influence the vector appearance and therefore must be considered during the evaluation of monthly capture results (Aguiar et al. 1996; Carneiro et al. 2004; Zeilhofer et al. 2008; Queiroz et al. 2012). Fruit trees help maintain an environment rich in organic matter, diminish insolation and provide an ecotope for vectors (Oliveira et al. 2012). The presence of domestic animals, especially chickens, is related to the occurrence of vectors (Barata et al. 2005; Teodoro et al. 2007; Fernandez et al. 2010). Precipitation interferes with phlebotomine appearance in both positive (*e.g.*, when it occurs at moderate levels and encourages breeding sites) and negative ways (*e.g.*, when it occurs at high levels and thus destroys breeding sites) (Missawa and Dias 2007; Silva et al. 2007; Oliveira et al. 2008).

In addition to environmental factors, the quantity and disposition of the sampling points, within a selected area for the entomological monitoring, may significantly influence data for the seasonality curve. The use of few sampling points may not generate sufficient data for a statistical analysis between areas, thus hindering the sampling period and spray area. The limitation regarding sampling points is normally related to the operational complexity of this activity, which—depending on the area extension—might require an additional team, vehicles and traps, among other supplies.

A meeting with the state entomology managers was held in 2016 to discuss entomological-surveillance activities. An instrument was used to diagnose the team status, demonstrated 40% of the Brazilian municipalities designated for entomological monitoring, defined the seasonality curve and used this information to guide the chemical control. Currently there is discussion about the feasibility and maintenance of this activity as an attribution of state and municipal entomology services with the proposal of joint activity to be carried out by academic and research groups.

The design of the entomological activities described previously must have sampling spatial representativeness so that the conclusions and activities are consistent with the results and to avoid decision-making based on average values from large areas. In contrast, even though a minimal number of consecutive nights is essential to normalize the climatic variation of the sampling, the frequency and duration of the activity must avoid unnecessary use of resources. Hence, the sampling effort should be defined based just on information required from the selected area to implement actions at a local level and not inferred from sampling in a different area.

Recommendation for an entomological activity depends on the classification of the municipality transmission status, such as conduct entomological survey in municipalities with no VL cases; conduct entomological investigation in municipalities presenting the first case and/or outbreaks; and conduct entomological survey and monitoring in municipalities with a moderate or high transmission (Brazil 2006).

For TL, the confirmation of *Lutzomyia* presence is based on entomological-surveillance activities, which consists of foci research and entomological monitoring. TL-foci research is like the VL-entomological investigation. It is recommended for new transmission areas to verify the presence and identify possible phlebotomine species and thus to assist in the identification of autochthonous cases and transmission occurrence within the home or work environment. Therefore, this supports the choice of which chemical-control measure, environmental management, or individual protection to employ—together or not—according to the transmission location. Considering the diversity of TL-transmitting species, during a foci research one should use the maximum possible number and types of traps (Shannon trap, light trap, and manual and/or electric Castro capturing devices).

TL-entomological monitoring is based on systematic captures and should be conducted monthly for at least 2 years at monitoring stations delineated from the natural vegetation and geomorphological region of the area to form homogenous capturing areas. Moreover, as in the foci research, one should use both light traps and Shannon traps (Brazil 2016).

Entomological surveillance in areas with low TL transmission is mainly based on orientation of individual and collective promotion and prevention actions. For areas with moderate or high transmission, the surveillance actions are the same, but the orientation is toward resource allocation and prioritization (Brazil 2016).

Indicators used for the analyses and evaluation of the data obtained in entomological surveillance activities are quantitative, qualitative, and standardized variables that can be compared in many situations. These indicators are based on four parameters: (1) characteristics of the vector-development stages (e.g., immature or mature); (2) characteristics of the vector capacity (e.g., anthropophily levels, activity period [diurnal or nocturnal] and activity location [intra-domicile or peri-domicile]); (3) vector competence with intrinsic information (*e.g.*, natural infection rate); and (4) non-biological environment factors (*e.g.*, temperature and humidity, which might influence the vector occurrence and density). Because gaps exist in the biology knowledge of immature forms of phlebotomies, indicators exclusively concern the adult form. Thus, the VL-infestation rate (IR) of an area is measured from the calculus of the domicile-infestation rate and the relative abundance of the vector, and the TL IR is measured from the monthly average, per species and per capture point, of captures with light and Shannon traps as well as manual capture (Gomes et al. 2002) (Table 2).

Scenarios faced by the leishmaniasis-surveillance program include localities at the cross-country border area. Brazil shares eco-regions and leishmaniasis-transmission areas with all of its bordering countries where animal and people transit is fluid and constant. Differences exist in the structures of the political system and urbanization, health-system coverage, and control programs across the countries, which demand an additional effort from the surveillance, especially those related to dispersion of parasitic species to areas with no transmission or outbreaks.

Table 2 Formula for domicile-infestation rates[1], relative abundance[2] and monthly average per species and per capture point for light traps[3] and Shannon traps[4]

1. Domicile-infestation rate	=	$\frac{\text{No. of positive domiciles / research site / technique}}{\text{No. of reaserch sites}}$
2. Relative abundance	=	$\frac{\text{No. of } \textit{Lu. longipalpis} \text{ collected using a domicile methodology (intra - or peri - domicile)}}{\text{Total investigated domiciles (intra or peri - domicile)}}$
3. Monthly average per specie and per capture point for light traps:		
3.1. Monthly average intra-domicile	=	• $\frac{\text{No. of captured samples of a determined species in the intra - domicile}}{\text{No. of worked days}}$
3.2. Monthly average peri-domicile	=	• $\frac{\text{No. of captured samples of a determined species in the peri - domicile}}{\text{No. of worked days}}$
3.3. Monthly average extra-domicile	=	• $\frac{\text{No. of captured samples of a determined sprcies in the extra - domicile(woods / edge of the woods)}}{\text{No. of worked days}}$
4. Monthly average per specie and per capture point for light Shannon traps:		
4.1. Monthly average intra-domicile	=	• $\frac{\text{No. of captured samples of a determined species in the intra - domicile}}{\text{No. of worked days}}$
4.2. Monthly average Peri	=	• $\frac{\text{No. of captured samples of a determined species in the peri - domicile}}{\text{No. of worked days}}$
4.3. Monthly average extra-domicile	=	• $\frac{\text{No. of captured samples of a determined species in the extra - domicile(woods / edge of the woods)}}{\text{No. of worked days}}$

These scenarios are aggravated by TL-transmission areas with social conflicts, massive migration processes or illegal activities.

To overcome these difficulties, the Latin American Leishmaniasis Programs and the PAHO/WHO are working together on a proposal of risk stratification and minimal recommendations for vector surveillance and control like those presented in this chapter: quality control of national reference inter-laboratories to obtain an adequate and comparable diagnostic evaluation; and a fast-alert system for cross-border municipalities to report the presence of new vectors and human (TL and VL) and canine cases. Furthermore, a research of specific border areas is being developed to characterize leishmaniasis transmission through a common methodology.

Vector Control

Chemical control is the main recommendation for vector control for the collective health scope (Brazil 2006). Information on the probable infection site and period of highest vector density, obtained from analysis of the collected data by entomological surveillance, are fundamental for the success and control of the phlebotomine population. Nevertheless, difficulties in the establishment of an entomological information flow in terms of the use of epidemiological parameters to determine the research areas—and the delay of data evaluation to guide actions recommended by leishmaniasis-control programs—lead to non-use of data generated by the entomological surveillance and, consequently, to discontinuity of the recommended control activities. In summary, limiting factors to the systematic and periodic execution of these activities result in difficulty producing reliable information on vectors (entomological indicators) and hence difficulties in the guidance of domicile spraying (Sena 2011).

In Brazil, vector chemical control is based on the use of alphacypermethrin 20% SC, and the federal government has the responsibility of acquiring and distributing this product to the states. This insecticide is part of the pyrethroid chemical class, has residual activity and is sprayed on all internal and external walls of the residence. The residual activity is variable and may last $\leq$3 months; however, studies exist that show a relation between the residual effect and the type of wall and environmental circumstances, such as precipitation and insolation (de Silans et al. 1998; Camargo-Neves et al. 2007). Spraying is performed according to a complex methodology because the person responsible for this activity must guarantee the necessary amount of insecticide on the wall of the residence. The uniform deposition of 40 mg of active element/m^2 built is indicated, and the application must be executed in the correct amount of time (Brazil 2006, 2016).

For VL control, spraying is indicated in specific circumstances, such as (1) area with the first record of a human and/or canine case; (2) residences within a 150-m radius from the first case; (3) in areas with moderate and high transmission where spraying should be executed in residences from pre-defined areas before and after the rainy season. TL recommendations for chemical control are spatially restricted because the transmission pattern of this disease, in its majority, occurs in non-modified, modified sylvatic, and rural areas, which makes the indication for chemical control more difficult. It is recommended in areas with occurrence of more than one human case of TL or the occurrence of cases in children younger than 10 years old—within a maximum of 6 months from the beginning of the symptoms—in new areas or during outbreaks associated with evidence that the transmission is occurring in the home environment, that is, confirmed the presence of a vector with competence inside the residence. Spraying must be exclusively executed on built areas and under no circumstance in forest areas (Ready et al. 2013; Brazil 2016).

Since 2003 in Brazil, leishmaniasis chemical control, particularly for VL, is based on the continuous use of alphacypermethrin 20% SC; thus, there is an evident necessity to establish population monitoring of *Lu. longipalpis* due to the possibility of this species to develop resistance to the used insecticides, especially in areas with overlapping vector-borne diseases—for which control is based on other pyrethroids—such as dengue, yellow fever and Chagas disease. Change in the susceptibility profile of the *Lu. longipalpis* population to alphacypermethrin has been described in Brazilian states, thus reinforcing the importance of vector-susceptibility monitoring as well as validation of other types of insecticide (Rocha 2016).

In addition to the aforementioned points, chemical control itself is complex with reports of its operation difficulty by municipal and state teams. Morais et al. (2015) showed great difficulty in the execution of vector-control activities in Belo Horizonte, which led to non-expansion of the spraying actions in the selected areas. A great percentage of the residents refused due to the required preparation of the residences (before and after spraying) and the need to exit the residence for some time after the spraying. Camargo-Neves et al. (2004) reported similar difficulties in the São Paulo state, mainly in urban areas with great demographic density, which makes the maintenance of activities over time unfeasible. Zuben and Donalísio (2016) applied a questionnaire to municipal managers from six Brazilian capitals (Fortaleza, Campo Grande, Campinas, Bauru, Goiânia and Belo Horizonte) to obtain their opinion on the control actions of the program. The results pointed that the execution difficulties were centered in the population's refusal of the vector- and reservoir-control actions, the high cost to maintain these actions associated with there being few resources available from the federal government and the weak involvement of other sectors from the city halls.

In addition to the adversities of chemical control, which is only adopted for the vector adult stage, valid tools are lacking to control the immature vector forms as well. Environmental management is currently recommended to decrease the presence and maintenance of phlebotomine breeding sites and human exposure to the vector. However, presently there exists no validated protocol in the national territory to adopt this activity in the daily routine of health services. Because the absence of a valid methodology is a concern, Brazilian health authorities are conducting a project—along with the Federal University of the Semiárido and the State Secretaries of Health—to validate the said protocol.

Other Tools

Due to the operational difficulties, quality of insecticides used, and change of the vector species-susceptibility profile, there is a need for alternative tools in the scope of vector control. In this regard, the repellent and insecticide effect of some plants, in the form of extracts and oils, has been widely researched and in some areas, such as agriculture, have been used to protect stored crop and grain (Matias et al. 2002; Boeke et al. 2004; Garcia and Azambuja 2004; Senthil et al. 2007).

Studies have demonstrated that plants from the Meliaceae family—especially *Azadiracta indica* and *Melia azedarach*, popularly known as "neem" and "amargoseira," respectively—repetitive insecticidal activity in the laboratory for all development stages of *Lu. longipalpis*. Laboratory bioassays of larvae feeding on leaves and fruits of *A. indica* and *M. azedarach*, in a restricted manner, have shown that both species of plants were capable to block molting between larvae stages. However, there was a difference with the *M. azedarach* leaves, which provoked a greater inhibition of molting compared with the fruits of this species and the leaves and fruits of *A.indica* (Andrade-Coelho et al. 2009).

Maciel et al. (2010a) demonstrated that contact with seed oil of *A. indica,* at a concentration of 100 mg/mL^{-1}, prevented egg hatching and larvae molting with an efficacy of 65% and 67%, respectively, for 10 days. Furthermore, at the same concentration, the study showed that the mortality rate of *Lu. longipalpis* remained >95% after 72 h from the initial spraying (Maciel et al. 2010a, 2010b).

Plants of the *Eucalipto* genus also showed evidence of insecticidal activity in the literature, and the species *E. staigeriana*, *E. citriodora*, and *E. globulus* demonstrated increased efficacy in preventing egg hatching and larvae molting as demonstrated by studies with *A. indica* e *M. azedarach*. However, the oil concentration needed for this blocking activity was different for each species of *Eucalipto* as well as for each development stage of the vector (egg, larvae or adult) (Maciel et al. 2010b).

In view of these results, it is possible to conclude that oil and extract of several plants have insecticidal activity against the three phases of *Lu. longipalpis* in the laboratory, thus indicating that some elements from these plants might be an alternative to chemical insecticide in VL-vector control. Nevertheless, there is still a need for further studies to confirm the activity of isolated substances against the vector and to evaluate the residual effect of these compounds in the field.

Another growing research field is the use of insecticide-treated nets (ITNs) to control the vector population of several diseases, such as malaria, VL, TL and Chagas disease (Wilson et al. 2014). This strategy is recommended by the WHO for malaria control in endemic areas, although its efficacy has not yet been proven in Brazil (Vieira et al. 2014).

When it comes to control of the phlebotomine population, the published studies with ITNs are for TL vectors only. In the systematic review and meta-analysis by Wilson et al. (2014), from a total of 21 articles only 4 studies on TL were included, which together presented an efficacy of 77% with the use of ITNs (95%CI: 39–91%).

Studies from six villages in the Northwest region of Cukurova, Turkey, using ITNs (permethrin 2% and piperonyl butoxide 1%) demonstrated an efficacy of 92.2% and a decrease from 4.78% to 0.37% of the TL-case incidence in the intervention areas (Gunay et al. 2014). These studies reveal a promising possibility of using ITNs in TL-endemic areas; still, further studies are needed to evaluate its efficacy against VL vectors in Brazil.

The use of synthetic pheromones, which are attractive to both female and male phlebotominae, is being studied to increase the efficacy of ITNs and domicile residual spraying in VL-endemic areas (Bray et al. 2010, 2014; Barbosa et al. 2016).

Four percent deltamethrin-impregnated collars (DICs) are also being evaluated for their efficacy, aiming at animal individual protection as well as effectiveness (David et al. 2001; Gavgani et al. 2002; Camargo-Neves et al. 2004; Reithinger et al. 2004). The repellent (or feeding-inhibition) effect was demonstrated in a study conducted in Fortaleza, Ceará, where feeding inhibition was confirmed in 99.3% of *Lu. longipalpis* at 4 weeks, 100% at 8–12 weeks, and 96% at 16–20 weeks after attachment of the DIC (David et al. 2001). A recent study revealed an efficacy of 61.8% and 88.3% for DICs impregnated with deltamethrin and flumethrin, respectively (Brianti et al. 2016). A multicentric study was conducted from 2012–2015 in 17 VL-endemic municipalities from Brazil to evaluate efficacy of the use of 4% DICs as a public-health tool. The study is still under analysis, although a preliminary result showed a decrease of the canine prevalence in intervention areas (Kazimoto 2016). It is necessary to evaluate the cost-efficiency of this strategy, after the final analysis of this study, to define the feasibility of adopting this measure as a control tool for public health seeing that the use of DICs must be done in a continuous and systematic manner to obtain the desired effect.

Moreover, it is necessary to evaluate *Lu. longipalpis* behavior, which is the main vector for VL, when faced with 4% DICs as a control tool because just the proof-of-concept of the active effect of the insecticide or repellent, or insecticide-based interventions, do not necessarily mean that the population wanting protection is actually protected by the desired effect. In this context, research is being performed to evaluate the behavior of *Lu. longipalpis* in two municipalities participating in the multicentric study mentioned previously.

Environment Management, Sanitary Education and Integrated Vector Management

Environment management and sanitary education are recognized by several publications as essential components of leishmaniasis surveillance and control. However, at the program level they are presented just as inter-sectorial management lines of action and general recommendations. These components are not applied in a systematic manner, and they lack standardized and controlled trials to support the evidence of their effectiveness and impact, as has been reported by reviews on other control activities (Alexander and Maroli 2003; Romero and Boelaert 2010; Bates et al. 2015; Salomón et al. 2015). Environment management and sanitary education are privileged areas to implement leishmaniasis inter-disciplinary management and integrated control of different vectors or "healthy housing/healthy environments" approaches (Rodrigues et al. 2011), in which there must be an evaluation of the resistance of the sympatric vectors with the largest impact in public health when using insecticide.

Proper environment management considers risk or protection factors for vector presence and abundance, parasite circulation associated with the landscape, and behavioral, social and economic factors, all of which should all be integrated with other surveillance and control activities and not considered as complementary or replacement activities. Generally, the recommendations are concentrated in the peri-domestic environment with there being no discrimination between rural, peri-urban, or urban leishmaniasis, except for urban canine VL and TL associated with occupational or recreational risk. The general recommendation usually includes (1) management of the peri-domicile, such as maintaining it free of refuge or potential breeding sites for vectors (plant or animal remains and garbage); (2) periodic rotation and distancing of the human sleeping area from animals, such as chickens and pigs; and (3) domicile improvement to reduce vector access (Marzochi et al. 2009; Coura-Vital et al. 2013; Curi et al. 2014).

Nonetheless, its implementation is difficult due to structural and economic problems (e.g. territory organization, sanitation, etc.); sustainability problems (e.g. resources, programmatic and community priorities when the incidence decreases); and the nature of the problem itself, which requires an adaptation of the guidelines to each particular scenario. Gouveia et al. (2012), in a successful integrated environment management for TL in Rio de Janeiro, identified factors on three different scales: (1) demographic scale (e.g. social-economic status and basic services); (2) domestic unit (e.g. domicile structure and animal refuge, proximity to woods, land use, garbage, vegetation, fruit trees and domestic and wild animals); and (3) micro-site: vector abundance. However, householder interventions are evaluated by the monthly entomological indicator in two domiciles, and each one of the interventions were different: One involves the reduction of livestock and animal dwellings, and the other involves decreasing the openness of the house structure to decrease vector indoor accessibility (Gouveia et al. 2012). Likewise, in a study conducted in the Paraná state, "cleaning" the peri-domicile—such as closing of piped sewer systems, ensuring animal refuge >100 m from the residence, increased insolation of land by tree trimming, removal of organic residues, and use of insecticides—decreased the intra-domicile vector population. However, domestic units being situated closer to the woods or to domestic animals, despite spraying, maintains the risk because the vector population recuperates over time (Teodoro et al. 2004; Legriffon et al. 2012).

Regarding sanitary education, despite its importance being repeatedly highlighted, this measure rarely has the support of professionals and specific financial resources in municipal programs, and it is limited to superficial informative messages by the same control agents (Zuben and Donalisio 2016). Thus, in many surveys of knowledge, attitudes, and practices (KAP [i.e. knowledge recorded on leishmaniasis among different actors of the community]), the education- and sanitary-system agents are found unsatisfactory or fragmented, thus reflecting the contradiction of rhetoric (especially for VL), in which the environment risks are not mentioned although they are high. Sanitary education should contain concrete recommendations for each social group both permanent or transitory, such as tourists (Carvalho et al. 2013), without containing unnecessary academic information and being adapted to the social context of those who will turn them into actions (Nieves et al. 2008; Margonari et al. 2012; Menezes et al. 2016). In addition, it should promote community participation and empowerment, complementary to the non-delegable responsibilities of the state at different levels, including social and economic promotion policies and territory organization, thus avoiding the delegitimization of public health speech by power, sectorial and jurisdictional conflicts (Dias 1998; Salomón et al. 2012).

Additional biases exist related to the methodological design and tools used in the generation of knowledge, which are used to develop environment-management and sanitary-education strategies. KAP studies usually have questions focused on the disease, vector identification, use of insecticide, animal reservoirs (such as dogs in the case of VL) and characterization of the environment and social status. Nonetheless, they contribute little to the knowledge on risk perception and disruption of a safe domicile or peri-domicile space, on the real capacity to modify the environment and on the conviction of doing it (where, when and what) by themselves or by external collective agents (Dobles-Ulloa et al. 1994; Gama et al. 1998; Santos et al. 2000; Borges et al. 2008; Margonari et al. 2012). Concerning TL, occupational, recreational, and extra-domestic subsistence activities (e.g. hunting, fishing and water and firewood collection), as well as the effect of the TL lesions on self-esteem and social exclusion, should be considered (Weigel et al. 1994; Moreira et al. 2002; Carrillo-Bonilla et al. 2014; Chahed et al. 2016). It is important to avoid blaming the individual and rather emphasise the role of the focal and peri-focal environment by defining concrete and feasible measures in time and space, promoting dialogue between sectors and fostering the participation of groups (Carmo et al. 2016).

In contrast, the evaluation of tools and information strategies also present difficulties of interpretation when transference of academic knowledge is privileged; therefore, information repetition is considered an impact indicator that only registers in the short- or medium-term memory (Uchôa et al. 2004; Luz et al. 2005; Rodrigues et al. 2011; Lobo et al. 2013), whereas sources of didactic but non-specific information on leishmaniasis usually contain useless or erroneous information (França et al. 2011, 2013). Regarding the methodology for KAP surveys, its paradox effect and limitation arises from the wide use of virtual platforms and social networks. Especially in the urban environment with canine VL, where cases might have more information than controls due to searches on the Internet about the diagnosis, such searching can result in the inference that information is a "risk factor" and is only an artefact of the methodology in the present time (López et al. 2016).

In conclusion, even if each household should find a particular solution toward environment management according to its scenario and risk, this fact implies (1) the need to register empowerment, education and permanent supervision to provide sustainable actions; (2) the need to contextualize recommendations to the real territory, thus guaranteeing necessary state follow-up and private-sector involvement (occupational risk and TL legal marks), (3) the need to promote joint actions in a spatial scale with biological consistency, which can overcome the domestic unit–dispersion area and vector colonization; and (4) the need to understand the real capacity (agency) of family groups to develop interventions that directly affect the economy, such as domicile intervention and the management of animals and crops for food or income.

Acknowledgements The authors thank all of the professionals working in the national and state leishmaniasis programs and in the epidemiological surveillance from endemic countries of the Americas; and Lia Puppim-Buzanovsky and Samantha Y.O.B. Valadas for their support in the maps and review, respectively.

References

Aguiar GM, Medeiros WM, Marco TS, Santos SC, Gambardella S (1996) Ecologia dos flebotomíneos da Serra do Mar, Itaguaí, estado do Rio de Janeiro, Brasil. I - A fauna flebotomínica e prevalência pelo local e tipo de captura (Diptera: Psychodidae: Phlebotominae). Cad Saude Publica 12(2):195–206. [Portuguese]

Alexander B, Maroli M (2003) Control of phlebotomine sandflies. Med Vet Entomol 17(1):1–18

Almeida AS, Werneck GL (2014) Prediction of high-risk areas for visceral leishmaniasis using socioeconomic indicators and remote sensing data. Int J Health Geogr 13:13

Alvar J, Yactayo S, Bern C (2006) Leishmaniasis and poverty. Trends Parasitol 22(12):552–557

Alvar J, Vélez ID, Bern C, Herrero M, Desjeux P, Cano J, Jannin J, den Boer M, and the WHO Leishmaniasis Control Team. Leishmaniasis Worldwide and Global Estimates of Its Incidence. PLoS One. 2012; 7(5): e35671

Andrade-Coelho C, Souza NA, Gouveia C, Silva VC, Gonzalez MS, Rangel EF (2009) Effect of fruit and leaves of meliaceae plants (*Azadirachta indica* and *Melia azedarach*) on the development of *Lutzomyia longipalpis* larvae (Diptera: Psychodidae: Phlebotominae) under experimental conditions. J Med Entomol 46(5):1125–1130

Antonialli SAC, Torres TG, Paranhos Filho AC, Tolezano JE (2007) Spatial analysis of American Visceral Leishmaniasis in Mato Grosso do Sul State, Central Brazil. J Infect 54(5):509–514. https://doi.org/10.1016/j.jinf.2006.08.004

Barata RA, França-Silva JC, Mayrink W, Silva JC, Prata A, Lorosa ES, Fiúza JA, Gonçalves CM, Paula KM, Dias ES (2005) Aspectos da ecologia e do comportamento de flebotomíneos em área endêmica de leishmaniose visceral, Minas Gerais. Rev Soc Bras Med Trop 38(5):421–425. [Portuguese]

Barbosa VA (2016) Avaliação de uma nova estratégia de controle de Lutzomyia longipalpis (Diptera: Psychodidae), vetor da Leishmania (Leishmania) infantum. Masters Dissertation, Instituto Oswaldo Cruz, Pós-Graduação em Medicina Tropical, Rio de Janeiro, pp 75

Bates PA, Depaquit J, Galati EAB, Kamhawi S, Maroli M, McDowell MA, Picado A, Ready PD, Salomón OD, Shaw JJ, Traub-Csekö YM, Warburg A (2015) Recent advances in phlebotomine sand fly research related to leishmaniasis control. Parasit Vectors 8:131

Belo VS, Werneck GL, Barbosa DS, Simões TC, Nascimento BWL et al (2013) Correction: factors associated with Visceral Leishmaniasis in the Americas: a systematic review and meta-analysis. PLoS Negl Trop Dis 7(5):e2182

Boeke SJ, Boersma MG, Alink GM, Loon JJA, Huis A, Dicke M, Rietjens IMCM (2004) Safety evaluation of neem (*Azadirachta indica*) derived pesticides. J Ethnopharmacol 94(1):25–41

Borges BKA, Silva JA, Haddad JPA, Moreira EC, Magalhães DF, Ribeiro LML, Fiúza VOP (2008) Avaliação do nível de conhecimento e de atitudes preventivas da população sobre a leishmaniose visceral em Belo Horizonte, Minas Gerais, Brasil. Cad Saúde Pública 24(4):777–784. [Portuguese]

Bray DP, Alves GB, Dorval ME, Brazil RP, Hamilton JG (2010) Synthetic sex pheromone attracts the leishmaniasis vector *Lutzomyia longipalpis* to experimental chicken sheds treated with insecticide. Parasit Vectors 3:16

Bray DP, Carter V, Alves GB, Brazil RP, Bandi KK, Hamilton JG (2014) Synthetic sex pheromone in a long-lasting lure attracts the visceral leishmaniasis vector, *Lutzomyia longipalpis*, for up to 12 weeks in Brazil. PLoS Negl Trop Dis 8(3):e2723

Brazil (2006) Ministério da Saúde. Secretaria de Vigilância em Saúde. Departamento de Vigilância Epidemiológica. Série A. Normas e Manuais Técnicos. Manual de vigilância e controle da leishmaniose visceral/Ministério da Saúde, Secretaria de Vigilância em Saúde, Departamento de Vigilância Epidemiológica. Ministério da Saúde. p 120. Available from: http://portal.saude.gov.br/portal/arquivos/pdf/manual_leish_visceral2006

Brazil (2007) Ministério da Saúde. Secretaria de Vigilância em Saúde. Manual de Vigilância da Leishmaniose Tegumentar Americana/Ministério da Saúde, Secretaria de Vigilância em Saúde. – 2. ed. atual. – Brasília: Editora do Ministério da Saúde (Série A. Normas e Manuais Técnicos). p 180. Available from: http://portalarquivos.saude.gov.br/images/pdf/2016/dezembro/14/manual_lta_2ed.pdf

Brazil (2016) Ministério da Saúde. Secretaria de Vigilância em Saúde. Casos de Leishmaniose Tegumentar. Brasil Grandes Regiões e Unidades Federadas. 1990 a 2015. [Internet] [cited 2017 February 28] Available from: http://portalarquivos.saude.gov.br/images/pdf/2016/novembro/07/LT-Casos.pdf

Brazil (2017) Ministério da Saúde. Secretaria de Vigilância em Saúde. Casos de Leishmaniose Visceral. Brazil Grandes Regiões e Unidades Federadas. 1990 a 2015. [Internet] [cited 2017 March 04] Available from:http://portalarquivos.saude.gov.br/images/pdf/2017/marco/03/LV-Casos.pdf

Brianti E, Napoli E, Gaglio G, Falsone L, Giannetto S, Solari Basano F (2016) Field evaluation of two different treatment approaches and their ability to control fleas and prevent canine leishmaniosis in a highly endemic area. PLoS Negl Trop Dis 10(9):e0004987. https://doi.org/10.1371/journal.pntd.0004987

Camargo-Neves VL (2004) A Leishmaniose Visceral Americana no Estado de São Paulo: situação atual. Boletim epidemiológico Paulista 4(48):12–14. [Portuguese]

Camargo-Neves VL, Colebrusco Rodas LA, Cabral G, Pauliquévis C (2007) Avaliação da eficácia Lambdacialotrina para o controle de *Lutzomyia longipalpis*/Evaluation of Lambdacialotrin efficacy for Lutzomyia longipalpis control. Boletim Epidemiologico Paulista 4(43):4–11

Canese A (1998) Focos de leishmaniasis visceral canina en las ciudades de Lambaré y Villa Elisa. Rev Parag Microbiol 18:18–24

Cardim MFM, Rodas LAC, Dibo MR, Guirado MM, Oliveira M, Chiaravalloti-Neto F (2013) Introdução e expansão da Leishmaniose visceral americana em humanos no estado de São Paulo, 1999-2011. Rev Saúde Pública 47(4):691–700. [Portuguese]

Cardim MFM, Guirado MM, Dibo MR, Chiaravalloti-Neto F (2016) Leishmaniose visceral no estado de São Paulo, Brazil: análise espacial e espaço-temporal. Rev Saúde Pública 50:48. [Portuguese]

Carmo RF, da Luz ZM, Bevilacqua PD (2016) Perceptions of the population and health professionals regarding visceral leishmaniasis. Cien Saude Colet 21(2):621–628

Carneiro D, Bavia ME, Rocha W, Lobão J, Madureira Filho C, Oliveira JB, Silva CE, Barbosa MG, Rios R (2004) Identificação de áreas de risco para a leishmaniose visceral americana através de estudos epidemiológicos e sensoriamento remoto orbital em Feira de Santana, Bahia, Brazil (2000-2002). Rev Baiana Saude Publica 28(1):19–32. [Portuguese]

Carrillo-Bonilla LM, Trujillo JJ, Álvarez-Salas L, Vélez-Bernal ID (2014) Estudio de los conocimientos, actitudes y prácticas de la leishmaniasis: evidencias del olvido estatal en el Darién Colombiano. Cad Saúde Pública Rio de Janeiro 30(10):2134–2144. [Spanish]

Carvalho MR, Valença HF, Silva FJ, Pitta-Pereira D, Pereira TA, Britto C, Brazil RP, Filho SB (2010) Natural Leishmania infantum infection in Migonemyia migonei (França, 1920) (Diptera:Psychodidae:Phlebotominae) the putative vector of visceral leishmaniasis in Pernambuco State, Brazil. Acta Trop 116(1):108–110

Carvalho BM, Maximo M, Costa WA, de Santana AL, da Costa SM, da Costa Rego TA, de Pita Pereira D, Rangel EF (2013) Leishmaniasis transmission in an ecotourism area: potential vectors in Ilha Grande, Rio de Janeiro State, Brazil. Parasit Vectors 6(1):325

Chahed MK, Bellali H, Ben Jemaa S, Bellaj T (2016) Psychological and Psychosocial Consequences of Zoonotic Cutaneous Leishmaniasis among Women in Tunisia: Preliminary Findings from an Exploratory Study. PLoS Negl Trop Dis 10(10):e0005090

Coura-Vital W, Reis AB, Reis LE, Braga SL, Roatt BM, Aguiar-Soares RD, Marques MJ, Veloso VM, Carneiro M (2013) Canine visceral leishmaniasis: incidence and risk factors for infection in a cohort study in Brazil. Vet Parasitol 197(3–4):411–417

Cousiño B (2006) Vigilancia y control de la leishmaniasis en Paraguay. In Panaftosa, Informe final de la reunión de expertos OPS/OMS sobre Leishmaniasis Visceral en las Américas, Panaftosa, Rio de Janeiro, pp 34–36. [Spanish]

Curi NH, Paschoal AM, Massara RL, Marcelino AP, Ribeiro AA, Passamani M, Demétrio GR, Chiarello AG (2014) Factors associated with the seroprevalence of leishmaniasis in dogs living around Atlantic Forest fragments. PLoS One 9(8):e104003

David JR, Stamm LM, Bezerra HS, Souza RN, Killick-Kendrick R, Lima JWO (2001) Deltamethrin-impregnated dog collars have a potent anti-feeding and insecticidal effect on *Lutzomyia longipalpis* and *Lutzomyia migonei*. Mem Inst Oswaldo Cruz Rio de Janeiro 96(6):839–847

De Silans LN, Dedet JP, Arias JR (1998) Field Monitoring of Cypermethrin Residual Effect on the Mortality Rates of the Phlebotomine Sand Fly *Lutzomyia longipalpis* in the State of Paraíba, Brazil. Mem Inst Oswaldo Cruz 93(3):339–344

De Vlas SJ, Stolk WA, le Rutte EA, Hontelez JAC, Bakker R, Blok DJ et al (2016) Concerted Efforts to Control or Eliminate Neglected Tropical Diseases: How Much Health Will Be Gained? PLoS Negl Trop Dis 10(2):e0004386. https://doi.org/10.1371/journal.pntd.0004386.

Desjeux P (2004) Leishmaniasis: current situation and new perspectives. Comp Immunol Microbiol Infect Dis 27(5):305–318

Dias JCP (1998) Problemas e possibilidades de participação comunitária no controle das grandes endemias do Brazil. Cad Saúde Pública 14(Supl. 2):19–37. [Portuguese]

Dobles-Ulloa A, Perriard C (1994) Representaciones, actitudes y practicas respecto a la leishmaniasis cutánea en la población del Cantón de Acosta, Provincia de San José, Costa Rica. Estudio Antropológico exploratório. Cad Saúde Pública 10(2):181–189. [Spanish]

Fernández MS, Salomón OD, Cavia R, Perez AA, Acardi SA, Guccione JD (2010) *Lutzomyia longipalpis* spatial distribution and association with environmental variables in an urban focus of cisceral leishmaniasis, Misiones, Argentina. Acta Trop 114(2):81–87

França VH, Margonari C, Schall VT (2011) Análise do conteúdo das Leishmanioses em livros didáticos de ciências e biologia indicados pelo Programa Nacional de Livros Didáticos (2008/2009). Ciência Educação 17(3):625–644. [Portuguese]

França VH, Margonari C, Schall VT (2013) Percepção de professores do ensino básico em relação as suas práticas educativas sobre leishmanioses: um estudo em área endêmica de Minas Gerais. Rev Ensaio 15(3):35–51. [Portuguese]

Franke CR, Ziller M, Staubach C, Latif M (2002) Impact of the El Niño- southern oscilation on visceral leishmaniasis, Brazil. Emerg Infect Dis 8(9):914–917

Galati EAB (2003) Morfologia, terminologia de adultos e identificac¸ ão dos táxonsda América. In: Rangel EF, Lainson R (eds) Flebotomíneos Do Brasil. Fiocruz, Rio de Janeiro, p 367

Galvis-Ovallos F et al (2017) Canine visceral leishmaniasis in the metropolitan area of São Paulo: *Pintomyia fischeri* as potential vector of *Leishmania infantum*. Parasite 24:2

Gama MEA, Barbosa JS, Pires B, Cunha AKB, Freitas AR, Ribeiro IR, Costa JML (1998) Avaliação do nível de conhecimento que populações residentes em áreas endêmicas têm sobre leishmaniose visceral, estado do Maranhão, Brazil. Cad Saúde Pública 14(2):3813–3890. [Portuguese]

Garcia ES, Azambuja P (2004) Lignoids in insects: chemical probes for the study of ecdysis, excretion and *Trypanosoma cruzi* triatomine interactions. Toxicon 44(4):431–440

Gavgani ASM, Hodjati MH, Mohite H, Davies CR (2002) Effect of insecticide-impregnated dog collars on incidence of zoonotic visceral leishmaniasis in iranian children: a matched-cluster randomised trial. Lancet 360(9330):374–379

Gomes AC (2002) Vigilância entomológica. Informe Epidemiológico Sus Brasília 11(2):79–90

Gouveia C, de Oliveira RM, Zwetsch A, Motta-Silva D, Carvalho BM, de Santana AF, Rangel EF (2012) Integrated tools for American cutaneous leishmaniasis surveillance and control:

intervention in an endemic area in Rio de Janeiro, RJ, Brazil. Interdiscip Perspect Infect Dis 2012:568312. https://doi.org/10.1155/2012/568312

Gramiccia M, Gradoni L (2005) The current status of zoonotic leishmaniasis and approaches to disease control. Int J Parasitol 35(11–12):1169–1180

Gunay F, Karakus M, Oguz G, Dogan M, Karakaya Y, Ergan G, Kaynas S, Kasap OE, Ozbel Y, Alten B (2014) Evaluation of the efficacy of Olyset® Plus in a village-based cohort study in the Cukurova Plain, Turkey, in an area of hyperendemic cutaneous leishmaniasis. J Vector Ecol 39(2):395–405

Kazimoto TA (2016) Uso de coleiras imregnadas com deltametrina 4% em cães no controle da leishmaniose visceral. [Masters Dissertation]. Universidade Federal do Semi Árido, Mossoró [Portuguese]

Koopman JS, Simon CP, Riolo CP (2005) When to control endemic infections by focusing on high-risk groups. Epidemiology 16(5):621–627

Legriffon CM, Reinhold-Castro KR, Fenelon VC, Neitzke-Abreu HC, Teodoro U (2012) Sandfly frequency in a clean and well-organized rural environment in the state of Paraná, Brazil. Rev Soc Brazil Med Trop 45(1):77–82

Lindoso JA, Cota GF, da Cruz AM, Goto H, Maia-Elkhoury ANS, Romero GA et al (2014) Visceral Leishmaniasis and HIV Coinfection in Latin America. PLoS Negl Trop Dis 8(9):e3136. https://doi.org/10.1371/journal.pntd.0003136

Lobo KS, Bezerra JMT, Brito LMO, Silva JS, Pinheiro VCS (2013) Conhecimentos de estudantes sobre Leishmaniose Visceral em escolas públicas de Caxias, Maranhão, Brazil. Ciênc Saúde Coletiva 18(8):2295–2300. [Portuguese]

López K, Tartaglino LC, Steinhorst II, Santini MS, Salomón OD (2016) Factores de riesgo en escenarios emergentes de leishmaniasis visceral urbana, Misiones, Argentina. Biomedica 36(S1):51–63. [Spanish]

Luz ZMP, Schall V, Rabello A (2005) Evaluation of a pamphlet on visceral leishmaniasis as a tool for providing disease information to healthcare professionals and laypersons. Cad Saude Publica 21(2):608–621

Maciel MV, Morais SM, Bevilaqua CML, Silva RA, Barros RS, Sousa RN, Sousa LC, Machado LKA, Brito ES, Souza-Neto MA (2010a) Atividade inseticida in vitro do óleo de sementes de nim sobre *Lutzomyia longipalpis* (Diptera: Psychodidae). Rev Bras Parasitol Vet Jaboticabal 19(1):7–11. [Portuguese]

Maciel MV, Morais SM, Bevilaqua CML, Silva RA, Barros RS, Sousa RN, Sousa LC, Brito ES, Souza-Neto MA (2010b) Chemical composition of *Eucalultus spp.* Essential oils and their insecticidal effects on *Lutzomyia longipalpis*. Vet Parasitol 167(1):1–7

Maia-Elkhoury ANS, Alves WA, Sousa-Gomes ML, Sena JM, Luna EA (2008) Visceral leishmaniasis in Brazil: trends and challenges. Cad Saúde Pública 24(12):2941–2947

Maia-Elkhoury ANS, Yadón ZE, Días MIS, Lucena FFA, Castellanos LG, Sanchez-Vazquez MJ (2016) Exploring Spatial and Temporal Distribution of Cutaneous Leishmaniasis in the Americas, 2001-2011. PLoS Negl Trop Dis 10(11):e0005086

Margonari C, Menezes JA, Rocha MN, Maia KN, Oliveira ME, Fonseca AL, Sousa FF, Ferreira EC, Madureira AP, Melo MN, Soares RP (2012) Public knowledge about and detection of canine visceral leishmaniasis in urban Divinópolis, Brazil. J Trop Med 2012:18

Marzochi MC, Fagundes A, Andrade MV, Souza MB, de Madeira MF, Mouta-Confort E, Schubach Ade O, Marzochi KB (2009) Visceral leishmaniasis in Rio de Janeiro, Brazil: eco-epidemiological aspects and control. Rev Soc Bras Med Trop 42(5):570–580

Matias R (2002) *Melia azedarach*, uso popular x estudos químico e farmacológico: breve revisão. Ensaios Ciência 6(1):91–121. [Portuguese]

Menezes JA, Luz TC, Sousa FF, Verne RN, Lima FP, Margonari C (2016) Peridomiciliary risk factors and knowledge concerning visceral leishmaniasis in the population of formiga, Minas Gerais, Brazil. Rev Bras Epidemiol 19(2):362–374

Mestre GLC, Fontes CJF (2007) A expansão da epidemia da leishmaniose visceral no Estado de Mato Grosso, 1998-2005. Rev Soc Brazil Med Trop 40(1):42–48. [Portuguese]

Missawa NA, Dias ES (2007) Phlebotomine sand flies (Diptera: Psychodidae) in the municipality of Várzea Grande: an area of transmission of visceral leishmaniasis in the state of Mato Grosso, Brazil. Mem Inst Oswaldo Cruz 102(8):913–918

Morais MHF, Fiuza VOP, Araújo VEM, Menezes FCM, Carneiro M (2015) Avaliação das atividades de controle da leishmaniose visceral em Belo Horizonte, Minas Gerais, 2006–2011. Epidemiol Serv Saúde Brasília 24(3):485–496. [Portuguese].

Moreira RCR, Rebelo JMM, Gama MEA, Costa JML (2002) Nível de conhecimento sobre leishmaniose tegumentar americana (LTA) e uso de terapias alternativas por populações de uma área endêmica da Amazônia do Maranhão, Brazil. Cad Saúde Pública 18(1):187–195. [Portuguese]

Nieves E, Villarrea N, Rondón M, Sánche M, Carrero J (2008) Evaluación de conocimientos y prácticas sobre la leishmaniasis tegumentaria en un área endémica de Venezuela. Biomedica 28(3):347–356. [Spanish]

Oliveira AL, Paniago AM, Dorval ME, Oshiro ET, Leal CR, Sanches M, Cunha RV, Bóia MN (2006) Foco emergente de leishmaniose visceral em Mato Grosso do Sul. Rev Soc Bras Med Trop 39(5):446–450. [Portuguese]

Oliveira AG, Galati EA, Fernandes CE, Dorval ME, Brazil RP (2008) Seasonal variation of *Lutzomyia longipalpis* (Lutz & Neiva, 1912) (Diptera:Psychodidae: Phlebotominae) in endemic area of visceral leishmaniasis, Campo Grande, state of Mato Grosso do Sul, Brazil. Acta Trop 105(1):55–61

Oliveira EF, Silva EA, Fernandes CE, Paranhos Filho AC, Gamarra RM, Ribeiro AA et al (2012) Biotic factors and occurrence of Lutzomyia longipalpis in endemic area of visceral leishmaniasis, Mato Grosso do Sul, Brazil. Mem Inst Oswaldo Cruz 107(3):396–401

Pan American Health Organization – World Health Organization (2016) Plan of action for the elimination of neglected infectious diseases and post-elimination actions 2016–2022. [Internet] [cited 2017 February 05] Available from: http://iris.paho.org/xmlui/bitstream/handle/123456789/31439/CD55-R9-e.pdf?sequence=1&isAllowed=y

Queiroz MFM, Varjão JR, Moraes SC, Salcedo GE (2012) Analysis of sandflies (Diptera: Psychodidae) in Barra do Garças, state of Mato Grosso, Brazil, and teh influence of environmental variables on the vector density of Lutzomyia longipalpis (Lutz & Neiva, 1912). Rev Soc Bras Med Trop 45(3):313–317

Ready PD (2013) Biology of phlebotomine sand flies as vectors of disease agents. Annu Rev Entomol 58(s.1):227–250

Reithinger R, Coleman PG, Alexander B, Vieira EP, Assis G, Davies CR (2004) Are insecticide-impregnated dog collars a feasible alternative to dog culling as a strategy for controlling canine visceral leishmaniasis in Brazil? Int J Parasitol 34(1):55–62

Rocha DA (2016) Caracterização de suscetibilidade de Phlebotominae (Diptera: Psychodidae) ao inseticida alfacipermetrina. [Masters Dissertation]. Universidade de Brasília, Brazilia, p 126

Rodrigues TO, Perri SHV, Nunes CM, Valladão GMR, Gallani SU, Pinheiro SR, Queiroz LH (2011) Ações Educativas para o Controle de Vetores da Dengue e Leishmaniose Visceral. Vet Zootec 18(3):462–472. [Portuguese]

Romero GA, Boelaert M (2010) Control of visceral leishmaniasis in Latin America - A systematic review. PLoS Negl Trop Dis 4(1):e584

Salomón OD, Orellano PW (2005) Lutzomyia longipalpis in Clorinda, Formosa province, an area of potential visceral leishmaniasis transmission in Argentina. Mem Inst Oswaldo Cruz 100(5):475–476

Salomón OD, Sinagra A, Nevot MC, Barberian G, Paulin P, Estevez JO, Riarte A, Estevez J (2008) First visceral leishmaniasis focus in Argentina. Mem Inst Oswaldo Cruz 103(1):109–111

Salomón, O.D.; Bezzi, G; Morán, M.L.; Betbeder, E.; Valdéz, D.V. *Lutzomyia migonei* as putative vector of visceral leishmaniasis in La Banda, Argentina Acta Trop v 113, p. 84–87, 2010

Salomón OD, Basmajdian Y, Fernández MS, Santini MS (2011) Lutzomyia longipalpis in Uruguay: the first report and the potential of visceral leishmaniasis transmission. Mem Inst Oswaldo Cruz 106(3):381–382

Salomón OD, Mastrángelo AV, Santini MS, Ruvinsky S, Orduna T, Sinagra A, Luna C, Riarte A, Casas N, Amiotti P (2012) Leishmaniasis visceral: senderos que confluyen, se bifurcan. Salud colectiva 8(Supl. 1):49–63. [Spanish]

Salomón OD, Feliciangeli MD, Quintana MG, Afonso MM, Rangel EF (2015) *Lutzomyia longipalpis* urbanisation and control. Mem Inst Oswaldo Cruz 110(7):831–846

Santos JB, Lauand L, Souza GS, Macêdo VO (2000) Fatores sócio-econômicos e atitudes em relação à prevenção domiciliar da leishmaniose tegumentar americana, em uma área endêmica do sul da Bahia, Brazil. Cad Saúde Pública 16(3):701–708. [Portuguese]

Satragno D, Faral-Tello P, Canneva B, Verger L, Lozano A, Vitale E et al (2017) Autochthonous outbreak and expansion of canine visceral leishmaniasis, uruguay. Emerg Infect Dis 23(3):536–538

Sena JM (2011) Vigilância Entomológica do Programa de Leishmaniose Visceral: limites e possibilidades para o monitoramento das ações. [Masters dissertation]. Fundação Oswaldo Cruz, Rio de Janeiro, p 159 [Portuguese]

Senthil NS, Choi MY, Paik CH, Seo HY, Kim JD, Kang SM (2007) The toxic effects of neem extract and azadirachtin on the brown plant hopper, Nilaparvata lugens (Stål) (BPH) (Homoptera:Delphacidae). Chemosphere 67(1):80–88

Silva EA, Andreotti R, Honer MR (2007) Comportamento de *Lutzomyia longipalpis*, vetor principal da leishmaniose visceral americana, em Campo Grande, Estado do Mato Grosso do Sul. Revista da Sociedade Brazileira de Medicina Tropical 40(4):420–425. [Portuguese]

Teodoro U, Thomaz-Soccol V, Kühl JB, Santos DR, Santos ES, Santos AR, Abbas M, Dias C (2004) Reorganization and cleanness of peridomiciliar area to control sand flies (Diptera, Psychodidae, Phlebotominae) in south Brazil. Braz Arch Biol Technol 47(2):205–212

Teodoro U, Valdrinez M, Lonardoni C, Silveira TGV, Dias AC, Abbas M, Alberton D, Santos DR (2007) Lights and hens as attraction factor of Nyssomyia whitmani in a rural área, Southern Brazil. Rev Saúde Pública 41(3):383–388

Uchôa CMA, Serra CMB, Magalhães CM, Silva RMM, Figliuolo LP, Leal CA, Madeira MF (2004) Educação em saúde: ensinando sobre a leishmaniose tegumentar americana. Cad Saúde Pública 20(4):935–941. [Portuguese]

Uniting to Combat Neglected Tropical Diseases. The London Declaration (2012) [Internet] [cited 2017 January 20]. Available from: http://unitingtocombatntds.org/sites/default/files/resource_file/london_declaration_on_ntds.pdf

Vieira GD, Basano SA, Katsuragawa TH, Camargo LMA (2014) Insecticide-Treated bed nets in Rondônia, Brazil: evaluation of their impact on malária control. Rev Insti Med Trop 56(6):493–497

Weigel MM, Armijos RX, Racines RJ, Zurita C, Izurieta R, Herrera E, Hinojsa E (1994) Cutaneous leishmaniasis in subtropical Ecuador: popular perceptions, knowledge, and treatment. Bull Pan Am Health Organ 28(2):142–155

Werneck G (2008) Forum: geographic spread and urbanization of visceral leishmaniasis in Brazil. Introduction Cad Saúde Pública 24(12):2937–2940

Wilson AL, Dhiman RC, Kitron U, Scott TW, Berg HVD, Lindasay SW (2014) Benefit of insecticide-treated nets, curtains and screening on vector borne diseases, excluding malaria: a systematic review and meta-analysis. PLoS Negl Trop Dis 8(10):3228–3241

World Health Organization (2010) WHO technical report series; no. 949 -Control of the leishmaniasis: report of a meeting of the WHO Expert Committee on the Control of Leishmaniasis, Geneva, 22–26 March

World Health Organization (2012) Accelerating work to overcome the global impact of neglected tropical diseases: a roadmap for implementation [Internet] [cited 2017 January 20]. Available from: http://www.who.int/neglected_diseases/NTD_RoadMap_2012_Fullversion.pdf

Young DG, Duncan MA (1994) Guide to the identification and geographic distribution of Lutzomyia sand flies in Mexico, the West Indies, Central and South America (Diptera: Psychodidae). Mem Amer Entomol Inst 54:1–881

Zeilhofer P, Kummer OP, Santos ES, Ribeiro ALM, Missawa NA (2008) Spatial modelling of Lutzomyia (Nyssomyia) whitmani s.l. (Antunes & Coutinho, 1939) (Diptera: Psychodidae: Phlebotominae) habitat suitability in the state of Mato Grosso, Brazil. Mem Inst Oswaldo Cruz 103(7):653–660

Zuben AP, Donalísio MR (2016) Dificuldades na execução das diretrizes do Programa de Vigilância e Controle da Leishmaniose Visceral em grandes municípios Brazileiros. Cad Saude Publica 32(6):e00087415. [Portuguese]

The manufacturer's authorised representative in the EU is Springer Nature Customer Service Centre GmbH, Europaplatz 3, 69115 Heidelberg, Germany. If you have any concerns regarding our products, please contact ProductSafety@springernature.com

Printed and bound by CPI Group (UK) Ltd, Croydon, CR0 4YY
22/09/2026
02221049-0002